App Available

The ARRL Repeater Directory® is powered by **RFinder—the Worldwide Repeater Directory**. Why wait for updates? Get an annual subscription to **RFinder**, and you'll have the latest repeater listings at your fingertips. Always updated! Get access to listings for 55,000+ repeaters in 175+ countries.

RFinder is integrated with EchoLink® on Android™ and iOS. An annual subscription includes access to the directory in the Apps, RT Systems, CHIRP, **web.rfinder.net**, **routes.rfinder.net**, and a growing list of third party applications that use repeater data!

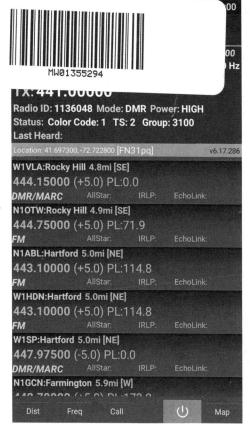

Screenshot of **RFinder** on the RFinder Android Radio.

To get the app...

- Go to **Subscribe.RFinder.net** on your phone's browser
- Click the Google Play™ or Apple® App Store logo [in most cases it detects the device and redirects to the app store]. Purchasing on Apple includes your first year subscription, or adds a year if you are already subscribed.
- If you do not have an Android or Apple device, just subscribe on that webpage!

Only $12.99* Annual Subscription. Visit **Subscribe.RFinder.net** for additional subscription options and pricing.

*Price subject to change without notice.

Hear weak signals clearly with.. ..a bhi DSP noise cancelling product!

bhi

ParaPro EQ20 Audio DSP Range

DSPKR
- 10W amplified DSP noise cancelling speaker
- Easy control of DSP filter
- 7 filter levels 9 to 35dB
- Filter select & store function
- Separate volume control
- Input overload LED
- Headphone socket
- Supplied with user manual and fused DC power lead

Boost the sound of your receive audio!

- 20W audio and parametric equalisation on all units
- Powerful high-performance audio processing system
- DSP noise cancelling and Bluetooth versions available
- Simple control of all DSP functions
- Two separate mono inputs or one stereo input - Use with passive speakers or head phones
- Fine-tune the audio to maximise your enjoyment
- Four models EQ20, EQ20B, EQ20-DSP, EQ20B-DSP

Precise audio adjustment to compensate for hearing loss

New HP-1 folding stereo headphones

Compact In-Line

Compact handheld DSP noise cancelling unit
- Easy to use rotary controls
- Use with a mono or stereo inputs
- 8 filter levels 9 to 40dB
- Ideal for portable use & DXing
- Use with headphones or speakers
- 12V DC power or 2 x AA batteries
- Over 40 hours battery life
- Size: 121mm x 70mm x 33mm
- Suitable for use with SDR, Elecraft K3 & KX3 plus FlexRadio products

Dual In-Line

Mono/stereo DSP noise eliminating module - Latest bhi DSP noise cancelling
- 8 Filter levels 8 to 40dB - 3.5mm Mono or stereo inputs - Line level in/out - 7 watts mono speaker output - 3.5mm stereo Headphone socket - Easy to adjust and setup - Ideal for DXing and club stations - Supplied with user manual and audio/power leads - Suitable for use with many radios and receivers including Elecraft K3, KX3 & FlexRadio products

DESKTOP

- 10W amplified DSP noise cancelling base station speaker
- Rotary volume and filter level controls
- 8 filter levels
- Speaker and line level audio inputs
- Headphone socket
- Size 200(H)x150(D) x160(W)mm, Wt 1.9 Kg
- For use with most radios, receivers & SDR including Elecraft & FlexRadio

DX ENGINEERING
DXEngineering.com -1-800-777-0703

www.bhi-ltd.com

The ARRL Repeater Directory® 2020 Edition

Published by:
ARRL The national association for AMATEUR RADIO®
225 Main Street, Newington, CT 06111-1400 USA
www.arrl.org

Repeater Directory® is a registered trademark of the American Radio Relay League, Inc.

Copyright © 2019 by
The American Radio Relay League, Inc.

Copyright secured under the Pan-American Convention

International Copyright secured

All rights reserved. No part of this work may be reproduced in any form except by written permission of the publisher. All rights of translation are reserved.

Printed in USA

Quedan reservados todos los derechos

ISBN: 978-1-62595-126-7

Advertising Contact Information

Advertising Department Staff

Janet Rocco, W1JLR, *Advertising Sales Manager*
Lisa Tardette, KB1MOI, *Account Executive*

Call Toll Free: 800-243-7768

Direct Line: 860-594-0207
Fax: 860-594-4285
E-mail: ads@arrl.org
Web: www.arrl.org/ads

Advertising Deadline:

Contact the Advertising Department in early October 2020 for advertising placements in the 2021 ARRL Repeater Directory.

If your company provides products or services of interest to ARRL Members, please contact the ARRL Advertising Department today for information on advertising in ARRL publications.

Please Note: All advertising in future editions of the ARRL Repeater Directory will be **Full Page** only.

Index of Advertisers

Advanced Specialties......................231

ASA, Inc. 7

BHI 2

Command Productions 10

Cushcraft 11

Dave's Hobby Shop/W5SWL 15

Hamcity ....Cover 2

Ham Radio Outlet.....................Cover 3

Hy-Gain... 11

ICOM America ....8, 9, Cover 4

Long Island Mobile Amateur
Radio Club.....................................245

MFJ Enterprises, Inc. 11, 12

Mirage 13

Radio Club of JHS 22 NYC 14

RFinder 1, 464

Sea Pac Ham Convention 369

TABLE OF CONTENTS

16   Labels and Abbreviations Used in Repeater DIrectory Listings

17   CTCSS and DCS Information

18   About the Repeater Directory

18   Errors and Corrections

19   Frequency Coordinators

23   Band Plans

37   Repeater Lingo/Hints

Repeater Listings

United States

Alabama: 39	Louisiana: 173	Oklahoma: 285
Alaska: 44	Maine: 177	Oregon: 291
Arizona: 46	Maryland: 180	Pennsylvania: 301
Arkansas: 55	Massachusetts: 183	Puerto Rico: 315
California: 60	Michigan: 189	Rhode Island: 319
Colorado: 104	Minnesota: 197	South Carolina: 320
Connecticut: 111	Mississippi: 204	South Dakota: 326
Delaware: 116	Missouri: 207	Tennessee: 327
Florida: 117	Montana: 214	Texas: 335
Georgia: 132	Nebraska: 217	U.S. Virgin Islands: 362
Guam: 142	Nevada: 220	Utah: 362
Hawaii: 142	New Hampshire: 227	Vermont: 369
Idaho: 146	New Jersey: 232	Virginia: 370
Illinois: 150	New Mexico: 238	Washington: 381
Indiana: 158	New York: 246	Washington DC: 394
Iowa: 159	North Carolina: 260	West Virginia: 395
Kansas: 164	North Dakota: 270	Wisconsin: 399
Kentucky: 168	Ohio: 273	Wyoming: 406

Canada

Alberta: 409	Nova Scotia: 430
British Columbia: 413	Nunavut: 431
Manitoba: 422	Ontario: 432
New Brunswick: 424	Prince Edward Island: 444
Newfoundland and Labrador: 423	Quebec: 444
Northwest Territories: 430	Saskatchewan: 457

ICOM

Handheld

IC-V86
2M Portable
More RF power,
more coverage.

ID-51A PLUS2
2M / 70CM Analog/Digital
D-STAR ready

ID-31A PLUS
70CM Analog / Digital
Terminal and Access
Point Modes
D-STAR ready

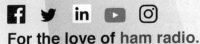

For the love of ham radio.

Mobile

ID-5100A
2M / 70CM Analog/Digital

D-STAR ready

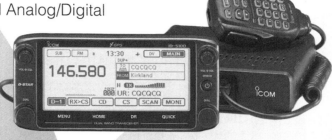

IC-7100
HF / 6M / 2M / 70CM

D-STAR ready

ID-4100A
2M / 70CM Analog / Digital
Terminal and Access
Point Modes

D-STAR ready

WWW.ICOMAMERICA.COM

©2020 Icom America Inc. The Icom logo is a registered trademark of Icom Inc.
All other trademarks remain the property of their respective owners.
All specifications are subject to change without notice or obligation. #31349

It's the highest class FCC Commercial License!

Get your "FCC General Radiotelephone Operator License with Radar Endorsement".

This impressive FCC License is your "ticket" to 1000's of high paying jobs in Communications, or start a business of your own!

Our proven Home-Study Course is easy, fast and low cost!

Our 40th Year

GUARANTEED PASS: You will get your FCC License or your money will be refunded.

www.LicenseTraining.com
or call now for FREE Info Kit
800-932-4268

COMMAND PRODUCTIONS • FCC License Training
480 Gate Five Road • PO Box 3000 • Sausalito, CA 94966-3000

Cushcraft Ringos

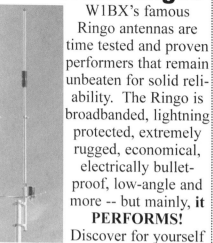

W1BX's famous Ringo antennas are time tested and proven performers that remain unbeaten for solid reliability. The Ringo is broadbanded, lightning protected, extremely rugged, economical, electrically bullet-proof, low-angle and more -- but mainly, *it* **PERFORMS!** Discover for yourself why you'll *love* this antenna!

ARX-2B, $139⁹⁵   135-160 MHz. 14 ft., 6 lbs. 1000 Watts.

Cushcraft
Amateur Radio Antennas
308 Ind'l Pk. Rd., Starkville, MS 39759
Toll-Free: 800-973-6572
www.cushcraftamateur.com

Hygain VHF FM Beam Antennas

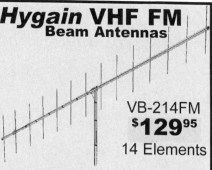

VB-214FM
$129⁹⁵
14 Elements

Hygain features tooled manufacturing: die-cast aluminum boom-to-mast bracket and element-to-boom compression clamps, tooled, swaged tubing that is easily and securely clamped in place and durable precision injection molded parts.

VB-23FM, $64⁹⁵
3 Elements.
VB-25FM, $74⁹⁵
5 Elements.
VB28FM, $119⁹⁵
8 Elements.

308 Ind'l Pk. Rd., Starkville, MS 39759
800-973-6572
www.hy-gain.com

MFJ-4230MV ... World's *Smallest*
30 Amp Metered Switching Power Supply

MFJ-4230MV
$99⁹⁵

Tiny 5Wx2¹/2Hx6D", weighs just 3 pounds! MFJ-4230MV gives you 25 Amps *continuously,* 30 Amps surge at 13.8 VDC. Voltage front-panel adjustable 4-16 Volts. *Selectable* input voltage 120 or 240 VAC at 47-63 Hz. Switch Amps or Volts for meter reading. Excellent 75% efficiency and extra low ripple and noise, <100 mV. Extremely quiet convection and heat-controlled fan cooling. Over-voltage, over-current protections. 5-way binding posts for DC output.

MFJ-1124, $79.95. DC Multi-Outlet with Anderson *PowerPole*⁽ᴿ⁾ connectors. Genderless, polarized and color coded connectors conform to ARES & RACES emergency 12V connector standard. 6 outlets. Installed fuses: one 1A, one 5A, two 10A, two 25A, one 40A. Outlets 1-4 are *PowerPole*⁽ᴿ⁾, 5-6 are binding posts.
MFJ-5512, $19.95. 14 ga., 15 ft. VHF/UHF power cable, in-line fuse.

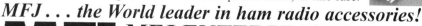

MFJ . . . the World leader in ham radio accessories!

MFJ ENTERPRISES, INC.
300 Industrial Park Road, Starkville, MS 39759
PH: (662) 323-5869 **FX:** (662) 323-6551
Free MFJ Catalog! Nearest Dealer or Order . . . 800-647-1800
www.mfjenterprises.com

MFJ VHF/UHF Headquarters
...the world leader in ham radio accessories!

$15⁹⁵
MFJ-281
MFJ speaker turns your HT into a super loud base!

$49⁹⁵
MFJ-1702C
2-Pos. coax switch, lightning surge protection, center ground.

$34⁹⁵
MFJ-916B
HF/144/440 Duplexer 200W, Low loss, SO-239s.

$69⁹⁵
MFJ-822
1.6-200 MHz SWR/Wattmeter. 30/300W FWD/REF ranges.

$114⁹⁵
MFJ-847
VHF/UHF digital SWR Wattmeter

$27⁹⁵
MFJ-108B
24/12 hour base station clock. Each side independently set

$49⁹⁵
MFJ-260C
MFJ dry dummy load handles 300W 30 to 650 MHz.

$24⁹⁵
MFJ-295K,I,Y,R Mini speaker mic for tiny new handheld radios

$79⁹⁵
MFJ-1868
Ultra wide band discone antenna receives 25-1300 MHz & transmits 50-1300 MHz, 200W, Ideal for 2/6/220/440.

$49⁹⁵
MFJ-1750
2-Meter, high-gain 5/8 Wave ground plane antenna. Strong, lightweight aluminum. Single U-bolt mount.

$34⁹⁵
MFJ-1717/S
15" very high gain 144/440 rubber duck antenna. Flexible and tough. Full 1/2 wave 440 BNC or SMA.

$34⁹⁵
MFJ-1724B
World's best selling 144/440 MHz mag mount antenna. Only 19" tall, 300W,15' coax, free BNC adapter.

$34⁹⁵
MFJ-1728B
Long range 5/8 Wave on 2M, 1/4 Wave on 6M. Mag mount. 12' coax, stainless radiator.

$49⁹⁵
MFJ-1729
Highest gain 144/440 MHz magnet mount antenna. Long 27.5" radiator. 300W,15' coax, PL-259/BNC.

$39⁹⁵
MFJ-1730
Pocket Roll-up 1/2 Wave 2-Meter antenna easily fits in pocket.

$24⁹⁵
MFJ-1722
UltraLite™ 144/440 MHz magnet mount mobile antenna, strong 1 1/8" dia. 2 oz. magnet, thin 20" whip, 12 ft. coax.

$59⁹⁵
MFJ-1734
Glass mount 144/440 MHz. High gain 26" stainless radiator, low SWR, 50W, 12 ft. coax.

$19⁹⁵
MFJ-1714
LongRanger™ 1/2 wave 2-M HT ant., 40" ext., 10.5" collapsed. Ultra long range -- outperforms a 5/8 Wave!

$44⁹⁵
MFJ-1422
RuffRider™ hi-gain 41.5 inch deluxe mobile antenna has super gain on 2M and ultra gain on 440. 150W, PL-259.

$44⁹⁵
MFJ-345S
Trunk lip SO-239 Mount with 14 ft. coax, rubber guard.

$19⁹⁵
MFJ-335BS
5" Magnet mount has 17' coax. For antennas with PL-259.

$44⁹⁵
MFJ-336S
Super-strong Tri-magnet mount, SO-239, 17 foot coax.

Free MFJ CATALOG!
Nearest Dealer or to Order... 800-647-1800

MFJ ENTERPRISES, INC.
300 Industrial Park Road,
Starkville, MS 39759
PH: (662) 323-5869 **FX:** (662) 323-6551
E-Mail:mfjcustserv@mfjenterprises.com

MFJ ... the world leader in ham radio accessories!

www.mfjenterprises.com

(c) 2018 MFJ Enterprises, Inc.

MIRAGE POWER!
Dual Band 144/440 MHz Amp
45 Watts on 2-Meters or 35 Watts on 440 MHz

FREE Catalog!

BD-35
$299.95
Suggested Retail

Call your dealer for your best price!

- 45 Watts on 2-Meters or . . . 35 Watts on 440 MHz
- Single connector for dual band radios and antennas
- Automatic band selection
- Reverse polarity protection
- Full Duplex Operation
- Auto RF sense T/R Switch
- D-Star/Fusion compatible
- Works with all Handhelds
- Compact 5Wx1½Hx5D"
- One year Mirage Warranty

35 Watts Out for 2M Handheld!

MIRAGE Rugged!

B-34-G
$199.95
Suggested Retail

- 35 Watts Output
- All Modes: FM, SSB, CW
- 18 dB GaAsFET preamp
- Reverse polarity protection
- Includes Mobile Bracket
- Auto RF sense T/R Switch
- Custom Heatsink runs cool
- Works with all HTs to 8 W
- Compact 5¼Wx1¾Hx4¾D"
- One year Mirage Warranty

Power Curve -- typical B-34-G output power for your HT									
Watts Out	8	12	18	30	33	35	35+	35+	35+
Watts In	.25	0.5	1	2	3	4	5	6	8

More Mirage Amplifiers!

B-34, $179.95.   35W out. For FM 2-Meter HTs, 3Wx1Hx4.25"
B-310-G, $359.95.   100W out. For 2M fm/ssb/cw HTs. Preamp
B-320-G, $599.95.   200W out. 2M all mode for HTs/Mobiles.

50 Watts In, 160W Out!
B-5018-G, $499.95.
FM, SSB, CW, Superb GaAsFET 20dB preamp, Over-drive and high SWR protection, remote w/RC-2 option.

MIRAGE
300 Industrial Park Road
Starkville, MS 39759 USA
Call Toll-Free: 800-647-1800

MIRAGE . . . the world's most rugged VHF/UHF Amplifiers

www.mirageamp.com (C) 2018 Mirage

The Radio Club of Junior High School 22

Bringing Communication to Education Since 1980

DONATE YOUR RADIO

Radios You Can Write Off - Kids You Can't

- Turn your excess Ham Radios and related items into a tax break for you and a learning tool for kids.

- Donate radios or related gear to an IRS approved 501(c)(3) charity. Get the tax credit and help a worthy cause.

- Equipment picked up anywhere or shipping arranged.

PO Box 1052
New York NY 10002
E-mail: crew@wb2jkj.org
www.wb2jkj.org

Call 516-674-4072
Fax 516-674-9600

RF Connectors and Adapters

DIN – BNC – C – FME
Low Pim – MC – MCX – MUHF
N – QMA – SMA – SMB – TNC
UHF & More

....

Attenuators
Loads & Terminations
Component Parts
Hardware
Mic & Headset Jacks
Mounts
Feet – Knobs
Speakers – Surge Protectors
Test Gear Parts
Gadgets – Tools

www.W5SWL.com

Labels and Abbreviations Used in *Repeater Directory* Listings

Location   The nearest city or geographic landmark

Mode   The repeater operating mode
- ATV – Amateur Television
- DMR – Digital Mobile Radio
- DMR/BM – DMR repeaters linked through the BrandMeister network
- DMR/MARC – DMR repeaters linked through the DMR-MARC network
- D STAR® – Digital Smart Technologies for Amateur Radio
- FM – Analog FM
- FUSION – Yaesu System Fusion® C4FM digital and analog FM
- NXDN – An NXDN repeater system
- P25 – APCO 25

Call Sign   The call sign of the repeater

Output   The repeater output frequency in MHz

Offset   Input frequency separation, plus or minus, in MHz. For example, –0.60000 equals minus 0.6 MHz or 600 kHz

When a listing shows "+" or "-", these common offsets typically apply:

Band (MHz)	Offset
50	500 kHz
144	600 kHz
222	1.6 MHz
430	5 MHz
1200	12 MHz

Access   The access method used by the repeater
- A frequency (such as 100 Hz) indicates analog CTCSS access
- CC followed by a number (such as CC25) is a DMR Color Code

Coordinator   The frequency coordination group that coordinated the repeater

CTCSS and DCS information

The purpose of CTCSS (PL)™ is to reduce co-channel interference during band openings. CTCSS (PL)™ equipped repeaters respond only to signals having the sub-audible CTCSS tone required for that repeater. These repeaters do not retransmit distant signals without the required tone, and congestion is minimized.

The standard Electronic Industries Association (EIA) tones, in hertz, with their Motorola alphanumeric designators are as follows:

67.0	- XZ	97.4	- ZB	141.3	- 4A	210.7	- M2
69.3	- WZ	100.0	- 1Z	146.2	- 4B	218.1	- M3
71.9	- XA	103.5	- 1A	151.4	- 5Z	225.7	- M4
74.4	- WA	107.2	- 1B	156.7	- 5A	233.6	- M5
77.0	- XB	110.9	- 2Z	162.2	- 5B	241.8	- M6
79.7	- WB	114.8	- 2A	167.9	- 6Z	250.3	- M7
82.5	- YZ	118.8	- 2B	173.8	- 6A		
85.4	- YA	123.0	- 3Z	179.9	- 6B		
88.5	- YB	127.3	- 3A	186.2	- 7Z		
91.5	- ZZ	131.8	- 3B	192.8	- 7A		
94.8	- ZA	136.5	- 4Z	203.5	- M1		

Some systems use tones not listed in the EIA standard. Motorola designators have been assigned to the most commonly used of these tones: 206.5 (8Z), 229.1 (9Z), and 254.1 (0Z). Some newer amateur transceivers support additional tones of 159.8, 165.5, 171.3, 177.3, 183.5, 189.9, 196.6 and 199.5 hertz.

Some newer amateur gear supports Digital Code Squelch (DCS), a similar form of access control less susceptible to false triggering than CTCSS. DCS codes are designated by three digit numbers and are enabled in a manner similar to CTCSS tones.

Those wishing to use a CTCSS or DCS equipped system should check equipment specifications prior to purchase to ensure capability for the specified tone(s) or code(s).

ABOUT THE *ARRL REPEATER DIRECTORY*®

For decades, *The ARRL Repeater Directory*® has served radio amateurs with an annual "snapshot" of repeater listings. With the *Repeater Directory*, you are only fingertips away from finding repeaters and the users who operate and support the repeaters for their community. Some repeaters are networked with other repeaters, extending your ability to contact radio amateurs across the world. Take the *Repeater Directory* with you when you're traveling for vacation and business trips. Keep a copy in your automobile glovebox and emergency Go Kit. You'll never be out of touch!

In recent years, a proliferation of digital repeaters and related communities maintaining lists of active repeaters has introduced new challenges to producing the annual Repeater Directory. Repeater users increasingly turn to online services such as RFinder for more-regularly-updated sources of repeater listings, and where listings are contributed and maintained by frequency coordinators, digital network databases, repeater owners, and users.

HAVE YOU FOUND AN ERROR?
SUBMIT CORRECTIONS ON-LINE

Since 2017, the listings included in the *Repeater Directory* have been supplied by RFinder. RFinder offers an online subscription service to its worldwide repeater database (sold separately). ARRL does not curate the listings included in the *Repeater Directory*. We encourage frequency coordinators and repeater owners to review listings published in the annual *Repeater Directory*, and to submit corrections directly to RFinder.

• If you believe an analog repeater is missing from the *Directory*, please add it at **http://add.rfinder.net** (or inside the RFinder app on Android or iOS devices). You do not need to be a registered user to add analog machines.

• If you are a registered RFinder user (or have the trial Android version), you can submit corrections to current analog repeater listings. Just log in at **https://www.rfinder.net/websearch.html**, search for the repeater in question, and then click the pencil icon that appears to the left of the call sign. If you are using the RFinder app, just click a repeater and press Submit Update, or request delete at the top of the Repeater Detail screen.

• RFinder does not accept corrections or additions for digital repeaters. Instead, RFinder obtains its digital repeater listings from network databases that support DMR, Yaesu Fusion, D-Star, Phoenix, UKRepeaters, etc. That information is supplied to the databases directly by the repeater owners and any corrections must be made by those owners. RFinder updates its digital listings automatically every day starting at 0500 UTC and typically ending by 0900 UTC.

Subscribe to RFinder by installing the app on iOS or Android! Just search RFinder in Google Play or the App Store on Apple devices. You can renew your subscription at **http://subscribe.rfinder.net**.

NEED MORE HELP?

Go to **www.arrl.org/repeaters**

FREQUENCY COORDINATORS

This book includes a listing of groups or individuals for the United States and Canada who are active in frequency coordination. Frequency coordinators are volunteers. The FCC Rules of the Amateur Radio Service define the frequency coordinator as "An entity, recognized in a local or regional area by amateur operators whose stations are eligible to be auxiliary or repeater stations, that recommends transmit/receive channels and associated operating and technical parameters for such stations in order to avoid or minimize potential interference" §97.3(a)(22).

A frequency coordinator will recommend frequencies for a proposed repeater in order to minimize interference with other repeaters and simplex operations. Therefore, anyone considering the installation of a repeater should check with the local frequency coordinator prior to such installation. The FCC Rules include the following provision: "Where the transmissions of a repeater cause harmful interference to another repeater, the two station licensees are equally and fully responsible for resolving the interference unless the operation of one station is recommended by a frequency coordinator and the operation of the other station is not. In that case, the licensee of the non-coordinated repeater has primary responsibility to resolve the interference" § 97.205(c).

Frequency coordinators keep extensive records of repeater input, output and control frequencies, including those not published in directories (at the owner's request). The frequency listings in this book are supplied by RFinder, the creator of an online directory of amateur radio repeaters worldwide. We encourage frequency coordinators and repeater owners to review listings published in the annual *Repeater Directory*, and to submit corrections to RFinder.

ARRL is not a frequency coordinator, nor does the ARRL organize or "certify" coordinators. Publication in the Repeater Directory *does not constitute nor imply endorsement or recognition of the authority of such coordinators, as coordinators derive their authority from the voluntary participation of the entire amateur community in the areas they serve.*

ALABAMA
Alabama Repeater Council
www.alabamarepeatercouncil.org

ALASKA
www.alaskarepeaters.kl7.net

ARIZONA
Amateur Radio Council of Arizona
www.azfreqcoord.org

ARKANSAS
Arkansas Repeater Council
www.arkansasrepeatercouncil.org

CALIFORNIA—NORTHERN
Northern Amateur Relay Council of
California
www.narcc.org

CALIFORNIA—SOUTHERN
(10 meters, 6 meters, 70 centimeters and
above)
Southern California Repeater and Remote
Base Association
www.scrrba.org

CALIFORNIA—SOUTHERN
(2 meters only)
Two-Meter Area Spectrum Management
Association
www.tasma.org

CALIFORNIA—SOUTHERN
(222 MHz only)
220 MHz Spectrum Management
Association
www.220sma.org

COLORADO
Colorado Council of Amateur Radio Clubs
www.ccarc.net/wordpress

CONNECTICUT
Connecticut Spectrum Management
www.ctspectrum.com

DELAWARE
The Mid Atlantic Repeater Council
www.tmarc.org

DISTRICT OF COLUMBIA
The Mid Atlantic Repeater Council
www.tmarc.org

FLORIDA
Florida Amateur Spectrum Management Association
www.fasma.org

GEORGIA
Southeastern Repeater Association
www.sera.org

HAWAII
Hawaii State Repeater Advisory Council
www.hawaiirepeaters.net

IDAHO—SOUTHEAST
Bill Wheeler, W7RUG
w7rug@arrl.net

IDAHO—SOUTHWEST
Larry Smith, W7ZRQ
smith_larry@hotmail.com

IDAHO—PANHANDLE
Ken Rau, K7YR
kenr@nwi.net

ILLINOIS
Illinois Repeater Association
www.ilra.net

INDIANA
Indiana Repeater Council
www.ircinc.org

IOWA
Iowa Repeater Council
www.iowarepeater.org

KANSAS
Kansas Amateur Repeater Council
www.ksrepeater.com

KENTUCKY
Southeastern Repeater Association
www.sera.org

LOUISIANA
Roger Farbe, N5NXL
n5nxl@bellsouth.net

MAINE
New England Spectrum Management Council
www.nesmc.org

MARYLAND
The Mid Atlantic Repeater Council
www.tmarc.org

MASSACHUSETTS
New England Spectrum Management Council
www.nesmc.org

MICHIGAN—LOWER PENNINSULA
Michigan Area Repeater Council
www.miarc.com

MICHIGAN—UPPER PENINSULA
Upper Peninsula Amateur Radio Repeater Association
www.uparra.org

MINNESOTA
Minnesota Repeater Council
www.mrc.gen.mn.us

MISSISSIPPI
Southeastern Repeater Association
www.sera.org

MISSOURI
Missouri Repeater Council
www.missourirepeater.org

MONTANA
Don Heide, W7MRI
w7mri@arrl.net

NEBRASKA
John Gebuhr, WB0CMC
wb0cmc@arrl.net

NEVADA—SOUTHERN
Southern Nevada Repeater Council
www.snrc.us

NEVADA—NORTHERN
Combined Amateur Relay Council of Nevada
www.carcon.org

NEW HAMPSHIRE
New England Spectrum Management Council
www.nesmc.org

NEW JERSEY
(All counties except Bergen, Essex, Hudson, Middlesex, Monmouth, Morris, Passaic, Somerset, and Union.)

Area Repeater Coordination Council
www.arcc-inc.org

NEW JERSEY
(Only Bergen, Essex, Hudson, Middlesex, Monmouth, Morris, Passaic, Somerset, and Union counties.)

Metropolitan Coordination Association
www.metrocor.net

NEW MEXICO
New Mexico Frequency Coordination Committee
www.qsl.net/nmfcc

NEW YORK—NORTHEAST
Saint Lawrence Valley Repeater Council
www.slvrc.org

NEW YORK—EASTERN AND CENTRAL UPSTATE
Upper New York Repeater Council
www.unyrepco.org

NEW YORK—NYC AND LONG ISLAND
Metropolitan Coordination Association
www.metrocor.net

NORTH CAROLINA
Southeastern Repeater Association
www.sera.org

NORTH DAKOTA
Joseph Ferrasa, N7IV
ferrara@srt.com

OHIO
Ohio Repeater Council
www.oarc.com

OKLAHOMA
Oklahoma Repeater Society
www.oklahomarepeatersociety.org

OREGON
Oregon Region Relay Council
www.orrc.org

PENNSYLVANIA—EASTERN
Area Repeater Coordination Council
www.arcc-inc.org

PENNSYLVANIA—WESTERN
Western Pennsylvania Repeater Council
www.wprcinfo.org

PUERTO RICO
Victor Madera, KP4PQ
wirci.pr@gmail.com

RHODE ISLAND
New England Spectrum Management Council
www.nesmc.org

SOUTH CAROLINA
Southeastern Repeater Association
www.sera.org

SOUTH DAKOTA
Richard Neish, W0SIR
neish@itctel.com

TENNESSEE
Southeastern Repeater Association
www.sera.org

TEXAS
Texas VHF/FM Society
www.txvhffm.org

UTAH
Utah VHF Society
utahvhfs.org/frqcoord.html

VERMONT
Vermont Independent Repeater Coordination Committee
www.ranv.org

VIRGINIA—SOUTH OF THE 38th PARALLEL AND US HIGHWAY 33
Southeastern Repeater Association
www.sera.org

VIRGINIA—NORTH OF THE 38th PARALLEL AND US HIGHWAY 33
The Mid Atlantic Repeater Council
www.tmarc.org

WASHINGTON—EASTERN
Ken Rau, K7YR
krau@nwi.net

WASHINGTON—WESTERN
Western Washington Amateur Relay Association
www.wwara.org

WEST VIRGINIA—EASTERN PANHANDLE
The Mid Atlantic Repeater Council
www.tmarc.org

WEST VIRGINA—ALL OTHER AREAS
Southeastern Repeater Association
www.sera.org

WISCONSIN
Wisconsin Association of Repeaters
www.wi-repeaters.org

WYOMING
Wyoming Repeater Coordinator Group
www.wyoham.com

CANADA

ALBERTA
Ken Oelke, VE6AFO
ve6afo@arrl.net

BRITISH COLUMBIA
British Columbia Amateur Radio Coordination Council
www.bcarcc.org

MANITOBA
Manitoba Amateur Repeater Coordination Council
www.winnipegarc.org/marcc/

MARITIME PROVINCES
Ron MacKay, VE1AIC
www.ve1cra.net

NEWFOUNDLAND AND LABRADOR
Ken Whalen, VO1ST
ken.vo1st@gmail.com

ONTARIO—EASTERN AND NORTHERN
Saint Lawrence Valley Repeater Council
www.slvrc.org

ONTARIO—SOUTHWEST
Western New York and Southern Ontario Repeater Council
www.wynsorc.net

QUEBEC
Radio Amateurs du Quebec
http://ccfq.ca/

SASKATCHEWAN
Saskatchewan Amateur Radio League
www.sarl.ca

BAND PLANS

Although the FCC rules set aside portions of some bands for specific modes, there's still a need to further organize our space among user groups by "gentlemen's agreements." These agreements, or band plans, usually emerge by consensus of the band occupants, and are sanctioned by a national body like ARRL. For further information on band planning, please contact your ARRL Division Director (see any issue of *QST*).

VHF-UHF BAND PLANS

When considering frequencies for use in conjunction with a proposed repeater, be certain that both the input and output fall within subbands authorized for repeater use, and do not extend past the subband edges. FCC regulation 97.205(b) defines frequencies which are currently available for repeater use.

For example, a 2-meter repeater on exactly 145.50 MHz would be "out-of-band," as the deviation will put the signal outside of the authorized band segment.

Packet-radio operations under automatic control should be guided by Section 97.109(d) of the FCC Rules.

REGIONAL FREQUENCY COORDINATION

The ARRL encourages regional frequency coordination efforts by amateur groups. Band plans published in the ARRL Repeater Directory are recommendations based on a consensus as to good amateur operating practice on a nationwide basis. In some cases, however, local conditions may dictate a variation from the national band plan. In these cases, the written determination of the regional frequency coordinating body shall prevail and be considered good amateur operating practice in that region.

28.000-29.700 MHZ

Please note that this bandplan is a general recommendation. Spectrum usage can be different depending upon local and regional coordination differences. Please check with your Frequency Coordinator for information.

28.000 – 28.070	CW
28.070 – 28.150	Data/CW
28.120 – 28.189	Packet/Data/CW
28.190 – 28.225	Foreign CW Beacons
28.200 – 28.300	Domestic CW Beacons (*)
28.300 – 29.300	Phone
28.680	SSTV
29.300 – 29.510	Satellites
29.510 – 29.590	Repeater Inputs
29.600	National FM Simplex Frequency
29.610 – 29.690	Repeater Outputs

*User note: In the United States, automatically controlled beacons may only operate on 28.2-28.3 MHz [97.203(d)].

In 1980, the ARRL Board of Directors adopted the following recommendations for CTCSS tones to be voluntarily incorporated by 10 meter repeaters:

Call Area	Tones	Call Area	Tones
W1	131.8/91.5	W7	162.2/110.9
W2	136.5/94.8	W8	167.9/114.8
W3	141.3/97.4	W9	173.8/118.8
W4	146.2/100.0	W0	179.9/123.0
W5	151.4/103.5	VE	127.3/88.5
W6	156.7/107.2	KP4	183.5/85.4
		KV4	186.2/82.5

The following band plan for 6 meters was adopted by the ARRL Board of Directors at its July, 1991 meeting.

50-54 MHZ

Please note that this bandplan is a general recommendation. Spectrum usage can be different depending on location and regional coordination differences. Please check with your Frequency Coordinator for information.

50.0-50.1	CW, beacons
50.060-50.080	beacon subband
50.1-50.3	SSB,CW
50.10-50.125	DX window
50.125	SSB calling
50.3-50.6	all modes
50.4	AM calling frequency
50.6-50.8	nonvoice communications
50.62	digital (packet) calling
50.8-51.0	radio remote control (20-kHz channels)

NOTE: Activities above 51.10 MHz are set on 20-kHz-spaced "even channels"

51.0-51.1	Pacific DX window
51.5-51.6	simplex (6 channels)
51.12-51.48	repeater inputs (19 channels)
51.12-51.18	digital repeater inputs
51.62-51.98	repeater outputs (19 channels)
51.62-51.68	digital repeater outputs
52.0-52.48	repeater inputs (except as noted; 23 channels)
52.02, 52.04	FM simplex
52.2	TEST PAIR (input)
52.5-52.98	repeater output (except as noted; 23 channels)
52.525	primary FM simplex
52.54	secondary FM simplex
52.7	TEST PAIR (output)
53.0-53.48	repeater inputs (except as noted; 9 channels)
53.0	base FM simplex
53.02	simplex
53.1, 53.2	radio remote control
53.3, 53.4	
53.5-53.98	repeater outputs (except as noted; 19 channels)
53.5, 53.6	radio remote control
53.7, 53.8	
53.52-53.9	simplex

Notes: The following packet radio frequency recommendations were adopted by the ARRL Board of Directors in July, 1987.

Duplex pairs to consider for local coordination for uses such as repeaters and meteor scatter:

50.62-51.62	50.68-51.68	50.76-51.76
50.64-51.64	50.72-51.72	50.78-51.78
50.66-51.66	50.74-51.74	

Where duplex packet radio stations are to be co-existed with voice repeaters, use high-in, low-out to provide maximum frequency separation from low-in, high-out voice repeaters.

144-148 MHZ

Please note that this bandplan is a general recommendation. Spectrum usage can be different depending on location and regional coordination differences. Please check with your Frequency Coordinator for information.

144.00-144.05	EME (CW)
144.05-144.10	General CW and weak signals
144.10-144.20	EME and weak-signal SSB
144.200	SSB calling frequency
144.20-144.275	General SSB operation
144.275-144.300	Propagation beacons
144.30-144.50	OSCAR subband
144.50-144.60	Linear translator inputs
144.60-144.90	FM repeater inputs
144.90-145.10	Weak signal and FM simplex (145.01,03,05,07,09 are widely used for packet radio)
145.10-145.20	Linear translator outputs
145.20-145.50	FM repeater outputs
145.50-145.80	Miscellaneous and experimental modes
145.80-146.00	OSCAR subband
146.01-146.37	Repeater inputs
146.40-146.58	Simplex (*)
146.52	National Simplex Calling Frequency
146.61-147.39	Repeater outputs
147.42-147.57	Simplex (*)
147.60-147.99	Repeater inputs

NOTES: (*) Due to differences in regional coordination plans the simplex frequencies listed may be repeater inputs/outputs as well. Please check with local coordinators for further information.

1) Automatic/unattended operations should be conducted on 145.01, 145.03, 145.05, 145.07 and 145.09 MHz.

a) 145.01 should be reserved for inter-LAN use.

b) Use of the remaining frequencies should be determined by local user groups.

2) Additional frequencies within the 2-meter band may be designated for packet radio use by local coordinators.

Footnotes

Specific VHF/UHF channels recommended above may not be available in all areas of the US.

Prior to regular packet radio use of any VHF/UHF channel, it is advisable to check with the local frequency coordinator. The decision as to how the available channels are to be used should be based on coordination between local packet radio users.

Some areas use 146.40-146.60 and 147.40-147.60 MHz for either simplex or repeater inputs and outputs.

States use differing channel spacings on the 146-148 MHz band. For further information on which states are currently utilizing which spacing structure see the Offset Map immediately following.

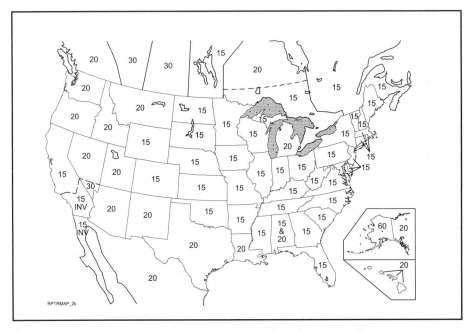

Note: This map shows channel spacing in the US and southern Canada. Spacing is in kHz unless otherwise specified. Please check with your Regional Frequency Coordinator for further information.

The following band plan for 222-225 MHz was adopted by the ARRL Board of Directors in July, 1991.

222-225 MHZ

222.00-222.15	Weak signal modes (No repeater operating)
222.00-222.025	EME
222.05-222.060	Propagation beacons
222.1	SSB & CW Calling
222.10-222.150	Weak signal CW & SSB
222.15-222.25	Local coordinator's option: weak signal, ACSB, repeater inputs and control
222.25-223.38	FM repeater inputs only
223.40-223.52	FM simplex
223.50	Simplex calling
223.52-223.64	Digital, packet
223.64-223.70	Links, control
223.71-223.85	Local coordinator's option; FM simplex, packet, repeater outputs
223.85-224.98	Repeater outputs only

Notes: Candidate packet simplex channels shared with FM voice simplex.
Check with your local fre-quency coordinator prior to use. Those channels are:

223.42	223.46
223.44	223.48

Footnotes

Specific VHF/UHF channels recommended above may not be available in all areas of the US.

Prior to regular packet radio use of any VHF/UHF channel, it is advisable to check with the local frequency coordinator. The decision as to how the available channels are to be used should be based on coordination between local packet radio users.

420-450 MHZ

Please note that this bandplan is a general recommendation. Spectrum usage can be different depending on location and regional coordination differences. Please check with your Frequency Coordinator for information.

420.00-426.00	ATV repeater or simplex with 421.25 MHz video carrier, control links and experimental
426.00-432.00	ATV simplex with 427.25 MHz video carrier frequency
432.00-432.07	EME (Earth-Moon-Earth)
432.07-432.10	Weak signal CW
432.10	Calling frequency
432.10-432.30	Mixed-mode and weak-signal work
432.30-432.40	Propagation beacons
432.40-433.00	Mixed-mode and weak signal work
433.00-435.00	Auxiliary/repeater links
435.00-438.00	Satellite only (internationally)
438.00-444.00	ATV repeater input with 439.250-MHz video carrier frequency and repeater links
442.00-445.00	Repeater inputs and outputs (local option)
445.00-447.00	Shared by auxiliary and control links, repeaters and simplex (local option)
446.00	National simplex frequency
447.00-450.00	Repeater inputs and outputs (local option)

The following packet radio frequency recommendations were adopted by the ARRL Board of Directors in January, 1988.

1) 100-kHz bandwidth channels

430.05	430.35	430.65
430.15	430.45	430.85
430.25	430.55	430.95

2)   25-kHz bandwidth channels

431.025	441.000	441.050
440.975	441.025	441.075

Footnotes

Specific VHF/UHF channels recommended above may not be available in all areas of the US.

Prior to regular packet radio use of any VHF/UHF channel, it is advisable to check with the local frequency coordinator. The decision as to how the available channels are to be used should be based on coordination between local packet radio users.

902-928 MHZ

Please note that this bandplan is a general recommendation. Spectrum usage can be different depending on location and regional coordination differences. Please check with your Frequency Coordinator for information.

The following band plan was adopted by the ARRL Board of Directors in January, 1991.

Frequency Range	Mode	Functional Use	Comments
902.000-902.075	FM / other including DV Or CW/SSB	Repeater inputs 25 MHz split paired with those in 927.000-927.075 or Weak signal	12.5 kHz channel spacing Note 2)
902.075-902.100	CW/SSB	Weak signal	
902.100	CW/SSB	Weak signal calling	Regional option
902.100-902.125	CW/SSB	Weak signal	
902.125-903.000	FM/other including DV	Repeater inputs 25 MHz split paired with those in 927.1250-928.0000	12.5 kHz channel spacing
903.000-903.100	CW/SSB	Beacons and weak signal	
903.100	CW/SSB	Weak signal calling	Regional option
903.100-903.400	CW/SSB	Weak signal	
903.400-909.000	Mixed modes	Mixed operations including control links	
909.000-915.000	Analog/digital	Broadband multimedia including ATV, DATV and SS	Notes 3) 4)
915.000-921.000	Analog/digital	Broadband multimedia including ATV, DATV and SS	Notes 3) 4)
921.000-927.000	Analog/digital	Broadband multimedia including ATV, DATV and SS	Notes 3) 4)
927.000-927.075	FM / other including DV	Repeater outputs 25 MHz split paired with those in 902.0000-902.0750	12.5 kHz channel spacing
927.075-927.125	FM / other including DV	Simplex	
927.125-928.000	FM / other including DV	Repeater outputs 25 MHz split paired with those in 902.125-903.000	12.5 kHz channel spacing Notes 5) 6)

Notes:
1) Significant regional variations in both current band utilization and the intensity and frequency distribution of noise sources preclude one plan that is suitable for all parts of the country. These variations will require many regional frequency coordinators to maintain band plans that differ in some respects from any national plan. As with all band plans, locally coordinated plans always take precedence over any general recommendations such as a national band plan.
2) May be used for either repeater inputs or weak-signal as regional needs dictate
3) Division into channels and/or separation of uses within these segments may be done regionally based on needs and usage, such as for 2 MHz-wide digital TV.
4) These segments may also be designated regionally to accommodate alternative repeater splits.
5) Simplex FM calling frequency 927.500 or regionally selected alternative.
6) Additional FM simplex frequencies may bedesignated regionally.

1240-1300 MHZ

Please note that this bandplan is a general recommendation. Spectrum usage can be different depending on location and regional coordination differences. Please check with your Frequency Coordinator for information.

Frequency Range	Suggested Emission Types	Functional Use
1240.000-1246.000	ATV	ATV Channel #1
1246.000-1248.000	FM, digital	Point-to-point links paired with 1258.000-1260.000
1248.000-1252.000	Digital	
1252.000-1258.000	ATV	ATV Channel #2
1258.000-1260.000	FM, digital	Point-to-point links paired with 1246.000-1248.000
1240.000-1260.000	FM ATV	Regional option
1260.000-1270.000	Various	Satellite uplinks, Experimental, Simplex ATV
1270.000-1276.000	FM, digital	Repeater inputs, 25 kHz channel spacing, paired with 1282.000-1288.000
1270.000-1274.000	FM, digital	Repeater inputs, 25 kHz channel spacing, paired with 1290.000-1294.000 (Regional option)
1276.000-1282.000	ATV	ATV Channel #3
1282.000-1288.000	FM, digital	Repeater outputs, 25 kHz channel spacing, paired with 1270.000-1276.000
1288.000-1294.000	Various	Broadband Experimental, Simplex ATV
1290.000-1294.000	FM, digital	Repeater outputs, 25 kHz channel spacing, paired with 1270.000-1274.000 (Regional option)
1294.000-1295.000	FM	FM simplex
	FM	National FM simplex calling frequency 1294.500
1295.000-1297.000		Narrow Band Segment
1295.000-1295.800	Various	Narrow Band Image, Experimental
1295.800-1296.080	CW, SSB, digital	EME
1296.080-1296.200	CW, SSB	Weak Signal
	CW, SSB	CW, SSB calling frequency 1296.100
1296.200-1296.400	CW, digital	Beacons
1296.400-1297.000	Various	General Narrow Band
1297.000-1300.000	Digital	

Note: The need to avoid harmful interference to FAA radars may limit amateur use of certain frequencies in the vicinity of the radars.

2300-2310 AND 2390-2450 MHZ

Please note that this bandplan is a general recommendation. Spectrum usage can be different depending on location and regional coordination differences. Please check with your Frequency Coordinator for information.

Frequency Range	Emission Bandwidth	Functional Use
2300.000-2303.000	0.05 - 1.0 MHz	Analog & Digital, including full duplex; paired with 2390 - 2393
2303.000-2303.750	< 50 kHz	Analog & Digital; paired with 2393 - 2393.750
2303.75-2304.000		SSB, CW, digital weak-signal
2304.000-2304.100	3 kHz or less	Weak Signal EME Band
2304.10-2304.300	3 kHz or less	SSB, CW, digital weak-signal (Note 1)
2304.300-2304.400	3 kHz or less	Beacons
2304.400-2304.750	6 kHz or less	SSB, CW, digital weak-signal & NBFM
2304.750-2305.000	< 50 kHz	Analog & Digital; paired with 2394.750 - 2395
2305.000-2310.000	0.05 - 1.0 MHz	Analog & Digital, paired with 2395 - 2400 (Note 2)
2310.000-2390.000	**NON-AMATEUR**	
2390.000-2393.000	0.05 - 1.0 MHz	Analog & Digital, including full duplex; paired with 2300- 2303
2393.000-2393.750	< 50 kHz	Analog & Digital; paired with 2303 - 2303.750
2393.750-2394.750		Experimental
2394.750-2395.000	< 50 kHz	Analog & Digital; paired with 2304.750 - 2305
2395.000-2400.000	0.05 - 1.0 MHz	Analog & Digital, including full duplex; paired with 2305- 2310
2400.000-2410.000	6 kHz or less	Amateur Satellite Communications
2410.000-2450.000	22 MHz max.	Broadband Modes (Notes 3, 4)

Notes:
1: 2304.100 is the National Weak-Signal Calling Frequency
2: 2305 - 2310 is allocated on a primary basis to Wireless Communications Services (Part 27). Amateur operations in this segment, which are secondary, may not be possible in all areas.
3: Broadband segment may be used for any combination of high-speed data (e.g. 802.11 protocols), Amateur Television and other high-bandwidth activities. Division into channels and/or separation of uses within this segment may be done regionally based on needs and usage.
4: 2424.100 is the Japanese EME transmit frequency

3300 – 3500 MHZ

Frequency (MHz)	Emission	Emission Bandwidth (Note1)	Functional Use
3300.000-3309.000	Analog, Digital	0.1 – 1.0 MHz	Paired with 3430.0-3439.0; 130 MHz Split
3309.000-3310.000			Experimental
3310.000-3330.000	Analog, Digital	>1.0 MHz	Paired with 3410.0-3430.0; 100 MHz Split
3330.000-3332.000			Experimental
3332.000-3339.000	**Radio Astronomy – Protected (Note 4)**		
3339.000-3345.800	Analog, Digital	0.1 – 1.0 MHz	Paired with 3439.0-3445.8; 100 MHz Split
3345.800-3352.500	**Radio Astronomy – Protected (Note 4)**		
3352.500-3355.000	Analog, Digital	0.05 - 0.2 MHz	Paired with 3452.5-3455.0; 100 MHz Split
3355.000-3357.000			Experimental
3357.000-3360.000	Analog, Digital	50 kHz or less	Paired with 3457.0-3460.0
3360.000-3400.000	OFDM, others	22 MHz max.	Broadband Modes (Note 3)
3360.000-3380.000	ATV	20 MHz	Television
3400.000-3410.000		CW, SSB, NBFM	6 kHz or less Amateur Satellite Communications
3400.000-3400.300	CW, SSB, Digital	3 kHz or less	Weak Signal EME
3400.300-3401.000	CW, SSB, Digital	3 kHz or less	Terrestrial Weak Signal Band - Future (Note 2)
3400.100	CW, SSB, Digital		EME Calling Frequency
3410.000-3430.000	Analog, Digital	>1.0 MHz	Paired with 3310.0-3330.0; 100 MHz Split
3430.000-3439.000	Analog, Digital	0.1 - 1.0 MHz	Paired with 3300.0-3309.0; 130 MHz Split
3439.000-3445.800	Analog, Digital	0.1 - 1.0 MHz	Paired with 3339.0-3345.8; 100 MHz Split
3445.800-3452.500			Experimental
3452.500-3455.000	Analog, Digital	0.05 - 0.2 MHz	Paired with 3352.5-3355.0; 100 MHz Split
3455.000-3455.500		100 kHz or less	Crossband linear translator (input or output)
3455.500-3457.000		CW, SSB, FM, Dig	6 kHz or less Terrestrial Weak Signal Band - Legacy (Note 2)
3456.100		6 kHz or less	Weak Signal Terrestrial Calling Frequency

Frequency (MHz)	Emission	Emission Bandwidth (Note1)	Functional Use
3456.300-3457.000		CW, Digital	1 kHz or less Propagation Beacons
3457.000-3460.000	Analog, Digital	50 kHz or less	Paired with 3357.0-3360.0; 100 MHz Split
3460.000-3500.000		OFDM, others	22 MHz max. Broadband Modes (Note 3)
3460.000-3480.000	20.0	ATV	Amateur Television

Note 1 – Includes all other emission modes authorized in the 9 cm amateur band whose necessary bandwidth does not exceed the suggested bandwidths listed.

Note 2 – Weak Signal Terrestrial legacy users are encouraged to move to 3400.3 to 3401.0 MHz as time and resources permit.

Note 3 – Broadband segments may be used for any combination of high-speed data (e.g. 802.11 protocols), Amateur Television and other high-bandwidth activities. Division into channels and/or separation of uses within these segments may be done regionally based on need and usage.

Note 4 – Per ITU RR 5.149 from WRC-07, these band segments are also used for Radio Astronomy. Amateur use of these frequencies should be first coordinated with the National Science Foundation (**esm@nsf.gov**).

5650 – 5925 MHZ

Frequency Range (MHz)	Emission Bandwidth	Functional Use
5650.0-5670.0		Amateur Satellite; Up-Link Only
5650.0-5675.0	0.05 - 1.0 MHz	Experimental
5675.0-5750.0	>= 1.0 MHz	Analog & Digital; paired with 5850-5925 MHz (Note 2)
5750.0-5756.0	>= 25 kHz and <1 MHz	Analog & Digital; paired with 5820-5826 MHz
5756.0-5759.0	<= 50 kHz	Analog & Digital; paired with 5826-5829 MHz
5759.0-5760.0	< 6 kHz	SSB, CW, Digital Weak-Signal
5760.0-5760.1	< 3kHz	EME
5760.1-5760.3	< 6 KHz	SSB, CW, Digital Weak-Signal (Note 1)
5760.3-5760.4	< 3 KHz	Beacons
5760.4-5761.0	< 6 KHz	SSB, CW, Digital Weak-Signal
5761.0-5775.0	<=50 kHz	Experimental
5775.0-5800.0	>=100 kHz	Experimental
5800.0-5820.0		Experimental
5820.0-5826.0	>=25 kHz and <1 MHz	Analog & Digital; paired with 5750-5756 MHz
5826.0-5829.0	<=50 kHz	Analog & Digital; paired with 5756-5759 MHz

Frequency Range (MHz)	Emission Bandwidth	Functional Use
5829.0-5850.0	0.05-1.0 MHz	Experimental
5830.0-5850.0		Amateur Satellite; Down-Link Only
5850.0-5925.0	>=1.0 MHz	Analog & Digital; paired with 5675-5750 MHz (Note 2)

Note 1: 5760.1 is the National Weak-Signal Calling Frequency.

Note 2: Broadband segment may be used for any combination of high-speed data (eg: 802.11 protocols), Amateur Television and other high-bandwidth activities. Division into channels and/or separation of uses within this segment may be done regionally based on needs and usage.

10.000 – 10.500 GHZ

Frequency Range (MHz)	Emission Bandwidth	Functional Use
10000.00-10050.000		Experimental
10050.000-10100.000	<=100 kHz	Analog & Digital; paired with 10300-10350
10100.000-10115.000	>=25 kHz and <1 MHz	Analog & Digital; paired with 10350-10365
10115.000-10117.000	<=50 kHz	Analog & Digital; paired with 10365-10367
10117.000-10120.000		Experimental
10120.000-10125.000	<=50 kHz	Analog & Digital; paired with 10370-10375
10125.000-10200.000	>=1 MHz	Analog & Digital; paired with 10375-10450 (Note 2)
10200.000-10300.000		Wideband Gunnplexers
10300.000-10350.000	<=100 kHz	Analog & Digital; paired with 10050-10100
10350.000-10365.000	>=25 kHz and <1 MHz	Analog & Digital; paired with 10100-10115
10365.000-10367.000	<=50 kHz	Analog & Digital; paired with 10115-10117
10367.000-10368.300	6 kHz or less	SSB, CW, Digital Weak-Signal & NBFM (Note 1)
10368.300-10368.400	6 kHz or less	Beacons
10368.400-10370.000	6 kHz or less	SSB, CW, Digital Weak-Signal & NBFM
10370.000-10375.000	<=50 kHz	Analog & Digital; paired with 10120-10125
10375.000-10450.000	>=1 MHz	Analog & Digital; paired with 10125-10200 (Note 2)
10450.000-10500.000		Space, Earth & Telecommand Stations

Note 1: 10368.100 is the National Weak-Signal Calling Frequency

Note 2: Broadband segment may be used for any combination of high-speed data (eg: 802.11 protocols), Amateur Television and other high-bandwidth activities. Division into channels and/or separation of uses within this segment may be done regionally based on needs and usage.

REPEATER LINGO / HINTS

This section covers a basic course in "repeater-speak" and explains many of the terms heard on your local repeater.

Definitions of the words and phrases commonly used on repeaters:

Autopatch – A device that interfaces the repeater system with the telephone system to extend ham communications over the telephone communications network.

Breaker – A ham who interjects his call sign during a QSO in an attempt to get a chance to communicate over a repeater.

Channel – The pair of frequencies (input and output) a repeater operates on.

Closed Repeater – A repeater whose use is limited to certain individuals. These are completely legal under FCC rules.

Control Operator – An individual ham designated to "control" the repeater, as required by FCC regulations.

COR – Carrier-Operated-Relay, a device that, upon sensing a received signal, turns on the repeater's transmitter to repeat the received signal.

Courtesy Tone – A short tone sounded after each repeater transmission to permit other stations to gain access to the repeater before the tone sounds.

Coverage – The geographical area in which the repeater may be used for communications.

CTCSS – Continuous Tone Coded Squelch System, a sub-audible tone system which operates the squelch (COR) of a repeater when the corresponding sub-audible tone is present on a transmitted signal. The squelch on a repeater which uses CTCSS will not activate if the improper CTCSS tone, or if no tone, is transmitted.

Crossband – Communications to another frequency band by means of a link interfaced with the repeater.

Desense – Degradation of receiver sensitivity caused by strong unwanted signals reaching the receiver front end.

Duplexer – A device that permits the use of one antenna for both transmitting and receiving with minimal degradation to either the incoming or outgoing signals.

Frequency Synthesis – A scheme of frequency generation in modern transceivers using digital techniques.

Full Quieting – Signal strength in excess of amount required to mask ambient noise.

Hand-Held – A portable FM transceiver that is small enough to use and carry in one hand.

Input – The frequency the repeater receiver is tuned to: The frequency that a repeater user transmits on.

Intermod – Interference caused by spurious signals generated by intermodulation distortion in a receiver front end or transmitter power amplifier stage.

Key-Up – Turning on a repeater by transmitting on its input frequency.

LiTZ – Long Tone Zero (LiTZ) Alerting system. Send DTMF zero (0) for at least three seconds to request emergency/urgent assistance.

Machine – The complete repeater system.

Mag-Mount – A mobile antenna with a magnetic base that permits quick installation and removal from the motor vehicle.

Offset – The spacing between a repeater's input and output.

Omnidirectional – An antenna system that radiates equally in all directions.

Output – The frequency the repeater transmits on; the frequency that a repeater user receives on.

Picket-Fencing – Rapid flutter on a mobile signal as it travels past an obstruction.

Polarization – The plane an antenna system operates in; most repeaters are vertically polarized.

Reverse Autopatch – A device that interfaces the repeater with the telephone system and permits users of the phone system to call the repeater and converse with on-the-air repeater users.

Reverse Split – A split-channel repeater operating in the opposite direction of the standard.

RPT/R – Abbreviation used after repeater call signs to indicate that the call sign is being used for repeater operation.

Simplex – Communication on one frequency, not via a repeater

Splinter Frequency – 2-meter repeater channel 15 kHz above or below the formerly standard 30 kHz-spaced channel.

Split Sites – The use of two locations for repeater operation (the receiver is at one site and the transmitter at another), and the two are linked by telephone or radio.

Squelch Tail – The noise burst that follows the short, unmodulated carrier following each repeater transmission.

Time-Out-Timer – A device that limits the length of a single repeater transmission (usually 3 minutes).

Tone Pad – A device that generates the standard telephone system tones used for controlling various repeater functions.

About Emergencies

Regardless of the band, mode, or your license class, FCC rules specify that, in case of emergency, the normal rules can be suspended. If you hear an emergency call for help, you should do whatever you can to establish contact with the station needing assistance, and immediately pass the information to the proper authorities. If you are talking with another station and you hear an emergency call for help, stop immediately and take the emergency call.

Location	Mode	Call sign	Output	Input	Access	Coordinator

ALABAMA   39

ALABAMA

Location	Mode	Call sign	Output	Input	Access	Coordinator
Alabaster	DMR/BM	N4FIV	444.75000	+	CC1	
	FM	N4PHP	444.55000	+	100.0 Hz	
Andalusia	FM	WC4M	147.26000	+	100.0 Hz	
Anniston	FM	KF4RGR	444.05000	+	131.8 Hz	
Anniston, Cheaha Mountain						
	DSTAR	WB4GNA	442.42500	+		
	DSTAR	WB4GNA	1285.00000	1265.00000		
	FM	WB4GNA	147.09000	+	131.8 Hz	
	FM	WB4GNA	444.75000	+	131.8 Hz	
Anniston, Oak Mountain						
	DSTAR	KJ4JGK	443.35000	+		
Arab	FM	AK4OV	443.22500	+	77.0 Hz	
Ashcraft Corner/Gordo						
	FM	KC4UG	444.40000	+	110.9 Hz	
Ashland	FM	KI4PSG	147.25500	+	131.8 Hz	
Athens	FM	KD4NTK	442.85000	+		
Auburn	DMR/BM	N4NQV	444.77500	+	CC1	
	FM	KA4Y	147.06000	+	123.0 Hz	
	FM	W4HOD	147.30000	+	123.0 Hz	
	FM	W4LEE	444.12500	+		
Auburn University, Haley Cente						
	FM	K4RY	444.80000	+	156.7 Hz	
Auburn, Auburn University						
	FM	K4RY	147.24000	+	156.7 Hz	
Battle Ground	FM	W4CFI	147.41500	146.41500	123.0 Hz	
Bay Minette, Raburn Tower Site						
	FM	WB4EMA	147.04500	+	123.0 Hz	
Bessemer, Tower Next To Water						
	FM	KA4KUN	444.62000	+	100.0 Hz	
Birmingham	DMR/BM	KA5GET	444.65000	+	CC1	
	DMR/MARC	W4RKZ	443.67500	+	CC1	
	DSTAR	KO4TM	442.07500	+		
	DSTAR	K4DSO	443.20000	+		
	DSTAR	KI4SBB	443.97500	+		
	DSTAR	KI4SBB	1282.50000	1262.50000		
	DSTAR	K4DSO	1283.40000	1263.40000		
	DSTAR	KI4SBB	1285.50000	1270.50000		
	FM	W4TPA	147.28000	+	100.0 Hz	
	FM	N4VSU	443.12500	+	146.2 Hz	
	FM	KK4BSK	443.70000	+	131.8 Hz	
	FM	KE4ADV	444.42500	+	131.8 Hz	
	FM	W4TPA	444.87500	+	85.4 Hz	
Birmingham, East Lake						
	FM	KK4BSK	443.17500	+		
Birmingham, Ruffner Mountain						
	FM	AG4ZV	444.82500	+	131.8 Hz	
Birmingham, UAB Campus						
	FM	KK4BSK	443.45000	+	131.8 Hz	
Boaz	FM	KN4UPN	444.20000	+	114.8 Hz	
Brewton	FM	KI4GGH	444.65000	+		
Brilliant, Pea Ridge						
	FM	KT4JW	147.04000	+	192.8 Hz	
Camden	FM	N5GEB	147.13000	+	123.0 Hz	
Cedar Bluff	FM	WA4OHM	444.40000	+	100.0 Hz	
Citronelle	FM	W4FRG	147.22500	+	203.5 Hz	
Clanton	DSTAR	W4AEC	444.37500	+		
	DSTAR	KF4LQK	1285.50000	1265.50000		
	FM	WB4UQT	147.10500	+	123.0 Hz	
	FM	KF4LQK	443.50000	+		

40   ALABAMA

Location	Mode	Call sign	Output	Input	Access	Coordinator
Columbiana	DMR/MARC	WA4CYA	444.70000	+	CC1	
Columbiana, Columbiana Mountai						
	FM	KC4EUA	444.60000	+	156.7 Hz	
Cullman	DMR/BM	N4UAI	444.90000	+	CC1	
Curry	FM	KI4GEA	441.80000	+		
Dadeville	DMR/BM	WA4TAL	145.27000	-	CC1	
	DMR/BM	WA4TAL	444.52500	+	CC1	
	FM	KB4MDD	224.24000	-	146.2 Hz	
Decatur	FM	W9KOP	442.35000	+		
	FM	W9KOP	442.67500	+		
	FM	W4ATD	443.85000	+		
Delta, Cheaha Mountain						
	FM	WX4ZAC	224.38000	-		
	FM	KF4RGR	443.67500	+	203.5 Hz	
Dixons Mills	FM	W4WTG	147.08000	+	210.7 Hz	
Dothan	FM	KC4JBF	147.14000	+	186.2 Hz	
	FM	N4RNU	147.34000	+		
	FM	KI4ZP	444.67500	+	156.7 Hz	
	FM	WB4ZPI	444.77500	+	186.2 Hz	
	FM	WA4MZL	444.90000	+		
Elba	DMR/BM	W4NQ	146.78000	-	CC1	
Elkmont	FM	WV4K	442.60000	+	203.5 Hz	
	FM	WV4K	443.45000	+	203.5 Hz	
Enterprise	FM	WD4ROJ	147.24000	+	100.0 Hz	
	FM	KJ4OTP	442.65000	+		
Eufaula	FM	W4EUF	147.28000	+	123.0 Hz	
	FM	N4TKT	442.07500	+	203.5 Hz	
	FM	WB4MIO	444.92500	+	151.4 Hz	
Falkville	FM	WR4JW	444.32500	+	107.2 Hz	
Fayette	DMR/BM	KK4QXJ	443.07500	+	CC2	
	FM	W4GLE	147.20000	+	110.9 Hz	
	FM	N4GRX	444.85000	+		
Florence	FM	KF4GZI	147.32000	+	100.0 Hz	
	FM	KF4GZI	444.00000	+	100.0 Hz	
	FM	KF4GZI	444.15000	+	100.0 Hz	
Florence, North Florence						
	FM	KQ4RA	444.65000	+	131.8 Hz	
Foley	FM	WA4MZE	147.24000	+		
Forestdale	FM	W4YMW	444.72500	+		
Fort Payne	FM	KF4FWZ	442.60000	+	100.0 Hz	
	FM	W4OZK	444.62500	+	141.3 Hz	
Fort Payne, Fischer Community						
	FM	KF4FWX	444.80000	+	100.0 Hz	
Fort Payne, Lookout Mountain						
	FM	W4DGH	147.27000	+	100.0 Hz	
Friendship	FM	KE4LTT	147.20000	+	107.2 Hz	
	FM	KE4LTT	444.57500	+	100.0 Hz	
Gadsden	DMR/MARC	KK4YOE	444.85000	+	CC1	
	FM	K4RBC	444.77500	+		
Gadsden, Hensley Mountain						
	FM	K4VMV	444.67500	+	100.0 Hz	
Gaylesville	FM	W4CCA	147.32000	+	100.0 Hz	
Geneva	FM	W4GEN	145.27000	139.27000	103.5 Hz	
Gold Hill	DMR/BM	KK4ICE	442.17500	+	CC1	
Greenville	DSTAR	K4TNS	442.22500	447.32500		
Grove Hill	FM	AB4BR	147.28000	+	210.7 Hz	
Gurley, Keel Moutain						
	FM	K4DED	442.97500	+	100.0 Hz	
Haleyville, Lakeland Community						
	FM	W4ZZA	442.22500	447.32500	203.5 Hz	
Hamilton	FM	KJ4I	147.02000	+	123.0 Hz	
Hanover	FM	K4GR	444.22500	+		

ALABAMA 41

Location	Mode	Call sign	Output	Input	Access	Coordinator
Heflin	FM	N4THM	444.17500	+		
Helena	DMR/BM	KC4SIG	444.92500	+	CC1	
Helena, Dearing Downs						
	FM	W4SHL	147.32000	+	88.5 Hz	
Helicon, Helicon Volunteer Fir						
	FM	W4FSH	442.72500	+	71.9 Hz	
Holtville	FM	K4IZN	444.95000	+	88.5 Hz	
Hoover	DMR/MARC	W4RKZ2	444.80000	+	CC1	
Hoover, Shades Mountain						
	FM	WA4CYA	147.14000	+	156.7 Hz	
	FM	KE4CAA	927.28750	902.28750	151.4 Hz	
Huntsville	DMR/BM	W4FMX	442.27500	+	CC1	
	DMR/BM	N4HSV	442.27500	+	CC1	
	DSTAR	KI4PPF	443.37500	+		
	DSTAR	W4WBC	443.42500	+		
	DSTAR	N4DTC	444.22500	+		
	DSTAR	W4WBC	1282.50000	1262.50000		
	DSTAR	KI4PPF	1284.00000	1264.00000		
	FM	W4VM	147.10000	+	103.5 Hz	
	FM	KB4CRG	147.24000	+	82.5 Hz	
	FM	W4QB	147.30000	+	103.5 Hz	
	FM	N4HSV	224.94000	-	100.0 Hz	
	FM	KE4BLC	442.00000	+	203.5 Hz	
	FM	KC4HRX	442.37500	+	156.7 Hz	
	FM	WA4NPL	443.00000	+		
	FM	WA1TDH	443.12500	+	107.2 Hz	
	FM	W4DYN	443.15000	+	103.5 Hz	
	FM	KE4LRX	443.25000	+	103.5 Hz	
	FM	W4VM	443.47500	+	103.5 Hz	
	FM	W4LDX	443.75000	+	186.2 Hz	
	FM	W4TCL	444.35000	+		
	FM	W4XE	444.37500	+	100.0 Hz	
	FM	WB4LTT	444.50000	+		
	FM	W4FMX	444.97500	+	103.5 Hz	
Huntsville, Brindley Mountain						
	FM	W4WLD	444.52500	+	DCS 143	
Huntsville, Drake Mountain						
	FM	KB4CRG	442.77500	+	107.2 Hz	
	FM	KB4CRG	444.57500	+	100.0 Hz	
Huntsville, Green Mountain						
	FM	WD4CPF	147.18000	+	100.0 Hz	
	FM	W4XE	444.30000	+	103.5 Hz	
Huntsville, Monte Sano						
	FM	W4TCL	147.50500	146.50500	123.0 Hz	
Huntsville, Monte Sano Mountai						
	FM	W4HMC	147.22000	+	136.5 Hz	
	FM	W4HSV	443.50000	+	110.9 Hz	
	FM	W4XE	443.62500	+	127.3 Hz	
	FM	KD4TFV	444.17500	+	151.4 Hz	
Huntsville, Redstone Arsenal						
	FM	N4WGY	444.75000	+	131.8 Hz	
Ider, Sand Mtn	FM	N2EMA	147.04500	+	100.0 Hz	
Jasper	DMR/MARC	N4MYI	443.92500	+	CC1	
	FM	KI4GEA	147.26000	+		
	FM	W4WCA	147.39000	+	110.9 Hz	
	FM	N4MYI	444.05000	+	123.0 Hz	
Jemison	FM	WB4UQT	444.47500	+	100.0 Hz	
Killen	FM	KS4QF	444.42500	+	203.5 Hz	
Killen, Greenhill	FM	AB4RC	442.47500	+	131.8 Hz	
Leighton	FM	AC4EG	147.34000	+	100.0 Hz	
Lineville	FM	WB4VBA	444.00000	+		
Madison	DMR/BM	KJ4NYH	443.02500	+	CC2	

42   ALABAMA

Location	Mode	Call sign	Output	Input	Access	Coordinator
Magnolia Springs	DSTAR	KI4SAZ	444.30000	+		
	DSTAR	KI4SAZ	1285.00000	1265.00000		
Marion	FM	KD4EXS	147.37500	+	123.0 Hz	
McCalla	DMR/MARC	W4RKZ	443.02500	+	CC1	
Mentone	DSTAR	KI4SAY	443.32500	+		
	DSTAR	KI4SAY	1285.00000	1265.00000		
	FM	W4OZK	224.72000	-	114.8 Hz	
Millry, Millry Communications						
	FM	KF4ZLK	147.18000	+	114.8 Hz	
Mobile	DMR/BM	N4FIV	444.78750	+	CC1	
	FM	W4IAX	444.90000	+		
Mobile, At Mobile EMA						
	FM	WB4BXM	147.15000	+	103.5 Hz	
Mobile, North Mobile County Hi						
	FM	N4RGJ	147.01500	+		
Mobile, Providence Hospital						
	FM	W4IAX	147.30000	+	203.5 Hz	
Mobile, USA Medical Center						
	FM	WX4MOB	444.52500	+	123.0 Hz	
Monroeville	FM	WB4UFT	147.16000	+	167.9 Hz	
	FM	K8IDX	444.77500	+	123.0 Hz	
Montgomery	DMR/BM	KK4AXA	443.97500	+	CC1	
	DMR/MARC	K4DJL	444.97500	+	CC1	
	FM	W4AP	147.18000	+	123.0 Hz	
	FM	WD4JRB	444.45000	+	100.0 Hz	
	FM	W4AP	444.50000	+	100.0 Hz	
	FM	W4OZK	444.62500	+	141.3 Hz	
Moody	DMR/MARC	KE4CAA	440.56250	+	CC1	
Moulton	FM	N4IDX	442.42500	+	107.2 Hz	
Moulton, Alabama	FM	N4IDX	444.77500	+	107.2 Hz	
Moundville	FM	K4CR	147.22000	+	77.0 Hz	
Muscle Shoals	FM	K4VFO	444.40000	+	79.7 Hz	
Nectar, Skyball Mountain						
	FM	N3AST	443.87500	+	123.0 Hz	
New Market	FM	W4TCL	442.17500	+	100.0 Hz	
Northport	DMR/MARC	KD9Q	444.17500	+	CC1	
Odenville, Bald Rock Mountain						
	FM	W4TCT	442.92500	+	103.5 Hz	
Opelika	DSTAR	W4LEE	147.37500	+		
	FM	WB4BYQ	147.06000	+	123.0 Hz	
	FM	WX4LEE	147.15000	+	123.0 Hz	
Opelika, East AL Medical Cente						
	FM	W4LEE	147.12000	+	123.0 Hz	
Opp, United States						
	FM	N4SYB	147.21000	+	100.0 Hz	
Parrish	FM	WR4Y	443.27500	+	123.0 Hz	
Parrish, On Parrish Water Tank						
	FM	W4WCA	442.30000	+	110.9 Hz	
Pelham	FM	W4TPA	444.15000	+		
Pell City, Bald Rock Mtn						
	FM	N4BRC	444.72500	+	156.7 Hz	
Phenix City	FM	WA4QHN	444.20000	+	123.0 Hz	
	FM	N4TKT	444.72500	+	123.0 Hz	
Pinson	FM	KA5GET	442.65000	+	131.8 Hz	
Quinton, Flint Ridge						
	FM	W4AI	444.20000	+	179.9 Hz	
Rainsville	FM	KF4BCR	442.55000	+	100.0 Hz	
Roanoke	DSTAR	KJ4JNX	1283.00000	1263.00000		
	FM	WD4KTY	147.04000	+		
	FM	KA4KBX	147.27000	+	141.3 Hz	
	FM	KA4KBX	224.92000	-		
	FM	KA4KBX	444.27500	+	141.3 Hz	

ALABAMA 43

Location	Mode	Call sign	Output	Input	Access	Coordinator
Roanoke	FM	KJ4JNX	444.90000	+		
Robertsdale, Baldwin County EM						
	FM	WB4EMA	147.09000	+	82.5 Hz	
Rogersville	FM	KF4GZI	442.02500	+	203.5 Hz	
	FM	KJ4LEL	442.50000	+		
Roxana	FM	W4KEN	224.84000	-		
Russellville	FM	NV4B	224.34000	-		
Russellville, Belgreen						
	FM	WX4FC	147.36000	+	103.5 Hz	
Russellville, Spruce Pine Moun						
	FM	WX4FC	147.21000	+	103.5 Hz	
	FM	WX4FC	444.67500	+		
Saginaw	FM	NR4J	224.50000	-	100.0 Hz	
Salem	FM	WA4QHN	444.10000	+	123.0 Hz	
Santuck	FM	W4KEN	224.88000	-	123.0 Hz	
Scottsboro	FM	K4NHA	224.58000	-		
Section	FM	K4SCO	147.36000	+	123.0 Hz	
Selma	FM	N4KTX	442.02500	+	100.0 Hz	
Sheffield, Water Tower						
	FM	N4GLE	444.05000	+	103.5 Hz	
Shelby Co	FM	N4PHP	442.00000	+	100.0 Hz	
Shelby, Lay Lake	FM	WB4CCQ	444.35000	+		
Shorter	FM	KK4ICE	442.07500	+	123.0 Hz	
Skipperville	FM	KD4KRP	147.03000	+	71.9 Hz	
Somerville	DMR/BM	WR4JW	440.61250	+	CC1	
Spanish Fort	DMR/BM	N4FIV	444.98750	+	CC1	
	FM	N4FIV	444.60000	+	203.5 Hz	
Springville, Simons Mountain						
	FM	KA5GET	443.65000	+	131.8 Hz	
Talladega	FM	W4PIG	224.22000	-	100.0 Hz	
Talladega, Cheaha Mountain						
	FM	WD4NOF	927.75000	902.75000	100.0 Hz	
Theodore	DMR/BM	W5TIN	443.50000	+	CC1	
Trafford	FM	W3NH	147.07500	+	67.0 Hz	
Trinity	FM	W4CFI	444.27500	+		
Troy	DMR/MARC	W4NQ	147.07500	+	CC1	
Trussville	DMR/BM	KK4YOE	442.10000	+	CC1	
Tuscaloosa	DMR/BM	W4XI	443.82500	+	CC1	
	DMR/MARC	KD9Q	444.90000	+	CC1	
	DSTAR	W4TTR	442.90000	+		
	DSTAR	W4KCQ	444.07500	+		
	DSTAR	W4KCQ	1284.40000	1264.40000		
	FM	KX4I	147.06000	-	179.9 Hz	
	FM	KR4ET	147.24000	+	186.2 Hz	
	FM	W4WYN	147.30000	+	131.8 Hz	
	FM	W4MD	442.15000	+		
	FM	K4CR	442.37500	+	DCS 411	
	FM	KX4I	442.55000	+	173.8 Hz	
	FM	W4MD	442.95000	+		
	FM	WS4I	443.30000	+	131.8 Hz	
	FM	KX4I	443.57500	+	192.8 Hz	
	FM	KX4I	444.02500	+	203.5 Hz	
	FM	KD9Q	444.57500	+	210.7 Hz	
	FM	N9YAY	444.90000	+		
Tuscumbia	FM	W4ZZK	444.87500	+	131.8 Hz	
Tuscumbia, S. Colbert County						
	FM	W4ZZK	443.72500	+	131.8 Hz	
Tuskegee	FM	N4LTX	444.87500	+	123.0 Hz	
Union Hill	FM	W5MEI	442.80000	+	162.2 Hz	
Vernon	DMR/BM	KC4UG	444.50000	+	CC1	
Vincent	FM	KK4OWL	224.40000	-	218.1 Hz	
Warrior	DMR/BM	KD4CIF	442.82500	+	CC1	

44   ALABAMA

Location	Mode	Call sign	Output	Input	Access	Coordinator
Warrior	DMR/MARC	N4UKE	443.55000	+	CC1	
	FM	W4GQF	224.12000	-		
Warrior, Corner	FM	KD4CIF	147.12000	+	100.0 Hz	
	FM	KD4CIF	224.44000	-		
Waterloo	FM	KF4GZI	442.12500	+	100.0 Hz	
Wetumpka	DMR/BM	KG4RCK	443.72500	+	CC1	
	FM	K4GR	444.22500	+		

ALASKA

Location	Mode	Call sign	Output	Input	Access	Coordinator
Anchorage	FM	WL7CWE	29.66000	-	110.9 Hz	
	FM	WL7CWE	51.65000	50.65000	103.5 Hz	
	FM	WL7CWA	146.79000	-	100.0 Hz	
	FM	KL3K	147.21000	+	123.0 Hz	
	FM	WL7CWE	444.85000	+	103.5 Hz	
	FM	KL3K	444.90000	+	123.0 Hz	
Anchorage, Elmandorf AFB						
	FM	KL7AIR	146.67000	-	103.5 Hz	
Anchorage, Flattop Mountain						
	FM	KL7AA	146.94000	-	100.0 Hz	
Anchorage, Flattop Mtn						
	FM	KL7AA	224.94000	-		
	FM	KL7AA	444.70000	+	100.0 Hz	
Anchorage, Mt Gordon Lyon						
	FM	KL7ION	147.30000	+	141.3 Hz	
Bethel	FM	AL7YK	146.10000	+	114.8 Hz	
	FM	AL7YK	444.10000	+	100.0 Hz	
Cantwell	FM	KL7KC	146.82000	-	103.5 Hz	
Central	FM	AL7FQ	146.82000	-	103.5 Hz	
Chicken	FM	KL7B	147.09000	+		
Delta Junction	DMR/MARC	KL2AVDMR	444.80000	+	CC1	ARC
	FM	KL2AV	146.76000	-	103.5 Hz	
	FM	KL2AV	444.60000	+	103.5 Hz	ARC
	FM	KL2AV	444.70000	+	103.5 Hz	
	FM	KL7KC	444.90000	+	103.5 Hz	
	FM	KL2AV	449.60000	-	103.5 Hz	
	FM	KL2AV	927.01250	902.01250	114.8 Hz	
Delta Junction, Donnelly Dome						
	FM	KL7KC	146.82000	-	103.5 Hz	
Denali Natl Park, Mt Mckinley						
	FM	KL7KC	146.76000	-	103.5 Hz	
Dot Lake	FM	KL7KC	146.88000	-	103.5 Hz	
Dot Lake, Knob Ridge						
	FM	KL7KC	146.82000	-		
Eilson AFB	FM	KL7KC	147.12000	+	103.5 Hz	
Fairbanks	FM	KL7EDK	147.30000	+	103.5 Hz	
	FM	KL3K	147.55000	+	123.0 Hz	
	FM	KL4BR	448.80000	-		
Fairbanks, Chena Dome						
	FM	KL7XO	146.79000	-	103.5 Hz	
Fairbanks, Cleary Summit						
	FM	KL7GNG	146.67000	-	103.5 Hz	
Fairbanks, Ester Dome						
	FM	KL7KC	146.88000	-	103.5 Hz	
	FM	AL7FG	224.88000	-	103.5 Hz	
	FM	KL7KC	444.80000	+	103.5 Hz	
Fairbanks, Porcupine Dome						
	FM	KL7KC	146.70000	-	103.5 Hz	
Fairbanks, U Of A	FM	KL7KC	146.94000	-	103.5 Hz	
Fort Richardson	FM	KL7GG	147.39000	+	100.0 Hz	
	FM	KL7GG	444.50000	+	100.0 Hz	
Galena	FM	AL2J	146.79000	-	103.5 Hz	
Healy, Mt Susitna	FM	KL3K	443.30000	+	103.5 Hz	

ALASKA 45

Location	Mode	Call sign	Output	Input	Access	Coordinator
Homer	DMR/MARC	KL4GR	443.40000	+	CC1	
Homer, Diamond Ridge Alascom S						
	FM	WL7PM	146.91000	-		
Houston	FM	KL3K	147.09000	+	123.0 Hz	
Juneau	FM	WA6AXO	146.88000	-	100.0 Hz	
	FM	KL7PF	147.00000	-		
	FM	WA6AXO	444.70000	+	141.3 Hz	
Juneau, Lena Point						
	FM	KL7PF	224.04000	-		
Juneau, Mendenhall Glacier						
	FM	KL2ZZ	147.12000	+	123.0 Hz	
Ketchikan	DSTAR	KL7FF	147.38000	+		
	FM	KL0RG	146.67000	-		
	FM	KL7GIH	146.79000	-		
Ketchikan, Ketchikan Internati						
	FM	WL7N	444.50000	+		
Kodiak	DMR/BM	KL1KE	441.87500	+	CC1	ARC
	FM	WL7CWZ	146.88000	-	141.3 Hz	
	FM	AL7LQ	444.85000	+	141.3 Hz	
Kodiak, Pillar Mt	FM	WL7AML	444.55000	+	136.5 Hz	
Kodiak, Spruce Cape						
	FM	KL4QY	145.31000	-	179.9 Hz	
Manley Hot Springs						
	FM	KL7KC	147.03000	+	103.5 Hz	
Mendenhall, Pederson Hill						
	FM	KL7IWC	147.30000	+		
Nenana	FM	WL7BDO	147.06000	+	103.5 Hz	
Ninilchik, Aurora Communicatio						
	FM	AL7Q	145.79000	+	91.5 Hz	
	FM	AL7Q	443.27500	+	91.5 Hz	
Nome	FM	KL0EF	147.21000	144.51000		
	FM	KL0EF	147.27000	144.57000		
	FM	KL7RAM	147.36000	+	103.5 Hz	
Nome, Anvil Mtn	FM	KL0EF	146.94000	-		
Nome, Downtown	FM	AL7X	144.64000	148.04000	103.5 Hz	
Nome, Newton Peak						
	FM	KL0EF	147.15000	144.55000		
Nome, Sinuk Mtn	FM	KL0EF	145.00000	+		
North Pole	FM	WL7TY	146.70000	-		
	FM	WL7LP	147.25000	+	123.0 Hz	
Northway	FM	KL7KC	146.82000	-	103.5 Hz	
Palmer	FM	KL3K	146.61000	-	123.0 Hz	
	FM	KL3K	444.30000	+	123.0 Hz	
	FM	KL7CC	447.57000	147.57000	103.5 Hz	
Palmer, Grubstake Mtn						
	FM	WL7CVG	147.33000	+	103.5 Hz	
	FM	WL7CVG	443.90000	+	103.5 Hz	
Petersburg, Duncan Canal						
	FM	KD7WN	444.50000	+		
Seward	FM	KL3K	146.76000	-	123.0 Hz	
Sitka	FM	KL7SRK	146.08000	+	114.8 Hz	
	FM	KL7SRK	146.82000	-	114.8 Hz	
	FM	KL7SRK	444.00000	+	114.8 Hz	
	FM	KL7SRK	446.00000	-	114.8 Hz	
Soldotna, Pickle Hill						
	FM	AL7LE	146.88000	-		
Talkeetna	FM	NL7E	147.12000	+	100.0 Hz	
	FM	KL2AV	444.60000	+	103.5 Hz	
Teller	FM	KL7RAM	146.73000	-	103.5 Hz	
Unalakleet	FM	KL7RAM	444.90000	+	103.5 Hz	
Valdez	FM	NL7R	146.94000	-		
Wasilla	DSTAR	WL7CWI	147.00000	+		

46   ALASKA

Location	Mode	Call sign	Output	Input	Access	Coordinator
Wasilla	DSTAR	WL7CWI	442.00000	+		
	FM	KL7DJE	147.09000	+	100.0 Hz	
	FM	NL7S	147.15000	+	107.2 Hz	
Wasilla, Wasilla High School						
	FM	KL7JFU	146.85000	-	103.5 Hz	
	FM	KL7JFU	444.60000	+	103.5 Hz	
Willow	FM	KL7DOB	146.64000	-		

ARIZONA

Location	Mode	Call sign	Output	Input	Access	Coordinator
Aguila, Smith Peak						
	FM	K7LKL	146.68000	-	162.2 Hz	
	FM	W7ARA	443.77500	+	100.0 Hz	
Ajo, Childs Mtn	FM	W7AJO	145.31000	-	100.0 Hz	
	FM	KL7DSI	448.10000	-	100.0 Hz	
Alpine, South Mountain						
	FM	K7EAR	145.27000	-	141.3 Hz	
	FM	N5IA	448.72500	-		
Anthem	FM	W6PAT	442.52500	+	114.8 Hz	
Anthem / Daisy Mtn. Fire Stati						
	FM	KE7KMI	448.37500	-	100.0 Hz	ARCA
Arizona City	FM	AE7RR	440.75000	+		
Benson	FM	WA7PIQ	448.82500	-	107.2 Hz	
Benson, Haystack Mtn						
	FM	K7SPV	145.37000	-	131.8 Hz	
Bisbee, Mule Mountains						
	FM	K7RDG	146.76000	-	162.2 Hz	
	FM	K7RDG	147.02000	+	162.2 Hz	
	FM	K7EAR	147.08000	+	141.3 Hz	
	FM	K7RDG	449.52500	-	100.0 Hz	
Black Canyon City						
	FM	KB7OCY	146.90000	-	118.8 Hz	
Blythe, Cunningham Mtn						
	FM	KR7AZ	147.06000	+	203.5 Hz	
Buckeye	DMR/BM	KE7CIU	446.57500	-	CC1	
Bullhead City	FM	K3MK	146.64000	-	123.0 Hz	
	FM	K7PFK	448.95000	-		
	FM	N7URK	449.50000	-		
Camp Verde, Squaw Peak						
	FM	W7EI	147.10000	+	131.8 Hz	
	FM	K6DLP	447.70000	-		
Casa Grande, Central Arizona C						
	FM	KA7TUR	446.82500	-	100.0 Hz	
Casa Grande, Sacaton Peak						
	FM	N7ULY	447.72500	-	100.0 Hz	
	FM	K7PNX	448.02500	-	136.5 Hz	
Catalina	FM	W7AI	147.32000	+	156.7 Hz	
Chandler	DMR/BM	N4SMG	436.00000		CC1	
	DMR/BM	N7YF	445.30000	-	CC1	
	DMR/MARC	KJ7CMR	438.80000	-	CC1	
	FM	WW7CPU	145.45000	-	162.2 Hz	
	FM	KE7JVX	224.92000	-	156.7 Hz	
	FM	KE7JVX	444.27500	+	100.0 Hz	
	FM	WB5DYG	447.50000	-	100.0 Hz	
	FM	WW7CPU	448.95000	-	100.0 Hz	
Chandler, Chandler Hospital						
	FM	W7MOT	927.43750	902.43750	151.4 Hz	
Chandler, Chandler Regional Me						
	FM	W7MOT	442.97500	+	100.0 Hz	
Chino Valley	FM	K7POF	449.25000	-	192.8 Hz	
Clarkdale	FM	K9FUN	441.77500	+	156.7 Hz	
	FM	W6NSA	443.90000	+	100.0 Hz	

ARIZONA 47

Location	Mode	Call sign	Output	Input	Access	Coordinator
Corona De Tucson, East OST & G						
	FM	K7RST	146.80000	-	156.7 Hz	
Cottonwood, Mingus Mountain						
	FM	K7YCA	145.29000	-	127.3 Hz	
	FM	W7ARA	146.82000	-	162.2 Hz	
	FM	K7MRG	147.00000	+	162.2 Hz	
	FM	W7EI	147.22000	+	162.2 Hz	
	FM	KI6FH	224.08000	-	156.7 Hz	
	FM	W7ARA	448.50000	-	100.0 Hz	
	FM	WA6LSE	449.42500	-	141.3 Hz	
	FM	N7CI	449.70000	-		
	FM	WA7JC	449.72500	-	110.9 Hz	
	FM	WB7BYV	927.11250	902.11250	DCS 432	
Crown King	FM	W7WHP	447.30000	-	88.5 Hz	
	FM	WA7ZZT	447.62500	-		
Crown King, Towers Mountain						
	FM	KF7EZ	927.28750	902.28750	151.4 Hz	
Crown King, Towers Mtn						
	FM	WB7EVI	448.57500	-	100.0 Hz	
	FM	K7STA	449.00000	-		
	FM	WB7EVI	449.17500	-	100.0 Hz	
Crown King, Wildflower Mountai						
	FM	WA7MKS	145.35000	-	162.2 Hz	
Dateland, Dateland Elementary						
	FM	N7ACS	147.22000	+	100.0 Hz	
Eagar, Greens Peak						
	FM	W7EH	145.31000	-	110.9 Hz	
	FM	N7QVU	146.62000	-	162.2 Hz	
	FM	K7EAR	146.70000	-	141.3 Hz	
	FM	W7ARA	448.37500	-	100.0 Hz	
	FM	N7QVU	449.35000	-	100.0 Hz	
	FM	W7ARA	927.16250	902.16250	151.4 Hz	
Flagstaff, Mormon Mtn						
	FM	KD7IC	145.27000	-	162.2 Hz	
Flagstaff, Mt. Elden						
	FM	NO7AZ	145.45000	-	103.5 Hz	
	FM	W7ARA	147.14000	+	162.2 Hz	
	FM	K6DLP	223.84000	-	107.2 Hz	
	FM	W7ARA	448.47500	-	100.0 Hz	
	FM	N7MK	448.62500	-		
	FM	W7ARA	448.87500	-	100.0 Hz	
	FM	NO7AZ	449.32500	-	103.5 Hz	
Fountain Hills	FM	N7MK	447.77500	-	110.9 Hz	
Gila Bend	FM	K7PO	145.29000	-	103.5 Hz	
Glendale	FM	K7CAE	446.60000	-	103.5 Hz	
	FM	KC7GHT	447.57500	-	151.4 Hz	
Glendale, Glendale City Hall						
	FM	KD7HJN	447.40000	-	100.0 Hz	
Globe	DMR/MARC	N7CI	445.86250	-	CC0	
Globe, Pinal Peak	FM	K7EAR	145.41000	-	141.3 Hz	
	FM	W7ARA	147.20000	+	162.2 Hz	
	FM	N7MK	224.10000	-	156.7 Hz	
	FM	N2QWF	445.86250	-		
	FM	N7TWW	448.42500	-	103.5 Hz	
	FM	W7ARA	448.47500	-	100.0 Hz	
	FM	WA7KUM	448.65000	-		
	FM	W7ARA	927.41250	902.41250	151.4 Hz	
	FM	N7TWW	927.83750	902.83750	151.4 Hz	
	FM	N7TWW	1283.65000	1263.65000	100.0 Hz	
Globe, Signal Peak						
	FM	WR7GC	146.74000	-	162.2 Hz	
	FM	WR7GC	449.65000	-	100.0 Hz	

48   ARIZONA

Location	Mode	Call sign	Output	Input	Access	Coordinator
Golden Valley	FM	N7FK	448.40000	-	123.0 Hz	
Goodyear, Estrella Mountain						
	FM	KD7YAT	446.55000	-	100.0 Hz	
Green Valley	FM	WE7GV	145.27000	-	107.2 Hz	
	FM	WB6TYP	145.43000	-		
	FM	AA7RP	449.22500	-	107.2 Hz	
Green Valley, Elephant Head Pe						
	FM	WE7GV	145.29000	-	107.2 Hz	
	FM	WE7GV	449.37500	-	107.2 Hz	
Green Valley, Keystone Peak						
	FM	KC0LL	444.87500	+	100.0 Hz	
Greens Peak	FM	N7QVU	146.62000	-	162.2 Hz	ARCA
Guadalupe	FM	KJ6KW	224.88000	-	156.7 Hz	
Henderson	DMR/BM	AF7JQ	449.62500	-	CC1	
Holbrook	FM	KA7ARZ	146.68000	-		
Huachuca City	FUSION	N7MUE	442.00000	+		
Jacob Lake, Jacob Lake Inn						
	FM	N7YSE	147.30000	+	100.0 Hz	
Kanab, TV Site	FM	N7YSE	449.10000	-	100.0 Hz	
Kingman	DSTAR	W7KDS	145.13500	-		
	DSTAR	W7KDS	1284.00000	1264.00000		
	DSTAR	W7KDS	1299.00000			
	FM	KA6NLS	51.94000	-	100.0 Hz	
	FM	KD7HVE	145.21000	-	123.0 Hz	
	FM	KD7HVE	147.12000	+	123.0 Hz	
	FM	KD7MIA	147.36000	+	123.0 Hz	
	FM	KD7MIA	224.96000	-	123.0 Hz	ARCA
	FM	K7RLW	446.22500	-		
	FM	KD7HVE	449.95000	-		
	FM	WB6TNP	927.47500	902.47500	DCS 532	
	FM	K7RLW	927.61250	902.61250		
Kingman AZ USA	FM	N7DPS	146.80000	-	100.0 Hz	ARCA
Kingman, Hualapai Mtn						
	FM	WB6RER	147.16000	+	131.8 Hz	
	FM	K7MPR	147.24000	+	123.0 Hz	
	FM	K6DLP	448.10000	-		
	FM	N7SKO	448.25000	-	131.8 Hz	
	FM	K7MPR	448.55000	-	123.0 Hz	
	FM	W6PNM	448.68000	-		
Kingman, Hualapai Peak						
	FM	KA6NLS	447.10000	-	100.0 Hz	
Kingman, Kingman Airport						
	FM	K7RLW	146.62000	-		
Kingman, Wind Turbine Site						
	FM	K7RLW	146.94000	-		
	FM	K7RLW	449.47500	-	94.8 Hz	
Lake Havasu	DMR/MARC	W7DXJ	448.45000	-	CC1	
Lake Havasu City	FM	W7DXJ	146.96000	-	162.2 Hz	
	FM	KE7ZIW	147.50000			
	FM	KF7X	224.24000	-	156.7 Hz	
	FM	KX7P	448.35000	-		
	FM	K6PNG	449.12500	-	67.0 Hz	
	FM	W7DXJ	449.95000	-	141.3 Hz	
Lake Havasu City, Dick Samp Me						
	FM	K7RLW	146.72000	-	123.0 Hz	
Lake Havasu City, Lake Havasu						
	FM	K7MPR	448.60000	-	123.0 Hz	
Lake Havasu City, Ram Peak						
	FM	WR7RAM	448.02000	-		
Laveen	FM	KC7QKS	449.25000	-	192.8 Hz	
Lukachukai, Roof Butte						
	FM	KB5ITS	145.25000	-	100.0 Hz	

ARIZONA 49

Location	Mode	Call sign	Output	Input	Access	Coordinator
Marana	FM	KC7CPB	448.00000	-	100.0 Hz	
Maricopa	FM	KC7KF	146.78000	-	100.0 Hz	
	FM	KE7JVX	927.02500	902.02500	151.4 Hz	
	FM	KE7JVX	927.65000	902.65000	151.4 Hz	
Maricopa, Central AZ College -						
	FM	WY7H	145.21000	-	162.2 Hz	
	FM	WY7H	449.12500	-	136.5 Hz	
Maricopa, Montycopa						
	FM	KE7JVX	224.92000	-	156.7 Hz	ARCA
McNary, Greens Peak						
	FM	W7ARA	146.72000	-	162.2 Hz	
Mesa	DMR/BM	K7DMK	440.40000	+	CC1	
	DMR/BM	KE7TR	440.72500	+	CC1	
	DMR/BM	K7EVR	440.75000	+	CC7	
	DMR/BM	K7DMK	441.75000	+	CC1	
	DMR/BM	K5EDJ	442.05000	+	CC3	
	DMR/MARC	KE7JFH	445.83750	-	CC1	
	DMR/MARC	N7DJZ	445.96250	-	CC1	
	FM	WB7TUJ	145.33000	-	114.8 Hz	
	FM	K7DAD	146.72000	-	100.0 Hz	
	FM	KF7EUO	440.42500	+		
	FM	K7DAD	449.37500	-	100.0 Hz	
	FM	KA7ZEM	449.55000	-	100.0 Hz	
	FM	WB7TJD	449.60000	-	100.0 Hz	
	FM	W5WVI	449.85000	-		
Mesa, 1201 S. Alma School Rd						
	FM	W7ARA	449.62500	-	100.0 Hz	
Mesa, Bank Of America						
	FM	W7ARA	224.68000	-	156.7 Hz	
Mesa, Dobson And University Dr						
	FM	N7OKN	440.67500	+	107.2 Hz	
Mesa, Usery Mountain						
	DMR/BM	N7DJZ	446.35000	-	CC1	
	DSTAR	KE7JFH	145.12500	-		
	DSTAR	KE7JFH	445.97500	-		
	DSTAR	KE7JFH	1285.65000	1265.65000		
	FM	WB7TJD	147.12000	+	162.2 Hz	
	FM	KE7JFH	448.72500	-		
	FM	N7DJZ	448.97500	-		
Mesa, Usery Pass	FM	KE7JFH	145.47000	-	79.7 Hz	
	FM	WB7TUJ	146.66000	-	162.2 Hz	
	FM	W7ARA	146.86000	-	162.2 Hz	
	FM	W7BSA	147.02000	+	162.2 Hz	
	FM	N7TWW	224.02000	-	151.4 Hz	
	FM	W7BSA	448.80000	-	100.0 Hz	
	FM	W7ARA	449.10000	-	100.0 Hz	
	FM	N7MK	927.46250	902.46250		
Morenci, Guthrie Peak						
	FM	N5IA	448.97500	-		
Mormon Mtn	FM	KD7IC	449.60000	-	162.2 Hz	
Mt Lemon	DMR/MARC	N7HND	447.87500	-	CC1	
Oracle	FM	WA7ELN	448.70000	-		
Oro Valley	DMR/BM	WD7ARC	445.80000	-	CC1	
	FM	WB7NUY	447.42500	-	107.2 Hz	
Overgaard	FM	W7RIM	146.80000	-	162.2 Hz	
Parker, Black Peak						
	FM	WA7RAT	146.85000	-	162.2 Hz	
	FM	K7ZEU	224.00000	-	100.0 Hz	
	FM	K6DLP	448.27500	-		
	FM	K7AY	448.65000	-		
Patagonia, Red Mtn						
	FM	W7JPI	146.64000	-	127.3 Hz	

50   ARIZONA

Location	Mode	Call sign	Output	Input	Access	Coordinator
Paulden, Paulden	FM	W7BNW	446.55000	-	100.0 Hz	
Payson	DMR/MARC	KE7TR	449.30000	-	CC1	
	FM	N7TAR	147.39000	+	100.0 Hz	
	FM	N7TAR	448.77500	-	77.0 Hz	
	FM	W7MOT	927.43750	902.43750	151.4 Hz	
Payson, Forest Lakes						
	FM	W7NAZ	447.47500	-	88.5 Hz	
Payson, Mt Ord	FM	WR7GC	146.96000	-	141.3 Hz	
Peoria	FM	WA7CBB	145.60000		162.2 Hz	
Peoria, Sunrise Mountain High						
	FM	W7DGL	446.57500	-	127.3 Hz	
Phoenix	DMR/BM	WA6PNP	444.20000	+	CC1	
	DMR/BM	N8NQP	445.25000	-	CC1	
	DMR/MARC	WA7KUM	440.10000	+	CC0	
	DMR/MARC	KD4IML	442.35000	+	CC1	
	DMR/MARC	KF6FM	447.87500	-	CC1	
	FM	K7JAX	146.84000	453.16000	123.0 Hz	
	FM	W7UXZ	147.06000	+	162.2 Hz	
	FM	WA7UID	147.28000	+	162.2 Hz	
	FM	K5VT	147.32000	+		
	FM	KA7RVV	440.77500	+		
	FM	KB7CGA	441.20000	+	77.0 Hz	
	FM	N1KQ	442.60000	+	100.0 Hz	
	FM	WW7B	442.67500	+	123.0 Hz	
	FM	KB7OBJ	447.32500	-		
	FM	N7TWB	447.95000	-	100.0 Hz	
	FM	WA7ZZT	448.77500	-		
Phoenix (ShawButte)						
	DMR/MARC	KE7JFH	440.03750	+	CC0	
Phoenix, Arizona National Guar						
	FM	WW7B	442.70000	+	123.0 Hz	
Phoenix, AZ	FM	W7ARA	147.24000	+	162.2 Hz	
	FM	WB7EVI	442.85000	+	100.0 Hz	
Phoenix, Chase Tower						
	FM	W7ARA	146.64000	-	162.2 Hz	
	FM	W7ARA	444.30000	+	100.0 Hz	
Phoenix, Downtown						
	FM	W7ATV	145.19000	-	162.2 Hz	
Phoenix, Far North Mountain						
	FM	K7PNX	448.85000	-	136.5 Hz	
Phoenix, Glendale						
	FM	W7TBC	147.04000	+	162.2 Hz	
Phoenix, Honor Health Hospital						
	FM	W7TBC	146.70000	-	162.2 Hz	
Phoenix, International Space S						
	FM	ARISS	145.80000	437.80000	100.0 Hz	
Phoenix, Metro Center						
	FM	W2MIX	445.27500	-		
Phoenix, North Mtn						
	FM	K7PNX	449.02500	-	136.5 Hz	
Phoenix, Shaw Butte						
	FM	W7ARA	224.90000	-	156.7 Hz	
	FM	N7ULY	449.35000	-		
	FM	W7ARA	449.52500	-	100.0 Hz	
	FM	KF7EZ	927.53750	902.53750	151.4 Hz	
Phoenix, South Mountain						
	FM	WA7ZZT	442.00000	+		
	FM	W7MOT	442.05000	+	100.0 Hz	
	FM	N7AUW	442.12500	+	100.0 Hz	
	FM	K7PNX	442.20000	+	136.5 Hz	
	FM	WA7ZZT	442.55000	+	100.0 Hz	
	FM	W7MOT	443.05000	+	100.0 Hz	

ARIZONA 51

Location	Mode	Call sign	Output	Input	Access	Coordinator
Phoenix, South Mountain						
	FM	AJ7T	444.82500	+	100.0 Hz	
	FM	W7MOT	927.21250	902.21250	151.4 Hz	
Phoenix, Twin Knolls						
	FM	K7PNX	447.10000	-	136.5 Hz	
Phoenix, Union Hills						
	FM	N6IME	224.56000	-	67.0 Hz	
Phoenix, Usery Pass						
	FM	W7BSA	147.02000	+	162.2 Hz	ARCA
Phoenix, White Tank Mountains-						
	P25	N9VJW	440.17500	+	100.0 Hz	
Phoenix, White Tanks						
	FM	W7TBC	147.04000	+	162.2 Hz	
	FM	W7DGL	223.98000	-	156.7 Hz	
Phoenix, White Tanks Mountain						
	DMR/BM	N8NQP	448.32500	-	CC1	
	DSTAR	W7MOT	145.13500	-		
	DSTAR	W7MOT	440.81250	+		
	DSTAR	K7PNX	1283.85000	1263.85000		
	DSTAR	W7MOT	1283.90000	1263.90000		
	DSTAR	K7PNX	1298.00000			
	DSTAR	W7MOT	1299.50000			
	FM	N7SKT	145.43000	-	100.0 Hz	
	FM	W7EX	146.94000	-	162.2 Hz	
	FM	N7ULY	147.38000	+	79.7 Hz	
	FM	KD7TKT	224.60000	-	156.7 Hz	
	FM	KE7JVX	224.84000	-	156.7 Hz	
	FM	W7EX	441.72500	+	100.0 Hz	
	FM	W1OQ	442.27500	+	100.0 Hz	
	FM	N7SKT	442.92500	+	100.0 Hz	
	FM	W7TBC	446.15000	-	100.0 Hz	
	FM	WA6LSE	447.17500	-		
	FM	WA7GBL	449.15000	-		
	FM	K7PNX	449.45000	-		
	FM	W7ARA	927.33750	902.33750	151.4 Hz	
Pinedale	FM	KB7ZIH	145.23000	-	110.9 Hz	
Pinetop-Lakeside	DMR/BM	AD7W	449.05000	-	CC1	
Pinetop-Lakeside, Porter Mount						
	FM	AD7W	146.76000	-	162.2 Hz	
Point Of Rocks	FM	WY7H	145.21000	-		
Potato Patch	FM	WB6TNP	447.60000	-	141.3 Hz	
	FM	K6DLP	448.10000	-		
	FM	W6PNM	448.68000	-		ARCA
	FM	K7RLW	927.61250	902.61250		ARCA
Prescott	DMR/MARC	KB6BOB	447.90000	-	CC1	
	DMR/MARC	KF7LRD	448.30000	-	CC1	
	FM	W7YRC	146.88000	-	100.0 Hz	
	FM	WB7BYV	449.67500	-	88.5 Hz	
	FM	WB7BYV	927.38750	902.38750	151.4 Hz	
Prescott, Airport	FM	WB7BYV	927.58750	902.58750	131.8 Hz	
Prescott, Mingus Mountain						
	FM	K7MRG	442.15000	+	100.0 Hz	
Prescott, Mt Francis						
	FM	K7YCA	145.37000	-	127.3 Hz	
Prescott, Mt Union						
	FM	N7NGM	52.56000	-	100.0 Hz	
	FM	K7YCA	147.26000	+	103.5 Hz	
	FM	WB7BYV	927.12500	902.12500	103.5 Hz	
Quartsite	DMR/MARC	WD6AML	448.95000	-	CC1	
Quartzite, Guadalupe Peak						
	FM	KJ6KW	224.88000	-	156.7 Hz	
Quartzsite	DMR/MARC	WD7FM	448.90000	-	CC1	

52   ARIZONA

Location	Mode	Call sign	Output	Input	Access	Coordinator
Quartzsite, Guadalupe Peak						
	FM	WB7FIK	145.31000	-	107.2 Hz	
	FM	K6TQM	147.36000	+	107.2 Hz	
	FM	WD6FM	448.90000	-		
	FM	WB7FIK	448.97500	-		
Quartzsite, Rice Ranch						
	FM	KG7HYF	145.19000	-	103.5 Hz	
Quartzsite, West Park						
	FM	W7AZQ	447.00000	-	123.0 Hz	
Red Rock, Roof Butte						
	FM	NM5SJ	146.82000	-	100.0 Hz	
Rio Rico	FM	KG7DNO	147.06000	+	127.3 Hz	
Safford	DMR/MARC	K7EAR	440.75000	+	CC1	
Safford, Guthrie Peak						
	FM	K7EAR	147.28000	+	141.3 Hz	
Safford, Heliograph Peak						
	FM	K7EAR	146.86000	-	141.3 Hz	
	FM	K7EAR	146.90000	-	141.3 Hz	
	FM	K7EAR	440.70000	+	141.3 Hz	
	FM	K7EAR	447.82500	-	100.0 Hz	
	FM	K7JEM	448.67500	-	100.0 Hz	
Saint Johns	FM	KE6GVK	145.23000	-	DCS 212	
San Tan Valley	FM	N7DJZ	449.32500	-	100.0 Hz	
Satellite, Space	FM	AO-85	145.98000	435.18000	67.0 Hz	
Scottsdale	DMR/MARC	KF7CUF	445.20000	-	CC1	
	FM	KB6POQ	145.31000	-	91.5 Hz	
	FM	W7MOT	443.15000	+	100.0 Hz	
	FM	KB6POQ	445.90000	-	91.5 Hz	
	FM	KC7WVE	446.67500	-	146.2 Hz	
	FM	WA7VEI	448.90000	-		
	FM	K0NL	449.42500	-		
	FM	N7KEG	449.65000	-		
	FM	K0NL	927.06250	902.06250	151.4 Hz	
	FM	KF7EUO	927.23750	902.23750	151.4 Hz	
Scottsdale, Airport						
	FM	W7ARA	146.76000	-	162.2 Hz	
	FM	W7ARA	441.62500	+	100.0 Hz	
	FM	W7ARA	927.16250	902.16250	151.4 Hz	
Scottsdale, Hayden Gd Site						
	FM	W7MOT	147.34000	+	162.2 Hz	
	FM	W7MOT	442.02500	+	100.0 Hz	
	FM	W7MOT	927.38750	902.38750	151.4 Hz	
Scottsdale, HonorHealth - 92nd						
	FM	W7UF	147.18000	+	162.2 Hz	
Scottsdale, HonorHealth Scotts						
	FM	W7UF	440.00000	+	100.0 Hz	
Scottsdale, Pinnacle Peak						
	FM	W0NWA	441.10000	+	103.5 Hz	
Scottsdale, Thompson Peak						
	FM	KG7UN	147.08000	+	162.2 Hz	
	FM	KG7UN	448.82500	-	100.0 Hz	
Seligman	FM	WA6GDF	147.12000	+	192.8 Hz	
Show Low	DMR/BM	W2MRA	446.57500	-	CC1	
	DMR/BM	N7OEI	448.85000	-	CC1	
	FM	W7EH	449.15000	-	110.9 Hz	
Sierra Vista	DMR/MARC	N2QWF	445.85000	-	CC1	
	FM	N0NBH	147.36000	+	100.0 Hz	
	FM	N0NBH	224.96000	-	100.0 Hz	
	FM	N0NBH	449.82500	-	100.0 Hz	
	FM	N0NBH	927.91250	902.91250	100.0 Hz	
	FM	N0NBH	1282.50000	1262.50000	100.0 Hz	

ARIZONA 53

Location	Mode	Call sign	Output	Input	Access	Coordinator
Sierra Vista, Green Acres						
	FM	K7RDG	447.95000	-	100.0 Hz	
Sierra Vista, Mustang Corners						
	FM	K7LTN	442.45000	+	100.0 Hz	
Solomon	DMR/MARC	K7EAR	440.07500	+	CC1	
St Johns, Red Sky Ranch						
	FM	NR7G	449.62500	-	136.5 Hz	
Sun City West	FM	KA7G	442.45000	+	91.5 Hz	
Sun City West, Dell Web Hospit						
	FM	NY7S	147.30000	+	162.2 Hz	
Sun City, Boswell Hospital						
	FM	W7JHQ	449.80000	-	100.0 Hz	
Sunflower	FM	W7ARA	146.92000	-	162.2 Hz	
	FM	W7ARA	147.36000	+	162.2 Hz	
Sunflower, Mt Ord	FM	W7ARA	444.50000	+	100.0 Hz	
Tempe	DMR/MARC	K7TMP	442.37500	+	CC1	
	FM	WA2DFI	145.27000	-	162.2 Hz	
	FM	WA2DFI	224.20000	-		
Tempe, A-Mountain ASU						
	FM	KE7EJF	927.03750	902.03750	DCS 023	
Tolleson	FM	AJ9Y	448.07500	-	100.0 Hz	
Tombstone	FM	NM2J	145.19000	-	88.5 Hz	
Tonopah	FM	WT9S	442.07500	+	123.0 Hz	
Tucson	DMR/BM	W7NFL	444.75000	+	CC1	
	DMR/BM	KG7PJV	445.13750	-	CC1	
	DMR/BM	W7AI	445.80000	-	CC1	
	DMR/BM	WN0EHE	449.50000	-	CC1	
	DMR/MARC	KG7GWN	440.76250	+	CC1	
	DMR/MARC	N7HND	444.25000	+	CC1	
	DMR/MARC	N7HND	445.87500	-	CC1	
	FM	K6PPT	145.17000	-		
	FM	KA7LVX	145.33000	-	127.3 Hz	
	FM	N7IQV	146.68000	-	173.8 Hz	
	FM	AG7H	146.85000	-		
	FM	WD7F	146.94000	-	110.9 Hz	
	FM	K7UAZ	146.96000	-	127.3 Hz	
	FM	N7OEM	147.30000	+	110.9 Hz	
	FM	KA3IDN	147.36000	+		
	FM	W7SA	224.70000	-	179.9 Hz	
	FM	K6PPT	446.57500	-	156.7 Hz	
	FM	W7RAP	448.30000	-		
	FM	KG7KV	448.90000	-		
	FM	N7CK	448.97500	-		
	FM	N7DQP	449.47500	-	107.2 Hz	
	FM	KW7RF	449.67500	-	77.0 Hz	
	FM	KA3IDN	927.07500	902.07500	218.1 Hz	
	FM	K6PYP	927.11250	902.11250	DCS 432	
	FM	N7OEM	927.97500	902.97500	DCS 606	
Tucson / Keystone Peak						
	FM	W7AI	146.62000	-	156.7 Hz	
Tucson, Bank Of America						
	DSTAR	K7RST	145.11500	-		
Tucson, East Tucson						
	FM	W7GV	146.66000	-	110.9 Hz	
Tucson, Keystone Peak						
	FM	W7AI	146.62000	-	156.7 Hz	
Tucson, Mt Lemmon						
	FM	K6PYP	51.86000	-	100.0 Hz	
	FM	KA7LFX	53.72000	52.72000	136.5 Hz	
	FM	N7OEM	146.88000	-	110.9 Hz	
	FM	KC0LL	147.14000	+	100.0 Hz	
	FM	K7EAR	147.16000	+	141.3 Hz	

54   ARIZONA

Location	Mode	Call sign	Output	Input	Access	Coordinator
Tucson, Mt Lemmon						
	FM	KA7LFX	224.06000	-	156.7 Hz	
	FM	K6PYP	224.50000	-	156.7 Hz	
	FM	KC0LL	444.97500	+	100.0 Hz	
	FM	W7ATN	445.22500	-	103.5 Hz	
	FM	N1DHS	448.35000	-	107.2 Hz	
	FM	N7OEM	448.55000	-	110.9 Hz	
	FM	N7HND	927.05000	902.05000	DCS 411	
Tucson, Mt Lemmon, Radio Ridge						
	FM	K6PYP	927.13750	902.13750	DCS 411	
	FM	K7SPV	1282.75000	-	131.8 Hz	
Tucson, Mt. Lemmon						
	FM	K7RST	29.64000	-	110.9 Hz	
Tucson, Oro Valley						
	DMR/BM	W7AI	444.10000	+	CC1	
Tucson, Raytheon	FM	W7SA	147.34000	+	179.9 Hz	
	FM	W7SA	448.77500	-	179.9 Hz	
Tucson, Rosemont And Broadway						
	FM	N7HND	147.00000	+	110.9 Hz	
Tucson, Saguaro National Park						
	FM	K6PYP	53.04000	52.04000	141.3 Hz	
Tucson, Sahuarita School						
	DSTAR	KR7ST	445.95000	-		
Tucson, Tucson Mtn						
	FM	KB7RFI	447.87500	-	88.5 Hz	
Tucson, Tucson Racquet And Fit						
	FM	K7RST	449.30000	-	156.7 Hz	
Tucson, West Tucson Mtn Range						
	DMR/BM	W7NFL	445.73750	-	CC1	
Vail	FM	KE7ULC	446.55000	-	100.0 Hz	
	FM	K7RST	448.32500	-	156.7 Hz	
	FM	K7LHR	449.55000	-	107.2 Hz	
Wilcox, West Peak						
	FM	K7EAR	145.35000	-	141.3 Hz	
Williams, Bill Williams Mounta						
	DSTAR	K7NAZ	145.11500	-		
	DSTAR	K7NAZ	445.78750	-		
	DSTAR	K7NAZ	1284.60000	1264.60000		
	FM	K7NAZ	146.78000	-	91.5 Hz	
	FM	K6JSI	449.75000	-	123.0 Hz	
	FM	WB7BYV	927.07500	902.07500	218.1 Hz	
Yarnell, Yarnell Hill						
	FM	N7LOQ	146.62000	-	162.2 Hz	
Yuma	DMR/MARC	KG7ARQ	440.95000	+	CC0	
	FM	N7ACS	224.72000	-	103.5 Hz	
	FM	KC7EQW	927.46250	902.46250	88.5 Hz	
Yuma, Filter Factory						
	FM	W7DIN	449.07500	-	88.5 Hz	
Yuma, Fire Station 4						
	FM	N7ACS	146.80000	-	162.2 Hz	
Yuma, Fortuna And I-8 Intersec						
	FM	KJ6IZQ	448.00000	-	100.0 Hz	
Yuma, Telegraph Pass						
	FM	N7ACS	146.78000	-	103.5 Hz	
	FM	N0RHZ	448.62500	-		
	FM	W7DIN	449.07500	-	88.5 Hz	ARCA
Yuma, YACS Comm Center						
	FM	N7ACS	146.62000	-	103.5 Hz	
Yuma, Yuma County Fairgrounds						
	FM	KJ7DLK	146.84000	-	88.5 Hz	

ARKANSAS 55

Location	Mode	Call sign	Output	Input	Access	Coordinator
ARKANSAS						
Adona, Perry Mtn	FM	W5SRE	447.15000	-	100.0 Hz	
Alexander	FM	N5YLE	145.29000	-	131.8 Hz	
Alpine	FM	KD5ARC	147.22500	+	114.8 Hz	
Arkadelphia	FM	W5RHS	444.67500	+	114.8 Hz	
	FM	KB5ILY	444.87500	+	114.8 Hz	
Ashdown	FM	KB5SSW	147.38000	+	100.0 Hz	
Athens, Tall Peak	FM	KD5NUP	146.92500	-	100.0 Hz	
	FM	N5THS	444.97500	+	88.5 Hz	
Batesville, Desha	FM	K5NES	444.25000	+		
Batesville, Jamestown Mountain						
	FM	N5TSC	224.50000	-	107.2 Hz	
Bauxite	FM	KJ5ZT	443.82500	+	141.3 Hz	
Bearden	FM	N5IOZ	147.33000	+	100.0 Hz	
	FM	N5IOZ	444.77500	+	100.0 Hz	
Benton, Mountain View Road						
	FM	W5RHS	147.12000	+	131.8 Hz	
Bentonville	FM	N5UFO	442.95000	+	97.4 Hz	
Berryville	FM	N6WI	443.80000	+	91.5 Hz	
Bismarck, Bismarck Mountain						
	FM	W5DI	147.27000	+	114.8 Hz	
Blytheville	FM	W5ENL	146.67000	-	107.2 Hz	
Bodcaw-Willisville	FM	W5UBS	444.75000	+	77.0 Hz	
Bono	DMR/BM	NZ5E	442.21250	+	CC1	
Bradford, Oakland	FM	W5BTM	146.74000	-	107.2 Hz	
Bryant	FM	N5TKG	444.30000	+	131.8 Hz	
Cabot	DMR/BM	W5STR	442.47500	+	CC1	
Cabot, Crace Ridge						
	FM	NL7RQ	145.41000	-	85.4 Hz	
Cabot, N Pulaski Water Tank						
	FM	AI5Z	53.75000	52.05000		
Caddo Valley, DeGray Lake						
	FM	KD5ARC	145.11000	-	88.5 Hz	
	FM	KD5ARC	444.95000	+	114.8 Hz	
Camden	FM	WA5OWG	146.91000	-	167.9 Hz	
Cash	FM	W5BE	443.87500	+	94.8 Hz	
Center Ridge, Woolverton Mount						
	FM	W5AUU	444.10000	+	114.8 Hz	
Cherokee Village	FM	KG5ICO	146.64000	-		
Clarkridge	DMR/MARC	N5EWC	443.22500	+	CC14	
Clarksville, Woods Mountain						
	FM	W5OI	147.28500	+	114.8 Hz	
Clinton, Evans Mountain						
	FM	N5UFC	146.91000	-	114.8 Hz	
Clinton, Holly Mountain Airpar						
	FM	W5DI	145.37000	-	114.8 Hz	
	FM	N5YU	442.00000	+	82.5 Hz	
Conway	DMR/BM	W5AUU	145.21000	-	CC1	
	DMR/BM	W5AUU	443.75000	+	CC1	
	FM	W5AUU	53.21000	51.51000	114.8 Hz	
Conway, Conway Landfill						
	FM	W5AUU	146.97000	-	114.8 Hz	
Conway, OEM Office						
	FM	W5AUU	147.03000	+	114.8 Hz	
Conway, Old Water Plant						
	FM	W5AUU	443.80000	+	114.8 Hz	
	FM	W5AUU	443.95000	+	114.8 Hz	
Coy	FM	W5STR	147.15000	+		
Dardanelle, Lake Dardanelle						
	FM	WD5B	146.68500	-	141.3 Hz	
	FM	WD5B	443.87500	+	79.7 Hz	

56   ARKANSAS

Location	Mode	Call sign	Output	Input	Access	Coordinator
Dardanelle, Mt Nebo						
	FM	K5PXP	146.82000	-	131.8 Hz	
	FM	K5PXP	443.40000	+	131.8 Hz	
De Queen, Lee's Summit						
	FM	WA5LTA	147.07500	+	100.0 Hz	
Decatur	FM	N5UXE	146.92500	-	114.8 Hz	
	FM	N5UXE	442.85000	+	123.0 Hz	
	FM	N5UXE	443.92500	+	114.8 Hz	
Decatur, Y City Rd And Hwy 59						
	FM	N5UXE	51.92500	52.92500	114.8 Hz	
Dell	FM	W5ENL	444.65000	+	186.2 Hz	
Dequeen, Beacon Hill						
	FM	WA5LTA	444.80000	+	88.5 Hz	
Dequeen, Hospital						
	FM	N5THR	145.13000	-	100.0 Hz	
	FM	N5THR	147.31500	+	100.0 Hz	
Edgemont	FM	W1ZM	145.18000	-		
El Dorado	FM	KC5AUP	146.74500	-		
Elkins	FM	WC5AR	146.70000	-	110.9 Hz	
Emerson	FM	N5PNB	146.95500	-		
Eureka Springs	DMR/BM	KG5JPK	443.42500	+	CC1	
	FM	K5AA	444.25000	+	100.0 Hz	
	FM	K5SRS	444.47500	+	110.9 Hz	
Eureka Springs, Highway 23 Tow						
	FM	N5UFO	146.86500	-	97.4 Hz	
Fayetteville	FM	WC5AR	442.05000	+	110.9 Hz	
	FM	K5SRS	444.92500	+	88.5 Hz	
Fayetteville , Kessler Mountia						
	FM	K5DVT	147.31500	+	97.4 Hz	
Fayetteville, Kessler Mountian						
	FM	W5KMP	442.00000	+	97.4 Hz	
Fayetteville, Wash Co Courthou						
	FM	WC5AR	147.03000	+	110.9 Hz	
Forrest City	FM	KD5DF	146.76000	-	100.0 Hz	
	FM	WA5CC	147.37500	+	107.2 Hz	
Fort Smith	FM	W5ANR	146.94000	-	88.5 Hz	
	FM	W5ANR	444.30000	+		
Fouke, Miller Co Tower						
	FM	N5MFI	147.28500	+	77.0 Hz	
Fox	IDAS	N5QT	145.11000	-	110.9 Hz	
Garfield, Mount Whitney						
	FM	KD5DMT	443.02500	+	110.9 Hz	
Gentry, Highway 12 Near Highwa						
	FM	K5SRS	146.67000	-	110.9 Hz	
Glenwood	FM	K5JSC	146.83500	-	114.8 Hz	
Green Forest, Bradshaw Mountai						
	FM	K5DVT	145.31000	-	97.4 Hz	
	FM	N5UFO	442.70000	+	97.4 Hz	
Green Forest, Bradshaw Mountia						
	FM	KJ6TQ	146.73000	-	136.5 Hz	
Greenbrier, Mountain Grove Bap						
	FM	W5AUU	146.62500	-	114.8 Hz	
Greers Ferry	FM	W5GFC	147.33000	+		
	FM	W5RHL	443.65000	+	88.5 Hz	
Gurdon, Bull Mountain						
	FM	KD5ARC	145.37000	-	88.5 Hz	
Harrisburg	FM	K5TW	146.83500	-	107.2 Hz	
	FM	K5TW	444.52500	+	94.8 Hz	
Harrison	DMR/BM	W5NWA	442.90000	+	CC1	
Harrison, Boat Mountain						
	FM	WB5CYX	53.15000	51.45000		
	FM	WB5CYX	147.00000	-	103.5 Hz	

ARKANSAS   57

Location	Mode	Call sign	Output	Input	Access	Coordinator
Harrison, OEM Building						
	FM	W5NWA	147.45000		103.5 Hz	
Hartford	FM	KC5JBX	146.89500	-	141.3 Hz	
Hartford, Poteau Mountain						
	FM	WD5MHZ	442.42500	+	88.5 Hz	
Heber Springs	FM	N5XUN	145.23000	-		
	FM	W5HSC	443.37500	+	91.5 Hz	
Heber Springs, Round Mountain						
	FM	W5HSC	145.43000	-	94.8 Hz	
Helena	FM	N5KGA	146.68500	-	100.0 Hz	
Holiday Island	FM	K5AA	146.83500	-	100.0 Hz	
Hollis	FM	K5KM	146.74500	-		
Hope	FM	N5OXP	146.68500	-	114.8 Hz	
Hot Springs	FM	KF5AF	145.27000	-	114.8 Hz	
	FM	W5YTR	224.16000	-		
	FM	WB5PIB	444.00000	+		
Hot Springs Village						
	FM	W5HSV	444.72500	+	114.8 Hz	
Hot Springs Village, Peral Way						
	FM	W5HSV	147.01500	+	114.8 Hz	
Hot Springs, St. Vincents Hosp						
	FM	WB5SPA	147.18000	+	114.8 Hz	
Hot Springs, West Mountain						
	FM	W5LVB	146.88000	-	114.8 Hz	
	FM	W5LVB	444.60000	+	114.8 Hz	
Huntsville	DMR/BM	K5DVT	444.07500	+	CC1	
Huntsville, Governor's Hill						
	FM	K5DVT	145.31000	-	103.5 Hz	
Huntsville, Huntsville Municip						
	FM	KG5OQF	443.62500	+	97.4 Hz	
Imboden	DMR/BM	W5WRA	442.40000	+	CC1	
Imboden, LCARC Tower						
	FM	W5WRA	147.04500	+		
Jacksonville	DMR/MARC	W5JWH	444.12500	+	CC1	
Jasper, Jasper Mountain						
	FM	WB5CYX	146.61000	-	103.5 Hz	
Jonesboro	FM	W5JBR	146.61000	-	107.2 Hz	
	FM	NI5A	147.24000	+	107.2 Hz	
Jonesboro, Crowleys Ridge						
	DSTAR	KG5NAU	443.93750	+		
Jonesboro, KIYS-FM Tower						
	FM	KF5OTW	444.15000	+	94.8 Hz	
Jonesboro, Liberty Bank Stadiu						
	FM	K5NEA	443.15000	+	107.2 Hz	
Ladelle, Lacey-Ladelle FD						
	FM	N5SEA	444.57500	+	127.3 Hz	
Lake City	FM	KC5TEL	444.32500	+	107.2 Hz	
Lamar, Red Oak Rd						
	FM	KE5SQC	442.12500	+	118.8 Hz	
Lepanto	FM	W5MLH	443.32500	+	107.2 Hz	
Little Rock	DMR/BM	K5NSX	145.17000	-	CC1	
	DMR/BM	K5NSX	442.65000	+	CC1	
	DMR/MARC	DMRXP	430.26250	439.66250	CC1	
	DMR/MARC	DMRXS	438.45000	430.85000	CC1	
	DMR/MARC	K5BRM	443.92500	+	CC1	
	FM	WA5LRU	146.85000	-		
	FM	W5FD	444.45000	+	123.0 Hz	
Little Rock, Baptist Med Ctr						
	FM	N5AT	145.13000	-	114.8 Hz	
Little Rock, Markham & Univers						
	FM	N5CG	443.00000	+	100.0 Hz	

58   ARKANSAS

Location	Mode	Call sign	Output	Input	Access	Coordinator
Little Rock, McClellan VA Hosp						
	DSTAR	N5DSD	147.48000	144.98000		
	DSTAR	N5DSD	444.01250	+		
Little Rock, Shinall Mountain						
	FM	N5CG	145.49000	-	114.8 Hz	
	FM	WA5PGB	146.73000	-	114.8 Hz	
	FM	N5CG	146.77500	-	162.2 Hz	
	FM	W5DI	146.94000	-		
	FM	W5DI	147.13500	+	114.8 Hz	
	FM	W5DI	147.30000	+	114.8 Hz	
	FM	N5CG	443.20000	+	114.8 Hz	
	FM	KG5DPK	443.47500	+	156.7 Hz	
	FM	W5DI	444.20000	+	114.8 Hz	
	FM	WA5OOY	444.40000	+	114.8 Hz	
	FM	K5XT	444.80000	+	146.2 Hz	
Little Rock, The Heights						
	FM	W5FD	147.06000	+	114.8 Hz	
Lowell	FM	K5SRS	147.22500	+	103.5 Hz	
Lynn, County Road 2645						
	DSTAR	K5CWR	444.80000	+		
Magazine Mountain						
	FM	N5XMZ	53.11000	51.41000	131.8 Hz	
	FM	N5XMZ	145.35000	-	151.4 Hz	
	FM	W5MAG	147.09000	+	151.4 Hz	
	FM	WD5MHZ	442.82500	+	88.5 Hz	
	FM	N5XMZ	443.25000	+	123.0 Hz	
Magnolia	FM	KC5OAS	147.10500	+	100.0 Hz	
Malvern	DMR/BM	W5DSD	444.52500	+	CC1	
	FM	KJ5YJ	145.31000	-	88.5 Hz	
	FM	W5BXJ	147.36000	+	136.5 Hz	
	FM	W5BXJ	147.39000	+	136.5 Hz	
	FM	KI5DLL	443.05000	+	131.8 Hz	
	FM	K5SZM	443.10000	+		
Malvern, Sulphur Springs Water						
	FM	W5BXJ	443.50000	+	136.5 Hz	
Marshall	FM	W5NWA	443.05000	+	103.5 Hz	
Mena	DMR/BM	KC5MMW	147.99000	144.51000	CC1	
	DMR/BM	KB5JBS	442.95000	+	CC1	
	FM	W5HUM	146.79000	-	100.0 Hz	
	FM	K5PS	444.67500	+		
	FM	KC5MMW	444.92500	+	88.5 Hz	
Monticello	FM	N5SEA	146.83500	-	DCS 023	
Monticello, CAP Tower						
	FM	W5GIF	444.82500	+	127.3 Hz	
Morrilton	FM	N5CG	145.33000	-	114.8 Hz	
Mount Ida, High Peak						
	FM	KA5WPC	52.91000	51.21000	100.0 Hz	
	FM	KA5WPC	146.71500	-	127.3 Hz	
	FM	KA5WPC	444.47500	+	114.8 Hz	
Mountain Home	DMR/BM	NA1MH	147.30000	+	CC1	
	FM	K5OZK	146.88000	-	103.5 Hz	
	FM	K5BAX	442.55000	+		
	FM	WB5NFC	444.00000	+	100.0 Hz	
Mountain Home, Old Lonnon Moun						
	FM	KC5RBO	147.07500	+		
Mountain Home, Wallace Knob						
	FM	WB5NFC	444.97500	+	100.0 Hz	
Mountain Home, Water Tower						
	FM	K5OZK	442.30000	+	103.5 Hz	
Mountain View	DMR/BM	KA0JTI	443.60000	+	CC1	
	DMR/MARC	KA0JTI	442.70000	+	CC1	
	FM	K5GNT	224.40000	-		

ARKANSAS 59

Location	Mode	Call sign	Output	Input	Access	Coordinator
Mountain View, KWOZ Tower						
	DSTAR	KF5BZP	442.63750	+		
	FM	KF5BZP	442.85000	+	97.4 Hz	
Mt. Nebo	DMR/BM	K5CS	442.07500	+	CC1	
Nashville	FM	N5THS	444.97500	+	88.5 Hz	
Nashville, Chapel Hill						
	FM	N5BAB	444.35000	+	88.5 Hz	
Nashville, Yates Tower						
	FM	KC5TSZ	147.04500	+	94.8 Hz	
North Little Rock	FM	W5RXU	444.65000	+	114.8 Hz	
North Little Rock, Camp Robins						
	FM	NT5LA	442.32500	+	85.4 Hz	
North Little Rock, Towbin VA H						
	FM	N5CG	444.70000	+	114.8 Hz	
Ola	FM	WA5YHN	147.21000	+		
	FM	WA5YHN	444.55000	+		
Ozone	FM	KC5LVW	442.62500	+		
Ozone, Cazort Knob						
	FM	K5OO	147.04500	+		
Paragould	FM	W5BJR	145.47000	-		
Pea Ridge	FM	K5SRS	146.95500	-	110.9 Hz	
Pine Bluff, Jefferson						
	FM	K5DAK	146.70000	-	82.5 Hz	
Pine Bluff, Watson Chapel Wate						
	FM	W5RHS	442.42500	+	114.8 Hz	
Prairie Grove	FM	WC5AR	146.76000	-	110.9 Hz	
Prattsville	FM	KD5RTO	145.19000	-	114.8 Hz	
Prattsville, Fire Station						
	FM	KD5RTO	442.87500	+	85.4 Hz	
Providence	FM	KG5S	146.92500	-	94.8 Hz	
Quitman, Miller Point						
	FM	KC5PLA	444.05000	+	127.3 Hz	
Redfield, White Bluff						
	FM	N5KWH	147.19500	+		
Rogers	DMR/BM	KG5JPJ	442.52500	+	CC1	
Rogers, Mercy Medical Center						
	FM	N5UFO	443.17500	+	97.4 Hz	
Rose Bud	FM	W4GXI	444.42500	+		
Royal	DMR/BM	KG5QYM	145.23000	-	CC2	
Royal`	DMR/BM	KG5QYM	442.10000	+	CC1	
Rudy, Landcaster Mountain						
	FM	KD5ZMO	147.16500	+	123.0 Hz	
	FM	KD5ZMO	443.72500	+	123.0 Hz	
Russell, Russell Mountain						
	FM	WA5OOY	147.31500	+	114.8 Hz	
Russellville	DMR/BM	K5CS	442.36250	+	CC1	
Searcy	DMR/BM	N5QZ	442.20000	+	CC1	
Searcy, Foster Chapel Road						
	FM	W1ZM	146.89500	-	85.4 Hz	
Searcy, Hill West Searcy						
	FM	AC5AV	147.39000	+	94.8 Hz	
Searcy, North Searcy						
	FM	AB5ER	146.65500	-	94.8 Hz	
	FM	N5LKE	442.10000	+		
	FM	N5QS	444.50000	+	192.8 Hz	
Sheridan	FM	K5BTM	146.98500	-		
	FM	K5BTM	444.90000	+	88.5 Hz	
Sherwood	DMR/BM	K5NSX	443.90000	+	CC1	
	FM	N1RQ	147.25500	+	114.8 Hz	
Siloam Springs	FM	N5YEI	444.32500	+	114.8 Hz	
Siloam Springs, North Water To						
	FM	WX5SLG	443.57500	+	114.8 Hz	

60   ARKANSAS

Location	Mode	Call sign	Output	Input	Access	Coordinator
Springdale, Beaver Lake Point						
	FM	N5UFO	442.20000	+	97.4 Hz	
Springdale, Carlock Mountain						
	IDAS	K5SRS	443.20000	+		
Springdale, Dodd Mtn.						
	FM	K5DVT	53.03000	51.33000	97.4 Hz	
	FM	KE5LXK	443.65000	+	97.4 Hz	
Star City	FM	W5DI	146.67000	-	114.8 Hz	
Stuttgart	FM	KB5LN	147.00000	-	114.8 Hz	
Summers	DMR/BM	KI5DWB	442.02500	+	CC1	
	DMR/BM	N9INK	442.02500	+	CC1	
Trumann, Quality Farm Supply						
	FM	NI5A	146.95500	-	107.2 Hz	
	FM	NI5A	443.50000	+	94.8 Hz	
Van Buren	FM	KC5YQB	145.19000	-	141.3 Hz	
Vilonia	FM	N5EWC	145.27000	-	100.0 Hz	
	FM	N5EWC	146.49000		100.0 Hz	
Vilonia, Cascade Mountain - Ne						
	FM	W5AMI	444.82500	+	85.4 Hz	
White Hall	FM	KJ5PE	442.17500	+	82.5 Hz	
	FM	AF5AR	443.70000	+		
White Hall, Hardin Water Tower						
	FM	K5DAK	147.24000	+		
Willisville	FM	N5ZAY	146.65500	-	100.0 Hz	
	FM	N5ZAY	444.92500	+		
Winslow, Bobby Hopper Tunnel						
	FM	K5SRS	444.77500	+	103.5 Hz	
Wynne	FM	WY5AR	146.86500	-	107.2 Hz	
Wynne AR	DMR/BM	AI5J	145.31000	-	CC1	
Y City, Buck Knob Mountain						
	FM	W5AWX	442.50000	+	79.7 Hz	
Yellville, Lee Mountain						
	FM	W5DHH	147.24000	+		
CALIFORNIA						
Acton	FM	K6ECS	147.70500	-	103.5 Hz	
Agoura, Castro Peak						
	FM	KE6HGO	447.50000	-		
	FM	N6IGG	1284.25000	-		
Ahwahnee	FM	WB6NIL	224.40000	-	123.0 Hz	
Alameda	DMR/BM	KG7NBL	442.30000	+	CC1	
	DMR/MARC	KA7QQV	443.50000	+	CC1	
	FM	KF6ALA	147.82500	-	88.5 Hz	
Alameda, Alameda Hospital						
	FM	K6QLF	444.57500	+	88.5 Hz	
Alhambra, Cal State LA						
	FM	KM6EON	445.06000	-	186.2 Hz	
Aliso Viejo	FM	KI6DB	445.10000	-	100.0 Hz	
Alleghany	FM	WR6ASF	444.92500	+	88.5 Hz	
Alpine	FM	N6LZR	441.55000	+		
	FM	K6KTA	447.58000	-	107.2 Hz	
Alpine, Monument Peak						
	FM	N6LVR	449.18000	-	88.5 Hz	
Altadena	FM	W6TOI	445.64000	-	156.7 Hz	
Altadena, Mt Disappointment						
	FM	K6CPT	145.30000	-	100.0 Hz	
	FM	K6CPT	224.30000	-	100.0 Hz	
	FM	K6VGP	446.24000	-		
	FM	K6VGP	1283.25000	-		
	FM	K6CPT	1285.30000	-	100.0 Hz	
Altadena, Mt Harvard						
	FM	N6LXX	53.62000	-	107.2 Hz	

CALIFORNIA 61

Location	Mode	Call sign	Output	Input	Access	Coordinator
Altadena, Mt Wilson						
	FM	K6VGP	147.36000	+	103.5 Hz	
	FM	WA6LWW	446.78000	-	107.2 Hz	
	FM	WA6TTL	449.72000	-		
Altadena, Mt. Disappointment						
	FM	K6VGP	224.56000	-	114.8 Hz	
	FM	WA6IBL	445.80000	-		
Altadena, Mt. Harvard						
	FM	WA6DVG	224.94000	-	94.8 Hz	
Altadena, Mt. Wilson						
	FM	WA6VLD	448.50000	-		
	FM	W6JYP	448.88000	-		
	FM	N6CIZ	449.70000	-	131.8 Hz	
Alturas	FM	N6KMR	147.36000	+	100.0 Hz	
	FM	N6KMR	444.25000	+	100.0 Hz	
Alturas, Likely Mountain						
	FM	WB6HMD	146.97000	-	100.0 Hz	
	FM	K6PRN	441.22500	+	100.0 Hz	
	FM	WB6HMD	442.35000	+	85.4 Hz	
Amboy	DMR/MARC	KF6FM	448.31250	-	CC1	
American Canyon	FM	K6ZRX	927.40000	902.40000	192.8 Hz	
Anaheim	DMR/BM	WI6Y	449.46250	-	CC3	
	FM	WI6Y	445.22000	-		
Anaheim, Disneyland						
	FM	KE6FUZ	146.94000	-	131.8 Hz	
Anaheim, HRO	DSTAR	W6HRO	147.57000	144.97000		
	DSTAR	W6HRO	449.84000	-		
Anchor Bay	FM	WA6RQX	147.27000	+	114.8 Hz	
Angels Camp	FM	NC6R	441.12500	+	156.7 Hz	
	FM	N6LZR	444.85000	+	103.5 Hz	
Angwin	DMR/MARC	K6LO	147.85500	-	CC1	
	DMR/MARC	K6LO	440.07500	+	CC1	
Antioch	FM	KC6WYA	440.65000	+	127.3 Hz	
Anza, Toro Peak	FM	W6YQY	224.10000	-		
	FM	WB6RHQ	224.18000	-	156.7 Hz	
	FM	WD6FZA	446.38000	-		
	FM	WV6H	1284.85000	-		
Apple Valley	FM	KB6BZZ	445.68000	-	141.3 Hz	
Arbuckle	FM	N6NHI	147.12000	+	118.8 Hz	
	FM	N6NHI	224.54000	-	118.8 Hz	
Arcadia	FM	N6AH	145.20000	-	103.5 Hz	
	FM	WA6SBH	449.40000	-		
	FM	KD6WLY	927.17500	902.17500	103.5 Hz	
Arcadia, Santa Anita Ridge						
	FM	WA6CGR	224.28000	-	107.2 Hz	
	FM	K6TEM	445.48000	-	131.8 Hz	
	FM	N6EX	445.50000	-	85.4 Hz	
	FM	KI6DYV	447.58000	-	100.0 Hz	
	FM	K6JSI	447.58000	-	100.0 Hz	
	FM	W6VHU	1283.30000	-		
	FM	WA6CGR	1283.90000	-		
Arnold	FM	KD6GIY	441.72500	+	162.2 Hz	
Arroyo Grande	FM	WB6FMC	146.70000	-	127.3 Hz	
	FM	N6CAV	147.03000	+	127.3 Hz	
Arroyo Grande, Lopez Hill						
	FM	W6YDZ	443.97500	+	127.3 Hz	
Arroyo Grande, Lopez Mountain						
	FM	W6SLO	146.94000	-	127.3 Hz	
Auberry	FM	KG6IBA	444.27500	+	127.3 Hz	
	P25	N6LYE	443.60000	+		
Auburn	DMR/BM	NG6D	442.96250	+	CC1	
	DSTAR	K6IOK	444.50000	+		

62   CALIFORNIA

Location	Mode	Call sign	Output	Input	Access	Coordinator
Auburn	FM	N6JSL	29.62000	-	156.7 Hz	
	FM	K6IOK	51.70000	-	77.0 Hz	
	FM	AK6OK	145.13000	-	114.8 Hz	
	FM	W6SAR	145.27000	-	156.7 Hz	
	FM	K6IOK	145.31000	-	114.8 Hz	
	FM	W6EK	145.43000	-	162.2 Hz	
	FM	N6JSL	146.76000	-	136.5 Hz	
	FM	N6NMZ	147.31500	+	77.0 Hz	
	FM	WB6ALS	223.82000	-	131.8 Hz	
	FM	W6EK	223.86000	-	110.9 Hz	
	FM	N6NMZ	223.90000	-	100.0 Hz	
	FM	W7FAT	224.02000	-	100.0 Hz	
	FM	W4WIL	224.58000	-	167.9 Hz	
	FM	W6EK	440.57500	+	162.2 Hz	
	FM	N6NMZ	444.60000	+	192.8 Hz	
	FM	N6NMZ	927.15000	902.15000		
	FM	KI6SSF	927.36250	902.36250		
	FM	N6NMZ	927.77500	902.77500	77.0 Hz	
	P25	N6LYE	443.60000	+		
Auburn, Stony Ridge						
	FM	K6IOK	444.47500	+	94.8 Hz	
Auburn, Sugar Pine Ridge						
	FM	K6JSI	444.95000	+	100.0 Hz	
Aukum, Mount Aukum						
	FM	W6HMT	52.64000	-	88.5 Hz	
Avalon, Buena Vista Pt.						
	FM	AA6DP	446.14000	-	110.9 Hz	
Avalon, Mt. Ada	FM	N6SCI	446.86000	-	141.3 Hz	
Baker	FM	N6BKL	53.94000	-	82.5 Hz	
Baker, Halloran Summit						
	FM	WR6TM	446.38000	-	136.5 Hz	
Baker, Shamrock Peak						
	FM	K6DK	147.15000	+	131.8 Hz	
Baker, Turquoise Mountain						
	FM	WD6DIH	446.96000	-	136.5 Hz	
	FM	W7DOD	448.16000	-	141.3 Hz	
Bakersfield	DMR/BM	WX6D	440.95000	+	CC1	NARCC
	DMR/BM	K6RET	443.00000	+	CC1	
	FM	KF6JOQ	145.21000	-	100.0 Hz	
	FM	W6LIE	146.91000	-	100.0 Hz	
	FM	KG6KKV	147.15000	+	100.0 Hz	
	FM	W6LIE	224.06000	-	100.0 Hz	
	FM	W6PVG	224.42000	-	156.7 Hz	
	FM	KG6KKV	224.52000	-	100.0 Hz	
	FM	N7BJD	443.27500	+	141.3 Hz	
	FM	W6LIE	443.90000	+	100.0 Hz	
	FM	KG6FOS	444.67500	+	107.2 Hz	
	FM	N6SMU	444.75000	+	141.3 Hz	
	FM	W6LIE	1285.45000	-	100.0 Hz	
Bakersfield, Breckenridge Moun						
	FM	W6LIE	145.15000	-	100.0 Hz	
	FM	K6JSI	447.64000	-	100.0 Hz	
Bakersfield, Cummings Mountain						
	FM	KG6KKV	444.42500	+	100.0 Hz	
Bakersfield, Getty	FM	K6RET	927.12500	902.12500	DCS 125	
Bakersfield, Mt Adelaide						
	FM	KC6EOC	147.21000	+	100.0 Hz	
Bakersfield, Pampa						
	FM	K6RET	147.27000	+	94.8 Hz	
Bakersfield, Round Mtn						
	P25	KE6YJC	440.97500	+		

CALIFORNIA 63

Location	Mode	Call sign	Output	Input	Access	Coordinator
Bakersfield, Shirley Peak						
	FM	W6LI	145.41000	-	103.5 Hz	
Baldwin Lake, Big Bear City, B						
	FM	KK6MOS	224.92000	-		
Banning	DMR/MARC	KF6FM	448.16250	-	CC3	
Banning, Mt Edna	FM	W6CDF	147.91500	-	123.0 Hz	
Banning, Snow Peak						
	FM	W6CTR	445.16000	-	67.0 Hz	
	FM	N6AJB	448.40000	-		
	FM	WB6TZC	448.66000	-		
Barstow	DMR/MARC	K6DEW	447.90000	-	CC1	
	FM	WA6TST	146.97000	-		
	FM	KE6JZS	447.58000	-	107.2 Hz	
	FM	KC6NKK	927.55000	902.55000	100.0 Hz	
Barstow, Flash II	FM	N6SLD	145.22000	-	114.8 Hz	
	FM	WA6TST	147.18000	+	151.4 Hz	
Barstow, Ord Mountain						
	FM	W6SCE	224.32000	-	100.0 Hz	
	FM	W6ATN	448.42000	-		
	FM	WR6OM	449.78000	-	136.5 Hz	
Bear Valley Springs, Cub Mount						
	FM	W6SLZ	146.70000	-	123.0 Hz	
Bell	FM	N6WZK	445.02000	-	167.9 Hz	
	FM	N6WZK	448.40000	-	151.4 Hz	
Belmont	FM	KR6WP	440.07500	+	114.8 Hz	
Belmont, City Hall	FM	WB6CKT	147.09000	+	100.0 Hz	
Ben Lomond	FM	WR6AOK	147.12000	+	94.8 Hz	
Benicia	FM	KR6BEN	441.25000	+	100.0 Hz	
	FM	KR6BEN	442.75000	+	100.0 Hz	
Berkeley	FM	K6GOD	440.17500	+	131.8 Hz	
	FM	WB6WTM	440.40000	+	127.3 Hz	
	FM	WB6IXH	440.82500	+		
	FM	WA2UNP	440.90000	+	131.8 Hz	
	FM	WA6ZTY	442.27500	+	100.0 Hz	
	FM	WB6UZX	442.67500	+		
	FM	KK6PH	1285.30000	-	88.5 Hz	
	FM	WA2UNP	1285.55000	-	114.8 Hz	
Berkeley, Grizzly Peak						
	FM	KB6LED	145.29000	-	131.8 Hz	
	FM	K6GOD	223.78000	-	110.9 Hz	
	FM	KB6LED	224.34000	-	100.0 Hz	
	FM	K6DJR	442.72500	+		
Berkely	FM	N6BRK	224.90000	-	131.8 Hz	
Beverly Hills, Hollywood Hills						
	FM	WA6PPS	144.50500	+	162.2 Hz	
Big Bear City	FM	WA6ITC	446.40000	-	162.2 Hz	
	FM	WA6ITC	446.42000	-	162.2 Hz	
Big Bear City, Bear Mtn.						
	FM	W6RRN	448.74000	-		
Big Bear Lake	FM	K6BB	147.33000	+	131.8 Hz	
Big Bear, E. Of Snow Summit						
	FM	K6BB	446.20000	-	103.5 Hz	
Bishop	FM	W6SCE	224.76000	-		
Bishop, Silver Peak						
	FM	N6OV	146.94000	-	103.5 Hz	
Black Mountain, Black Mountain						
	FM	WA6LAW	146.88000	-	162.2 Hz	
	FM	WM6Z	147.12000	+	103.5 Hz	
Blythe	FM	K6JRM	446.05000	-	88.5 Hz	
Blythe, Big Maria	FM	W6SCE	224.32000	-		
Blythe, Black Rock						
	FM	W6CDF	147.09000	+	123.0 Hz	

64   CALIFORNIA

Location	Mode	Call sign	Output	Input	Access	Coordinator
Bodega Bay	DMR/BM	KJ6QBM	440.32500	+	CC1	
Bonny Doon	DMR/BM	WB6ECE	440.58750	+	CC2	
	FM	N6NAC	224.06000	-	110.9 Hz	
Boulder Creek	FM	KI6YDR	145.35000	-	94.8 Hz	
Brawley, Black Mountain						
	FM	WD6AWP	448.04000	-	136.5 Hz	
Brawley, Superstition Mountain						
	FM	N6LVR	146.67000	-	103.5 Hz	
Brea, Brea Civic Center						
	FM	WB6DNX	446.54000	-	192.8 Hz	
Brentwood	DMR/BM	W6APX	443.70000	+	CC1	
Brisbane	FM	K6CV	440.70000	+	123.0 Hz	
Buena Park, Knotts Berry Farm						
	FM	K6KBF	445.52000	-	85.4 Hz	
Burbank	DMR/BM	KE6ZRP	439.42000		CC1	
Burbank, Verdugo Mountain						
	FM	WA6PPS	147.30000	+	110.9 Hz	
	FM	N6MQS	224.20000	-	123.0 Hz	
	FM	KA6AZB	1283.95000	-		
Burney, Hatchet Mountain						
	FM	KE6CHO	444.65000	+	94.8 Hz	
Cactus City	FM	NR6P	146.02500	+	107.2 Hz	
	FM	W6SCE	224.32000	-		
	FM	W6CDF	445.02000	-	186.2 Hz	
Calexico	DMR/BM	N2IX	449.02500	-	CC5	
	DMR/BM	K6CLX	449.70000	-	CC1	
Calexico, Desert View Tower						
	FM	K6CLX	147.00000	+	162.2 Hz	
Calistoga	FM	K6ZRX	52.62000	-	114.8 Hz	
	FM	N6PMF	444.15000	+		
	FM	N6TKW	444.17500	+	151.4 Hz	
	FM	K6IRC	444.47500	+		
	FM	WZ6X	1283.90000	-	88.5 Hz	
Calsbad	DMR/MARC	N6ZEK	449.90000	-	CC4	
Camarillo Springs	FM	K6ERN	147.91500	-	127.3 Hz	
Camarillo, Camarillo Hills						
	FM	K6BBK	147.61500	-		
	FM	WD6EBY	445.60000	-	141.3 Hz	
Camarillo, Camarillo Springs						
	FM	K6ERN	447.00000	-	103.5 Hz	
Cambria	FM	KC6TOX	147.27000	+	127.3 Hz	
	FM	WB6JWB	224.68000	-		NARCC
Cameron Park	FM	N6RDE	147.03000	+	77.0 Hz	
	FM	N6RDE	440.12500	+		
	FM	WA6NHC	441.00000	+	100.0 Hz	
Camino	FM	W6MPD	224.06000	-		
Camino, Mt Danaher						
	FM	AG6AU	147.82500	-	82.5 Hz	
Campbell	FM	K6KMT	1284.85000	-	100.0 Hz	
Campbell, Crystal Peak						
	FM	K9GVF	442.17500	+	100.0 Hz	
Canyon Country	FM	KC6TKA	445.90000	-	107.2 Hz	
	FM	KI6JL	1282.00000	-		
	FM	KI6JL	1284.30000	-		
Canyon Lake	FM	KI6IGR	147.91500	-	100.0 Hz	
Carlsbad	DMR/BM	AI6BX	445.30000	-	CC3	
	DMR/BM	K4DJN	447.60000	-	CC7	
	FM	W6THC	147.91500	-		
	FM	K6VST	446.86000	-	103.5 Hz	
Carmichael	FM	KJ6JD	224.88000	-	162.2 Hz	
Carson	FM	KK6BE	1285.37500	-		
Casmalia	FM	WA6VPL	52.60000	-	82.5 Hz	

CALIFORNIA 65

Location	Mode	Call sign	Output	Input	Access	Coordinator
Catalina	DMR/BM	AA6DP·	147.09000	+	CC1	
Catalina Island, Airport						
	FM	AA6DP	224.42000	-	110.9 Hz	
	FM	KR6AL	927.93750	902.93750	DCS 311	
Catalina Island, Catalina Isla						
	FM	N6KNW	51.86000	-	82.5 Hz	
Cathedral City, Edom Hill						
	FM	WD6RAT	146.94000	-	107.2 Hz	
	FM	K6JR	449.24000	-	131.8 Hz	
Cathedral City, Palm Springs						
	FM	KD6QLT	146.76000	-	107.2 Hz	
Cazadero	FM	K6ACS	147.97500	-	88.5 Hz	
Cedarpines Park, Jobs Peak						
	FM	W6CDF	224.26000	-	110.9 Hz	
	FM	W6CDF	445.02000	-	107.2 Hz	
Channel Islands Harbor, Ventur						
	FM	KI6HHU	445.76000	-	141.3 Hz	
Chatsworth	DMR/MARC	KF6FM	447.52000	-	CC2	
Chatsworth Peak	FM	K6LAM	445.46000	-		
Chatsworth, Chatsworth Peak						
	FM	WD6EBY	145.24000	-	127.3 Hz	
	FM	W6RRN	445.80000	-		
	FM	WD6EBY	445.84000	-	141.3 Hz	
	FM	W6XC	446.38000	-		
	FM	WA6TTL	446.66000	-		
	FM	WA6TTL	447.16000	-		
Chatsworth, Magic Mountain						
	FM	WA6DVG	224.58000	-	156.7 Hz	
Chatsworth, Oat Mountain						
	FM	K6LRB	53.76000	-	82.5 Hz	
	FM	KB6C	147.73500	-	100.0 Hz	
	FM	KF6JWT	147.94500	-	136.5 Hz	
	FM	WD6FZA	224.40000	-		
	FM	K6LRB	224.62000	-	82.5 Hz	
	FM	W6SCE	224.74000	-		
	FM	WD6AWP	445.10000	-	82.5 Hz	
	FM	K7FY	445.88000	-		
	FM	WA6LWW	446.68000	-		
	FM	N6LXX	446.80000	-	103.5 Hz	
	FM	WB6LST	447.56000	-	136.5 Hz	
	FM	KE6PGN	447.82000	-	67.0 Hz	
	FM	N6LXX	927.58750	902.58750	131.8 Hz	
Chester	FM	N6TZG	441.37500	+		
	FM	KF6CCP	444.50000	+	103.5 Hz	
Chico	DMR/BM	KI6ND	440.40000	+	CC1	
	DMR/MARC	K6CHO	146.89500	-	CC0	
	FM	KI6ND·	145.49000	-	110.9 Hz	
	FM	W6RHC	146.85000	-	110.9 Hz	
	FM	W6ECE	146.94000	-	123.0 Hz	
	FM	K6NP	147.30000	+	141.3 Hz	
	FM	WA6UHF	224.28000	-	110.9 Hz	
	FM	KI6ND	224.62000	-	88.5 Hz	
	FM	KI6PNB	440.00000	+	100.0 Hz	
	FM	WA6UHF	440.50000	+		
	FM	WB6RHC	440.55000	+		
	FM	W6RHC	440.65000	+	110.9 Hz	
	FM	N6EJX	440.67500	+		
	FM	WB6RHC	441.40000	+	110.9 Hz	
	FM	W6ECE	442.37500	+	100.0 Hz	
	FM	KI6ND	927.07500	902.07500	88.5 Hz	
Chico, Paradise	FM	K6JSI	224.44000	-	100.0 Hz	

66   CALIFORNIA

Location	Mode	Call sign	Output	Input	Access	Coordinator
Chico, Platte Mountain						
	FM	N6TZG	147.97500	-	110.9 Hz	
Chino	FM	K6OPJ	445.56000	-	136.5 Hz	
	FM	K0JPK	447.50000	-		
Chino Hills	FM	W6AJP	445.84000	-		
Chula Vista	FM	KK6KD	145.26000	-	107.2 Hz	
Chula Vista, Sharp Hospital						
	FM	KK6KD	224.94000	-	107.2 Hz	
Cisco Grove, Cisco Butte						
	FM	N6MVT	443.47500	+	100.0 Hz	
Citrus Heights	FM	KA6FTY	444.72500	+		
Claremont	DMR/BM	K6IRF	448.26000	-	CC1	
Claremont, Sunset Ridge						
	FM	WA1IRS	224.90000	-	103.5 Hz	
	FM	KD6AFA	445.92000	-	186.2 Hz	
	FM	N6LXX	446.10000	-	100.0 Hz	
	FM	N6SIM	446.30000	-	127.3 Hz	
	FM	K4ELE	446.48000	-		
	FM	N6DKA	447.72000	-		
	FM	W7BF	448.36000	-		
	FM	N6ENL	448.52000	-		
	FM	W6RRN	448.76000	-		
	FM	WH6NZ	448.78000	-		
	FM	WB6TZA	449.00000	-		
	FM	W6OY	449.14000	-	100.0 Hz	
	FM	K6UFX	449.52000	-	100.0 Hz	
	FM	N6LXX	927.55000	902.05000	123.0 Hz	
Clayton	FM	N6AMG	927.11250	902.11250		
Clearlake	FM	WR6COP	442.82500	+		
Cloverdale	FM	KI6B	146.97000	-	103.5 Hz	
	FM	WB6QAZ	449.70000	-	88.5 Hz	
Clovis	DMR/MARC	WX6D	145.45000	-	CC1	
	DMR/MARC	WX6D	440.05000	+	CC1	NARCC
	FM	NI6M	440.35000	+	141.3 Hz	
	FM	N6IB	443.82500	+	141.3 Hz	
Clovis, Meadow Lakes Ridge						
	FM	K6ARP	147.67500	-	141.3 Hz	
	FM	N6JXL	224.38000	-	141.3 Hz	
Clovis, Mile High	FM	K6ARP	444.72500	+	141.3 Hz	
Coalinga	DMR/BM	N6VYT	145.02500	147.50000	CC1	NARCC
	FM	N6LEX	440.52500	+	146.2 Hz	
	FM	K6JKL	440.75000	+	114.8 Hz	
	FM	KF6FM	442.42500	+		
	FM	NC9RS	927.66250	902.66250	146.2 Hz	
Coalinga, Joaquin Ridge						
	P25	K6NOX	443.72500	+	107.2 Hz	
Coalinga, Santa Rita Peak						
	FM	W6VFZ	224.44000	-	100.0 Hz	
	FM	W6EMS	440.67500	+	146.2 Hz	
	FM	K6JSI	441.67500	+	100.0 Hz	
	FM	N6OA	441.90000	+	100.0 Hz	
Coarsegold	FM	W6HMH	146.64000	-	127.3 Hz	
	FM	W6HMH	442.90000	+	127.3 Hz	
	FM	KE6YMW	444.40000	+	82.5 Hz	
	FM	WB6NIL	444.50000	+	131.8 Hz	
Coarsegold, Trabuco Mountain						
	FM	K6MXZ	444.37500	+	123.0 Hz	
Cohasset	FM	KH8AF	444.12500	+		
Cohassett	FM	N6YCK	224.36000	-	110.9 Hz	
Columbia	FM	W6FEJ	147.94500	-	100.0 Hz	
	FM	K6DEL	440.85000	+	146.2 Hz	
	FM	K6TUO	440.97500	+	103.5 Hz	

CALIFORNIA 67

Location	Mode	Call sign	Output	Input	Access	Coordinator
Columbia	FM	K6NOX	927.40000	902.40000	107.2 Hz	
Concord	FM	W6YOP	223.98000	-	85.4 Hz	
	FM	W6YOP	224.92000	-	85.4 Hz	
	FM	WB6FRM	440.32500	+	127.3 Hz	
	FM	N6BLA	440.77500	+		
	FM	WA6HAM	440.87500	+		
	FM	W6YOP	441.75000	+	127.3 Hz	
	FM	WB6BDD	441.82500	+	114.8 Hz	
	FM	WA6ZTY	442.65000	+	100.0 Hz	
	FM	K6JJC	443.50000	+		
	FM	K6IRC	443.57500	+		
	FM	K6FJ	444.87500	+	123.0 Hz	
	FM	N6OLD	927.55000	902.55000	141.3 Hz	
Concord, Kregor Peak						
	FM	WA6HAM	147.73500	-	107.2 Hz	
Concord, Mt Diablo						
	DSTAR	W6CX	145.00000	147.50000		
	FM	K6POU	145.33000	-	100.0 Hz	
	FM	AB6CR	145.35000	-	88.5 Hz	
	FM	W6CX	224.78000	-	77.0 Hz	
	FM	W6CX	441.32500	+	100.0 Hz	
	FM	K6POU	443.80000	+	100.0 Hz	
Contractors Point	FM	KC6JAR	445.34000	-	103.5 Hz	
Copperopolis	FM	KG6TXA	29.66000	-	141.3 Hz	
	FM	KB6RYU	224.34000	-	141.3 Hz	
	FM	KD6FVA	441.65000	+	156.7 Hz	
	FM	N6MAC	442.37500	+	186.2 Hz	
Copperopolis, Gopher Ridge						
	FM	N6GKJ	147.01500	+	114.8 Hz	
Cordelia	FM	WZ6X	1282.40000	-	88.5 Hz	
Corning	FM	NC9RS	927.63750	902.63750	123.0 Hz	
Corona	FM	W6CPD	147.22500	+	156.7 Hz	
	FM	W6CPD	147.88500	-	88.5 Hz	
Corona, Lake Elsinore						
	FM	KI6ITV	146.76000	-	136.5 Hz	
Corona, Norco Hills						
	FM	W6PWT	147.06000	+	162.2 Hz	
Corona, Pleasants Peak						
	FM	W7BF	224.80000	-		
	FM	W7BF	448.36000	-		
Corona, Sierra Peak						
	FM	KD6DDM	146.61000	-	103.5 Hz	
	FM	W6NUT	147.45000	146.45000	127.3 Hz	
	FM	W6KRW	223.76000	-	110.9 Hz	
	FM	WB6ORK	447.10000	-	136.5 Hz	
	FM	K6ARN	448.28000	-	107.2 Hz	
	FM	W6KRW	1282.27500	-	88.5 Hz	
Coronado	DSTAR	W6SH	447.20000	-		
	FM	W6SH	147.18000	+	110.9 Hz	
Coronado, Point Loma						
	FM	W6RDF	145.38000	-	107.2 Hz	
Corralitos	FM	KJ6FFP	146.70000	-	94.8 Hz	
Costa Mesa	FM	AD6HK	224.32000	-	151.4 Hz	
Costa Mesa, City Hall						
	FM	WB6HRO	147.06000	+	100.0 Hz	
Cottonwood	FM	K6JSI	147.30000	+	123.0 Hz	
Covelo	FM	WB6TCS	147.21000	+	103.5 Hz	NARCC
Covelo, Anthony Pk						
	FM	WB6TCS	147.21000	+	103.5 Hz	
Covina	DMR/MARC	K9KAO	145.36000	-	CC1	
	FM	WA6NJJ	223.84000	-	151.4 Hz	
Crescent City	FM	W6HY	146.88000	-	136.5 Hz	

68   CALIFORNIA

Location	Mode	Call sign	Output	Input	Access	Coordinator
Crescent City	FM	K6JSI	147.06000	+	100.0 Hz	
	FM	KA7PRR	224.62000	-	91.5 Hz	
	FM	KA7PRR	224.72000	-	103.5 Hz	
	FM	KA7PRR	442.52500	+		
	FM	KD6GDZ	443.05000	+	100.0 Hz	
	FM	KF6QBW	444.82500	+	100.0 Hz	
Crescent City, Flynn Center						
	FM	W6HY	147.39000	+	136.5 Hz	
Crestline	FM	N6QCU	224.20000	-		
	FM	W6JBT	224.86000	-	77.0 Hz	
	FM	W6JJR	445.36000	-		
	FM	K6JTH	448.70000	-		
	FM	K6DLP	449.86000	-	107.2 Hz	
	FM	K6DLP	449.92000	-		
Crestline, Jobs Peak						
	FM	W6CTR	146.91000	-	151.4 Hz	
	FM	W6ATN	1253.25000	-		
Crestline, Skyland Peak						
	FM	W6JBT	146.85000	-	146.2 Hz	
Crestline, Strawberry Peak						
	FM	K6IOJ	448.42000	-		
Culver City	DMR/MARC	KJ6YQW	446.43000	-	CC1	
	FM	K6CCR	445.60000	-	131.8 Hz	
Culver City, Baldwin Hills						
	FM	WA6TFD	146.92500	-	114.8 Hz	
	FM	WA6MDJ	445.32000	-	88.5 Hz	
Cupertino	FM	W6AMT	440.12500	+		
	FM	W6TDM	440.15000	+	100.0 Hz	
	FM	W6VB	441.55000	+		
	FM	N6MBB	1284.00000	-	100.0 Hz	
	FM	W6MOW	1285.65000	-	110.9 Hz	
Cupertino, Montebello Ridge						
	FM	KA2FND	224.62000	-	110.9 Hz	
Cypress	DMR/BM	N0CSW	444.50000	+	CC1	
	DSTAR	N6CYP	144.89500	+		
Daly City	DMR/MARC	W6BUR	442.75000	+	CC0	
	DSTAR	KN6CDC	144.95000	147.45000		
	FM	WD6INC	146.83500	-	123.0 Hz	
	FM	K6TEA	440.67500	+		
	FM	K6JDE	442.37500	+	114.8 Hz	
	FM	KC6PGV	442.37500	+	156.7 Hz	
	FM	KF6REK	442.47500	+	114.8 Hz	
Dana Point	FM	N6OCS	927.17500	902.17500	123.0 Hz	
Danville	DMR/BM	N6TRB	440.65000	+	CC1	
Davis, UC Davis	FM	K6JRB	145.45000	-	203.5 Hz	
Desert Center	FM	KA6GBJ	147.03000	+	107.2 Hz	
Desert Center, Chuckwalla Moun						
	FM	W6DRA	145.38000	-	100.0 Hz	
	FM	WC6MRA	224.04000	-		
	FM	W6YQY	224.70000	-		
	FM	W6SCE	224.76000	-		
Diamond Bar	FM	W7BF	146.64000	-	167.9 Hz	
	FM	N6XPG	147.03000	+	100.0 Hz	
	FM	WR6JPL	224.70000	-	114.8 Hz	
	FM	NO6B	445.08000	-	103.5 Hz	
	FM	NO6B	446.28000	-		
	FM	WA6TTL	447.14000	-		
Diamond Springs	FM	WA6EUZ	444.07500	+		
Dillon Beach	FM	KI6SUD	146.86500	-	88.5 Hz	
Dinsmore	FM	K6FWR	146.98000	-	103.5 Hz	
Dinuba	FM	N6SGW	147.30000	+	94.8 Hz	
	FM	N6SGW	444.82500	+	141.3 Hz	

CALIFORNIA   69

Location	Mode	Call sign	Output	Input	Access	Coordinator
Dixon	DMR/BM	K6JWN	444.52500	+	CC9	
	FM	K6JWN	441.88750	+	94.8 Hz	
Dobbins	FM	N6NMZ	147.04500	+	77.0 Hz	
	FM	N6ICW	444.30000	+		
Duarte	FM	KA6AMR	146.08500	+	110.9 Hz	
Dublin	FM	KQ6RC	224.40000	-		
Dunnigan	FM	NA6DF	224.72000	-	77.0 Hz	
Dunsmuir	FM	K6SIS	146.82000	-	100.0 Hz	
Dunsmuir, Mt Bradley						
	FM	W7PRA	146.67000	-	136.5 Hz	
El Cajon	DMR/BM	KF6YB	449.90000	-	CC1	
	DMR/MARC	W6HDC	446.06250	-	CC1	
	FM	KN6NA	146.35500	+	123.0 Hz	
	FM	WA6BGS	147.42000	146.52000	107.2 Hz	
	FM	WA6BGS	445.90000	-	107.2 Hz	
	FM	W6HDC	447.52000	-	127.3 Hz	
El Cajon, Crest Water Tank						
	FM	WA6BGS	224.08000	-	107.2 Hz	
El Cajon, Lyons Peak						
	FM	W6SS	146.26500	+	107.2 Hz	
	FM	K6JCC	223.80000	-		
	FM	WV6H	223.94000	-	141.3 Hz	
El Cerrito	DMR/MARC	AH6KD	441.87500	+	CC0	
	FM	N6GVI	444.70000	+	100.0 Hz	
El Dorado	FM	W6OIU	52.56000	-	107.2 Hz	
	FM	WA6JQV	927.02500	902.02500		
	FM	WA6JQV	927.47500	902.47500		
El Dorado Hill	FM	N6QDY	441.10000	+	123.0 Hz	NARCC
El Dorado Hills	DMR/BM	W6JMP	145.01250	147.51250	CC1	
	FM	W7CCE	52.82000	-	110.9 Hz	
	FM	KD6CQ	145.37000	-	100.0 Hz	
	FM	N6QDY	441.10000	+	123.0 Hz	
	FM	NC9RS	927.01250	902.01250	100.0 Hz	
El Mirage	FM	AA7SQ	147.99000	-		
El Monte	DMR/BM	W1REI	443.72500	+	CC1	
El Paso Peak	DMR/MARC	WI6RE	447.82500	-	CC15	
El Segundo	FM	WB6VMV	445.24000	-	88.5 Hz	
El Segundo, Raytheon						
	FM	W6HA	445.62000	-	127.3 Hz	
Elk Creek	FM	N6YCK	147.10500	+	110.9 Hz	
Elsinore Peak, Ca	DMR/BM	WD6FZA	445.72000	-	CC1	
Elverta	DMR/BM	K9WWV	441.85000	+	CC1	
	FM	N6ICW	927.95000	902.95000		
Engineer Springs, Hi-Pass						
	FM	K6GAO	145.28000	-	107.2 Hz	
Escondido	DMR/MARC	W6PSA	446.14000	-	CC1	
Escondido, Palomar Mountain						
	DSTAR	KI6MGN	147.57000	144.97000		
	DSTAR	KI6MGN	1282.67500	-		
	FM	W6JLL	51.72000	-	82.5 Hz	
	FM	W6NWG	52.68000	-	107.2 Hz	
	FM	W6ZN	145.28000	-	74.4 Hz	
	FM	N6NIK	145.44000	-	186.2 Hz	
	FM	W6NWG	146.70000	-		
	FM	K6RIF	147.03000	+	103.5 Hz	
	FM	W6NWG	147.07500	+	107.2 Hz	
	FM	W6NWG	147.13000	+	107.2 Hz	
	FM	K6JCC	147.19500	+	203.5 Hz	
	FM	KK6KD	224.38000	-	107.2 Hz	
	FM	WD6HFR	224.90000	-	107.2 Hz	
	FM	W6NWG	447.00000	-	107.2 Hz	
	FM	K6RIF	447.80000	-	88.5 Hz	

70   CALIFORNIA

Location	Mode	Call sign	Output	Input	Access	Coordinator
Escondido, Palomar Mountain						
	FM	K6JXY	448.54000	-	91.5 Hz	
	FM	KE6VK	448.96000	-	131.8 Hz	
	FM	K6JSI	449.08000	-	123.0 Hz	
	FM	W6YJ	449.32000	-		
	FM	WA6TTL	449.72000	-		
	FM	W6YJ	927.11250	902.11250		
Esparto	FM	NC6R	441.12500	+	127.3 Hz	
Eureka	FM	W6ZZK	145.47000	-	103.5 Hz	
	FM	K6FWR	146.70000	-	103.5 Hz	
	FM	WA6RQX	442.22500	447.32500		
Eureka, Bunker Hill						
	FM	WB6HII	147.44500	-	103.5 Hz	
	FM	WA6RQX	444.75000	+	100.0 Hz	
Eureka, Horse Mountain						
	FM	AE6R	442.00000	+	100.0 Hz	
Exeter	FM	WA7HRG	147.19500	+	156.7 Hz	
Fair Oaks	FM	W6HIR	146.79000	-	100.0 Hz	
Fairbanks Ranch	DMR/BM	K6RYA	449.90000	-	CC10	
Fairfield	DMR/BM	WG6R	443.40000	+	CC1	
	FM	W6ER	224.38000	-	77.0 Hz	
	FM	K6SOL	441.15000	+	77.0 Hz	
	FM	KC6UJM	442.77500	+	77.0 Hz	
	FM	WL3DZ	443.40000	+		
	FM	KE3RQ	444.12500	+		
Fallbrook	FM	KG6HSQ	446.80000	-	127.3 Hz	
Fallbrook, Red Mountain						
	FM	N6FQ	146.17500	+	107.2 Hz	
	FM	KF6ATL	445.22000	-		
	FM	N6FQ	445.60000	-	107.2 Hz	
Felton, Empire Grade						
	FM	W6JWS	146.74500	-	94.8 Hz	
Fiddletown	FM	NC6R	146.88000	-	156.7 Hz	
	FM	W6SF	147.16500	+	107.2 Hz	
	FM	KG6TXA	224.86000	-	141.3 Hz	
	FM	W6SF	442.25000	+	107.2 Hz	
	FM	K6SZQ	443.87500	+	156.7 Hz	
Folsom	FM	KS6HRP	146.61000	-	136.5 Hz	
	FM	KS6HRP	224.72000	-	136.5 Hz	
	FM	AB6LI	440.35000	+	156.7 Hz	
	FM	KS6HRP	442.35000	+	136.5 Hz	
	FM	K6MFM	442.52500	+	77.0 Hz	
	FM	W6YDD	1283.75000	-	88.5 Hz	
Folsom, Carpenter Hill						
	DSTAR	KS6HRP	147.67500	-		
Fontana	FM	KA6GRF	447.32000	-	136.5 Hz	
Foresthill	FM	W6YDD	146.35500	+	94.8 Hz	
	FM	W6SAR	146.74500	-	156.7 Hz	
	FM	W6SAR	223.76000	-	100.0 Hz	
	FM	KA6ZRJ	442.70000	+	114.8 Hz	
	FM	KA6EBR	442.87500	+	131.8 Hz	
Fort Bragg	FM	K6MHE	147.03000	+	103.5 Hz	
Fort Jones, Oro Fino Valley						
	FM	KK6OAH	146.92000	-	DCS 023	
Fortuna	FM	KA6ROM	147.09000	+	103.5 Hz	
Fountain Valley	FM	WA6FV	145.26000	-	136.5 Hz	
	FM	WA6FV	447.32000	-	94.8 Hz	
Frazier Park	DMR/MARC	WD6FM	448.16250	-	CC1	
	FM	K6NYB	445.64000	-	67.0 Hz	
Frazier Park, Frazier Mountain						
	FM	N6BKL	52.56000	-	82.5 Hz	
	FM	N6XKI	224.72000	-	156.7 Hz	

CALIFORNIA   71

Location	Mode	Call sign	Output	Input	Access	Coordinator
Frazier Park, Frazier Mtn.						
	FM	WB6TZH	448.68000	-		
	FM	W6HWW	448.86000	-		
	FM	N6ENL	449.12000	-		
Fremont	DMR/MARC	KI6KGN	440.02500	+	CC12	NARCC
	FM	KU6V	224.18000	-	94.8 Hz	
	FM	WA6GG	224.84000	-	127.3 Hz	
	FM	WA6FSP	440.00000	+	110.9 Hz	
	FM	KI6KGN	440.90000	+	141.3 Hz	
	FM	KC6WXO	441.12500	+	100.0 Hz	
	FM	WA6PWW	442.60000	+	107.2 Hz	
	FM	N6IGF	443.40000	+		
	FM	K6JJC	443.70000	+	136.5 Hz	
Fremont, Mission Peak						
	FM	K6GOD	223.78000	-	110.9 Hz	
	FM	K6GOD	440.17500	+	131.8 Hz	
Fremont, Mt Allison						
	FM	NT6S	224.66000	-	110.9 Hz	
	FM	WB6ECE	441.30000	+	110.9 Hz	
	FM	N6HWI	443.72500	+	127.3 Hz	
Fresno	DMR/MARC	KE6YJC	435.10000		CC1	
	DMR/MARC	KE6YJC	440.06250	+	CC1	NARCC
	DMR/MARC	WX6D	442.23750	+	CC1	NARCC
	DMR/MARC	W6EDX	442.80000	+	CC1	
	FM	WA6IPZ	52.84000	-	82.5 Hz	
	FM	K6WGJ	145.43000	-	141.3 Hz	
	FM	W7POR	145.47000	+	141.3 Hz	
	FM	KE6JZ	146.82000	-	141.3 Hz	
	FM	WQ6CWA	146.85000	-	141.3 Hz	
	FM	N6HEW	147.15000	+	141.3 Hz	
	FM	N6VRC	147.16500	+	141.3 Hz	
	FM	W6DXW	147.30000	+	94.8 Hz	
	FM	KE6JZ	224.70000	-	156.7 Hz	
	FM	KJ6HUP	224.74000	-	156.7 Hz	
	FM	N6VRC	440.00250	+	141.3 Hz	
	FM	N6AMG	440.37500	+		
	FM	N6LYE	441.80000	+		
	FM	N6IB	442.52500	+	141.3 Hz	
	FM	WQ6CWA	443.25000	+	107.2 Hz	
	FM	KE6JZ	443.37500	+	123.0 Hz	
	FM	NA6MM	443.40000	+	141.3 Hz	
	FM	W6WYT	443.42500	+	141.3 Hz	
	FM	KE6SHK	443.65000	+	141.3 Hz	
	FM	K6SRA	443.97500	+		
	FM	W6TO	444.20000	+	141.3 Hz	
	FM	KJ6HUP	927.03750	902.03750	146.2 Hz	
	FM	N6VRC	927.05000	902.05000	141.3 Hz	
	FM	W6YEP	1283.45000	-	100.0 Hz	
	P25	KF6FGL	443.77500	+		
	P25	KF6FGL	443.87500	+		
Fresno, Bald Mountain						
	FM	K6VAU	927.06250	902.06250	141.3 Hz	
Fresno, Bear Mountain						
	FM	W6FSC	145.23000	-	141.3 Hz	
	FM	W6FSC	443.60000	+	141.3 Hz	
Fresno, Black Mountain						
	FM	WR6VHF	51.82000	-	162.2 Hz	
Fresno, Meadow Lakes						
	FM	KF6FGL	440.12500	+	110.9 Hz	
	FM	W6NIF	444.10000	+	100.0 Hz	
Frink	FM	W6SCE	224.32000	-		
Fullerton	FM	N6ME	145.40000	144.60000	103.5 Hz	

72   CALIFORNIA

Location	Mode	Call sign	Output	Input	Access	Coordinator
Fullerton	FM	N6ME	224.18000	-	103.5 Hz	
Fullerton, Fullerton Hills						
	FM	KK6HS	445.12000	-	100.0 Hz	
Fullerton, Raytheon Bldg 606						
	FM	K6QEH	146.97000	-	136.5 Hz	
	FM	K6QEH	446.44000	-	114.8 Hz	
Galt	FM	K6SAL	443.45000	+	107.2 Hz	
Garberville	FM	W6CLG	146.79000	-	103.5 Hz	
	FM	KA6ROM	147.15000	+	103.5 Hz	
Garberville, Pratt Mountain						
	FM	N6VA	146.61000	-	103.5 Hz	
Gardena	DMR/MARC	KM6BXA	449.62500	-	CC2	
Gasquet	FM	W6HY	147.18000	+	136.5 Hz	
Georgetown	FM	W6YDD	146.62500	-	123.0 Hz	
	FM	W6HIR	224.84000	-	100.0 Hz	
	FM	K6IRC	441.57500	+		
	FM	K6IRC	443.17500	+		
	FM	K6SRA	443.55000	+		
	FM	WA6APX	443.85000	+		
	FM	K6JJC	444.02500	+	107.2 Hz	
	FM	K6YC	927.85000	902.85000	123.0 Hz	
Georgetown, Bald Mountain						
	FM	AG6AU	52.78000	-	107.2 Hz	
Georgetown, Rubicon Trail						
	FM	KA6GWY	444.98750	+	156.7 Hz	
Geyserville	FM	WA6OYK	442.05000	+	100.0 Hz	
Gibraltar	FM	WB9KMO	224.86000	-	131.8 Hz	
Gilroy	DMR/BM	K6SIA	440.13750	+	CC3	NARCC
	DMR/BM	KJ6WZS	443.52500	+	CC0	
Glendale	DMR/MARC	N6JLY	445.68000	-	CC1	
	FM	K6VGP	445.26000	-	100.0 Hz	
Glendale, Flint Peak						
	FM	WB6FYR	224.92000	-	94.8 Hz	
	FM	K6CCC	445.38000	-		
Glendale, Mount Lukens						
	FM	WA6ZRB	448.56000	-		
Glendale, Mt Lukens						
	FM	W6AM	145.48000	-	100.0 Hz	
	FM	KD6AFA	146.67000	-	192.8 Hz	
	FM	AA6TL	147.49500	146.49500	186.2 Hz	
	FM	W6CPA	223.90000	-	136.5 Hz	
	FM	WA6TTL	224.04000	-		
	FM	N6ENL	224.78000	-	151.4 Hz	
	FM	KI6QK	446.22000	-	123.0 Hz	
	FM	WA6UZS	447.24000	-	100.0 Hz	
	FM	WR6TWE	448.02000	-	100.0 Hz	
	FM	WB6YMH	449.22000	-	131.8 Hz	
	FM	WA6DPB	449.80000	-		
	FM	KO6TD	1282.07500	-	100.0 Hz	
	FM	WA6DPB	1282.47500	-	77.0 Hz	
	FM	KB6SUA	1284.12500	-		
	FM	AA6TL	1285.32500	-		
	FM	AB6BX	1285.60000	-		
	FM	KC6MQP	1286.32500	-		
Glendale, Mt Thom						
	FM	WB6ZTY	146.02500	+	136.5 Hz	
	FM	KC6ZQR	448.54000	-	136.5 Hz	
Glendale, Mt. Lukens						
	FM	WA6DYX	447.06000	-		
	FM	K6UHF	448.20000	-		
	FM	KF6JBN	448.58000	-	118.8 Hz	
	FM	WB6LVZ	448.62000	-		

CALIFORNIA 73

Location	Mode	Call sign	Output	Input	Access	Coordinator
Glendale, Mt. Lukens						
	FM	N6CIZ	449.26000	-		
Glendora	FM	KK6JYT	446.86000	-	123.0 Hz	
Glennville	FM	W6SCE	224.32000	-		
Gold Run	FM	WB6OHV	440.95000	+	192.8 Hz	
Goleta, Brush Peak						
	FM	N6HYM	224.00000	-	156.7 Hz	
Gorman, Burnt Peak						
	FM	KF6BXW	448.66000	-		
	FM	W6RRN	448.72000	-		
Gorman, Frazier Mountain						
	FM	N6SMU	447.04000	-	136.5 Hz	
	FM	WB6ORK	447.10000	-	203.5 Hz	
	FM	WR6FM	447.36000	-		
	FM	KK6AC	447.86000	-	141.3 Hz	
	FM	W6RLW	1282.97500	-	88.5 Hz	
	FM	WA6RLW	1285.15000	-		
Grass Valley	FM	W6YDD	146.62500	-	151.4 Hz	
	FM	W6DD	147.01500	+	151.4 Hz	
	FM	W6DD	147.28500	+	151.4 Hz	
	FM	KD6GVO	224.90000	-	151.4 Hz	
	FM	KO6CW	440.10000	+	151.4 Hz	
	FM	KB6LCS	440.52500	+	192.8 Hz	
	FM	W6RCA	441.02500	+	151.4 Hz	
	FM	KF6GLZ	442.42500	+		
	FM	W6AI	442.62500	+		
	FM	N6VYQ	442.95000	+	107.2 Hz	
	FM	WA6WER	443.02500	+	114.8 Hz	
	FM	KG6BAJ	443.65000	+	114.8 Hz	
	FM	K6NP	444.05000	+		
	FM	K6RTL	444.75000	+	167.9 Hz	
	FM	WB4YJT	927.05000	902.05000		
	FM	N6NMZ	927.13750	902.13750	77.0 Hz	
Grass Valley, Banner Mountain						
	FM	N1OES	52.72000	-	100.0 Hz	
Grass Valley, Wolf Mountain						
	FM	KF6GLZ	52.76000	-	131.8 Hz	
Green Valley	FM	KI6BKN	147.64500	-		
Guadalupe	FM	KA6BFB	146.17500	+	100.0 Hz	
Gualala	FM	W6ABR	147.82500	-	103.5 Hz	
Guerneville, McCray Mtn						
	FM	KM6XU	51.80000	-	114.8 Hz	
Guerneville, Mt Jackson						
	FM	K6CDF	145.19000	-	88.5 Hz	
Hagador Peak	FM	W6CTR	445.94000	-	151.4 Hz	
Half Moon Bay	FM	WR6HMB	147.28500	+	114.8 Hz	
	FM	N6IMS	927.70000	902.70000		
Hamilton Branch, Dyer Mtn						
	FM	K6PLU	145.37000	-	123.0 Hz	
Hanford	DMR/BM	WX6D	442.53750	+	CC1	
	FM	N6VRC	147.28500	+	141.3 Hz	
	FM	W6FBW	444.66000	+	123.0 Hz	
	FM	N6CVC	444.95000	+	100.0 Hz	
Hanford, Hospital	FM	N6CVC	145.11000	-	100.0 Hz	
Happy Camp, Slater Butte						
	FM	K6SIS	146.91000	-	100.0 Hz	
Harbor City	DMR/BM	N9QE	449.46000	-	CC15	
Hartley	DMR/BM	K6LNK	440.21250	+	CC1	
Hawthorne	FM	AC6FB	1283.72500	-		
Hayward	DMR/BM	KQ6RC	145.25000	-	CC2	
	DMR/BM	KB6LED	442.25000	+	CC1	
	DMR/BM	KB6LED	444.85000	+	CC1	

74   CALIFORNIA

Location	Mode	Call sign	Output	Input	Access	Coordinator
Hayward	FM	K6EAG	145.13000	-	127.3 Hz	
	FM	KQ6YG	440.05000	+	156.7 Hz	
	FM	KB6LED	440.95000	+	100.0 Hz	
	FM	K6DDR	442.87500	+	100.0 Hz	
Hayward, Walpert Ridge						
	FM	K6EAG	52.76000	-	114.8 Hz	
Healdsburg	FM	NN6J	224.36000	-	88.5 Hz	
Hemet	DMR/MARC	WA6HXG	446.86000	-	CC1	
	DMR/MARC	WA6HXG	446.88000	-	CC1	
	FM	K6JRM	446.86000	-	100.0 Hz	
	FM	KB6JAG	446.88000	-	88.5 Hz	
Hemet, Hemet Valley Med Ctr						
	FM	N7OD	145.42000	-	88.5 Hz	
Hemet, Simpson Park						
	FM	W6COH	144.50500	+	100.0 Hz	
	FM	W6COH	224.12000	-	97.4 Hz	
Hesperia	FM	W6ECS	146.02500	+	DCS 205	
	FM	WA6AV	146.17500	+	97.4 Hz	
High Lakes, Spring Valley Moun						
	FM	K6FHL	146.70000	-	110.9 Hz	
High Pass	FM	W6JAM	147.99000	-		
Hollister	FM	W6KRK	145.11000	-	94.8 Hz	
	FM	N6SBC	147.31500	+	94.8 Hz	
	FM	W6MOW	441.90000	+	110.9 Hz	
	FM	W6MOW	443.60000	+	110.9 Hz	
	FM	W6MOW	1286.22500	-	110.9 Hz	
Hollywood	FM	WB6BJM	147.03000	+		
	FM	WD8CIK	446.56000	-	127.3 Hz	
Hollywood Beach	FM	WD5B	443.87500	+		
Hollywood, Hollywood Hills						
	FM	WB6BJM	147.00000	+		
	FM	KD6JTD	147.07500	+	100.0 Hz	
	FM	WB6BJM	446.28000	-		
	FM	KB6IBB	449.62000	-		
	FM	N6LXX	927.47500	901.97500	110.9 Hz	
Hopland	FM	WA6RQX	145.47000	-	103.5 Hz	
Huntington Beach	FM	W6BRP	447.94000	-	100.0 Hz	
Huntington Beach, Boeing						
	FM	W6VLD	147.46500	146.46500	103.5 Hz	
	FM	W6VLD	445.58000	-	94.8 Hz	
Idyllwild	FM	KD6OI	146.89500	-	118.8 Hz	
Independence, Mazourka Peak						
	FM	W6TD	146.76000	-	103.5 Hz	
Indian Wells, Bird Spring Pass						
	FM	KF6FM	146.08500	+	141.3 Hz	
Indio	DMR/MARC	KF6FM	448.81250	-	CC1	
	FM	N6BKL	52.90000	52.50000	82.5 Hz	
Indio, Chiriaco Smt						
	FM	W6KSN	447.58000	-	100.0 Hz	
Indio, Indio Hills	FM	WA6MDJ	51.92000	-	82.5 Hz	
Inglewood	FM	WS6C	224.46000	-	131.8 Hz	
Inglewood, Loyola	FM	W6LMU	147.85500	-	127.3 Hz	
Inverness	FM	KI6B	145.17000	-	88.5 Hz	
Ione	FM	K6KBE	224.00000	-	107.2 Hz	
Irvine	DMR/MARC	WB6SRC	448.13750	-	CC1	
Irvine, Sierra Peak						
	FM	N6FFI	52.80000	-	82.5 Hz	
Irvine, Signal Peak						
	DMR/BM	WD6DIH	445.40000	-	CC1	
	FM	N6OCS	927.18750	902.18750	103.5 Hz	
	FM	W6KRW	1282.52500	-	88.5 Hz	
Johnson Valley	DMR/MARC	KF6FM	448.13750	-	CC1	

CALIFORNIA   75

Location	Mode	Call sign	Output	Input	Access	Coordinator
Juniper Hills	FM	KD6KTQ	145.20000	-	114.8 Hz	
	FM	WA6GDF	449.50000	-	192.8 Hz	
Kelseyville	FM	KG6UFR	441.35000	+	100.0 Hz	
	FM	N6GJM	441.42500	+	100.0 Hz	
Kelseyville, Mount Konocti						
	FM	N1PPP	146.77500	-	103.5 Hz	
Kensington	DMR/MARC	AH6KD	441.88750	+	CC1	
King City	FM	K6TAZ	443.97500	+	131.8 Hz	
	FM	W6FM	444.55000	+		
King City, Williams Hill						
	FM	N6SPD	145.37000	-	94.8 Hz	
	FM	N6SPD	441.65000	+	156.7 Hz	
	FM	WA6VPL	444.27500	+		
Kings Canyon NP, Park Ridge						
	P25	WA6BAI	146.88000	-	103.5 Hz	
Kingsburg	FM	KB6RHD	444.60000	+	141.3 Hz	
Klamath	FM	KA7PRR	224.86000	-		
La Canada, JPL Mesa						
	FM	WR6JPL	147.15000	+	103.5 Hz	
La Canada, Mt. Wilson						
	FM	W6ATN	448.42000	-		
La Conchita, Rincon Mountain						
	FM	W6SCE	224.76000	-		
La Crescenta	FM	NW6B	447.96000	-		
La Honda	FM	W6SCF	146.73000	-	114.8 Hz	
	FM	KC6ULT	146.80500	-	114.8 Hz	
	FM	WA6DQP	440.10000	+	114.8 Hz	
La Jolla, UCSD	FM	KK6UC	927.46250	902.46250	100.0 Hz	
La Mesa	DMR/BM	N0MIS	448.26000	-	CC1	
	FM	WA6HYQ	145.24000	-	123.0 Hz	
	FM	N6QWD	146.67000	-	156.7 Hz	
	FM	WA6ZFT	446.88000	-	131.8 Hz	
La Mesa, Nebo Hill						
	FM	WA6HYQ	223.88000	-	107.2 Hz	
La Presa, San Miguel Mountain						
	FM	W6HDC	145.12000	-	123.0 Hz	
La Quinta	FM	WR7NV	145.36000	-	100.0 Hz	
La Quinta, Santa Rosa Mountain						
	FM	WA6HYQ	223.88000	-	110.9 Hz	
Laguna Beach	FM	K6SOA	147.64500	-	110.9 Hz	
Laguna Beach, Temple Hill						
	DSTAR	K6SOA	146.11500	+		
	DSTAR	K6SOA	445.70500	-		
	DSTAR	K6SOA	1282.60000	-		
	FM	K6SOA	224.10000	-	110.9 Hz	
	FM	K6SOA	445.66000	-	110.9 Hz	
Laguna Seca	FM	W6JSO	927.28250	902.08250	DCS 423	
Laguna Woods	FM	W6LY	147.61500	-	136.5 Hz	
Lake Arrowhead	FM	NO6B	445.46000	-		
Lake Arrowhead, Heaps Peak						
	FM	WW6Y	445.10000	-	179.9 Hz	
	FM	K6ZXZ	447.08000	-	136.5 Hz	
	FM	WB6TZU	448.18000	-		
	FM	W6YJ	449.66000	-		
	FM	WA6RQD	449.68000	-	156.7 Hz	
Lake Elsinore, Elsinore Peak						
	FM	W6CDW	144.89500	+	156.7 Hz	
	FM	W6CDW	445.62000	-	173.8 Hz	
Lake Forest, Santiago Peak						
	FM	W6KRW	52.62000	-	103.5 Hz	
	FM	KA6EEK	145.16000	-	136.5 Hz	
	FM	N6SLD	145.22000	-	103.5 Hz	

76   CALIFORNIA

Location	Mode	Call sign	Output	Input	Access	Coordinator
Lake Forest, Santiago Peak						
	FM	K6MWT	147.43500	146.43500	103.5 Hz	
	FM	WA6SVT	223.82000	-		
	FM	WR6AAC	224.22000	-	151.4 Hz	
	FM	K6SOA	224.64000	-	151.4 Hz	
	FM	W6SCE	224.76000	-		
	FM	K8BUW	224.82000	-	156.7 Hz	
	FM	KB6TRD	224.88000	-	107.2 Hz	
	FM	WB6MIE	446.12000	-		
	FM	KI6QK	446.22000	-		
	FM	N6SIM	446.30000	-	127.3 Hz	
	FM	AA4CD	446.32000	-		
	FM	KF6PHX	446.64000	-	77.0 Hz	
	FM	WD6DIH	446.90000	-	110.9 Hz	
	FM	KG6GI	447.18000	-	131.8 Hz	
	FM	W6YJ	447.38000	-		
	FM	K6UHF	447.68000	-	118.8 Hz	
	FM	W6YQY	447.70000	-		
	FM	WD6AWP	448.04000	-		
	FM	K6JSI	448.06000	-	100.0 Hz	
	FM	KD6ZLZ	448.08000	-	162.2 Hz	
	FM	KA6JRG	448.12000	-		
	FM	WB6SRC	448.14000	-	131.8 Hz	
	FM	W6KRW	448.32000	-	141.3 Hz	
	FM	W6ZOJ	448.82000	-		
	FM	N6SLD	448.92000	-	91.5 Hz	
	FM	WA6SBH	449.40000	-		
	FM	WA6TTL	449.74000	-		
	FM	WB6BWU	449.80000	-	110.9 Hz	
	FM	WR6SGO	449.94000	-		
	FM	KB6SUA	1184.27500	-		
	FM	W6KRW	1282.02500	-	88.5 Hz	
	FM	WA6MDJ	1282.15000	-	114.8 Hz	
	FM	K6ARN	1282.62500	-	82.5 Hz	
	FM	W6KRW	1282.72500	-	88.5 Hz	
	FM	KB6KZA	1284.17500	-		
	FM	WA6SVT	1286.15000	-		
	P25	N6OCS	927.12500	902.12500		
Lake Hughes, Burnt Peak						
	FM	K6DK	147.15000	+	131.8 Hz	
Lake Isabella, Sawmill Road						
	FM	N6SR	145.45000	-	156.7 Hz	
Lake Isabella, Shirley Peak						
	FM	WB6RHQ	224.64000	-	156.7 Hz	
Lakehead	FM	N0ASA	440.32500	+	100.0 Hz	
	FM	KH8AF	442.17500	+		
Lakside	DMR/BM	KJ6KHI	147.22500	146.44500	CC1	
Lancaster	FM	WB6RSM	146.67000	-		
	FM	WA6YVL	223.92000	-	100.0 Hz	
	FM	KI6CHH	445.22000	-		
Landers, Goat Mountain						
	FM	WB6CDF	447.58000	-	173.8 Hz	
Laytonville	FM	WA6RQX	145.43000	-	103.5 Hz	
	FM	K7BUG	146.65500	-	103.5 Hz	
	FM	WA6RQX	444.80000	+		
Laytonville, Cahto Peak						
	FM	K6JSI	443.00000	+	100.0 Hz	
Leemore	FM	KM6OU	146.80500	-	118.8 Hz	
Lemon Grove	FM	W6YEC	223.98000	-	107.2 Hz	
Lemoore	FM	W6BY	145.27000	-	88.5 Hz	
	FM	KM6OU	145.33000	-	146.2 Hz	
	FM	N6BEN	442.02500	+	146.2 Hz	

CALIFORNIA 77

Location	Mode	Call sign	Output	Input	Access	Coordinator
Leona Valley	FM	AF6TG	445.84000	-	100.0 Hz	
Lewiston	FM	K6SDD	145.11000	-	85.4 Hz	
Lincoln	FM	K6PAC	147.33000	+	123.0 Hz	
	FM	KU6V	224.04000	-	123.0 Hz	
	FM	W6LHR	443.22500	+	167.9 Hz	
Littlerock	FM	K6SRT	145.38000	-	151.4 Hz	
	FM	KN6RW	147.07500	-	100.0 Hz	
Livermore	DMR/BM	K6LRG	441.82500	+	CC1	
	FM	KO6PW	145.43000	-	100.0 Hz	
	FM	W6LLL	146.77500	-	100.0 Hz	
	FM	KO6PW	224.74000	-	100.0 Hz	
	FM	K6LRG	224.88000	-		
	FM	K7FED	444.12500	+	100.0 Hz	
	FM	WA6JQV	927.02500	902.02500		
	FM	WA6JQV	927.43750	902.43750		
	FM	WA6JQV	927.47500	902.47500		
	FM	W6RLW	1282.22500	-	88.5 Hz	
Lodi	DMR/BM	N6GKJ	444.22500	+	CC1	
	FM	WB6ASU	444.25000	+	114.8 Hz	
	FM	WB6ASU	927.07500	902.07500	100.0 Hz	
	FM	N6GKJ	927.08750	902.08750	100.0 Hz	
	FM	N6GKJ	927.10000	902.10000	100.0 Hz	
Lodi, Bear Mountain						
	FM	WB6ASU	147.09000	+	114.8 Hz	
Loma Linda, LLUMC						
	FM	K6LLU	445.60000	-	118.8 Hz	
Lomita	FM	WA6TTL	447.16000	-		
	FM	N6CA	927.65000	902.65000	DCS 411	
Lompoc	DMR/BM	WA6VPL	440.57500	+	CC1	
	DMR/BM	K7AZ	443.35000	+	CC3	
	DMR/BM	K7AZ	443.42500	+	CC1	NARCC
	FM	W6AB	145.36000	-	131.8 Hz	
	FM	WA6VPL	147.12000	+	131.8 Hz	
	FM	K7AZ	443.27500	+	91.5 Hz	
	FM	WA6VPL	444.80000	+		
Lompoc, Gravel Peak						
	FM	WA6VPL	145.12000	-	100.0 Hz	
Lompoc, Tranquillon Peak						
	FM	W6AB	224.50000	-	131.8 Hz	
	FM	W6AB	449.14000	-	131.8 Hz	
Lompoc, Vandenberg AFB						
	FM	WA6VPL	444.27500	+	88.5 Hz	
Lone Pine, Alabama Hills						
	FM	W6PH	147.21000	+	103.5 Hz	
	FM	N6AZY	447.70000	-	100.0 Hz	
Lone Pine, Owens Valley						
	FM	N6BKL	52.90000	-	100.0 Hz	
Long Beach	FM	KE6HE	146.80500	-		
	FM	KD6CIX	1283.97500	-		
Long Beach, VA Hospital						
	FM	K6SYU	146.79000	-	103.5 Hz	
	FM	K6SYU	224.50000	-	156.7 Hz	
Loomis	FM	N6ZN	927.21250	902.21250		
Loop Canyon	FM	WA6WLZ	446.98000	-		
Loop, Leviathan Peak						
	FM	KB6DWO	146.65500	-	131.8 Hz	
Los Altos	FM	KH6N	440.87500	+	100.0 Hz	
	FM	WB6WTM	441.25000	+		
	FM	K6MSR	443.67500	+		
	FM	W6SRI	1283.15000	-	100.0 Hz	
Los Altos Hills	FM	W6LAH	146.74500	-	110.9 Hz	
	FM	K6AIR	146.94000	-	123.0 Hz	

78   CALIFORNIA

Location	Mode	Call sign	Output	Input	Access	Coordinator
Los Altos Hills	FM	K6AIR	441.52500	+	123.0 Hz	
	FM	W6BUG	443.85000	+		
	FM	KE6JTK	444.22500	+	131.8 Hz	
	FM	K6AIR	1282.60000	-	100.0 Hz	
Los Angeles	DMR/BM	W6DVI	438.00000		CC1	
	DMR/BM	KI6KQU	449.38000	-	CC1	
	DMR/BM	W6DVI	449.46250	-	CC1	
	FM	W6JUN	224.04000	-		
	FM	KK6QY	1285.02500	-		
Los Angeles, Contract Point						
	FM	W6NUT	147.45000	146.45000	127.3 Hz	
Los Angeles, Downtown						
	FM	WR6JPL	224.70000	-	114.8 Hz	
Los Banos	FM	K6TJS	146.92500	-	123.0 Hz	
	FM	K6TJS	147.21000	+	123.0 Hz	
	FM	K6TJS	444.00000	+	123.0 Hz	
Los Gatos	FM	KU6V	51.92000	-	114.8 Hz	
	FM	KB6LCS	223.82000	-	156.7 Hz	
	FM	K6INC	224.48000	-		
	FM	NU6P	224.80000	-	118.8 Hz	
	FM	N6DVC	224.88000	-	88.5 Hz	
	FM	KC6TYG	440.65000	+	94.8 Hz	
	FM	W6RCA	441.62500	+	100.0 Hz	
	FM	K6UB	441.70000	+	127.3 Hz	
	FM	WA6ABB	443.02500	+		
	FM	K9GVF	443.75000	+	100.0 Hz	
	FM	WB6LPZ	444.12500	+		
	FM	KB6LCS	444.92500	+	151.4 Hz	
	FM	WA6JQV	927.02500	902.02500		
	FM	N6NMZ	927.15000	902.15000	156.7 Hz	
	FM	K6INC	927.25000	902.25000	114.8 Hz	
	FM	WA6JQV	927.46250	902.46250		
	FM	WA6JQV	927.47500	902.47500		
	P25	KI6KGN	440.52500	+		
	P25	N6MM	440.52500	+	110.9 Hz	
Los Gatos, Castle Rock						
	FM	K6FB	145.45000	-	100.0 Hz	
	FM	K6FB	223.88000	-	100.0 Hz	
Los Gatos, Crystal Peak						
	FM	K6DND	927.91250	902.91250	167.9 Hz	
Los Gatos, Santa Cruz Mountain						
	FM	WB6KHP	224.54000	-	100.0 Hz	
	FM	WB6KHP	444.97500	+	127.3 Hz	
Los Molinos	FM	KI6PNB	145.39000	-	110.9 Hz	
Los Osos	FM	W6SLO	146.86000	-	127.3 Hz	
Los Osos, Clark Valley / Irish						
	FM	W6SLO	444.97500	+	127.3 Hz	
Lotus, Bakers Mountain						
	FM	AG6AU	441.72500	+	82.5 Hz	
Loyalton	FM	N5TEN	440.10000	+		
Lucerne Valley	FM	KC6JTN	145.18000	-	123.0 Hz	
	FM	K6DLP	224.52000	-		
Ludlow	FM	WA6TST	147.88500	-	151.4 Hz	
Madera	P25	N6LYE	443.31250	+		
Magalia	DMR/MARC	KB6FEC	147.15000	+	CC0	
	FM	KC6USM	51.94000	-	114.8 Hz	
Malibu, Castro Peak						
	FM	K6DCS	147.22500	+	94.8 Hz	
Malibu, Paradise Cove						
	FM	N6FDR	145.26000	-	100.0 Hz	
Malibu, Saddle Peak						
	FM	K6VGP	224.98000	-		

CALIFORNIA   79

Location	Mode	Call sign	Output	Input	Access	Coordinator
Malibu, Saddle Peak						
	FM	WB6ZTR	445.18000	-		
	FM	N6CIZ	446.26000	-		
	FM	W6YJ	446.42000	-		
	FM	WA6ZPS	448.24000	-		
	FM	WR6BRN	448.70000	-		
Malibu, Santa Monica Mountains						
	FM	W6XC	449.74000	-		
Mammoth Lakes, Mammoth Lakes						
	FM	WA6TTL	444.72500	+		
Mammoth Lakes, Mammoth Mtn						
	FM	NW6C	146.73000	-	100.0 Hz	
Mammoth, Conway Summit						
	FM	KB6DWO	146.88000	-	131.8 Hz	
Manteca	DMR/BM	KM6IRY	443.02500	+	CC14	
Maricopa	FM	KK6PHE	145.29000	-	94.8 Hz	
Maricopa, Cerro Noroeste						
	FM	KC6WRD	224.98000	-	94.8 Hz	
	FM	KC6WRD	443.07500	+	94.8 Hz	
Mariposa	DMR/BM	AB6BP	442.70000	+	CC1	NARCC
	FM	W6HHD	145.13000	-		
	FM	W6MPA	147.25500	+	146.2 Hz	
	FM	KF6CLR	224.16000	-	74.4 Hz	
	FM	KF6CLR	224.30000	-	74.4 Hz	
	FM	W6BRB	224.50000	-	123.0 Hz	
	FM	KF6CLR	441.42500	+	74.4 Hz	
	FM	N6IB	442.35000	+	141.3 Hz	
	FM	W6HHD	444.80000	+		
	FM	W6BXN	1284.30000	-	88.5 Hz	
Mariposa, Mt Bullion						
	DSTAR	W6HHD	1284.10000	-		
	FM	W6PPM	146.74500	-	146.2 Hz	
	FM	W6BXN	147.03000	+	100.0 Hz	
	FM	N6LYE	440.83750	+		
	FM	K6SIX	441.35000	+	107.2 Hz	
	FM	KI6HHU	442.67500	+	107.2 Hz	
	FM	K6IXA	443.07500	+	107.2 Hz	
	FM	W6BXN	444.70000	+	107.2 Hz	
	FM	K6RDJ	927.15000	902.15000	100.0 Hz	
	P25	N6LYE	440.80000	+	114.8 Hz	
Martinez	FM	KF6HTE	444.45000	+	107.2 Hz	
Maxwell, Coast Range Foothills						
	FM	N6NMZ	147.04500	+	156.7 Hz	
	FM	N6NMZ	442.27500	+	100.0 Hz	
Mcfarland	FM	N6RDN	443.27500	+	141.3 Hz	
Meadow Lakes	DMR/BM	AB6BP	145.45000	-	CC1	
	FM	KJ6CE	145.25000	-	141.3 Hz	
	FM	WA6OIB	146.61000	-	141.3 Hz	
	FM	K6JSI	146.79000	-	100.0 Hz	
	FM	W6TO	146.94000	-	141.3 Hz	
	FM	N6VQL	147.09000	+		
	FM	N6IB	147.39000	+	141.3 Hz	
	FM	K6TVI	441.40000	+	107.2 Hz	
	FM	K6JSI	444.25000	+	100.0 Hz	
	FM	W6BJ	444.97500	+	136.5 Hz	
	P25	N6VQL	147.31500	+		
Meadow Lakes, CA						
	DMR/MARC	KF6FGL	440.03750	+	CC1	
Mendocino	FM	WD6HDY	146.82000	-	103.5 Hz	
Menifee	DMR/MARC	KN6CLM	433.25000		CC1	
Menlo Park	DMR/BM	W6FBK	440.02500	+	CC1	
	DSTAR	W6OTX	1284.15000	-		

80   CALIFORNIA

Location	Mode	Call sign	Output	Input	Access	Coordinator
Menlo Park	FM	KB7IP	51.78000	-	114.8 Hz	
	FM	W6FBK	441.65000	+	162.2 Hz	
	FM	W6OTX	1272.15000	-		
Merced	DMR/BM	AB6BP	145.37000	-	CC1	NARCC
	DMR/MARC	KF6FM	442.40000	-	CC1	
	FM	N6WEB	441.60000	+		
	FM	W6TCD	444.52500	+	107.2 Hz	
Merritt	DMR/BM	K6LNK	145.01250	147.51250	CC9	
Meyers	FM	NC9RS	927.67500	902.67500	156.7 Hz	
Middletown	FM	AC6VJ	29.64000	-	156.7 Hz	
	FM	WA6JQV	927.02500	902.02500		
	FM	WA6JQV	927.48750	902.48750		
	FM	WA6JQV	927.93750	902.93750		
Mill Valley	FM	W6GHZ	1285.05000	-	88.5 Hz	
Mill Valley, Mt Tamalpais						
	FM	K6GWE	443.25000	+	179.9 Hz	
	FM	KJ6RA	444.67500	+	100.0 Hz	
Millbrae	FM	K6HSV	442.10000	+	107.2 Hz	
Milpitas	FM	W6MLP	145.43000	-	85.4 Hz	
	FM	N6QDY	147.94500	-	77.0 Hz	
	FM	W6MLP	224.72000	-	100.0 Hz	
	FM	W6KCS	442.02500	+	162.2 Hz	
	FM	K6GOD	442.35000	+	100.0 Hz	
	FM	K6EXE	443.02500	+		
	FM	K6ATF	927.06250	902.06250	100.0 Hz	
	FM	WB6PHE	927.16250	902.16250	127.3 Hz	
Milpitas, Mission Peak						
	FM	K6GOD	444.72500	+	162.2 Hz	
Miramar	FM	K6ARN	51.80000	-	127.3 Hz	
Mission Hills	FM	WD6APP	145.32000	-	107.2 Hz	
	FM	WD6APP	223.80000	-		
	FM	KJ6HVS	446.88000	-	123.0 Hz	
Mission Viejo	FM	KA6TBF	146.16000	+	107.2 Hz	
Mobile	DMR/BM	K6BIV	144.93750	147.43750	CC8	
Mobile NorCal / USA						
	DMR/BM	K6BIV	144.52000	147.99000	CC1	
Moccasin	FM	K6DPB	145.29000	-	100.0 Hz	
Modesto	DMR/BM	K6ACR	442.17500	+	CC1	
	FM	N6APB	442.55000	+		
	FM	K6JJC	443.17500	+	107.2 Hz	
	FM	WA6JQV	927.02500	902.02500		
	FM	WA6JQV	927.03750	902.03750		
Modesto, Double Tree Hotel						
	FM	WD6EJF	145.11000	-	136.5 Hz	
Modesto, Memorial Med Ctr						
	FM	WA6OYF	146.35500	+	156.7 Hz	
Modesto, Mt Oso	FM	K6GTO	51.94000	-	114.8 Hz	
	FM	WD6EJF	145.39000	-	136.5 Hz	
	FM	N5FDL	146.89500	-	114.8 Hz	
	FM	WD6EJF	224.14000	-	136.5 Hz	
	FM	WD6EJF	440.22500	+	136.5 Hz	
	FM	K6RDJ	441.27500	+	77.0 Hz	
	FM	N6QOP	442.07500	+	123.0 Hz	
	FM	K6JSI	443.52500	+	107.2 Hz	
	FM	K6TJS	444.00000	+	123.0 Hz	
	P25	N6LYE	440.80000	+		
Moffett Field	DMR/BM	W6CMU	443.82500	+	CC1	NARCC
Monrovia	FM	WA6DVG	224.58000	-	100.0 Hz	
Montclair	DMR/MARC	WA6FM	446.03750	-	CC1	
Monterey	FM	WE6R	51.76000	-	114.8 Hz	
	FM	WE6R	146.08500	+		
	FM	N6FNP	147.67500	-		

CALIFORNIA 81

Location	Mode	Call sign	Output	Input	Access	Coordinator
Monterey	FM	WA6YBD	224.90000	-	107.2 Hz	
	FM	WB6ECE	441.30000	+	136.5 Hz	
	FM	WE6R	441.32500	+		
	FM	N6SPD	441.65000	+	123.0 Hz	
	FM	WH6KA	442.22500	447.32500		
	FM	N6SPD	444.27500	+	123.0 Hz	
	FM	N6AMO	444.47500	+	110.9 Hz	
	FM	W6JSO	927.28750	902.28750	DCS 423	
	FM	WE6R	927.97500	902.97500		
	P25	N6AMO	443.65000	+	110.9 Hz	
Monterey Park	FM	KF6YLB	146.35500	+	71.9 Hz	
Monument Peak	FM	W6ZN	145.42000	-	74.4 Hz	
Moraga	FM	KI6O	442.45000	+		
	FM	KB7IP	443.42500	+		
	FM	K6SJH	927.32500	902.32500	100.0 Hz	
Moreno Valley, Box Springs Mtn						
	FM	KI6REC	449.30000	-	103.5 Hz	
Moreno Valley, Civic Center						
	FM	WA6HYQ	146.67000	-	123.0 Hz	
Moreno Valley, March AFB						
	FM	K6AFN	223.96000	-	100.0 Hz	
Moreno Valley, Reche Peak						
	FM	KJ6QFS	445.18000	-		
Morgan Hill	FM	K7DAA	147.33000	+	103.5 Hz	
	FM	W6GGF	147.82500	-	100.0 Hz	
	FM	WA6YBD	223.80000	-	107.2 Hz	
	FM	KA6ZRJ	440.57500	+	114.8 Hz	
Morgan Hill, Crystal Peak						
	FM	K6GOD	440.47500	+		
Morgan Hill, Henry Coe Park						
	FM	K7DAA	442.97500	+	100.0 Hz	
Mount San Miguel	FM	N6LXX	927.57500	902.57500	151.4 Hz	
Mount Shasta	DMR/BM	NR6J	442.27500	+	CC1	
	DMR/MARC	W6BML	441.27500	+	CC1	
Mountain Ranch	FM	N6GVI	440.90000	+		
Mountain View	FM	N6SGI	1284.25000	-	88.5 Hz	
Mountain View, El Camino Hospi						
	FM	W6ASH	145.27000	-	100.0 Hz	
	FM	W6ASH	224.14000	-	88.5 Hz	
	FM	W6ASH	440.80000	+	100.0 Hz	
Mt Laguna, Stephenson Pk						
	FM	K6KTA	446.42000	-	107.2 Hz	
Mt Shasta City	FM	W6BML	146.88000	-	123.0 Hz	
	FM	W6PRN	444.35000	+	100.0 Hz	
	FM	AB6MF	444.82500	+	100.0 Hz	
Mt Shasta City, Gray Butte						
	FM	W6PRN	52.72000	-	110.9 Hz	
	FM	KI6WJP	440.27500	+	118.8 Hz	
Mt Shasta City, Mt Bradley						
	FM	AB7BS	444.47500	+	141.3 Hz	
Mt. Allison	DMR/BM	W6TCP	144.98750	147.48750	CC1	
	DMR/BM	W6TCP	440.12500	+	CC3	
Mt. Bullion	DMR/BM	W6BXN	144.93750	147.43750	CC1	
	DMR/BM	W6BXN	444.78750	+	CC1	
Mt. Diablo	DMR/BM	K6MDD	144.97500	147.47500	CC1	
Mt. Pinos	FM	K6NYB	147.37500	149.87500	67.0 Hz	
	FM	K6NYB	445.07500	-	67.0 Hz	
Mt. Umunhum	DMR/MARC	WA6YCZ	444.40000	+	CC1	
Mt. Wilson	FM	K6JTH	448.10000	-		
Murrieta	DMR/BM	KE6UPI	442.40000	+	CC1	
	DMR/BM	KE6UPI	447.40000	-	CC2	
Napa	DMR/BM	KE6O	442.48750	+	CC2	

82　CALIFORNIA

Location	Mode	Call sign	Output	Input	Access	Coordinator
Napa	FM	N6TKW	51.72000	-	114.8 Hz	
	FM	N6TKW	146.11500	+	123.0 Hz	
	FM	N6TKW	146.65500	-	88.5 Hz	
	FM	W6BYS	146.82000	-	151.4 Hz	
	FM	N6NAR	440.85000	+	173.8 Hz	
	FM	W6CO	441.80000	+	151.4 Hz	
	FM	N6TKW	442.25000	+	151.4 Hz	
	FM	K6ZRX	444.52500	+	151.4 Hz	
	FM	W6FMG	1285.70000	+	173.8 Hz	
Napa, Mt St Helena						
	DSTAR	W6CO	440.05000	+		
	FM	WR6VHF	51.82000	-	151.4 Hz	
	FM	W6CO	147.18000	+	151.4 Hz	
	FM	W6CO	441.90000	+	151.4 Hz	
	FM	K6ZRX	927.52500	902.52500	173.8 Hz	
Nevada City	FM	KG6TZT	145.31000	-	151.4 Hz	
	FM	W6JP	444.95000	+	100.0 Hz	
Newberry Springs	FM	WA6MTZ	146.70000	-		
Newbury Park	FM	N6JMI	146.67000	-	127.3 Hz	
Newbury Park, Rasnow Peak						
	FM	N6JMI	147.88500	-	127.3 Hz	
	FM	N6CFC	223.96000	-	141.3 Hz	
	FM	W6RRN	448.60000	-		
Newhall	FM	WB6DZO	223.98000	-	110.9 Hz	
Newport Beach	DMR/BM	KC6AGL	447.40000	-	CC1	
	FM	K6NBR	145.42000	-	136.5 Hz	
	FM	K6NBR	445.22000	-		
Newport Beach, Signal Peak						
	FM	WB6TZD	448.64000	-		
	FM	N6EX	927.13750	902.13750	DCS 411	
Nipomo	DMR/MARC	K6DOA	444.65000	+	CC4	
	FM	WA6VPL	52.58000	-	82.5 Hz	
	FM	WB6MIA	147.99000	-	127.3 Hz	
	FM	KB6Q	444.70000	+	100.0 Hz	
Northridge, Northrop ISD						
	FM	WA6AQQ	146.26500	+	103.5 Hz	
Novato	DMR/BM	KG6MZV	440.65000	+	CC2	
	FM	N6GVI	440.25000	+		
	FM	KI6B	443.60000	+		
	FM	KM6PA	927.35000	902.35000	131.8 Hz	
Novato, Big Rock Ridge						
	FM	K6GWE	146.70000	-	203.5 Hz	
Oak Flat, Kern Canyon						
	FM	KK6AC	145.19000	-	141.3 Hz	
Oak Glen	FM	N6LIZ	53.56000	-	107.2 Hz	
Oak Glen, Oak Glen Div						
	FM	N6AJB	927.37500	902.37500	DCS 023	
Oakhurst	FM	W6PPM	146.74500	-	123.0 Hz	
Oakhurst, Deadwood Peak						
	FM	W6WGZ	147.18000	+	146.2 Hz	
	FM	W6WGZ	441.17500	+	146.2 Hz	
Oakhurst, Goat Mountain						
	FM	WB6BRU	224.90000	-	156.7 Hz	
Oakland	DMR/BM	W6UUU	144.95000	147.45000	CC1	
	DMR/MARC	KB6FEC	443.82500	+	CC0	
	FM	W6EBW	53.72000	-	118.8 Hz	
	FM	W6MTF	146.62500	-		
	FM	W6BUR	146.67000	-	85.4 Hz	
	FM	WB6NDJ	146.88000	-	77.0 Hz	
	FM	W6JMX	224.00000	-	123.0 Hz	
	FM	KC6LHL	224.16000	-	156.7 Hz	
	FM	W6MTF	224.68000	-	114.8 Hz	

CALIFORNIA 83

Location	Mode	Call sign	Output	Input	Access	Coordinator
Oakland	FM	W6YOP	224.76000	-	85.4 Hz	
	FM	W6WOP	224.92000	-	100.0 Hz	
	FM	KM6EF	440.35000	+	123.0 Hz	
	FM	W6EBW	440.57500	+	118.8 Hz	
	FM	W6RCA	441.22500	+	100.0 Hz	
	FM	W6MTF	441.42500	+	156.7 Hz	
	FM	W6YOP	441.47500	+	127.3 Hz	
	FM	KH8AF	442.20000	+		
	FM	WB6NDJ	442.40000	+	77.0 Hz	
	FM	N6GVI	443.20000	+		
	FM	WB6SHU	443.37500	+	114.8 Hz	
	FM	W6MTF	443.87500	+		
	FM	WW6BAY	443.97500	+	100.0 Hz	
	FM	W6PUE	444.65000	+		
	FM	W6YOP	444.77500	+	127.3 Hz	
	FM	KD6GLT	444.80000	+	110.9 Hz	
	FM	WA6JQV	927.03750	902.03750		
	FM	WA6JQV	927.42500	902.42500		
	FM	N6SSB	927.57500	902.57500	100.0 Hz	
	FM	NC9RS	927.65000	902.65000	131.8 Hz	
	FM	WA6JQV	927.93750	902.93750		
	FM	KD6GLT	1284.45000	-		
Oakland, Calistoga						
	FM	W6MTF	52.62000	-	114.8 Hz	
Oakland, Grizzly Peak						
	FM	K6LNK	146.85000	-	162.2 Hz	
Oakland, Oakland Hills						
	FM	WB6TCS	444.25000	+	100.0 Hz	
Oceanside	DMR/MARC	K6JSI	448.80000	-	CC1	
	FM	WF6OCS	144.50500	+	107.2 Hz	
Oildale	FM	KG6KKV	51.88000	-	114.8 Hz	
Ojai	DMR/BM	N6BMW	445.70000	-	CC1	
	DMR/BM	N6BMW	445.72000	-	CC3	
Ojai, Black Mountain						
	FM	N6FL	145.40000	-	114.8 Hz	
Ojai, Sulphur Mountain						
	FM	WD6EBY	145.20000	-	127.3 Hz	
	FM	WD6EBY	445.56000	-	141.3 Hz	
Onyx, Onyx Peak	FM	N6LXX	53.82000	-	107.2 Hz	
	FM	N6LXX	446.88000	-	110.9 Hz	
Orange	DMR/MARC	KF6FM	448.15000	-	CC3	
	FM	K6COV	147.91500	-	136.5 Hz	
	FM	KB6CJZ	446.14000	-	94.8 Hz	
	FM	KB6CJZ	1283.15000	-	85.4 Hz	
Orange Cove	FM	KC6QIT	146.89500	-	100.0 Hz	
Orange, Pleasants Peak						
	FM	W6HBR	145.14000	-	110.9 Hz	
Orcutt	FM	WB9STH	927.42500	902.42500	82.5 Hz	
Orcutt, Graciosa Ridge						
	FM	W6AB	145.14000	-	131.8 Hz	
Orinda	DMR/BM	K6LNK	144.93750	147.43750	CC5	
	FM	K6CHA	52.68000	-	162.2 Hz	
	FM	WA6HAM	440.62500	+		
	FM	W6CBS	441.97500	+	100.0 Hz	
	FM	K6JJC	443.82500	+	136.5 Hz	
	FM	KE6PTT	444.00000	+	100.0 Hz	
Orinda, Bald Peak	FM	WA6HAM	145.49000	-	107.2 Hz	
Orinda, Grizzly Peak						
	FM	N6QOP	443.05000	+	114.8 Hz	
Orland	DMR/BM	W6GRC	144.96250	147.46250	CC1	
Orland Low Level	DMR/BM	W6GRC	144.98750	147.48750	CC2	
Oroville	DSTAR	KJ6LVV	444.27500	+		

84   CALIFORNIA

Location	Mode	Call sign	Output	Input	Access	Coordinator
Oroville	FM	W6AF	146.65500	-	136.5 Hz	
	FM	WA6UHF	224.50000	-	110.9 Hz	
	FM	W6SCR	440.90000	+	110.9 Hz	
	FM	W6YOP	441.47500	+	114.8 Hz	NARCC
	FM	WA6UHF	442.35000	+	110.9 Hz	
Oroville, Downtown						
	FM	WA6CAL	442.17500	+	123.0 Hz	
Oxnard	FM	W6XC	146.73000	-		
	FM	KJ6HCX	146.80500	-	127.3 Hz	
	FM	K6BBK	147.85500	-		
	FM	W6XC	448.34000	-	141.3 Hz	
	FM	K6BBK	448.80000	-	131.8 Hz	
Oxnard, Laguna Peak						
	FM	WD6EBY	445.64000	-	141.3 Hz	
Oxnard, Near Oxnard Airport						
	FM	WB6YQN	146.97000	-	127.3 Hz	
Oxnard, Plains Knoll						
	FM	W6XC	51.90000	-	82.5 Hz	
Pacific Grove, Forest Hill						
	FM	K6CQX	444.60000	+	151.4 Hz	
Pacific Palisades	FM	K6BDE	445.52000	-	123.0 Hz	
Pacifica	DMR/MARC	K6HN	440.67500	+	CC1	
	FM	WA6AFT	440.72500	+		
	FM	K6HN	441.72500	+		
Pacifica, Montara Mountain						
	FM	WA6TOW	441.07500	+	114.8 Hz	
Pacifica, North Peak						
	FM	WA6TOW	146.92500	-	114.8 Hz	
Pacines, Call Mountain						
	FM	W6KRK	442.85000	447.95000	100.0 Hz	
Paicines	FM	N6SBC	145.41000	-	118.8 Hz	
	FM	N6SBC	146.62500	-	94.8 Hz	
Palm Desert	FM	W6DRA	447.32000	-	107.2 Hz	
Palm Desert, Santa Rosa Peak						
	FM	N6MRN	145.34000	-	107.2 Hz	
Palm Springs	DMR/MARC	K6IFR	446.58000	-	CC1	
	FM	KD6QLT	144.93000	147.63000	107.2 Hz	
Palm Springs, Indio Hills						
	FM	K6IFR	445.64000	-	131.8 Hz	
Palm Springs, Tram 1						
	FM	W6DRA	145.48000	-	107.2 Hz	
Palm Springs, Tram 2						
	FM	W6DRA	145.20000	-	131.8 Hz	
Palmar Mountain	FM	W6WNG	147.13000	+	107.2 Hz	
	FM	WB6FMT	446.14000	-	123.0 Hz	
Palmdale	FM	WA6YVL	52.66000	-	82.5 Hz	
	FM	K6LMA	445.52000	-	162.2 Hz	
Palmdale, Hauser Peak						
	FM	W6CLA	146.73000	-	100.0 Hz	
	FM	N6ND	445.48000	-	77.0 Hz	
	FM	WB6FYR	927.37500	902.37500	114.8 Hz	
Palmdale, Mt McDill						
	FM	WB6BFN	147.27000	+	156.7 Hz	
Palmdale, Ten-Hi	FM	KJ6W	445.60000	-	100.0 Hz	
	FM	K6VGP	446.74000	-	107.2 Hz	
Palo Alto	DMR/BM	K6OTR	441.85000	+	CC1	NARCC
	DMR/BM	WW6BAY	444.35000	+	CC1	
	FM	N6NFI	145.23000	-	100.0 Hz	
	FM	N6BDE	440.20000	+	123.0 Hz	
	FM	K6IRC	441.57500	+		
	FM	WW6HP	442.00000	+	151.4 Hz	
	FM	KJ6VU	442.45000	449.35000		

CALIFORNIA 85

Location	Mode	Call sign	Output	Input	Access	Coordinator
Palo Alto	FM	K6OTR	442.80000	+	114.8 Hz	
	FM	W6OOL	443.00000	+		
	FM	KJ6K	443.22500	+		
	FM	WB6NNY	443.75000	+	100.0 Hz	
	FM	KB6LED	444.95000	+	162.2 Hz	
	FM	K6BAM	1284.95000	-	88.5 Hz	
	FM	W6RLW	1285.15000	-	88.5 Hz	
Palo Alto, Black Mountain						
	DSTAR	WW6BAY	444.07500	+		
	FM	WA6FUL	52.64000	-	114.8 Hz	
	FM	WW6BAY	145.39000	-	100.0 Hz	
	FM	W6TI	147.36000	+	110.9 Hz	
	FM	W6PMI	442.12500	+	162.2 Hz	
	FM	K6FB	442.57500	+	100.0 Hz	
	FM	WW6BAY	443.22500	+	100.0 Hz	
	FM	WA6FUL	927.13750	902.13750	DCS 506	
	FM	KJ6VU	927.86250	902.86250	DCS 023	
	FM	W6YX	1282.50000	-	88.5 Hz	
Palo Alto, Gunn High School						
	FM	KK6KPQ	443.60000	+		
Palomar Mountain	DMR/BM	K6VZK	446.04000	-	CC1	
Palos Verdes	DMR/MARC	K6EH	446.06000	-	CC1	
	FM	WD6FZA	446.70000	-		
	FM	W6RRN	448.60000	-		
Palos Verdes Estates						
	FM	W6PVE	447.80000	-		
Palos Verdes, San Pedro Hill						
	FM	K6VGP	445.28000	-		
Panorama City	FM	K6ARN	51.74000	-	146.2 Hz	
Paradise	FM	K6JSI	147.33000	+	123.0 Hz	
Pasadena	FM	W6MPH	145.18000	-	156.7 Hz	
Pasadena, JPL Mesa						
	FM	WR6JPL	224.08000	-	156.7 Hz	
	FM	WR6JPL	445.20000	-	103.5 Hz	
Pasadena, Mirador Peak						
	FM	W6UE	445.44000	-		
	FM	N6EX	927.15000	902.15000	DCS 411	
Pasadena, Mt. Harvard						
	FM	N6LIZ	29.66000	-	107.2 Hz	
Paso Robles	DMR/BM	AG6VS	446.85000	-	CC15	
	FM	W6HD	51.82000	-	100.0 Hz	
	FM	KK6ATA	146.98000	-	127.3 Hz	
	FM	WB6JWB	224.68000	-	91.5 Hz	
	FM	WB6JWB	441.05000	+		
Patterson	FM	NC6R	146.77500	-	156.7 Hz	
	FM	KK6AT	927.62500	902.62500	100.0 Hz	
Pebble Beach, Huckleberry Hill						
	FM	K6LY	146.97000	-	94.8 Hz	
	FM	K6LY	444.70000	+	123.0 Hz	
Pescadero	FM	KE6MNJ	146.62500	-	114.8 Hz	
	FM	N6QZH	442.32500	+		
Petaluma	FM	WB6TMS	146.91000	-	88.5 Hz	
	FM	NI6B	444.22500	+		
	FM	W6GHZ	1286.25000	-	88.5 Hz	
Phelan	FM	N6RPG	445.90000	-	146.2 Hz	
	FM	N6LXX	927.47500	902.47500	DCS 532	
Phelan, Blue Ridge						
	FM	WA6KXK	448.98000	-		
	FM	K6UHF	449.02000	-		
Pine Cove	DMR/MARC	K6TMD	446.06250	-	CC1	
Pine Grove, Mount Zion						
	FM	K6ARC	146.83500	-	100.0 Hz	

86   CALIFORNIA

Location	Mode	Call sign	Output	Input	Access	Coordinator
Pine Grove, Mount Zion						
	FM	K6ARC	441.52500	+	100.0 Hz	
Pine Mountain Club						
	FM	K6SS	147.76500	-	123.0 Hz	
Pine Valley, Stephenson Peak						
	FM	N6JAM	147.99000	-	156.7 Hz	
Pinecrest	FM	K6TUO	147.97500	-	100.0 Hz	
Pioneer	FM	K6MSR	443.62500	+		
Pismo Beach	DMR/BM	KB6BF	444.60000	+	CC2	
Pittsburg	DMR/BM	K6PIT	440.13750	+	CC2	
	DMR/BM	K6BIV	440.60000	+	CC2	
Place Holder	DMR/MARC	N6IB	446.00000	+	CC1	
Placentia	FM	WA6YNT	147.85500	-	100.0 Hz	
Placerville	FM	KA6GWY	146.80500	-	123.0 Hz	
	FM	WA6BTH	440.70000	+	123.0 Hz	
	FM	N6UUI	441.05000	+	127.3 Hz	
	FM	W6LOA	441.25000	+	94.8 Hz	
	FM	W6RCA	441.62500	+		
	FM	WA6BTH	442.47500	+	110.9 Hz	
Placerville, El Dorado Hills						
	FM	N6QDY	147.25500	+	123.0 Hz	
	FM	N6QDY	443.92500	+	179.9 Hz	
Pleasanton	DMR/BM	N6LDJ	443.51250	+	CC1	
	DMR/BM	N6LDJ	444.27500	+	CC1	
	DMR/BM	N6LDJ	444.58750	+	CC1	
	FM	W6SRR	147.04500	+	94.8 Hz	
	FM	W6SRR	442.62500	+	94.8 Hz	
	FM	W6RGG	442.92500	+		
	FM	K6TEA	443.65000	+		
	FM	N6QL	927.37500	902.37500	88.5 Hz	
	FM	N6QL	1284.72500	-	88.5 Hz	
	FM	N6QL	1284.75000	-	88.5 Hz	
Pleasanton, Sunol Ridge						
	FM	W6SRR	29.68000	-	94.8 Hz	
	FM	W6SRR	927.18750	902.18750	94.8 Hz	
Pleasonton	FM	N6AKK	1283.55000	-	88.5 Hz	
Point Arena	FM	W6ABR	146.61000	-	88.5 Hz	
Point Reyes Station						
	FM	WB6TMS	145.47000	-	88.5 Hz	
Point Sur, Lighthouse						
	FM	KI6PAU	146.94000	-	94.8 Hz	
Pollock Pines	FM	WB6DAX	52.98000	-	141.3 Hz	
	FM	WA6BTH	146.86500	-	146.2 Hz	
Pomona	FM	WB6RSK	146.02500	+	103.5 Hz	
	FM	K6CPP	445.58000	-	156.7 Hz	
Pomona, ACS Enterprises, Inc.						
	DSTAR	KC6ACS	446.16000	-		
Pomona, Sunset Ridge						
	FM	N6USO	145.44000	-	136.5 Hz	
	FM	K4ELE	146.70000	-		
	FM	KC6FMX	147.07500	+	110.9 Hz	
	FM	K6JSI	147.21000	+	100.0 Hz	
	FM	K6JSI	224.16000	-	71.9 Hz	
	FM	KF6FM	446.04000	-	131.8 Hz	
	FM	W6GAE	447.52000	-		
	FM	N6DD	447.62000	-		
	FM	N6XPG	447.92000	-		
	FM	K9KAO	449.50000	-	100.0 Hz	
	FM	K6TEM	449.88000	-	146.2 Hz	
	FM	K6DLP	927.61250	902.61250	100.0 Hz	
	FM	K6TEM	1282.82500	-	88.5 Hz	
	FM	KB6MQQ	1284.05000	-		

CALIFORNIA 87

Location	Mode	Call sign	Output	Input	Access	Coordinator
Pomona, Sunset Ridge						
	FM	KM6NP	1284.37500	-		
	FM	WH6NZ	1285.27500	-		
Porterville	FM	AB6MJ	440.25000	+	186.2 Hz	
	FM	WC6HP	441.52500	+	67.0 Hz	
	FM	AC6KT	442.27500	+	100.0 Hz	
	FM	KE6WDX	443.05000	+	123.0 Hz	
Porterville, Blue Ridge						
	FM	KE6WDX	145.31000	-	100.0 Hz	
Porterville, Jordan Peak						
	FM	W6XC	440.82500	+		
Porterville, Republican Hill						
	FM	KE6WDX	146.65500	-	123.0 Hz	
Portola Valley	FM	WB5NVN	146.08500	+	100.0 Hz	
	FM	KC6ULT	440.97500	+	114.8 Hz	
Poway	DMR/BM	KB6TWT	445.42000	-	CC1	
	DMR/MARC	W6CRC	446.05000	-	CC2	
	FM	K6JCC	147.19500	+	110.9 Hz	
Poway, Mt Woodson						
	FM	K6KTA	145.18000	-	107.2 Hz	
Pozo, Black Mountain						
	FM	W6SLO	146.83500	-	127.3 Hz	
Prunedale	FM	W6OPI	146.91000	-	94.8 Hz	
	FM	KC6UDC	441.12500	+	123.0 Hz	
	FM	W6DXW	442.77500	+	110.9 Hz	
Quartz Hill	FM	KD6PXZ	446.40000	-	100.0 Hz	
Quartz Hill, Water Tank						
	FM	KG6SLC	445.56000	-	110.9 Hz	
Quincy	FM	W7OWC	51.90000	-	103.5 Hz	
	FM	AF6AP	147.94500	-	123.0 Hz	
	FM	W6RCA	441.62500	+	100.0 Hz	
Quintette	FM	AG6AU	927.27500	902.27500	127.3 Hz	
Ramona	FM	KD6RSQ	145.30000	-	88.5 Hz	
	FM	KD6RSQ	445.76000	-	88.5 Hz	
Rancho Bernardo, Carmel Mounta						
	FM	NG6ST	146.79000	-	107.2 Hz	
Rancho Cordova	FM	W6AK	224.10000	-	100.0 Hz	
Rancho Cucamonga						
	FM	K6ONT	447.20000	-	114.8 Hz	
Rancho Palos Verdes						
	DMR/BM	K6PV	447.12000	-	CC1	
	DMR/MARC	N6RPV	445.72000	-	CC1	
	FM	WA6LA	145.38000	-	100.0 Hz	
	FM	KA6TSA	146.23500	+	127.3 Hz	
	FM	K6VGP	147.36000	+	97.4 Hz	
	FM	AA6RJ	147.42000	146.52000	131.8 Hz	
	FM	WA6LA	223.78000	-	100.0 Hz	
	FM	W6SBA	224.38000	-	192.8 Hz	
	FM	WZ6A	445.98000	-		
	FM	K6VGP	446.74000	-	100.0 Hz	
	FM	W6TRW	447.00000	-	100.0 Hz	
	FM	KE6LE	449.82000	-		
	FM	WA6DPB	1282.47500	-	131.8 Hz	
	FM	N6YKE	1283.32500	-		
	FM	N6UL	1283.55000	-		
	FM	K6MOZ	1284.50000	-		
	FM	KV6D	1285.15000	-		
Rancho Palos Verdes, San Pedro						
	FM	K6IUM	449.98000	-	173.8 Hz	
Randsburg	FM	N6BKL	52.68000	-	82.5 Hz	
Red Bluff	FM	KF6KDD	145.45000	-	88.5 Hz	
	FM	KH8AF	444.15000	+		

88 CALIFORNIA

Location	Mode	Call sign	Output	Input	Access	Coordinator
Redding	DMR/BM	W6GRC	145.00000	147.50000	CC1	
	DMR/BM	WA6IO	443.70000	+	CC1	
	FM	KE6CHO	145.11000	-	94.8 Hz	
	FM	WB6CAN	146.76000	-	107.2 Hz	
	FM	NC6I	147.09000	+	88.5 Hz	
	FM	KK6JP	147.36000	+	162.2 Hz	
	FM	NA0SA	440.05000	+		
	FM	NA0SA	440.17500	+		
	FM	WR6TV	927.12500	902.12500	107.2 Hz	
Redding, Bully Choop Mountain						
	FM	N4SMF	146.92500	-	85.4 Hz	
Redding, Shasta Bally						
	FM	WR6TV	52.66000	-	107.2 Hz	
	FM	AB7BS	444.32500	+	100.0 Hz	
Redding, South Fork Mountain						
	FM	W6STA	146.64000	-	88.5 Hz	
	FM	NC6SV	444.55000	+	100.0 Hz	
	FM	KE6CHO	927.22500	902.22500		
Redlands	DMR/BM	N6LKA	446.88000	-	CC10	
	DMR/BM	AD5MT	449.08000	-	CC7	
	FM	AI6BX	145.26000	-	123.0 Hz	
	FM	AI6BX	147.18000	+	88.5 Hz	
Redlands, Sunset Hills						
	FM	AI6BX	445.34000	-	88.5 Hz	
Redlands, Tremont Ranch						
	FM	N6AJB	447.02000	-		
Redondo Beach	DMR/MARC	W6TRW	449.46250	-	CC2	
	FM	W6TRW	145.32000	-	114.8 Hz	
Redwood City	DMR/BM	W6BSD	443.50000	+	CC1	
	FM	K6MPN	53.68000	-	114.8 Hz	
	FM	KC6ULT	146.86500	-	114.8 Hz	
	FM	WD6GGW	441.40000	+	114.8 Hz	
	FM	K6MPN	444.50000	+	100.0 Hz	
	FM	WI6H	927.88750	902.88750	192.8 Hz	
	FM	WD6GGW	1284.70000	-	114.8 Hz	
	FM	KE6UIE	1285.25000	-	88.5 Hz	
Reedley	FM	K6NOX	927.40000	902.40000	114.8 Hz	
Rescue	FM	AG6AU	224.06000	-	127.3 Hz	
Rescue, Pine Hill	FM	AG6AU	927.23750	902.23750	127.3 Hz	
Rialto, Fire Station						
	FM	K6RIA	147.64500	-	127.3 Hz	
Richmond	FM	K6LOU	440.97500	+		
	FM	WA6DUR	442.15000	+	100.0 Hz	
Ridgecrest	FM	WA6YBN	147.00000	+	107.2 Hz	
	FM	K6RFO	147.21000	+		
	FM	W5HMV	447.02000	-	123.0 Hz	
	FM	NC9RS	927.01250	902.01250	88.5 Hz	
Ridgecrest, El Paso Peak						
	FM	WI6RE	147.97500	-	100.0 Hz	
	FM	W5WH	224.90000	-		
	FM	WI6RE	448.80000	-	100.0 Hz	
Ridgecrest, Randsburg Peak						
	FM	WA6YBN	145.34000	-	100.0 Hz	
Rio Linda	DMR/BM	K9WWV	441.61250	+	CC1	
Riverside	DMR/BM	N6WZK	449.36000	-	CC1	
	DMR/BM	KC7NP	449.46250	-	CC1	
	FM	KB6OZX	445.06000	-	162.2 Hz	
Riverside, Box Springs Mountai						
	FM	W6CDF	224.46000	-	110.9 Hz	
Riverside, County Admin Bldg.						
	FM	W6TJ	146.88000	-	146.2 Hz	
Rocklin	DMR/BM	K6IOK	442.52500	+	CC1	

CALIFORNIA   89

Location	Mode	Call sign	Output	Input	Access	Coordinator
Rocklin	FM	N6UG	442.00000	+	179.9 Hz	
Rohnert Park	FM	WD6FTB	223.90000	-	88.5 Hz	
Rolling Hills	DMR/BM	K6RH	445.30000	-	CC1	
Rosamond	FM	KK6KU	224.66000	-	110.9 Hz	
Rosamond, Taco Bell						
	FM	WA6CAM	444.57500	+		
Rosemead	FM	KB6MRC	445.90000	-	123.0 Hz	
Roseville	FM	W6SAR	146.64000	-	156.7 Hz	
	FM	KA6OIJ	1282.65000	-	100.0 Hz	
Rowland Heights	FM	NO6B	446.28000	-		
Running Springs	FM	WA6ISG	145.12000	-	131.8 Hz	
	FM	KI6JVF	147.61500	-	186.2 Hz	
	FM	WA6DVG	224.06000	-	94.8 Hz	
	FM	K6AMS	445.70000	-	151.4 Hz	
	FM	WR6HP	447.28000	-	136.5 Hz	
	FM	KV6D	447.98000	-		
	FM	KA6MEP	449.98000	-	151.4 Hz	
	FM	AA6QO	1285.07500	-		
	FM	KA6RWW	1286.00000	-		
Running Springs , Keller Peak						
	DSTAR	KI6WZX	446.34000	-		
Running Springs, Heaps Peak						
	FM	K6ECS	147.70500	-	167.9 Hz	
	FM	KE6PCV	447.74000	-		
	FM	WA6VAW	448.86000	-		
	FM	N6CIZ	449.76000	-		
Running Springs, Keller Peak						
	DSTAR	KI6WZX	147.55000	144.95000		
	FM	KE6TZG	146.38500	+	146.2 Hz	
	FM	WA6MTN	224.00000	-		
Sacramento	FM	KC6MHT	145.23000	-	162.2 Hz	
	FM	N6NA	145.25000	-	162.2 Hz	
	FM	K6INC	146.70000	-		
	FM	KF6SQL	147.12000	+	162.2 Hz	
	FM	K6NP	147.30000	+	179.9 Hz	
	FM	KC6MHT	224.22000	-	123.0 Hz	
	FM	WA6ZZK	224.56000	-	94.8 Hz	
	FM	AA6IP	224.70000	-	107.2 Hz	
	FM	KJ6JD	224.82000	-	77.0 Hz	
	FM	KU6P	440.20000	+	131.8 Hz	
	FM	KJ6KO	441.45000	+		
	FM	NA6DF	441.85000	+	77.0 Hz	
	FM	K6YC	441.95000	+	114.8 Hz	
	FM	KF6BIK	442.32500	+		
	FM	WB6GWZ	442.40000	+		
	FM	WA6ZZK	442.50000	+	151.4 Hz	
	FM	W6AK	442.80000	+	100.0 Hz	
	FM	K6NP	442.90000	+	136.5 Hz	
	FM	KJ6JD	443.27500	+	127.3 Hz	
	FM	N0RM	443.45000	+		
	FM	W6YDD	443.90000	+	136.5 Hz	
	FM	K6YC	927.20000	902.20000	100.0 Hz	
	FM	N6ICW	927.96250	902.96250		
	FM	KD6GFZ	1284.85000	-	88.5 Hz	
Sacramento, Folsom Lake						
	FM	K6IS	145.19000	-	162.2 Hz	
	FM	K6IS	224.40000	-	162.2 Hz	
Sacramento, KOVR/KXTV Tower						
	FM	K6DTV	147.39000	+	146.2 Hz	
Sacramento, Pilot Hill						
	FM	W6PRN	444.42500	+	100.0 Hz	
Salinas	DMR/BM	W6JSO	444.52500	+	CC1	

90 CALIFORNIA

Location	Mode	Call sign	Output	Input	Access	Coordinator
Salinas	FM	W6CER	145.41000	-		
	FM	W6RTF	145.49000	-	100.0 Hz	
	FM	KC6UDC	146.08500	+	100.0 Hz	
	FM	N6LEX	224.32000	-	146.2 Hz	
	FM	N6LEX	442.02500	+	146.2 Hz	
	FM	KG6UYZ	442.60000	+	110.9 Hz	
	FM	KG6UYZ	927.95000	902.95000		
Salinas, Fremont Pk						
	FM	K6JE	145.47000	-	94.8 Hz	
	FM	K6JE	441.45000	+	123.0 Hz	
Salinas, Mt Toro	FM	N6SPD	145.43000	-	94.8 Hz	
	FM	WB6ECE	147.27000	+	94.8 Hz	
San Andreas	FM	W6ALL	441.70000	+		
	FM	NC6R	443.35000	+	156.7 Hz	
	FM	N6GKJ	927.07500	902.07500	100.0 Hz	
San Anselmo	FM	K6BW	146.77500	-	127.3 Hz	
	FM	W6RV	440.55000	+	100.0 Hz	
San Ardo, Williams Hill						
	FM	WR6VHF	51.82000	-	136.5 Hz	
	FM	W6FM	146.73000	-	127.3 Hz	
San Bernardino	DMR/BM	AE6TV	449.16000	-	CC1	
San Bernardino, Onyx Peak						
	FM	N6LXX	224.44000	-	94.8 Hz	
San Bernardino, San Bernardino						
	FM	WA6JBD	927.95000	902.95000		
San Bernardino, Twin Peaks						
	FM	WA6TJQ	224.56000	-	100.0 Hz	
San Bruno	DMR/MARC	AH6KD	145.01250	147.51250	CC0	
	FM	W6JMX	224.24000	-	141.3 Hz	
	FM	KM6EF	1286.05000	-	123.0 Hz	
San Carlos	FM	W6CBS	441.61250	+	100.0 Hz	
San Clemente	DMR/MARC	KK6TTJ	144.65000	147.55000	CC1	
	FM	K6SOA	146.02500	+	110.9 Hz	
	FM	W6KRW	146.89500	-	136.5 Hz	
	FM	W6KRW	1282.77500	-	88.5 Hz	
San Diego	DMR/BM	W6RDX	444.46250	+	CC8	
	DMR/BM	W2NOR	445.20000	-	CC1	
	DMR/BM	KI6MGN	445.86000	-	CC1	
	DMR/BM	K6RRR	446.35000	-	CC1	
	DMR/BM	K6RRR	446.36500	-	CC1	
	DMR/BM	K6RRR	446.72000	-	CC1	
	DMR/BM	KK6KD	448.46000	-	CC1	
	DMR/BM	N6VVY	449.36000	-	CC3	
	DMR/BM	AI6DZ	449.46250	-	CC6	
	DMR/BM	KB6TWT	449.74000	-	CC1	
	DMR/BM	N7OEI	449.85000	-	CC1	
	DMR/BM	N6RVI	449.96000	-	CC3	
	DMR/MARC	WA6NVL	445.62000	-	CC0	
	DMR/MARC	KI6KQU	448.52000	-	CC1	
	DMR/MARC	WA6YVX	449.86250	-	CC1	
	FM	N6WYF	147.94500	-	107.2 Hz	
	FM	W6GIC	224.74000	-	107.2 Hz	
	FM	K6ERN	445.58000	-	100.0 Hz	
	FM	KE6VK	448.94000	-		
	FM	KJ6GRS	449.00000	-		
	FM	N6VCM	449.14000	-		
	FM	N6OEI	449.84000	-		
San Diego, Black Mtn						
	FM	WR7NV	145.36000	-	100.0 Hz	
San Diego, Kearney Mesa						
	FM	W6UUS	447.32000	-	107.2 Hz	

CALIFORNIA 91

Location	Mode	Call sign	Output	Input	Access	Coordinator
San Diego, Mount Laguna						
	FM	W6HDC	145.12000	-	107.2 Hz	
	FM	WB6WLV	147.15000	+	107.2 Hz	
	FM	KA6DAC	446.75000	-	107.2 Hz	
	FM	K6KTA	449.58000	-	107.2 Hz	
San Diego, Mount Otay						
	FM	W6XC	145.34000	0.00000		
	FM	K6XI	449.44000	-	107.2 Hz	
San Diego, Mt San Miguel						
	FM	N6LXX	446.10000	-	123.0 Hz	
	FM	N6LXX	927.57500	901.97500	151.4 Hz	
San Diego, Mt. Laguna/Cuyamaca						
	FM	K6JCC	52.60000	-	107.2 Hz	
San Diego, Mt. Otay						
	FM	WD6FZA	224.40000	-	103.5 Hz	
	FM	WR6BLU	448.52000	-		
	FM	KF6BYB	448.98000	-		
	FM	N6JOJ	449.82000	-		
San Diego, Mt. San Miguel						
	FM	N6LXX	224.72000	-	94.8 Hz	
San Diego, Mt. Soledad						
	P25	KI6KHB	447.28000	-	91.5 Hz	
San Diego, Mt. Woodson						
	FM	WA6JAF	448.72000	-		
San Diego, Old Town						
	FM	KM6RPT	448.28000	-	82.5 Hz	
San Diego, Otay Mountain						
	FM	WB6WLV	146.64000	-	107.2 Hz	
	FM	N6VVY	147.21000	+	91.5 Hz	
	FM	K6XI	223.84000	-	107.2 Hz	
	FM	WA6HYQ	224.16000	-	107.2 Hz	
	FM	N6VVZ	224.26000	-	107.2 Hz	
	FM	WR6MO	448.02000	-	91.5 Hz	
	FM	K6RRR	449.06000	-	88.5 Hz	
	FM	K6XI	449.12000	-	100.0 Hz	
	FM	WB6CYT	449.22000	-	91.5 Hz	
	FM	KW6HRO	449.38000	-	100.0 Hz	
	FM	W6XC	449.74000	-		
	FM	WB6DTR	449.78000	-	123.0 Hz	
	FM	K6RIF	449.98000	-	88.5 Hz	
	FM	WB6WLV	1282.30000	-	107.2 Hz	
San Diego, Otay Mtn						
	FM	KK6BAD	447.04000	-	123.0 Hz	
	FM	K6JSI	447.64000	-	100.0 Hz	
	FM	WA6OSB	449.70000	-	151.4 Hz	
San Diego, Paradise Hills						
	FM	W6JVA	145.48000	-	127.3 Hz	
San Diego, Rattlesnake Peak						
	FM	AA6WS	147.76500	-	79.7 Hz	
	FM	N6JOJ	446.18000	-		
San Diego, San Miguel						
	DMR/BM	KB6PLH	449.84000	443.85000	CC1	
San Diego, San Miguel Mountain						
	FM	N6LXX	53.58000	-	103.5 Hz	
	FM	KR6FM	53.66000	-	103.5 Hz	
San Diego, Sharp Hospital						
	FM	K6AIL	147.88500	-	107.2 Hz	
San Diegto	DMR/MARC	W6SS	448.16000	-	CC1	
San Dimas	FM	K6VGP	447.62000	-	114.8 Hz	
San Dimas, Johnstone Peak						
	FM	W6FNO	146.82000	-		
	FM	W6NRY	223.98000	-	103.5 Hz	

92   CALIFORNIA

Location	Mode	Call sign	Output	Input	Access	Coordinator
San Dimas, Johnstone Peak						
	FM	K6VGP	224.84000	-		
	FM	W6FNO	446.02000	-		
	FM	WA6FZH	446.40000	-	103.5 Hz	
	FM	KM6RW	446.80000	-	131.8 Hz	
	FM	W6NRY	447.30000	-		
	FM	K6OES	448.34000	-	114.8 Hz	
	FM	K6MVH	448.84000	-		
San Fernando	DMR/BM	N6JVH	447.26000	-	CC1	
	FM	N6BKL	51.80000	-	82.5 Hz	
	FM	W6IN	146.91000	-	136.5 Hz	
San Fernando, Loop Canyon						
	FM	WA6TTL	446.62000	-		
San Francisco	FM	W6TP	146.79000	-	114.8 Hz	
	FM	WA6GG	224.22000	-	100.0 Hz	
	FM	W6TP	224.50000	-	114.8 Hz	
	FM	KA6TGI	224.52000	-	67.0 Hz	
	FM	W6TP	443.10000	+	DCS 114	
	FM	K6MSR	443.67500	+		
	FM	KB6LCS	444.92500	+	136.5 Hz	
	FM	KA6TGI	1284.90000	-	67.0 Hz	
San Francisco, Downtown						
	FM	N6MVT	442.07500	+	100.0 Hz	
San Francisco, Twin Peaks						
	FM	WA6GG	442.05000	+	127.3 Hz	
San Francsico	DMR/BM	W6PW	444.22500	+	CC1	
San Jose	DMR/BM	W6OTX	144.96250	147.46250	CC3	
	DMR/BM	KC6IAU	147.10500	+	CC1	
	DMR/BM	N6JET	441.87500	+	CC1	
	DMR/BM	K6HLE	444.02500	+	CC1	
	DMR/BM	W6OTX	444.47500	+	CC1	
	DMR/MARC	TI0RHU	145.10000	+	CC1	
	DMR/MARC	W6YYY	440.03750	+	CC1	
	DMR/MARC	K6LLC	440.22500	+	CC1	
	DMR/MARC	AD1U	441.82500	+	CC1	
	DMR/MARC	AD1U	443.31250	+	CC1	
	FM	KD6AOG	52.66000	-	127.3 Hz	
	FM	KG6HAT	52.94000	-	100.0 Hz	
	FM	WA2IBM	145.19000	-	151.4 Hz	
	FM	KB6FEC	145.31000	-	162.2 Hz	
	FM	KE6MON	146.35500	+	123.0 Hz	
	FM	K6INC	146.82000	-	123.0 Hz	
	FM	KB6FEC	146.89500	-	110.9 Hz	
	FM	AD1U	146.97000	-		
	FM	WB6KHP	147.03000	+	107.2 Hz	
	FM	KB6FEC	147.16500	+	100.0 Hz	
	FM	K6GOD	223.86000	-	107.2 Hz	
	FM	KU6V	224.04000	-	100.0 Hz	
	FM	W6IOS	430.30000	+		
	FM	W6SMQ	440.10000	+	127.3 Hz	
	FM	WA6YOP	440.27500	+	127.3 Hz	
	FM	N6TNR	440.37500	+		
	FM	KC6BJO	441.15000	+	100.0 Hz	
	FM	KF6FWO	441.17500	+	103.5 Hz	
	FM	K6BEN	441.27500	+		
	FM	W6PIY	441.35000	+	88.5 Hz	
	FM	WA6QDP	441.72500	+		
	FM	KG6KCL	441.85000	+		
	FM	N1UFD	442.17500	+		
	FM	K6INC	442.30000	+	114.8 Hz	
	FM	WB6KHP	442.45000	+	100.0 Hz	
	FM	WB6ZVW	442.50000	+	100.0 Hz	

CALIFORNIA 93

Location	Mode	Call sign	Output	Input	Access	Coordinator
San Jose	FM	N6MNV	442.70000	+	100.0 Hz	
	FM	KU6V	442.77500	+	131.8 Hz	
	FM	WR6COP	442.82500	+		
	FM	WA6YLV	442.87500	+	100.0 Hz	
	FM	K6YZS	442.95000	+	85.4 Hz	
	FM	N6MVT	443.07500	+	123.0 Hz	
	FM	W6AMT	443.27500	+	107.2 Hz	
	FM	KB5JR	443.30000	+	136.5 Hz	
	FM	K6MF	443.45000	+		
	FM	K6TAZ	443.55000	+		
	FM	K6MSR	443.62500	+		
	FM	WA6GFY	443.77500	+	100.0 Hz	
	FM	K6JJC	444.02500	+	136.5 Hz	
	FM	KD6CUC	444.05000	+		
	FM	N6TLQ	444.10000	+		
	FM	WB6RNH	444.30000	+	162.2 Hz	
	FM	WA6INC	444.32500	+	114.8 Hz	
	FM	WB6OQS	444.60000	+	141.3 Hz	
	FM	N6NAC	444.62500	+	110.9 Hz	
	FM	WA6GEL	444.80000	+		
	FM	W6RLW	1282.00000	-	88.5 Hz	
	FM	W6RLW	1282.20000	-	88.5 Hz	
	FM	N6AKK	1283.10000	-	88.5 Hz	
	FM	N6EEZ	1283.40000	-	94.8 Hz	
	FM	WA6GFY	1283.70000	-	100.0 Hz	
	FM	N6AKB	1284.30000	-	100.0 Hz	
	FM	W6RLW	1285.00000	-	88.5 Hz	
	FM	KD6AOG	1285.80000	-	127.3 Hz	
	FM	KU6V	1285.95000	-	100.0 Hz	
	FM	N6NAC	1286.00000	-	110.9 Hz	
	FM	WB6OCD	1286.07500	-	88.5 Hz	
	FM	KD6AOG	1286.15000	-	127.3 Hz	
	FM	W6PIY	1286.20000	-	100.0 Hz	
	P25	KE6STH	443.57500	+		
San Jose, Almaden Valley						
	FM	WB6KHP	224.68000	-	107.2 Hz	
San Jose, Alum Rock Park						
	FM	K6GOD	441.10000	+	203.5 Hz	
San Jose, Central San Jose						
	FM	NT6S	224.98000	-	110.9 Hz	
San Jose, Eagle Rock						
	FM	W6UU	146.98500	-	114.8 Hz	
San Jose, Good Samaritan Hosp						
	FM	W6PIY	52.58000	-	151.4 Hz	
	FM	W6PIY	147.39000	+	151.4 Hz	
	FM	W6PIY	223.96000	-	156.7 Hz	
San Jose, Good Samaritan Hospi						
	FM	WB6KHP	444.70000	+	127.3 Hz	
San Jose, Loma Prieta						
	FM	WR6ABD	146.64000	-	162.2 Hz	
	FM	WB6OQS	146.76000	-	151.4 Hz	
	FM	WB6OQS	224.26000	-	123.0 Hz	
	FM	K6JSI	442.90000	+	162.2 Hz	
San Jose, Mount Umunhum						
	DMR/BM	WA6YCZ	442.53750	+	CC1	
San Jose, Mt Hamilton						
	FM	KB6ABM	224.60000	-	156.7 Hz	
San Jose, Mt Umunhum						
	FM	WA6YCZ	147.15000	+	110.9 Hz	
	FM	WB6ECE	441.30000	+	146.2 Hz	
	FM	K6LNK	443.47500	+	123.0 Hz	

94   CALIFORNIA

Location	Mode	Call sign	Output	Input	Access	Coordinator
San Jose, Regional Medical Cen						
	FM	W6UU	442.42500	+	107.2 Hz	
San Jose, San Jose State Univ.						
	FM	W6YL	443.90000	+	100.0 Hz	
San Jose, Santa Cruz Mountains						
	FM	KU6V	444.90000	+	110.9 Hz	
San Jose, Santa Cruz Mts.						
	FM	N6NAC	224.64000	-	110.9 Hz	
San Leandro	FM	W6RGG	147.24000	+	107.2 Hz	
	FM	KB6NCL	442.77500	+		
	FM	K6KBL	443.37500	+	156.7 Hz	
	FM	W6RGG	444.20000	+	107.2 Hz	
San Lorenzo	FM	KQ6YG	224.70000	-	156.7 Hz	
	FM	KM6HJA	441.45000	+	162.2 Hz	
San Luis Obispo	DMR/BM	K7AZ	443.30000	+	CC3	NARCC
	DMR/BM	KK6DJ	444.93750	+	CC1	
	DMR/MARC	WX6D	443.43750	+	CC1	
	FM	KC6WRD	145.29000	-		
	FM	WB6JWB	441.07500	+	94.8 Hz	
	FM	WB6FMC	442.87500	+		
	FM	W6FM	443.57500	+		
	FM	N6HYM	443.80000	+		
	FM	KC6WRD	444.90000	+	127.3 Hz	
San Luis Obispo, Black Butte						
	FM	WB6NYS	224.58000	-	151.4 Hz	
San Luis Obispo, Cal Poly						
	FM	W6BHZ	146.76000	-	91.5 Hz	
	FM	W6BHZ	442.30000	+	127.3 Hz	
San Luis Obispo, Cuesta Peak						
	FM	W6SLO	146.80000	-	127.3 Hz	
	FM	W6SLO	442.70000	+	127.3 Hz	
San Luis Obispo, Mt Lowe						
	FM	W6FM	147.36000	+	127.3 Hz	
San Luis Obispo, Rocky Butte						
	FM	W6SLO	146.62000	-	127.3 Hz	
San Luis Obispo, Tassajara Pea						
	FM	W6SLO	444.10000	+	127.3 Hz	
	FM	W6FM	444.52500	+	127.3 Hz	
San Luis Obispo, Tassajera Pea						
	FM	W6SLO	146.67000	-	127.3 Hz	
San Marcos	DMR/BM	AA4CD	447.52000	-	CC7	
	FM	WD6FZA	446.58000	-	156.7 Hz	
	FM	K6VGP	447.36000	-	91.5 Hz	
San Marcos, Ca	DMR/BM	AA4CD	445.42000	-	CC1	
San Martin	FM	KU6V	223.92000	-	100.0 Hz	
San Mateo	FM	N6MPX	147.30000	+	100.0 Hz	
San Miguel	DMR/BM	KB6CIO	444.17500	+	CC1	
San Pablo	FM	WA6KQB	145.11000	-	82.5 Hz	
	FM	WA6KQB	224.30000	-	82.5 Hz	
	FM	WA6JQV	927.02500	902.02500		
	FM	WA6JQV	927.45000	902.45000		
	FM	WA6JQV	927.47500	902.47500		
San Rafael	FM	KH8AF	440.92500	+		
	FM	KH8AF	442.17500	+		
San Rafael, San Pedro Ridge						
	FM	K6GWE	147.33000	+	173.8 Hz	
	FM	K6GWE	443.52500	+	82.5 Hz	
San Ramon	FM	WA6HAM	440.42500	+	79.7 Hz	
San Simeon, Rocky Butte						
	FM	W6SLO	444.10000	+	127.3 Hz	
Sanel Mountain	DMR/BM	NN6J	440.13750	+	CC1	NARCC
Santa Barbara	DMR/BM	KK6GFX	446.98000	-	CC1	

CALIFORNIA 95

Location	Mode	Call sign	Output	Input	Access	Coordinator
Santa Barbara	DMR/MARC	KA6SOX	445.38000	-	CC1	
	FM	AF6VU	147.07500	+	131.8 Hz	
	FM	WC6MRA	224.04000	-		
Santa Barbara, Gibraltar Peak						
	FM	K6RCL	224.04000	-		
	FM	K6RCL	1286.20000	-		
Santa Barbara, Gibralter Peak						
	FM	KD6OVS	146.70000	-	131.8 Hz	
Santa Barbara, La Cumbre Peak						
	FM	K6TZ	224.08000	-	131.8 Hz	
	FM	WA6TTL	446.20000	-		
Santa Barbara, La Vigia Hill						
	FM	K6TZ	51.82000	-	82.5 Hz	
	FM	K6TZ	146.79000	-	131.8 Hz	
	FM	K6TZ	446.40000	-	131.8 Hz	
Santa Barbara, Santa Ynez Peak						
	FM	W6XC	447.16000	-		
	FM	W6RRN	448.76000	-		
	FM	K6JSI	448.90000	-	123.0 Hz	
	FM	WB6BBE	449.56000	-		
	FM	K6RCL	1284.05000	-		
Santa Barbara, UCSB						
	FM	K6TZ	224.16000	-	131.8 Hz	
	FM	KG6MNB	927.46250	902.46250	131.8 Hz	
Santa Barbara, UCSB - Storke T						
	FM	W6RFU	145.48000	-	136.5 Hz	
Santa Clara	DMR/BM	KQ6RC	444.87500	+	CC2	
	FM	K6SNC	442.02500	+		
	FM	K6SNC	927.73750	902.73750	100.0 Hz	
	FM	N6MEF	927.83750	902.83750	100.0 Hz	
Santa Clarita	FM	N6KNW	51.86000	-	82.5 Hz	
	FM	KI6AIT	145.20000	-		
	FM	N6NMC	448.34000	-	67.0 Hz	
Santa Clarita, Duck Mountain						
	FM	W6MEP	147.24000	+	67.0 Hz	
Santa Clarita, Mad Mountain						
	FM	W6JW	146.79000	-	123.0 Hz	
	FM	W6JW	445.30000	-	100.0 Hz	
Santa Clarita, Oat Mtn.						
	FM	WA6EQU	448.22000	-		
	FM	KB6C	448.48000	-		
	FM	WA6TXY	449.24000	-		
Santa Cruz	DSTAR	K6DRI	145.00000	147.50000		
	FM	K6HJU	29.66000	-	156.7 Hz	
	FM	KE6IEL	52.68000	-	94.8 Hz	
	FM	K6BJ	146.79000	-	94.8 Hz	
	FM	KA6TGI	224.52000	-	136.5 Hz	
	FM	W6JWS	440.85000	+	94.8 Hz	
	FM	N7WG	441.67500	+	123.0 Hz	
	FM	WB6PHE	443.47500	+	127.3 Hz	
Santa Cruz , Near Loma Prieta						
	FM	WA6FUL	442.55000	+		
Santa Cruz Island, Diablo Peak						
	FM	W6XC	446.60000	-		
	P25	W6XC	146.65500	-		
Santa Cruz Island, Diablo Pk						
	FM	K6TZ	223.92000	-	131.8 Hz	
Santa Cruz, Empire Grade						
	FM	W6WLS	147.18000	+		
	P25	W6DXW	442.75000	+	110.9 Hz	
Santa Cruz, Loma Prieta						
	FM	AE6KE	146.83500	-	94.8 Hz	

96   CALIFORNIA

Location	Mode	Call sign	Output	Input	Access	Coordinator
Santa Cruz, UCSC Baskin School						
	FM	W6SLG	145.31000	-	94.8 Hz	
Santa Maria	FM	N6UE	146.64000	-		
	FM	W6NO	147.91500	-	103.5 Hz	
Santa Maria, Marian Med Ctr						
	FM	W6AB	147.30000	+	131.8 Hz	
Santa Monica	DMR/BM	N6JKF	446.08000	-	CC1	
	FM	K6FCC	145.28000	-	127.3 Hz	
Santa Monica, Brentwood / San						
	FM	K6CYC	147.03000	+	127.3 Hz	
Santa Monica, Saddle Peak						
	FM	WA6TTL	224.34000	-		
	FM	WD6FZA	445.42000	-	127.3 Hz	
	FM	K6VGP	445.82000	-		
	FM	W6XC	447.78000	-		
Santa Monica, Saddle Pk						
	FM	WA6JQB	447.84000	-		
	FM	WR6SP	447.86000	-		
Santa Paula	DMR/BM	WB6VKR	447.36000	-	CC1	
Santa Paula, South Mountain						
	FM	K6ERN	51.84000	-	103.5 Hz	
	FM	WA6ZSN	146.38500	+	127.3 Hz	
	FM	WA6ZSN	224.10000	-	127.3 Hz	
Santa Rosa	DMR/BM	K6ACS	442.11250	+	CC1	
	DSTAR	K6ACS	145.04000	-		
	FM	WA6YGD	145.35000	-	88.5 Hz	
	FM	K6ACS	146.73000	-	88.5 Hz	
	FM	KE6EAQ	146.83500	-	88.5 Hz	
	FM	KE6N	223.76000	-	85.4 Hz	
	FM	K6ACS	224.82000	-	103.5 Hz	
	FM	KD6CJQ	440.20000	+	88.5 Hz	
	FM	K6EAR	440.45000	+	88.5 Hz	
	FM	KD6RC	441.30000	+	88.5 Hz	
	FM	WB7ABP	443.82500	+	100.0 Hz	
	FM	WB6RUT	444.37500	+		
	FM	KC6REK	1283.20000	-	88.5 Hz	
Santa Rosa, Sanel Mountain						
	FM	WA6RQX	444.75000	+	123.0 Hz	
Santa Ynez	DSTAR	KK6GFX	447.84000	-		
Santa Ynez, Broadcast Peak						
	FM	WB6OBB	147.00000	+	131.8 Hz	
Santa Ynez, Santa Ynez Peak						
	FM	K6TZ	145.16000	-	127.3 Hz	
	FM	W6YJO	145.18000	-	131.8 Hz	
	FM	K6BVA	145.44000	-		
	FM	K6TZ	224.12000	-	131.8 Hz	
	FM	WA6TZE	449.00000	-	131.8 Hz	
Santee	FM	WA6OSB	147.91500	-	151.4 Hz	
Santee, Kearney Mesa						
	FM	KD6GNB	224.92000	-	107.2 Hz	
Santiago Peak	ATV	W6ATN	1253.25000	2441.55000		SCRRBA
	DMR/BM	KB6CRE	447.54000	-	CC1	
	FM	W6KRW	29.64000	-	107.2 Hz	
Saratoga	DMR/BM	KK6USZ	441.95000	+	CC1	NARCC
	DMR/MARC	NU6P	440.43750	+	CC1	
	FM	K6SA	146.65500	-	114.8 Hz	
	FM	K6BEN	224.46000	-		
	FM	K6BEN	443.12500	+		
	FM	W6RLW	1283.00000	-	88.5 Hz	
Sausalito, Wolfback Rdg						
	FM	K6ER	442.52500	+	114.8 Hz	
Scotia	FM	N7HQZ	51.84000	-	114.8 Hz	

CALIFORNIA 97

Location	Mode	Call sign	Output	Input	Access	Coordinator
Scotia	FM	K6FWR	443.25000	+	103.5 Hz	
Scotia, Monument Peak						
	FM	K6FWR	146.76000	-	103.5 Hz	
Scotia, Pt Reyes	FM	WB6TMS	145.17000	-	103.5 Hz	
Sea Ranch	FM	W6ABR	147.94500	-	88.5 Hz	
Seal Beach	FM	K6NX	445.36000	-	100.0 Hz	
	FM	WA6FZH	446.40000	-	88.5 Hz	
	FM	KC6YNQ	449.30000	-	141.3 Hz	
Sebastapol	FM	WA6FUL	52.64000	-	114.8 Hz	
Sebastopol	DMR/BM	W6SON	444.82500	+	CC1	
	DMR/BM	KC6SOT	444.98750	+	CC1	
	FM	WA6FUL	440.47500	+	97.4 Hz	
	FM	WA6FUL	440.95000	+	123.0 Hz	
	FM	WA6FUL	443.42500	+	88.5 Hz	
Sebastopol, English Hill						
	FM	W6SON	147.31500	+	88.5 Hz	
	FM	W6SON	224.48000	-	88.5 Hz	
Seigler	FM	KI6QCU	145.15000	-	103.5 Hz	
Sequoia, Buck Rock						
	FM	KK6AC	442.95000	+	141.3 Hz	
Shasta Lake	FM	K7JKL	442.07500	+	114.8 Hz	
Sherman Oaks	FM	NK6S	145.24000	-	91.5 Hz	
Shingle Springs	FM	KG6HAT	52.90000	-	100.0 Hz	
	FM	N6NA	145.25000	-	162.2 Hz	
	FM	WO3B	146.94000	-	136.5 Hz	
	FM	N6NA	441.30000	+		
Shingle Springs, Mt Aukum						
	FM	NC6R	146.67000	-	156.7 Hz	
	FM	NC6R	442.05000	+	156.7 Hz	
Shingletown	FM	WO6P	147.03000	+	88.5 Hz	
	FM	WO6P	147.16500	+	88.5 Hz	
Sierra City	FM	W7FEH	145.17000	-	114.8 Hz	
Sierra Madre, Santa Anita Ridg						
	FM	W6QFK	147.76500	-	131.8 Hz	
Signal Hill, Reservoir Hill						
	FM	K6CHE	146.14500	+	156.7 Hz	
Signal Hill, Signal Hill Tower						
	FM	K6CHE	223.80000	-	156.7 Hz	
	FM	K6CHE	449.78000	-	131.8 Hz	
	FM	K6CHE	1286.30000	-	156.7 Hz	
Silicon Valley	FM	KA6DWN	443.75000	+	100.0 Hz	
Simi Valley	FM	WA6FGK	146.64000	-	127.3 Hz	
	FM	K6ERN	146.80500	-	100.0 Hz	
	FM	AD6SV	224.06000	-	156.7 Hz	
Simi Valley, Chatsworth Peak						
	FM	WA6TTL	145.34000	-		
	FM	N6XPG	224.44000	-	100.0 Hz	
Simi Valley, Mellow Lane						
	FM	K6ERN	445.58000	-	100.0 Hz	
Smith River	FM	K6SLS	443.10000	+	100.0 Hz	
Soledad	FM	N6HU	145.21000	-		
	FM	WA6RQX	444.37500	+	100.0 Hz	
Solvang	FM	K6SYV	146.89500	-	131.8 Hz	
Sonoma	DMR/BM	W6AJF	440.01250	+	CC1	
Sonoma Mtn	DMR/BM	NN6J	444.03750	+	CC2	NARCC
Sonora	DMR/BM	KJ6NRO	442.97500	+	CC1	NARCC
	FM	WB6PHE	441.47500	+	110.9 Hz	
	FM	K6LNK	443.47500	+	103.5 Hz	
	FM	K6KVA	444.65000	+	114.8 Hz	
	FM	NC9RS	927.61250	902.61250	107.2 Hz	
Soulsbyville	FM	N6HUH	146.11500	+	100.0 Hz	

98   CALIFORNIA

Location	Mode	Call sign	Output	Input	Access	Coordinator
South Lake Tahoe						
	FM	W6SUV	144.89500	-		
	FM	N6ICW	145.15000	-	123.0 Hz	
	FM	KA6GWY	145.35000	-	110.9 Hz	
	FM	W6SUV	146.11500	+	192.8 Hz	
	FM	WA6EWV	146.85000	-	123.0 Hz	
	FM	NR7A	224.02000	-		
	FM	N3KD	1285.00000	-	88.5 Hz	
South Lake Tahoe, East Peak						
	FM	NR7A	147.24000	+	123.0 Hz	
South Lake Tahoe, Heavenly Val						
	FM	W6SUV	442.82500	+	88.5 Hz	
South San Francisco						
	FM	K6HN	440.60000	+		
	FM	K6DNA	441.25000	+	141.3 Hz	
	FM	N6MNV	442.70000	+	173.8 Hz	
Springville, Blue Ridge						
	FM	KE6WDX	441.97500	+	100.0 Hz	
St. George	DMR/MARC	K6IB	448.47500	-	CC1	
St. Helena	DMR/BM	K6LO	440.10000	+	CC1	
Stockton	FM	KD6ITH	223.94000	-		
	FM	WA6TCG	224.62000	-	192.8 Hz	
	FM	KE6DXF	440.07500	+	131.8 Hz	
	FM	K6GTO	442.27500	+	103.5 Hz	
	FM	N6GVI	443.10000	+		
	FM	KI6KGN	443.12500	+	107.2 Hz	
	FM	KI6KGN	444.87500	+	107.2 Hz	
	FM	K6NOX	927.41250	902.41250	107.2 Hz	
Stockton, Coronado Tower						
	FM	N5FDL	147.21000	+	114.8 Hz	
	FM	K6TRK	444.50000	+	114.8 Hz	
Stockton, Stockton Police Dept						
	FM	K6KJQ	444.32500	+	94.8 Hz	
Stonyford	FM	K6LNK	146.11500	-	123.0 Hz	
	FM	N6MVT	443.07500	+	114.8 Hz	
	FM	K6BIQ	443.87500	+	100.0 Hz	
Sun City	FM	KB6SSB	146.70000	-	103.5 Hz	
Sun Valley	FM	KC6HUR	445.22000	-	110.9 Hz	
Sunnyvale	FM	K6SNY	145.17000	-	94.8 Hz	
	FM	WA6DY	147.10500	+	100.0 Hz	
Sunnyvale, Moffett Field						
	FM	NA6MF	145.25000	-	123.0 Hz	
Sunol	DMR/MARC	W6SRR	443.51250	+	CC1	
Susanville	DMR/MARC	KE6NDG	444.57500	+	CC1	
	DMR/MARC	KE6NDG	444.77500	+	CC1	
	DMR/MARC	KE6NDG	444.87500	+	CC1	
Susanville, Fredonyer Peak						
	FM	K6LRC	146.91000	-	91.5 Hz	
Susanville, Hamilton Mountain						
	FM	K6LRC	146.83500	-	91.5 Hz	
Susanville, Shaffer Mountain						
	FM	K6LRC	146.88000	-	91.5 Hz	
Swingle	DMR/BM	K6LNK	927.28750	902.28750	CC3	
Sylmar, Contractor's Point						
	FM	WA6TTL	224.36000	-		
	FM	W6CTR	445.16000	-	100.0 Hz	
Sylmar, Contractors Point						
	DSTAR	WA6IRC	147.56000	144.96000		
	DSTAR	WA6IRC	447.04000	-		
	DSTAR	WA6IRC	1286.10000	-		
	FM	KC6PXL	145.12000	-	103.5 Hz	
	FM	W6WAX	224.24000	-	162.2 Hz	

CALIFORNIA 99

Location	Mode	Call sign	Output	Input	Access	Coordinator
Sylmar, Contractors Point						
	FM	K6VE	224.48000	-	110.9 Hz	
	FM	KC6PXL	224.52000	-	103.5 Hz	
	FM	KC6JAR	445.34000	-	71.9 Hz	
	FM	W6CPA	1285.90000	-	123.0 Hz	
Sylmar, Contractors Pt						
	FM	KF6HKM	224.26000	-	103.5 Hz	
	FM	K6HOG	445.04000	-	107.2 Hz	
	FM	KC6PXL	447.22000	-		
	FM	W6FRT	447.34000	-	162.2 Hz	
	FM	K6LRB	927.48750	902.48750	DCS 311	
Sylmar, Loop Canyon						
	FM	K6VGP	449.64000	-		
Taft	DMR/BM	KC6KGE	441.57500	+	CC1	
	DMR/MARC	KC6KGE	441.12500	+	CC1	
	FM	KC6KGE	927.08750	902.08750	100.0 Hz	
Tahoe City	FM	WA6FJS	146.94000	-	100.0 Hz	
	FM	W6AV	224.76000	-	123.0 Hz	
	FM	K1BMW	440.27500	+	114.8 Hz	
	FM	KH8AF	440.92500	+		
	FM	N7VXB	441.17500	+	107.2 Hz	
	FM	KH8AF	442.17500	+		
	FM	WA6FJS	442.95000	+	131.8 Hz	
	FM	K6SRA	443.97500	+		
	FM	W6PUE	444.95000	+		
Tancred	FM	N6QDY	147.25500	+	123.0 Hz	
Tassajara	DMR/MARC	W6FM	444.47500	+	CC1	NARCC
Tassajera Peak	DMR/BM	K6DOA	444.35000	+	CC2	
Tehachapi	FM	W6PVG	147.06000	+		
	FM	KI6HHU	442.92500	+	141.3 Hz	
	FM	WA6CAM	927.02500	902.02500	146.2 Hz	
Tehachapi, Cub Peak						
	FM	WA6CGR	927.72500	902.72500	DCS 411	
Tehachapi, Double Mt						
	FM	K6RET	224.78000	-	100.0 Hz	
Tehachapi, Double Mtn.						
	FM	KI6HHU	446.32000	-	141.3 Hz	
	FM	KI6HHU	447.92000	-	141.3 Hz	
Tehachapi, Oak Creek Pass						
	FM	WB6FYR	224.92000	-	94.8 Hz	
Tehachapi, Tehachapi Mountains						
	FM	W6SLZ	440.62500	+	100.0 Hz	
Temecula	FM	KE6UPI	445.08000	-	82.5 Hz	
Templeton	FM	W6YDZ	146.88000	-	127.3 Hz	
Thousand Oaks	DMR/MARC	KF6GOI	445.56000	-	CC1	
	FM	K0AKS	147.15000	+	127.3 Hz	
	FM	KC6IJM	447.02000	-	127.3 Hz	
	FM	K6JSI	448.94000	-	100.0 Hz	
Thousand Oaks, Grissom Pk						
	FM	N6EVC	146.85000	-	94.8 Hz	
	FM	K6HB	224.70000	-	156.7 Hz	
Thousand Oaks, Rasnow Peak						
	FM	WB6RHQ	223.94000	-	156.7 Hz	
	FM	WA6TTL	449.10000	-		
	FM	W6AMG	449.44000	-	131.8 Hz	
Thousand Oks	DMR/MARC	WB6YES	446.36000	-	CC5	
Thousand Palms	FM	KA6GBJ	447.20000	-	107.2 Hz	
Thousand Palms, Indio Hills						
	FM	KA6GBJ	51.84000	-	107.2 Hz	
Tiki Mobile Village	FM	W6YJ	449.32000	-		
Timber Cove	FM	K6CHG	444.03750	+		
Topanga Canyon	FM	W6DRT	146.11500	+		

100   CALIFORNIA

Location	Mode	Call sign	Output	Input	Access	Coordinator
Torrance	DMR/BM	K6DAN	448.26000	-	CC3	
	DMR/MARC	K6UD	448.26000	-	CC1	
	FM	K6TPD	223.86000	-	100.0 Hz	
	FM	KE6LDM	1284.07500	-		
Trabuco Canyon	FM	K6SOA	145.24000	-	110.9 Hz	
Tracy	FM	W6LLL	146.65500	-	100.0 Hz	
	FM	KB6LED	224.34000	-	141.3 Hz	
Tracy, Mt Delux	FM	WA6SEK	145.21000	-	100.0 Hz	
Tranquillion Peak VAFB						
	DMR/BM	K7AZ	440.50000	+	CC3	
Trona	FM	K6YYJ	146.97000	-	123.0 Hz	
Truckee	DMR/BM	K1BMW	446.01250	433.01250	CC1	
	FM	K1BMW	441.75000	+	123.0 Hz	
	FM	WA6JQV	444.27500	+		
Truckee, Donner Peak						
	FM	W6SAR	146.64000	-	131.8 Hz	
	FM	W6SAR	223.82000	-	100.0 Hz	
	FM	W6SAR	440.70000	+	131.8 Hz	
Tujunga	FM	W7JAM	146.16000	+	146.2 Hz	
	FM	NW6B	447.98000	-		
	FM	NW6B	1285.25000	-		
Tulare	DMR/MARC	WX6D	442.47500	+	CC2	NARCC
Tuolumne	DMR/BM	KJ6NRO	144.96250	147.46250	CC2	
	DMR/BM	KJ6NRO	442.47500	+	CC1	
Turlock	DMR/BM	W6BXN	444.35000	+	CC2	
	FM	WB6PBN	442.17500	+	110.9 Hz	
Tustin, Loma Ridge						
	FM	W6KRW	146.89500	-	136.5 Hz	
	FM	W6KRW	449.10000	-		
Twain Harte	FM	KE6KUA	440.55000	+	114.8 Hz	
	FM	W6YOP	441.92500	+	123.0 Hz	
Twentynine Palms						
	DSTAR	KJ6KTV	147.57000	144.97000		
	FM	W6BA	147.06000	+	136.5 Hz	
Ukiah	FM	WA6RQZ	440.02500	+	141.3 Hz	
Union City	FM	KM6EF	146.61000	-	123.0 Hz	
Upland	FM	WB6QHB	147.30000	+	123.0 Hz	
	FM	K6PQN	224.58000	-	88.5 Hz	
	FM	K6PQN	445.86000	-	114.8 Hz	
	FM	K6PQN	927.32500	902.32500	114.8 Hz	
Vacavila	FM	KI6SSF	927.26250	902.26250		
Vacaville	FM	WA6CAX	51.98000	-	114.8 Hz	
	FM	W6YDD	146.62500	-	100.0 Hz	
	FM	KM6KW	223.92000	-	85.4 Hz	
	FM	KJ6MB	224.12000	-	141.3 Hz	
	FM	WV6F	224.20000	-	127.3 Hz	
	FM	KB6SJG	224.24000	-	136.5 Hz	
	FM	WV6F	440.02500	+	127.3 Hz	
	FM	KB6LCS	440.52500	+	136.5 Hz	
	FM	KH8AF	440.92500	+		
	FM	W6RCA	441.60000	+	100.0 Hz	
	FM	W6RCA	441.97500	+	94.8 Hz	
	FM	W6KCS	442.02500	+	179.9 Hz	
	FM	KH8AF	442.22500	447.32500		
	FM	W6NQJ	442.30000	+		
	FM	N6APB	442.55000	+		
	FM	AB6CQ	442.85000	+	146.2 Hz	
	FM	K6JJC	442.97500	+	136.5 Hz	
	FM	WA6KBP	443.75000	+		
	FM	W6SEL	444.12500	+		
	FM	WA6RTL	444.75000	+	107.2 Hz	
	FM	W6PUE	444.85000	+		

CALIFORNIA   101

Location	Mode	Call sign	Output	Input	Access	Coordinator
Vacaville	FM	WA6JQV	927.02500	902.02500		
	FM	WA6JQV	927.03750	902.03750		
	FM	N6ICW	927.05000	902.05000	77.0 Hz	
	FM	WV6F	927.08750	902.08750	127.3 Hz	
	FM	W6NQJ	927.12500	902.12500	131.8 Hz	
	FM	WA6JQV	927.37850	902.37850		
	FM	WA6JQV	927.47500	902.47500		
	FM	WA6JQV	927.93750	902.93750		
	FM	N6ICW	927.97500	902.97500		
	FM	K6HEW	1282.90000	-	88.5 Hz	
	FM	W6YDD	1285.85000	-	100.0 Hz	
	FM	KD6ZNG	1285.90000	-	156.7 Hz	
Vacaville, Blue Ridge						
	FM	N6NMZ	52.86000	-	136.5 Hz	
	FM	N6NMZ	224.42000	-	100.0 Hz	
Vacaville, Mt Vaca						
	FM	WR6VHF	51.82000	-	141.3 Hz	
	FM	W6VVR	145.47000	-	127.3 Hz	
	FM	W6SAR	146.74500	-		
	FM	K6MVR	147.00000	-	136.5 Hz	
	FM	N6ICW	147.19500	+	123.0 Hz	
	FM	W6AEX	147.27000	+	77.0 Hz	
	FM	KB6ABM	223.84000	-	141.3 Hz	
	FM	K6LNK	440.75000	+	173.8 Hz	
	FM	N6ICW	441.77500	+		
	FM	K6MVR	443.95000	+	136.5 Hz	
	FM	WA6EUZ	444.57500	+		
	FM	W6KCS	927.06250	902.06250	167.9 Hz	
	FM	KI6SSF	927.33750	902.33750	162.2 Hz	
Vallejo	FM	K6LI	145.31000	-	88.5 Hz	
	FM	K6LI	442.42500	+	88.5 Hz	
	FM	KC6PGV	1284.35000	-	131.8 Hz	
Valley Center	DMR/BM	N6VCC	449.92000	-	CC1	
Valley Center , Paradise Mtn						
	FM	N6VCC	146.23500	+		
Valley Springs	FM	KC6TTZ	146.92500	-	94.8 Hz	
	FM	W6EBW	441.07500	+	114.8 Hz	
Valyermo, Blue Ridge						
	FM	N6GMS	1285.12500	-		
Various	DMR/BM	AA1HD	442.00000	+	CC1	
Ventura	FM	N6VUY	147.97500	-		
	FM	KB6LJQ	445.60000	-	114.8 Hz	
	FM	WA6TTL	447.14000	-		
Ventura, Red Mountain						
	FM	K6ERN	52.98000	-	82.5 Hz	
	FM	K6ERN	448.18000	-	100.0 Hz	
	FM	K6ERN	927.87500	902.87500	103.5 Hz	
	FM	KO6TD	1282.10000	-	127.3 Hz	
Ventura, South Mtn						
	FM	WA6ZSN	447.32000	-	100.0 Hz	
Verdugo Peak, Verdugo Peak						
	FM	K6VGP	445.26000	-		
Victorville	DMR/MARC	N6GGS	446.06000	-	CC2	
	FM	AA7SQ	147.39000	+	100.0 Hz	
Victorville, Blue Ridge						
	FM	W6NVY	51.96000	-	82.5 Hz	
Victorville, Fairgrounds						
	FM	K7GIL	146.11500	+	91.5 Hz	
Victorville, Quartzite Mountai						
	FM	W6NVY	147.19500	+	141.3 Hz	
	FM	W6NVY	147.85500	-	186.2 Hz	
	FM	K7GIL	223.84000	-	156.7 Hz	

102   CALIFORNIA

Location	Mode	Call sign	Output	Input	Access	Coordinator
Victorville, Quartzite Mountai						
	FM	WR6QZ	446.10000	-	136.5 Hz	
	FM	K6PNG	448.90000	-	67.0 Hz	
	FM	KK6SVL	449.20000	-	146.2 Hz	
	FM	W6NVY	449.70000	-	141.3 Hz	
Victorville, Quartzite Mtn.						
	FM	K7GIL	147.12000	+	91.5 Hz	
Visalia	DSTAR	K6VIS	145.01250	147.51250		
	DSTAR	K6VIS	145.45000	-		
	DSTAR	K6VIS	442.30000	+		
	DSTAR	K6VIS	1286.32500	-		
	FM	WA6BAI	223.88000	-	103.5 Hz	
	FM	N6BYH	440.45000	+	141.3 Hz	
	FM	WA6BLB	443.02500	+	88.5 Hz	
	FM	WA6MSN	444.45000	+	127.3 Hz	
	FM	WA6BAI	1286.30000	-	103.5 Hz	
	P25	N6VQL	147.39000	+	127.3 Hz	
	P25	WA6YLB	443.35000	+	141.3 Hz	
	P25	N6IB	927.02500	902.02500	141.3 Hz	
Visalia, Bear Mountain						
	DSTAR	WX6D	1286.35000	+		
Visalia, Blue Ridge						
	FM	KM6OR	146.73000	-	141.3 Hz	
	FM	KM6OR	443.20000	+	141.3 Hz	
Visalia, Downtown Visalia						
	FM	KE6WDX	146.97000	-	100.0 Hz	
Visalia, Kings Canyon National						
	FM	N6BYH	146.76000	-	141.3 Hz	
Visalia, Park Ridge						
	FM	WA6BAI	440.40000	+	103.5 Hz	
Vista	DMR/MARC	KA3AJM	447.90000	-	CC0	
	FM	KI6AZQ	146.67000	-	156.7 Hz	
	FM	KA3AJM	146.97000	-	107.2 Hz	
	FM	KA3AJM	446.30000	-		
	FM	WI6RE	448.80000	-	100.0 Hz	
Vista View Resort Mobile Home						
	FM	W6SEL	444.12500	+		
Volcano	FM	W6KAP	440.45000	+	127.3 Hz	
	FM	W6KAP	927.90000	902.90000	127.3 Hz	
Walnut Creek	DMR/BM	WB6PQM	444.15000	+	CC1	
	DMR/BM	WB6PQM	927.50000	902.50000	CC3	
	DMR/MARC	W6CX	147.06000	+	CC1	
	FM	N6MVT	443.47500	+	114.8 Hz	
	FM	KK6BSN	443.82500	+	167.9 Hz	
	FM	K6NOX	927.71250	902.71250		
Walnut Grove	FM	WA6JIV	443.70000	+		
	FM	K6YC	927.30000	902.30000	100.0 Hz	
Watsonville	FM	W6UNI	145.17000	-	131.8 Hz	
	FM	W6TUW	145.29000	-	94.8 Hz	
	FM	W6DNC	145.33000	-	123.0 Hz	
	FM	KB6MET	146.77000	-	123.0 Hz	
	FM	W6NAD	146.95500	-		
	FM	KB6MET	224.38000	-		
	FM	KB6MET	224.84000	-	156.7 Hz	
	FM	KB6MET	443.35000	+	123.0 Hz	
	FM	N6NAC	1286.20000	-	110.9 Hz	
Watsonville, Fire Station						
	FM	KI6EH	147.94500	-	94.8 Hz	
Watsonville, Hospital						
	FM	K6RMW	147.00000	+	94.8 Hz	
Watsonville, Watsonville Commu						
	FM	K6RMW	443.05000	+	94.8 Hz	

CALIFORNIA 103

Location	Mode	Call sign	Output	Input	Access	Coordinator
Weaverville, Hayfork Bally						
	FM	N6TKY	146.73000	-	85.4 Hz	
Weott	FM	KM6TE	147.33000	+	103.5 Hz	
West Covina	DMR/BM	KF6ITC	449.66000	-	CC13	
	FM	WB6QZK	1282.87500	-		
West Los Angeles	FM	N6GLA	447.32000	-	103.5 Hz	
	FM	K6PYP	1282.57500	-	103.5 Hz	
West Point	FM	WB6LZV	441.37500	+	123.0 Hz	
West Sacramento	FM	W6AK	146.91000	-	162.2 Hz	
	FM	K7QDX	927.10000	902.10000	103.5 Hz	
Westley	FM	K6ACR	147.12000	+	77.0 Hz	
	FM	K6MSR	443.62500	+		
	FM	WA6RQX	444.17500	+	141.3 Hz	
	FM	K6RDJ	1282.80000	-		
Westley, Mt Oso	FM	K6JJC	443.82500	+	107.2 Hz	
Westwood, UCLA	FM	W6YRA	448.54000	-	82.5 Hz	
Whitmore, Blue Mountain						
	DMR/BM	KD6MTU	145.07500	147.57500	CC1	
Whitmore/Redding						
	DMR/BM	KD6MTU	444.50000	+	CC1	
Whittier	FM	N6WZK	449.56000	-		
	FM	KA6VHA	1283.05000	-		
Whittier, Rio Hondo Peak						
	FM	K0JPK	146.73000	-	103.5 Hz	
Whittier, Whittier Hills						
	FM	N6CRG	224.12000	-	151.4 Hz	
Wildomar	FM	KI6ITV	145.40000	144.60000	146.2 Hz	
Williams	FM	N6NMZ	146.76000	-	131.8 Hz	
	FM	N6NMZ	224.26000	-	100.0 Hz	
Willits	FM	WA6RQX	440.07500	+		
Willits, Laughlin Ridge						
	FM	K7WWA	51.74000	-	114.8 Hz	
	FM	K7WWA	147.12000	+	103.5 Hz	
	FM	WA6RQX	147.39000	+	103.5 Hz	
	FM	K7WWA	444.92500	+	100.0 Hz	
	P25	K7WWA	442.10000	+		
Willow Creek, Horse Mountain						
	FM	K6FWR	147.00000	+	103.5 Hz	
Wilson	FM	K6JP	1284.10000	-		
Winchester	FM	WR6AAC	224.36000	-	151.4 Hz	
Windsor	FM	W6IBC	146.98500	-	88.5 Hz	
Winterhaven, Black Mountain						
	FM	K6CKS	1284.60000	-		
Winters	DMR/BM	N6UTX	144.93750	147.43750	CC2	
Wofford Heights	FM	KB6DJT	224.54000	-	77.0 Hz	
Woodland, Berryessa Peak						
	FM	KE6YUV	146.97000	-	123.0 Hz	
Woodside	FM	W6MTF	52.62000	-	192.8 Hz	
	FM	KC6ZIS	224.44000	-	107.2 Hz	
	FM	KB6LED	224.56000	-	107.2 Hz	
	FM	WB6ECE	441.30000	+	123.0 Hz	
	FM	KF6JEE	446.28000	-	131.8 Hz	
Wrightwood	DMR/BM	K6UHF	449.74000	-	CC7	
	FM	KW6WW	147.24000	+	91.5 Hz	
	FM	KW6WW	224.40000	-	91.5 Hz	
	FM	KW6WW	445.24000	-	127.3 Hz	
	FM	KB6BZZ	927.22500	902.22500		
Wrightwood, Blue Ridge						
	DSTAR	W6CPA	147.54000	144.94000		
	DSTAR	W6CPA	446.87000	-		
	FM	N6LXX	53.58000	-	107.2 Hz	
	FM	KC6JAR	224.02000	-	110.9 Hz	

104   CALIFORNIA

Location	Mode	Call sign	Output	Input	Access	Coordinator
Wrightwood, Blue Ridge						
	FM	N6LXX	224.14000	-	94.8 Hz	
	FM	KD6AFA	224.74000	-	186.2 Hz	
	FM	N6LXX	446.86000	-	110.9 Hz	
	FM	KD6OFD	447.44000	-		
	FM	WB6SLR	449.74000	-		
Wrightwood, Table Mountain						
	FM	WR6AZN	145.28000	-	131.8 Hz	
	FM	WR6AZN	223.96000	-	156.7 Hz	
Yermo	FM	N6LXX	53.84000	-	107.2 Hz	
	FM	N6LXX	446.80000	-	110.9 Hz	
Yermo, Mount Rodman						
	FM	WB6TNP	448.26000	-		
Yermo, Mt. Rodman						
	FM	KF6FM	448.14000	-	123.0 Hz	
Yorba Linda	DMR/BM	W0PE	449.44000	-	CC1	
	DMR/BM	W6ELL	449.46000	-	CC1	
Yosemite, Turtle Back Dome						
	FM	W6BXN	147.00000	+	100.0 Hz	
Yountville, Mt Veeder						
	FM	N6TKW	444.72500	+	151.4 Hz	
Yreka	FM	K6SIS	443.75000	+	100.0 Hz	
	FM	W6PRN	444.42500	+	100.0 Hz	
Yreka, Gunsight Peak						
	FM	K6SIS	146.79000	-	100.0 Hz	
	FM	AB7BS	147.12000	+	136.5 Hz	
Yuba City	FM	N6IQY	145.21000	-	127.3 Hz	
	FM	W6GNO	224.96000	-	100.0 Hz	
Yuba City, Sutter Butte						
	FM	WD6AXM	146.08500	+	127.3 Hz	
Yuba City, Sutter Buttes						
	FM	KN6TED	927.73750	902.03750	127.3 Hz	
Yucaipa, Little San Gorgonio P						
	FM	W6DXX	445.14000	-		
Yucaipa, Snow Peak						
	FM	W7DOD	448.16000	-	131.8 Hz	
Yucca Valley	FM	W6BA	146.79000	-	136.5 Hz	
Yucca Valley, Yucca Mesa						
	FM	WB6CDF	145.12000	-		
Yuma, Quartz Peak						
	FM	WB6YFG	224.96000	-		
Zambales	DMR/BM	AB6BP	145.33000	-	CC1	
Zamora	FM	W6OF	440.15000	+		

COLORADO

Location	Mode	Call sign	Output	Input	Access	Coordinator
Akron	DMR/MARC	W0FT	448.17500	-	CC7	CCARC
	FM	KB0VJJ	145.40000	-	88.5 Hz	
Allenspark, AFPD Station #5						
	FM	KI0HG	147.03000	+	100.0 Hz	
Aspen, Aspen Mountain						
	FM	K0VQ	447.05000	-	136.5 Hz	CCARC
Aurora	FM	N0OBA	145.19000	-	151.4 Hz	
	FM	W0BG	448.27500	-	156.7 Hz	CCARC
	FM	N0ZUQ	448.40000	-	94.8 Hz	CCARC
	FM	W9SL	449.95000	-	77.0 Hz	CCARC
Aurora, Smokey Hill						
	FM	KB0UDD	448.50000	-	100.0 Hz	CCARC
Aurora, Smoky Hill						
	FM	KB0UDD	145.40000	-	100.0 Hz	
Bailey, Dick Mountain						
	FM	AB0PC	146.89500	-	100.0 Hz	
Boulder	DSTAR	KC0DS	145.38750	-		

COLORADO 105

Location	Mode	Call sign	Output	Input	Access	Coordinator
Boulder	DSTAR	KC0DS	446.86250	-		CCARC
	DSTAR	KC0DS	1283.86250	1263.86250		CCARC
	DSTAR	W0DK	1299.50000	1279.50000		CCARC
	FM	W0DK	146.61000	-	100.0 Hz	
	FM	W0IA	449.55000	-	100.0 Hz	CCARC
Boulder 1	DMR/MARC	N0SZ	440.07500	+	CC1	
Boulder North	DMR/MARC	K2AD	445.05000	-	CC1	
Boulder South	DMR/MARC	K7PFJ	446.98750	-	CC1	CCARC
Boulder, Blue Mountain						
	FM	KE0SJ	145.47500	-	100.0 Hz	
Boulder, Eldorado Mountain						
	FM	W0CRA	145.46000	-	107.2 Hz	
	FM	W0CRA	447.97500	-	107.2 Hz	CCARC
Boulder, Gunbarrel Hill						
	FM	W0IA	146.76000	-	100.0 Hz	
Boulder, Lee Hill	FM	W0IA	224.02000	-		
	FM	N0SZ	446.98750	-		
	FM	N0SZ	447.75000	-	141.3 Hz	CCARC
Boulder, Table Mesa						
	FM	W0DK	448.90000	-	100.0 Hz	CCARC
Boulder, Table Mountain						
	FM	W0DK	146.70000	-	100.0 Hz	
Boulder, Thorodin Mountain						
	FM	KB0VJJ	145.31000	-	88.5 Hz	
Breckenridge	DMR/MARC	N0SZ	445.08750	-	CC7	
	FM	WB0QMR	146.70000	-	107.2 Hz	
Breckenridge, Bald Mountain						
	FM	KB0VJJ	147.39000	+	88.5 Hz	
Canon City	DMR/MARC	N0SZ	438.00000	-	CC15	
	DMR/MARC	K0JSC	446.73750	-	CC1	
Canon City, Colorado						
	FM	K0JSC	447.97500	-	100.0 Hz	CCARC
Canon City, Eightmile Mountain						
	FM	WB0WDF	449.00000	-	67.0 Hz	CCARC
Canon City, Fremont Peak						
	FM	WB0WDF	53.03000	51.33000	88.5 Hz	
	FM	WD0EKR	145.49000	-	103.5 Hz	
	FM	WD0EKR	223.96000	-	103.5 Hz	
	FM	K0JSC	447.25000	-	100.0 Hz	CCARC
	FM	WD0EKR	447.75000	-	103.5 Hz	CCARC
	FM	KB0TUC	927.70000	902.70000	DCS 114	CCARC
Carbondale, Four Mile Ridge						
	FM	K0SNO	147.39000	+	107.2 Hz	
	FM	K0ELK	449.72500	-	179.9 Hz	CCARC
Carbondale, Missouri Heights						
	FM	K0VQ	447.15000	-	136.5 Hz	
Castle Rock	DMR/BM	WA0DE	446.82500	-	CC1	
Cedar Point	FM	K0UPS	449.25000	-	123.0 Hz	CCARC
Cedaredge	FM	W0ALC	147.36000	+	100.0 Hz	
	FM	W0ALC	449.82500	-	100.0 Hz	CCARC
Cedaredge, Cedar Mesa						
	FM	KC0QXX	147.19500	+	107.2 Hz	
Centennial	DMR/MARC	K0PWO	448.45000	-	CC7	
Centennial, Warren Mountain						
	FM	WB0TUB	146.88000	-	100.0 Hz	
	FM	WB0TUB	449.60000	-	100.0 Hz	CCARC
Center	FM	N0KM	146.64000	-	77.0 Hz	
	FM	N0KM	447.10000	-	77.0 Hz	
Colorado Springs	DMR/BM	K0HYT	440.00000	+	CC1	
	DMR/BM	K0HYT	449.47500	-	CC1	CCARC
	FM	AD0TP	146.85000	-	156.7 Hz	
	FM	K0IRP	146.91000	-	151.4 Hz	

106   COLORADO

Location	Mode	Call sign	Output	Input	Access	Coordinator
Colorado Springs	FM	AA0L	147.13500	+	100.0 Hz	
	FM	W0MOG	147.39000	+	103.5 Hz	
	FM	KB0SRJ	224.06000	-		
	FM	W0MOG	224.72000	-	103.5 Hz	
	FM	K0IRP	447.35000	-	151.4 Hz	CCARC
	FM	NX0G	447.47500	-	107.2 Hz	CCARC
	FM	WA6IFI	447.55000	-	123.0 Hz	CCARC
	FM	AA0L	448.30000	-	100.0 Hz	CCARC
	FM	W0MOG	448.60000	-	114.8 Hz	CCARC
	FM	W0MOG	927.80000	902.80000	DCS 116	CCARC
Colorado Springs, Almagre Moun						
	FM	AA0SP	147.18000	+	100.0 Hz	
	FM	WA6IFI	927.72500	902.02500		CCARC
Colorado Springs, Cheyenne Mou						
	DSTAR	KC0CVU	145.38500	-		
	FM	KC0CVU	146.76000	-	107.2 Hz	
	FM	KC0CVU	147.34500	+	107.2 Hz	
	FM	N0SZ	445.06250	-		
	FM	KC0CVU	448.00000	-	107.2 Hz	CCARC
	FM	KC0CVU	448.10000	-	107.2 Hz	CCARC
	FM	KC0CVU	927.85000	902.85000	DCS 114	CCARC
Colorado Springs, Fillmore Hil						
	FM	KA0TTF	448.72500	-	123.0 Hz	CCARC
Colorado Springs, Pikes Peak						
	WX	KB0SRJ	146.97000	-	100.0 Hz	
	WX	KB0SRJ	448.45000	-	100.0 Hz	CCARC
Colorado Springs, ~2 Mi ENE US						
	FM	KB0SRJ	448.80000	-	100.0 Hz	CCARC
Conifer, Critchell Mountain						
	FM	N0OWY	447.50000	-	88.5 Hz	CCARC
Cortez, Downtown						
	FM	KD5LWU	146.79000	-	127.3 Hz	
Craig	FM	WD0HAM	145.26500	-	107.2 Hz	
	FM	KB0VJJ	146.97000	-	88.5 Hz	
Creede, Bristol Head						
	FM	N0PKT	449.52500	-	100.0 Hz	
Cripple Creek	FM	WB0WDF	145.46000	-	67.0 Hz	
	FM	NX0G	147.01500	+	107.2 Hz	
	FM	WB0WDF	224.94000	-	67.0 Hz	
	FM	WB0WDF	447.40000	-	67.0 Hz	CCARC
	FM	WB0WDF	1287.70000	1267.70000	67.0 Hz	CCARC
Critchell, Critchell Mountain						
	FM	N0ARA	147.12000	+	88.5 Hz	
Deer Trail	FM	N6LXX	447.87500	-	107.2 Hz	
Delta, Uncompahgre Butte						
	FM	KB0YNA	449.40000	-	131.8 Hz	CCARC
Denver	DMR/BM	KF4TNP	445.23750	-	CC7	
	DMR/BM	WR0AEN	449.00000	-	CC1	CCARC
	DMR/BM	WR0AEN	449.15000	-	CC1	CCARC
	DMR/BM	WR0AEN	449.65000	-	CC1	CCARC
	DMR/BM	WR0AEN	449.67500	-	CC1	CCARC
	DMR/MARC	N0SZ	145.36250	-	CC1	
	DMR/MARC	KD0NQA	146.92500	-	CC1	
	DMR/MARC	N0SZ	440.06250	+	CC1	
	DMR/MARC	KD0WHB	444.90000	+	CC2	
	DMR/MARC	N0SZ	446.80000	-	CC1	CCARC
	DMR/MARC	WA2YZT	446.83750	-	CC1	CCARC
	DMR/MARC	N0SZ	446.93750	-	CC1	CCARC
	FM	N0JXN	146.71500	-	123.0 Hz	
	FM	W0CRA	147.22500	+	107.2 Hz	CCARC
	FM	K0FEZ	447.92500	-	100.0 Hz	CCARC
	FM	N5EHP	448.07500	-	123.0 Hz	CCARC

COLORADO 107

Location	Mode	Call sign	Output	Input	Access	Coordinator
Denver, Broadway Plaza Lofts						
	FM	W0JRL	447.17500	-	100.0 Hz	CCARC
Denver, Green Valley Ranch						
	FM	WE0FUN	448.15000	-	141.3 Hz	CCARC
Denver, Thorodin Mountain						
	FM	N0SZ	449.22500	-	141.3 Hz	CCARC
Deployable	DMR/MARC	N0SZ	445.01250	-	CC0	
Devils Head	DMR/MARC	N0ESQ	446.92500	-	CC1	CCARC
Dolores, Summit Ridge						
	FM	KB5ITS	145.19000	-	100.0 Hz	
	FM	KB5ITS	447.27500	-	100.0 Hz	
Durango	DMR/MARC	N0SZ	445.13750	-	CC7	
	FM	KC5EVE	448.62500	-	131.8 Hz	CCARC
	FM	K0EP	449.85000	-	100.0 Hz	CCARC
Durango, Eagle Pass						
	FM	K0EP	146.67000	-	100.0 Hz	
Durango, Missionary Ridge						
	FM	K0EP	146.70000	-	100.0 Hz	
	FM	KB0VJJ	147.34500	+	88.5 Hz	
Durango, Smelter Mountain						
	FM	KB5ITS	147.13500	+	100.0 Hz	
Eads	DMR/MARC	KZ0DEM	446.75000	-	CC7	
Englewood	DMR/MARC	N0SZ	438.22500	-	CC7	
Estes Park	DMR/BM	N0WAR	147.24000	+	CC2	
	FM	N0FH	146.68500	-	123.0 Hz	
	FM	N0FH	449.80000	-	123.0 Hz	CCARC
Evans	DMR/BM	WD0HDR	445.57500	-	CC2	
Evergreen	DMR/MARC	N0SZ	145.17500	-	CC1	
	FM	KE4GUQ	145.34000	-	103.5 Hz	
Evergreen, Squaw Mountain						
	FM	N0PYY	147.30000	+	103.5 Hz	
Fairplay	DMR/MARC	K7PFJ	446.76250	-	CC7	CCARC
Fairplay, Badger	FM	N0SZ	446.76250	-		
Fairplay, Sacramento Mountain						
	FM	AA0BF	447.12500	-	103.5 Hz	CCARC
Falcon, Black Forest						
	FM	KD0MDP	447.62500	-	100.0 Hz	
Firestone	DMR/MARC	K7PFJ	445.06250	-	CC7	
Fort Collins	DMR/BM	W1VAN	147.16500	747.16500	CC2	
	DMR/BM	W1VAN	446.90000	-	CC2	
	DMR/MARC	KT0L	145.20500	-	CC10	
	DMR/MARC	W0DMR	147.39000	+	CC2	
	DMR/MARC	N7VDR	445.02500	-	CC1	
	DMR/MARC	N0AOL	446.73750	-	CC2	
	DMR/MARC	W0DMR	446.77500	-	CC2	
	FM	KC0RBT	447.45000	-	123.0 Hz	CCARC
	FM	W7RF	447.72500	-	131.8 Hz	CCARC
	FM	K1TJ	1283.55000	1263.55000	100.0 Hz	CCARC
Fort Collins, Buckhorn Mountai						
	FM	W0UPS	146.62500	-	100.0 Hz	
	WX	W0UPS	447.70000	-	100.0 Hz	
Fort Collins, CSU Durward Hall						
	DSTAR	W0QEY	446.81250	-		CCARC
	FM	W0QEY	147.36000	+	100.0 Hz	
	FM	W0QEY	449.85000	-	100.0 Hz	CCARC
	FM	W0QEY	927.95000	902.95000	100.0 Hz	CCARC
Fort Collins, Horsetooth Mount						
	WX	W0UPS	447.27500	-	100.0 Hz	CCARC
Fort Lupton	FM	AC0KC	447.30000	-	107.2 Hz	CCARC
Fort Morgan	DMR/MARC	W0FT	448.20000	-	CC1	CCARC
Franktown	DSTAR	K0PRA	446.85000	-		CCARC
Ft Collins	DMR/BM	KB0VGD	145.26500	-	CC2	

108 COLORADO

Location	Mode	Call sign	Output	Input	Access	Coordinator
Genoa	FM	KE0AE	147.06000	+	103.5 Hz	
Glenwood Springs						
	FM	KI0G	224.02000	-		
	FM	KI0G	447.60000	-		CCARC
Glenwood Springs, Lookout Moun						
	FM	WA4HND	449.85000	-	131.8 Hz	CCARC
Glenwood Springs, Sunlight Pea						
	FM	KB0VJJ	146.85000	-	88.5 Hz	
	FM	KI0G	146.88000	-	107.2 Hz	
	FM	KI0G	447.10000	-	107.2 Hz	CCARC
	FM	N0XLI	449.60000	-	107.2 Hz	CCARC
Gold Hill	FM	W0JZ	146.91000	+	123.0 Hz	
Golden	DMR/BM	KB0VGD	446.97500	-	CC2	
	DMR/BM	W0WYX	449.75000	-	CC1	
	FM	N0PYY	448.12500	-	107.2 Hz	CCARC
	FM	W0GV	448.97500	-	123.0 Hz	CCARC
	FM	N0PYY	927.78750	902.78750	156.7 Hz	
Golden, Centennial Cone						
	FM	KE0SJ	145.28000	-	100.0 Hz	
	FM	WB0TUB	146.64000	-	100.0 Hz	
	FM	K0FEZ	146.98500	-	100.0 Hz	
	FM	W0TX	448.62500	-	100.0 Hz	CCARC
	FM	KE0SJ	449.52500	-		CCARC
	FM	KI0HC	927.83750	902.03750		CCARC
Golden, Guy Hill	DMR/BM	KI0GO	449.75000	-	CC1	CCARC
	FM	W0CBI	147.15000	+	100.0 Hz	
	FM	N0MHU	224.00000	-	103.5 Hz	
	FM	N0POH	224.74000	-	88.5 Hz	
	FM	K0IBM	448.85000	-	88.5 Hz	CCARC
	P25	WT0C	145.22000	-	103.5 Hz	
Golden, Lookout Mountain						
	DSTAR	W0CDS	145.25000	-		
	DSTAR	W0CDS	446.96250	-		CCARC
	DSTAR	W0CDS	1283.96250	1263.96250		CCARC
	DSTAR	W0CDS	1299.90000			
	FM	N0SZ	145.37000	-		
	FM	KE0VH	449.62500	-	141.3 Hz	CCARC
	FM	W0SKY	927.97500	902.07500	67.0 Hz	
Granby	FM	KA0YDW	146.82000	-	123.0 Hz	
Grand Junction	DSTAR	KD0RED	446.77500	-		
	FM	KE0TY	145.22000	-	107.2 Hz	
	FM	K0SSI	147.10500	+		
	FM	KE0TY	147.39000	+	107.2 Hz	
	FM	KC0ARV	447.50000	-	114.8 Hz	CCARC
	FM	W0GJT	448.15000	-	100.0 Hz	CCARC
	FM	KB0SW	449.00000	-		CCARC
	FM	KD0SMZ	449.77500	-	173.8 Hz	CCARC
Grand Junction, 9 Mile, Hwy 14						
	FM	W0GJT	146.82000	-	107.2 Hz	
Grand Junction, Black Ridge						
	FM	WA4HND	449.57500	-	131.8 Hz	CCARC
Grand Junction, Black Ridge (A						
	FM	W0RRZ	146.94000	-	107.2 Hz	
Grand Junction, GJT						
	FM	W0RRZ	145.17500	-	107.2 Hz	
Grand Junction, Grand Mesa						
	FM	KB0VJJ	145.35500	-	88.5 Hz	
	FM	KB0VJJ	147.35500	-	88.5 Hz	
	FM	WA4HND	449.30000	-	107.2 Hz	CCARC
	WX	KB0YNA	449.65000	-	151.4 Hz	CCARC
Grand Lake	FM	WC3W	449.42500	-	103.5 Hz	
Greeley	DMR/MARC	W0DMR	445.13750	-	CC2	

COLORADO 109

Location	Mode	Call sign	Output	Input	Access	Coordinator
Greeley	DMR/MARC	K0AEN	449.32500	-	CC1	
	FM	KC0KWD	448.47500	-	100.0 Hz	CCARC
	FM	WR0AEN	449.32500	-		CCARC
Greeley, UNC	WX	W0UPS	146.85000	-	100.0 Hz	
	WX	K0OJ	449.72500	-	127.3 Hz	CCARC
Gunnison	FM	W0VTL	147.12000	+		
	FM	K5GF	449.95000	-		CCARC
Gunnison, W Mountain Or Tender						
	FM	KB0YNA	447.65000	-	151.4 Hz	CCARC
Hesperus	FM	KB0VIU	145.37000	-	100.0 Hz	
Highlands Ranch	DMR/BM	WA0DE	445.08750	-	CC1	
Holyoke	DMR/BM	N0FON	146.95500	-	CC1	
Hudson	FM	K0EB	449.92500	-	114.8 Hz	CCARC
Idaho Springs, Centennial Cone						
	FM	W0TX	145.49000	-	100.0 Hz	
Idaho Springs, Squaw Mountain						
	FM	KB0UDD	146.67000	-	100.0 Hz	
	FM	W0WYX	146.94000	-	103.5 Hz	
	FM	W0CRA	447.57500	-	107.2 Hz	CCARC
	FM	N0SZ	448.22500	-	141.3 Hz	CCARC
	FM	W0CFI	448.67500	-	100.0 Hz	CCARC
	FM	WG0N	449.05000	-	107.2 Hz	CCARC
	FM	W0TX	449.35000	-	100.0 Hz	CCARC
	FM	K1DUN	449.45000	-	103.5 Hz	CCARC
Ignacio, Spring Creek						
	FM	KB5ITS	147.15000	+	100.0 Hz	
Indian Hills, Chief Mtn						
	FM	KI0HC	927.93750	902.03750		CCARC
Kremmling, Santoy Peak						
	FM	KB0VJJ	147.07500	+	88.5 Hz	
La Veta	FM	N0JPX	145.35500	-	100.0 Hz	
La Veta, Cordova Pass						
	FM	N0ZSN	449.75000	-	100.0 Hz	CCARC
Lake City, Hill 71	FM	KB5ITS	147.13500	-	123.0 Hz	
Lake George	FM	WZ0N	448.57500	-	103.5 Hz	CCARC
Lake George, Badger Mountain						
	FM	NX0G	146.68500	-	107.2 Hz	
	FM	KC0CVU	147.36000	+	107.2 Hz	
Lakewood, Green Mountain						
	FM	W0TX	147.33000	+	100.0 Hz	
	FM	W0TX	224.38000	-	100.0 Hz	
Lakewood, Green Mtn						
	FM	W0TX	449.77500	-		CCARC
Lakewood, Moffat Treatment Pla						
	FM	KD0SSP	147.21000	+	100.0 Hz	
Lakewood, St Anthony Medical C						
	FM	W0TX	447.82500	-	DCS 073	CCARC
Lamar	FM	N0LAR	146.61000	-	123.0 Hz	
	FM	KC0HH	449.50000	-	123.0 Hz	CCARC
Larkspur, Westcreek						
	FM	W0CRA	147.22500	+	107.2 Hz	
	FM	W0CRA	448.42500	-	107.2 Hz	
	FM	K0JSC	449.97500	-	100.0 Hz	CCARC
Leadville	FM	KB0VJJ	145.44500	-	123.0 Hz	CCARC
Leadville, Quail Mountain						
	FM	N0ZSN	147.24000	+	156.7 Hz	
Limon	DMR/MARC	K0RTS	446.73750	-	CC8	
Lochbuie	FM	KE0SJ	447.62500	-	100.0 Hz	
Loma, Baxter Pass						
	FM	KB0SW	447.00000	-	107.2 Hz	CCARC
Longmont	FM	W0ENO	147.27000	+	100.0 Hz	
	FM	N0EPA	448.52500	-	151.4 Hz	CCARC

110   COLORADO

Location	Mode	Call sign	Output	Input	Access	Coordinator
Longmont	FM	W0ENO	448.80000	-	88.5 Hz	CCARC
Louisville	DMR/BM	W0RMT	445.70000	-	CC7	
Loveland, Namaqua Hill						
	FM	W0LRA	147.19500	+	100.0 Hz	
	FM	W0LRA	449.57500	-	100.0 Hz	CCARC
Mancos	DMR/MARC	N5UBJ	446.73750	-	CC6	
Mancos, Caviness Mountain						
	FM	KB5ITS	442.37500	+	100.0 Hz	
Mancos, Menefee Mountain						
	FM	KB5ITS	145.32500	-	100.0 Hz	
Montrose	FM	KD5OPD	146.79000	-	107.2 Hz	
	FM	KC0UUX	146.91000	-	107.2 Hz	
	FM	KB0YNA	448.65000	-	151.4 Hz	CCARC
Montrose, Waterdog Peak						
	FM	WA4HND	447.20000	-	107.2 Hz	CCARC
	FM	KB0YNA	448.65000	-		
Monument, Monument Hill						
	FM	K0NR	447.72500	-	100.0 Hz	CCARC
Morrison, Mt. Morrison						
	FM	N6LXX	447.87500	-	107.2 Hz	CCARC
Nathrop	FM	W0LSD	146.74500	-	100.0 Hz	
Northglenn	FM	K0ML	147.04500	+	123.0 Hz	
Ouray, Engineer Mountain						
	FM	KB5ITS	147.27000	+	127.3 Hz	
Pagosa Springs, Oakbrush Hill						
	WX	N0JSP	146.61000	-	123.0 Hz	
Palmer Divide	FM	N0PWZ	449.72500	-	100.0 Hz	CCARC
Palmer Lake, Monument Hill						
	FM	N0XLF	145.19000	-	131.8 Hz	
Paonia	FM	KI0MR	147.33000	+	107.2 Hz	
Parker	DMR/BM	N0KEG	147.31500	+	CC1	
	DMR/BM	N0KEG	448.05000	-	CC1	
	DMR/MARC	W9CN	440.01250	+	CC7	
	DMR/MARC	W9CN	442.00000	+	CC7	
	DMR/MARC	K0PRA	445.07500	-	CC1	
	FM	WQ8M	448.70000	-	146.2 Hz	
	P25	KI0HC	927.83750	902.03750		CCARC
Parker, Hess Reservoir						
	FM	N0ESQ	447.52500	-	203.5 Hz	CCARC
Perry Park	FM	N0OBA	145.19000	-		
Pueblo	DMR/MARC	KF0KR	446.86250	-	CC1	
Pueblo West	FM	K0JSC	448.97500	-	100.0 Hz	CCARC
Pueblo West, Liberty Point						
	FM	NA0PW	447.45000	-	123.0 Hz	
Pueblo, Baculite Mesa						
	FM	W0PHC	146.79000	-	88.5 Hz	
Pueblo, Cedarwood						
	FM	K0JSC	449.97500	-	100.0 Hz	CCARC
Pueblo, CSU	FM	NE0Z	147.00000	+		
Pueblo, Deer Peak						
	FM	KC0CVU	449.62500	-	107.2 Hz	CCARC
Pueblo, Metro	FM	K0JSC	447.27500	-	100.0 Hz	CCARC
Pueblo, Pueblo	FM	K0ST	146.65500	-	123.0 Hz	
Punkin Center	FM	N1FSX	449.90000	-	107.2 Hz	CCARC
Rangely	DMR/MARC	N0SZ	445.26250	-	CC7	
	DMR/MARC	N0SZ	445.27500	-	CC7	
Ridgway	DMR/BM	AD0RM	447.80000	-	CC1	
Rollinsville	FM	W0RM	224.60000	-	100.0 Hz	
Salida	DMR/MARC	N0SZ	446.81250	-	CC7	
Salida, Methodist Mountain						
	DSTAR	KD0QPG	446.97500	-		CCARC
	FM	KC0CVU	145.29500	-	107.2 Hz	

COLORADO   111

Location	Mode	Call sign	Output	Input	Access	Coordinator
Salida, Methodist Mountain						
	FM	KB0VJJ	147.28500	+	88.5 Hz	
	FM	KC0CVU	449.65000	-	107.2 Hz	CCARC
	FM	WZ0N	449.92500	-	103.5 Hz	CCARC
	FM	K0JSC	449.97500	-	100.0 Hz	CCARC
Severance, Cactus Hill						
	FM	N6RFI	447.20000	-	82.5 Hz	CCARC
Silverton	FM	KB5ITS	447.52500	-	127.3 Hz	CCARC
Silverton, Buffalo Boy						
	FM	KB5ITS	145.32500	-	127.3 Hz	
Silverton, Engineer Mountain						
	FM	KB5ITT	147.27000	+	127.3 Hz	
Silverton, Hazelton						
	FM	KB5ITS	444.00000	+	67.0 Hz	
Silverton, Kendall Mountain						
	FM	KB5ITS	147.37500	+	156.7 Hz	
Simla	FM	WA0DE	147.10500	+	107.2 Hz	
Snowmass Village						
	FM	K0CL	146.67000	-	107.2 Hz	
Springfield	DMR/MARC	KZ0DEM	446.76250	-	CC7	
	FM	KZ0DEM	147.09000	+	118.8 Hz	
	WX	KZ0DEM	449.20000	-	118.8 Hz	
Steamboat Springs						
	FM	KB0VJJ	147.16500	+	88.5 Hz	
	FM	KB0VJJ	449.62500	-	123.0 Hz	
	WX	KD0H	147.21000	+	107.2 Hz	
Sterling, CO	FM	WA0JTB	145.29500	-		
Stratton	FM	KE0AE	146.89500	-	103.5 Hz	
Trinidad, Raton Pass						
	FM	WA6IFI	145.43000	-	107.2 Hz	
Vail	DMR/MARC	N0SZ	445.07500	-	CC7	
	FM	KB0VJJ	147.34500	+	88.5 Hz	
Vail, Bald Mountain						
	FM	K0RV	146.61000	-	107.2 Hz	
Walsenburg, N Rattlesnake Butt						
	FM	W0PHC	146.73000	-	88.5 Hz	
Wellington	FM	WB7UPS	448.32500	-	110.9 Hz	CCARC
Westcliffe	FM	KB0TUC	147.06000	+	77.0 Hz	
	FM	KB0TUC	448.15000	-	103.5 Hz	CCARC
	FM	K0JSC	448.32500	-	100.0 Hz	
	FM	KB0TUC	927.77500	902.77500	DCS 116	CCARC
Westcreek	DMR/MARC	N0SZ	446.87500	-	CC6	CCARC
Westminster	ATV	W0ATV	1253.25000	1257.25000		
	FM	N1UPS	449.30000	-	100.0 Hz	CCARC
Wiley	DMR/BM	W0CTS	444.20000	+	CC1	
Winter Park	FM	WA4CCC	447.45000	-	103.5 Hz	CCARC
Winter Park, Winter Park						
	FM	KB0VJJ	147.28500	+	88.5 Hz	
Woodland Park	FM	KA0WUC	145.41500	-	179.9 Hz	
	FM	NX0G	146.82000	-	107.2 Hz	
	FM	KA0WUC	447.67500	-	179.9 Hz	CCARC
	FM	NX0G	448.65000	-	107.2 Hz	CCARC
	FM	KA0WUC	449.02500	-	141.3 Hz	CCARC
	FM	KA0WUC	927.90000	902.90000	DCS 205	CCARC
Woodland Park, Paradise Pines						
	FM	KA4EPS	449.32500	-	103.5 Hz	CCARC

CONNECTICUT

Location	Mode	Call sign	Output	Input	Access	Coordinator
Ansonia	DMR/MARC	K1EIR	444.35000	+	CC2	CSMA
	FM	WK1M	145.19000	-	77.0 Hz	CSMA
Avon	DMR/MARC	W1HDN	442.05000	+	CC1	
	FM	W1JNR	224.94000	-		CSMA

112   CONNECTICUT

Location	Mode	Call sign	Output	Input	Access	Coordinator
Barkhamsted	FM	W1RWC	147.27000	+	77.0 Hz	CSMA
Bethel	FM	KA1KD	147.03000	+	100.0 Hz	CSMA
Bloomfied	DMR/MARC	W1SP	446.43750	-	CC1	CSMA
Bloomfield	DMR/BM	N1AJW	426.00000	+	CC1	
	DMR/BM	N1AJW	449.80000	-	CC1	
	FM	W1CWA	146.82000	-		CSMA
	FM	W1SP	449.12500	-		
Bloomfield ,CT	DMR/BM	N1AJW	449.50000	-	CC1	
Branford	DMR/MARC	N1HUI	449.32500	-	CC1	CSMA
	FM	N1HUI	927.81250	902.81250	DCS 311	CSMA
Bridgeport	DMR/MARC	N1TGE	440.76250	+	CC1	CSMA
	DMR/MARC	AG2K	442.20000	+	CC1	
	FM	WA1RJI	146.44500	147.44500	77.0 Hz	
	FM	N1MUC	146.89500	-	77.0 Hz	CSMA
	FM	N1LXV	224.96000	-	77.0 Hz	CSMA
	FM	KA1HCX	449.40000	-	110.9 Hz	CSMA
Bridgeport, Bridgeport Hospita						
	FM	N1KGN	441.70000	+	77.0 Hz	CSMA
Bristol	DMR/BM	W1IXU	448.87500	-	CC1	
	FM	KB1CDI	29.64000	-	88.5 Hz	CSMA
	FM	WA1IXU	53.05000	52.05000	162.2 Hz	CSMA
	FM	KB1CDI	53.39000	52.39000	88.5 Hz	
	FM	K1DII	145.31000	-	110.9 Hz	
	FM	K1IFF	146.59250	147.59250		
	FM	W1DHT	146.68500	-	77.0 Hz	CSMA
	FM	K1CRC	146.88000	-	77.0 Hz	CSMA
	FM	KB1AEV	224.16000	-	77.0 Hz	CSMA
	FM	WA1IXU	224.22000	-	118.8 Hz	CSMA
	FM	KB1CDI	224.82000	-	88.5 Hz	
	FM	K1CRC	442.85000	+		CSMA
	FM	KB1AEV	444.65000	+	151.4 Hz	CSMA
	FM	WE1SPN	448.72500	-	77.0 Hz	
	FM	W1DHT	448.72500	-	141.3 Hz	CSMA
Bristol, South Mountain						
	DSTAR	W1IXU	145.14000	-		CSMA
	DSTAR	W1IXU	448.37500	-		
Brooklyn	DMR/MARC	KB1NTA	445.73750	-	CC3	CSMA
Burlington	FM	K1CRC	147.15000	+	77.0 Hz	
Colchester	DMR/MARC	K1JCL	440.80000	+	CC1	
	DMR/MARC	WH6SW	445.98750	-	CC1	CSMA
	DSTAR	KB1YPL	442.95000	+		CSMA
Cornwall	DMR/MARC	W1SP	444.35000	+	CC1	CSMA
Coventry	DMR/MARC	K1JCL	449.87500	-	CC1	CSMA
CT Portable Rptr	DMR/MARC	W1SP	442.00000	+	CC3	
Danbury	DMR/MARC	KX1EOC	445.73750	-	CC1	CSMA
Danbury, Danbury Hospital						
	FM	W1HDN	147.12000	+	141.3 Hz	
	FM	W1HDN	443.65000	+	114.8 Hz	CSMA
Danbury, Spruce Mountain						
	FM	W1QI	147.30000	+	100.0 Hz	CSMA
	FM	W1QI	447.77500	-	100.0 Hz	CSMA
Durham	DSTAR	KB1UHS	444.55000	+		
	FM	KB1MMR	927.83750	902.83750		
Durham/Middlefield						
	FM	KB1MMR	446.92500	-		CSMA
East Haddam	FM	K1IKE	147.01500	+	110.9 Hz	CSMA
East Hampton	DSTAR	KB1CDI	147.13500	+	88.5 Hz	CSMA
	FM	KB1CDI	53.11000	52.11000	88.5 Hz	
	FM	KC1AJR	444.32500	+	77.0 Hz	
East Hartford	FM	W1EHC	443.25000	+	141.3 Hz	CSMA
East Hartland	FM	W1XOJ	53.19000	52.19000	162.2 Hz	CSMA
	FM	K1YON	145.23000	-		CSMA

CONNECTICUT   113

Location	Mode	Call sign	Output	Input	Access	Coordinator
East Haven	DMR/MARC	KA1MJ	449.82500	-	CC1	CSMA
East Killingly	FM	K1MUJ	147.22500	+	156.7 Hz	CSMA
Enfield	FM	N1XDN	146.43000	+		
	FM	K1ENF	442.40000	+	94.8 Hz	
Fairfield	DMR/MARC	KA1HCX	444.50000	+	CC3	
	FM	WB1CQO	146.62500	-	100.0 Hz	CSMA
	FM	N3AQJ	224.10000	-	77.0 Hz	CSMA
	FM	N1ZU	440.45000	447.45000		
	FM	N1LXV	441.50000	+	77.0 Hz	CSMA
	FM	N3AQJ	446.82500	-	110.9 Hz	CSMA
Farmington	FM	N1GCN	442.70000	+	173.8 Hz	CSMA
	FM	WA1ARC	448.57500	-	146.2 Hz	
Glastonbury	FM	W1EDH	147.09000	+	110.9 Hz	CSMA
	FM	W1EDH	449.62500	-	110.9 Hz	CSMA
Goshen, Goshen Fairgrounds						
	FM	KU1Q	440.25000	+	123.0 Hz	CSMA
Groton	FM	W1NLC	146.67000	-	156.7 Hz	CSMA
	FM	W1CGA	146.86500	-	156.7 Hz	
	FM	KB1CJP	448.42500	-	127.3 Hz	
Guilford	DMR/MARC	W1SP	441.26250	+	CC1	CSMA
	FM	NI1U	53.75000	52.75000	110.9 Hz	
Hamden	FM	N1GTL	444.45000	+	100.0 Hz	
	FM	WA1MIK	927.41250	902.41250	DCS 311	CSMA
Hartford	DMR/MARC	W1SP	447.97500	-	CC1	CSMA
	FM	W1HDN	146.64000	-	141.3 Hz	CSMA
	FM	N1ABL	443.10000	+	114.8 Hz	CSMA
Hartland	FM	W1OBQ	448.00000	-	162.2 Hz	CSMA
Hebron	FM	N1CBD	224.70000	-	156.7 Hz	CSMA
	FM	K1PTI	449.37500	-		CSMA
Killingly	DMR/MARC	KZ1M	444.80000	+	CC1	
	DMR/MARC	K1JCL	444.85000	+	CC1	
	DSTAR	N1GAU	444.10000	+		CSMA
Killingworth	FM	W1BCG	145.29000	-	110.9 Hz	CSMA
	FM	KB1MMR	146.41500	+		
Lebanon	FM	NA1RC	147.30000	-	77.0 Hz	
Ledyard	DMR/MARC	W1SP	449.27500	-	CC1	CSMA
	FM	W1DX	224.38000	-	103.5 Hz	CSMA
Ledyard Center	FM	W1DX	145.39000	-	156.7 Hz	CSMA
Manchester	FM	WA1VOA	145.33000	-	88.5 Hz	CSMA
	FM	N1SPI	147.04500	+	88.5 Hz	CSMA
	FM	WA1YQB	449.22500	-	77.0 Hz	CSMA
Meriden	DSTAR	W1ECV	145.49000	-		CSMA
	DSTAR	W1ECV	444.25000	+		CSMA
	FM	K1HSN	224.80000	-	77.0 Hz	CSMA
	FM	N1SZM	442.45000	+		
	FM	KB1AEV	444.20000	+	151.4 Hz	
	FM	W1OBQ	448.00000	-	162.2 Hz	CSMA
Middlebury	DMR/MARC	W1SP	445.83750	-	CC2	
Middletown	FM	N1SFE	446.32500	-	203.5 Hz	
	FM	K1IKE	446.87500	-	156.7 Hz	CSMA
Milford	DMR/BM	KA1FAI	147.22500	+	CC1	CSMA
	DMR/BM	N1EG	445.92500	-	CC1	
	FM	KA1OYS	53.27000	52.27000	77.0 Hz	CSMA
	FM	N1JKA	223.88000	-		CSMA
	FM	N1LUF	443.55000	+	77.0 Hz	CSMA
	FM	KA1FAI	446.60000	-	136.5 Hz	
Milford, Eels Hill	FM	KB1CBD	146.92500	-	67.0 Hz	
Monroe	DMR/MARC	KA1HCX	444.40000	+	CC3	CSMA
Montville	FM	K1IKE	53.41000	52.41000	156.7 Hz	CSMA
	FM	WA1IXU	224.82000	-	110.9 Hz	CSMA
Morris	FM	KB1TIF	146.95500	-	100.0 Hz	
Morris, Firehouse	FM	KB1CDI	224.32000	-	88.5 Hz	CSMA

114 CONNECTICUT

Location	Mode	Call sign	Output	Input	Access	Coordinator
Mystic	DSTAR	KB1TMO	448.97500	-		
	FM	KB1CJP	147.27000	+	127.3 Hz	
	FM	KB1JCP	446.57500	-	127.3 Hz	CSMA
Naugatuck	FM	WA1NQP	224.46000	-		CSMA
New Britain	FM	WA1VRP	446.72500	-	123.0 Hz	
New Canaan	FM	N1LLL	146.77500	-	100.0 Hz	CSMA
	FUSION	N1LLL	447.27500	-	123.0 Hz	CSMA
New Haven	DMR/MARC	KB1TTN	441.46250	+	CC1	CSMA
	FM	WA1UFC	224.18000	-		CSMA
New Haven, Yale New Haven						
	FM	KB1CDI	224.08000	-		
New London	FM	W1NLC	224.26000	-	156.7 Hz	CSMA
New Milford	DMR/MARC	NA1RA	440.96250	+	CC1	CSMA
	FM	KA1RFM	224.72000	-		CSMA
Newington	DSTAR	W1HQ	1284.12500	1264.12500		
	FM	W1AW	145.45000	-	127.3 Hz	CSMA
	FM	W1HQ	147.39000	+		
	FM	N1OGB	224.02000	-	100.0 Hz	CSMA
	FM	W1AW	224.84000	-	127.3 Hz	CSMA
	FM	W1OKY	443.05000	+	100.0 Hz	CSMA
	FM	WA1UTQ	449.57500	-	79.7 Hz	CSMA
	FM	WA1UTQ	1292.15000	1272.15000	88.5 Hz	CSMA
Newtown	FM	WA1SOV	145.23000	-		CSMA
North Guilford, Bluff Head						
	FM	NI1U	927.48750	902.48750	DCS 311	CSMA
Northford	DMR/MARC	N1OFJ	449.47500	-	CC1	
Northford, Bluff Head						
	FM	NI1U	449.47500	-		CSMA
Norwalk	DMR/MARC	W1NLK	448.07500	-	CC1	CSMA
	DSTAR	W1NLK	441.60000	+		
	FM	W1NLK	146.47500	147.47500	100.0 Hz	CSMA
Norwich	FM	N1NW	449.72500	-	156.7 Hz	CSMA
	FM	N1NW	927.43750	902.43750	156.7 Hz	CSMA
Norwich, Plain Hill	FM	N1NW	146.73000	-	156.7 Hz	CSMA
Orange	FM	KC1LVB	447.67500	-		CSMA
Plainville	FM	AA1WU	447.07500	-	110.9 Hz	CSMA
Portable	DMR/MARC	KA1HCX	448.62500	-	CC1	
Portland	FM	W1EDH	147.03000	+	110.9 Hz	CSMA
	FM	N1JML	147.19500	+	71.9 Hz	CSMA
	FM	N1JML	443.45000	+	100.0 Hz	
Prospect	DMR/MARC	W1LAS	448.17500	-	CC1	
	FM	W1HDN	147.18000	+	141.3 Hz	
Prospect, Fire Dept						
	FM	W1LAS	146.83500	-	100.0 Hz	
Ridgefield	FM	KR1COM	145.47000	-	100.0 Hz	CSMA
Rocky Hill	DMR/MARC	W1VLA	444.15000	+	CC3	CSMA
	DSTAR	KB1CDI	444.40000	+	88.5 Hz	
	FM	N1OTW	145.19000	-	71.9 Hz	
	FM	KB1CDI	147.37500	+	88.5 Hz	CSMA
	FM	KB1CDI	224.30000	-	88.5 Hz	CSMA
	FM	N1JBS	224.68000	-	123.0 Hz	
	FM	KB1CDI	224.78000	-	88.5 Hz	CSMA
	FM	N1OTW	444.75000	+	71.9 Hz	
Rocky Hill, Fire Station #2						
	DSTAR	W1VLA	145.27000	-		CSMA
Salem	DSTAR	KD1STR	443.40000	+		
	FM	W1DX	224.14000	-	103.5 Hz	CSMA
Seymour	DMR/BM	W0JAY	145.16000	-	CC1	
	DMR/BM	W0JAY	444.50000	+	CC1	
	DMR/MARC	W0JAY	146.16000	290.72000	CC1	
	DMR/MARC	KA1HCX	442.90000	+	CC1	
Sharon	FM	W1BAA	147.28500	+	77.0 Hz	CSMA

CONNECTICUT   115

Location	Mode	Call sign	Output	Input	Access	Coordinator
Shelton	FM	W1VAR	446.97500	-	77.0 Hz	
Shelton, White Hills Rec. Cent						
	FM	W1VAR	146.98500	-	141.3 Hz	
Somers	DMR/MARC	W1SP	445.88750	-	CC1	CSMA
	FM	N1TUP	441.80000	+	77.0 Hz	CSMA
Southington	DSTAR	WD1STR	446.45000	-		
	FM	W1ECV	145.17000	-	77.0 Hz	CSMA
Stamford	DMR/MARC	W1EE	447.12500	-	CC2	CSMA
	FM	W1EE	146.65500	-	100.0 Hz	CSMA
Sterling	DMR/MARC	N1CLV	445.83750	-	CC1	CSMA
Storrs	DMR/MARC	W1JLZ	441.46250	+	CC2	CSMA
Terryville	FM	KB1CDI	147.31500	+	88.5 Hz	CSMA
	FM	K1MKY	224.30000	-	88.5 Hz	
	FM	KB1CDI	442.30000	+	88.5 Hz	
Tolland	FM	N1PAH	53.29000	52.29000	127.3 Hz	CSMA
Torrington	DMR/BM	W1HDN	441.55000	+	CC1	CSMA
	DSTAR	N1GAU	444.10000	+		
	FM	W1RWC	145.37000	-	77.0 Hz	CSMA
	FM	W1HDN	146.85000	-	141.3 Hz	CSMA
	FM	W1RWC	147.24000	+	141.3 Hz	CSMA
	FM	W1HDN	443.60000	+	82.5 Hz	CSMA
	FM	KB1AEV	447.22500	-	77.0 Hz	CSMA
	FM	K1KGQ	449.77500	-		CSMA
Union	DMR/MARC	KB1NTA	443.80000	+	CC1	CSMA
Vernon	DMR/MARC	W1HDN	443.95000	+	CC1	CSMA
	DMR/MARC	K1IIG	444.95000	+	CC1	
	DSTAR	AA1HD	442.15000	+		CSMA
	FM	W1HDN	53.45000	52.45000	82.5 Hz	CSMA
	FM	W1HDN	145.41000	-	141.3 Hz	CSMA
	FM	W1HDN	146.79000	-	82.5 Hz	CSMA
	FM	KB1AEV	147.34500	+	151.4 Hz	CSMA
	FM	W1HDN	224.12000	-	82.5 Hz	CSMA
	FM	KB1AEV	224.36000	-	77.0 Hz	CSMA
	FM	K1WMS	224.60000	-	123.0 Hz	CSMA
	FM	KB1AEV	442.60000	+	77.0 Hz	CSMA
	FM	W1HDN	443.30000	+	114.8 Hz	CSMA
	FM	W1BRS	443.75000	+	77.0 Hz	CSMA
	FM	K1IIG	449.95000	-		
Vernon, Box Mountain						
	DSTAR	AA1HD	145.26000	-		CSMA
	FM	W1BRS	145.11000	-	77.0 Hz	CSMA
Wallingford	DMR/MARC	K1IIG	448.62500	-	CC1	
	FM	W1KKF	147.36000	+	162.2 Hz	CSMA
Warren	FM	NA1RA	53.97000	52.97000	110.9 Hz	
Washington, Fenn Hill						
	FM	NA1RA	441.85000	+	203.5 Hz	CSMA
Waterbury	FM	KB1EPA	442.45000	+	88.5 Hz	
	FM	K1IFF	449.98750	-		
Waterford	DMR/MARC	W1NLC	448.97500	-	CC1	CSMA
	FM	W1NLC	146.97000	-		CSMA
Watertown	DSTAR	KB1AEV	441.65000	+		CSMA
	FM	KB1ALU	224.04000	-	151.4 Hz	CSMA
West Hartford	FM	N1XLU	146.70000	-	162.2 Hz	
	FM	W1HDN	146.74500	-	141.3 Hz	CSMA
	FM	N1XLU	224.28000	-	114.8 Hz	CSMA
	FM	N1CRS	449.37500	-		
West Haven	DMR/MARC	KB1TTN	449.92500	-	CC1	
	FM	W1GB	146.61000	-	110.9 Hz	CSMA
	FM	K1SOX	224.50000	-	77.0 Hz	CSMA
Westbrook	DMR/MARC	WB1EOC	445.73750	-	CC2	
	DSTAR	W1BCG	444.00000	+		

116   CONNECTICUT

Location	Mode	Call sign	Output	Input	Access	Coordinator
Westbrook, High School						
	FM	W1BCG	146.77500	-	110.9 Hz	CSMA
Wethersfield	FM	KA1BQO	145.35000	-		CSMA
	FM	KA1DFH	224.68000	-		CSMA
Wilton	DMR/MARC	W1SP	440.75000	+	CC1	
Winsted	FM	N1ZCW	146.80000	-	100.0 Hz	
	FM	W1EOO	147.33000	+	141.3 Hz	CSMA
	FM	W1ECR	447.47500	-	100.0 Hz	
Wolcott	DMR/MARC	KB1TTN	440.81250	+	CC1	
	DMR/MARC	KB1TTN	446.47500	-	CC1	
Woodbridge	FM	K1SOX	147.50500	146.50500	77.0 Hz	
	FM	W1WPD	442.50000	+	DCS 073	
Woodbury, Upper Grassy Hill Ro						
	FM	NA1RA	444.80000	+	192.8 Hz	

DELAWARE

Location	Mode	Call sign	Output	Input	Access	Coordinator
Dagsboro	DMR/MARC	N3YMS	443.10000	+	CC1	T-MARC
Delaware City	FM	N3JLH	448.82500	-	131.8 Hz	T-MARC
Dover	DMR/MARC	N3YMS	146.79000	-	CC1	T-MARC
	DMR/MARC	N3YMS	449.07500	-	CC1	T-MARC
	FM	N3IOC	444.50000	-	114.8 Hz	T-MARC
Dover, DMV Complex						
	FM	KC3ARC	146.97000	-	123.0 Hz	T-MARC
Frederica	DMR/MARC	N3YMS	449.72500	-	CC1	
Greenwood	FM	W3WMD	224.44000	-		T-MARC
Harrington	FM	KB3IWV	442.45000	+	127.3 Hz	T-MARC
Hazlettville	DMR/MARC	N3YMS	448.07500	-	CC2	
	FM	N3YMS	147.30000	+	77.0 Hz	T-MARC
Laurel	DMR/MARC	N3KNT	442.05000	+	CC1	
Lewes	FM	W3LRS	147.33000	+	156.7 Hz	T-MARC
Lewes, WGMD-FM Tower						
	FM	W4ALT	443.55000	+	156.7 Hz	T-MARC
Middletown, Christiana Care Un						
	FM	W3CER	442.50000	+	100.0 Hz	T-MARC
Millsboro	FM	WS3ARA	147.09000	+	156.7 Hz	T-MARC
	FM	WS3ARA	224.84000	-	156.7 Hz	T-MARC
	FM	WS3ARA	449.82500	-	156.7 Hz	T-MARC
Newark	DSTAR	W3CER	145.11000	-		T-MARC
	FM	KB3MEC	224.00000	-	100.0 Hz	T-MARC
	FM	N3JCR	224.72000	-		T-MARC
Newark , University Of Delawar						
	FM	W3UD	145.31000	+	141.3 Hz	
Newark, Christiana Hospital						
	FM	W3CER	444.95000	+	100.0 Hz	T-MARC
Newark, Windy Hills Water Towe						
	FM	W3DRA	146.70000	-	131.8 Hz	T-MARC
	FM	W3DRA	449.02500	-		T-MARC
Roxana	DMR/MARC	W3BXW	448.72500	-	CC1	T-MARC
Seaford	DMR/MARC	N3YMS	442.81250	+	CC1	
	FM	N3KNT	146.71500	-	156.7 Hz	T-MARC
Seaford, WSUX	FM	W3TBG	145.21000	-	156.7 Hz	T-MARC
Selbyville	FM	WS3ARA	147.01500	+	156.7 Hz	T-MARC
Smyrna	FM	K3CRK	146.65500	-	146.2 Hz	T-MARC
	FM	K3CRK	443.05000	+	156.7 Hz	T-MARC
Wilmington	DMR/MARC	WR3IRS	448.42500	-	CC1	T-MARC
	FM	W3DRA	146.73000	-	131.8 Hz	T-MARC
	FM	WA3UYJ	146.95500	-	131.8 Hz	T-MARC
	FM	W3DRA	224.52000	-	131.8 Hz	T-MARC
	FM	N3KZ	442.00000	+	131.8 Hz	
	FM	W3DRA	448.37500	-	131.8 Hz	T-MARC
Wilmington, Wilmington Hospita						
	FM	W3CER	444.40000	+	100.0 Hz	T-MARC

DELAWARE 117

Location	Mode	Call sign	Output	Input	Access	Coordinator
Woodside	FM	KC3ARC	146.91000	-	77.0 Hz	T-MARC
Wyoming	FM	N3IOC	449.77500	-	114.8 Hz	T-MARC

FLORIDA

Location	Mode	Call sign	Output	Input	Access	Coordinator
Aguadilla	DMR/BM	WP3JJ	447.97500	-	CC1	
Altamonte Springs						
	FM	N1FL	147.09000	+	103.5 Hz	FASMA
	FM	N4EH	147.28500	+	107.2 Hz	FASMA
	FM	N1FL	442.75000	+	103.5 Hz	FASMA
	FM	N4EH	442.97500	+	103.5 Hz	FASMA
Andytown	FM	N0LO	442.82500	+	110.9 Hz	
Anthony	FM	KA4WJA	224.10000	-	103.5 Hz	FASMA
	FM	KA2MBE	444.32500	+	123.0 Hz	FASMA
Arcadia	FM	W4MIN	147.07500	+	100.0 Hz	
	FM	W4MIN	147.18000	+	100.0 Hz	FASMA
	FM	W4MIN	444.20000	+	100.0 Hz	
Auburndale	DMR/MARC	NP2OL	441.60000	+	CC1	
Ave Maria	FM	N4DJJ	444.07500	+	103.5 Hz	
Aventura	FM	K4PAL	147.21000	+	103.5 Hz	FASMA
	FM	K4PAL	442.25000	+	114.8 Hz	FASMA
	FM	K4PAL	443.82500	+	114.8 Hz	FASMA
Avon Park	FM	N4EMH	145.29000	-	127.3 Hz	FASMA
	FM	W4HCA	442.35000	+		FASMA
Avon Park, Bombing Range						
	FM	W4HEM	444.82500	+	100.0 Hz	FASMA
Avon Park, National Guard Armo						
	FM	W4HEM	145.33000	-	100.0 Hz	FASMA
Bartow	DMR/MARC	W4VCO	444.85000	+	CC1	FASMA
	FM	NI4CE	442.82500	+	100.0 Hz	
Bell	FM	KE4HDG	147.28500	+	123.0 Hz	
Belle Glade	FM	AB4BE	147.12000	+		FASMA
Big Pine Key	DMR/BM	N2GKG	145.37000	-	CC1	FASMA
	FM	KC2CWC	145.23000	-	94.8 Hz	FASMA
	FM	NQ2Z	442.37500	+		FASMA
Bithlo	FM	N4FL	145.23000	-	103.5 Hz	FASMA
	FM	N4FL	442.62500	+	103.5 Hz	
Boca Raton	DMR/BM	N4BRF	442.87500	+	CC1	FASMA
	DMR/MARC	N4DES	442.00000	+	CC8	
	FM	N4BRF	145.29000	-	110.9 Hz	FASMA
	FM	KC4GH	444.70000	+	123.0 Hz	
	FM	KF4LZA	444.75000	+	110.9 Hz	
	FM	KI4LJM	927.62500	902.62500	100.0 Hz	FASMA
	P25	N4DES	147.39000	+		
Boca Raton, Boca Raton Regiona						
	FM	W4BUG	146.82000	-	110.9 Hz	FASMA
Bonifay	FM	N4LMI	145.11000	-	100.0 Hz	FASMA
	FM	KF4KQE	146.91000	-	100.0 Hz	FASMA
Bonita Springs	DMR/BM	W9LP	442.67500	+	CC1	
	DMR/BM	NP2DL	443.60000	+	CC1	
	FM	KM4OWA	444.75000	+		
Boynton Beach	DMR/MARC	WX3C	442.30000	+	CC7	FASMA
	DMR/MARC	WX3C	444.25000	+	CC7	FASMA
	FM	NR4P	444.65000	+	127.3 Hz	FASMA
Boynton Beach, Bethesda West						
	P25	WX3C	442.12500	+	DCS 265	
Boynton Beach, Florida's Turnp						
	FM	NR4P	147.22500	+	107.2 Hz	FASMA
Bradenton	FM	K4GG	146.82000	-	100.0 Hz	FASMA
	FM	KF4MBN	147.19500	+	103.5 Hz	FASMA
	FM	K4CVL	224.62000	-		
	FM	KF4MBN	442.12500	+	100.0 Hz	FASMA
	FM	K4MPX	444.27500	+		FASMA

118   FLORIDA

Location	Mode	Call sign	Output	Input	Access	Coordinator
Bradenton, Mixons Fruit Farm						
	FM	K4TAP	444.87500	+	82.5 Hz	
Brandon	FM	W4HSO	146.61000	-	141.3 Hz	FASMA
	FM	K4TN	147.16500	+	136.5 Hz	FASMA
	P25	WA6KDW	927.60000	902.60000		
Brandon, Fire Station 92						
	P25	WA6KDW	444.37500	+		
Bristol	FM	KG4ITD	146.71500	-	94.8 Hz	FASMA
Brooksville	DMR/BM	K4WZV	222.22200	222.82200	CC1	
	FM	K4BKV	146.71500	-		FASMA
	FM	K4BKV	442.12500	+		FASMA
	FM	W4RPT	443.82500	+	103.5 Hz	FASMA
	P25	W4RPT	147.07500	+		FASMA
Bryceville, Near Baldwin						
	FM	W4NAS	146.83500	-	127.3 Hz	FASMA
Bunnell	DMR/MARC	KC2CWT	444.97500	+	CC1	
	FM	KB4JDE	145.45000	-	94.8 Hz	FASMA
	FM	KB4JDE	224.02000	-		FASMA
	FM	KB4JDE	444.00000	+	123.0 Hz	FASMA
Bushnell	FM	KI4DYE	145.49000	-	123.0 Hz	
Callahan	FM	W4NAS	145.31000	-		
Callahan, County Landfill						
	FM	W4NAS	147.00000	-	127.3 Hz	FASMA
Cape Coral	DMR/MARC	KN4EOF	443.80000	+	CC1	
	FM	KN2R	146.61000	-		
	FM	KN2R	444.72500	+	136.5 Hz	
Chattahoochee	FM	K4GFD	444.97500	+	94.8 Hz	
Chiefland	FM	W4DAK	147.39000	+	123.0 Hz	FASMA
Chipley	FM	N4PTW	146.62500	-	100.0 Hz	FASMA
	FM	WA4MN	444.75000	+	100.0 Hz	FASMA
Chuluota	FM	N1FL	147.16500	+	103.5 Hz	FASMA
Clearwater	DMR/BM	KD4YAL	441.95000	+	CC1	
	DMR/MARC	KN4GVY	443.37500	+	CC1	
	DSTAR	KA9RIX	147.28500	+		
	FM	KK4EQF	145.47000	-		
	FM	WD4SCD	147.03000	+	156.7 Hz	
	FM	WD0DIA	443.05000	+	141.3 Hz	
	FM	KA9RIX	444.46750	+		
Clermont	DMR/MARC	W4VCO	442.45000	+	CC1	
	FM	KA0OXH	223.94000	-		FASMA
	FM	KG4RPH	442.45000	+	103.5 Hz	FASMA
	FM	K4VJ	442.47500	+	103.5 Hz	FASMA
	FM	KA0OXH	444.97500	+	103.5 Hz	FASMA
Clermont, South Lake Hospital						
	FM	WA2UPK	442.60000	+	103.5 Hz	FASMA
Clermont, Sugarloaf Mountain						
	FM	KR4Q	224.82000	-		
Clewiston	FM	WB4TWQ	145.35000	-	127.3 Hz	
	FM	WB4TWQ	444.40000	+	127.3 Hz	FASMA
Clewiston, Clewiston Inn						
	FM	WA4PAM	146.76000	-	97.4 Hz	
Cocoa	ATV	K4ATV	427.25000	428.25000		FASMA
	DMR/BM	K4DJN	444.57500	+	CC3	FASMA
	FM	W2SDB	145.37000	-	156.7 Hz	FASMA
	FM	W4NLX	444.65000	+	107.2 Hz	FASMA
Cocoa Beach, KSC						
	FM	AA4CD	444.40000	+	103.5 Hz	FASMA
Cocoa, Kennedy Space Center						
	FM	AJ4IR	224.12000	-	123.0 Hz	FASMA
Coral Gables	FM	KD4BBM	146.76000	-		FASMA
	FM	K4AG	147.15000	+	94.8 Hz	FASMA
Coral Springs	FM	WR4AYC	145.27000	-	110.9 Hz	

FLORIDA 119

Location	Mode	Call sign	Output	Input	Access	Coordinator
Coral Springs	FM	N4RQY	146.65500	-	131.8 Hz	FASMA
	FM	WR4AYC	443.85000	+	110.9 Hz	
Crawfordville	FM	K4WAK	145.45000	-	94.8 Hz	FASMA
	FM	KN4NN	147.25500	+	94.8 Hz	FASMA
	FM	K4WAK	444.45000	+	94.8 Hz	FASMA
Crawfordville, Bethel						
	FM	K4TLH	442.85000	+	94.8 Hz	FASMA
Crescent City	FM	KJ4UOP	145.19000	-	127.3 Hz	
Cresent City	FM	KJ4UOP	53.73000	52.73000		
Crestview	DMR/MARC	KB4LSL	443.00000	+	CC1	
	DSTAR	KO4EOC	145.15000	-		
	DSTAR	KO4EOC	444.60000	+		
	DSTAR	KO4EOC	1291.30000	1271.30000		
	FM	W4AAZ	147.36000	+	100.0 Hz	FASMA
	FM	N4NID	444.95000	+	DCS 023	FASMA
	FM	KB4LSL	927.76250	902.76250	100.0 Hz	
Crestview, DOT Tower Hwy 85 An						
	FM	KC4YBZ	444.90000	+	100.0 Hz	
Cudjoe Key	DMR/BM	K4HG	441.52500	+	CC1	
	FM	AK3ML	147.06000	+		FASMA
Cutler Bay	DMR/MARC	KD4NYC	442.12500	+	CC1	
Dade City	DMR/MARC	N3OS	442.37500	+	CC1	
	FM	W4PEM	147.13500	+		FASMA
Dade City, Exit 293 On I-75						
	FM	K4EX	146.88000	-	146.2 Hz	FASMA
Dania Beach	FM	W6BXQ	146.64000	-	103.5 Hz	
	FM	KG4UGK	441.52500	+	77.0 Hz	FASMA
Davenport	FM	WC4PEM	444.62500	+	127.3 Hz	
Daytona	DMR/MARC	W2BFD	443.21250	+	CC1	
Daytona Beach	DMR/MARC	NY4Z	442.12500	+	CC1	
	FM	K4BV	147.15000	+	127.3 Hz	FASMA
	FM	N4ZKF	147.37500	+	103.5 Hz	FASMA
	FM	W4TAM	444.85000	+		
	FM	KE4NZG	927.65000	902.65000	107.2 Hz	FASMA
Deerfield Beach	FM	N4ZUW	444.42500	+	110.9 Hz	FASMA
Deerfield Beach, Watertower						
	FM	KA4EPS	444.92500	+	110.9 Hz	
DeFuniak Springs	FM	WF4X	147.28500	+	100.0 Hz	
	FM	KJ4JAH	147.37500	+	100.0 Hz	FASMA
DeLand	FM	KV4EOC	147.24000	+	123.0 Hz	FASMA
	FM	WV4ARS	147.31500	+		FASMA
	FM	K4HEK	444.15000	+	103.5 Hz	
Delray Beach	DMR/MARC	W2GGI	443.30000	+	CC1	
Deltona	FM	NP4ND	146.95500	-	107.2 Hz	
	FM	NP4ND	444.25000	+	103.5 Hz	FASMA
Destin	FM	W4RH	147.00000	+	100.0 Hz	FASMA
Destin, On Top Of Silver Beach						
	FM	N4NID	145.29000	-	DCS 023	FASMA
Drew Park	DMR/BM	KJ4SHL	442.61250	+	CC15	
Duke Field	FM	W4ZBB	147.22500	+	100.0 Hz	FASMA
Dundee	FM	WC4PEM	146.98500	-	127.3 Hz	FASMA
	FM	K4LKW	442.42500	+	127.3 Hz	
Dundee, Dundee Water Tower						
	FM	K4LKW	442.42500	+	127.3 Hz	FASMA
Dunedin	DMR/MARC	KN4GVY	145.11000	-	CC1	
	DMR/MARC	KN4GVY	444.35000	+	CC1	
	FM	K4LK	145.23000	-	146.2 Hz	FASMA
	FM	KE4EMC	146.70000	-	146.2 Hz	FASMA
	FM	K4JMH	444.15000	+	146.2 Hz	FASMA
	FM	KE4EMC	444.97500	+	100.0 Hz	FASMA
	FM	KJ4JBO	1285.50000	1265.50000	103.5 Hz	FASMA
Dunnellon	FM	KI4LOB	147.36000	+		FASMA

120 FLORIDA

Location	Mode	Call sign	Output	Input	Access	Coordinator
Durbin	FM	KK4BD	146.80500	-	127.3 Hz	FASMA
	FM	AJ4FR	444.57500	+	127.3 Hz	FASMA
Eglin AFB	FM	W4NN	147.12000	+	100.0 Hz	FASMA
	FM	W4NN	444.80000	+	100.0 Hz	FASMA
Englewood	DMR/BM	W4MO	444.10000	+	CC1	
	DMR/MARC	W4AC	444.10000	+	CC1	
	FM	K0DGF	146.77500	-	77.0 Hz	
	FM	W4AC	146.80500	-	100.0 Hz	
	FM	K0DGF	444.62500	+	77.0 Hz	FASMA
Englewood, WENG Radio Station						
	FM	AA4FB	145.25000	-		FASMA
Espanola, 1000 Foot Tower						
	FM	KD4QOF	146.74500	-	123.0 Hz	FASMA
Estero	FM	N5ICT	442.12500	+		FASMA
Eustis	DMR/MARC	W4ALR	444.55000	+	CC1	
	DSTAR	KK4KYK	443.02500	+		
	FM	KD4MBN	53.23000	52.23000	103.5 Hz	FASMA
	FM	W4ALR	146.89500	-	103.5 Hz	FASMA
	FM	KD4MBN	147.18000	+	103.5 Hz	FASMA
	FM	K4AUS	444.55000	+		FASMA
	FM	N4ZSN	444.87500	+	103.5 Hz	FASMA
Fanning Springs	FM	KB8BSO	53.17000	52.17000	107.2 Hz	FASMA
Fll	DMR/MARC	KK4ZMG	145.40000	-	CC1	
Fort Lauderdale	DMR/BM	N4MOT	442.42500	+	CC1	
	DMR/BM	K4ABB	444.25000	+	CC7	
	DMR/MARC	KK4ZMG	440.00000	+	CC1	
	DSTAR	W4AB	442.45000	+		
	FM	N4THW	146.73000	-	103.5 Hz	
	FM	W4AB	146.91000	-	110.9 Hz	
	FM	K4FK	147.33000	+	103.5 Hz	FASMA
	FM	KF4LZA	224.40000	-	110.9 Hz	
	FM	W4BEN	224.76000	-	110.9 Hz	
	FM	KD4CPG	443.80000	+	131.8 Hz	
	FM	W4AB	444.82500	+	110.9 Hz	
	FM	KF4LZA	927.05000	902.05000		
	FM	KB2TZ	927.67500	902.67500		FASMA
	FM	KF4LZA	927.70000	902.70000	110.9 Hz	
Fort Lauderdale, 110 Tower Dow						
	FM	W4AB	442.85000	+	110.9 Hz	
Fort Lauderdale, Points Of Ame						
	DSTAR	W4BUG	442.20000	+	110.9 Hz	FASMA
Fort Lauderdale, Port Everglad						
	P25	N4MOT	146.79000	-	88.5 Hz	
Fort Meyers	DMR/BM	W4LCO	443.45000	+	CC1	
Fort Myers	DMR/MARC	KF5IW	442.17500	+	CC1	
	DMR/MARC	NQ6U	442.50000	+	CC9	
	FM	AA4JS	445.92500	-	136.5 Hz	
Fort Pierce	DMR/MARC	K4SRN	927.70000	902.70000	CC1	
Frostproof	FM	WC4PEM	444.95000	+	127.3 Hz	FASMA
Ft Lauderdale	DMR/MARC	W2GGI	443.62500	+	CC1	
Ft McCoy	FM	N4STP	147.16500	+	123.0 Hz	
	FM	N4STP	444.37500	+	123.0 Hz	
Ft Myers	FM	KG4VDS	145.17000	-	136.5 Hz	
	FM	K4QCW	145.39000	-	136.5 Hz	
	FM	WA4PIL	224.52000	-	136.5 Hz	
	FM	NG2F	443.05000	+	141.3 Hz	
	FM	NX4Y	444.22500	+	136.5 Hz	FASMA
	FM	WB4FOW	444.45000	+	77.0 Hz	
	FM	WX4L	444.77500	+	136.5 Hz	
Ft Myers, FGCU Campus						
	FM	AC4TM	146.82000	-	136.5 Hz	

FLORIDA 121

Location	Mode	Call sign	Output	Input	Access	Coordinator
Ft Myers, Lee County EOC						
	FM	W4LCO	147.16500	+	127.3 Hz	
	FM	NT4TS	147.34500	+	136.5 Hz	
	FM	W4WJN	444.67500	+		
Ft Myers, North Ft Myers						
	FM	KI4ODC	146.88000	-	136.5 Hz	
Ft Pierce	DSTAR	KB4DD	145.44000	-		
	DSTAR	W4AKH	444.50000	+		
	FM	AF4CN	146.77500	-	107.2 Hz	
	FM	W4SLC	444.60000	+	107.2 Hz	
	FM	K4NRG	927.60000	902.60000	100.0 Hz	FASMA
	FM	K4NRG	927.61250	902.61250	100.0 Hz	FASMA
Ft Pierce, EOC	FM	W4SLC	147.24000	+	107.2 Hz	
Ft Pierce, WQCS-FM Tower						
	FM	W4AKH	147.34500	+		
Ft Walton Beach	FM	W4ZBB	146.79000	-		
	FM	K4FWB	444.45000	+	100.0 Hz	
Ft. Lauderdale	DMR/MARC	N4MOT	443.00000	+	CC1	
Ft. Pierce	DMR/MARC	W4AKH	444.80000	+	CC1	
	DMR/MARC	KC4DEA	927.61250	902.61250	CC1	
Gainesville	DMR/MARC	W4DFU	444.81250	+	CC1	FASMA
	FM	KD4MGR	146.85000	-		FASMA
	FM	W4DFU	146.91000	-	123.0 Hz	FASMA
Gainesville, Devil's Milhopper						
	FM	K4GNV	146.82000	-	123.0 Hz	
	FM	K4GNV	146.98500	-	123.0 Hz	
	FM	K4GNV	444.92500	+	123.0 Hz	
Gasden	FM	W4EAF	147.16500	+	94.8 Hz	FASMA
Golden Gate Estates						
	FM	KC4SSD	444.87500	+	67.0 Hz	
Grand Island	FM	KD4WOV	440.05000	+	103.5 Hz	FASMA
Greensboro	DMR/MARC	NX4DN	444.98750	+	CC1	
	FM	NX4DN	147.39000	+	94.8 Hz	
Greensboro, 170 Mile Marker In						
	FM	NX4DN	444.12500	+	94.8 Hz	
Groveland	DMR/MARC	K4AUS	444.05000	+	CC1	FASMA
	FM	KD4MBN	147.34500	+	103.5 Hz	FASMA
Gulf Hammock	FM	K4NCA	147.33000	+	123.0 Hz	
Hawthorne	FM	K3YAN	147.10500	+		FASMA
Hialeah	DMR/BM	N2GKG	145.37000	-	CC1	
	DMR/MARC	W4PHR	441.85000	+	CC4	FASMA
	FM	KC2CWC	145.23000	-	110.9 Hz	FASMA
	FM	AE4EQ	145.25000	-	110.9 Hz	FASMA
	FM	KB4AIL	145.33000	-		FASMA
	FM	KF4ZCL	145.43000	-		FASMA
	FM	KC2CWC	146.70000	-	156.7 Hz	FASMA
	FM	N2GKG	442.37500	+	103.5 Hz	FASMA
	FM	WD4DPS	442.92500	+	110.9 Hz	
	FM	WB4ESB	444.00000	+	110.9 Hz	
	FM	KA4EPS	444.35000	+	103.5 Hz	FASMA
Hickory Flat	FM	KD4Z	442.25000	+		
High Springs	FM	KB4MS	145.47000	-	123.0 Hz	
Hines	FM	K4PRY	146.74500	-	94.8 Hz	FASMA
Hobe Sound	FM	W4JUP	146.62500	-	110.9 Hz	FASMA
Holiday	FM	KP4PC	145.15000	-	146.2 Hz	FASMA
Hollister	FM	KF4CWI	147.06000	+	123.0 Hz	FASMA
	FM	W4OBB	443.90000	+	94.8 Hz	FASMA
Holly Hill	FM	KI4RF	29.66000	-	100.0 Hz	FASMA
	FM	KI4RF	53.05000	52.05000	100.0 Hz	FASMA
	FM	KI4RF	146.65500	-	103.5 Hz	
	FM	KI4RF	223.85000	-	131.8 Hz	FASMA
Hollywood	DMR/BM	K4ABB	442.90000	+	CC7	

122   FLORIDA

Location	Mode	Call sign	Output	Input	Access	Coordinator
Hollywood	FM	AC4XQ	145.21000	-	107.2 Hz	FASMA
	FM	W4RCC	147.03000	+	103.5 Hz	FASMA
	FM	AC4XQ	444.17500	+	107.2 Hz	FASMA
	FM	KC4MNI	444.55000	+	88.5 Hz	FASMA
Hollywood, Memorial Regional H						
	FM	WF2C	147.18000	+	91.5 Hz	
	FM	WF2C	444.15000	+	88.5 Hz	
Homestead	FM	W4MBU	147.00000	-		
	FM	W4MBU	224.00000	-		
	FM	W4MBU	444.00000	+	103.5 Hz	
HotSpots	DMR/MARC	KN4GVY	444.44440	+	CC1	
Hudson	DMR/BM	KD4ACG	442.58750	+	CC1	
Inverness, Citrus Memorial Hos						
	FM	KC4EOC	442.05000	+	103.5 Hz	FASMA
Jacksonville	DMR/BM	KF4EOK	147.36000	+	CC1	
	DMR/BM	K7BEN	442.00000	+	CC1	FASMA
	DMR/BM	N2XDA	444.42500	+	CC7	FASMA
	DMR/MARC	KM4SM	444.11500	+	CC2	
	DMR/MARC	K4QHR	444.27500	+	CC7	FASMA
	FM	W4IZ	146.70000	-	127.3 Hz	FASMA
	FM	W4RNG	146.76000	-	127.3 Hz	FASMA
	FM	WJ4EOC	146.95500	-	127.3 Hz	FASMA
	FM	W4EMN	147.13500	+	127.3 Hz	FASMA
	FM	W4IZ	444.40000	+	127.3 Hz	FASMA
Jacksonville , CSX Building						
	FM	W4RNG	444.67500	+	127.3 Hz	
Jacksonville Beach						
	DMR/BM	KM4CTB	442.42500	+	CC1	FASMA
	FM	K2LSF	147.39000	+	127.3 Hz	
	FM	KB4ARS	444.87500	+		FASMA
Jacksonville, Dames Point Brid						
	FM	W4RNG	147.31500	+	127.3 Hz	FASMA
	FM	K4QHR	444.20000	+		FASMA
Jupiter	DMR/MARC	KD4SJF	442.60000	+	CC10	
	FM	KA4EPS	443.82500	+	110.9 Hz	FASMA
Jupiter Farms, Florida Turnpik						
	FM	AG4BV	145.17000	-	110.9 Hz	FASMA
Kathleen, Socrum Loop Rd.						
	FM	WC4PEM	443.90000	+	127.3 Hz	FASMA
Kendall	FM	KC2CWC	145.17000	-	141.3 Hz	FASMA
	FM	KJ4OBN	444.62500	+	97.4 Hz	FASMA
Kendall Lake	FM	KI4BCO	145.45000	-	94.8 Hz	FASMA
Kendall, AT&T Mobile Switch Ce						
	FM	KA4EPS	444.12500	+	94.8 Hz	FASMA
Key Largo	FM	WX4SFL	146.61000	-	94.8 Hz	FASMA
	FM	KC4SFA	147.16500	+	94.8 Hz	FASMA
Key West	FM	N2GKG	145.17000	-		FASMA
Key West, NAS	FM	KA4EPS	146.55000		94.8 Hz	
Keystone Heights	FM	WB4EN	145.13000	-		
	FM	KI4UWC	147.22500	+	156.7 Hz	FASMA
Kings Point	FM	W4KPR	927.01250	952.01250	162.2 Hz	
Kissimmee	DMR/MARC	N4GUS	444.97500	+	CC1	
	FM	N4ARG	147.21000	+		FASMA
	FM	N4ARG	224.00000	-	DCS 411	
	FM	N4OTC	442.10000	+	103.5 Hz	FASMA
	FM	NO9S	444.45000	+	DCS 411	FASMA
LaBelle, EOC	FM	WA4PAM	145.47000	-	97.4 Hz	
Lady Lake	FM	K9GMZ	442.55000	+	146.2 Hz	
Lake Buena Vista	DMR/MARC	WD4WDW	444.00000	+	CC3	FASMA
Lake City	FM	N4SVC	53.39000	52.39000		
	FM	NF4CQ	145.49000	-		FASMA
	FM	WA4ZFQ	146.94000	-	123.0 Hz	

FLORIDA 123

Location	Mode	Call sign	Output	Input	Access	Coordinator
Lake City	FM	NF4CQ	147.15000	+	110.9 Hz	FASMA
	FM	NF4CQ	444.90000	+	110.9 Hz	FASMA
Lake Placid, Happiness Tower						
	FM	W4HCA	147.04500	+	100.0 Hz	FASMA
	FM	NI4CE	443.95000	+	100.0 Hz	FASMA
Lake Placid, Lake Placid						
	FM	W4HEM	145.21000	-	100.0 Hz	FASMA
Lake Wales	FM	K4LKW	147.33000	+	127.3 Hz	FASMA
Lake Worth	DMR/BM	WX3C	443.37000	+	CC7	
	DMR/MARC	KG4GOQ	444.45000	+	CC10	
	FM	KA4EPS	444.85000	+	110.9 Hz	FASMA
	FM	K9EE	445.52500	-	110.9 Hz	
Lakeland	DMR/MARC	W4CLL	146.65500	-	CC6	
	DMR/MARC	W4CLL	442.02500	+	CC1	FASMA
	DMR/MARC	W4VCO	444.27500	+	CC1	
	DMR/MARC	N4KEG	444.66250	+	CC1	
	DSTAR	KJ4ACN	1293.00000	1273.00000		FASMA
	FM	WP3BC	145.27000	-	127.3 Hz	FASMA
	FM	N4KEG	146.65500	-		
	FM	K4LKL	146.68500	-	127.3 Hz	FASMA
	FM	N4AMC	442.27500	+	82.5 Hz	FASMA
	P25	WP3BC	444.30000	+	127.3 Hz	FASMA
Lakewood Ranch	FM	NI4MX	443.87500	+	100.0 Hz	FASMA
Land O Lakes	DMR/MARC	KJ4SHL	439.90000	-	CC0	
Land O Lakes, Land O' Lakes Hi						
	FM	WA4GDN	145.33000	-	146.2 Hz	FASMA
Lantana	FM	WV4I	147.04500	+	110.9 Hz	FASMA
	FM	WV4I	443.97500	+	110.9 Hz	FASMA
Largo	FM	W4ACS	145.17000	-	156.7 Hz	FASMA
	FM	KO4CR	224.22000	-		
	FM	KJ4RUS	442.92500	+	146.2 Hz	FASMA
Laurel	DMR/MARC	N4SER	444.70000	+	CC1	
Lecanto	FM	W1XJ	146.95500	-	103.5 Hz	FASMA
Lecanto, Key Training Center						
	FM	W4CIT	146.77500	-	146.2 Hz	FASMA
Lee , SOUTH OF I-10 ON CR-255						
	FM	W1JXG	145.11000	-	123.0 Hz	
Lee, Lee Water Tower						
	FM	W4FAO	145.19000	-	123.0 Hz	FASMA
Leesburg	DSTAR	KJ4TJD	444.50000	+		FASMA
	FM	K4FC	147.00000	+	103.5 Hz	FASMA
Lehigh Acres	FM	N2FSU	146.71500	-	74.4 Hz	FASMA
	FM	WB4TWQ	442.80000	+	82.5 Hz	
	FM	WB4TWQ	444.50000	+	67.0 Hz	
Lehigh Acres, Lehigh Medical C						
	FM	W1RP	147.28500	+	136.5 Hz	
Lithia, Fire Station 92						
	P25	WA6KDW	927.10000	902.10000		
Little Torch Key	FM	K4CIO	146.64000	-	94.8 Hz	FASMA
	FM	K4CIO	444.77500	+	94.8 Hz	FASMA
Live Oak	DMR/MARC	KC4GOL	443.60000	+	CC7	
	FM	N4SVC	145.41000	-	100.0 Hz	
	FM	N4SVC	442.92500	+	127.3 Hz	FASMA
Longwood	DMR/MARC	KE4GLA	444.00000	893.00000	CC1	
Lox	DMR/MARC	KK4ZMG	144.00000		CC1	
Loxahachee	FM	KA4EPS	444.30000	+	110.9 Hz	
Loxahatche	DMR/BM	KK4ZMG	444.70000	+	CC7	
Loxahatchee	DMR/BM	N4KMM	444.27500	+	CC1	
	DMR/MARC	N4KMM	443.72500	+	CC1	
	FM	WB2SNN	444.35000	+	110.9 Hz	
Loxahatchee, 20 Mile Bend/Lion						
	P25	AK4JQ	147.36000	+		FASMA

124   FLORIDA

Location	Mode	Call sign	Output	Input	Access	Coordinator
Macclenny	DMR/BM	W4DNQ	442.87500	+	CC1	
	FM	AB4GE	147.09000	+	100.0 Hz	
	FM	AB4GE	444.07500	+	100.0 Hz	FASMA
Madeira Beach	DSTAR	KA9RIX	444.46250	+		
Madison	DMR/BM	W4FAO	442.00000	+	CC1	
	FM	K4NRD	444.30000	+	94.8 Hz	FASMA
Magnonia Park	P25	AG4BV	444.32500	+		FASMA
Maitland	FM	AG4YD	442.80000	+	103.5 Hz	FASMA
Mangonia Park	FM	W4ESA	146.94000	-	91.5 Hz	
Marathon	FM	KA4EPS	444.02500	+	94.8 Hz	FASMA
	FM	W1FXX	444.32500	+		
Marco Island	DMR/MARC	K5MI	444.81250	+	CC1	
Marco Island, Mainsail Drive H						
	FM	K5MI	146.85000	-	141.3 Hz	
Marco Island, South Shore						
	DSTAR	K5MI	146.98500	-		FASMA
Margate, ATT Microwave Tower						
	FM	KA4EPS	444.02500	+	107.2 Hz	FASMA
Marianna	FM	W4BKD	146.67000	-	123.0 Hz	
	FM	W4BKD	444.95000	+	123.0 Hz	
Mas Verde Mobile Home Estates						
	FM	W2SDB	145.37000	-		
Melbourne	DMR/BM	K4MRG	442.02500	+	CC1	FASMA
	FM	AF4Z	146.68500	-		FASMA
	FM	K4YWC	147.00000	-	167.9 Hz	FASMA
	FM	W4MLB	444.42500	+	107.2 Hz	FASMA
	IDAS	KI4SWB	444.90000	+	118.8 Hz	FASMA
Melbourne, Holmes Reg Med Ctr						
	FM	W4MLB	146.61000	-	107.2 Hz	FASMA
Melbourne, I95 & US-192						
	FM	KI4SWB	147.48000		107.2 Hz	
Melbourne, Sun Trust Building						
	FM	K4HRS	145.47000	-	107.2 Hz	FASMA
Melbourne, Trinity Towers (Eas						
	FM	K4DCS	444.32500	+	107.2 Hz	
Merritt Island	DMR/BM	KC2UFO	444.77500	+	CC1	
Merritt Island, Cocoa Beach, K						
	FM	KC2UFO	444.87500	+	107.2 Hz	FASMA
Metcalf	DMR/MARC	NX4DN	444.05000	+	CC1	
Miami	DMR/MARC	K7HJE	441.10000	+	CC1	
	DMR/MARC	W2GGI	444.98750	+	CC1	
	DSTAR	WD4ARC	145.33000	-		
	FM	AC4XQ	53.03000	52.03000	107.2 Hz	FASMA
	FM	W4HN	53.25000	52.25000	110.9 Hz	FASMA
	FM	KB4AIL	145.28000	-		FASMA
	FM	KI4IJQ	146.70000	-	156.7 Hz	FASMA
	FM	AE4EQ	146.80500	-	110.9 Hz	FASMA
	FM	K4PAL	146.85000	-	91.5 Hz	
	FM	AE4WE	146.89500	-	100.0 Hz	FASMA
	FM	KF4ACN	146.92500	-	103.5 Hz	FASMA
	FM	W4NR	146.95500	-	110.9 Hz	FASMA
	FM	KB4MBU	147.06000	+	103.5 Hz	FASMA
	FM	KR4DQ	147.30000	+	88.5 Hz	FASMA
	FM	K4AG	147.36000	+	94.8 Hz	FASMA
	FM	K4AG	442.15000	+	94.8 Hz	FASMA
	FM	KI4IJQ	442.32500	+	156.7 Hz	FASMA
	FM	KR4DQ	442.52500	+	88.5 Hz	FASMA
	FM	KB4ELI	442.65000	+	94.8 Hz	FASMA
	FM	KC4MND	442.72500	+		FASMA
	FM	N4CR	443.05000	+		FASMA
	FM	AE4EQ	443.92500	+	110.9 Hz	FASMA
	FM	KA4HLO	444.20000	+	94.8 Hz	FASMA

FLORIDA 125

Location	Mode	Call sign	Output	Input	Access	Coordinator
Miami	FM	K4PCS	444.27500	+	94.8 Hz	FASMA
	FM	KI4ZYV	444.32500	+	156.7 Hz	
	FM	KC4MNE	444.37500	+	94.8 Hz	FASMA
	FM	KC2CWC	444.50000	+	127.3 Hz	FASMA
	FM	KD4IMM	444.52500	+	114.8 Hz	FASMA
	FM	K4AG	444.60000	+	94.8 Hz	FASMA
	FM	KS4WF	444.77500	+		FASMA
Milton	DSTAR	KI4WZA	147.33000	+		
	DSTAR	KI4WZA	444.92500	+		
	FM	W4VIY	145.49000	-	100.0 Hz	FASMA
	FM	WA5HC	146.44000		100.0 Hz	
	FM	K4SRC	146.70000	-	100.0 Hz	FASMA
Milton, NAS Whiting Field						
	FM	N3CMH	145.25000	-	100.0 Hz	FASMA
Milton, NAS WHTG FLD						
	FM	N3CMH	443.97500	+	DCS 114	FASMA
Mims	FM	KE4NUZ	146.62500	-	100.0 Hz	FASMA
Mims, Hog Valley	FM	K4KSC	146.77500	-	100.0 Hz	
Minneola	FM	W4ALR	147.22500	+	103.5 Hz	FASMA
Miramar	DMR/BM	KA2ZPN	442.30000	+	CC3	
Monticello	FM	WX4JEF	145.43000	-	94.8 Hz	FASMA
Moore Haven	FM	KJ4FJD	147.30000	+	100.0 Hz	FASMA
Murdock	FM	N9OJ	444.60000	+	82.5 Hz	
Naples	DMR/MARC	AB4NP	438.38750	430.78750	CC1	
	DMR/MARC	KC4RPP	443.10000	+	CC0	
	DSTAR	AA4PP	441.50000	+		
	DSTAR	AB4FL B	448.80000	-		
	FM	WA3JGC	146.64000	-	136.5 Hz	
	FM	KC4RPP	147.10500	+	136.5 Hz	
	FM	KC1AR	147.50500	-	67.0 Hz	
	FM	KC1AR	224.38000	-	67.0 Hz	FASMA
	FM	KC1AR	442.75000	+		
	FM	KF4YEN	443.60000	+	114.8 Hz	
	FM	N5ICT	443.90000	+	67.0 Hz	
	FM	WA1QDP	444.90000	+	67.0 Hz	
Naples, Airport	FM	AB4NP	444.72500	+		
Naples, County Barn Road, BES						
	DSTAR	AA4PP	145.49000	-		FASMA
Naples, Golden Gate Blvd West						
	FM	K4YHB	147.03000	+		FASMA
Naples, Marbella	DSTAR	AB4NP	145.27000	-		
Nassauville	FM	W4NAS	444.47500	+		
Navarre	FM	KC4ERT	444.20000	+	100.0 Hz	FASMA
New Port Richey	DMR/MARC	KN4GVY	444.20000	+	CC1	FASMA
	FM	WA4T	145.35000	-		
New Port Richey, Pasco County						
	FUSION	WA4GDN	146.67000	-	146.2 Hz	FASMA
New Smyrna , Hospital						
	FM	K4BO	145.33000	-	127.3 Hz	
North Bay County	FM	KE4FD	146.94000	-	100.0 Hz	
North Dade	DMR/MARC	AC4XQ	443.12500	+	CC10	
North Miami Beach						
	FM	K4PAL	147.37500	+	118.8 Hz	FASMA
North Naples, Marbella						
	FM	WB2QLP	146.67000	-	136.5 Hz	
North Port	FM	K4NPT	147.12000	+	136.5 Hz	FASMA
	FM	KA2LAL	442.15000	+	94.8 Hz	
Ocala	DSTAR	KK4DFC	146.79000	-		FASMA
	DSTAR	KK4DFC	147.19500	+		
	FM	KA2MBE	29.68000	-		FASMA
	FM	KG4NXO	145.17000	-	123.0 Hz	FASMA
	FM	KD4GME	145.43000	-	141.3 Hz	

126   FLORIDA

Location	Mode	Call sign	Output	Input	Access	Coordinator
Ocala	FM	K4GSO	146.61000	-	123.0 Hz	FASMA
	FM	NX4Y	444.02500	+	123.0 Hz	
	FUSION	K2ADA	147.21000	+	123.0 Hz	FASMA
Ocala, MRMC	FM	WA3YOX	145.27000	-	123.0 Hz	FASMA
Okeechobee	FM	K4OKE	147.09000	+	100.0 Hz	FASMA
	FM	K4OKE	147.19500	+	100.0 Hz	
Orange City	DMR/BM	N2NEI	442.47500	+	CC1	
Orange Park	DMR/MARC	KK4ECR	443.08750	+	CC1	
	FM	K4SIX	53.19000	52.19000		
	FM	K4TB	146.67000	-	127.3 Hz	FASMA
	FM	K4BT	444.50000	+	127.3 Hz	FASMA
Orange Park, Fleming Island						
	FM	KI4UWC	146.92500	-	156.7 Hz	FASMA
Orange Park, Orange Park Hospi						
	FM	W4NEK	147.25500	+	103.5 Hz	FASMA
Orlando	DMR/BM	W4MCO	443.16250	+	CC11	
	DMR/MARC	KJ4OVA	443.13750	+	CC1	
	DSTAR	K1XC	146.82000	-		FASMA
	DSTAR	K1XC	1285.00000	1265.00000		FASMA
	FM	WD4IXD	145.13000	-	103.5 Hz	FASMA
	FM	W4MCO	146.73000	-	103.5 Hz	FASMA
	FM	KB4UT	146.76000	-	103.5 Hz	FASMA
	FM	WD4MRR	147.06000	+	103.5 Hz	FASMA
	FM	N4LGH	147.12000	+	103.5 Hz	FASMA
	FM	WD4WDW	147.30000	+	103.5 Hz	FASMA
	FM	KR4KZ	224.46000	-		
	FM	W4LOV	442.07500	+	103.5 Hz	FASMA
	FM	KD4Z	442.25000	+	103.5 Hz	
	FM	W4MCO	442.70000	+		FASMA
	FM	W4MCO	443.05000	+	103.5 Hz	FASMA
	FM	N4ATS	443.85000	+	103.5 Hz	FASMA
	FM	KR4KZ	443.97500	+	103.5 Hz	FASMA
	FM	K4HEK	444.15000	+	88.5 Hz	FASMA
	FM	K4HEK	444.80000	+	88.5 Hz	FASMA
Orlando, Disney World						
	DSTAR	WD4WDW	442.00000	+		
Orlando, Downtown						
	FM	NN4TT	442.37500	+	103.5 Hz	FASMA
Orlando, Orange County Convent						
	FM	W4AES	444.12500	+	103.5 Hz	FASMA
Ormond Beach	FM	KA2AYR	145.27000	-	156.7 Hz	FASMA
	FM	N4GOA	146.86500	-	127.3 Hz	FASMA
	FM	N4GOA	147.27000	+	127.3 Hz	FASMA
	FM	N4JRF	442.40000	+		
	FM	KA2AYR	442.65000	+	127.3 Hz	FASMA
	FM	KE4NZG	443.82500	+	118.8 Hz	FASMA
	FM	N4GOA	443.87500	+	127.3 Hz	FASMA
Oviedo	FM	WD4DSV	442.95000	+	103.5 Hz	FASMA
Palatka	FM	KF4PXZ	145.37000	-		
Palm Bay	DSTAR	N4OTC	444.47500	+		
	FM	WW4AL	145.25000	-	82.5 Hz	
	FM	W4MLB	146.85000	-	107.2 Hz	FASMA
	FM	K4EOC	146.89500	-		FASMA
	FM	K4MER	927.60000	902.60000	100.0 Hz	
Palm Bay, Palm Bay South Tower						
	FM	K4DCS	147.25500	+	107.2 Hz	FASMA
Palm Beach Gardens						
	DMR/MARC	KD4SJF	442.10000	+	CC10	FASMA
Palm Beach Gardens, I-95/PGA B						
	FM	W4JUP	444.22500	+	110.9 Hz	
Palm Beach Gardens, PGA Blvd./						
	P25	AG4BV	444.40000	+		FASMA

FLORIDA   127

Location	Mode	Call sign	Output	Input	Access	Coordinator
Palm City	DMR/MARC	N4IRS	927.62500	902.62500	CC1	
Palm Coast	DMR/BM	KF4I	443.30000	+	CC1	
	DMR/MARC	KG4IDD	442.20000	+	CC1	
	DSTAR	W4SRT	442.32500	+		
	DSTAR	KG4TCC	1293.40000	1313.40000		
	FM	KG4IDD	145.47000	-	123.0 Hz	FASMA
	FM	W4FPC	146.71500	-	123.0 Hz	FASMA
	FM	W4FPC	147.07500	+	123.0 Hz	FASMA
Palm Harbor	DMR/MARC	KN4GVY	444.57500	+	CC1	
	DSTAR	W4AFC	442.50000	+		FASMA
	FM	W4AFC	147.12000	+	100.0 Hz	
Palm Harbor, Florida						
	FM	W4AFC	442.70000	+		
Panama City	FM	AC4QB	53.05000	52.05000		
	FM	W4RYZ	145.21000	-		FASMA
	FM	KV4ATV	145.25000	-	100.0 Hz	
	FM	AC4QB	145.33000	-	100.0 Hz	FASMA
	FM	KF4JMM	146.74500	-		
	FM	KM4OAR	147.06000	+		FASMA
	FM	KV4ATV	224.25000	-	100.0 Hz	
	FM	N1HQ	444.10000	+		FASMA
	FM	KV4ATV	444.25000	+	100.0 Hz	
	FM	KF4JMM	444.50000	+	103.5 Hz	
	FM	KV4ATV	919.25000	433.95000		
Parkland	FM	WR4AYC	145.11000	-	110.9 Hz	
Pembroke Pines	FM	K2HXC	441.90000	+		FASMA
Pensacola	DMR/BM	WB4OQF	443.70000	+	CC1	
	DSTAR	KB4TXY	144.00000			
	FM	W4UC	146.76000	-	100.0 Hz	FASMA
	FM	WB4OQF	146.85000	-	100.0 Hz	
Pensacola, Naval Hospital						
	FM	WA4ECY	145.45000	+	100.0 Hz	
Pensacola, Naval Hospital Pens						
	FM	WA4ECY	443.85000	+	100.0 Hz	FASMA
Perry	FM	K4PRY	145.35000	-	123.0 Hz	FASMA
	FM	K4PRY	146.97000	-	123.0 Hz	
	FM	K4III	444.10000	+	94.8 Hz	FASMA
Pinellas Park	DMR/MARC	KN4GVY	443.82500	+	CC1	
Plant City	DMR/MARC	W4CLL	442.67500	+	CC1	
Plantation	DMR/MARC	W4MOT	442.40000	+	CC1	
	FM	K4GET	147.48000		110.9 Hz	
	FM	N4RQY	224.18000	-	131.8 Hz	FASMA
	FM	K4GET	441.42500	+	103.5 Hz	FASMA
Poinciana	FM	K9YCG	146.71500	-	103.5 Hz	FASMA
Polk City	DSTAR	AD1I	442.78750	+	67.0 Hz	FASMA
Pompano Beach	DMR/BM	W4BUG	443.35000	+	CC11	
	FM	W5BUG	927.55000	902.55000	100.0 Hz	FASMA
Pompano Beach, John Knox Build						
	FM	W4BUG	146.61000	-	110.9 Hz	FASMA
	FM	W4BUG	442.50000	+	110.9 Hz	FASMA
Ponte Vedra	FM	KX4EOC	147.01500	+	127.3 Hz	FASMA
Port Charlotte	FM	KB0EVM	147.01500	+	136.5 Hz	FASMA
	FM	N4FOB	442.70000	+	136.5 Hz	
Port Charlotte, Charlotte Coun						
	FM	K8ONV	146.86500	-	136.5 Hz	FASMA
Port Charlotte, County Tower B						
	FM	W4DUX	147.25500	+	136.5 Hz	FASMA
Port Richey	DMR/MARC	KJ4LXT	444.93750	+	CC1	
	FM	KG4YZY	442.65000	+		FASMA
Port Salerno	DMR/MARC	N4IRS	444.97500	+	CC1	
Port St Joe	FM	W4WEB	147.30000	+	103.5 Hz	
Port St Lucie	DMR/MARC	K2DMR	440.85000	+	CC1	

128 FLORIDA

Location	Mode	Call sign	Output	Input	Access	Coordinator
Port St Lucie	FM	W4SLC	147.01500	+	107.2 Hz	FASMA
	FM	K4NRG	444.00000	+		FASMA
	FM	K4NRG	927.66250	902.66250	100.0 Hz	FASMA
Port St Lucie, WAVW Tower						
	FM	K4PSL	146.95500	-	107.2 Hz	FASMA
Port St. Lucie	DMR/MARC	W4SLC	442.02500	+	CC1	
Port St. Lucie, Tradition Medi						
	FM	W4SLC	442.00000	+	107.2 Hz	
Port St.Lucie	DMR/MARC	K4SRN	444.00000	+	CC1	
Portable 1	DMR/MARC	KN4GVY	444.10000	+	CC2	
Portable 2	DMR/MARC	KN4GVY	444.37500	+	CC2	
Portable Rpt	DMR/MARC	NX4DN	444.17500	+	CC1	
Princeton	FM	KF4ACN	146.83500	-	94.8 Hz	FASMA
	FM	KF4ACN	442.35000	+	103.5 Hz	FASMA
	FM	KF4ACN	442.68750	+		FASMA
Punta Gorda	DMR/MARC	W4DUX	442.92500	+	CC1	FASMA
	DMR/MARC	KA2OOR	443.00000	+	CC9	
	FM	KF4QWC	146.68500	-	136.5 Hz	FASMA
	FM	WX4E	146.74500	-	136.5 Hz	FASMA
	FM	WX4E	444.97500	+	136.5 Hz	
Riverview	FM	K4SIP	441.90000	+	146.2 Hz	
	FM	NI4CE	442.55000	+	100.0 Hz	
Riverview, TB Central						
	FM	NI4CE	444.42500	+		FASMA
Riviera Beach	FM	AK4JQ	146.88000	-		FASMA
	FM	KF4ACN	443.92500	+		FASMA
	FM	KF4ACN	927.52500	902.52500		FASMA
Rockledge	FM	W4NLX	146.88000	-	107.2 Hz	FASMA
	FM	K4GCC	146.94000	-		FASMA
	FM	K4EOC	147.13500	+	107.2 Hz	FASMA
	FM	K4EOC	444.52500	+	103.5 Hz	FASMA
Safety Harbor	DSTAR	KJ4ARB	1292.00000	1272.00000		FASMA
Saint Lucie	DMR/MARC	K2DMR	440.75000	+	CC10	
Samsula	FM	KB4GW	147.21000	+		
Sanford	DSTAR	W4PLB	145.16000	-		FASMA
	DSTAR	W4PLB	442.30000	+		
	DSTAR	KK4CQQ	1291.30000	1271.30000		
	FM	N4EH	146.80500	-	103.5 Hz	FASMA
Sanibel Island	FM	W4SBL	146.79000	-	136.5 Hz	
Sarasota	FM	W4IE	146.91000	-	100.0 Hz	FASMA
	FM	N4SER	147.39000	+	100.0 Hz	FASMA
	FM	W4IE	444.92500	+	100.0 Hz	FASMA
	P25	NX4Y	444.80000	+		
Sarasota, Comcast Tower						
	FM	N4SER	146.73000	-	100.0 Hz	FASMA
Sebastian	DMR/BM	KB1YBB	442.60000	+	CC1	
Sebring	DMR/BM	K4DRZ	441.95000	+	CC1	
Sebring, Highlands County EOC						
	FM	W4HEM	147.27000	+	100.0 Hz	FASMA
Seminole	FM	KA4CNP	444.40000	+	192.8 Hz	FASMA
South Vero Beach	FM	AB4AZ	145.13000	-	107.2 Hz	FASMA
Spring Hill	FM	KC4MTS	146.76000	-	123.0 Hz	FASMA
	FM	KF4IXU	146.80500	-		FASMA
Spring Hill, Cell Tower Insect						
	FM	KF4IXU	443.80000	+		
St Augustine	DMR/BM	KE4LF	447.70000	-	CC13	
	FM	AB4EY	145.17000	-	107.2 Hz	FASMA
	FM	KF4MX	146.62500	-		
St Augustine, EOC						
	FM	KX4EOC	145.21000	-	127.3 Hz	
St Augustine, Summer Haven						
	FM	KC5LPA	442.80000	+	127.3 Hz	FASMA

FLORIDA 129

Location	Mode	Call sign	Output	Input	Access	Coordinator
St Cloud	FM	KG4EOC	145.35000	-	103.5 Hz	
	FM	W4SIE	146.79000	-	103.5 Hz	
	FM	KG4EOC	444.10000	+	123.0 Hz	
St Petersburg	FM	KA9RIX	145.39000	-	141.3 Hz	
	FM	W4ORM	146.85000	-	146.2 Hz	
	FM	N4BSA	147.31500	+		FASMA
	FM	W4MRA	147.36000	+	127.3 Hz	FASMA
	FM	WA4AKH	224.66000	-		
	FM	W4ABC	443.92500	+		FASMA
St Petersburg, North East Park						
	FM	N3FU	444.02500	+	146.2 Hz	
St Petersburg, Pinellas Park						
	FM	W4GAC	147.06000	+		
St Petersburg, Skyway Bridge						
	FM	AG4UU	442.25000	+	146.2 Hz	FASMA
St. Petersburg	DMR/MARC	KJ4SHL	444.96250	+	CC1	FASMA
	IDAS	KJ4SHL	443.76250	+		
St. Petersburgh #1						
	DMR/MARC	KN4GVY	443.97500	+	CC1	
St. Petersburgh #2						
	DMR/MARC	KN4GVY	444.62500	+	CC1	
Starke, Bradford County EOC						
	FM	K4BAR	145.15000	-		FASMA
Stuart	FM	KA3COZ	444.15000	+	107.2 Hz	
Stuart, EOC Tower						
	FM	WX4MC	145.15000	-	107.2 Hz	
Stuart, Martin Memorial Hospit						
	FM	K4ZK	147.06000	+	107.2 Hz	
Stuart, Stuart Memorial Hospit						
	DMR	WX4MC	444.90000	+	CC0	
Summerfield, The Villages						
	FM	K4HOG	147.03000	+	123.0 Hz	FASMA
Sumterville	FM	KS4EOC	146.92500	-	123.0 Hz	FASMA
Sun City Center	FM	KE4ZIP	147.22500	+	146.2 Hz	
	FM	W4KPR	440.10000	+	162.2 Hz	
	FM	W4KPR	442.45000	+		FASMA
	FM	KE4ZIP	443.25000	+	146.2 Hz	
Sun City Center, Clubhouse						
	FM	W4KPR	145.45000	-	162.2 Hz	FASMA
Sun City Center, Kings Point R						
	DSTAR	W1SCC	147.26250	+		FASMA
	DSTAR	W1SCC	442.22500	447.32500		FASMA
Sun City Center, Sun Towers						
	P25	W4KPR	927.01250	902.01250	162.2 Hz	FASMA
Sunny Isles	DMR/MARC	N4LJQ	442.95000	+	CC7	
Sunrise	FM	K4BRY	443.05000	+	110.9 Hz	FASMA
Tallahassee	DMR/BM	W4JMF	441.00000	+	CC1	
	DMR/MARC	NX4DN	443.13750	+	CC1	
	FM	K4TLH	29.66000	-	94.8 Hz	FASMA
	FM	N4PG	146.61000	-	203.5 Hz	
	FM	K4TLH	146.91000	-	94.8 Hz	
	FM	K4TLH	147.03000	+	94.8 Hz	
	FM	K4TLH	442.10000	+	94.8 Hz	FASMA
	FM	AE4S	443.95000	+	94.8 Hz	
	FM	K4TLH	444.80000	+		FASMA
Tallahassee, Capital Regional						
	FM	N4NKV	444.40000	+	131.8 Hz	FASMA
Tallahassee, State EOC						
	FM	KA4EOC	147.28500	+	94.8 Hz	FASMA
Tallahassee, Tallahassee Memor						
	FM	KD4MOJ	444.00000	+	94.8 Hz	FASMA
Tampa	DMR/MARC	W4CLL	145.41000	-	CC5	FASMA

130 FLORIDA

Location	Mode	Call sign	Output	Input	Access	Coordinator
Tampa	DMR/MARC	YO6NAM	439.85000	430.45000	CC1	
	DMR/MARC	W4PJT	443.11250	+	CC1	
	DMR/MARC	W4CLL	443.11250	+	CC1	
	DMR/MARC	W4CLL	443.77500	+	CC1	
	DMR/MARC	KJ4SHL	444.00000	+	CC1	
	DSTAR	KJ4ARB	147.01000	+		FASMA
	DSTAR	KJ4ARB	443.98750	+		FASMA
	DSTAR	W4RNT	444.81250	+		FASMA
	FM	W4EFK	145.49000	-	88.5 Hz	FASMA
	FM	KC4LSQ	146.83500	-	131.8 Hz	FASMA
	FM	W3YT	146.86500	-	123.0 Hz	
	FM	NI4M	146.94000	-	146.2 Hz	FASMA
	FM	W4AQR	147.00000	+	107.2 Hz	FASMA
	FM	N4TP	147.10500	+	146.2 Hz	
	FM	KD4HVC	147.24000	+	88.5 Hz	FASMA
	FM	W4RNT	442.72500	+	146.2 Hz	
	FM	NX4Y	442.85000	+		FASMA
	FM	W4RNT	444.25000	+	146.2 Hz	
	FM	W4EFK	444.60000	+	88.5 Hz	FASMA
	FM	W4AQR	444.67500	+	103.5 Hz	FASMA
	FM	W4HSO	444.90000	+	141.3 Hz	FASMA
	FM	W9CR	927.05000	902.05000		
	FM	N1CDO	927.31250	902.31250	162.2 Hz	
	P25	N1CDO	440.10000	+		FASMA
Tampa Airport	DMR/MARC	KJ4SHL	442.60000	+	CC1	FASMA
Tampa Downtown	DMR/MARC	KJ4SHL	443.75250	+	CC1	
Tampa, BCI Communications						
	FM	W4BCI	146.79000	-	146.2 Hz	
Tampa, Downtown						
	FM	W9CR	224.28000	-	146.2 Hz	
	P25	WA6KDW	927.20000	902.20000		
Tampa, Hillsborough EOC Tower						
	FM	N4TP	147.10500	+	146.2 Hz	FASMA
Tampa, Near Tampa Internationa						
	FM	KK4AFB	147.34500	+	146.2 Hz	FASMA
Tampa, Old County EOC						
	FM	W4RNT	224.74000	-		
Tampa, Raymond James Stadium						
	FM	N4TP	443.02500	+	146.2 Hz	FASMA
Tampa, Regions Building, Downt						
	FM	W4BCI	444.22500	+	146.2 Hz	FASMA
Tampa, Rocky Point						
	FM	KB4ABE	444.52500	+	141.3 Hz	FASMA
Tampa, St Joseph's Hospital						
	FM	N4TP	444.75000	+	146.2 Hz	FASMA
Tarpon Springs	FM	K4JMH	444.45000	+	146.2 Hz	FASMA
Tavares	DMR/MARC	K4AUS	444.96250	+	CC1	
	FM	N4FLA	147.25500	+	103.5 Hz	FASMA
	FM	K4AUS	147.39000	+	103.5 Hz	FASMA
	FM	K4FC	442.90000	+		FASMA
The Villages	FM	K4LFK	145.21000	-	110.9 Hz	FASMA
	FM	K4LFK	146.67000	-	103.5 Hz	
	FM	WA1UTQ	444.25000	+	110.9 Hz	FASMA
	FM	WA1UTQ	1292.15000	1272.15000	91.5 Hz	FASMA
	P25	WN4AMO	146.85000	-	103.5 Hz	FASMA
Titusville	DMR/BM	K4DJN	444.67500	+	CC3	FASMA
	DMR/MARC	K2JO	444.15000	+	CC1	
	FM	WN3DHI	145.49000	-		
	FM	KJ4VEH	145.60000		107.2 Hz	
	FM	K4EOC	147.07500	+	107.2 Hz	FASMA
	FM	KJ4VEH	445.60000	-	107.2 Hz	
	FUSION	N4TDX	442.85000	+		

FLORIDA 131

Location	Mode	Call sign	Output	Input	Access	Coordinator
Titusville, Jess Parrish Hospi						
	FM	K4NBR	147.33000	+	107.2 Hz	FASMA
Titusville, Parrish Medical Ce						
	FM	N4TDX	147.36000	+	107.2 Hz	FASMA
	FM	N4TDX	442.85000	+	107.2 Hz	FASMA
Titusville, The Great Outdoors						
	FM	N4TDX	444.75000	+	156.7 Hz	FASMA
Titusville, Titusville Towers						
	FM	K4KSC	146.97000	-	107.2 Hz	FASMA
Titusville, Vehicle Assembly B						
	FM	N1KSC	444.92500	+	131.8 Hz	FASMA
Trenton	FM	N4TSV	921.20000	-		FASMA
Turkey Foot	FM	NX4Y	444.95000	+	100.0 Hz	
Valkaria	FM	K4HV	444.70000	+		FASMA
Valparaiso	FM	K4DTV	146.73000	-		FASMA
Venice	FM	KB2WVY	442.05000	+	100.0 Hz	
Venice, Laurel Rd/I-75						
	FM	N4SER	145.13000	-	100.0 Hz	FASMA
Verna	FM	NI4CE	442.95000	+		FASMA
	FM	NI4CE	444.31250	+		FASMA
Verna, Cox Radio Tower						
	FM	NI4CE	145.43000	-	100.0 Hz	FASMA
Vero Beach	DMR/BM	KJ4YZI	444.35000	+	CC1	
	DMR/MARC	KJ4YZI	444.32500	+	CC1	
	FM	K4CPJ	444.72500	+	107.2 Hz	FASMA
	FM	KA4EPS	444.85000	+	107.2 Hz	FASMA
	FUSION	AB4AZ	145.13000	-		
Vero Beach, Indian River Medic						
	FM	W4IRC	145.31000	-	107.2 Hz	
Vero Beach, North County						
	FM	W4PHJ	146.64000	-	107.2 Hz	FASMA
Wacissa	FM	K4TLH	53.03000	52.03000	94.8 Hz	FASMA
	FM	K4TLH	147.00000	+	94.8 Hz	FASMA
Wacssia	DMR/MARC	NX4DN	443.02500	+	CC1	
Wagon Wheel	DMR/MARC	KC4RPP-R	147.10500	+	CC0	
Wauchula	FM	N4EMH	146.62500	-	127.3 Hz	
Weeki Wachee	DSTAR	KK4ONE	445.44500	-		
	FM	KF4CIK	53.13000	52.13000	100.0 Hz	FASMA
Weeki Wachee, Weeki Wachee Spr						
	FM	KB4SYU	147.04500	+		FASMA
Wellborn	FM	W1QBI	145.27000	-	123.0 Hz	FASMA
Wellington	DMR/BM	WX3C	443.32500	+	CC7	
	DMR/MARC	K4PKT	442.93750	+	CC1	
	FM	K4WRC	147.28500	+		FASMA
Wesley Chapel	DSTAR	W4SRT	1293.40000	1273.40000		
West Palm Beach	DMR/MARC	W2GGI	444.91250	+	CC1	
	DSTAR	K4WPB	145.32000	-		FASMA
	FM	K4LJP	146.97000	-	162.2 Hz	
	FM	AK4JQ	147.36000	+	110.9 Hz	
West Palm Beach, 20 Mile Bend.						
	FM	KA4EPS	444.12500	+	107.2 Hz	
West Palm Beach, Mangonia Park						
	FM	KC4UDZ	53.21000	52.21000		
West Palm Beach, Palm Beach Co						
	FM	AK4JQ	145.39000	-		FASMA
	FM	W4JUP	146.71500	-	110.9 Hz	FASMA
West Palm Beach, Sewage Proces						
	FM	WR4AKX	146.67000	-		FASMA
West Palm Beach, WPB Christian						
	FM	WD4CQH	444.37500	+	118.8 Hz	
Westly Chapel	DMR/MARC	KN4GVY	444.77500	+	CC1	
Wewahitchka	FM	W4FFC	146.86500	-	100.0 Hz	FASMA

32   FLORIDA

Location	Mode	Call sign	Output	Input	Access	Coordinator
Wildwood	FM	WA1UTQ	224.98000	-	91.5 Hz	FASMA
Wildwood , The Villages						
	FUSION	WA1UTQ	444.57500	+	91.5 Hz	FASMA
Wildwood, The Villages						
	FM	WA1UTQ	146.94000	-	103.5 Hz	FASMA
Winter Park	P25	W4MCO	442.52500	+	103.5 Hz	FASMA
Yulee, FDOT Tower						
	FM	KC5LPA	145.23000	-	127.3 Hz	FASMA
	FM	KC5LPA	442.90000	+	127.3 Hz	FASMA
Zephyrhills	FM	NI4M	145.19000	-	146.2 Hz	FASMA
	FM	W1PB	146.91000	-	146.2 Hz	FASMA

GEORGIA

Location	Mode	Call sign	Output	Input	Access	Coordinator
Acworth	DMR/BM	KO4ACQ	443.97500	+	CC13	
	FM	KC4YNF	441.80000	+	77.0 Hz	
Adairsville	FM	WB4AEG	443.72500	+	167.9 Hz	
Albany	DSTAR	W4MM	440.70000	+		
	FM	W4MM	146.73000	-		
	FM	W4MM	146.82000	-	110.9 Hz	
	FM	W4MM	444.50000	+		
Albany, Phoebe Putney Memorial						
	DSTAR	W4MM	144.96000	+		
Allenhurst	FM	KF4ZUR	147.27000	+	162.2 Hz	
Alma	FM	KM4GFV	145.25000	-	167.9 Hz	
Alpharetta	FM	K9RFA	224.58000	-	100.0 Hz	
Americus	FM	W4VIR	147.27000	+	131.8 Hz	
Appling	FM	K4KNS	145.19000	-	71.9 Hz	SERA
	FM	K4KNS	442.90000	+	71.9 Hz	SERA
Ashburn	FM	N4OME	145.35000	-	141.3 Hz	SERA
Athens	DSTAR	KJ4PXY	144.98000	+		
	DSTAR	KJ4PXY	440.63250	+		
	FM	KD4QHB	146.74500	-	123.0 Hz	
	FM	KD4AOZ	146.95500	-	123.0 Hz	
	FM	K4TQU	147.37000	+	127.3 Hz	
Atlanta	DMR/BM	W4JEW	438.80000	+	CC12	
	DMR/BM	W4DOC	444.82500	+	CC10	SERA
	DMR/BM	W4KIP	444.93750	+	CC1	
	DMR/BM	N4YCI	927.57500	902.57500	CC1	
	DMR/MARC	W7QO	442.45000	+	CC1	
	DMR/MARC	W7QO	443.02500	+	CC1	SERA
	DMR/MARC	KC6OVD	444.12500	+	CC0	
	DMR/MARC	K1DMR	444.43750	+	CC1	
	DSTAR	WX4ATL	444.56700	+		
	FM	K4CLJ	146.97000	-	100.0 Hz	SERA
	FM	WA4GBT	147.03000	+		
	FM	N4NFP	224.12000	-	151.4 Hz	
	FM	N4MTA	224.32000	-	123.0 Hz	
	FM	N4NEQ	224.44000	-	151.4 Hz	
	FM	N4MTA	224.50000	-	123.0 Hz	
	FM	K4RFL	224.96000	-	100.0 Hz	
	FM	W4CML	442.02500	+	127.3 Hz	SERA
	FM	W4CML	443.65000	+	123.0 Hz	
	FM	N4NEQ	444.05000	+	151.4 Hz	
	FM	W4PME	444.15000	+	100.0 Hz	
	FM	N4NEQ	444.77500	+	151.4 Hz	
	FM	WA4NNO	444.92500	+		
	FM	K5TEX	927.11250	902.11250	DCS 432	
	FM	KD4GPI	927.51250	902.51250		
	FM	KB4KIN	1292.00000	1272.00000		
Atlanta, Bank Of America Tower						
	DSTAR	W4DOC	145.35000	-		
	DSTAR	W4DOC	1282.60000	1262.60000		

GEORGIA 133

Location	Mode	Call sign	Output	Input	Access	Coordinator
Atlanta, Bank Of America Tower						
	FM	W4DOC	146.82000	-	146.2 Hz	
	FM	W4DOC	224.34000	-	146.2 Hz	
Atlanta, Buckhead	FM	WB4RTH	147.10500	+	110.9 Hz	SERA
	FM	WB4RTH	444.97500	+	100.0 Hz	SERA
Atlanta, Georgia Tech						
	FM	W4AQL	145.15000	-	167.9 Hz	SERA
Augusta	FM	K4KNS	53.03000	52.03000		
	FM	W4DV	145.11000	-	71.9 Hz	SERA
	FM	W4DV	145.29000	-	71.9 Hz	
	FM	W4QK	145.41000	-	71.9 Hz	
	FM	W4DV	147.18000	+	71.9 Hz	
	FM	K4KNS	444.67500	+	71.9 Hz	SERA
	FM	W4DV	444.95000	+	71.9 Hz	
Augusta, Bon Air Hotel						
	FM	KT4N	146.94000	-	146.2 Hz	
Austell	FM	WA4YUR	442.82500	+		SERA
Bainbridge	FM	W4DXX	443.00000	+	100.0 Hz	
Baldwin	FM	WD4NHW	147.18000	+		
	FM	WD4NHW	442.35000	+		SERA
Barnesville	FM	W8JI	147.22500	+	131.8 Hz	SERA
	FM	N4GWO	443.67500	+		
Between	DMR/MARC	N4TAW	443.73750	+	CC3	
	FM	WC4RG	147.27000	+		SERA
Blackshear	FM	KI4LDE	145.37000	-	141.3 Hz	
Blairsville	FM	KF4SKT	442.20000	+		
	FM	W6IZT	444.60000	-	100.0 Hz	
Blairsville, Rocky Top Mtn						
	FM	K5PRE	146.95500	-	100.0 Hz	
Blueridge	FM	KD4GRU	442.12500	+	146.2 Hz	
Bogart	FM	W4EEE	147.00000	+	85.4 Hz	
	FM	WW4GA	443.30000	+		
	FM	W4EEE	443.47500	+		
Boston	FM	W4UCJ	147.24000	+	141.3 Hz	
Bowlingbrook	FM	WB4NFG	146.83500	-	77.0 Hz	
Braselton	FM	WX4TC	146.62500	-	127.3 Hz	
	FM	W4CLE	441.82500	+	88.5 Hz	
Brent	FM	KJ4ZZF	146.50000	+		
Brunswick	DMR/BM	KG4PXG	442.20000	+	CC7	
Brunswick , Sidney Lanier Brid						
	FM	WX4BWK	145.33000	-	131.8 Hz	
Buckhead	FM	K5TEX	927.06250	902.06250	151.4 Hz	
Buford	FM	W4DDM	147.00000	-	123.0 Hz	SERA
	FM	W4DDM	443.40000	+	123.0 Hz	SERA
	FM	N4GJF	927.62500	902.62500		
Butler	FM	KC4TVY	145.31000	-		
Byron, Peach County Fire Depar						
	FM	WX4PCH	145.29000	-	82.5 Hz	SERA
Calhoun	FM	K4WOC	146.74500	-	100.0 Hz	
	FM	K4WOC	443.67500	+	100.0 Hz	
Canton	DMR/MARC	KD4KHO	434.40000	+	CC1	
	FM	WA4EOC	443.07500	+	107.2 Hz	SERA
Carnesville	FM	N4VNI	146.89500	-	100.0 Hz	
Carrollton	FM	W4FWD	146.64000	-	131.8 Hz	
	FM	WR4VR	442.77500	+	127.3 Hz	SERA
Cartersville, Pine Mt.						
	FM	W4CLM	443.17500	+	100.0 Hz	SERA
Cartersville, Ponder's Mtn						
	FM	W4CLM	147.24000	+	103.5 Hz	
Cedar Grove	FM	WA4FRI	147.15000	+	123.0 Hz	
Cedar Grove, Biskey Mtn						
	FM	KC4JNN	53.05000	52.05000	100.0 Hz	

134 GEORGIA

Location	Mode	Call sign	Output	Input	Access	Coordinator
Cedartown	FM	W4CMA	147.12000	+		SERA
Chatsworth	FM	N4YYD	224.24000	-	71.9 Hz	
	FM	KJ4SPI	443.80000	+	141.3 Hz	
	FM	N4DMX	444.85000	+	141.3 Hz	
	FM	NS4U	927.61250	902.61250		SERA
Chatsworth, Fort Mountain						
	FM	W4DRC	145.23000	-	141.3 Hz	
	FM	W4DRC	443.00000	+		
Chickamauga	DMR/MARC	K4KR	438.00000	+	CC1	
Clarkesville	FM	K4HCA	147.12000	+	123.0 Hz	
Claxton	FM	W4CLA	147.07500	+	123.0 Hz	
Clayton	FM	W1CP	442.82500	+	162.2 Hz	
Clayton, Rainey Mountain						
	FM	KK4BSA	444.50000	+		
Clermont	DMR/MARC	KA3JIJ	444.81250	+	CC0	
Cleveland	DSTAR	K4GAR	440.51250	+		
	FM	K4VJM	442.62500	+	100.0 Hz	
Cleveland, Long Mountain						
	FM	K4GAR	146.91000	-	100.0 Hz	
	FM	K4GAR	443.55000	+		
Cochran	FM	W4MAZ	53.01000	52.01000	77.0 Hz	
Colbert	FM	N4ALE	147.30000	+	123.0 Hz	
College Park	DMR/MARC	KK4EQB	442.92500	+	CC4	
Columbus	DMR/BM	WB4ULK	441.97500	+	CC1	SERA
	FM	W4CVY	146.88000	-	123.0 Hz	
Columbus, WXTX CH54 Tower						
	FM	WX4RUS	146.74500	-	123.0 Hz	
	FM	W9TVM	442.10000	+	123.0 Hz	
Columbus, WXTX Tower						
	FM	W4CVY	442.20000	+		
Concord	DMR/BM	WB4GWA	442.31250	+	CC1	
	FM	WB4GWA	145.25000	-	110.9 Hz	
	FM	WB4GWA	224.46000	-	110.9 Hz	
	FM	WB4GWA	443.40000	+	110.9 Hz	
Conley	FM	N4MNA	53.65000	52.65000	100.0 Hz	
	FM	N4MNA	224.28000	-		
Conyers	FM	WB4JEH	51.55000	50.55000	151.4 Hz	
	FM	WB4JEH	53.55000	52.55000	151.4 Hz	
	FM	K1KC	146.61000	-	103.5 Hz	SERA
	FM	KF4GHF	147.21000	+	162.2 Hz	
	FM	WB4JEH	442.55000	+	151.4 Hz	
	FM	WX4RCA	444.30000	+	131.8 Hz	
	FM	K1KC	444.55000	+	103.5 Hz	SERA
	FM	KC4ELV	444.75000	+	162.2 Hz	
	FM	WX4RCA	449.30000	-	131.8 Hz	SERA
Cornelia	FM	WB4VAK	444.27500	+	100.0 Hz	
Covington	DMR/BM	WA4ASI	444.80000	+	CC2	SERA
Covington, Fibervisions Buildi						
	FM	WA4ASI	443.35000	+	88.5 Hz	SERA
Covington, Foresty Fire Tower						
	FM	WA4ASI	146.92500	-	88.5 Hz	SERA
Crawford	FM	KD4FVI	53.33000	52.33000	88.5 Hz	
Cumming	DMR/MARC	W4OO	440.52500	+	CC1	
	DMR/MARC	W9SH	444.12500	+	CC2	
	DMR/MARC	W4CBA	444.62500	+	CC1	SERA
	DSTAR	W4CBA	145.20000	-		SERA
	DSTAR	W4CBA	444.35000	+		SERA
Cumming, Sawnee Mountain						
	FM	WB4GQX	147.15000	+	141.3 Hz	SERA
	FM	WB4GQX	441.90000	+	141.3 Hz	SERA
Dacula	DMR/BM	KN4OAJ	443.92500	+	CC5	SERA

GEORGIA   135

Location	Mode	Call sign	Output	Input	Access	Coordinator
Dacula, Hog Mountain						
	FM	KD4YDD	145.33000	-	103.5 Hz	
Dahlonega, Bisky Mountain						
	FM	N4KHQ	146.83500	+	100.0 Hz	SERA
Dahlonega, Black Mountain						
	FM	N4KHQ	443.10000	+	100.0 Hz	
Dahlonega, High House						
	FM	N4KHQ	224.48000	-	100.0 Hz	
Dallas	DMR/MARC	KJ4KKB	442.95000	+	CC3	
	DMR/MARC	KJ4KKB	443.80000	+	CC3	
	FM	KJ4ZZF	146.50000		88.5 Hz	
	FM	N4YDX	224.18000	-	71.9 Hz	
	FM	N4YEA	224.54000	-		
	FM	K4CGA	442.67500	+	DCS 411	
Dallas, Macland	FM	WB4QOJ	146.95500	-	77.0 Hz	SERA
Dalton	DMR/BM	W4DMM	442.17500	+	CC1	
	FM	N4BZJ	147.13500	+	141.3 Hz	
	FM	N4BZJ	224.46000	-	141.3 Hz	
	FM	N4BZJ	224.68000	-	141.3 Hz	
	FM	N4KVC	224.74000	-	141.3 Hz	
Dalton, Dug Gap Mountain						
	DSTAR	KA4RVT	145.33000	-		
	DSTAR	KA4RVT	444.50000	+		
Darien	FM	KT4J	146.98500	-	100.0 Hz	
Decatur	DMR/BM	N4MPC	444.90000	+	CC1	
Decatur, Exchange Park						
	FM	W4BOC	145.45000	-	107.2 Hz	
	FM	W4BOC	224.76000	-		
	FM	W4BOC	444.25000	+	131.8 Hz	SERA
Doerun	FM	KG4ABK	147.22500	+		
Doraville	DSTAR	WB4HRO	440.71250	+		
	FM	WB4HRO	1284.45000	1264.45000		
Douglas	DMR/BM	W4SCD	443.13750	+	CC3	
	FM	KE4ZRT	147.04500	+	141.3 Hz	
	FM	W4JSF	147.16500	+	141.3 Hz	SERA
	FM	AD4EQ	147.31500	+	141.3 Hz	
	FM	KE4ZRT	443.00000	+	141.3 Hz	
Douglasville	FM	K4NRC	145.11000	-	88.5 Hz	SERA
Dover Bluff	FM	KG4PXG	53.11000	52.11000	100.0 Hz	
	FM	KG4PXG	146.68500	-	100.0 Hz	
Dry Branch	FM	KC4TVY	444.65000	+	77.0 Hz	
Dublin	DSTAR	KJ4YNR	145.18000	-		
	FM	KD4IEZ	147.33000	+	77.0 Hz	
	FM	WA4HZX	147.36000	+		
Eastman	FM	KC4YNB	145.21000	-	103.5 Hz	
Eastman, EMA Office Water Tank						
	FM	KB4MQ	444.85000	+		
Eatonton	FM	NZ2X	53.19000	52.19000	156.7 Hz	
	FM	K4EGA	146.65500	-	186.2 Hz	SERA
	FM	K4PAR	443.17500	+	186.2 Hz	
	FM	K4EGA	444.42500	+	186.2 Hz	
Eden	FM	K4VYX	145.19000	-		
Elberton	FM	KI4CCZ	145.21000	-	118.8 Hz	SERA
	FM	KI4CCZ	444.70000	+	118.8 Hz	SERA
Ellerslie	DMR/BM	WB4ULK	444.61250	+	CC1	SERA
Ellijay	FM	W4HHH	145.17000	-	100.0 Hz	SERA
	FM	KC4ZGN	146.98500	-	77.0 Hz	SERA
	FM	KC4ZGN	442.70000	+	77.0 Hz	SERA
Ellijay, Coosawattee River Res						
	DSTAR	W4HHH	443.98750	+		SERA
Emerson	FM	N4RSW	443.30000	+		
	FM	AE4JO	443.42500	+	103.5 Hz	

136　GEORGIA

Location	Mode	Call sign	Output	Input	Access	Coordinator
Emerson, Lake Point						
	FM	N4RSW	146.46000		88.5 Hz	
	FM	N4RSW	446.17500	-		
Evans	FM	K4KNS	146.98500	-	71.9 Hz	
	FM	W4QK	444.90000	+	71.9 Hz	
Fairmount	FM	AB4LZ	146.68500	-	167.9 Hz	
Fayetteville	DMR/BM	W0WHS	442.50000	+	CC1	
	DMR/BM	KN4RQL	446.00000	-	CC15	
	DSTAR	KK4GQ	442.56200	+		
	DSTAR	KK4GQ	444.43750	+		
	FM	KK4GQ	145.21000	-	131.8 Hz	
	FM	KK4GQ	146.68500	-	131.8 Hz	
	FM	AG4ZR	224.56000	-	131.8 Hz	
	FM	W4PSZ	444.60000	+	77.0 Hz	
Folkston	FM	KD4GEY	146.79000	-	141.3 Hz	
Forsyth	DMR/BM	KK4JPG	444.75000	+	CC1	
	DMR/BM	KK4JPG	444.97500	+	CC1	
	DSTAR	KK4JPG	146.48000	145.08000		
	FM	KK4JPG	449.75000	-	100.0 Hz	
Fort Benning	DMR/BM	W8JVF	441.80000	+	CC1	SERA
Freehome	DMR/BM	N5FL	927.95000	902.95000	CC11	
Gainesville	DMR/BM	KA3JIJ	145.31000	-	CC1	SERA
	DMR/MARC	KA3JIJ	444.95000	+	CC1	SERA
	FM	AA4BA	145.08000	146.48000		
	FM	AA4BA	441.86200	+		
	FM	AA4BA	444.07500	+	100.0 Hz	
	FM	W4CBA	444.35000	+	131.8 Hz	
Gainesville, Wauka Mountain						
	FM	W4ABP	146.67000	-	131.8 Hz	SERA
Gray	FM	WB4JOE	145.37000	-	88.5 Hz	
	FM	N5BI	443.70000	+	100.0 Hz	SERA
Gray, Round Oak	FM	WB4JOE	145.37000	-	88.5 Hz	
Grayson	FM	W4GR	224.58000	-	100.0 Hz	
Graysville	FM	N4YAV	444.52500	+	131.8 Hz	
Griffin	FM	WB4GWA	145.39000	-	110.9 Hz	
	FM	K4HYB	146.91000	-	88.5 Hz	
	FM	NQ4AE	443.55000	+		SERA
Griffin, Spalding Hospital						
	FM	W4AMI	145.48000	-		
Guyton	FM	W4ECA	146.74500	-	97.4 Hz	
Hannah's Mill	FM	K4NRC	147.36000	+	88.5 Hz	SERA
Hawkinsville	FM	WR4MG	224.82000	-	107.2 Hz	
Helen	FM	K4PE	444.42500	+	127.3 Hz	
Hiawassee, Bell Mtn						
	FM	KI4ENN	146.86500	+	151.4 Hz	
High Point	FM	KB4VAK	443.45000	+	77.0 Hz	
Hinesville	FM	KG4OGC	147.01500	+		
	FM	KG4OGC	444.85000	+		
Hiram	FM	W4TIY	224.70000	-	100.0 Hz	
Howardville	FM	W4CLM	147.24000	+		
Irwington	FM	WB4NFG	443.27500	+	77.0 Hz	SERA
Irwinton	FM	WB4NFG	147.24000	+	77.0 Hz	
	FM	WB4NFG	444.92500	+	77.0 Hz	
Jackson	FM	WX4BCA	443.32500	+	131.8 Hz	
Jackson, Sylvan Grove Hospital						
	FM	WX4BCA	147.28500	+	131.8 Hz	
Jasper	DMR/MARC	N8WHG	145.37000	-	CC3	
	DSTAR	KJ4PYC	145.08000	146.48000		
	DSTAR	KJ4PYC	440.66250	+		
	FM	KB4IZF	145.37000	-	103.5 Hz	
	FM	K4UFO	146.70000	-	123.0 Hz	
	FM	KC4AQS	224.40000	-	100.0 Hz	

GEORGIA 137

Location	Mode	Call sign	Output	Input	Access	Coordinator
Jasper	FM	KC4AQS	224.60000	-	100.0 Hz	SERA
	FM	KC4AQS	443.37500	+	100.0 Hz	
	FM	KB4IZF	444.37500	+		
Jasper, Biskey Mountain						
	FM	KC4JNN	224.16000	-	100.0 Hz	SERA
	FM	KC4AQS	443.50000	+	100.0 Hz	
Jasper, Burnt Mountain						
	FM	W4RRG	224.98000	-		
Jasper, Little Hendricks Mount						
	FM	N3DAB	147.19500	+	77.0 Hz	
Jasper, Mount Oglethorpe						
	P25	K4SJR	441.67500	+	100.0 Hz	SERA
	P25	W3CP	927.02500	902.02500	DCS 073	SERA
Jasper, Mt. Oglethorpe						
	FM	KC4AQS	146.80500	+	100.0 Hz	
Jasper, Sassafras Mtn						
	FM	KB3KHP	443.95000	+	131.8 Hz	
Jesup	DMR/BM	AB4KK	145.75000		CC3	
	DMR/BM	AB4KK	438.80000		CC3	
	DMR/BM	AB4KK	446.87500	-	CC3	
	DMR/MARC	KD4GGY	441.81250	+	CC1	
	FM	N4ZON	146.86500	-	141.3 Hz	
	FM	N4PJR	146.92500	-	141.3 Hz	
	FM	KE4ZFR	441.67500	+	131.8 Hz	
	FM	N4PJR	444.92500	+	141.3 Hz	
Kingsland	DMR/BM	N9USN	442.85000	+	CC3	
	DMR/MARC	K4QHR	442.11250	+	CC7	
	DMR/MARC	W1KFR	444.62500	+	CC3	SERA
	FM	W4ULB	146.50000			
	FM	W4NAS	146.83500	-	127.3 Hz	
	FM	K4QHR	146.89500	-	127.3 Hz	SERA
Kingsland, Fire Station 10						
	FM	N6EMA	147.19500	+	118.8 Hz	SERA
Kingston	FM	AF4PX	444.12500	+		
Knoxville	FM	N4PQR	443.52500	+	114.8 Hz	
Lagrange	DMR/MARC	WB4BXO	147.33000	747.33000	CC1	
	FM	WB4BXO	224.72000	-		
Lake Park	FM	WR4SG	147.13500	+	141.3 Hz	
Lakemont	FM	W4WCR	443.15000	+	127.3 Hz	
	FM	N4ZRF	444.75000	+		
Lavonia	FM	K4NVG	146.71500	-	100.0 Hz	
	FM	K4NVG	442.47500	+	203.5 Hz	
	FM	N4VNI	443.20000	+	151.4 Hz	SERA
Lawrenceville	DMR/BM	AI1U	442.70000	+	CC1	
	DMR/BM	N5JMD	442.90000	+	CC1	
	DMR/MARC	AI1U	442.53750	+	CC7	
	DSTAR	WD4STR	145.06000	+		
	DSTAR	WD4STR	440.55000	+		
	DSTAR	WD4STR	1282.55000	1262.55000		
	FM	W4GR	53.11000	52.11000	82.5 Hz	
	FM	WB4HJG	442.85000	+	82.5 Hz	
	FM	N5JMN	442.90000	+	100.0 Hz	
	FM	K4HQV	444.20000	+	100.0 Hz	
Lawrenceville, Georgia Gwinnet						
	FM	W4GGC	442.22500	447.32500		
Lookout Mountain	DMR/MARC	N4LMC	145.35000	-	CC1	
	DMR/MARC	N4LMC	441.87500	+	CC1	
	DMR/MARC	N4LMC	442.65000	+	CC1	
Lookout Mountain, Lookout Moun						
	FM	W4GTA	145.35000	-	100.0 Hz	SERA
Lookout Mountain, Lookout Mt						
	FM	W4RRG	224.56000	-	100.0 Hz	SERA

138   GEORGIA

Location	Mode	Call sign	Output	Input	Access	Coordinator
Lula	FM	WB4HJG	53.89000	52.89000	82.5 Hz	
Mableton	FM	W4JLG	442.32500	+		
Macon	DSTAR	WX4EMA	145.34000	-		
	FM	AA4RI	145.43000	-	88.5 Hz	
	FM	WA4DDI	146.77500	-		
	FM	K4PDQ	146.80500	-	77.0 Hz	SERA
	FM	KD4UTQ	146.89500	-	88.5 Hz	
	FM	WX4EMA	147.01500	+	88.5 Hz	SERA
	FM	WA4DDI	147.06000	+	141.3 Hz	
	FM	WA4DDI	224.64000	-	88.5 Hz	
	FM	WA4DDI	442.27500	+	123.0 Hz	
	FM	WX4EMA	443.07500	+	88.5 Hz	
	FM	W4OCL	444.27500	+	123.0 Hz	
	FM	WA4DDI	444.70000	+	103.5 Hz	
Macon, Round Oak						
	FM	AA4RI	53.43000	52.43000	88.5 Hz	SERA
Madison, Confederate Rd & Brow						
	FM	WB4DKY	146.86500	-	179.9 Hz	
Madison, Morgan County Public						
	FM	WR4MC	443.75000	+	118.8 Hz	
Madras / Newnan	FM	K4NRC	147.16500	+	131.8 Hz	SERA
	FM	K4NRC	442.25000	+		SERA
Mansfield, South Newton County						
	FM	K4IO	443.12500	+	186.2 Hz	SERA
Mansfield, White Pine Ln						
	FM	K4IO	145.23000	-	151.4 Hz	SERA
Marietta	DMR/BM	W4KIP	146.73000	-	CC1	SERA
	DMR/BM	W4KIP	444.77500	+	CC1	
	DMR/MARC	W4OO	442.97500	+	CC1	
	DMR/MARC	KD4Z	442.97500	+	CC1	SERA
	DMR/MARC	N4IRR	927.70000	902.70000	CC1	
	FM	WC4RAV	147.53500	146.63500	103.5 Hz	SERA
	FM	KE4QFG	224.26000	-	110.9 Hz	
	FM	N1KDO	442.00000	+	91.5 Hz	SERA
	FM	WK4E	442.42500	+	107.2 Hz	
	FM	WC4RAV	443.45000	+	103.5 Hz	SERA
Marietta, Fort Hill Tower						
	FM	KK4OIO	145.49000	-	107.2 Hz	SERA
Marietta, Sweat Mountain						
	DSTAR	W4BTI	440.68750	+		
	FM	W4BTI	146.88000	-		
Marietta, Sweat Mtn						
	FM	W4PME	224.62000	-	100.0 Hz	SERA
Marlow	FM	W1MED	442.47500	+	114.8 Hz	
Martinez	FM	WE4GW	441.90000	+	123.0 Hz	
McDonough	DMR/MARC	KE4UAS	444.87500	+	CC1	SERA
	FM	KI4FVI	146.71500	-	146.2 Hz	SERA
	FM	W4NOC	927.13750	902.13750	103.5 Hz	SERA
McIntyre	FM	KC4TVY	53.73000	52.73000	77.0 Hz	
Metcalf	FM	W4UCJ	147.19500	+	141.3 Hz	
Midland	FM	WB4ULK	224.44000	-		SERA
Midway	FM	KA4CID	443.30000	+	97.4 Hz	SERA
Milledgeville	FM	W4PCF	146.70000	-	67.0 Hz	SERA
	FM	W4PCF	147.13500	+	123.0 Hz	SERA
Milton	FM	K5TEX	927.65000	902.65000		
Monroe	FM	WC4RG	442.05000	+	88.5 Hz	
Montezuma	FM	KI4BEO	146.64000	-	97.4 Hz	
Morganton, Brawley Mountain						
	DSTAR	KM4MAD	145.40000	-		
Morganton, Brawley Mtn						
	FM	W4HBS	443.22500	+	100.0 Hz	
Moultrie	FM	WD4KOW	146.79000	-	141.3 Hz	

GEORGIA   139

Location	Mode	Call sign	Output	Input	Access	Coordinator
Moultrie	FM	N4JMD	443.32500	+	114.8 Hz	
	FM	W1BPP	444.62500	+	123.0 Hz	
	FM	N4JMD	927.01250	902.01250	100.0 Hz	SERA
Newnan	DMR/MARC	WX4SKY	442.25000	+	CC1	
Newnan, Georgia	FM	K4NRC	145.13000	-	156.7 Hz	SERA
Nicholls	DMR/BM	W4JSF	444.41250	+	CC3	SERA
Norcross	FM	W4GR	442.10000	+	100.0 Hz	SERA
Ola	FM	K1KC	444.55000	+	103.5 Hz	
Omaha	FM	KI4VDP	443.72500	+	77.0 Hz	
Parrott	FM	WG4JOE	147.36000	+	173.8 Hz	
Peachtree City	FM	W4PSZ	442.50000	+	77.0 Hz	
Pembroke	DSTAR	KJ4GGV	145.28000	-		SERA
	DSTAR	KJ4GGV	440.70000	+		SERA
	DSTAR	KJ4GGV	1282.70000	1302.70000		SERA
Pembroke, WVAN Tower						
	FM	KF4DG	147.10500	+		SERA
Perry	FM	WR4MG	146.95500	-	107.2 Hz	SERA
Pine Mountain	DSTAR	KJ4KLE	144.92000	147.42000		
	FM	WB4ULJ	145.19000	-	110.9 Hz	
	FM	WB4ULJ	224.36000	-		
	FM	WB4ULJ	224.66000	-		
	FM	WB4ULJ	444.40000	+		
Powder Springs	DMR/BM	N8WHG	440.80000	+	CC3	
Quitman	FM	WA4NKL	146.88000	-		
	FM	WA4NKL	444.60000	+	141.3 Hz	
Ray City	FM	WR4SG	224.22000	-	141.3 Hz	
Reno	FM	KE4URL	145.17000	-	141.3 Hz	
Riceboro	FM	KG4OGC	145.47000	-		
Ringgold	FM	W4ABZ	146.71500	-	67.0 Hz	
	FM	KK4LPW	442.35000	+		SERA
	FM	W4BAB	443.92500	+		
Rockmart	FM	WX4PCA	443.47500	+	77.0 Hz	SERA
Rome	FM	N4EBY	147.30000	+	100.0 Hz	
	FM	WB4LRA	224.64000	-	141.3 Hz	
Rome, Mount Alto	FM	W4VO	146.94000	-	88.5 Hz	
Rome, Mt Alto	FM	W4VO	147.39000	+		SERA
Rome, Mt. Alto	DSTAR	W4VO	145.38000	-		
	DSTAR	W4VO	440.62500	+		SERA
	FM	WA4OKJ	443.20000	+	100.0 Hz	
Roswell	FM	NF4GA	147.06000	+	100.0 Hz	SERA
	FM	W4PME	443.15000	+	100.0 Hz	
Sandersville	FM	K4GK	145.27000	-	77.0 Hz	
Sandy Springs	DMR/MARC	KE4OKD	441.95000	+	CC0	SERA
	FM	NF4GA	145.47000	-	100.0 Hz	
Savannah	DMR/MARC	K4VYX	442.81250	+	CC3	
	FM	K3SRC	146.85000	-	100.0 Hz	
	FM	W4HBB	146.97000	-	123.0 Hz	
	FM	W4LHS	147.21000	+		
	FM	W4LHS	147.33000	+	203.5 Hz	
	FM	W4LHS	442.70000	+		
	FM	WD4AFY	444.00000	+		
Sharpsburg	FM	AG4ZR	29.64000	-	146.2 Hz	
Snellville	DMR/MARC	W8RED	442.60000	+	CC3	
	FM	W4GR	147.25500	+	107.2 Hz	
	FM	W4CSX	224.92000	-	100.0 Hz	
	FM	W4GR	444.52500	+	82.5 Hz	
Snellville, Lanier Mountain						
	FM	W4CML	444.02500	+	127.3 Hz	
Snellville, Lanier Mountain, B						
	FM	W4GR	147.07500	+	82.5 Hz	SERA
Sparta	DMR/BM	KC4YAP	147.19500	+	CC4	
Split Silk	FM	KD4HLV	444.10000	+	77.0 Hz	

140   GEORGIA

Location	Mode	Call sign	Output	Input	Access	Coordinator
Statesboro, WWNS Tower						
	FM	KF4DG	147.39000	+		
Statham, Hwy 53/Hebron Church						
	FM	WW4T	442.30000	-		
Stockbridge	DMR/MARC	KJ4KPX	443.22500	+	CC1	
Stockbridge, Piedmont Henry Ho						
	FM	KJ4KPY	443.22500	+	146.2 Hz	SERA
Stockbridge, Piedmont Henry Me						
	FM	KJ4KPY	145.17000	-	146.2 Hz	SERA
Stone Mountain	DMR/MARC	W4BOC	441.81250	+	CC1	SERA
	FM	W4BOC	146.76000	-	107.2 Hz	
	FM	KG4LMT	927.71250	902.71250	DCS 223	
Stone Mountain, Stone Mountain						
	DSTAR	WX4GPB	145.35000	-		
	DSTAR	WX4GPB	440.60000	+		
Summerville	FM	K4PS	53.75000	52.75000	127.3 Hz	
	FM	W4RLP	147.22500	+	100.0 Hz	
Sumner	FM	W4CCS	444.90000	+	141.3 Hz	
Sumner , WOBB Tower						
	FM	W4PVW	147.28500	+	141.3 Hz	SERA
Swainsboro	FM	K4VYX	146.79000	-		
Sycamore	FM	KF4BI	444.92500	+	141.3 Hz	
Thomaston, Hagans Mountain						
	FM	W4OHH	147.39000	+	131.8 Hz	
Thomaston, URMC Hospital						
	FM	W4OHH	444.45000	+	131.8 Hz	
Thomasville	DSTAR	KJ4PYB	440.65000	+		
	DSTAR	KJ4PYB	1248.75000	1268.75000		
	FM	W4UCJ	145.37000	-	141.3 Hz	SERA
	FM	W4UCJ	147.06000	+		
	FM	WR4SG	224.32000	-	141.3 Hz	
	FM	W4UCJ	442.60000	+	141.3 Hz	SERA
Tifton	FM	W4PVW	444.56250	+		
Tifton, MRS Warehouse						
	FM	KE4RJI	444.87500	+	141.3 Hz	SERA
Tifton, Tiftnet Tower						
	DSTAR	W4PVW	145.12000	-		SERA
	DSTAR	W4PVW	1282.65000	1262.65000		SERA
Toccoa	FM	K4TRS	145.25000	-	71.9 Hz	
	FM	KR4CW	147.33000	+	127.3 Hz	
	FM	W4BNG	442.50000	+	88.5 Hz	
Trenton	FM	K4GC	441.85000	+		
Trenton, Sand Mountain Water T						
	FM	K4SOD	146.76000	-		SERA
Twin City	FM	N4SFU	146.71500	-		
	FM	N4SFU	147.00000	+	156.7 Hz	SERA
	FM	N4SFU	444.25000	+		
Union	FM	WD4LUQ	224.88000	-	77.0 Hz	
Valdosta	DMR/MARC	WR4SG	442.71250	+	CC1	
	DSTAR	W4VLD	145.15000	-		
	DSTAR	W4VLD	1282.82500	1262.82500		
	FM	W4VLD	146.76000	-	141.3 Hz	SERA
	FM	WR4SG	224.46000	-	141.3 Hz	
	FM	W4VLD	443.71250	+		
	FM	W4VLD	444.70000	+	141.3 Hz	
Vidalia	DMR/MARC	KG4BKO	444.98750	+	CC1	
	FM	K4HAO	146.62500	-	88.5 Hz	SERA
Vidalia , Hwy 280 Water Tank						
	FM	KE4UHF	444.57500	+	88.5 Hz	SERA
Vienna	FM	K4WDN	147.37500	+	131.8 Hz	
Villa Rica	DSTAR	WR4VR	145.24000	-		SERA
	FM	WR4VR	147.18000	+	127.3 Hz	SERA

GEORGIA 141

Location	Mode	Call sign	Output	Input	Access	Coordinator
Villa Rica	FM	WR4VR	224.30000	-	127.3 Hz	SERA
Villanow	FM	N2YYP	443.52500	+		SERA
Waleska	DMR/BM	W4KIP	444.05000	+	CC1	SERA
	FM	KD4ALC	224.14000	-		
	FM	KR4FN	224.20000	-	100.0 Hz	SERA
	FM	KD4DXR	441.35000	+		
Waleska, Pine Log						
	P25	K9APD	145.43000	-		SERA
Waleska, Pine Log Mountain						
	FM	K4PLM	53.29000	52.29000	192.8 Hz	
	FM	WD4OVN	53.45000	52.45000		
	FM	KG4VUB	145.27000	-	100.0 Hz	SERA
	FM	K4AIS	224.52000	-		
	FM	KJ4JJX	224.94000	-		
	FM	K4PLM	443.85000	+	192.8 Hz	
Waleska, Pine Log Mtn						
	DSTAR	KI4GOM	145.02000	146.42000		
Waleska, Pine Log Mtn.						
	FM	KK4YLX	147.01500	+	100.0 Hz	SERA
Warm Springs	FM	KD4BDB	53.23000	52.23000	97.4 Hz	
	FM	N4UER	146.98500	-		
	FM	N4UER	442.40000	+		
Warner Robins	FM	WR4MG	29.66000	-		
	FM	WR4MG	53.79000	52.79000		
	FM	WB4BDP	147.18000	+	107.2 Hz	
	FM	WR4MG	442.90000	+	107.2 Hz	
Warner Robins, Houston County						
	FM	WR4MG	147.30000	+	107.2 Hz	SERA
Warner Robins, Houston Medical						
	FM	WM4B	146.67000	-	82.5 Hz	SERA
	FM	WM4B	443.15000	+	82.5 Hz	SERA
Warner Robins, Water Tower						
	FM	WA4ORT	146.85000	-		SERA
Watkinsville	FM	KD4AOZ	53.57000	52.57000	123.0 Hz	
	FM	KD4AOZ	53.71000	52.71000		
	FM	KD4AOZ	224.42000	-	123.0 Hz	
	FM	KD4AOZ	444.72500	+	123.0 Hz	
Watkinsville, GA	FM	KD4AOZ	147.04500	+	123.0 Hz	
Waycross	DMR/BM	KK4GXP	448.07500	-	CC1	
	DMR/MARC	KM4DND	444.02500	+	CC3	SERA
	FM	AE4PO	53.41000	52.41000	141.3 Hz	
	FM	KM4DND	145.27000	-	141.3 Hz	
	FM	KF4SUY	146.64000	-	141.3 Hz	
	FM	KM4DND	146.82000	-		
Waynesboro	DSTAR	K4BRK	145.23000	-	71.9 Hz	SERA
	DSTAR	K4BRK	444.10000	+	71.9 Hz	SERA
	FM	K4BRK	146.64000	-	71.9 Hz	SERA
	FM	K4BRK	442.80000	+		SERA
Willow Wind	FM	W4FLQ	443.32500	+		
Winder	FM	WR4BC	145.13000	-	100.0 Hz	
Winder, Winder Fire Station #1						
	FM	W4WYI	443.52500	+	100.0 Hz	SERA
Woodbury	FM	WB4GWA	443.80000	+	110.9 Hz	
Woodstock	FM	KF4RMB	442.27500	+	67.0 Hz	
	FM	KE4SJO	444.22500	+		
Wrens	FM	KT4N	147.12000	+		
Wrightsville	FM	WA4RVB	146.94000	-		
	FM	WA4RVB	443.02500	+	156.7 Hz	
Young Harris	FM	W4NGT	147.21000	+	100.0 Hz	
	FM	NP2Y	224.66000	-	100.0 Hz	

142   GUAM

Location	Mode	Call sign	Output	Input	Access	Coordinator
GUAM						
Asan, Nimitz Hill	FM	AH2G	146.94000	-		
Barragada Heights						
	FM	AH2G	146.91000	-		
HAWAII						
Aiea	DMR/BM	NH7QH	444.80000	+	CC1	HSRAC
Aiea, Waiau District Park						
	FM	WH6PD	147.16000	+	103.5 Hz	
Central	DMR/MARC	KH6DT	444.95000	+	CC0	
Ewa	FM	WH6MK	147.38000	+		
	FM	WH6PD	224.50000	-	103.5 Hz	
Ewa By Gentry	FM	WH6PD	146.60000	-	103.5 Hz	HSRAC
Ewa, Ewa By Gentry						
	FM	WH6PD	442.02500	+		
Ewa, Leahi Hospital						
	FM	KH6CY	443.10000	+	103.5 Hz	
Glenwood	FM	AH6GG	52.20000	51.20000	141.3 Hz	HSRAC
	FM	AH6GG	147.00000	+	141.3 Hz	HSRAC
	FM	AH6GG	442.02500	+	156.7 Hz	
Hakalau	FM	AF7DX	443.20000	+	100.0 Hz	
Haleakala	FM	KH6COM	146.94000	-	110.9 Hz	HSRAC
	FM	KH6H	147.02000	+	103.5 Hz	HSRAC
	FM	NH6XO	147.08000	+	123.0 Hz	HSRAC
	FM	N6HPQ	147.26000	+		
	FM	AH6GR	442.10000	+	136.5 Hz	HSRAC
	FM	AH6GR	444.22500	+	110.9 Hz	HSRAC
Haleiwa	FM	KH6LJ	146.90000	-		
	FM	KH6LJ	224.02000	-		
Hana	DMR/BM	AH6GR	442.30000	+	CC1	HSRAC
Hanalei, Waimea Canyon						
	FM	NH6HF	147.10000	+	100.0 Hz	
Hanamaulu	FM	KH6BFU	223.98000	-		
Hilo	DMR/MARC	WH6FM	444.90000	+	CC4	HSRAC
	FM	AH6GG	146.66000	-	141.3 Hz	
	FM	AF7DX	147.28000	+		
	FM	AH6JA	444.15000	+		HSRAC
	FM	WH6FM	444.72500	+		HSRAC
	FM	AH6JA	444.75000	+		HSRAC
	FM	WH6FM	444.77500	+	123.0 Hz	
	FM	KH6KL	444.92500	+		HSRAC
Hilo, Hilo Shopping Ctr						
	FM	KH6QAI	145.35000	-		HSRAC
Hilo, Waiakea High School						
	FM	WH6HQ	147.10000	+		HSRAC
Honaunau	DMR/BM	KH7MS-1	444.95000	+	CC1	
Honolulu	DMR/BM	WH6CZB	441.10000	+	CC1	
	DMR/BM	AH6OO	443.20000	+	CC1	
	DMR/BM	WH6CZB	444.50000	+	CC5	HSRAC
	DMR/MARC	KH6OCD	146.86000	-	CC1	
	DMR/MARC	KH6HPZ	147.06000	+	CC1	
	DMR/MARC	AH6CP	147.30000	+	CC5	HSRAC
	DMR/MARC	AH6BZ	443.20000	+	CC1	
	DMR/MARC	KH6FV	443.42500	+	CC2	
	DMR/MARC	KH6FV	443.82500	+	CC1	
	DMR/MARC	AH6KD	444.20000	+	CC0	
	DMR/MARC	AH6CP	444.30000	+	CC0	
	DMR/MARC	AH6CP	444.40000	+	CC0	
	DMR/MARC	AH6HI	444.42500	+	CC0	HSRAC
	DMR/MARC	AH6Q	444.57500	+	CC1	
	DMR/MARC	AH6CP	444.90000	+	CC0	HSRAC

HAWAII   143

Location	Mode	Call sign	Output	Input	Access	Coordinator
Honolulu	FM	WH6F	53.03000	52.03000		HSRAC
	FM	KH7EC	145.19000	-		
	FM	AH7HI	145.21000	-		
	FM	WH6CZP	145.39000	-		
	FM	WH6F	145.47000	-		HSRAC
	FM	KH6JUU	146.84000	-		HSRAC
	FM	WH6DIG	146.92000	-		HSRAC
	FM	WH6DIG	147.18000	+		HSRAC
	FM	WH7MN	147.34000	+		HSRAC
	FM	KH6OJ	223.94000	-		HSRAC
	FM	NH7ZD	224.74000	-		HSRAC
	FM	KH7NM	224.94000	-		
	FM	KH6MEI	442.30000	+	103.5 Hz	HSRAC
	FM	KH6AZ	442.80000	+		HSRAC
	FM	KH7EC	443.00000	+	103.5 Hz	
	FM	KH7EC	443.02500	+		
	FM	KH7TX	443.20000	+		
	FM	KH6FV	443.25000	+	114.8 Hz	
	FM	KH6OJ	443.45000	+	103.5 Hz	
	FM	WH6HR	443.62500	+		
	FM	WH6UG	443.67500	+		HSRAC
	FM	KH7TK	443.95000	+	118.8 Hz	HSRAC
	FM	NH6WP	444.00000	+		HSRAC
	FM	KH6OCD	444.05000	+		HSRAC
	FM	WH6CZB	444.20000	+		HSRAC
	FM	WH6FM	444.72500	+	123.0 Hz	HSRAC
	FM	AH7HI	444.82500	+		HSRAC
	FM	KH6MP	444.87500	+		HSRAC
	FM	WH6DIG	444.92500	+		HSRAC
	FM	AH6CP	925.60000	922.00000		HSRAC
Honolulu, Diamond Head						
	FM	WH6CZB	146.88000	-	88.5 Hz	HSRAC
	FM	AH6RH	147.06000	+	103.5 Hz	HSRAC
	FM	WR6AVM	147.26000	+		HSRAC
Honolulu, Municipal Building						
	FM	WH6CZB	146.98000	-	88.5 Hz	HSRAC
Honolulu, Pearl Harbor						
	FM	KH6BB	146.78000	-		HSRAC
Honolulu, Round Top						
	FM	KH6MEI	147.10000	+		
	FM	KH6HFD	443.42500	+	114.8 Hz	HSRAC
	FM	NH6XO	443.77500	+	123.0 Hz	HSRAC
	FM	KH6HFD	443.82500	+	114.8 Hz	HSRAC
	FM	KH6MEI	444.70000	+	100.0 Hz	HSRAC
	FM	KH6HFD	444.85000	+	114.8 Hz	
Honolulu, St Francis Liliha Me						
	FM	KH6ICX	147.22000	+		HSRAC
Honolulu, St. Francis Medical						
	FM	NH6WP	147.28000	+		HSRAC
Ka'u	FM	WH6FC	145.29000	-	100.0 Hz	HSRAC
Kaaawa	DMR/MARC	KH7HO	443.47500	+	CC0	
Kaala	FM	NH6XO	146.68000	-	123.0 Hz	HSRAC
	FM	NH6XO	147.36000	+		
	FM	NH6XO	444.77500	+	123.0 Hz	HSRAC
Kaanapali	DMR/BM	AH6GR	442.50000	+	CC1	
Kahalui	FM	AH6GR	147.18000	+		HSRAC
Kahanamoku	DMR/MARC	AH6CP	444.52500	+	CC0	
Kahua	FM	AH6GR	442.27500	+	136.5 Hz	HSRAC
Kahuku	DMR/MARC	KH7HO	443.72500	+	CC0	
Kailua	DMR/MARC	KH7HO	443.97500	+	CC0	
	FM	WR6AVM	147.00000	+		HSRAC
	FM	WH6CZB	444.15000	+		HSRAC

144   HAWAII

Location	Mode	Call sign	Output	Input	Access	Coordinator
Kailua Kona	DMR/BM	KH6BFD	444.67500	+	CC1	
	FM	WB6EGR	927.13750	902.13750	DCS 411	HSRAC
Kailua Kona, Koloko Peak						
	FM	KH6BFD	443.65000	+	100.0 Hz	HSRAC
Kailua, Mount Olomana						
	FM	WH6CZB	146.66000	-	88.5 Hz	HSRAC
Kaimuki, Diamond Head						
	FM	WH6ARC	147.36000	+		HSRAC
Kalaheo	DMR/BM	NH6HI	442.22500	+	CC1	
Kalaheo, Mt. Kahili						
	FM	KH6E	146.92000	-	100.0 Hz	
Kalaoa	FM	KH7MS	145.41000	-		HSRAC
Kalaoa, Hualalai	FM	WH6DEW	147.16000	+		HSRAC
Kaneohe	DMR/BM	AH6PR	442.65000	+	CC7	
	DMR/MARC	KH7HO	444.60000	+	CC0	
	FM	WH6CZP	145.35000	-		
	FM	KH6BFZ	147.20000	+		
Kapaa	DMR/BM	WH7J	442.45000	+	CC1	
	DMR/MARC	WH6TF	444.30000	+	CC0	
	FM	KH6KWS	147.34000	+	100.0 Hz	HSRAC
Kapaa, Wilcox Hospital						
	FM	NH6JC	147.08000	+	100.0 Hz	
Kapalua	DMR/MARC	KH6DT	444.92500	+	CC0	
	FM	KH6RS	442.00000	+	110.9 Hz	HSRAC
Kapolei	DMR/MARC	KH6FV	443.47500	+	CC1	
Kau	FM	KH6EJ	444.60000	+		HSRAC
Kau, Police	FM	KH6EJ	146.92000	-		HSRAC
Kaunakakai Molokai						
	DMR/BM	AH6GR-5	442.30000	+	CC1	
Kea'au	FM	KH6EJ	146.68000	-		HSRAC
Keaau	FM	NH6HT	147.28000	+		HSRAC
	FM	NH6HT	442.57500	+		
Keaau, HPP	FM	KH6EJ	442.50000	+		HSRAC
Keaau, Kaumana	FM	K2FFT	146.64000	-	100.0 Hz	HSRAC
Keanae	FM	AH6GR	146.90000	-	110.9 Hz	HSRAC
Kilauea	FM	KH6E	146.70000	-		HSRAC
Kohala, Parker Ranch						
	FM	KH6EJ	444.45000	+	88.5 Hz	HSRAC
Kona	DMR/MARC	KH6FV	444.55000	+	CC4	
	FM	WB6EGR	927.21250	902.21250	DCS 411	HSRAC
Kualapuu	FM	W6KAG	145.37000	-		HSRAC
Kukuilono	DMR/MARC	AH6CP	444.35000	+	CC1	HSRAC
Kula	DMR/MARC	KH6HPZ	147.02000	+	CC1	
Kurtistown	FM	K1ENT	146.84000	-	100.0 Hz	HSRAC
Lahaina	FM	AL4A	146.64000	-	136.5 Hz	HSRAC
Laie, BYU	FM	KH6BYU	145.29000	-	123.0 Hz	HSRAC
Lanai	DMR/BM	AH6GR	442.07500	+	CC1	
Lanai City	FM	KH6HC	146.74000	+		HSRAC
	FM	KH6RS	442.22500	447.32500	110.9 Hz	HSRAC
Leeward	DSTAR	WH6DHT	145.45000	-		HSRAC
	DSTAR	WH6DHT	442.70000	-		
	DSTAR	WH6DHT	1293.00000	1273.00000		HSRAC
	FM	KH6AZ	145.13000	-		HSRAC
	FM	KH7TK	145.43000	-	77.0 Hz	
	FM	KH7INC	145.49000	-		HSRAC
	FM	AH6IH	147.32000	+		HSRAC
	FM	NH7QH	224.92000	-		HSRAC
	FM	KH6NYC	442.60000	+		
Lihue	DMR/MARC	KH6HPZ	147.04000	+	CC1	
	DMR/MARC	NH7YS	444.32500	+	CC0	
	FM	NH6HF	147.04000	+		HSRAC
	FM	KH6S	442.25000	+		

HAWAII 145

Location	Mode	Call sign	Output	Input	Access	Coordinator
Lihue	FM	KH6CVJ	444.67500	+	100.0 Hz	HSRAC
Manawahua	FM	KH6HFD	443.47500	+	114.8 Hz	HSRAC
	FM	KH6HFD	443.55000	+	114.8 Hz	HSRAC
Maui	DMR/BM	AH6GR	442.85000	+	CC1	
	DMR/BM	AH6GR-1	442.85000	+	CC1	
	DMR/MARC	KH6DT	443.15000	+	CC0	HSRAC
Mauna Kapu, Palihua Ridge						
	FM	KH6OCD	147.12000	+	88.5 Hz	HSRAC
Mauna Loa	FM	WH6FM	146.82000	-	100.0 Hz	
	FM	KH6EJ	146.82000	-		
	FM	AH6JA	147.04000	+		
Moana Loa	DMR/MARC	AH6CP	444.75000	+	CC1	
Mokuleia	DMR/MARC	KH6FV	443.60000	+	CC0	
Mokuleia, Peacock Flats						
	FM	KH6FV	146.76000	-	114.8 Hz	HSRAC
Mountain View	FM	KH6QAJ	444.35000	+		HSRAC
Mt. Halekala	DMR/MARC	KH6FV	444.85000	+	CC0	HSRAC
Na'alehu, Discovery Harbour Go						
	FM	KH6VFD	145.41000	-		HSRAC
North Shore	DMR/MARC	KH7HO	442.62500	+	CC0	
	FM	WR6AVM	147.00000	-		HSRAC
Oahu	FM	AH6GR	442.27500	+	136.5 Hz	
Ocean View	DMR/BM	KH7MS-2	145.41000	-	CC1	
	DMR/BM	KH7MS	443.40000	+	CC1	HSRAC
	DMR/BM	KH7MS-5	443.40000	+	CC1	
	DMR/BM	KH7MS	444.95000	+	CC1	HSRAC
	DMR/BM	KH7MS-3	444.95000	+	CC1	
Paauilo, Mauna Kea						
	FM	KH6EJ	146.72000	-		HSRAC
Pahala	DMR/BM	KH7MS-4	444.95000	+	CC1	
Pahoa	FM	NH6P	147.14000	+		
	FM	NH6P	442.25000	+	114.8 Hz	HSRAC
Pahoa, Hawaii	FM	WH6DYN	147.12000	+		
Pepeekeo	FM	KH6EJ	146.88000	-		HSRAC
Portable	DMR/MARC	KH6FV	444.65000	+	CC5	
Puu Hoku	FM	AH6GR	442.12500	+	110.9 Hz	HSRAC
Puu Kilea	FM	KH6RS	442.07500	+	110.9 Hz	HSRAC
Puu Mahoe	FM	KH6RS	442.05000	+	136.5 Hz	HSRAC
Salt Lake	FM	KH6MEI	442.17500	+	103.5 Hz	HSRAC
Volcano	FM	WH6ECW	442.15000	+		
Volcano, Kulani	FM	KH6EJ	146.76000	-		HSRAC
Waianae, Mauna Kapu						
	DMR/BM	WH6CZB	444.10000	+	CC4	HSRAC
	FM	KH7O	146.62000	-	103.5 Hz	HSRAC
	FM	WH6CZB	146.80000	-		HSRAC
	FM	KH7O	442.40000	+		
	FM	WR6AVM	442.47500	+		HSRAC
Waikiki	DMR/MARC	WB6EGR	443.65000	+	CC0	
Waikiki Beach	DSTAR	WH6DWF	444.70000	+		
	FM	WH6DWF	442.27500	+	118.8 Hz	
Waikoloa	FM	NH6EE	147.24000	+	100.0 Hz	HSRAC
	FM	NH6EE	444.97500	+		HSRAC
Wailuku	DMR/BM	AH6GR	442.75000	+	CC1	
	DMR/BM	AH6GR-2	442.75000	+	CC1	
	DMR/BM	WH6AV	443.02500	+	CC1	
	DMR/BM	WH6AV	444.10000	+	CC1	
	FM	KH6DT	146.76000	-	100.0 Hz	HSRAC
	FM	AH6GR	443.22500	+	110.9 Hz	HSRAC
Wailuku, Haleakala						
	DMR/BM	AH6GR	442.75000	+	CC1	HSRAC
Waimanalo	FM	WH6CXI	145.23000	-		
	FM	KH6HFD	443.40000	+	114.8 Hz	

146   HAWAII

Location	Mode	Call sign	Output	Input	Access	Coordinator
Waimanalo	FM	KH6HFD	443.60000	+	100.0 Hz	HSRAC
	FM	KH6CB	444.02500	+		HSRAC
	FM	AH6CP	444.32500	+	103.5 Hz	HSRAC
	FM	KH6HFD	444.37500	+	114.8 Hz	HSRAC
Waimea	FM	KH7T	147.38000	+		HSRAC
Waimea, Hospital	FM	NH7HI	147.32000	+	100.0 Hz	HSRAC
Waimea, Mauna Kea						
	FM	WB6EGR	443.62500	+	100.0 Hz	
Waiohinu	FM	AH6DX	145.49000	-		
Windward	FM	KH6BS	147.14000	+		HSRAC

IDAHO

Location	Mode	Call sign	Output	Input	Access	Coordinator
Blackfoot	FM	WB6EVM	447.95000	-		
Blacks Creek	FM	K7ZZL	443.25000	+	110.9 Hz	UVHFS
Boise	DMR/BM	WA7GSK	444.07500	+	CC1	
	DSTAR	W7VOI	444.35000	+		W7ZRQ
	FM	W7VWR	146.78000	-	100.0 Hz	
	FM	N7DJX	147.30000	+	100.0 Hz	
	FM	N7FYZ	147.32000	+	100.0 Hz	UVHFS
	FM	KG6GCQ	147.50000		107.2 Hz	
	FM	KB7ZD	223.94000	-	100.0 Hz	UVHFS
	FM	KA7EWN	224.50000	-	100.0 Hz	UVHFS
	FM	N7DJX	443.80000	+	100.0 Hz	W7ZRQ
	FM	W7VWR	444.72500	+	100.0 Hz	W7ZRQ
Boise, Blacks Creek						
	FM	K3ZFF	145.25000	-	100.0 Hz	W7ZRQ
Boise, Boise Bench						
	FM	W7VOI	145.29000	-	100.0 Hz	W7ZRQ
Boise, Deer Point	FM	KE7YD	444.47500	+	DCS 245	W7ZRQ
Boise, Downtown	FM	W7VWR	444.27500	+		
Boise, Hewlett-Packard						
	FM	AB7HP	147.26000	+	100.0 Hz	UVHFS
Boise, Idaho	FM	WA7FDR	145.13000	-	100.0 Hz	
Boise, Seaman's Gulch						
	FM	N7BOI	147.38000	+	DCS 026	W7ZRQ
Boise, Shafer Butte						
	FM	N7BOI	145.15000	-	DCS 026	W7ZRQ
Boise, Wilderness Ridge						
	FM	WA9WSJ	52.62000	51.62000	110.9 Hz	UVHFS
	FM	KA7EWN	444.65000	+	110.9 Hz	UVHFS
Bonners Ferry	FM	W7BFI	147.04000	+	100.0 Hz	IACC
	FM	W7BFI	443.02500	+	100.0 Hz	IACC
Bonners Ferry, Black Mountain						
	FM	W7BFI	146.96000	-	123.0 Hz	IACC
Buhl	DMR/BM	W6RNK	447.60000	-	CC1	
Burley	ATV	K6ZVA	434.00000	1253.30000		
Burley, Mt Harrison						
	FM	WA7FDR	145.27000	-	100.0 Hz	UVHFS
	FM	K7ACA	145.33000	-	123.0 Hz	W7ZRQ
	FM	KC7SNN	147.00000	-	100.0 Hz	UVHFS
	FM	KC7SNN	449.20000	-		UVHFS
Cascade	DMR/BM	W7CIA	146.96000	-	CC1	UVHFS
	DMR/BM	W7CIA	441.92500	+	CC1	
	FM	K7ZZL	443.30000	+	110.9 Hz	UVHFS
	FM	NB7C	927.16250	902.16250	151.4 Hz	W7ZRQ
Cascade, Snowbank Mountain						
	FM	W7VOI	146.62000	-	100.0 Hz	W7ZRQ
Challis, Grouse Peak						
	FM	AA7WG	146.78000	-	110.9 Hz	
Cleft	FM	KD7RMB	145.41000	-	114.8 Hz	
Cocolalla	FM	K7ZOX	442.00000	+	110.9 Hz	IACC
Coeur D'Alene	DMR/MARC	WA7DMR	440.92500	+	CC1	IACC

IDAHO 147

Location	Mode	Call sign	Output	Input	Access	Coordinator
Coeur D'Alene, Canfield Mounta						
	FM	K7ID	443.97500	+	136.5 Hz	IACC
Coeur D'Alene, Mica Peak/RX Bl						
	FM	KB6UMY	53.39000	51.69000	100.0 Hz	IACC
Coolin	FM	K7KAM	145.41000	-	77.0 Hz	IACC
Coolin, Priest Lake						
	FM	N7KAM	145.41000	-	77.0 Hz	
Cottonwood, Cottonwood Butte						
	FM	K7EI	444.95000	+	100.0 Hz	UVHFS
Council, Council Mtn						
	FM	W7KAU	145.37000	-	100.0 Hz	
Deary	FM	N7WEE	145.35000	-	114.8 Hz	UVHFS
Driggs	DMR/BM	W7RAC	449.80000	-	CC1	
Driggs, Relay Ridge						
	FM	K7ENE	146.94000	-	123.0 Hz	UVHFS
	FM	W7RAC	147.14000	+	123.0 Hz	
	FM	KB7ITU	447.87500	-		
Eagle	FM	WV7I	449.85000	-		UVHFS
Elk River, Elk Butte						
	FM	KF7WOR	145.19000	-	67.0 Hz	IACC
Emida	FM	KB7SIJ	145.31000	-	88.5 Hz	IACC
Emmett	FM	N7UBO	147.22000	+		
	FM	K7WIR	224.88000	-	100.0 Hz	UVHFS
Emmett, Squaw Butte						
	FM	K7WIR	147.20000	+	100.0 Hz	UVHFS
Filer, Everton Building						
	FM	KB7SQS	444.60000	+		W7ZRQ
Fruitland	FM	KC7BSA	443.65000	+	100.0 Hz	UVHFS
Grace, Sedgwick Peak						
	FM	AE7TA	146.80000	-	88.5 Hz	UVHFS
	FM	AE7T	449.37500	-	88.5 Hz	UVHFS
Grangeville	FM	KC7MGR	146.68000	-	100.0 Hz	UVHFS
Howe, Jumpoff Peak						
	FM	WA7FDR	146.85000	-	100.0 Hz	UVHFS
	FM	W7RUG	447.62000	-	88.5 Hz	
Idaho City, Grimes Creek						
	FM	K7WIR	443.35000	+		
Idaho City, Schaefer Butte						
	FM	WI7ATV	145.47000	-	100.0 Hz	W7ZRQ
Idaho City, Shafer Butte						
	FM	W7VOI	146.84000	-	100.0 Hz	W7ZRQ
	FM	K7BSE	146.94000	-	100.0 Hz	UVHFS
	FM	WR7ID	444.50000	+	100.0 Hz	
	FM	W7VOI	444.90000	+	100.0 Hz	W7ZRQ
	FM	NB7C	927.11250	902.11250	DCS 432	UVHFS
Idaho Falls	DMR/BM	KB7SJZ	443.40000	+	CC0	
	DSTAR	KG7WZG	448.90000	-		
Idaho Falls, Peterson Hill						
	FM	K7EFZ	146.74000	-		UVHFS
Idaho Falls, Quality Inn						
	FM	WA4VRV	443.00000	+	100.0 Hz	UVHFS
Iona, Iona Hill	FM	K7EFZ	146.64000	-		UVHFS
	FM	KD7SUF	448.17500	-	100.0 Hz	UVHFS
Island Park	FM	WA7FDR	145.23000	-	123.0 Hz	EIFC
Island Park, Sawtell Peak						
	FM	WA7FDR	145.23000	-	100.0 Hz	UVHFS
Jerome	DSTAR	KF7VTM	444.80000	+		W7ZRQ
Jerome, Flat Top Butte						
	FM	K7MVA	146.66000	-	100.0 Hz	UVHFS
	FM	K7MVA	442.30000	+	100.0 Hz	
Kamiah	FM	KD6MNA	443.92500	+	100.0 Hz	IACC

148 IDAHO

Location	Mode	Call sign	Output	Input	Access	Coordinator
Kellogg, Wardner Peak						
	FM	N7SZY	146.94000	-	127.3 Hz	IACC
	FM	N7SZY	444.00000	+	127.3 Hz	
Ketchum, Bald Mountain						
	FM	WX7XX	147.18000	+	100.0 Hz	W7ZRQ
Kimberly, Hansen Butte						
	FM	K7MVA	146.76000	-	100.0 Hz	UVHFS
	FM	K7MVA	147.10000	+	100.0 Hz	
Kooskia	FM	KK3ARC	146.62000	-	88.5 Hz	IACC
Kootenai, Schweitzer Mountain						
	FM	KD7WPQ	442.77500	+	100.0 Hz	
Lewiston	FM	KB7RKY	223.96000	-		UVHFS
	FM	W7TRO	444.40000	+	162.2 Hz	UVHFS
	FM	K7EI	444.92500	+	100.0 Hz	UVHFS
Lewiston, Craig Mountain						
	FM	K7EI	53.35000	51.65000	100.0 Hz	UVHFS
	FM	K7EI	146.92000	-	110.9 Hz	UVHFS
	FM	K7EI	442.10000	+	103.5 Hz	UVHFS
Lewiston, Lewiston Hill						
	FM	KK6RYR	145.21000	-	203.5 Hz	IACC
	FM	W7VJD	146.96000	-		UVHFS
	FM	K7EI	444.90000	+	100.0 Hz	UVHFS
Lowman, Pilot Peak						
	FM	KA7ERV	145.31000	-	100.0 Hz	
Malad City	DMR/BM	KI7WQR	447.98750	-	CC1	
Marsing	FM	K7ZZL	146.88000	-	100.0 Hz	W7ZRQ
	FM	K7ZZL	443.55000	+	100.0 Hz	W7ZRQ
Marsing, French John Hill						
	FM	K7TRH	147.36000	+	100.0 Hz	UVHFS
	FM	K7TRH	442.90000	+	100.0 Hz	W7ZRQ
Mccall	DMR/BM	W7ELE	146.64000	-	CC1	
	DMR/BM	N7IBC	441.92500	+	CC1	
	DMR/BM	N7IBC	444.12500	+	CC1	W7ZRQ
McCall, Brundage Mtn						
	FM	KC7MCC	146.90000	-	123.0 Hz	
McCall, No Business Montain						
	FM	KC7MCC	147.02000	+	100.0 Hz	UVHFS
Melba, Hat Butte	FM	K7LCD	145.35000	-		W7ZRQ
	FM	K7LCD	444.17500	+		W7ZRQ
Menan, Menan Buttes						
	FM	K7ENE	146.88000	-	100.0 Hz	UVHFS
Meridian, Eagle Road And Ustic						
	FM	KC7LHV	442.60000	+		W7ZRQ
Meridian, Meridian						
	FM	KC9CJ	147.08000	+	100.0 Hz	W7ZRQ
Middleton	FM	K6LOR	146.50000			
	FM	K6LOR	444.27500	+		
Montpelier	FM	AC7TJ	147.38000	+	88.5 Hz	UVHFS
Montpelier, M-Hill	FM	AG7BL	147.12000	+	123.0 Hz	UVHFS
Moore, Mike Betts Home						
	FM	N7GJV	146.96000	-	100.0 Hz	
Moscow	FM	K9GRZ	443.00000	+	88.5 Hz	IACC
	FM	W9GRZ	443.30000	+	88.5 Hz	IACC
Moscow, Moscow Mountain						
	FM	WA7HWD	146.82000	-	127.3 Hz	UVHFS
	FM	K7EI	444.97500	+	100.0 Hz	UVHFS
Moscow, Moscow Mtn						
	FM	KA7FVV	147.32000	+	103.5 Hz	
Mountain Home	FM	K7ECI	145.19000	-	110.9 Hz	
	FM	WA7ZAF	224.34000	-	123.0 Hz	W7ZRQ
	FM	K7ECI	442.82500	+	110.9 Hz	
	FM	K7ECI	444.70000	+	110.9 Hz	

IDAHO   149

Location	Mode	Call sign	Output	Input	Access	Coordinator
Mountain Home, Rattlesnake But						
	FM	K7ECI	147.34000	+	100.0 Hz	W7ZRQ
Moyie Springs	DSTAR	KF7MJA	145.12500	-		IACC
	FM	AF7LJ	145.45000	-		
Nampa	FM	NG7O	444.10000	+	156.7 Hz	W7ZRQ
Nampa, Hat Butte	FM	K7OVG	927.18750	902.18750	DCS 432	W7ZRQ
Naples, Peterson Hill						
	FM	NK7I	147.32000	+	167.9 Hz	
New Meadows	DMR/BM	KC7MCC	442.50000	+	CC1	
Orofino	FM	KD7ALJ	145.27000	-	100.0 Hz	UVHFS
	FM	KC7VBT	145.49000	-		UVHFS
	FM	K7EI	444.87500	+	100.0 Hz	UVHFS
Orofino, Wells Bench						
	FM	K7NDX	146.76000	-	131.8 Hz	UVHFS
Payette	FM	NB7C	443.05000	+	114.8 Hz	UVHFS
	FM	NB7C	927.12500	902.12500	103.5 Hz	UVHFS
Peck, Teakean Butte						
	FM	KK6RYR	145.21000	-	206.5 Hz	IACC
Pocatello	FM	AD7UI	146.82000	-	100.0 Hz	UVHFS
	FM	KA7MLM	147.30000	+		
	FM	K9GP	147.34000	+		UVHFS
	FM	KF7FY	147.52000		123.0 Hz	
	FM	WB6EVM	449.12500	-		
Pocatello, Kinport Peak						
	FM	N7PI	147.36000	+	100.0 Hz	UVHFS
Pocatello, Scout Montain						
	FM	N7PI	147.06000	+	100.0 Hz	UVHFS
Post Falls, Blossom Peak						
	FM	KB6UMY	147.28000	+	100.0 Hz	IACC
	FM	KB6UMY	442.97500	+	100.0 Hz	IACC
Post Falls, Idaho Mica Peak						
	FM	K7ID	443.27500	+		
Post Falls, Mica Peak						
	FM	K7ID	146.98000	-	127.3 Hz	IACC
	FM	KC7ODP	147.08000	+	100.0 Hz	IACC
Preston	DMR/BM	KE7EYY	448.75000	-	CC1	
Priest River	FM	K7MEE	444.40000	+	107.2 Hz	IACC
Rexburg, BYU-Idaho Campus						
	FM	K7BYI	145.41000	-	100.0 Hz	W7ZRQ
	FM	K7WIP	448.60000	-	100.0 Hz	W7ZRQ
Rexburg, Water Tower						
	FM	N7UNY	146.70000	-	100.0 Hz	UVHFS
Roberts	FM	KE7JFA	448.80000	-	100.0 Hz	
Saint Maries	DMR/BM	KB7SIJ	443.75000	+	CC1	
Salmon, Baldy Mountain						
	FM	AA7WG	442.10000	+	100.0 Hz	
	FM	AA7WG	442.22500	447.32500	100.0 Hz	
Salmon, Baldy Summit						
	FM	AA7WG	146.98000	-	100.0 Hz	
Salmon, Old Dump Hill Above Sa						
	FM	AA7WG	147.03000	+	100.0 Hz	
Sandpoint	FM	K7BNR	442.50000	+	131.8 Hz	
	FM	KD7WPQ	442.77500	+	100.0 Hz	IACC
	FM	N7JCT	446.02500	-		
	FM	K7JEP	448.37500	+		
Sandpoint, Hoodoo Mountain						
	FM	K7JEP	145.49000	-	123.0 Hz	IACC
Sawtooth City, Galena Peak						
	FM	AE6DX	147.14000	+	100.0 Hz	UVHFS
Shafer Butte	FM	N7KNL	444.67500	+	156.7 Hz	
Silver City, War Eagle						
	FM	KJ7GGR	147.04000	+	123.0 Hz	

150　IDAHO

Location	Mode	Call sign	Output	Input	Access	Coordinator
Spirit Lake	FM	K7ZOX	442.00000	+	110.9 Hz	IACC
St. Marie's, St. Joe Baldy						
	FM	KB7SIJ	147.26000	+	88.5 Hz	IACC
SW Idaho	FM	NB7C	927.11250	902.11250	DCS 432	
Twin Falls	FM	W7CTH	442.60000	+	100.0 Hz	
	FUSION	KB7SQS	443.75000	+		W7ZRQ
Twin Falls, Mt Harrison						
	ATV	K6ZVA	1253.25000	426.25000		W7ZRQ
Wallace, Goose Peak						
	FM	KB7BYR	147.18000	+	118.8 Hz	
	FM	KB7BTU	224.76000	-		IACC
Weiser, Sheep Creek						
	FM	K7OJI	145.39000	-		
	FM	K7OJI	147.12000	+		UVHFS
Wendell	FUSION	K8MPW	443.95000	+		

ILLINOIS

Location	Mode	Call sign	Output	Input	Access	Coordinator
Aledo	FM	KC9HDD	145.31000	-	100.0 Hz	IRA
	FM	KC9HDD	443.25000	+	100.0 Hz	IRA
Algonquin	DMR/MARC	WD9BBE	443.95625	+	CC15	IRA
	FM	N9IVM	444.02500	+	103.5 Hz	IRA
Allerton	FM	K9LOF	147.28500	+	146.2 Hz	IRA
Alton	FM	KB9GPF	927.60000	902.60000	79.7 Hz	IRA
Alton, OSF Saint Anthony's Hea						
	FM	K9HAM	224.64000	-	123.0 Hz	
	FM	K9HAM	442.22500	+	123.0 Hz	IRA
	FM	K9HAM	927.01250	902.01250	123.0 Hz	
Alton, Verizon Cell Tower - Ya						
	FM	K9HAM	145.23000	-	79.7 Hz	IRA
	FM	K9HAM	442.90000	+	123.0 Hz	IRA
Anna	FM	KD9EVL	145.25000	-		IRA
	FM	WA9LM	442.85000	+	88.5 Hz	IRA
Antioch	FM	KA9VZD	145.29000	-	107.2 Hz	IRA
	FM	N9EMS	927.65000	902.65000	192.8 Hz	IRA
Arcola	FM	WA9WOB	444.37500	+	192.8 Hz	IRA
Arlington Heights	FM	N9IVM	444.02500	+	100.0 Hz	IRA
Athens	FM	W9DUA	147.04500	+	103.5 Hz	IRA
Aurora	DMR/MARC	W9LSL	443.42500	+	CC1	IRA
Ava	FM	W9RNM	147.09000	+	88.5 Hz	IRA
Aviston	FM	KT9TR	147.21000	+	79.7 Hz	IRA
	FM	KT9TR	443.17500	+	79.7 Hz	IRA
Batavia	DSTAR	W9CEQ	147.22500	+		IRA
	DSTAR	W9CEQ	442.10625	+		IRA
	FM	W9ZGP	147.06000	+	103.5 Hz	IRA
	FM	W9XA	224.40000	-	110.9 Hz	IRA
	FM	KA9LFU	444.10000	+	100.0 Hz	IRA
	FM	WB9IKJ	444.22500	+	114.8 Hz	IRA
	FM	W9CEQ	444.30000	+	114.8 Hz	IRA
	FM	W9XA	1292.00000	1272.00000	88.5 Hz	IRA
Beardstown	FM	W9ACU	443.95000	+	103.5 Hz	IRA
Belleville, Turkey Hill						
	FM	K9GXU	147.12000	+		IRA
	FM	K9GXU	224.12000	-	127.3 Hz	IRA
	FM	K9GXU	444.62500	+	127.3 Hz	IRA
Bellwood	FM	KC9ZI	444.57500	+	114.8 Hz	IRA
Belvidere	DMR/MARC	K9VO	442.75000	+	CC10	IRA
	FM	K9ORU	147.37500	+	100.0 Hz	IRA
	FM	N9KUX	442.82500	+	114.8 Hz	IRA
Berwyn	FM	WA9HIR	444.15000	+	146.2 Hz	IRA
Bethalto, St. Louis Regional A						
	FM	K9HAM	145.13000	-	123.0 Hz	IRA
Bloomingdale	FM	K9NB	224.22000	-	110.9 Hz	IRA

ILLINOIS   151

Location	Mode	Call sign	Output	Input	Access	Coordinator
Bloomington	DMR/BM	WX9WX	444.23750	+	CC12	
	FM	N9BXI	147.01500	+	156.7 Hz	IRA
	FM	W9NUP	147.19500	+	97.4 Hz	IRA
	FM	W9NUP	444.95000	+	97.4 Hz	IRA
Bloomington, BMI	FM	K9MBS	224.46000	-	107.2 Hz	IRA
Bloomington, BMI Airport						
	DSTAR	KJ9P	444.58125	+		IRA
Blue Island	FM	W9SRC	442.67500	+	131.8 Hz	IRA
Bolingbrook	DMR/MARC	K9BAR	443.70000	+	CC1	IRA
	FM	K9BAR	147.33000	+	107.2 Hz	IRA
	FM	K9BAR	224.54000	-	110.9 Hz	IRA
	FM	K9BAR	443.52500	+	114.8 Hz	IRA
Bridgeview	FM	KB9EPL	224.48000	-	110.9 Hz	IRA
Brookfield	FM	K9SAD	224.16000	-	110.9 Hz	IRA
Buffalo Grove	FM	WB9TAL	224.58000	-	110.9 Hz	IRA
Burnt Prairie	FM	W9KXP	147.33000	+	DCS 023	IRA
Cadwell	DSTAR	W9BIL	145.19500	-		IRA
	DSTAR	W9BIL	440.64375	+		IRA
Canton	FM	K9ILS	147.28500	+	103.5 Hz	IRA
	FM	K9ILS	444.72500	+	103.5 Hz	IRA
Carbondale	FM	W9UIH	442.02500	+	88.5 Hz	IRA
Carbondale, SIU Morris Library						
	DSTAR	W9UIH	442.65625	+		
Carlinville	FM	N9OWS	443.27500	+	100.0 Hz	IRA
Carthage	FM	KC9LMF	147.10500	+	103.5 Hz	IRA
Cary	DSTAR	KO9H	224.96000	-		IRA
Champaign	DSTAR	W9YR	443.48125	+		IRA
	FM	K9SI	444.10000	+	162.2 Hz	IRA
	FM	W9YH	444.52500	+	162.2 Hz	
Channahon	FM	W9CHI	444.60000	+	136.5 Hz	IRA
Cherry Valley	FM	W9FT	442.62500	+	123.0 Hz	IRA
Chicago	DMR/BM	KB9NTX	442.45000	+	CC15	IRA
	DMR/BM	KP4EOP	443.97500	+	CC1	
	DMR/MARC	K9QI	440.00000	+	CC1	
	DMR/MARC	N9OZR	440.20000	+	CC1	
	DMR/MARC	W9BMK	440.30000	+	CC1	IRA
	DMR/MARC	KC9MNL	440.40000	445.00000	CC1	
	DMR/MARC	W9DIG	440.85625	+	CC1	IRA
	DMR/MARC	K9QI	440.90000	+	CC1	
	DMR/MARC	AA9VI	441.21875	+	CC1	
	DMR/MARC	K9TOW	441.33125	+	CC4	
	DMR/MARC	NS9RC	442.72500	+	CC1	IRA
	DMR/MARC	WA9VGI	442.97500	443.47500	CC4	IRA
	DMR/MARC	KC9DTO	444.12500	5444.12500	CC10	
	DSTAR	NS9RC	442.09375	+		IRA
	DSTAR	NS9RC	1292.20000	1272.20000		IRA
	FM	W9SRO	147.15000	+	107.2 Hz	IRA
	FM	KC9EBB	223.88000	-	110.9 Hz	IRA
	FM	W9TMC	224.02000	-	103.5 Hz	IRA
	FM	WD9GEH	224.06000	-	110.9 Hz	IRA
	FM	WA9ORC	224.10000	-	110.9 Hz	IRA
	FM	W9RA	224.34000	-	103.5 Hz	IRA
	FM	WM9W	224.52000	-	110.9 Hz	IRA
	FM	AB9OV	442.17500	+	114.8 Hz	IRA
	FM	K9NBC	442.40000	+	114.8 Hz	IRA
	FM	K9CDW	442.57500	+	131.8 Hz	IRA
	FM	K9QKW	443.37500	+	114.8 Hz	IRA
	FM	NS9RC	443.60000	+	156.7 Hz	IRA
	FM	WA9ORC	443.75000	+	114.8 Hz	IRA
	FM	K9SAD	443.80000	+	114.8 Hz	IRA
	FM	K9GFY	444.37500	+	114.8 Hz	IRA
	FM	N9SHB	444.62500	+	110.9 Hz	IRA

152   ILLINOIS

Location	Mode	Call sign	Output	Input	Access	Coordinator
Chicago	FM	N9JDZ	927.71250	902.71250	151.4 Hz	IRA
	FM	WA9ORC	1291.10000	1271.10000	114.8 Hz	IRA
Chicago, AON Building (Old Amo						
	FM	KC9DFK	443.67500	+		IRA
Clinton	FM	KA9KEI	442.37500	+	91.5 Hz	IRA
Collinsville	FM	KD6TVP	442.17500	+	103.5 Hz	IRA
	FM	K9HAM	442.77500	+	123.0 Hz	IRA
Congerville, WYZZ Tower						
	FM	KE9HB	443.32500	+	107.2 Hz	IRA
Crescent City	FM	AD9L	147.03000	+	103.5 Hz	IRA
Crystal Lake	DMR/MARC	K9QI	439.22500	431.62500	CC6	
	DMR/MARC	K9VI	444.80625	+	CC4	IRA
	FM	K9VI	224.70000	-	100.0 Hz	IRA
	FM	N9EAO	443.20000	+	131.8 Hz	IRA
	FM	N9HEP	443.47500	+	114.8 Hz	IRA
Dakota	FM	N9WSQ	147.30000	+	88.5 Hz	IRA
Dallas City	FM	KA9JNG	444.92500	+	123.0 Hz	IRA
Danville	DMR/BM	NE9RD	443.72500	+	CC1	IRA
	DMR/MARC	N9WEW	443.82500	+	CC1	
	DMR/MARC	KC9DTN	443.82500	+	CC10	
	FM	W9MJW	29.66000	-	100.0 Hz	IRA
	FM	NU9R	52.97000	51.27000	88.5 Hz	IRA
	FM	W9MJW	927.62500	902.62500	100.0 Hz	IRA
Darien	FM	W9ANL	145.19000	-	114.8 Hz	IRA
Decatur	DSTAR	KC9YFX	147.24000	+		IRA
	DSTAR	KC9YFX	442.64375	+		
	FM	K9HGX	29.62000	-	103.5 Hz	IRA
	FM	K9HGX	53.23000	51.53000	103.5 Hz	IRA
	FM	WA9RTI	147.10500	+	103.5 Hz	IRA
	FM	WA9RTI	442.25000	+	103.5 Hz	IRA
	FM	K9HGX	443.80000	+	123.0 Hz	IRA
	FM	K9MCA	444.17500	+	100.0 Hz	IRA
Deer Park	FM	KP4EOP	444.00000	+		IRA
Deerfield	FM	KA9REN	224.24000	-	110.9 Hz	IRA
Dekalb	FM	KB9FMU	444.45000	+	114.8 Hz	IRA
Downers Grove	FM	W9DUP	224.68000	-	110.9 Hz	IRA
	FM	KC9WPR	442.25000	+	DCS 072	
	FM	W9DUP	442.55000	+	114.8 Hz	IRA
	FM	W9YRC	442.87500	+	114.8 Hz	IRA
	FM	N9ATO	443.90000	+	114.8 Hz	IRA
Downers Grove, Butterfield And						
	FM	W9CCU	444.47500	+	114.8 Hz	IRA
Dunlap	FM	N9BBO	224.08000	-	156.7 Hz	IRA
	FM	N9BBO	443.12500	+	156.7 Hz	IRA
Eagleton	FM	W9IMP	224.84000	-	82.5 Hz	IRA
	FM	KB2MAU	443.40000	+		
East Dundee	P25	W9DWP	927.62500	902.62500	151.4 Hz	IRA
Edwardsville, Edwardsville Wat						
	FM	W9AIU	224.06000	-	127.3 Hz	IRA
Edwardsville, SIUE Water Tower						
	FM	W9AIU	442.40000	+	127.3 Hz	IRA
Effingham	FM	K9UXZ	444.12500	+	110.9 Hz	IRA
Elburn	DMR/MARC	W9XA	443.64375	+	CC6	IRA
	FM	W9DWP	145.27000	-		IRA
	FM	W9CEQ	147.21000	+	103.5 Hz	IRA
	FM	W9DWP	443.02500	+		
Elgin	DMR/MARC	N9NLE	444.52500	+	CC15	IRA
	FM	WR9ABQ	52.95000	51.95000	114.8 Hz	IRA
	FM	WR9ABQ	444.95000	+	114.8 Hz	IRA
Elizabeth	FM	W9SBA	147.33000	+	250.3 Hz	IRA
Elk Grove Village	FUSION	KB9L	147.01500	+	107.2 Hz	IRA
Elmhurst	DMR/MARC	KB9UUU	440.85000	448.85000	CC0	

ILLINOIS 153

Location	Mode	Call sign	Output	Input	Access	Coordinator
Fairfield	FM	KC9TON	145.13000	-	88.5 Hz	IRA
	FM	N9BRG	444.82500	+		IRA
Forest City	FM	WI9MP	223.94000	-	110.9 Hz	IRA
Frankfort	FM	WD9HSY	443.32500	+	114.8 Hz	IRA
	FM	W9WIL	444.55000	+	114.8 Hz	IRA
Freeport	FM	KB9RNT	147.39000	+	114.8 Hz	IRA
	FM	W9FN	224.92000	-	74.4 Hz	IRA
	FM	W9SBA	442.00000	+	146.2 Hz	IRA
	FM	KB9RNT	443.27500	+	114.8 Hz	IRA
Galesburg, Hawthorne Center						
	FM	KA9QMT	147.21000	+	107.2 Hz	IRA
Galesburg, St. Mary's Hospital						
	FM	W9GFD	147.00000	-	103.5 Hz	IRA
	FM	W9GFD	444.45000	+	103.5 Hz	IRA
Galva	FM	WA9BA	443.30000	+	103.5 Hz	IRA
	FM	W9YPS	443.35000	+	225.7 Hz	IRA
Geff	FM	KC9GMX	444.40000	+		IRA
Geneseo	FM	W9MVG	444.87500	+	136.5 Hz	IRA
Gillespie	FM	K9MCE	444.25000	+	103.5 Hz	IRA
Glendale Heights	DSTAR	KC9PWC	440.10625	+		IRA
	FM	K9XD	444.87500	+	114.8 Hz	IRA
	FM	KD9AUP	927.55000	902.55000	151.4 Hz	IRA
Glenview	FM	W9AP	224.60000	-	110.9 Hz	IRA
Grand Detour, IL Route 2						
	FM	W9DXN	444.80000	+	114.8 Hz	IRA
Grant Park	FM	WA9WLN	441.30000	+	114.8 Hz	IRA
	FM	W9LEO	441.87500	+		IRA
Greenup, Old Water Tower						
	FM	W9GWF	147.03000	+	107.2 Hz	IRA
Greenville	DMR/MARC	K7QLL	443.43125	+	CC7	IRA
	FM	AD9OV	442.92500	+	103.5 Hz	IRA
Greenville, Water Tower						
	FM	AD9OV	147.16500	+	103.5 Hz	IRA
Gridley	FM	KE9HB	444.35000	+	107.2 Hz	IRA
Groveland	DSTAR	W9PIA	145.10500	-		IRA
	FM	W9UVI	147.07500	+	156.7 Hz	IRA
Gurnee	FM	N9OZB	443.15000	+	114.8 Hz	IRA
Hardin	FM	ND2D	147.30000	+		IRA
Hawthorne Woods						
	FM	W9AP	147.09000	+	107.2 Hz	IRA
Herald	FM	W9KXP	147.30000	+	DCS 023	IRA
Herod, Williams Hill						
	FM	KB9JNO	145.13000	-		
	FM	K9OWU	444.80000	+	88.5 Hz	IRA
Heyworth	FM	KG9DW	442.82500	+	141.3 Hz	IRA
Hinsdale	FM	KB9OYP	444.20000	+	114.8 Hz	IRA
Hoffman Estates, St Alexus Hos						
	FM	N9RJV	444.12500	+		
Homer Glen	DMR/MARC	WB9PHK	423.30625	420.30625	CC3	
	DMR/MARC	KC9NCS	442.83125	+	CC3	IRA
Homewood	FM	WA9WLN	442.37500	+	114.8 Hz	IRA
Hoopeston	FM	KB9YZI	444.82500	+	127.3 Hz	IRA
Huntley	DMR/MARC	W9IV	440.04375	+	CC1	IRA
	FM	KC9ONA	441.62500	+	DCS 072	IRA
	FM	AB9OU	927.72500	902.72500	114.8 Hz	IRA
Ina, Rend Lake College - Theat						
	FM	W9RLC	145.19000	-	71.9 Hz	IRA
Ingleside, In The Chain Of Lak						
	FM	K5TAR	440.81875	+	114.8 Hz	IRA
Inverness	DMR/MARC	WX9NC	441.95625	+	CC1	IRA
	DSTAR	WX9NC	147.39000	+		
Jacksonville	FM	K9JX	444.67500	+	103.5 Hz	IRA

154   ILLINOIS

Location	Mode	Call sign	Output	Input	Access	Coordinator
Joliet	FM	KC9PLK	145.25000	-	156.7 Hz	IRA
	FM	W9OFR	223.82000	-		IRA
	FM	W9OFR	442.30000	+	114.8 Hz	IRA
	FM	WA9VGI	442.95000	+		IRA
	FM	N9WYS	927.52500	902.52500	151.4 Hz	IRA
Joliet, Joshua Arms Retirement						
	FM	WD9AZK	147.30000	+	94.8 Hz	IRA
Jonesboro, Union County Emerge						
	FM	WA9LM	443.57500	+	192.8 Hz	
Kankakee	FM	WD9HSY	147.16500	+	107.2 Hz	IRA
	FM	W9AZ	444.80000	+	100.0 Hz	IRA
Kansas	FM	W9COD	53.29000	51.59000	162.2 Hz	IRA
	FM	W9COD	147.37500	+	162.2 Hz	IRA
	FM	W9COD	443.62500	+	162.2 Hz	IRA
Kewanee	FM	N9ZK	442.17500	+	225.7 Hz	IRA
Kickapoo, Near Exit						
	FM	K9WRA	444.20000	+	103.5 Hz	IRA
La Grange Park	FM	K9ONA	443.30000	+	114.8 Hz	IRA
Lake Villa	FM	WB9RKD	147.03000	+	107.2 Hz	IRA
	FM	N9FJS	442.32500	+	107.2 Hz	IRA
	FUSION	WB9RKD	444.40000	+	114.8 Hz	IRA
Lake Zurich	DSTAR	KC9OKW	441.23125	+		IRA
	FM	K9SA	223.84000	-	110.9 Hz	IRA
	FM	W9SRO	224.86000	-	110.9 Hz	IRA
	FM	K9SA	443.25000	+	114.8 Hz	IRA
	FM	KD9GY	443.85000	+	114.8 Hz	IRA
	FM	W9JEM	927.68750	902.68750	151.4 Hz	IRA
Libertyville	FM	K9IQP	147.18000	+	127.3 Hz	IRA
	FM	N9KTW	441.12500	+	114.8 Hz	IRA
	FM	K9IQP	442.52500	+	114.8 Hz	IRA
Lincoln	FM	K9ZM	147.34500	+	103.5 Hz	IRA
Lincoln, ALMH	FM	KC9WMV	442.80000	+	203.5 Hz	IRA
Lisle	FM	WA9WSL	224.36000	-	110.9 Hz	IRA
	FM	W9AEK	224.62000	-	110.9 Hz	IRA
	FM	W9AUX	442.05000	+		
	FM	WA9WSL	442.22500	+	114.8 Hz	IRA
	FM	W9AEK	442.70000	+	203.5 Hz	IRA
	FM	W9AEK	1293.10000	1273.10000	114.8 Hz	IRA
Litchfield	FM	W9BXR	444.45000	+	94.8 Hz	IRA
Lockport	DMR/MARC	N2BJ	443.22500	+	CC2	IRA
	FM	N2BJ	224.94000	-	114.8 Hz	IRA
	FM	NC9T	442.02500	+	100.0 Hz	IRA
	FM	N9OWR	927.58750	902.58750	DCS 411	IRA
Loda	FM	K9UXC	442.42500	+	179.9 Hz	IRA
Loves Park	FM	K9RFD	147.19500	+	114.8 Hz	IRA
Lovington, Grain Elevator						
	FM	KR9X	223.86000	-	103.5 Hz	IRA
	FM	WC9V	444.27500	+	103.5 Hz	IRA
Macomb, WIU	FM	WB9TEA	444.30000	+	103.5 Hz	IRA
Macomb, WIUM Tower						
	FM	W9SSP	147.06000	+	103.5 Hz	IRA
Markham	FM	W9YPC	147.13500	+		IRA
Marseilles	DMR/MARC	KA9FER	146.74500	-	CC1	
	DMR/MARC	KA9FER	442.60000	+	CC1	
Maryville	FM	KG9OV	224.70000	-	151.4 Hz	IRA
	FM	KB9KLD	443.20000	+	103.5 Hz	IRA
McCook	DMR/MARC	N9CWM	440.75625	+	CC1	
McHenry	FM	WA9VGI	442.92500	+	114.8 Hz	
	FM	KB9I	444.07500	+	88.5 Hz	IRA
Melrose Park	FM	W9FT	442.62500	+	114.8 Hz	IRA
	FM	K9VMP	443.87500	+	114.8 Hz	IRA
Metamora	DMR/MARC	KC9GQR	444.00000	+	CC1	

ILLINOIS   155

Location	Mode	Call sign	Output	Input	Access	Coordinator
Monmouth	FM	KD9J	444.32500	+	173.8 Hz	IRA
Monticello	IDAS	KB9ZAM	442.72500	+	103.5 Hz	IRA
Morris	DMR/MARC	KC9KKO	146.71500	-	CC1	
	DMR/MARC	KC9KKO	441.50000	+	CC1	
	DMR/MARC	KC9KKO	442.00000	442.60000	CC1	IRA
	DMR/MARC	KB9SZK	442.32500	+	CC0	IRA
	FM	KB9SZK	147.27000	+	107.2 Hz	IRA
Morrison	FM	KA9QYS	145.21000	-	114.8 Hz	IRA
Mount Carmel	FM	KC9MAK	147.25500	+	151.4 Hz	IRA
	FM	W9KTL	442.15000	+	203.5 Hz	
	FM	AI9H	442.32500	+	114.8 Hz	IRA
	FM	KC9MAK	443.87500	+	151.4 Hz	IRA
	FM	AI9H	444.77500	+	114.8 Hz	IRA
Mount Morris	FM	K9AMJ	145.13000	-	114.8 Hz	IRA
	FM	K9AMJ	147.10500	+	114.8 Hz	IRA
	FM	K9AMJ	224.12000	-	118.8 Hz	IRA
	FM	K9AMJ	224.84000	-	118.8 Hz	IRA
	FM	K9AMJ	442.67500	+	114.8 Hz	IRA
Mount Pulaski	FM	N9NWI	443.82500	+	94.8 Hz	
Mount Vernon	FM	KB9KDE	147.13500	+	88.5 Hz	IRA
Mt. Pulaski	FM	N9NWI	223.90000	-	94.8 Hz	
Mulberry Grove, TNG Wireless T						
	FM	W9KXQ	224.14000	-	103.5 Hz	IRA
Mundelein	DMR/MARC	WB9PHK	423.29375	420.29375	CC2	
Naperville	FM	WA9WSL	145.17000	-	103.5 Hz	IRA
	FM	W9NPD	224.20000	-	110.9 Hz	IRA
	FM	NE9MA	443.05000	+	114.8 Hz	IRA
New Lenox	DMR/MARC	N2BJ	444.40000	+	CC2	IRA
	FM	WB9IRL	145.21000	-	107.2 Hz	IRA
Niles, Leaning Tower YMCA						
	FM	W9FO	147.31500	+	107.2 Hz	IRA
Noble	FM	KC9RHH	442.37500	+		
Normal	FM	WB9UUS	442.70000	+	107.2 Hz	IRA
North Riverside	FM	K9ONA	224.82000	-	110.9 Hz	IRA
Northbrook	FM	NS9RC	224.32000	-	110.9 Hz	IRA
O'Fallon	FM	K9AIR	443.10000	+	127.3 Hz	IRA
Oak Brook	FM	N9XKY	443.12500	+		IRA
Oak Forest	FM	N9ZD	443.27500	+	114.8 Hz	IRA
Oak Lawn	FM	W9OAR	444.90000	+	114.8 Hz	IRA
Oblong	DMR/BM	W9DJF	444.87500	+	CC1	IRA
Olive Branch	FM	K9IM	147.25500	+	118.8 Hz	IRA
Oregon	FM	N9AUW	147.16500	+	146.2 Hz	IRA
Orland Park	FM	WA9PAC	444.77500	+	114.8 Hz	IRA
	FM	WD9HGO	444.85000	+	71.9 Hz	IRA
Oswego	FM	NK9M	224.92000	-	110.9 Hz	IRA
Palatine	FM	KA9ORD	443.00000	+	114.8 Hz	IRA
Pana	DMR/MARC	KB9TZQ	444.72500	+	CC1	IRA
	FM	KB9TZQ	145.15000	-	94.8 Hz	IRA
Park Forest	FM	WB9UAR	223.96000	-	110.9 Hz	IRA
Park Ridge	DSTAR	WA9ORC	441.90625	+		
	FM	WA9ZMY	224.78000	-		IRA
Pawnee	FM	N9RYR	442.60000	+	94.8 Hz	IRA
Pekin	FM	W9FED	147.39000	+	110.9 Hz	
Peoria	DMR/BM	KB9YVN	442.22500	+	CC12	
	DMR/BM	WX9PIA	442.50000	+	CC12	IRA
	FM	W9UVI	443.87500	+	156.7 Hz	IRA
	FM	N9BBO	444.37500	+	156.7 Hz	IRA
Peoria, Bradley University - G						
	FM	W9JWC	444.47500	+	103.5 Hz	IRA
Peoria, Office Of Emergency Ma						
	FM	WX9PIA	147.33000	+	103.5 Hz	IRA

156   ILLINOIS

Location	Mode	Call sign	Output	Input	Access	Coordinator
Peoria, Proctor Hospital						
	FM	K9WRA	443.17500	+	156.7 Hz	IRA
Plano	DMR/MARC	W9XA	443.65625	+	CC6	IRA
Plato Center	DMR/MARC	WR9ABQ	444.97500	+	CC1	
	FM	W8ZS	223.92000	-	114.8 Hz	IRA
	FM	W9ZS	444.97500	+		IRA
Princeton, 911 Tower						
	FM	KD9ABX	444.92500	+	118.8 Hz	IRA
Quincy	DSTAR	W9AWE	147.19500	+		IRA
	DSTAR	W9AWE	441.89375	+		
	FM	W9AWE	147.03000	+	103.5 Hz	IRA
	FM	W9AWE	443.90000	+	103.5 Hz	IRA
Roanoke, Woodford County EMA B						
	FM	K9WRA	444.75000	+	DCS 464	IRA
Roanoke, Woodford EMA Building						
	FM	K9WRA	147.25500	+	103.5 Hz	IRA
Robinson, Water Tower						
	FM	WA9ISV	147.36000	+	107.2 Hz	IRA
	FM	WA9ISV	442.80000	+	107.2 Hz	IRA
Rock Falls	FM	KB9LRT	444.70000	+		
Rock Island	DSTAR	W9QCR	440.83125	+		IRA
Rock Island, Trinity Hospital						
	FM	W9QCR	444.90000	+	100.0 Hz	IRA
Rockford	DMR/BM	WX9MCS	147.25500	+	CC1	IRA
	DMR/MARC	WW9P	443.32500	+	CC1	IRA
	DMR/MARC	WX9MCS	443.45000	+	CC1	
	FM	W9AXD	147.00000	+	114.8 Hz	IRA
	FM	K9AMJ	224.04000	-	118.8 Hz	IRA
	FM	K9AMJ	224.44000	-	118.8 Hz	IRA
	FM	K9RFD	442.65000	+	192.8 Hz	IRA
	FM	WW9P	442.77500	+	118.8 Hz	IRA
	FM	K9AMJ	444.35000	+	114.8 Hz	IRA
Rockton	FM	N9JTA	440.87500	+	88.5 Hz	IRA
Rolling Meadows	FM	N9EP	442.80000	+		IRA
	FM	N9EW	444.92500	+		IRA
Round Lake	FM	W9GWP	444.60000	+	114.8 Hz	IRA
Round Lake Beach						
	FM	N9VUD	443.10000	+	114.8 Hz	
	FM	N9JSF	443.77500	+		IRA
	FUSION	KC9NSA	440.57500	+	114.8 Hz	IRA
Salem, I-57 And US-50						
	FM	W9CWA	147.27000	+	103.5 Hz	IRA
	FM	W9CWA	442.20000	+	103.5 Hz	IRA
Sandwich	FM	KA9QPN	444.42500	+	131.8 Hz	IRA
Sandwich, Sandwich High School						
	FM	N9EF	443.50000	+	123.0 Hz	IRA
Savanna	FM	N9FID	147.13500	+	107.2 Hz	IRA
Schaumburg	DMR/MARC	WB9PHK	146.70000	-	CC1	
	DMR/MARC	K9PW	439.87500	430.47500	CC1	
	DMR/MARC	WB9PHK	443.06875	+	CC0	IRA
	DMR/MARC	WB9PHK	443.08125	+	CC1	IRA
	DMR/MARC	K9MOT	443.57500	+	CC1	IRA
	DMR/MARC	K9MOT	444.79375	+	CC1	IRA
	DMR/MARC	WB9PHK	927.66250	902.66250	CC0	IRA
	FM	K9IIK	145.23000	-	107.2 Hz	IRA
	FM	N9CXQ	147.28500	+	107.2 Hz	IRA
	FM	K9EL	224.56000	-	110.9 Hz	IRA
	FM	WB9YBM	224.66000	-		IRA
	FM	N9CXQ	224.76000	-	110.9 Hz	IRA
	FM	KB2MAU	224.88000	-	110.9 Hz	IRA
	FM	K9MOT	442.17500	+		IRA
	FM	K9IIK	442.27500	+	114.8 Hz	IRA

ILLINOIS   157

Location	Mode	Call sign	Output	Input	Access	Coordinator
Schaumburg	FM	WB9PHK	443.10000	+	114.8 Hz	IRA
	FM	N9CXQ	443.62500	+	114.8 Hz	IRA
	FM	N9KNS	443.72500	+		IRA
	FM	K9PW	444.50000	+	114.8 Hz	IRA
	FM	KB2MAU	444.80000	+	203.5 Hz	IRA
Schiller Park	DMR/MARC	K9TOW	441.33750	+	CC4	IRA
	FM	WB9AET	224.98000	-	110.9 Hz	IRA
Shiloh	DMR/BM	WS9IDG	444.30000	+	CC7	IRA
	FM	AA9RT	145.11000	-		IRA
Skokie	DMR/BM	KB9JRC	440.45000	+	CC1	
Skokie, Old Orchard Shopping C						
	FM	KB9TAP	443.17500	+	127.3 Hz	IRA
Springfield	DSTAR	W9DUA	443.78125	+		IRA
	FM	W9DUA	443.00000	+	94.8 Hz	IRA
	FM	WA9KRL	443.37500	+	94.8 Hz	IRA
	FM	K9CZ	444.32500	+		IRA
	FM	KB9TZS	444.40000	+	103.5 Hz	IRA
Sterling	FM	N9JWI	444.02500	+	82.5 Hz	IRA
Stockton	FM	N9NIX	443.97500	+	127.3 Hz	IRA
Sugar Grove	DMR/MARC	K9NRO	442.42500	+	CC1	IRA
	FM	KA9HPL	442.47500	+	103.5 Hz	IRA
Swansea	DSTAR	KC9WKE	444.17500	+	DCS 025	IRA
Tallula	FM	W9DUA	442.67500	+	151.4 Hz	IRA
	FM	W9DUA	444.90000	+	151.4 Hz	IRA
Taylorville	FM	N9FU	442.05000	+	79.7 Hz	IRA
Tinley Park	FM	W9IC	441.80000	+	107.2 Hz	IRA
Towerhill, Williamsburg Hill						
	FM	WB9QPM	147.39000	+	203.5 Hz	IRA
Tremont	DMR/BM	W6PC	444.97500	+	CC12	
Tremont / Peoria	DMR/MARC	W6PC	444.15000	+	CC1	
Tremont, Groveland Tower						
	FM	W9TAZ	444.55000	+		IRA
Trivoli	DMR/BM	KT9Y	442.10000	+	CC12	
Troy	FM	KT9R	442.65000	+	79.7 Hz	IRA
Tunnel Hill	FM	W9WG	147.34500	+	88.5 Hz	IRA
	FM	WB9F	224.86000	-	88.5 Hz	IRA
Urbana	FM	KD9FDD	147.06000	+	162.2 Hz	IRA
Utica	FM	KC9CFU	145.29000	-	103.5 Hz	IRA
Versailles	FM	KB9JVU	29.68000	-	103.5 Hz	IRA
	FM	KB9JVU	443.92500	+	88.5 Hz	
Washington / Morton						
	DMR/BM	KB9YVN	442.20000	+	CC12	IRA
Waterloo	FM	KC0TPS	147.25500	+		IRA
	FM	N9OMD	444.70000	+	127.3 Hz	IRA
Watseka	FM	AD9L	444.62500	+	103.5 Hz	IRA
Wauconda	DMR/MARC	N9CWM	441.75625	+	CC1	
	FM	K9SGR	442.50000	+	114.8 Hz	IRA
Waukegan	FM	N9IJ	441.37500	+	88.5 Hz	
	FM	AA9RA	442.17500	+		IRA
Wayne	DSTAR	W9AUX	440.26875	+		IRA
	DSTAR	W9AUX	1292.60000	1272.60000		IRA
West Chicago	DMR/MARC	WD9BBE	442.96250	+	CC5	
	FM	N9XP	224.64000	-	110.9 Hz	IRA
	FM	W9DMW	927.70000	902.70000	114.8 Hz	IRA
West Frankfort	FM	AB9ST	147.04500	+	97.4 Hz	IRA
Westmont	FM	N9TO	223.86000	-	110.9 Hz	IRA
Westville	DMR/BM	W9MJW	443.08750	+	CC1	
Wheaton	FM	W9BZW	147.36000	+	136.5 Hz	IRA
	FM	W9CCU	224.14000	-	110.9 Hz	IRA
	FM	KA9KDC	444.27500	+	114.8 Hz	IRA
Wheeling	FM	WB9OUF	444.32500	+	114.8 Hz	IRA
Winnebago	DMR/MARC	W9TMW	440.05625	+	CC3	IRA

158   ILLINOIS

Location	Mode	Call sign	Output	Input	Access	Coordinator
Winnebago	FM	W9TMW	442.35000	+	88.5 Hz	IRA
Worth	FM	WA9ORC	224.18000	-	110.9 Hz	IRA
Yorkville	FM	WX9KRC	145.15000	-	103.5 Hz	IRA
	FM	WA9BSA	443.55000	+	114.8 Hz	IRA
Zion	FM	KA9VMV	444.82500	+	74.4 Hz	IRA

INDIANA

Location	Mode	Call sign	Output	Input	Access	Coordinator
Atlanta	DMR/MARC	KB9PFM	442.47500	+	CC1	
Attica	DMR/BM	W9ABH	442.97500	+	CC1	
Bedford	DMR/MARC	N9UMJ	442.47500	+	CC1	
	DMR/MARC	W9QYQ	444.05000	+	CC1	
Bloomington	DMR/BM	K9IU	147.18000	+	CC1	
	DMR/MARC	K9IU	444.90000	+	CC1	
Brookville	DMR/BM	KD9COF	441.93125	+	CC1	
Columbia CIty	DMR/BM	N9MTF	442.80000	+	CC1	
Connersville	DMR/MARC	W2NAP	441.97500	+	CC1	
Covington	DMR/BM	W9ABH	443.98750	+	CC1	
Crown Point	DMR/MARC	N9IAA	444.35000	+	CC1	
Culver	DMR/MARC	N9GPY	443.92500	+	CC1	
Danville	DMR/BM	WX9HC	145.13000	-	CC1	
	DMR/MARC	WX9HC	444.57500	+	CC1	
Edinburgh	DMR/MARC	KC9TKJ	441.91250	+	CC1	
Elkhart	DMR/MARC	K9DEW	444.05000	+	CC1	
Evansville	DMR/BM	W9OG	442.18750	+	CC1	
Ferdinand	DMR/MARC	KC9CFM	444.17500	+	CC1	
Floyds Knobs	DMR/MARC	KC9TKJ	443.11250	+	CC1	
Frankfort	DMR/MARC	W9SMJ	441.37500	+	CC1	
Freetown	DMR/BM	NA9VY	145.78500		CC1	
Ft. Wayne	DMR/MARC	K9MMQ	443.10000	+	CC1	
Galveston	DMR/MARC	W9SMJ	441.85000	+	CC1	
Gary	DMR/MARC	W9CTO	146.91000	-	CC1	
	DMR/MARC	W9CTO	442.75000	+	CC1	
Indianapolis	DMR/BM	W9AMT	145.41000	-	CC1	
	DMR/BM	W9AMT	441.20000	+	CC1	
	DMR/MARC	W9AMT	927.93750	902.93750	CC1	
Kokomo	DMR/MARC	W9SMJ	444.60000	+	CC1	
La Porte	DMR/MARC	N9IAA	444.67500	+	CC1	
Lawrenceburg	FM	KB9GPS	443.87500	+	146.2 Hz	
Lowell	FM	KA9OOL	443.95000	892.90000	131.8 Hz	
Lynn	DMR/MARC	N9CZV	441.17500	+	CC1	
Marion	DMR/BM	KB9CRA	442.75000	+	CC1	
Mishawaka	DMR/MARC	K9DEW	442.05000	+	CC1	
Mobile	DMR/MARC	KB8SWR	444.70000	+	CC1	
Monrovia	DMR/MARC	NF9K	443.47500	+	CC1	
Muncie	DMR/MARC	W9AMT	441.28750	+	CC1	
	DMR/MARC	N9CZV	441.30000	+	CC1	
New Castle	DMR/MARC	N9CZV	441.56750	+	CC1	
	DMR/MARC	K9NZF	441.95500	+	CC1	
Noblesville	DMR/MARC	NF9K	441.57500	+	CC5	
	DMR/MARC	NF9K	443.10000	+	CC1	
	DMR/MARC	K3HTK	444.41250	+	CC1	
Osgood	DMR/MARC	N9ICV	434.60000		CC2	
Peru	DMR/MARC	W9SMJ	441.26250	+	CC1	
Plainfield	DMR/BM	N9ALD	441.50000	+	CC1	
Plymouth	DMR/MARC	N9GPY	444.92500	+	CC1	
Portable	DMR/MARC	N9DMR	441.21250	+	CC1	
Portage	FM	W9BUZ	445.95000	-	131.8 Hz	
Reynolds	DMR/MARC	KC9PQA	441.95625	+	CC1	
Richmond	DMR/MARC	N9CZV	444.55000	+	CC1	
Roanoke	DMR/MARC	K9MMQ	442.92500	+	CC1	
Saint Paul	DMR/MARC	KD9GUD	441.98125	+	CC1	
Scipio	FM	N9JWR	443.02500	+	103.5 Hz	IRC

INDIANA 159

Location	Mode	Call sign	Output	Input	Access	Coordinator
Shelbyville	DMR/MARC	W9NTP	441.48750	+	CC1	
	DMR/MARC	W9NTP	441.84375	+	CC1	
South Bend	DMR/MARC	N9IAA	443.42500	+	CC1	
Terre Haute	DMR/BM	K9IKQ	442.08750	+	CC1	
Upland	DMR/BM	KB9TTX	442.40000	+	CC1	
Valparaiso	DMR/MARC	N9IAA	441.57500	+	CC1	
West Lafayette	DMR/BM	W9YB	443.60000	+	CC1	
Winchester	DMR/MARC	N9CZV	441.80000	+	CC1	

IOWA

Location	Mode	Call sign	Output	Input	Access	Coordinator
Ackley	FM	WB0EMJ	145.11000	-		IRC
	FM	WB0EMJ	147.25500	+	136.5 Hz	IRC
	FM	WB0EMJ	223.85000	-	136.5 Hz	
	FM	KB0EMJ	443.75000	+	136.5 Hz	
Afton	FM	AC0IK	442.40000	+	151.4 Hz	IRC
Algona	DMR/BM	K0HU	444.82500	+	CC2	
	FM	KC0MWG	147.21000	+	110.9 Hz	IRC
	FM	KC0MWG	444.82500	+	110.9 Hz	IRC
Alleman, Broadcast Antenna Far						
	FM	W0DM	145.31000	-	114.8 Hz	IRC
Ames	DMR/BM	W0YR	443.97500	+	CC1	
	FM	W0YL	147.24000	+	114.8 Hz	IRC
	FM	KI0Q	443.25000	+	114.8 Hz	IRC
	FM	W0ISU	443.37500	+	114.8 Hz	IRC
	FM	W0DP	444.25000	+		IRC
Ames, ISU	FM	W0ISU	147.37500	+	114.8 Hz	IRC
Anamosa	DMR/BM	K7PEM	444.63750	+	CC1	
	FM	W0CWP	145.39000	-	77.0 Hz	IRC
Atlantic	FM	N0DYB	147.15000	+	151.4 Hz	IRC
Audubon	FM	WA0GUD	147.12000	+		IRC
Avoca	FM	N0DYB	147.25500	+	151.4 Hz	IRC
	FM	N0DYB	443.95000	+	151.4 Hz	IRC
Baxter	FM	KC0NFA	444.22500	+	151.4 Hz	IRC
Bedford	DMR/BM	KA0RDE	442.25000	+	CC1	
	FM	KA0RDE	443.70000	+	136.5 Hz	
	FM	KA0RDE	443.75000	+	136.5 Hz	IRC
Bedford, Iowa	FM	KA0RDE	147.13500	+	127.3 Hz	IRC
Beetown	DMR/BM	KD9HAE	443.20000	+	CC2	
Bondurant	DMR/BM	N0MB	443.83750	+	CC1	
Boone, Hospital	FM	KB0TLM	146.85000	-		IRC
	FM	KB0TLM	443.90000	+		IRC
Breda	FM	N0NAF	147.28500	+	110.9 Hz	IRC
Bridgewater	DMR/MARC	WD0FIA	443.80000	+	CC10	IRC
	FM	WD0FIA	145.23000	-	136.5 Hz	IRC
	FM	WD0FIA	224.82000	-	136.5 Hz	IRC
Brooklyn	FM	K0TSK	442.65000	+		
Brunsville	FM	KD0XD	444.22500	+	110.9 Hz	IRC
Burlington	FM	W0LAC	146.79000	-	100.0 Hz	IRC
Calrinda, KMA-FM Tower						
	FM	N0NHB	145.35000	-		IRC
Carroll	FM	KC0UIO	146.80500	-	110.9 Hz	IRC
Castana	FM	K0BVC	145.47000	-	136.5 Hz	IRC
Cedar Falls	DMR/BM	N0CF	444.65000	+	CC1	
	DMR/BM	N0CF	927.08750	902.08750	CC1	
	FM	NK0T	224.90000	-	136.5 Hz	IRC
Cedar Rapids	DMR/BM	NO3R	442.00000	+	CC1	
	DMR/BM	W0GQ	443.60000	+	CC1	IRC
	DSTAR	N0CXX	147.40500	146.40500		
	DSTAR	N0CXX	442.05000	+		
	FM	W0WSV	145.15000	-	192.8 Hz	
	FM	N0DX	145.19000	-	192.8 Hz	
	FM	W0GQ	146.74500	-	192.8 Hz	IRC

160   IOWA

Location	Mode	Call sign	Output	Input	Access	Coordinator
Cedar Rapids	FM	W0WSV	147.09000	+	192.8 Hz	IRC
	FM	W0WSV	224.94000	-	192.8 Hz	IRC
	FM	W0VCK	443.00000	+	192.8 Hz	
	FM	N0MA	443.80000	+	192.8 Hz	
	FM	NN0V	444.30000	+	192.8 Hz	
	FM	W0IY	927.11250	902.11250	DCS 432	
Chariton	FM	KB0AJ	146.83500	-	123.0 Hz	IRC
Chelsea	FM	WD0GAT	442.12500	+	151.4 Hz	IRC
Clarinda	FM	K0SKU	146.97000	-	136.5 Hz	
Clear Lake	DMR/BM	KK6RQ	147.30000	144.30000	CC1	
Clinton	DMR/MARC	KD0WY	147.31500	+	CC23	
	FM	W0CS	145.43000	-	100.0 Hz	
	FM	KN0BS	224.18000	-	136.5 Hz	IRC
Coralville	FM	W0FDA	147.15000	+	192.8 Hz	IRC
	FM	K0GH	444.75000	+	151.4 Hz	IRC
Council Bluffs	FM	WB0GXD	442.52500	+		IRC
Council Bluffs, I	FM	K0SWI	146.82000	-		IRC
Cresco, Hawkeye Tri County REC						
	FM	W0CYY	146.92500	-	103.5 Hz	IRC
Creston	FM	N0GMH	146.79000	-	136.5 Hz	IRC
Davenport	FM	WB0BIZ	421.25000	910.25000		IRC
Davenport, KWQC						
	FM	W0BXR	146.70000	-		IRC
Davenport, Saint Ambrose Unive						
	FM	W0BXR	146.94000	-		
Denison	FM	K0CNM	147.09000	+		IRC
	FM	KC0LGI	147.33000	+		IRC
	FM	K0FZZ	147.42000		110.9 Hz	
	FM	KC0LGI	444.00000	+		IRC
Des Moines	DMR/BM	N0VPR	443.10000	+	CC1	
	DMR/BM	W0KWM	443.50000	+	CC1	
	DMR/BM	W0RAY	444.20000	+	CC1	
	DMR/BM	W0RAY	444.87500	+	CC1	IRC
	DSTAR	KD0IAN	147.10500	-		IRC
	FM	WA0QBP	147.30000	+	114.8 Hz	IRC
	FM	WA0QBP	444.62500	+	151.4 Hz	
	FM	K0SXY	444.67500	+	114.8 Hz	IRC
Des Moines, Alleman						
	FM	W0KWM	444.57500	+	151.4 Hz	
Des Moines, Broadlawn's Hospit						
	FM	K0DSM	444.05000	+	151.4 Hz	
Des Moines, IA	FM	W0AK	145.13000	-	114.8 Hz	IRC
Des Moines, Lucas State Office						
	FM	W0KWM	146.82000	-	203.5 Hz	IRC
	FM	WD0FIA	224.98000	-	114.8 Hz	IRC
Des Moines, Methodist Hospital						
	FM	K0DSM	444.10000	+	151.4 Hz	IRC
Des Moines, Park And Fleur						
	FM	W0AK	146.94000	-	114.8 Hz	IRC
Des Moines, Sherman Hills						
	FM	KD0WPK	29.67000	-	103.5 Hz	IRC
Des Moines, VA Hospital						
	FM	W0AK	146.98500	-	114.8 Hz	IRC
Dubuque, Iowa	FM	W0OIC	146.79000	-		
Dumont	FM	N0RJJ	145.43000	-	136.5 Hz	IRC
Early	FM	W0DOG	146.61000	-		
Eldridge	FM	W0BXR	146.88000	-	77.0 Hz	IRC
Estherville, Airport	FM	W0MDM	146.70000	-		IRC
Fairport	FM	N2AM	444.95000	+	100.0 Hz	
Fayette	DMR/BM	W0OEL	443.95000	+	CC1	IRC
	FM	W0OEL	147.34500	+	103.5 Hz	IRC
Forest City	FM	WB0URC	147.27000	+	103.5 Hz	IRC

IOWA 161

Location	Mode	Call sign	Output	Input	Access	Coordinator
Fort Dodge	FM	K0RJV	146.68500	-	110.9 Hz	IRC
Fort Madison	FM	WF0RT	146.86500	-	100.0 Hz	IRC
Frankville	FM	K0RTF	146.67000	-	103.5 Hz	IRC
	FM	K0RTF	444.10000	+	103.5 Hz	IRC
Garrison	FM	K0DKS	145.23000	-	141.3 Hz	IRC
Gilman	FM	NF0T	53.03000	51.33000	151.4 Hz	IRC
Glenwood	FM	N0WKF	145.29000	-		IRC
	FM	N0WKF	444.32500	+		IRC
Greenfield	DMR/BM	N0BKB	444.70000	+	CC1	IRC
	FM	N0BKB	146.86500	-	146.2 Hz	IRC
Grimes	DMR/BM	N0INX	444.72500	+	CC1	IRC
	DSTAR	KD0IAN	443.17500	+		IRC
	FM	N0INX	53.25000	51.55000	110.9 Hz	IRC
	FM	N0INX	146.61000	-	114.8 Hz	IRC
	FM	N0INX	224.54000	-	114.8 Hz	IRC
	FM	N0INX	443.40000	+	151.4 Hz	IRC
	FM	N0INX	927.02500	902.02500	156.7 Hz	IRC
	FM	N0INX	927.43500	902.43500		
Grundy Center	DMR/BM	W0RBK	443.87500	+	CC1	
	FM	N0MXK	444.32500	+	110.9 Hz	IRC
Grundy Center, High School						
	FM	W0RBK	146.65500	-	136.5 Hz	IRC
Herndon	FM	N2RDP	444.27500	+		IRC
Homestead	FM	WC0C	442.42500	+	151.4 Hz	IRC
Honey Creek	FM	AB0VX	145.41000	-	97.4 Hz	IRC
	FM	AB0VX	444.80000	+	97.4 Hz	IRC
Humboldt	DMR/BM	K0HU	145.50000	-	CC3	
	DMR/BM	K0HU	147.18000	+	CC3	
	DMR/BM	K0HU	442.40000	+	CC3	IRC
	FM	NI0A	147.39000	+		IRC
Independence	DMR/BM	KC0RMS	442.90000	+	CC1	IRC
	FM	KC0RMS	145.33000	-	103.5 Hz	IRC
Indianola	FM	N0FAM	444.12500	+	114.8 Hz	IRC
Indianola, Indianola Fire Stat						
	FM	KD0FGV	146.64000	-	114.8 Hz	IRC
Indianola, Pickard Recreation						
	FM	N0FAM	147.19500	+	114.8 Hz	IRC
Iowa City	DMR/BM	KD0MVJ	443.47500	+	CC1	
	DMR/BM	K0GH	443.77500	+	CC1	
	FM	KE0BX	145.47000	-	100.0 Hz	
	FM	W0JV	146.85000	-	192.8 Hz	IRC
Johnston	DMR/BM	N0VPR	442.50000	+	CC1	
	FM	KC0MTI	147.16500	+	114.8 Hz	IRC
	FM	KC0MTI	442.80000	+	151.4 Hz	IRC
Johnston, Camp Dodge						
	FM	KC0MTI	146.70000	-	114.8 Hz	IRC
Kelly	FM	KC0MTI	444.42500	+	151.4 Hz	IRC
Keokuk	FM	KC0TPI	147.30000	+	146.2 Hz	IRC
Lamoni	FM	AA0OS	147.36000	+	114.8 Hz	IRC
Laurel	FM	WC0C	444.80000	+	151.4 Hz	IRC
Le Mars	DMR/BM	KI0EO	147.01500	+	CC1	IRC
	FM	KI0EO	444.67500	+	110.9 Hz	IRC
	FM	KD0VYD	446.12500	-	110.9 Hz	
Le Mars, Iowa	FM	KD0PMM	147.19500	+	110.9 Hz	IRC
LeMars	DMR/BM	KD0PMM	444.92500	+	CC1	
Lenox	DMR/BM	K9ADL	444.95000	+	CC1	
	FM	KD0TWE	146.88000	-	136.5 Hz	IRC
Leon	FM	K0FFX	444.97500	+	114.8 Hz	
Madrid	FM	N0QIX	145.25000	-	114.8 Hz	IRC
	FM	N0SFF	442.60000	+	151.4 Hz	IRC
	FM	KC0MTE	443.85000	+	151.4 Hz	IRC
Magnolia	DMR/MARC	WB0QQK	444.65000	+	CC1	

162   IOWA

Location	Mode	Call sign	Output	Input	Access	Coordinator
Manchester	FM	W0II	147.30000	+		IRC
Manilla	FM	N0JRX	147.22500	+	151.4 Hz	IRC
Marshalltown	DMR/BM	K0MIW	444.52500	+	CC2	
	FM	NF0T	147.13500	+	141.3 Hz	IRC
	FM	N0MXK	443.32500	+	110.9 Hz	IRC
	FM	NF0T	927.82500	902.82500	151.4 Hz	IRC
Mason City, Interstate Grain E						
	FM	W0MCW	146.76000	-	103.5 Hz	IRC
Mason City, Red & White Tower						
	FM	KB0JBF	147.31500	+	103.5 Hz	IRC
Mediapolis	FM	N0GES	443.10000	+	151.4 Hz	
Menlo	FM	N0BKB	147.04500	+	127.3 Hz	IRC
Minden	FM	N0GR	442.07500	+	136.5 Hz	IRC
Mineola	FM	N0WKF	442.02500	+		IRC
	FM	N0WKF	444.02500	+		IRC
Missouri Valley	DMR/BM	N0ZHX	444.07500	+	CC1	
Mondamin	FM	K0BVC	53.39000	51.69000	136.5 Hz	IRC
	FM	K0BVC	145.13000	-	136.5 Hz	IRC
	FM	K0BVC	444.92500	+	136.5 Hz	IRC
Moravia	DMR/BM	W0ALO	444.47500	+	CC1	IRC
	FM	W0ALO	146.92500	-	146.2 Hz	IRC
	FM	W0ALO	927.33750	902.33750	136.5 Hz	IRC
Mount Pleasant	DSTAR	KD0NJC	444.35000	+		
	FM	KE0IHD	146.45000		100.0 Hz	
	FM	W0MME	147.16500	+	156.7 Hz	
	FM	W0MME	147.39000	+		IRC
Mount Pleasant, Iowa						
	FM	KE0IHD	444.62500	+	100.0 Hz	
Mt. Ayr	FM	KA0RDE	443.30000	+	136.5 Hz	IRC
Mt. Pleasant	DMR/BM	WB0VHB	444.35000	+	CC1	
	DMR/BM	WB0VHB	444.52500	+	CC1	IRC
Muscatine	DMR/BM	AC0EC	444.12500	+	CC2	
	DMR/BM	KC0WWV	444.40000	+	CC1	
	FM	WA0VUS	444.27500	+	192.8 Hz	IRC
Muscatine, Muscatine Community						
	FM	KC0AQS	146.91000	-	192.8 Hz	IRC
Newton	FM	W0WML	147.03000	+	114.8 Hz	IRC
	FM	KC0NFA	442.30000	+	151.4 Hz	IRC
Ogden	FM	N0AN	144.39000			
	FM	N0AN	438.50000			IRC
Osceola	FM	KC0UNH	147.21000	+	136.5 Hz	
Oskaloosa	FM	KB0VXL	145.49000	-	146.2 Hz	IRC
Otho	DMR/BM	K0HU	443.57500	+	CC5	
Ottumwa	DMR/BM	KE0BX	444.85000	+	CC1	
	DMR/MARC	WA0DX	443.97500	+	CC1	
	FM	KE0BX	146.97000	-	100.0 Hz	
Pacific Junction	FM	N0WKF	443.02500	+		IRC
Paullina	FM	N0OYK	147.13500	+	110.9 Hz	IRC
Pella	FM	KE0SQA	145.17000	-	114.8 Hz	IRC
Perry	FM	KD0NEB	444.37500	+	151.4 Hz	IRC
Perry, L Ave. & RRVT Bike Trai						
	FM	KD0NEB	145.19000	-	114.8 Hz	IRC
Portsmouth	FM	K0BVC	146.74500	-	136.5 Hz	IRC
Prescott	FM	N0DTS	145.15000	-	127.3 Hz	IRC
Primghar	FM	KC0TQU	444.87500	+		
Red Oak	FM	N0NHB	146.65500	-	146.2 Hz	IRC
Reinbeck	DMR/BM	W0RBK	444.87500	+	CC1	IRC
Rock Valley, Perkins Corner (J						
	FM	W0VHQ	147.30000	+	110.9 Hz	IRC
Rockwell City	FM	K0FBP	145.49000	-	110.9 Hz	IRC
Sac City	FM	WD0CLO	146.92500	-		IRC
Saint Ansgar	FM	KC0VII	147.19500	+	103.5 Hz	IRC

IOWA   163

Location	Mode	Call sign	Output	Input	Access	Coordinator
Sanborn, On City Water Tower						
	FM	W0VHQ	145.31000	-	110.9 Hz	IRC
	FM	W0VHQ	443.90000	+		
Saylor Township, Saylor FD						
	FM	KD0QED	145.39000	-	114.8 Hz	IRC
Saylorville	FM	KB0NFF	443.67500	+		
Saylorville, East Mixmaster						
	FM	KB0NFF	444.50000	+		
Saylorville, Fire Dept Comm Ct						
	FM	KB0NFF	444.17500	+		IRC
Saylorville, I-35 & I-80 Cross						
	FM	KB0NFF	146.89500	-	114.8 Hz	IRC
Sheldahl	FM	N0QFK	53.09000	51.39000		IRC
	FM	W0QFK	147.07500	+	114.8 Hz	IRC
Shenandoah	FM	KB0NUR	145.21000	-		IRC
Sioux City	DMR/BM	KI0EO	444.62500	+	CC1	
	FM	K0AAR	146.91000	-	110.9 Hz	IRC
	FM	K0TFT	146.97000	-	110.9 Hz	IRC
	FM	K0TFT	147.06000	+	110.9 Hz	IRC
	FM	KC0DXD	147.27000	+	110.9 Hz	IRC
	FM	KS0F	443.57500	+	110.9 Hz	
	FM	K0NH	444.72500	+	110.9 Hz	IRC
Spencer	DMR/BM	KG0CK	146.82000	-	CC1	IRC
	DMR/BM	KG0CK	444.70000	+	CC1	IRC
	FM	WA0DOY	444.97500	+	110.9 Hz	IRC
Spirit Lake	FM	W0DOG	146.61000	-	110.9 Hz	IRC
Springbrook	FM	W0DBQ	147.06000	+	114.8 Hz	
Storm Lake	DMR/BM	WB0FNA	444.52500	+	CC1	
	FM	WA0UZI	146.77500	-		IRC
	FM	WB0FNA	444.75000	+		
Stratford	FM	K0KQT	146.62500	-		IRC
Thurman	DMR/MARC	WB0YLA	444.50000	+	CC1	IRC
	FM	WB0YLA	145.39000	-	136.5 Hz	IRC
Truro	FM	W0AK	146.98500	-	114.8 Hz	
Twin Lakes	FM	K0FBP	442.50000	+		IRC
Urbandale	FM	N2RDP	444.27500	+	151.4 Hz	
Van Meter	FM	N0XD	443.10000	+	151.4 Hz	IRC
Victor	DMR/BM	K0TSK	444.95000	+	CC1	
Wapello, Louisa County Sheriff						
	FM	KC0AQS	146.98500	-	192.8 Hz	
Washington	FM	W0ARC	147.04500	+	146.2 Hz	
	FM	W0ARC	443.70000	+	146.2 Hz	IRC
Waterloo	DMR/BM	W0GEN	442.82500	+	CC1	
	DMR/BM	W0GEN	443.27500	+	CC1	
	DMR/BM	W0GEN	443.72500	+	CC1	
	DMR/BM	W0ALO	444.90000	+	CC1	IRC
	DMR/MARC	W0EL	444.70000	+	CC1	
	FM	W0ALO	146.82000	-	136.5 Hz	
	FM	W0MG	444.97500	+	136.5 Hz	IRC
	P25	W0ALO	444.92500	+	136.5 Hz	IRC
Waterloo, Cedar Valley TechWor						
	FM	KB0VGG	444.55000	+	103.5 Hz	IRC
Waterloo, Kimball Ridge Center						
	FM	N0CF	927.06250	902.06250		IRC
Waterloo, Water Tower						
	FM	W0MG	146.94000	-	136.5 Hz	IRC
Webster	FM	N0PSF	146.91000	-		IRC
Webster City	FM	K0KWO	147.01500	+	103.5 Hz	IRC
West Bend	FM	N0DOB	145.17000	-	110.9 Hz	IRC
	FM	N0QQS	444.77500	+	110.9 Hz	IRC
West Des Moines, KB0SL QTH						
	FM	KB0SL	927.90000	902.90000	DCS 023	IRC

164   IOWA

Location	Mode	Call sign	Output	Input	Access	Coordinator
Williams	FM	W0MCW	444.50000	+	151.4 Hz	IRC
Winterset	FM	WA0O	147.27000	+	114.8 Hz	IRC
Woodbine	FM	K0BVC	444.35000	+	136.5 Hz	IRC

KANSAS

Location	Mode	Call sign	Output	Input	Access	Coordinator
Alma, K-Link	FM	W0KHP	444.52500	+	162.2 Hz	Kansas RC
Arkansas City	FM	WA0JBW	147.00000	+	97.4 Hz	Kansas RC
Basehor	FM	K0HAM	145.39000	-	88.5 Hz	Kansas RC
	FM	N0GRQ	443.55000	+	151.4 Hz	Kansas RC
Beaumont	FM	KS0KE	145.13000	-	156.7 Hz	Kansas RC
	FM	KD5IMA	443.52500	+	156.7 Hz	Kansas RC
Beloit, K-Link	FM	K0KSN	442.80000	+	DCS 162	Kansas RC
Carbondale	FM	KB0WTH	147.30000	+	88.5 Hz	Kansas RC
	FM	WB0PTD	443.12500	+		Kansas RC
Chanute	FM	KZ0V	147.10500	+	91.5 Hz	Kansas RC
Chanute, Walnut	FM	AI0E	146.74500	-	91.5 Hz	Kansas RC
Clay Center	DMR/MARC	N0XRM	442.75000	+	CC1	Kansas RC
	FM	N0XRM	147.16500	+	162.2 Hz	Kansas RC
Clay Center, Water Tower						
	FM	KD7QAS	145.15000	-	100.0 Hz	Kansas RC
Coffeyville	FM	WR0CV	146.61000	-	91.5 Hz	Kansas RC
	FM	NU0B	147.30000	+	91.5 Hz	Kansas RC
	FM	N0TAP	224.52000	-	91.5 Hz	Kansas RC
	FM	WR0MG	442.87500	+	91.5 Hz	Kansas RC
	FM	N0TAP	444.55000	+	91.5 Hz	Kansas RC
Colby	DMR/MARC	NV8Q	444.75000	+	CC1	Kansas RC
	FM	NW0K	146.82000	-	156.7 Hz	Kansas RC
Concordia	FM	K0KSN	146.86500	-		Kansas RC
Derby	FM	W0UUS	146.85000	-	156.7 Hz	Kansas RC
Dodge City	FM	K0ECT	147.03000	+	123.0 Hz	
	FM	K0ECT	442.37500	+	123.0 Hz	Kansas RC
	FM	KY0J	443.35000	+	123.0 Hz	Kansas RC
	FM	K0ECT	444.55000	+	141.3 Hz	Kansas RC
Dodge City, K-Link						
	FM	K0HAM	443.67500	+	162.2 Hz	Kansas RC
Edgerton	FM	WB0OUE	224.82000	-	151.4 Hz	Kansas RC
Effingham	FM	K0DXY	146.61000	-	100.0 Hz	
El Dorado	DMR/MARC	K0JWH	441.97500	+	CC1	
	FM	KB0VAC	147.15000	+		Kansas RC
El Dorado, K-Link	FM	K0JWH	443.10000	+	162.2 Hz	Kansas RC
Elkader	FM	K0ECT	444.30000	+	141.3 Hz	Kansas RC
Ellsworth, K-Link	FM	K0HAM	444.77500	+	162.2 Hz	Kansas RC
Emporia	DMR/BM	K0ESU	443.00000	+	CC1	Kansas RC
	FM	N0OFG	145.31000	-	103.5 Hz	Kansas RC
	FM	K0HAM	146.98500	-	88.5 Hz	Kansas RC
Flush	DMR/MARC	K0USY	444.80000	+	CC1	
Fort Scott	FM	K0EFJ	146.71500	-	91.5 Hz	Kansas RC
	FM	K0EFJ	444.17500	+	88.5 Hz	Kansas RC
Garden City	FM	WA0OQA	52.87000	51.17000	141.3 Hz	Kansas RC
	FM	WA0OQA	146.91000	-	141.3 Hz	Kansas RC
Garden City, Tennis						
	FM	K0ECT	442.50000	+	141.3 Hz	Kansas RC
Gardner	FM	K0NK	224.78000	-		Kansas RC
Gas	FM	WI0LA	147.37500	+	179.9 Hz	Kansas RC
Girard	FM	K0SEK	147.24000	+	91.5 Hz	Kansas RC
Great Bend	DMR/BM	N0VPX	442.07500	+	CC1	
	FM	KI0NN	146.76000	-	103.5 Hz	Kansas RC
	FM	KE0PTS	444.10000	+	100.0 Hz	Kansas RC
Hays	DMR/BM	N7JYS	147.04500	+	CC7	Kansas RC
	DMR/BM	N7JYS	224.28000	-	CC7	
	DMR/MARC	K0USY	444.82500	+	CC1	MACC-KARC
	FM	KE0KIY	147.18000	+	100.0 Hz	Kansas RC

KANSAS 165

Location	Mode	Call sign	Output	Input	Access	Coordinator
Hays	FM	N0ECQ	444.82500	+	114.8 Hz	Kansas RC
Hays, K-Link	FM	N7JYS	442.45000	+	162.2 Hz	Kansas RC
Haysville	FM	KA0RT	147.10500	+	151.4 Hz	Kansas RC
	FM	W0SY	444.97500	+	103.5 Hz	Kansas RC
Hiawatha	FM	WA0W	147.18000	+	88.5 Hz	Kansas RC
Hill City, K-Link	FM	N0NM	443.70000	+	162.2 Hz	Kansas RC
Hillsboro, K-Link	FM	WX0RG	442.50000	+	162.2 Hz	Kansas RC
Hoisington	FM	N7JYS	224.00000	-	131.8 Hz	
Hoisington, Beaver						
	FM	KE0PTS	444.92500	+	100.0 Hz	Kansas RC
Hoisington, KS	DMR/MARC	K0HAM	147.13500	+	CC1	Kansas RC
Holton, Courthouse						
	FM	AA0MM	146.77500	-		Kansas RC
Hoyt	FM	W0CET	145.27000	-	88.5 Hz	
	FM	K0HAM	444.72500	+	88.5 Hz	Kansas RC
Hutchinson	FM	W0WR	147.12000	+		Kansas RC
	FM	KA0HN	443.72500	+		Kansas RC
Hutchinson, KWCH-12						
	FM	W0UUS	146.82000	-	103.5 Hz	Kansas RC
	FM	W0UUS	444.65000	+	103.5 Hz	Kansas RC
Hutchinson, Partridge						
	FM	W0WR	146.67000	-		Kansas RC
Independence	FM	N0ID	147.01500	+	91.5 Hz	Kansas RC
	FM	KW0I	442.65000	+	91.5 Hz	Kansas RC
Independence, Elk City State L						
	FM	N0ID	145.49000	-	91.5 Hz	Kansas RC
Iola	FM	KS0AL	444.82500	+	91.5 Hz	Kansas RC
Isabel	FM	K0HPO	442.02500	+	103.5 Hz	Kansas RC
	FM	W5ALZ	442.40000	+	103.5 Hz	Kansas RC
Itinerant	DMR/MARC	WA0EDA	444.98750	+	CC1	
Junction City	FM	N0VGY	146.88000	-	DCS 162	Kansas RC
Junction City, K-Link						
	FM	KS1EMS	147.31500	+	162.2 Hz	Kansas RC
Kansas City	DMR/BM	WA0SLL	442.57500	+	CC1	
	DMR/BM	WD0GQA	442.85000	+	CC1	Kansas RC
	DMR/BM	WD0GQA	443.85000	+	CC1	Kansas RC
	FM	WB0NSQ	53.85000	52.15000		Kansas RC
	FM	WA0NQA	145.13000	-	151.4 Hz	Kansas RC
	FM	K0HAM	146.94000	-	88.5 Hz	Kansas RC
	FM	W0LB	147.15000	+	151.4 Hz	Kansas RC
Kansas City, Foxridge Towers						
	FM	WB0NSQ	444.85000	+	151.4 Hz	Kansas RC
Kansas City, KU Med						
	FM	KU0MED	442.32500	+	151.4 Hz	Kansas RC
Kansas City, Providence Medica						
	FM	W0KCK	147.21000	+	151.4 Hz	
Kingman	FM	KD0SLE	442.12500	+	103.5 Hz	Kansas RC
Lakin	FM	N0OMC	146.98500	-	156.7 Hz	Kansas RC
Lathrop	DMR/BM	KC0DMR	442.35000	+	CC1	Kansas RC
Lawrence	DMR/BM	KU0JHK	444.50000	+	CC7	Kansas RC
	FM	W0UK	146.76000	-	88.5 Hz	Kansas RC
	FM	KE0QNS	147.03000	+	88.5 Hz	Kansas RC
	FM	K0HAM	444.90000	+	88.5 Hz	MACC-KARC
Leavenworth	FM	N0KOA	145.33000	-		Kansas RC
	FM	KS0LV	147.00000	+	151.4 Hz	Kansas RC
Leavenworth, St. John Hospital						
	FM	W0ROO	444.80000	+	151.4 Hz	Kansas RC
Lenora, K-Link	FM	N0KOM	146.88000	-	162.2 Hz	Kansas RC
Liberal	FM	W0KKS	146.80500	-	103.5 Hz	Kansas RC
	FM	K0ECT	443.10000	+	141.3 Hz	Kansas RC
Logan	DMR/MARC	N0KOM	444.47500	+	CC1	Kansas RC
Louisburg	FM	K0HAM	53.13000	52.13000	88.5 Hz	MACC-KARC

166   KANSAS

Location	Mode	Call sign	Output	Input	Access	Coordinator
Louisburg	FM	K0HAM	145.41000	-		Kansas RC
	FM	K0HAM	147.31500	+	88.5 Hz	MACC-KARC
	FM	K0HAM	442.12500	+		Kansas RC
Manhattan	FM	W0QQQ	145.41000	-	151.4 Hz	Kansas RC
	FM	KS0MAN	147.25500	+	88.5 Hz	Kansas RC
	FM	KS0MAN	442.00000	+	88.5 Hz	Kansas RC
	FM	W0QQQ	444.17500	+	88.5 Hz	
Matfield Green, K-Link						
	FM	K0HAM	147.04500	+	88.5 Hz	Kansas RC
McPherson	DMR/BM	N5NIQ	444.80000	+	CC1	Kansas RC
	FM	W0TWU	147.33000	+	162.2 Hz	Kansas RC
McPherson, K-Link						
	FM	N0SGK	444.60000	+	162.2 Hz	Kansas RC
Medicine Lodge, Gyp Hills						
	FM	K0UO	146.88000	-	103.5 Hz	Kansas RC
Merriam	DMR/BM	KC0DMR	442.30000	+	CC4	
Merriam, Shawnee Mission Medic						
	FM	K0KN	927.71250	902.71250	151.4 Hz	Kansas RC
Minneapolis	DMR/MARC	NV8Q	147.22500	+	CC1	Kansas RC
	DMR/MARC	N7KLR	444.75000	+	CC1	Kansas RC
Minneapolis, K-Link						
	FM	KS0LNK	444.85000	+	162.2 Hz	Kansas RC
Mission	DMR/BM	WB0KIA	147.16500	+	CC1	Kansas RC
	DMR/BM	WB0YRG	442.10000	+	CC1	
	DMR/BM	WB0YRG	443.10000	+	CC1	Kansas RC
	DMR/MARC	WB0YRG	927.58750	902.58750	CC1	Kansas RC
	FM	WB0KIA	224.10000	-		Kansas RC
	FM	KC1WIZ	442.10000	+	167.9 Hz	Kansas RC
Montezuma	FM	K0ECT	444.25000	+	141.3 Hz	Kansas RC
Mound City	FM	WA0PPN	147.28500	+	91.5 Hz	Kansas RC
	FM	W0PT	444.42500	+		Kansas RC
Mulvane	FM	N0KTA	146.71500	-	100.0 Hz	Kansas RC
	FM	N0KTA	443.55000	+		Kansas RC
Neodesha	FM	KC0QYD	146.77500	-	91.5 Hz	Kansas RC
	FM	KC0QYD	442.20000	+	146.2 Hz	Kansas RC
Newton	FM	W0BZN	146.61000	-	103.5 Hz	Kansas RC
Norway, K-Link	FM	K0KSN	146.92500	-	162.2 Hz	Kansas RC
Oberlin	FM	KB0DZB	145.19000	-		Kansas RC
Olathe	DMR/BM	KD0JWD	145.23000	-	CC1	Kansas RC
	FM	N0CRD	52.97000	51.27000	91.5 Hz	Kansas RC
	FM	K0ECS	145.47000	-	151.4 Hz	
	FM	W0QQ	224.94000	-		
	FM	K0HAM	444.25000	+	88.5 Hz	Kansas RC
	FM	K0HCV	444.40000	+		
	FM	KD0JWD	927.03750	902.03750	151.4 Hz	Kansas RC
Olathe, Garmin Building						
	DSTAR	WA0RC	442.52500	+		Kansas RC
Olathe, I-35 And 135th St.						
	FM	KE5BR	442.20000	+	151.4 Hz	Kansas RC
Olathe, Sheridan Bridge						
	FM	N0CRD	442.62500	+	91.5 Hz	Kansas RC
Osawatomie	FM	AA0X	145.25000	-	151.4 Hz	Kansas RC
	FM	N0SWP	442.05000	+	151.4 Hz	Kansas RC
Osborne, K-Link	FM	NZ0M	147.37500	+	162.2 Hz	Kansas RC
Ottawa	DMR/BM	W0QW	147.39000	+	CC1	Kansas RC
Overland Park	DMR/MARC	W0WJB	146.83500	-	CC1	Kansas RC
	FM	W0ERH	145.29000	-	151.4 Hz	Kansas RC
	FM	K0HAM	146.91000	-		Kansas RC
	FM	W0LHK	442.15000	+	82.5 Hz	Kansas RC
	FM	W0ERH	443.72500	+		Kansas RC
Paola	FM	WS0WA	147.36000	+	151.4 Hz	Kansas RC
Paola, K-Link	FM	N0SWP	442.47500	+	151.4 Hz	Kansas RC

KANSAS 167

Location	Mode	Call sign	Output	Input	Access	Coordinator
Parsons	FM	AA0PK	146.68500	-	91.5 Hz	Kansas RC
Pawnee Rock	DMR/BM	N0VPX	444.32500	+	CC1	
	FM	KE0PTS	146.83500	-	100.0 Hz	Kansas RC
Phillipsburg	FM	AA0HJ	443.27500	+	100.0 Hz	Kansas RC
Pittsburg	DMR/MARC	K0PRO	146.94000	-	CC1	Kansas RC
	DMR/MARC	K0PRO	444.80000	+	CC1	Kansas RC
	FM	K0SEK	442.67500	+	91.5 Hz	Kansas RC
Plains	FM	WK0DX	147.18000	+		Kansas RC
	FM	WK0DX	443.50000	+	141.3 Hz	Kansas RC
Ransom, K-Link	FM	K0HAM	443.57500	+	162.2 Hz	Kansas RC
Riley, K-Link	FM	KS1EMS	146.68500	-	162.2 Hz	Kansas RC
Russell	DMR/BM	KB0SJR	442.00000	+	CC1	
	DMR/BM	N7JYS	442.47500	+	CC7	Kansas RC
	FM	AB0UO	147.28500	+	162.2 Hz	Kansas RC
	FM	KC0HFA	442.85000	+	141.3 Hz	Kansas RC
Russell, K-Link	FM	AB0UO	444.95000	+	162.2 Hz	Kansas RC
Saint Marys	FM	K0HAM	146.95500	-	88.5 Hz	Kansas RC
Salina	DMR/BM	NV8Q	444.37500	+	CC1	
	FM	W0CY	147.03000	+		Kansas RC
	FM	N0KSC	147.27000	+	118.8 Hz	Kansas RC
	FM	W0CY	443.90000	+	118.8 Hz	Kansas RC
Salina, K-Link	FM	N0KSC	442.20000	+	162.2 Hz	Kansas RC
Scott City	FM	WA0OQA	52.87000	51.17000		MACC-KARC
	FM	WA0OQA	146.70000	-	141.3 Hz	Kansas RC
Sedan	FM	WX0EK	146.95500	-	100.0 Hz	
Shawnee	FM	WB0HAC	145.21000	-	151.4 Hz	Kansas RC
	FM	W0ERH	223.94000	-	151.4 Hz	Kansas RC
	FM	W0ERH	442.60000	+		Kansas RC
Shawnee Mission	FM	K0GXL	443.52500	+	167.9 Hz	Kansas RC
Smith Center	FM	N0LL	146.61000	-		Kansas RC
Smolan	FM	WD0GAH	146.62500	-		Kansas RC
St. John	FM	W5ALZ	146.70000	-	103.5 Hz	Kansas RC
Sterling, K-Link	FM	WB0LUN	444.45000	+	162.2 Hz	Kansas RC
Syracuse	FM	KB0CKE	146.77500	-		Kansas RC
	FM	K0ECT	444.00000	+	141.3 Hz	Kansas RC
Topeka	DMR/BM	KC0DMR	442.87500	+	CC1	Kansas RC
	DMR/MARC	WV0S	146.80500	-	CC1	Kansas RC
	FM	K0HAM	52.91000	51.21000	88.5 Hz	Kansas RC
	FM	W0CET	145.45000	-	88.5 Hz	Kansas RC
	FM	WA0VRS	146.67000	-	88.5 Hz	Kansas RC
	FM	WA0VRS	224.84000	-	88.5 Hz	
	FM	W0SIK	442.02500	+		Kansas RC
	FM	W0CET	442.22500	447.32500	88.5 Hz	
	FM	W0CET	442.42500	+	88.5 Hz	Kansas RC
	FM	WA0VRS	443.92500	+	88.5 Hz	Kansas RC
	FM	N0CBG	444.40000	+	88.5 Hz	
Topeka, NEKSUN	FM	K0HAM	444.90000	+	88.5 Hz	Kansas RC
Tribune	FM	K0WPM	442.17500	+	156.7 Hz	
Udall	FM	KD0HNA	147.16500	+	97.4 Hz	
	FM	KD0HNA	444.30000	+	97.4 Hz	Kansas RC
Ulysses	FM	K0ECT	147.06000	+		Kansas RC
Ulysses, Wagon Bed Springs						
	FM	K0ECT	444.52500	+	141.3 Hz	Kansas RC
Wallace	FM	WA0VJR	444.60000	+	146.2 Hz	
Wichita	DMR/BM	W0SOE	145.17000	-	CC3	Kansas RC
	DMR/BM	W0SOE	442.17500	+	CC3	Kansas RC
	DSTAR	W0SOE	442.52500	+		Kansas RC
	FM	KC0AHN	146.89500	-		Kansas RC
Wichita, College Hill						
	FM	N0EQS	444.57500	+	100.0 Hz	Kansas RC
Wichita, Colwich	FM	WA0RJE	146.94000	-	103.5 Hz	Kansas RC
	FM	WA0RJE	444.00000	+		Kansas RC

168　KANSAS

Location	Mode	Call sign	Output	Input	Access	Coordinator
Wichita, K-Link	FM	W0VFW	443.32500	+	162.2 Hz	Kansas RC
Wichita, VFW 3115						
	FM	W0VFW	145.27000	-	103.5 Hz	Kansas RC
Wichita, WMC	FM	W0SOE	146.79000	-	103.5 Hz	Kansas RC
	FM	W0SOE	442.25000	+		Kansas RC
	FM	W0SOE	442.32500	+	103.5 Hz	
Wilson	FM	K0BHN	146.97000	-	118.8 Hz	Kansas RC
Winfield	FM	WA0JBW	145.19000	-		Kansas RC
	FM	WA0JBW	444.02500	+	97.4 Hz	Kansas RC
Winfield, Downtown						
	FM	N5API	442.10000	+	97.4 Hz	Kansas RC

KENTUCKY

Location	Mode	Call sign	Output	Input	Access	Coordinator
Agnes	FM	W4RRA	145.45000	-	77.0 Hz	
Ashland	DMR/MARC	K4AHS	145.41000	-	CC1	
	DMR/MARC	K4AHS	440.61250	+	CC1	
	FM	KC4QK	147.24000	+	107.2 Hz	
	FM	KC4QK	223.94000	-	107.2 Hz	
Ashland, Tarpin Ridge						
	DMR/BM	KY4TVS	145.41000	-	CC1	SERA
	FM	KG4DVE	146.94000	-	107.2 Hz	SERA
Bardstown	DMR/BM	K4KTR	443.00000	+	CC1	
	FM	KB4KY	145.47000	-	151.4 Hz	
	FM	W4CMY	442.05000	+	91.5 Hz	SERA
Beaver Dam	FM	KI4HEC	145.17000	-	136.5 Hz	SERA
Berea	FM	KF4OFT	146.71500	-	100.0 Hz	
Big Hill, Big Hill Mountain						
	FM	AJ4AJ	224.36000	-	77.0 Hz	
Blackey	DMR/BM	KY4RDB	446.50000	+	CC1	
Bonnieville	FM	KY4X	146.89500	-	114.8 Hz	
	FM	KY4X	444.85000	+	114.8 Hz	SERA
Bowling Green	DMR/BM	W4WSM	146.62500	-	CC11	SERA
	DMR/BM	W4WSM	444.56250	+	CC1	
	DMR/MARC	W4WSM	444.70000	+	CC1	SERA
	FM	W4WSM	147.16500	+		SERA
	FM	KY4BG	147.33000	+	107.2 Hz	
	FM	W4WSM	444.10000	+	100.0 Hz	SERA
Brooks, WAMZ Tower						
	FM	KY4KY	146.70000	-	79.7 Hz	SERA
Buckhorn Lake	FM	K4XYZ	147.37500	+	103.5 Hz	
Burkesville	FM	KI4NTU	147.37500	+	100.0 Hz	
Burnside, Alien Grave Mountain						
	ATV	KY6MTR	53.03000	52.03000		
Buttonsberry	FM	KY4MA	146.73000	-	82.5 Hz	
Campbellsville	DSTAR	WA4UXJ	146.81000	-		
Cane Valley, WGRB TV						
	FM	WA4UXJ	146.64000	-		
Carrollton	FM	KI4CER	146.71500	-		
Cerulean	FM	KY4KEN	147.19500	+	136.5 Hz	
Corbin	DSTAR	KK4RQX	442.11250	+		
	FM	WD4KWV	146.61000	-	100.0 Hz	
	FM	WB4IVB	444.90000	+	100.0 Hz	
Crestwood	FM	KY4OC	147.39000	+	151.4 Hz	SERA
Cumberland, Kindom Come State						
	FM	KK4WH	146.76000	-	103.5 Hz	SERA
Danville	DSTAR	KM4OON	444.23750	+		
Danville, Locklin Lane						
	FM	W4CDA	440.80000	+		
Danville, Persimmon Knob						
	FM	W4CDA	145.31000	-		SERA
Dawson Springs	FM	KF4CWK	146.77500	-		
	FM	W4WKY	147.31500	+		

KENTUCKY 169

Location	Mode	Call sign	Output	Input	Access	Coordinator
Dewdrop	FM	KD4DZE	147.03000	+	107.2 Hz	
Dixon	FM	AJ4SI	145.35000	-	71.9 Hz	
Dorton	DMR/BM	W4VJE	440.60000	+	CC1	
Dorton, Flatwoods	FM	KD4KZT	442.15000	+	167.9 Hz	
Dorton, Flatwoods Mountain						
	FM	KS4XL	224.52000	-		SERA
Draffenville	FM	KI4HUS	145.39000	-	118.8 Hz	
Drakesboro	FM	KF4DKJ	146.82000	-	107.2 Hz	
Edgewood	DMR/BM	K4TCD	440.60000	+	CC1	
	FM	K4CO	147.25500	+	123.0 Hz	
Edgewood, St. Elizabeth Hospit						
	FM	K8SCH	146.62500	-	123.0 Hz	
Elizabethtown	DMR/BM	KG4LHQ	444.76250	+	CC1	
	DMR/BM	KG4LHQ	444.91250	+	CC1	
	FM	WX4HC	145.35000	-	103.5 Hz	SERA
	FM	W4BEJ	146.98000	-		SERA
	FM	W4BEJ	444.80000	+		
Elkton	FM	KI4CJT	145.17000	-	97.4 Hz	
Eminence	FM	NG0O	443.42500	+	203.5 Hz	
Fairdale	FM	KK4CZ	53.41000	52.41000	100.0 Hz	SERA
	FM	KK4CZ	145.41000	-		SERA
	FM	KK4CZ	444.41000	+		
Flemingsburg	FM	KF4BRO	146.95500	-	107.2 Hz	
Frankfort	FM	K4TG	147.10500	+	107.2 Hz	SERA
	FM	K4TG	147.24000	+	100.0 Hz	SERA
Franklin	FM	KE4SZK	147.13500	+	136.5 Hz	
Georgetown	DMR/BM	W4VJE	440.51250	+	CC1	
	DMR/BM	KY4RCN	440.51250	+	CC1	SERA
	FM	NE4ST	146.68500	-	141.3 Hz	SERA
	FM	NE4ST	443.62500	+		
Glasgow	FM	KY4X	146.94000	-	114.8 Hz	
	FM	KY4X	444.92500	+		
Goshen	FM	KC4ZMZ	145.28000	-	77.0 Hz	
Graefenburg	FUSION	N4HZX	443.55000	+	123.0 Hz	SERA
Grayson	FM	KD4DZE	146.70000	-	107.2 Hz	
Halls Gap	FM	AG4TY	146.79000	-	79.7 Hz	
Hamlin	FM	W4GZ	147.24000	+	91.5 Hz	
Hardinsburg	FM	KG4LHQ	443.32500	+	107.2 Hz	SERA
Harlan	FM	KK4KCQ	147.10500	+	103.5 Hz	
Harrodsburg	DMR/BM	KK4FJO	440.53750	+	CC1	SERA
	IDAS	KK4FJO	145.39000	-	107.2 Hz	SERA
Hawesville	FM	KY4HC	146.71500	-	136.5 Hz	SERA
Hazard	DMR/BM	KY4MT	444.82500	+	CC1	
	FM	K4TDO	146.85000	-	77.0 Hz	SERA
	FM	WR4AMS	224.72000	-	203.5 Hz	
Hazard, Buffalo Mountain						
	FM	KY4MT	146.67000	-	103.5 Hz	SERA
Henderson	FM	W4KVK	145.49000	-	103.5 Hz	
	FM	WA4GDU	146.97000	-		SERA
Henderson, WEHT TV Tower						
	FM	WA4GDU	444.72500	+	82.5 Hz	
Highland Heights	FM	AD4CC	53.33000	52.33000	123.0 Hz	
	FM	W4YWH	146.79000	-	123.0 Hz	
	FM	K4CO	146.89500	-	123.0 Hz	
Hopkinsville	FM	WD9HIK	224.78000	-	179.9 Hz	
	FM	N1YKT	442.45000	-	192.8 Hz	
Hopkinsville, Trace Industries						
	FM	K4ULE	147.03000	+	103.5 Hz	SERA
	FUSION	KG4CKR	442.50000	+	103.5 Hz	
Hueysville	FM	K4NLT	145.47000	-	79.7 Hz	
Inez	DMR/BM	W4VJE	440.55000	+	CC3	
	FM	N4KJU	145.27000	-	127.3 Hz	

170 KENTUCKY

Location	Mode	Call sign	Output	Input	Access	Coordinator
Ingle	FM	AC4DM	29.68000	-	146.2 Hz	
Irvine	FM	W4CMR	146.82000	-	192.8 Hz	
	FM	AD4RT	147.01500	+	100.0 Hz	
	FM	AD4RT	224.94000	-		
	FM	AD4RT	444.00000	+	100.0 Hz	
Irvine, Sandhill Road Irvine						
	FM	KA4PND	443.47500	+		
Irvington	DMR/BM	KG4LHQ	444.51250	+	CC1	
Jonathan Creek	FM	N4SEI	146.98500	-	123.0 Hz	
Lancer	FM	KB8QEU	53.41000	52.41000	127.3 Hz	SERA
	FM	KB8QEU	224.86000	-	127.3 Hz	SERA
	FM	KB8QEU	442.80000	+	127.3 Hz	SERA
Lawrenceburg	DMR/BM	KG4LHQ	444.81250	+	CC1	
	FM	K4TG	145.11000	-	107.2 Hz	
	FM	K4TG	146.83500	-	107.2 Hz	
	FM	K4TG	444.37500	+	107.2 Hz	SERA
	FM	K4TG	444.50000	+	107.2 Hz	
Lebanon	FM	KN4HWX	145.21000	-	146.2 Hz	
Lebanon, Marion/Taylor County						
	FM	KN4HWX	145.21000	-	146.2 Hz	
Leitchfield	FM	KY4RFE	147.22500	+	179.9 Hz	
Lexington	DMR/MARC	KC9TKJ	440.71250	+	CC1	
	FM	AD4YJ	145.25000	-	110.9 Hz	
	FM	K4KJQ	147.16500	+		
	FM	AD4YJ	444.72500	+		
	FM	KK4PQU	444.95000	+	88.5 Hz	SERA
Lexington, Lexington Financial						
	DSTAR	W4DSI	145.46000	-		
	DSTAR	W4DSI	441.81250	+		
	FM	KY4K	146.94000	-	88.5 Hz	SERA
	FUSION	KY4K	444.12500	+		
Lexington, UK Hospital						
	FM	K4UKH	147.12000	+	141.3 Hz	
Lexington, WKYT Tower						
	FM	K4KJQ	146.76000	-		
Liberty	FM	AG4TY	442.97500	+		
London	DMR/BM	KE4GJG	443.20000	+	CC1	SERA
	FM	KI4FRJ	442.90000	+	77.0 Hz	
London, Cold Hill	FM	KI4FRJ	147.28500	+		
Louisa	DMR/BM	KJ4GRJ	440.65000	+	CC1	
	FM	WA4SWF	147.39000	+	127.3 Hz	
Louisville	DMR/BM	KK4JPB	441.90000	+	CC1	
	DMR/MARC	KK4PGE	147.46000	144.96000	CC1	SERA
	DMR/MARC	KK4JPB	443.10000	+	CC1	
	FM	N7BBW	53.43000	52.43000	167.9 Hz	
	FM	KE4JVM	145.23000	-		SERA
	FM	W4PF	146.88000	-	100.0 Hz	
	FM	W4PJZ	147.03000	+	151.4 Hz	
	FM	WB4EJK	147.27000	+	151.4 Hz	
	FM	KQ9Z	147.36000	+		
	FM	KA4MKT	224.30000	-		
	FM	KK4CZ	441.35000	+	DCS 654	SERA
	FM	N4ORL	442.00000	+	146.2 Hz	
	FM	N4KWT	443.97500	+	100.0 Hz	
	FM	N4MRM	444.60000	+	151.4 Hz	
	FM	W4PF	444.90000	+	100.0 Hz	
Louisville, Audibon Hospital						
	FM	W4CN	147.18000	+	79.7 Hz	
Louisville, Bardstown Rd., Fer						
	FM	WB4UMR	444.42500	+	DCS 071	
Lynch	DSTAR	KK4BSG	145.40000	-		
	DSTAR	KK4BSG	444.46250	+		

KENTUCKY 171

Location	Mode	Call sign	Output	Input	Access	Coordinator
Lynch, Black Mountain						
	FM	K4BKR	147.21000	+	103.5 Hz	
	FM	WB4IVB	442.35000	+	100.0 Hz	
Lynch, Black Mtn	FM	KK4WH	444.25000	+	107.2 Hz	SERA
Madisonville	DMR/BM	KC4FIE	444.60000	+	CC1	
	FM	KC4FRA	146.61000	-	100.0 Hz	
	FM	WB4JRO	147.27000	+	77.0 Hz	
	FM	KK4RYS	442.42500	+	82.5 Hz	
	FM	KC4FIE	442.77500	+	82.5 Hz	
Magnolia	FM	WA4FOB	146.67000	-	77.0 Hz	
	FM	WA4FOB	443.67500	+	77.0 Hz	
Manchester	FM	WR4AMS	224.98000	-	100.0 Hz	
	FM	KF4IFC	444.27500	+	79.7 Hz	
	FM	KD4GMH	444.50000	+		
Manchester, Pilot Mountain						
	FM	KG4LKY	146.92500	-	79.7 Hz	SERA
Mayfield	FM	WA6LDV	145.11000	-		SERA
	FM	WA6LDV	224.82000	-	179.9 Hz	
	FM	WA6LDV	441.87500	+	179.9 Hz	
Maysville	FM	KF4BRO	145.47000	-		
Meta	FM	WR4AMS	224.38000	-		SERA
Middlesboro	DSTAR	AJ4G	145.30000	-		
	DSTAR	AJ4G	443.46250	+		
	DSTAR	AJ4G	1282.40000	1262.40000		
	FM	KA4OAK	146.77500	-	79.7 Hz	SERA
	FM	AJ4G	224.12000	-	100.0 Hz	
Middlesboro, White Oak Spur						
	FM	AJ4G	442.32500	+	100.0 Hz	
Monticello	FM	AC4DM	145.15000	-	100.0 Hz	
	FM	WB9SHH	146.99500	-	88.5 Hz	
Morehead	DMR/BM	KY4HS	442.50000	+	CC1	
	FM	KJ4VF	145.13000	-	100.0 Hz	
Morehead, Triangle Mountain						
	FM	KY4HS	146.91000	-	123.0 Hz	
Morgantown	DMR/BM	W4WSM	443.98750	+	CC1	
Morgantown, County Courthouse						
	FM	W4WSM	146.65500	-	100.0 Hz	SERA
Mount Sterling, Montgomery Co.						
	FM	KD4ADJ	147.31500	+	100.0 Hz	SERA
	FM	KD4ADJ	442.05000	+		SERA
Murray	FM	K4MSU	443.80000	+	91.5 Hz	
Murray, Price Doyle Fine Arts						
	FM	K4MSU	146.94000	-	91.5 Hz	
Nancy	DSTAR	NN4H	444.23750	+		
	FM	AC4DM	53.27000	52.27000	100.0 Hz	
	FM	AC4DM	224.10000	-	100.0 Hz	
	FM	AC4DM	443.60000	+	100.0 Hz	
Nicholasville	FM	K4HH	145.49000	-		
	FM	KC4UPE	444.77500	+	167.9 Hz	
	FM	K4HH	444.97500	+		
Owensboro	DMR/BM	KG4LHQ	444.96250	+	CC1	SERA
	FM	N4WJS	146.69000	-	110.9 Hz	
	FM	WA4GDU	146.86500	-	82.5 Hz	
	FM	K4HY	147.21000	+	146.2 Hz	
	FM	N4WJS	443.65000	+	110.9 Hz	
	FM	KI4JXN	444.55000	+	103.5 Hz	
Owensboro, Greenwood Cemetery						
	FM	K4HY	145.33000	-	103.5 Hz	
Owingsville	FM	W4WOO	442.00000	+	100.0 Hz	SERA
Owingsville, Gateway Health De						
	FM	N4EWW	147.07500	+	123.0 Hz	
Paducah	FM	KY4OEM	146.76000	-	179.9 Hz	

172   KENTUCKY

Location	Mode	Call sign	Output	Input	Access	Coordinator
Paducah	FM	W4NJA	147.06000	+	179.9 Hz	
	FM	KD4DVI	147.12000	+	179.9 Hz	
	FM	KD4DVI	443.00000	+	179.9 Hz	
Paintsville	DMR/BM	W4VJE	145.48000	-	CC1	SERA
	FM	N4KJU	441.52500	+	127.3 Hz	
Paintsville, Starfire Hill						
	FM	KY4ARC	147.22500	+	127.3 Hz	SERA
Payneville	FM	K4ULW	146.62500	-	151.4 Hz	
Phelps	DMR/BM	W4VJE	440.62500	+	CC4	
Phelps, Dick's Knob						
	FM	N4MWA	147.09000	+	100.0 Hz	
Pikeville	DMR/BM	W4VJE	440.70000	+	CC2	
	FM	K4PDM	145.15000	-	127.3 Hz	
	FM	AD4BI	444.47500	+		
Pikeville, Poor Farm Ridge						
	FM	KD4DAR	224.62000	-		SERA
Pineville, Chained Rock						
	FM	WA4YZY	146.83500	+	100.0 Hz	SERA
Pleasure Ridge Park						
	FM	N4MRM	147.37500	+	151.4 Hz	
Prestonburg, Abbott Hill						
	FM	KY4ARC	145.31000	-	127.3 Hz	
Prestonsburg	FM	WR4AMS	224.72000	-	203.5 Hz	
Princeton, Princeton Vocationa						
	FM	W4KBL	444.17500	+		SERA
Princeton, US62 Near Caldwell/						
	FM	W4KBL	145.23000	-	179.9 Hz	
Providence	FM	AG4BT	147.25500	+		
Radcliff	FM	W4BEJ	146.92500	-		SERA
	FM	WX4HC	147.15000	+	103.5 Hz	
Richmond	DSTAR	KE4YVD	442.81250	+		SERA
	FM	KE4YVD	145.37000	-	192.8 Hz	
	FM	KM4EOC	146.86500	-	192.8 Hz	
	FM	KE4ISW	444.62500	+	192.8 Hz	SERA
Rockport	FM	KD4BOH	444.07500	+	77.0 Hz	
Russell Springs	FM	KV4D	444.40000	+	179.9 Hz	
Russell Springs, Lake Cumberla						
	FM	N4SQV	146.95500	-	100.0 Hz	
Salvisa	FM	KC4UPE	444.87500	+	167.9 Hz	
Salyersville	FM	KY4CM	146.62500	-	127.3 Hz	
Sandy Hook	FM	KD4DZE	147.13500	+	107.2 Hz	
Shelbyville	FM	KE4LR	147.00000	+	173.8 Hz	SERA
Shelbyville, Jeptha Knob						
	FM	KB4PTJ	444.05000	+	91.5 Hz	SERA
Somerset	FM	AC4DM	146.88000	-	77.0 Hz	SERA
	FM	N4AI	224.30000	-		
	FM	N4AI	224.88000	-		
	FM	KY4TB	443.40000	+	136.5 Hz	
Stamping Ground	FUSION	W4IOD	442.10000	+		SERA
Stanton	FM	KC4TUK	145.29000	-		
	FM	KC4TUK	442.07500	+		
Taylor Mill	FM	WB8CRS	147.03000	+	123.0 Hz	SERA
Taylorsville	FM	KX4DN	443.82500	+		
Tompkinsville	FM	KJ4OG	146.77500	-	151.4 Hz	
Union	DSTAR	WW4KY	147.39000	+		SERA
	FM	AD4CC	441.80000	+	179.9 Hz	
Uniontown	FM	KJ4HNC	145.29000	-	77.0 Hz	SERA
Versailles	FM	KY4WC	224.22000	-	107.2 Hz	SERA
	FM	AD4CR	444.25000	+	107.2 Hz	
Versailles, United Bank Buildi						
	FM	KY4WC	443.77500	+	107.2 Hz	SERA
Vine Grove	DMR/BM	KM4LNN	444.91250	+	CC1	

KENTUCKY 173

Location	Mode	Call sign	Output	Input	Access	Coordinator
Vine Grove	DSTAR	KM4LNN	145.12000	-		
Walton	FM	K4CO	147.37500	+	123.0 Hz	
Waynesburg	FM	AC4DM	53.39000	52.39000	100.0 Hz	
Westview	FM	KY3O	147.06000	+	136.5 Hz	SERA
Whitesburg, Little Shephard Tr						
	FM	KK4WH	224.96000	-	203.5 Hz	SERA
Whitesburg, Pine Mtn						
	FM	KM4IAL	145.35000	-	186.2 Hz	SERA
Williamsburg	FM	KB4PTJ	444.05000	+	100.0 Hz	
	FUSION	KB4PTJ	146.70000	-		
Williamstown	FM	K4KPN	444.42500	+	107.2 Hz	
Winchester	FM	W4PRC	145.43000	-	203.5 Hz	SERA
	FM	W4PRC	441.90000	+	203.5 Hz	

LOUISIANA

Location	Mode	Call sign	Output	Input	Access	Coordinator
Abbeville	IDAS	KD5QYV	147.06000	+	103.5 Hz	LCARC
Alexandria	DSTAR	KC5ZJY	145.15000	-		LCARC
	DSTAR	KF5PIE	147.21000	+		LCARC
	FM	KC5ZJY	145.47000	-	173.8 Hz	LCARC
	FM	KC5ZJY	444.97500	+	173.8 Hz	LCARC
Alexandria, Downtown						
	FM	KC5ZJY	53.23000	52.23000	173.8 Hz	LCARC
Alexandria, I-49 South						
	FM	KC5ZJY	147.33000	+	173.8 Hz	LCARC
	FM	KC5ZJY	443.30000	+	173.8 Hz	LCARC
Amelia	FM	W5BMC	146.74500	-		LCARC
Amite	FM	W5TEO	444.32500	+	107.2 Hz	LCARC
Bastrop	FM	N5EXS	146.92500	-	127.3 Hz	LCARC
Baton Rouge	DMR/BM	W5SJL	145.45000	454.55000	CC1	
	DMR/BM	W5RAR	443.10000	+	CC1	LCARC
	DMR/BM	W5SJL	444.45000	+	CC1	
	DMR/BM	KF5PLQ	444.45000	-	CC2	
	DSTAR	KD5CQB	146.88000	-		LCARC
	DSTAR	KD5CQB	442.92500	+		LCARC
	FM	W5GQ	145.23000	-	107.2 Hz	
	FM	WA5TQA	145.45000	-	107.2 Hz	LCARC
	FM	W5GIX	146.79000	-	107.2 Hz	LCARC
	FM	KD5QZD	443.37500	+		LCARC
	FM	W5DJA	443.45000	+		
	FM	W5GIX	444.40000	+		LCARC
	FM	WB5LHS	444.62500	+	156.7 Hz	LCARC
	FM	KC5BMA	444.85000	+	107.2 Hz	LCARC
	FM	KE5QJQ	444.95000	+	107.2 Hz	LCARC
Baton Rouge, Downtown						
	FM	KD5CQB	443.92500	+	107.2 Hz	LCARC
Baton Rouge, LSU						
	FM	K5LSU	442.80000	+		LCARC
Bayou L'Ourse	FM	W5LMR	145.21000	-	114.8 Hz	LCARC
Belle Chasse	FM	KA5EZQ	444.17500	+	114.8 Hz	
Belle Chasse, Near Bridge/Tunn						
	FM	KA5EZQ	146.89500	-	114.8 Hz	LCARC
Belle Chasse, Port Sulphur Fir						
	FM	KA5EZQ	444.17500	+	114.8 Hz	LCARC
Bernice	FM	W5JC	147.07500	+	127.3 Hz	LCARC
	FM	W5JC	444.07500	+		LCARC
Bossier	FM	K5BMO	147.15000	+	186.2 Hz	LCARC
Bossier City	DMR/BM	AF6BZ	442.02500	+	CC1	LCARC
	FM	N5FJ	444.30000	+	186.2 Hz	LCARC
Bryceland	FM	N5RD	147.30000	+	186.2 Hz	LCARC
Calhoun, Portable Tower						
	FM	W5KGT	145.17000	-		LCARC
	FM	W5KGT	444.10000	+		LCARC

174   LOUISIANA

Location	Mode	Call sign	Output	Input	Access	Coordinator
Carencro	DMR/BM	K5LPD	144.98000	147.48000	CC1	LCARC
	DMR/MARC	K5LPD	144.00000	+	CC1	
	DSTAR	WD5TR	444.90000	+		LCARC
	FM	W5NB	145.29000	-	103.5 Hz	LCARC
	FM	W5NB	443.80000	+	103.5 Hz	LCARC
Centerville	FM	N5KWW	444.02500	+	103.5 Hz	LCARC
Central	FM	KD5CQB	146.94000	-	107.2 Hz	LCARC
	FM	WB5BTR	442.40000	+	156.7 Hz	LCARC
	FM	KD5CQB	443.55000	+	107.2 Hz	LCARC
Chalmette	FM	W5MCC	146.86000	-	114.8 Hz	LCARC
	FM	W5MCC	444.80000	+	114.8 Hz	LCARC
Convent	DMR/BM	W5RAR	442.07500	+	CC1	LCARC
	FM	WB5BTR	442.50000	+		LCARC
	FM	K5ARC	443.27500	+	107.2 Hz	LCARC
	FM	K5ARC	444.72500	+	107.2 Hz	LCARC
Convent, Sunshine Bridge						
	FM	K5ARC	146.98500	-	107.2 Hz	LCARC
Coushatta	FM	K5EYG	145.27000	-	186.2 Hz	LCARC
Covington	DMR/BM	KD5KNZ	444.87500	+	CC5	
Crowley, Falcon Rice Mill						
	FM	N9QO	145.19000	-	103.5 Hz	LCARC
Denham Springs	FM	WZ5A	442.70000	+	107.2 Hz	LCARC
DeRidder	FM	KE5PFA	146.85000	-		LCARC
	FM	K5LJO	147.24000	+	203.5 Hz	LCARC
Des Allemands	FM	WX5RLT	442.10000	+	114.8 Hz	LCARC
Duson	FM	W5EXI	147.04000	+	103.5 Hz	LCARC
Farmerville	FM	KC5DR	145.23000	-	DCS 023	LCARC
Folsom	FM	W5NJJ	146.71500	-	114.8 Hz	LCARC
	FM	W5NJJ	147.37500	+	114.8 Hz	LCARC
	FM	W5NJJ	444.05000	+	114.8 Hz	LCARC
Franklin	FM	W5BMC	147.12000	+	103.5 Hz	LCARC
Gonzales	FM	K5ARC	147.22500	+	107.2 Hz	LCARC
Gonzales, Praireville						
	FM	K5ARC	145.31000	-	107.2 Hz	LCARC
Gray	FM	W5YL	147.39000	+	114.8 Hz	LCARC
Greensburg	FM	WB5BTR	442.27500	+	156.7 Hz	LCARC
Gretna	DSTAR	KF5SKU	145.25000	-		LCARC
	DSTAR	KF5SKU	444.47500	+		LCARC
	DSTAR	KF5SKU	1295.30000	1275.30000		LCARC
	FM	W5MCC	147.07500	+	114.8 Hz	
	FM	W5UK	444.20000	+	114.8 Hz	LCARC
Hammond	FM	WB5NET	147.00000	-	107.2 Hz	LCARC
Hammond, North Oaks Medical Ce						
	FM	WB5NET	145.13000	-	107.2 Hz	LCARC
	FM	WB5NET	444.25000	+	107.2 Hz	LCARC
Haughton	FM	W5JLH	53.05000	52.05000	186.2 Hz	LCARC
	FM	N5RD	145.43000	-	186.2 Hz	LCARC
	FM	WN5AIA	147.03000	+	186.2 Hz	LCARC
	FM	KC5UCV	147.24000	+	123.0 Hz	LCARC
	FM	KG5RWA	444.65000	+	186.2 Hz	LCARC
Holden	FM	W5LRS	146.73000	-	107.2 Hz	LCARC
Hornbeck	FM	KE5RRA	146.62500	-	173.8 Hz	LCARC
Houma	FM	W5YL	147.33000	+	114.8 Hz	
Houma, Bayou Cane						
	FM	W5YL	444.50000	+	114.8 Hz	LCARC
Jackson	FM	KD5UZA	53.83000	52.83000	107.2 Hz	LCARC
	FM	KD5UZA	146.83500	-	114.8 Hz	
	FM	WB5BTR	443.62500	+	156.7 Hz	LCARC
Jefferson	DSTAR	W5GAD	146.92500	-		LCARC
	DSTAR	W5GAD	444.92500	+		LCARC
	DSTAR	W5GAD	1285.00000	1265.00000		LCARC
Jonesboro	FM	WB5NIN	444.80000	+		LCARC

LOUISIANA 175

Location	Mode	Call sign	Output	Input	Access	Coordinator
Jonesboro, Water Tower						
	FM	WB5NIN	146.79000	-		LCARC
Jonesville	FM	N5TZH	146.96000	-	127.3 Hz	
	FM	N5TZH	444.57500	+	127.3 Hz	
Kentwood, Lewiston Water Tank						
	FM	WB5ERM	442.05000	+	107.2 Hz	LCARC
Kinder	FM	W5ELM	146.92500	-	203.5 Hz	LCARC
LaCombe	DMR/BM	W5SLA	145.29000	-	CC1	LCARC
	DMR/BM	W5SLA	147.27000	+	CC1	LCARC
	DMR/BM	W5SLA	444.10000	+	CC1	LCARC
	FM	W5MCC	146.62000	-	114.8 Hz	LCARC
	FM	N5UK	146.64000	-	114.8 Hz	LCARC
Lafayette	FM	WA5TNK	145.37000	-	103.5 Hz	LCARC
	FM	W5DDL	146.82000	-	103.5 Hz	LCARC
	FM	W5DDL	443.00000	+	103.5 Hz	LCARC
Lake Charles	DSTAR	KJ4QKD	444.67500	+		LCARC
	FM	W5BII	146.73000	-	173.8 Hz	LCARC
	FM	W5SUL	147.36000	+		LCARC
	FM	W5BII	444.22500	+	103.5 Hz	LCARC
	FM	W5BII	444.30000	+	88.5 Hz	LCARC
LaPlace	DMR/BM	W5RAR	442.67500	+	CC1	LCARC
	FM	W5RAR	53.73000	52.73000	114.8 Hz	LCARC
	FM	W5RAR	146.80500	-	114.8 Hz	LCARC
	FM	W5RAR	443.82500	+	114.8 Hz	LCARC
	FM	W5RAR	444.67500	+	114.8 Hz	LCARC
Leesville	DSTAR	KE5PFA	147.34500	+		LCARC
	DSTAR	KE5PFA	442.05000	+		LCARC
	FM	W5LSV	145.31000	-	203.5 Hz	LCARC
	FM	W5TMP	147.39000	+	203.5 Hz	LCARC
	FM	KE5PFA	442.62500	+	203.5 Hz	LCARC
	FM	W5LSV	444.70000	+	118.8 Hz	LCARC
Livingston	FM	W5GQ	145.23000	-	107.2 Hz	LCARC
	FM	WB5LIV	147.16500	+	107.2 Hz	LCARC
	FM	WB5BTR	147.25500	+		LCARC
	FM	WB5LIV	442.35000	+	156.7 Hz	LCARC
	FM	WB5BTR	444.35000	+	156.7 Hz	LCARC
Lockport	DMR/BM	KD5JFE	443.32500	+	CC1	LCARC
	FM	W5XTR	147.19500	+	114.8 Hz	LCARC
Loranger	FM	K5WDH	443.87500	+	107.2 Hz	LCARC
Lydia	FM	K5BLV	145.41000	-		
Madisonville	FM	W5NJJ	224.14000	-	114.8 Hz	LCARC
Many	DSTAR	N5MNY	146.80500	-		LCARC
	FM	K5MNY	147.28000	+	173.8 Hz	LCARC
	FM	K5MNY	444.20000	+	173.8 Hz	LCARC
Marrero	DMR/BM	W5MCC	147.03000	+	CC1	
Mathews	DMR/BM	W5LPG	442.62500	+	CC2	
Metairie	FM	W5GAD	147.24000	+	114.8 Hz	LCARC
	FM	W5GAD	444.00000	+		LCARC
	FM	AE5BZ	444.62500	+	114.8 Hz	LCARC
Minden	DSTAR	N5MAD	144.92000	147.42000		LCARC
	DSTAR	N5MAD	442.51250	+		LCARC
Monroe	FM	N5DMX	146.85000	-		LCARC
Morgan City	FM	W5BMC	443.75000	+	103.5 Hz	LCARC
Morgan City, Bayou L'Ourse						
	FM	W5MCC	147.01500	+	103.5 Hz	LCARC
Morgan City, Rig Museum						
	FM	WA5MC	444.62500	+		LCARC
Natchitoches	FM	KC5ZJY	146.88000	-	173.8 Hz	LCARC
New Iberia	FM	K5BLV	145.41000	-	123.0 Hz	LCARC
	FM	K5ARA	146.68000	-	103.5 Hz	LCARC
New Orleans	DMR/BM	N5UXT	444.22500	+	CC5	
	DMR/BM	KB5OZE	444.82500	+	CC1	LCARC

176 LOUISIANA

Location	Mode	Call sign	Output	Input	Access	Coordinator
New Orleans	FM	N5OZG	146.82000	-	114.8 Hz	
	FM	W4NDF	147.12000	+	114.8 Hz	LCARC
	FM	W5RU	147.36000	+		
	FM	KB5AVY	444.15000	+	114.8 Hz	LCARC
	FM	N5OZG	444.57500	+	114.8 Hz	LCARC
	FM	W5MCC	444.70000	+	114.8 Hz	LCARC
	FM	N5UXT	444.95000	+	114.8 Hz	LCARC
New Orleans, Canal St, Downtow						
	DMR/BM	K5LZP	146.77500	146.71500	CC1	LCARC
New Orleans, LA	FM	W5MCC	224.12000	-	114.8 Hz	LCARC
Opelousas, N/NW Of Opelousas						
	FM	N5TBU	444.87500	+	103.5 Hz	LCARC
Parks	FM	WR5U	443.20000	+	103.5 Hz	LCARC
Paulina	FM	WB5GCL	443.67500	+	114.8 Hz	LCARC
Pineville, Red River						
	FM	KC5ZJY	147.37500	+	173.8 Hz	LCARC
Port Sulphur	FM	KA5EZQ	444.07500	+	114.8 Hz	LCARC
Port Sulphur, Port Sulpher Fir						
	FM	KA5EZQ	146.65500	-	114.8 Hz	LCARC
Rayville	FM	WA5KNV	145.49000	-	100.0 Hz	LCARC
	FM	WA5KNV	444.95000	+	100.0 Hz	LCARC
Ruston	DSTAR	N5APB	145.14000	-		LCARC
	FM	WC5K	147.12000	+	94.8 Hz	LCARC
	FM	W5MCH	444.35000	+	94.8 Hz	LCARC
Sheridan	DSTAR	KF5BSZ	147.44000	144.94000		LCARC
	DSTAR	KF5BSZ	444.58750	+		LCARC
	DSTAR	KF5BSZ	1293.00000	1273.00000		LCARC
	FM	WB5BTR	442.42500	+	156.7 Hz	
Sheridan, WP 911 Center						
	FM	WA5ARC	145.43000	-	107.2 Hz	LCARC
Shreveport	DSTAR	W5SHV	147.36000	+		LCARC
	DSTAR	W5SHV	442.00000	+		LCARC
	FM	K5SAR	145.11000	-	186.2 Hz	LCARC
	FM	K5EMU	145.60000			
	FM	K5SAR	146.82000	-	186.2 Hz	LCARC
	FM	N5SYV	224.26000	-	100.0 Hz	LCARC
	FM	KD0KMT	444.77500	+	186.2 Hz	LCARC
Shreveport, Downtown						
	FM	N5SHV	146.76000	-	186.2 Hz	LCARC
	FM	WB5QFM	147.09000	+		LCARC
Shreveport, Schumpert Hospital						
	FM	K5SAR	444.90000	+	186.2 Hz	LCARC
Shreveport, VA Hospital						
	FM	K5SAR	146.70000	-	186.2 Hz	
Slidell, Water Tower On Front						
	FM	W5SLA	444.42500	+	114.8 Hz	
Springhill	DMR/BM	N5IGN	443.75000	+	CC1	LCARC
	FM	W5KJN	146.73000	-		
	FM	AF5P	147.16500	+		LCARC
St Martinville	DSTAR	KF5ZUZ	147.00000	+		LCARC
	DSTAR	KF5ZUZ	443.85000	+		LCARC
	DSTAR	KF5ZUZ	1292.10000	1272.10000		LCARC
Stanley	FM	WI5M	146.92500	-	186.2 Hz	LCARC
Sulphur	FM	W5BII	145.35000	-	103.5 Hz	LCARC
	FM	AF5XP	444.62500	+		LCARC
Tallulah	FM	KC5GIB	147.24000	+	94.8 Hz	
Thibodaux	DMR/BM	W5XTR	442.10000	+	CC3	
	FM	W5XTR	442.00000	+	114.8 Hz	
Walker	FM	W5LRS	444.52500	+	107.2 Hz	LCARC
West Monroe	DMR/BM	KC5DR	443.80000	+	CC3	LCARC
	DMR/MARC	KC5DR	443.70000	+	CC3	
	FM	WB5SOT	146.97000	-		LCARC

LOUISIANA   177

Location	Mode	Call sign	Output	Input	Access	Coordinator
West Monroe	FM	KC5DR	147.13500	+		LCARC
	FM	WB5SOT	444.30000	+		LCARC
Wilmer	FM	KE5ILT	147.34500	+		LCARC
Winnfield, Red Hill						
	FM	WB5NIN	147.06000	+	173.8 Hz	LCARC
Winnsboro, LA	FM	KC5DR	146.70000	-	127.3 Hz	LCARC

MAINE

Location	Mode	Call sign	Output	Input	Access	Coordinator
Alfred	FM	WJ1L	448.72500	-	103.5 Hz	NESMC
	FM	KB1PRG	449.82500	-	103.5 Hz	NESMC
Alfred, Brackett Hill						
	FM	WJ1L	145.41000	-	103.5 Hz	NESMC
Alfred, York County EMA EOC						
	FM	W1BHR	147.34500	+	123.0 Hz	
Allagash, Rocky Mountain						
	FM	N1SJV	146.71500	-	100.0 Hz	
Arundel, Arundel Hill						
	FM	W6BZ	146.92500	-	103.5 Hz	NESMC
Auburn	FM	KA1SHU	146.95500	-	107.2 Hz	NESMC
Auburn, Goff Hill	FM	W1NPP	146.61000	-	88.5 Hz	NESMC
Augusta	DMR/MARC	KQ1L	145.17000	-	CC12	
	DMR/MARC	KQ1L	145.30000	-	CC12	
	FM	KA1SHU	146.95500	-	100.0 Hz	NESMC
	FM	KQ1L	224.72000	-		
Augusta, Sand Hill						
	FM	KQ1L	146.67000	-	100.0 Hz	NESMC
Bar Harbor, Ireson Hill						
	FM	W1TU	147.03000	+	100.0 Hz	NESMC
Belfast, Waldo Co EMA						
	FM	W1EMA	147.16500	+		
Belgrade Lakes	FM	W1PIG	449.27500	-	88.5 Hz	
Biddeford	DMR/BM	KC1ALA	146.75000	-	CC1	
Brownville, Stickney Hill						
	FM	N1BUG	147.10500	+	103.5 Hz	NESMC
	FM	K1PQ	444.95000	+	103.5 Hz	
Brunswick, Growston Hill						
	FM	WZ1J	147.13500	+	103.5 Hz	NESMC
Brunswick, Oak Hill						
	FM	K1MNW	1284.00000	1264.00000	88.5 Hz	NESMC
Brunswick, Woodward Cove						
	FM	K0LDO	147.33000	+	100.0 Hz	NESMC
Buckfield	DMR/MARC	K1YFY	145.32000	-	CC12	
Buckfield, Streaked Mtn						
	FM	KQ1L	146.88000	-	100.0 Hz	NESMC
Calais	DMR/MARC	W1LH	147.04500	+	CC1	
Calais, Magurrewock Mtn						
	FM	K1QA	145.15000	-	100.0 Hz	NESMC
Camden	DMR/MARC	K1XI	145.37000	-	CC12	
Camden, Ragged Mtn						
	FM	KQ1L	146.82000	-	100.0 Hz	NESMC
Caribou	FM	N1RTX	444.40000	+	241.8 Hz	NESMC
Carrabasset, Sugarloaf Mountai						
	FM	KQ1L	146.97000	-	100.0 Hz	NESMC
Cooper, Cooper Hill						
	FM	W1LH	444.30000	+	100.0 Hz	NESMC
	WX	W1LH	146.98500	-		NESMC
	WX	W1LH	147.33000	+	118.8 Hz	NESMC
Corinna	DMR/MARC	KB1UAS	145.22000	-	CC0	
	FM	N1ME	146.94000	-	100.0 Hz	
	FM	KB1UAS	224.84000	-	103.5 Hz	
	FM	N1GNN	449.72500	-	103.5 Hz	

178   MAINE

Location	Mode	Call sign	Output	Input	Access	Coordinator
Cornish, Hessian Hill						
	FM	N1KMA	145.21000	-	156.7 Hz	NESMC
Dedham, Bald Mtn						
	FM	KC1FRJ	146.80500	-	82.5 Hz	NESMC
	FM	KC1FRJ	224.90000	-	103.5 Hz	
	FM	KC1FRJ	444.00000	+	100.0 Hz	
Dixmont, Mt Harrison						
	FM	KQ1L	146.85000	-	100.0 Hz	NESMC
Dresden	DMR/MARC	N1UGR	145.43000	-	CC12	
Ellsworth, Beckwith Hill						
	FM	KB1NEB	146.91000	-	151.4 Hz	NESMC
Exeter	FM	AA1PN	224.24000	-	103.5 Hz	
Falmouth	DMR/BM	N1XBM	448.82500	-	CC1	
Falmouth, Blackstrap Hill						
	FM	W1QUI	147.09000	+	100.0 Hz	NESMC
Falmouth, Blackstrap Hill (PAW						
	FM	W1KVI	146.73000	-	100.0 Hz	NESMC
Farmington	DMR/MARC	W1BHR	442.40000	+	CC11	
	FM	W1PIG	145.39000	-		NESMC
	FM	KY1C	147.18000	+	123.0 Hz	NESMC
	FM	W1IMD	449.07500	-	114.8 Hz	
	FM	W1BHR	449.92500	-		
Fort Kent	FM	N1SJV	146.64000	-	100.0 Hz	
Franklin , Martins Ridge						
	FM	WB5NKJ	146.61000	-	107.2 Hz	NESMC
Franklin, Martins Ridge						
	FM	N1DP	444.80000	+		
Frenchville	FM	N1CHF	224.18000	-		
Gardiner, Libby Hill						
	FM	KB1RAI	147.25500	+	114.8 Hz	NESMC
Glastonbury	DMR/BM	KA1VSC	449.82500	-	CC1	NESMC
Gouldsboro	DMR/MARC	KC1FRJ	145.21000	-	CC12	
Gray, NWS Forecast Office						
	FM	K1MV	147.04500	+	103.5 Hz	NESMC
Hampden	FM	W1GEE	147.30000	+	100.0 Hz	NESMC
Hebron	FM	W1IF	224.62000	-	103.5 Hz	
Hiram	DMR/BM	W1IMD	448.57500	+	CC2	
	FM	K1AAM	53.37000	52.37000	136.5 Hz	
	FM	W6BZ	442.20000	+	82.5 Hz	
Hiram, Peaked Pass Mtn						
	FM	K1AAM	147.01500	+	103.5 Hz	NESMC
Holden	DMR/MARC	N1ME	145.31000	-	CC10	
Holden, Riders Bluff						
	FM	N1ME	444.40000	+		
Hope	FM	WA1ZDA	224.00000	-		
Hope, Hatchet Mtn						
	FM	WA1ZDA	147.24000	+	110.9 Hz	NESMC
	FM	K1EHO	449.52500	-	110.9 Hz	
Houlton, Resevoir Hill						
	WX	W1BC	146.79000	-	110.9 Hz	NESMC
Island Falls	FM	KB1JVQ	145.17000	-	123.0 Hz	
Kents Hill	FM	W1PIG	147.00000	+	100.0 Hz	NESMC
Knox	DMR/MARC	W1EMA	145.42000	-	CC12	NESMC
Knox, Aborn Hill	FM	W1EMA	147.27000	+	136.5 Hz	NESMC
	FM	KD1KE	443.50000	+	103.5 Hz	
Levant, Pember Ridge						
	FM	W1DLO	147.36000	+	77.0 Hz	NESMC
Lincoln	DMR/MARC	KC1FRJ	145.35000	-	CC12	
	FM	K1AQ	449.27500	-		
Lincoln , Bagley Mtn						
	FM	KQ1L	147.00000	+	100.0 Hz	NESMC
Litchfield	FM	K1AAM	53.05000	52.05000	136.5 Hz	

MAINE 179

Location	Mode	Call sign	Output	Input	Access	Coordinator
Livermore Falls, Moose Hill						
	FM	W1BHR	147.22500	+	123.0 Hz	
Long A Township, Lincoln Ridge						
	FM	KA1EKS	146.74500	-	100.0 Hz	NESMC
Machias	DMR/MARC	KC1KPC	444.50000	+	CC1	
Madawaska	FM	W1ACP	449.82500	-		
Madison	FM	KA1C	449.62500	-	91.5 Hz	
Madison, Blackwell Hill						
	FM	KA1C	146.73000	-	91.5 Hz	NESMC
Marshfield	FM	K1HF	146.77500	-		NESMC
Millinocket, South Twin Lake						
	FM	KA1EKS	145.25000	-	100.0 Hz	NESMC
Milo, Sargent Hill	FM	KB1ZQY	147.15000	+	123.0 Hz	NESMC
Naples	FM	K1AAM	146.83500	-	103.5 Hz	NESMC
New Sharon	DMR/MARC	W1BHR	145.14000	-	CC11	
New Sharon, York Hill						
	FM	N1UGR	147.37500	-	131.8 Hz	
Nobleboro	FM	W1AUX	224.32000	-	103.5 Hz	
North Wade, Near Presque Isle						
	FM	K1FS	146.73000	-		
Norway, Pike Hill	FM	W1OCA	147.12000	+		
Palermo, Marden Hill						
	FM	K1XI	145.27000	-	100.0 Hz	NESMC
Phippsburg , Near Fuller Mtn						
	FM	KS1R	147.21000	+	100.0 Hz	NESMC
Phippsburg , Nr. Fuller Mt						
	DSTAR	KS1R	447.57500	-		
Poland Spring, White Oak Hill						
	FM	W1NPP	147.31500	+	103.5 Hz	NESMC
Portland	DMR/MARC	W1IMD	145.34000	-	CC12	
	FM	KA1SHU	146.76000	-	100.0 Hz	NESMC
Portland, Mitchell Hill						
	FM	K1SA	147.36000	+	100.0 Hz	NESMC
Presque Isle	DMR/MARC	KC1FRJ	145.18000	-	CC12	
Rangeley Saddleback Mtn						
	DMR/MARC	K1XI	145.20000	-	CC12	
Rumford, Black Mtn						
	FM	N1BBK	146.91000	-	100.0 Hz	
Sanford	FM	N1ROA	146.80500	-	103.5 Hz	
	FM	N1KMA	223.82000	-	DCS 023	
	FM	W1LO	441.60000	+	203.5 Hz	
Sanford, Mt Hope	FM	KQ1L	147.18000	+	131.8 Hz	NESMC
Scarborough	FM	W1KVI	444.10000	+	82.5 Hz	
Shapleigh	DMR/MARC	K1DQ	145.11000	-	CC4	
Sidney	DMR/MARC	KQ1L	145.24000	-	CC12	
Skowhegan	FM	W1LO	446.32500	-	203.5 Hz	
Skowhegan, Bigelow Hill						
	FM	KA1ZGC	147.34500	+		NESMC
Smyrna	DMR/MARC	KC1FRJ	147.09000	+	CC12	
Springfield, Almanac Mtn						
	FM	WA1ZJL	147.37500	+	100.0 Hz	NESMC
Sweden	DMR/MARC	SK3WH	145.57500	-	CC3	
Topsfield, Musquash Mtn						
	FM	W1LH	146.67000	-	100.0 Hz	NESMC
Topsham	DMR/MARC	N1IPA	145.19000	-	CC13	
	FM	KS1R	444.40000	+	88.5 Hz	
Waldoboro	FM	N1PS	224.78000	-	107.2 Hz	
Waldoboro, RW Glidden Auto Bod						
	FM	K1NYY	147.39000	+	179.9 Hz	NESMC
Wales, Oak Hill	FM	W1PIG	145.29000	-	100.0 Hz	NESMC
Warren	FM	WA1ZDA	224.10000	-		
Washington	FM	KC1CG	53.55000	52.55000	91.5 Hz	

180   MAINE

Location	Mode	Call sign	Output	Input	Access	Coordinator
Washington	FM	KC1CG	224.28000	-	91.5 Hz	NESMC
	FM	WZ1J	444.90000	+	91.5 Hz	
Washington, Benner Hill						
	FM	W1PBR	147.06000	+	91.5 Hz	NESMC
Washington, Lenfest Mountain						
	FM	KC1CG	145.49000	-	91.5 Hz	NESMC
Waterville	DMR/MARC	KQ1L	146.92500	-	CC12	
Wells	DMR/MARC	KY1C	442.40000	+	CC11	
Westbrook	FM	N1ULM	444.25000	+	82.5 Hz	
	FM	W1CKD	444.60000	+	82.5 Hz	
Westbrook, Rocky Hill						
	FM	W1CKD	147.27000	+	103.5 Hz	NESMC
Wilton, Voter Hill	FM	W1PIG	145.39000	-	100.0 Hz	NESMC
Windham Hill	FM	N1FCU	29.68000	-	173.8 Hz	NESMC
Winslow	FM	KD1MM	146.76000	-	103.5 Hz	NESMC
Wiscasset, Blinn Hill						
	FM	K1LX	146.98500	-	136.5 Hz	
Woodstock	FM	W1IMD	53.09000	52.09000	71.9 Hz	
	FM	W1IMD	223.94000	-	103.5 Hz	
	FM	W1IMD	449.02500	-	82.5 Hz	
Yarmouth	FM	K1JW	146.94000	-		
York	FM	W1XOJ	448.00000	-	173.8 Hz	

MARYLAND

Location	Mode	Call sign	Output	Input	Access	Coordinator
Accokeek	FM	W3TOM	444.50000	+	127.3 Hz	T-MARC
Adelphi	FM	W3ARL	145.37000	-		T-MARC
Annapolis	DMR/MARC	W4ATN	443.78750	+	CC3	T-MARC
Annapolis, Broad Creek Park						
	FM	KB3CMA	442.30000	+	107.2 Hz	T-MARC
Ashton	DMR/MARC	K3UCB	442.23750	+	CC1	T-MARC
	FM	N3AGB	53.25000	52.25000	100.0 Hz	T-MARC
	FM	K3WX	147.00000	+		T-MARC
	FM	K3WX	224.54000	-	156.7 Hz	T-MARC
	FM	K3WX	443.15000	+		T-MARC
	FM	K3WX	927.72500	902.72500	156.7 Hz	T-MARC
Baltimore	DMR/MARC	N3CDY	927.06250	902.06250	CC1	
	FM	WB3DZO	147.03000	+		T-MARC
	FM	N3HIA	224.48000	-		T-MARC
	FM	KS3L	224.68000	-		T-MARC
	FM	W3WCQ	439.25000	426.25000		T-MARC
	FM	WA3KOK	442.25000	+	107.2 Hz	T-MARC
	FM	K3CUJ	448.27500	-	156.7 Hz	T-MARC
	FM	W3PGA	449.57500	-	123.0 Hz	T-MARC
	FM	N3ST	449.67500	-	167.9 Hz	T-MARC
	FM	W3WCQ	911.25000	426.25000		T-MARC
Baltimore, Kenwood High School						
	FM	W3PGA	147.24000	+		T-MARC
Bel Air	DMR/MARC	KB3WHR	449.61250	-	CC1	T-MARC
	DSTAR	KC3FHC	145.12000	-		T-MARC
	DSTAR	KC3FHC	447.98750	-		T-MARC
	DSTAR	KC3FHC	1282.30000	1262.30000		T-MARC
	FM	KC3FHC	146.77500	-	146.2 Hz	T-MARC
	FM	N3EKQ	147.12000	+		T-MARC
	FM	N3EKQ	223.96000	-		T-MARC
	FM	KC3FHC	449.77500	-	162.2 Hz	T-MARC
Bethesda, NIH	FM	K3YGG	145.29000	-	156.7 Hz	T-MARC
Bladensburg	FM	K3GMR	146.61000	-		T-MARC
Boonsboro	FM	KD3SU	442.95000	+	94.8 Hz	T-MARC
Bowie	FM	W3XJ	442.15000	+		T-MARC
Brandywine	FM	W3SMR	147.15000	+	114.8 Hz	T-MARC
Burtonsville	FM	WA3KOK	444.05000	+		T-MARC
Centerville	FM	KE3AO	146.94000	-	107.2 Hz	T-MARC

MARYLAND 181

Location	Mode	Call sign	Output	Input	Access	Coordinator
Centreville	FM	N8ADN	448.22500	-	107.2 Hz	T-MARC
Charles Town	FM	KD8DWU	927.58750	902.58750	123.0 Hz	
Charlestown	FM	N3RCN	145.47000	-	94.8 Hz	T-MARC
Charlestown, Water Tower						
	FM	N3RCN	442.95000	+	103.5 Hz	T-MARC
Charlotte Hall	DMR/MARC	K3OCM	443.18750	+	CC6	T-MARC
Chestertown	DMR/MARC	K3ARS	449.17500	+	CC1	T-MARC
Clear Spring	FM	K3MAD	147.34500	+	123.0 Hz	T-MARC
	FM	N3UHD	442.65000	+	79.7 Hz	T-MARC
Clear Spring, Fairview Mountai						
	FM	W3CWC	146.94000	-	100.0 Hz	T-MARC
Cockeysville	FM	K3NXU	145.19000	-	110.9 Hz	T-MARC
College Park, UMD						
	FM	W3EAX	145.49000	-		T-MARC
Columbia	FM	K3CUJ	147.13500	+	156.7 Hz	T-MARC
	FM	W3CAM	224.86000	-		T-MARC
	FM	K3CUJ	449.47500	-	156.7 Hz	T-MARC
Cooksville	FM	K3CUJ	147.39000	+		
Cumberland	FM	AB3FE	442.30000	+	167.9 Hz	T-MARC
	FM	KK3L	444.00000	+	123.0 Hz	T-MARC
Curtis Bay	FM	W3VPR	147.07500	+	107.2 Hz	T-MARC
Damascus	DMR/MARC	KB3LYL	442.21250	+	CC1	
	FM	N3VNG	145.25000	-	146.2 Hz	T-MARC
Damascus, Smokehill BBQ						
	FM	KB3LYL	441.88750	+		T-MARC
Davidsonville	FM	W3VPR	147.10500	+	107.2 Hz	T-MARC
	FM	W3VPR	223.88000	-	107.2 Hz	T-MARC
	FM	W3VPR	444.40000	+	107.2 Hz	T-MARC
Dayton	FM	W3YVV	443.95000	+		T-MARC
	FM	W3YVV	927.53750	902.53750	156.7 Hz	T-MARC
District Heights	FM	K3ERA	145.23000	-	110.9 Hz	T-MARC
Dundalk	FM	N3HIA	147.34500	+	192.8 Hz	T-MARC
Dundalk, Key Bridge						
	FM	KC3APF	448.67500	-	91.5 Hz	T-MARC
Easton, UM/Shore Regional Medi						
	FM	K3EMD	442.20000	+		T-MARC
Elkton	DMR/BM	K3DRF	145.25000	139.25000	CC1	
	FM	KX3B	448.77500	-	131.8 Hz	
Ellicott City	FM	N3EZD	224.32000	-		T-MARC
Fallston	DMR/BM	N3CNJ	145.27000	-	CC1	
Flintstone	FM	KB3VJO	147.31500	+		T-MARC
	FM	KB3VJO	443.90000	+		T-MARC
Frederick	FM	N3KTX	53.47000	52.47000	107.2 Hz	
	FM	N3IGM	53.75000	52.75000	100.0 Hz	T-MARC
	FM	K3ERM	146.64000	-	156.7 Hz	T-MARC
	FM	W3ICF	146.73000	-	141.3 Hz	T-MARC
	FM	K3MAD	147.06000	+	146.2 Hz	T-MARC
	FM	K3MAD	224.20000	-	123.0 Hz	T-MARC
	FM	WA3KOK	443.40000	+	136.5 Hz	T-MARC
	FM	N3ST	444.10000	+	167.9 Hz	T-MARC
	P25	N3ITA	442.80000	+		T-MARC
	P25	K3MAD	448.12500	-		T-MARC
Frederick, Braddock Mountain						
	FM	W3ARK	444.35000	+	100.0 Hz	T-MARC
Frederick, Gambril Mountain						
	FM	K3ERM	448.42500	-	100.0 Hz	T-MARC
Frederick, Gambrill Mountain						
	DSTAR	KB3YBH	444.80000	+		T-MARC
Frederick, Gambrill State Park						
	DMR/BM	K3DO	441.96250	+	CC1	T-MARC
	DSTAR	KB3YBH	145.17000	-		T-MARC

182 MARYLAND

Location	Mode	Call sign	Output	Input	Access	Coordinator
Frederick, Gambrils State Park						
	FM	N3KTX	53.47000	52.47000	107.2 Hz	T-MARC
Frostburg	DMR/MARC	K3YDA	147.10500	+	CC1	
Gaithersburg	FM	KV3B	53.27000	52.27000	156.7 Hz	T-MARC
Galena	FM	KB3MEC	224.00000	-	131.8 Hz	T-MARC
Germantown	DSTAR	KV3B	444.20000	+		T-MARC
	FM	WA3KOK	147.27000	+	156.7 Hz	T-MARC
Glen Burnie	FM	N3MIR	224.60000	-		T-MARC
	FM	N3MIR	442.60000	+	127.3 Hz	T-MARC
Greenbelt	FM	WA3NAN	146.83500	-		T-MARC
	FM	W3GMR	146.88000	-		T-MARC
Hagerstown	FM	K3UMV	147.37500	+		T-MARC
	FM	W3ARK	447.12500	-	123.0 Hz	T-MARC
Havre De Grace	FM	N3KZ	444.15000	+	131.8 Hz	T-MARC
Hollywood	DSTAR	N3PX	147.19500	+		T-MARC
Hughesville	FM	W3ZO	145.39000	-	186.2 Hz	T-MARC
Hyattsville	DMR/MARC	N3LHD	444.65000	+	CC6	T-MARC
Jarrettsville	FM	N3UR	53.93000	52.93000		T-MARC
	FM	N3UR	448.47500	-		T-MARC
Jefferson	FM	K3LMS	443.30000	+	100.0 Hz	T-MARC
Jessup	FM	WA3DZD	146.76000	-	107.2 Hz	T-MARC
	FM	WA3DZD	444.00000	+	107.2 Hz	T-MARC
Joppa, Joppa Magnolia Fire Com						
	DMR/BM	N3RQP	145.23000	-	CC1	
La Plata	FM	KA3GRW	443.70000	+		T-MARC
Lanham, Doctors Community Hosp						
	FM	K3ERA	447.32500	-	110.9 Hz	
Laurel	DMR/BM	W3SQN	442.90000	+	CC1	
	FM	W3LRC	442.50000	+	156.7 Hz	T-MARC
	FM	WA3GPC	444.70000	+	167.9 Hz	T-MARC
	FM	K3UQQ	923.25000	1265.25000		T-MARC
Lexington Park	FM	K3HKI	146.64000	-	146.2 Hz	T-MARC
	FM	WA3UMY	443.05000	+		T-MARC
Littlestown	FM	N3ST	449.67500	-	167.9 Hz	
Manchester	FM	N3KZS	146.89500	-		
Midland	FM	WX3M	147.10500	+	192.8 Hz	T-MARC
	FM	AB3FE	442.75000	+	167.9 Hz	T-MARC
Midland, Dan's Mountain						
	FM	W3YMW	146.88000	-	123.0 Hz	T-MARC
Midland, Dans Mountain						
	FM	KK3L	147.24000	+		T-MARC
Midland, Dans Rock						
	FM	W3YMW	444.50000	+	118.8 Hz	T-MARC
Millersville	DMR/BM	KP4IP	442.40000	+	CC1	T-MARC
	FM	W3CU	146.80500	-	107.2 Hz	T-MARC
	FM	W3VPR	224.56000	-	107.2 Hz	T-MARC
	FM	W3CU	449.12500	-	107.2 Hz	T-MARC
Nanos	DMR/BM	S55DKP	438.70000	431.10000	CC1	
Oakland	FM	KB3AVZ	146.70000	-	173.8 Hz	T-MARC
	FM	KB8NUF	146.80500	-	123.0 Hz	T-MARC
	FM	KB3AVZ	444.27500	+	110.9 Hz	T-MARC
Ocean City	DMR/MARC	N3HF	444.01250	+	CC1	T-MARC
Ocean Pines	FM	NW2M	146.95500	-		
Oldtown, Warrior Mountain WMA						
	FM	W3YMW	145.45000	-	123.0 Hz	T-MARC
Olney	DMR/MARC	K3UCB	442.90000	+	CC1	
Owings Mills	DMR/MARC	N3CDY	449.07500	-	CC1	
	FM	N3CDY	927.51250	902.51250	156.7 Hz	T-MARC
Pen Mar, High Rock						
	FM	W3CWC	147.09000	+	100.0 Hz	T-MARC
	FM	W3CWC	447.97500	-	100.0 Hz	T-MARC
Perry Hall	FM	W3JEH	223.84000	-		T-MARC

MARYLAND 183

Location	Mode	Call sign	Output	Input	Access	Coordinator
Port Deposit	FM	WA3SFJ	53.83000	52.83000	94.8 Hz	T-MARC
	FM	WA3SFJ	146.85000	-	107.2 Hz	T-MARC
	FM	WA3SFJ	449.82500	-	167.9 Hz	T-MARC
Prince Frederick	FM	W3PQS	53.17000	52.17000	100.0 Hz	T-MARC
Prince Frederick, Barstow						
	FM	N3PX	145.35000	-	156.7 Hz	T-MARC
	FM	W3SMD	223.90000	-		T-MARC
Princess Anne	DMR/MARC	N3YMS	441.88750	+	CC2	T-MARC
	DMR/MARC	N3NRL	442.03750	+	CC1	
	FM	KA3MRX	146.62500	-	156.7 Hz	T-MARC
Rockville	DMR/BM	N3JFW	442.48750	+	CC1	
	FM	KV3B	146.95500	-		T-MARC
	FM	K3ATV	224.94000	-	156.7 Hz	T-MARC
	FM	KV3B	442.75000	+	156.7 Hz	T-MARC
	FM	K3ATV	923.25000	1277.25000		T-MARC
Rockville, Mongomery County Ex						
	FM	WA3YOO	443.90000	+	156.7 Hz	T-MARC
Salisbury	DMR/MARC	K3OCM	442.65000	+	CC6	
	DSTAR	W3PRO	145.28000	-		T-MARC
	DSTAR	W3PRO	444.20000	+		T-MARC
	FM	W3PRO	146.92500	-	156.7 Hz	T-MARC
	FM	N3HQJ	442.65000	+	156.7 Hz	T-MARC
	FM	K3RIC	444.05000	+		
Salisbury, WMDT Tower						
	FM	K3DRC	146.82000	-	156.7 Hz	T-MARC
Shawsville	FM	K3HT	145.33000	-		T-MARC
	FM	N3UR	224.92000	-		T-MARC
	FM	W3EHT	449.37500	-		T-MARC
Silver Spring	FM	N3AUY	29.66000	-	141.3 Hz	T-MARC
	FM	KA3LAO	147.18000	+		T-MARC
	FM	WB3GXW	147.22500	+	156.7 Hz	T-MARC
	FM	N3HF	443.45000	+	156.7 Hz	T-MARC
	FM	WB3GXW	444.25000	+	156.7 Hz	T-MARC
	FM	N3AUY	449.02500	29.52500	141.3 Hz	T-MARC
Starr	FM	WA3NAN	146.83500	-		
Suitland	FM	N3ST	448.92500	-	167.9 Hz	T-MARC
Sunderland	FM	K3CAL	146.98500	-	156.7 Hz	T-MARC
Thurmont	DMR/MARC	N3EJT	147.19500	+	CC1	T-MARC
	FM	K3KMA	448.02500	-	103.5 Hz	T-MARC
Towson	DMR/MARC	WR3IRS	441.40000	+	CC1	T-MARC
	DMR/MARC	K3OCM	443.85000	+	CC6	T-MARC
	DSTAR	W3DHS	145.14000	-		T-MARC
	DSTAR	W3DHS	442.11250	+		T-MARC
	DSTAR	W3DHS	1282.70000	1262.70000		T-MARC
	FM	W3FT	146.67000	-	107.2 Hz	T-MARC
	FM	N3CDY	449.27500	-	107.2 Hz	T-MARC
	FM	N3CDY	927.48750	902.48750	156.7 Hz	T-MARC
Trappe	DMR/MARC	N3YMS	442.53750	+	CC1	
	FM	K3EMD	147.04500	+	156.7 Hz	T-MARC
Tyaskin	DMR/MARC	N3HQJ	146.86500	-	CC1	
West Laurel	FM	K3WS	447.92500	-	123.0 Hz	T-MARC
Westminster	FM	K3PZN	53.09000	52.09000	107.2 Hz	T-MARC
	FM	K3PZN	145.41000	-	114.8 Hz	T-MARC
	FM	K3PZN	449.87500	-	127.3 Hz	T-MARC
White Oak	FM	N3HF	443.45000	+	151.4 Hz	
Woodbine	DMR/MARC	KA3LAO	442.46250	+	CC1	T-MARC
Worton	FM	K3ARS	147.37500	+	156.7 Hz	T-MARC

MASSACHUSETTS

Location	Mode	Call sign	Output	Input	Access	Coordinator
Abington	FM	WG1U	927.45000	902.45000	DCS 031	
	FM	W1EHT	927.62500	902.62500	131.8 Hz	
Acton	DMR/BM	NO1A	442.35000	+	CC1	

184   MASSACHUSETTS

Location	Mode	Call sign	Output	Input	Access	Coordinator
Acton	DMR/MARC	NO1A	146.42000	144.92000	CC1	
Adams, Mount Greylock						
	DMR/BM	K1FFK	449.42500	-	CC1	
	FM	K1FFK	53.23000	52.23000	162.2 Hz	NESMC
	FM	K1FFK	146.91000	-		NESMC
Adams, Mt Greylock						
	FM	KB1EXR	145.21000	-	77.0 Hz	
	FM	K1FFK	927.87500	902.87500	100.0 Hz	NESMC
Agawam	DSTAR	W1KK	449.17500	-		
	FM	KA1JJM	449.77500	-		
	FM	W1KK	927.80000	902.80000	DCS 244	
Agawam, Provin Mtn						
	FM	W1TOM	146.67000	-	127.3 Hz	
Andover	FM	N1LHP	146.83500	-	77.0 Hz	NESMC
Ashland	FM	N3HFK	927.88750	902.88750	131.8 Hz	
Assonet	IDAS	N1KIM	145.43750	-		
	IDAS	WG1U	449.99695	-		
Attleboro	WX	K1SMH	147.19500	+	127.3 Hz	
Auburn	FM	K1WPO	443.90000	+	100.0 Hz	
	FM	K1WPO	448.12500	-	88.5 Hz	
Barnstable	FM	N1YHS	53.01000	52.01000	173.8 Hz	
	FM	W1SGL	146.73000	-	67.0 Hz	NESMC
	FM	W1SGL	927.82500	902.02500	82.5 Hz	NESMC
Belchertown	FM	N1SIF	443.70000	+	71.9 Hz	
Belmont	FM	KC1CLA	145.43000	-	146.2 Hz	NESMC
	FM	KB1FX	223.86000	-	100.0 Hz	
Beverly	FM	WA1PNW	147.39000	+		
	FM	WA1PNW	442.85000	+	103.5 Hz	
Billerica	FM	W1DC	147.12000	+	103.5 Hz	
Boston	DMR/BM	KB1VKI	442.05000	+	CC2	
	DMR/MARC	N1PA	146.49000	144.99000	CC1	
	DMR/MARC	W1BOS	449.17500		CC1	
	FM	KB1BEM	145.21000	-		
	FM	W1BOS	145.23000	-	88.5 Hz	
	FM	W1KBN	145.31000	-	123.0 Hz	
	FM	K1BOS	146.82000	-	127.3 Hz	NESMC
	FM	W1KRU	444.70000	+		NESMC
	FM	K1RJZ	927.06250	902.06250	DCS 244	NESMC
Bourne	DMR/MARC	K1RK	145.20000	-	CC10	
	FM	N1YHS	224.22000	-	100.0 Hz	
Braintree	FM	K1GUG	53.03000	52.03000		
	FM	AE1TH	53.39000	52.39000	71.9 Hz	
	FM	AE1TH	442.50000	+	118.8 Hz	
Bridgewater	FM	W1MV	444.55000	+	88.5 Hz	
	FM	W1WCF	927.42500	902.42500	131.8 Hz	
	WX	W1MV	147.18000	+	67.0 Hz	
Brookline	DSTAR	K1MRA	145.16000	-		
	FM	K1DVD	146.98500	-	88.5 Hz	
	FM	W1CLA	446.32500	-	146.2 Hz	
Burlington	FM	W1DYJ	446.77500	-	88.5 Hz	
	FM	W1CLA	447.02500	-	146.2 Hz	
Cambridge, Green Building						
	FM	W1XM	449.72500	-	114.8 Hz	NESMC
Canton	FM	K1BFD	146.74500	-	146.2 Hz	
	FM	K1BFD	449.42500	-	88.5 Hz	
Chelmsford	DMR/MARC	N1IW	145.18000	-	CC2	
Clinton	FM	N1ZUZ	146.65500	-	74.4 Hz	
	FM	N1KUB	442.30000	+	74.4 Hz	
Concord	FM	N1CON	447.57500	-	110.9 Hz	
Concord, Annursnac Hill						
	FM	N1CON	145.11000	-	110.9 Hz	
Dalton	FM	N1FZH	224.40000	-		

MASSACHUSETTS 185

Location	Mode	Call sign	Output	Input	Access	Coordinator
Danvers	DMR/MARC	NS1RA	442.80000	+	CC4	
	FM	N1UEC	53.85000	52.85000	71.9 Hz	
	FM	NS1RA	145.47000	-	136.5 Hz	
	FM	NS1RA	223.88000	-	136.5 Hz	
Darthmouth	FM	W1AEC	224.80000	-	67.0 Hz	
Dartmouth	DSTAR	NN1D	145.29000	-		
	FM	W1AEC	147.00000	+	67.0 Hz	
	FM	W1AEC	927.83750	902.03750	77.0 Hz	
Deerfield	FM	AB1RS	145.13000	-	DCS 054	
	FM	AB1RS	443.45000	+	173.8 Hz	
Dennis	DMR/MARC	W1MLL	146.47000	144.97000	CC11	
	FM	K1PBO	146.95500	-	88.5 Hz	
Dorchester	DMR/BM	KC1AVD	442.40000	+	CC1	
Egremont	FM	WB2BQW	145.25000	-	100.0 Hz	
Fairhaven	FM	W1SMA	145.49000	-	67.0 Hz	
Fall River	DSTAR	K1RFI	145.42000	-		
	DSTAR	K1RFI	449.52500	-		
	FM	WA1DGW	145.15000	-	123.0 Hz	
	FM	NN1D	146.80500	-	67.0 Hz	
	FM	KB1NYT	224.18000	-	67.0 Hz	
	FM	NN1D	927.03750	902.03750	DCS 031	
	FM	N1JBC	927.65000	902.65000	91.5 Hz	
Falmouth	DSTAR	KB1ZEG	145.21000	-		
	FM	N1YHS	147.37500	+	110.9 Hz	
	FM	K1RK	927.85000	902.05000	88.5 Hz	
Falmouth, Hospital						
	FM	K1RK	146.65500	-	88.5 Hz	
Feeding Hills	DSTAR	W1KK	145.15000	-		
	DSTAR	W1KK	1282.50000	1262.50000		
Fitchburg	FM	WB1EWS	53.83000	52.83000	71.9 Hz	
	FM	WB1EWS	147.31500	+	100.0 Hz	
	FM	WB1EWS	442.95000	+	88.5 Hz	
Fitchburg, Burbank Hospital						
	FM	W1GZ	145.45000	-	74.4 Hz	
Florence	FM	AA1AK	53.35000	52.35000	71.9 Hz	
Framingham	FM	W1FRS	53.27000	52.27000	71.9 Hz	
	FM	WB1CTO	224.24000	-	103.5 Hz	NESMC
	FM	N1OMJ	444.75000	+	88.5 Hz	
	FM	W1FRS	448.17500	-	88.5 Hz	
	FM	W1FRS	927.01250	902.01250	131.8 Hz	
Framingham, MEMA Headquarters						
	FM	W1FY	147.15000	+	100.0 Hz	
Freetown	DSTAR	KB1WUW	146.41000	144.91000		
	DSTAR	KB1WUW	147.57500	-		
	DSTAR	KB1WUW	449.77500	-		
Gardner	DMR/MARC	N1WW	145.34000	-	CC3	
	FM	W1GCD	145.37000	-	136.5 Hz	
	FM	W1GCD	442.10000	+	88.5 Hz	
Gloucester	FM	W1GLO	145.13000	-	107.2 Hz	
	FM	W1GLO	224.90000	-		
	FM	W1GLO	443.70000	+	107.2 Hz	
	FM	W1RAB	447.52500	-	114.8 Hz	
Granville	DMR/MARC	KB1AEV	442.75000	+	CC1	
	WX	W1TOM	147.00000	+	127.3 Hz	
Great Barrington	FM	KC1AJX	146.71500	-	77.0 Hz	
Great Barrington, Monument Mou						
	FM	KC1GLK	442.65000	+	162.2 Hz	NESMC
Great Barrington, Ski Butternu						
	IDAS	KA1OA	145.27000	-		
Greenfield	FM	KB1BSS	448.27500	-	136.5 Hz	
Harvard	FM	W1DVC	145.41000	-	74.4 Hz	
Harwich	FM	WA1YFV	145.27000	-	67.0 Hz	

186   MASSACHUSETTS

Location	Mode	Call sign	Output	Input	Access	Coordinator
Harwich	FM	K1KEK	224.34000	-	100.0 Hz	NESMC
Haverhill	FM	KT1S	145.35000	-	136.5 Hz	
	FM	N1IRS	224.12000	-	103.5 Hz	
Hingham	DMR/MARC	K1GAS	146.43000	144.93000	CC1	
Holliston	DMR/MARC	W1DSR	145.14000	-	CC1	NESMC
Holyoke	DSTAR	AA1KK	447.37500	-		
	FM	K1ZJH	146.71500	-	100.0 Hz	NESMC
	FM	W1TOM	443.20000	+	127.3 Hz	
	FM	AA1KK	927.83750	902.83750	DCS 244	
Holyoke, Mt Tom	FM	W1TOM	146.94000	-	127.3 Hz	
Hopkinton	FM	K1KWP	223.94000	-	103.5 Hz	
	FM	W1BRI	449.57500	-	88.5 Hz	NESMC
Lawrence	DMR/BM	KB1SFG	446.37500	-	CC1	
	DMR/MARC	KB1SFG	441.37500	+	CC1	
	FM	N1EXC	146.65500	-	107.2 Hz	
	FM	N1EXC	224.30000	-		
	FM	N1EXC	447.62500	-	88.5 Hz	
Leominster	FM	AA1JD	224.76000	-	85.4 Hz	
Leyden	FM	KB1BSS	146.98500	-	136.5 Hz	
Littleton	FM	K1PRE	927.43750	902.43750	131.8 Hz	NESMC
Lowell	FM	K1LVF	442.25000	+	88.5 Hz	NESMC
	FM	KB2KWB	444.96250	+	179.9 Hz	
Lynn	FM	W1DVG	147.01500	+	88.5 Hz	
Mansfield	FM	KB1CYO	147.01500	+	67.0 Hz	
	FM	KB1JJE	446.92500	-	100.0 Hz	
	FM	N1UEC	449.67500	-	146.2 Hz	
Marblehead	DMR/MARC	K1XML	145.37000	-	CC0	
	DMR/MARC	K1XML	445.87500	-	CC1	
Marlborough	FM	W1MRA	29.68000	-	131.8 Hz	
	FM	W1BRI	53.81000	52.81000	71.9 Hz	NESMC
	FM	WA1NPN	147.24000	+	71.9 Hz	
	FM	W1MRA	147.27000	+	146.2 Hz	NESMC
	FM	W1MRA	224.88000	-	103.5 Hz	NESMC
	FM	N1EM	446.67500	-	88.5 Hz	
	FM	K1IW	447.87500	-	136.5 Hz	
	FM	W1MRA	449.92500	-	88.5 Hz	NESMC
	FM	W1MRA	927.70000	902.70000	DCS 244	NESMC
Marshfield	DMR/BM	W1ATD	145.39000	-	CC9	
Marshfield, WATD-FM Tower						
	P25	W1ATD	927.47500	902.07500	131.8 Hz	NESMC
Medfield	FM	N1KUE	441.50000	+	88.5 Hz	
Medford	DMR/MARC	W1DSR	146.94000	-	CC1	
Medway	FM	W1KG	147.06000	+		
	FM	W1KG	224.66000	-		
Mendon	FM	K1KWP	146.61000	-	146.2 Hz	NESMC
Methuen	DMR/BM	N1EXC	147.01500	+	CC1	
	FM	WB1CXB	223.92000	-	141.3 Hz	
	FM	N1LHP	224.68000	-	88.5 Hz	
	FM	N1WPN	448.32500	-	88.5 Hz	
Milford	FM	WA1QGU	446.82500	-	100.0 Hz	NESMC
	FM	W1NAU	448.72500	-	DCS 343	
Milton	FM	N1MV	224.36000	-		
Mt. Greylock	FM	K1FFK	224.10000	+	162.2 Hz	
N Oxford	DMR/BM	KC1ACI	447.27500	-	CC1	
Nantucket	FM	W1TUK	146.79000	-	107.2 Hz	NESMC
Natick	FM	W1STR	449.12500	-	146.2 Hz	
Natick, Police Station						
	FM	KB1DFN	447.67500	-	203.5 Hz	
New Bedford	FM	W1RJC	145.11000	-	67.0 Hz	NESMC
Newton	FM	W1TKZ	147.03000	+	123.0 Hz	
	FM	W1LJO	147.36000	+	67.0 Hz	
	FM	WA1GPO	442.75000	+	141.3 Hz	

MASSACHUSETTS 187

Location	Mode	Call sign	Output	Input	Access	Coordinator
Newton	FM	W1TKZ	444.60000	+	88.5 Hz	
North Adams	FM	KC1EB	145.49000	-	100.0 Hz	
North Andover	FM	N1LHP	444.10000	+		
North Attleboro	DMR/BM	NA1HS	447.97500	-	CC3	
North Reading, Tower						
	FM	KC1US	146.71500	-	146.2 Hz	
Northampton	DMR/MARC	KA1QFE	145.18000	-	CC1	
	DMR/MARC	KA1OAN	449.52500	-	CC1	
Northborough	FM	K1WPO	441.60000	+	88.5 Hz	
Norwell	DMR/BM	AC1M	145.25000	-	CC5	
	DMR/BM	N1ZZN	224.06000	-	CC6	NESMC
	FM	KC1HO	53.33000	52.33000	71.9 Hz	
Norwood	FM	W1JLI	147.21000	+	100.0 Hz	
Oakham	FM	KA1OXQ	53.67000	52.67000	123.0 Hz	
Orange	DMR/MARC	W1WWX	447.15000	446.65000	CC1	
	FM	NA1P	146.62500	-	110.9 Hz	
Orleans	FM	N5API	145.67000		141.3 Hz	
Otis	DMR/BM	N1ATP	443.93750	+	CC1	
Oxford	FM	K1AOI	147.25500	+	88.5 Hz	
Paxton	FM	W1BIM	146.97000	-	114.8 Hz	
	FM	WR1O	224.38000	-		
	FM	W1XOJ	447.98750	-	136.5 Hz	
Pelham	FM	N1PAH	53.09000	52.09000	162.2 Hz	
	FM	WA1VEI	224.74000	-	88.5 Hz	
Pepperell	FM	N1MNX	145.07000			
	FM	N1MNX	147.34500	+	100.0 Hz	
	FM	WA1VVH	224.64000	-		
	FM	N1MNX	442.90000	+	100.0 Hz	
	FM	WA1VVH	446.52500	-		
	FM	WA1VVH	927.46250	902.46250	88.5 Hz	
Pepperrell	FM	N1MNX	53.89000	52.89000	100.0 Hz	
Pittsfield	DMR/BM	N1ATP	449.93750	-	CC1	
	FM	KD2NSA	146.70000	-	69.3 Hz	
	FM	K1FFK	147.03000	+		
Plymouth	FM	N1ZIZ	146.68500	-	82.5 Hz	
	FM	WG1U	147.31500	+	67.0 Hz	
Princeton	WX	W3DEC	448.62500	-	88.5 Hz	NESMC
Princeton, Mount Wachusett						
	FM	WC1MA	53.31000	52.31000	71.9 Hz	
Quincy	FM	W1BRI	146.67000	-	146.2 Hz	NESMC
	FM	N1KUG	224.40000	-	103.5 Hz	NESMC
Reading	FM	WA1RHN	446.52500	-	151.4 Hz	
Rehoboth	FM	N1UMJ	927.57500	902.57500	DCS 051	
Salem	FM	NS1RA	146.88000	-	118.8 Hz	
	FM	NS1RA	446.62500	-	88.5 Hz	
	FM	NS1RA	927.75000	902.75000	131.8 Hz	
Sharon	FM	K1CNX	146.86500	-	103.5 Hz	
South Deerfield	FM	N1PMA	147.16500	+	118.8 Hz	
Southboro	DMR/MARC	AE1C	448.37500	-	CC1	NESMC
Spencer	FM	N1VOR	224.54000	-		
Springfield	FM	K4PR	147.30000	+	127.3 Hz	
	FM	KP4R	447.82500	-	127.3 Hz	
Stoneham	FM	WA1HUD	53.25000	52.25000	71.9 Hz	NESMC
Sutton	DMR/BM	KC1AZZ	442.85000	+	CC1	NESMC
Swansea	FM	KC1JET	145.32000	-		
Taunton	FM	KA1GG	147.13500	+	67.0 Hz	
Topsfield	FM	W1VYI	147.28500	+	100.0 Hz	
Truro	FM	WA1YFV	147.25500	+	67.0 Hz	
Uxbridge	FM	KB1MH	53.43000	52.43000	118.8 Hz	NESMC
	FM	KB1MH	147.39000	+		
	FM	W1WNS	447.32500	-	DCS 244	
	FM	KB1MH	447.47500	-	118.8 Hz	

188   MASSACHUSETTS

Location	Mode	Call sign	Output	Input	Access	Coordinator
Wakefield	FM	WA1RHN	147.07500	+	151.4 Hz	
	FM	WA1RHN	223.80000	-		
	FM	WA1WYA	224.26000	-	67.0 Hz	
Walpole	DMR/MARC	W1JFR	145.38000	-	CC12	
	DSTAR	WA1PLE	446.43750	-		
	FM	W1ZSA	224.32000	-	118.8 Hz	
	FM	W1ZSA	448.97500	-	141.3 Hz	
	WX	K1HRV	146.89500	-	123.0 Hz	
Waltham	FM	W1MHL	224.94000	-		
	FM	W1MHL	449.07500	-	88.5 Hz	
	FM	W1MHL	927.13750	902.13750	131.8 Hz	
	P25	W1MHL	146.64000	-	136.5 Hz	
Warren	FM	K1QVR	147.21000	+	88.5 Hz	
Webster	DMR/MARC	N1PFC	446.00000	-	CC0	
West Bridgewater	DMR/BM	KA1GG	449.98750	-	CC4	
	IDAS	WG1U	145.28000	-		
West Bridgwater	FM	AB1CQ	146.77500	-	DCS 244	
West Millbury	FM	KA1AQP	444.90000	+	100.0 Hz	
West Newbury	FM	K1KKM	146.62500	-	131.8 Hz	
West Tisbury	FM	KB1QL	147.34500	+	88.5 Hz	
Westborough	FM	W1WNS	448.77500	-	DCS 244	
Westfield	FM	W1JWN	147.07500	+	88.5 Hz	
	FM	W1MBT	446.77500	-	77.0 Hz	
	FM	N1PAH	449.82500	-	110.9 Hz	NESMC
Westford	DMR/BM	KB1SBJ	443.05000	+	CC1	
	FM	WB1GOF	442.45000	+		NESMC
Westford, Prospect Hill						
	FM	WB1GOF	146.95500	-	74.4 Hz	NESMC
Westford, Prospect Hill Water						
	FM	WB1GOF	145.33000	-		
Weston	FM	N1BE	146.79000	-	146.2 Hz	
	FM	N1NOM	224.70000	-	103.5 Hz	
	FM	W1MRA	442.70000	+	88.5 Hz	
Weymouth	FM	N1BGT	147.30000	+		
Weymouth, South Shore Hospital						
	FM	W1SSH	147.34500	+	110.9 Hz	NESMC
Whitman	FM	WA1NPO	147.22500	+	67.0 Hz	
	P25	KC1EFG	927.68750	902.68750		NESMC
Wilbraham	FM	W1XOJ	147.10500	+	162.2 Hz	
Wilmington	DMR/MARC	K1KZP	146.47000	144.97000	CC10	
	FM	K1KZP	224.16000	-	67.0 Hz	
	FM	K1KZP	441.90000	+		
Winchendon	FM	AA1JD	224.44000	-		
	FM	WC1P	447.20000	-	88.5 Hz	
Woburn	FM	N1LHP	449.82500	-	136.5 Hz	
Worcester	DMR/MARC	W1DSR	147.37500	+	CC1	
	DMR/MARC	N1PFC	442.20000	+	CC1	
	FM	N1EKO	224.48000	-		
	FM	N1OHZ	443.30000	+	100.0 Hz	
	FM	W1WPI	449.02500	-	88.5 Hz	
	FM	WE1CT	927.73750	902.73750	DCS 244	
	P25	WE1CT	146.48000	144.98000	107.2 Hz	NESMC
	P25	W1YK	146.92500	-	100.0 Hz	
	P25	N1PFC	449.87500	-		NESMC
Worcester, WPI	FM	W1WPI	145.31000	-	100.0 Hz	
Wrentham	FM	K1LBG	147.09000	+	146.2 Hz	
	FM	N1UEC	224.78000	-		
	FM	K1LBG	444.45000	+	127.3 Hz	
	FM	K1LBG	448.57500	-	88.5 Hz	
	FM	N1UEC	927.48750	902.48750	131.8 Hz	NESMC

MICHIGAN 189

Location	Mode	Call sign	Output	Input	Access	Coordinator
MICHIGAN						
49024	DMR/BM	KM8CC	438.87500		CC1	
Ada	DMR/BM	K8SN	444.16250	+	CC1	
Adrian	FM	W8TQE	145.37000	-	85.4 Hz	MiARC
	FM	K8ADM	443.37500	+	107.2 Hz	MiARC
Alanson	FM	KC8YGT	443.20000	+	151.4 Hz	MiARC
Allegan	FM	AC8RC	147.24000	+	94.8 Hz	MiARC
Alma, Alma College						
	FM	KC8MUV	145.37000	-	100.0 Hz	MiARC
Alpena	FM	N8BIT	442.47500	+	100.0 Hz	MiARC
Alpena, Manning Hill						
	FM	K8PA	146.76000	-	88.5 Hz	MiARC
Ann Arbor	DMR/BM	W8RP	443.50000	+	CC1	MiARC
	DMR/MARC	N8LBV	443.05000	+	CC1	MiARC
	FM	N8DUY	145.15000	-	100.0 Hz	MiARC
	FM	WB8TKL	146.96000	-	100.0 Hz	MiARC
	FM	W8PGW	224.38000	-		MiARC
Ann Arbor, University Of Michi						
	FM	W8UM	145.23000	-	100.0 Hz	MiARC
Aurelius	FM	KC8LMI	443.87500	+	136.5 Hz	MiARC
Avoca, Greenwood Energy Center						
	FM	K8DD	147.30000	+	192.8 Hz	MiARC
Bad Axe	FM	N8LFR	145.47000	-	110.9 Hz	MiARC
	FM	KA8PZP	146.88000	-		MiARC
Baldwin, Wolf Lake						
	FM	AF8U	146.90000	-	94.8 Hz	MiARC
Bancroft	DMR/MARC	W8FSM	443.31250	+	CC1	
Bangor	FM	K8BRC	147.36000	+	94.8 Hz	MiARC
Barton City	FM	WB8ZIR	145.49000	-	100.0 Hz	
Battle Creek	DSTAR	W8DF	146.79000	-		MiARC
	DSTAR	W8DF	442.76250	+		MiARC
	FM	W8IRA	145.15000	-	94.8 Hz	MiARC
	FM	W8DF	146.66000	-	94.8 Hz	MiARC
	FM	KD8PVK	147.12000	+	186.2 Hz	MiARC
	FM	W8DF	224.24000	-		MiARC
	FM	W8DF	443.95000	+	94.8 Hz	MiARC
Bay City	FM	N8BBR	147.36000	+	131.8 Hz	MiARC
	FM	KB8YUR	444.50000	+	123.0 Hz	MiARC
Belle River - East China						
	FM	KD8GRU	147.32000	+	131.8 Hz	
Benzonia	FM	W8BNZ	442.60000	+	100.0 Hz	MiARC
Bessemer, Blackjack Mountain S						
	FM	K8ATX	146.76000	-		UPARRA
Beverly Hills	FM	W8HP	443.22500	+	107.2 Hz	MiARC
Big Rapids	FM	W8IRA	145.29000	-	94.8 Hz	MiARC
Big Rapids, Ferris State Unive						
	FM	KB8QOI	443.90000	+		MiARC
Big Rapids, WDEE Tower						
	FM	KB8QOI	146.74000	-		MiARC
Bloomfield Township						
	FM	AA8GK	443.82500	+		
Breckenridge	FM	W8QPO	442.65000	+	100.0 Hz	MiARC
Bridgeport	FM	KC8BXI	443.40000	+		MiARC
Bridgman	FM	W8MAI	442.77500	+	88.5 Hz	MiARC
Britton	DMR/BM	W8ATE	444.05000	+	CC1	
Brooklyn	DMR/BM	N8GY	443.90000	+	CC1	
	DMR/MARC	KB8POO	145.12000	-	CC1	
Brooklyn, Clark Lake						
	FM	N8URX	443.75000	+	146.2 Hz	
Buchanan	FM	N8NIT	443.65000	+		MiARC
Burnside	DMR/MARC	W8CMN	443.11250	+	CC1	

190   MICHIGAN

Location	Mode	Call sign	Output	Input	Access	Coordinator
Burt	FM	KC8ELQ	442.20000	+	103.5 Hz	MiARC
Burton	FM	N8NE	147.38000	+	88.5 Hz	MiARC
	FM	W8JDE	224.72000	-		MiARC
Byron Center	DMR/BM	KD8RXD	442.92500	+	CC2	
	DMR/MARC	KD8RXD	444.62500	+	CC1	MiARC
Cadillac	FM	K8CAD	146.98000	-		
	FM	W8IRA	147.16000	+	103.5 Hz	MiARC
Calumet, Mount Horace Greeley						
	FM	K8MDH	147.31500	+	100.0 Hz	UPARRA
Calumet, Old Calumet Air Base						
	FM	K8MDH	443.15000	+	100.0 Hz	UPARRA
Caro	FM	KC8CNN	146.66000	-	100.0 Hz	MiARC
	FM	WA8CKT	146.82000	-	100.0 Hz	MiARC
	FM	KC8CNN	442.55000	+	103.5 Hz	MiARC
Cassopolis	DMR/MARC	KD8UJM	443.55000	+	CC1	
Cedar Springs	FM	NW8J	52.72000	-	136.5 Hz	MiARC
	FM	NW8J	146.88000	-	141.3 Hz	MiARC
	FM	NW8J	224.14000	-		MiARC
	FM	NW8J	443.07500	+	94.8 Hz	MiARC
	FUSION	AB8DT	445.50000	-	100.0 Hz	
Centreville, Glen Oaks Communi						
	FM	K8SJC	145.31000	-	123.0 Hz	MiARC
	FM	KC8BRO	442.15000	+	94.8 Hz	MiARC
Charlotte	DMR/BM	N8HEE	442.26250	+	CC1	MiARC
	FM	K8CHR	147.08000	+	103.5 Hz	MiARC
	FM	N8HEE	443.62500	+	100.0 Hz	MiARC
Charlotte, County EOC						
	DSTAR	K8ETN	145.20000	-		MiARC
	DSTAR	K8ETN	443.43750	+		MiARC
Cheboygan	FM	W8IPQ	146.74000	-	103.5 Hz	MiARC
	FM	WB8DEL	444.85000	+	100.0 Hz	MiARC
Chelsea	DMR/BM	KB8POO	444.75000	+	CC1	
	FM	WD8IEL	145.45000	-	100.0 Hz	MiARC
	FM	KC8LMI	443.57500	+	100.0 Hz	MiARC
Chelsea, Sylvan Township Water						
	FM	KC8LMI	145.31000	-	136.5 Hz	
Clarkston	DMR/MARC	KD8VIV	444.83750	+	CC2	
	FM	W8JWB	146.84000	-	100.0 Hz	MiARC
Clinton Township	DMR/BM	N8PBX	443.67500	+	CC2	MiARC
Clio	FM	KB5TOJ	443.40000	+	156.7 Hz	MiARC
	FM	W8JDE	444.37500	+		MiARC
Coldwater	DSTAR	KD8JGF	442.96250	+		MiARC
	FM	WD8KAF	147.30000	+	100.0 Hz	MiARC
	FM	WD8KAF	443.30000	+	123.0 Hz	MiARC
Commerce Township						
	DMR/BM	N4KCD	442.97500	+	CC1	
Commerce Twp	DMR/MARC	WB8SFY	444.93750	+	CC1	
Commerce Twp.	DMR/MARC	WB8SFY	446.50000	-	CC5	
Cooks	FM	WA8WG	146.70000	-	110.9 Hz	UPARRA
Copper Harbor	FM	KE8IL	146.97000	-		
Dansville	DMR/MARC	N8OBU	444.68750	5444.68750	CC2	
	DMR/MARC	N8OBU	444.70000	+	CC1	
	FM	N8OBU	444.57500	+	107.2 Hz	MiARC
Davisburg	DMR/MARC	N8UE	444.83750	+	CC2	MiARC
Dearborn	FM	K8UTT	145.27000	-		MiARC
	FM	WR8DAR	147.16000	+	100.0 Hz	MiARC
	FM	K8UTT	224.52000	-	100.0 Hz	MiARC
	FM	WR8DAR	442.80000	+	107.2 Hz	MiARC
	FM	K8UTT	443.42500	+	107.2 Hz	MiARC
Decatur	FM	KF8ZF	52.94000	-	94.8 Hz	MiARC
Detroit	DMR/MARC	W8FSM	444.00000	+	CC1	
	DMR/MARC	W8CMC	444.67500	+	CC2	MiARC

MICHIGAN 191

Location	Mode	Call sign	Output	Input	Access	Coordinator
Detroit	DMR/MARC	K9DPD	444.87500	+	CC1	
	FM	K8PLW	51.84000	-	100.0 Hz	MiARC
	FM	W8DET	145.11000	-	100.0 Hz	MiARC
	FM	WR8DAR	145.33000	-	100.0 Hz	MiARC
	FM	KC8LTS	147.33000	+		MiARC
	FM	KC8LTS	224.36000	-	103.5 Hz	MiARC
	FM	K8PLW	442.10000	+	107.2 Hz	MiARC
	FM	KC8LTS	442.17500	+	123.0 Hz	MiARC
	FM	KD8IFI	443.02500	+	107.2 Hz	MiARC
	FM	WR8DAR	443.47500	+	88.5 Hz	MiARC
Detroit, AT&T Michigan Headqua						
	FM	KE8HR	146.76000	-	100.0 Hz	MiARC
Detroit, Renaissance Center -						
	FM	WW8GM	443.07500	+	123.0 Hz	MiARC
Dexter	FM	W8SRC	446.15000	-	100.0 Hz	MiARC
Dimondale	FM	N9UV	442.05000	+	100.0 Hz	MiARC
Dowagiac	FM	KU8Y	145.21000	-	94.8 Hz	MiARC
Dowagiac, Building						
	FM	N9QID	927.68750	902.68750	94.8 Hz	
Dundee	FM	K8RPT	442.82500	+	100.0 Hz	MiARC
Dunken	FM	W8CDZ	146.67000	-	100.0 Hz	UPARRA
Durand	FM	N8IES	145.29000	-	100.0 Hz	MiARC
	FM	N8IES	224.86000	-	100.0 Hz	MiARC
	FM	N8IES	442.62500	+	100.0 Hz	MiARC
Eagle	DMR/BM	KB8SXK	403.02500		CC2	
	DMR/BM	KB8SXK	444.71250	+	CC2	
Eagle / Portland	FM	K8VEB	443.35000	+	100.0 Hz	MiARC
Eagle/Portland	FM	K8VEB	224.66000	-	100.0 Hz	MiARC
East Jordan	FM	W8COL	147.28000	+	103.5 Hz	MiARC
East Lansing, MSU						
	FM	W8MSU	442.90000	+		MiARC
Edmore	FM	WB8VWK	146.80000	-	103.5 Hz	MiARC
	FM	KC8LEQ	444.70000	+	103.5 Hz	MiARC
Elmira	DSTAR	NM8ES	145.32000	-		MiARC
	DSTAR	NM8ES	444.11250	+		MiARC
Escanaba	FM	KB9BQX	145.13000	-		UPARRA
	FM	N8JWT	147.15000	+	123.0 Hz	
	FM	N8JWT	147.24000	+	123.0 Hz	
	FM	W8JRT	442.40000	+		
	FM	WD8RTH	444.30000	+		UPARRA
Farmington Hills	FM	WA8SEL	442.70000	+	100.0 Hz	MiARC
Fenton	DMR/MARC	W8FSM	443.20000	+	CC2	
	FM	W8CMN	146.78000	-	151.4 Hz	MiARC
	FM	KB8PGF	443.97500	+	67.0 Hz	MiARC
	FM	N8VDS	927.53750	902.53750	131.8 Hz	MiARC
Flint	DMR/BM	W8CMN	443.80000	+	CC1	MiARC
	DSTAR	N8UMW	442.00000	+	100.0 Hz	MiARC
	FM	KC8KGZ	147.10000	+	100.0 Hz	MiARC
	FM	KC8KGZ	147.26000	+	100.0 Hz	MiARC
	FM	W8ACW	147.34000	+	100.0 Hz	MiARC
	FM	KC8KGZ	224.48000	-	100.0 Hz	MiARC
	FM	W8ACW	444.20000	+	107.2 Hz	MiARC
	FM	W8JDE	444.60000	+		MiARC
	FM	KC8KGZ	1253.25000	439.25000		MiARC
Fort Gratiot	FM	KG8OU	146.80000	-	100.0 Hz	MiARC
Frankenmuth	DMR/MARC	W8FSM	444.25000	+	CC1	
	DMR/MARC	KB8SWR	444.72500	+	CC1	
	FM	KB8SWR	444.02500	+	100.0 Hz	MiARC
Frankfort	FM	W8BNZ	442.20000	+	114.8 Hz	MiARC
Fremont	FM	KC8MSE	146.92000	-	94.8 Hz	MiARC
Garden City	DSTAR	N8RTS	442.12500	+		MiARC
	FM	KK8GC	146.86000	-	100.0 Hz	MiARC

192 MICHIGAN

Location	Mode	Call sign	Output	Input	Access	Coordinator
Gaylord	DSTAR	KD8QCC	444.03750	+		
	FM	NM8RC	146.82000	-	118.8 Hz	MiARC
	FM	W1WRS	147.12000	+	151.4 Hz	MiARC
Gaylord, Wilderness Valley						
	FM	K7IOU	224.74000	-	136.5 Hz	
Gladwin	FM	W8GDW	147.18000	+		MiARC
	FM	K8EO	442.45000	+		
Gladwin, Lake Lancer						
	FUSION	W8CSX	145.25000	-	103.5 Hz	
Glen Arbor	FM	WI0OK	52.92000	-	146.2 Hz	MiARC
	FM	WI0OK	444.72500	+	114.8 Hz	MiARC
Glendale	FM	W8GDS	224.84000	-	94.8 Hz	MiARC
Glenwood	FM	W8GDS	224.84000	-	94.8 Hz	
Grand Haven	FM	W8CSO	145.49000	-	94.8 Hz	MiARC
	FM	W8CSO	443.77500	+	94.8 Hz	MiARC
Grand Marais	FM	KC8BAN	147.19500	+	100.0 Hz	UPARRA
Grand Rapids	DMR/BM	KD8RXD	446.50000	-	CC2	
	DMR/MARC	KD8RXD	444.25000	+	CC1	
	DSTAR	WX8GRR	147.29000	+		MiARC
	DSTAR	WX8GRR	442.55000	+		MiARC
	FM	NW8J	145.11000	-	94.8 Hz	MiARC
	FM	W8DC	146.76000	-	94.8 Hz	MiARC
	FM	W8IRA	147.16000	+	94.8 Hz	MiARC
	FM	W8DC	147.26000	+	94.8 Hz	MiARC
	FM	K8DMR	421.25000	439.25000		MiARC
	FM	K8EFK	442.00000	+	141.3 Hz	MiARC
	FM	KA8YSM	443.80000	+	94.8 Hz	MiARC
	FM	N8NET	444.10000	+		MiARC
	FM	N8WKN	444.32500	+	82.5 Hz	MiARC
	FM	W8DC	444.40000	+	94.8 Hz	MiARC
	FM	W8USA	444.45000	+	94.8 Hz	
	FM	K8WM	444.77500	+	94.8 Hz	MiARC
	FM	N8WKM	927.26250	902.26250		MiARC
Grand Rapids, NW Grand Rapids						
	FM	K8WM	145.41000	-	94.8 Hz	MiARC
Grass Lake	DMR/MARC	W8CMN	443.81250	+	CC1	MiARC
	FM	KC8LMI	224.16000	-	100.0 Hz	MiARC
Grayling	FM	N8AHZ	145.13000	-	107.2 Hz	MiARC
Greenville	DMR/BM	KB8ZGL	440.27500	+	CC1	
	DMR/MARC	KD8RXD	443.38750	+	CC1	
	FM	KB8ZGL	927.48750	902.48750	131.8 Hz	MiARC
Greilickville	DSTAR	WI0OK	145.36000	-		MiARC
Grosse Ile	FM	N8ZPJ	444.90000	+	107.2 Hz	MiARC
Grosse Pointe Farms						
	FM	N8XN	444.22500	+	107.2 Hz	MiARC
Gwinn	FM	K8LOD	146.64000	-	100.0 Hz	UPARRA
Hancock	FM	W8CDZ	146.88000	-	100.0 Hz	UPARRA
Hanover	FM	K8WBG	52.62000	-	123.0 Hz	MiARC
Harrison	FM	KA8DCJ	147.20000	+	103.5 Hz	MiARC
Harrisville	FM	W8HUF	147.04000	+	123.0 Hz	MiARC
Hart	FM	N8UKH	146.64000	-	94.8 Hz	MiARC
	FM	W8VTM	443.67500	+	94.8 Hz	MiARC
Hastings	FM	K8YPW	146.84000	-	94.8 Hz	
Hell	DSTAR	K8LCD	147.21000	+		MiARC
	DSTAR	K8LCD	444.06250	+		MiARC
	DSTAR	K8LCD	1294.48000	-		MiARC
Hemlock	FM	N8ERL	145.33000	-	88.5 Hz	MiARC
Hesperia	DMR/MARC	KC8MSE	442.01250	+	CC1	
Holland	DMR/BM	K8DAA	443.88750	+	CC4	
	FM	K8DAA	146.50000	147.50000	94.8 Hz	MiARC
	FM	K8DAA	147.06000	+	94.8 Hz	MiARC
	FM	K8DAA	443.82500	+	94.8 Hz	MiARC

MICHIGAN   193

Location	Mode	Call sign	Output	Input	Access	Coordinator
Holland	FUSION	N8XPQ	444.80000	+		MiARC
Holly	FM	W8FSM	224.62000	-	100.0 Hz	MiARC
	FM	W8FSM	442.35000	+	107.2 Hz	MiARC
Holton, M-120 Highway						
	FM	WD8MKG	147.32000	+	91.5 Hz	MiARC
Houghton	DMR/MARC	N8WAV	147.12000	+	CC1	
	DMR/MARC	N8WAV	435.00000	+	CC1	
	DMR/MARC	W8YY	444.50000	+	CC1	UPARRA
	FM	N8WAV	147.39000	+	100.0 Hz	UPARRA
Houghton, Wadsworth Hall						
	FM	W8YY	444.65000	+	100.0 Hz	UPARRA
Howell	DMR/BM	W8LRK	442.57500	+	CC1	
	DSTAR	W8LIV	145.32000	-		MiARC
	DSTAR	W8LIV	444.03750	+		MiARC
	FM	K8JBA	145.41000	-	162.2 Hz	
	FM	W8LRK	146.68000	-	162.2 Hz	MiARC
Hudsonville	FM	K8TB	442.25000	+	94.8 Hz	MiARC
	FM	K8IHY	444.90000	+	94.8 Hz	MiARC
Ida	FM	K8RPT	146.72000	-	100.0 Hz	MiARC
	FM	K8RPT	442.65000	+	100.0 Hz	MiARC
Iron Mountain	FM	WA8FXQ	146.85000	-	100.0 Hz	UPARRA
	FM	WA8FXQ	444.85000	+	100.0 Hz	UPARRA
Iron River, Ski Brule Moutain						
	FM	N8LVQ	145.17000	-	107.2 Hz	UPARRA
Ironwood	DMR/MARC	N8JJB	441.43750	440.93750	CC1	
	FM	N8JJB	146.80500	-	141.3 Hz	UPARRA
	FM	WA1MAR	441.43750	+	107.2 Hz	
Ishpeming, Cliffs Shaft Museum						
	FM	K8LOD	146.91000	-		UPARRA
Ithaca	DSTAR	KD8IEK	147.15000	+		MiARC
	DSTAR	KD8IEK	443.13750	+		MiARC
Jackson	FM	W8IRA	145.47000	-	114.8 Hz	MiARC
	FM	W8JXN	146.88000	-	100.0 Hz	MiARC
	FM	KA8HDY	147.36000	+	100.0 Hz	MiARC
	FM	WD8EEQ	443.17500	+	77.0 Hz	MiARC
	FM	KA8YRL	444.17500	+	100.0 Hz	MiARC
	FM	K4KWQ	444.90000	+	131.8 Hz	
James Twp.	DMR/MARC	N8VDS	443.60000	+	CC1	MiARC
Jerome	FM	KC8QVX	444.82500	+	107.2 Hz	MiARC
Jonesville	FM	KC8QVX	147.06000	+	179.9 Hz	MiARC
Kalamazoo	DMR/BM	KE8EVF	442.67500	+	CC1	
	FM	K8KZO	51.72000	-	94.8 Hz	MiARC
	FM	N8FYZ	145.17000	-	94.8 Hz	MiARC
	FM	W8VY	147.00000	+	94.8 Hz	MiARC
	FM	K8KZO	147.04000	+	94.8 Hz	MiARC
	FM	W8VY	444.65000	+	131.8 Hz	MiARC
	FM	K8KZO	444.87500	+	94.8 Hz	MiARC
Kalamazoo, Borgess Hospital						
	DSTAR	NK8X	444.50000	+		
Kalkaska	FM	W8KAL	52.82000	-		MiARC
Kent City	DMR/MARC	KD8RXD	442.21250	+	CC1	MiARC
Kincheloe	FM	W8EUP	444.90000	+	107.2 Hz	
La Salle	FM	K8RPT	224.78000	-	100.0 Hz	MiARC
Lake Angelus	FM	NE9Y	53.94000	-	131.8 Hz	MiARC
Lake City	FM	KG8QY	145.21000	-		MiARC
	FM	KA8ABM	444.52500	+	100.0 Hz	MiARC
Lake Leelanau	FM	N8JKV	146.92000	-	114.8 Hz	MiARC
Lake Orion	DSTAR	K8DXA	145.13000	-		MiARC
	DSTAR	K8DXA	444.26250	+		MiARC
Lake Orion, Great Lakes Shoppi						
	FM	WW8GM	145.61000		123.0 Hz	
Lansing	DMR/BM	KB8SXK	444.78750	+	CC2	MiARC

194   MICHIGAN

Location	Mode	Call sign	Output	Input	Access	Coordinator
Lansing	DMR/MARC	KB8SXK	442.08750	+	CC1	MiARC
	DMR/MARC	W8JTT	446.50000	-	CC1	
	FM	KD8PA	52.96000	-	100.0 Hz	MiARC
	FM	KB8LCY	147.28000	+	100.0 Hz	MiARC
	FM	KD8PA	442.42500	+	100.0 Hz	MiARC
	FM	KD8IFI	443.00000	+	107.2 Hz	MiARC
	FM	KD8PA	927.97500	902.97500		MiARC
Lansing, Ingham Regional Medic						
	FM	W8BCI	146.70000	-	107.2 Hz	MiARC
	FM	W8BCI	146.94000	-	100.0 Hz	MiARC
	FM	W8BCI	224.98000	-	100.0 Hz	MiARC
Lapeer	DSTAR	W8LAP	442.75000	+		MiARC
	FM	W8LAP	146.62000	-	100.0 Hz	MiARC
Leland	FM	W8SGR	145.39000	-	103.5 Hz	
Lewiston	FM	N8SCY	145.19000	-		MiARC
Lincoln	DMR/MARC	W8JJR	442.01250	+	CC1	MiARC
Livonia	FM	K8UNS	145.35000	-	100.0 Hz	MiARC
	FM	K8PLW	224.84000	-	100.0 Hz	MiARC
	FM	K8UNS	444.87500	+	123.0 Hz	MiARC
Lowell	DMR/MARC	KD8RXD	443.11250	+	CC1	MiARC
	FM	W8LRC	145.27000	-	94.8 Hz	MiARC
	FM	AA8JR	443.85000	+	94.8 Hz	MiARC
Ludington	FM	W8IRA	145.31000	-	94.8 Hz	MiARC
	FM	K8DXF	145.47000	-	103.5 Hz	MiARC
	FM	WB8ERN	146.62000	-	94.8 Hz	MiARC
Lupton	FM	N8RSH	147.08000	+		MiARC
Mackinaw City, Icebreaker Mack						
	FM	W8AGB	444.37500	+	107.2 Hz	
Mancelona	FM	K8WQK	51.98000	-		MiARC
	FM	K8WQK	147.38000	+	107.2 Hz	
Manchester	DMR/MARC	KB8POO	146.60000	-	CC1	
	FM	WD8IEL	146.98000	-	100.0 Hz	MiARC
Manistee	FM	W8GJX	146.78000	-	94.8 Hz	MiARC
	FM	KB8BIT	224.12000	-	100.0 Hz	MiARC
Marcellus	FM	KD8UJM	442.22500	+	94.8 Hz	MiARC
Marine City	DMR/MARC	KB8VLL	443.88750	+	CC2	
Marquette	FM	K8LOD	147.27000	+	100.0 Hz	UPARRA
	FM	KB0P	442.20000	+	100.0 Hz	UPARRA
	FM	K8LOD	443.45000	+	100.0 Hz	UPARRA
Marquette, Marquette Mountain						
	FM	KE8IL	146.97000	-	100.0 Hz	UPARRA
Mason	DMR/MARC	N8URW	442.50000	+	CC0	MiARC
	FM	WB8RJY	51.70000	-	192.8 Hz	MiARC
	FM	WB8RJY	443.70000	+		MiARC
Mayville	DMR/MARC	KB8SWR	443.85000	+	CC1	MiARC
	FM	KB8ZUZ	51.82000	-	131.8 Hz	MiARC
	FM	KB8ZUZ	443.77500	+	100.0 Hz	MiARC
Menominee	FM	AB9PJ	53.11000	51.41000	114.8 Hz	UPARRA
	FM	W8PIF	147.00000	+	107.2 Hz	UPARRA
	FM	W8PIF	444.07500	+	107.2 Hz	UPARRA
Midland	DSTAR	WB8WNF	444.35000	+	131.8 Hz	MiARC
	FM	W8KEA	147.00000	+	103.5 Hz	MiARC
	FM	W8QN	443.32500	+	103.5 Hz	MiARC
Milan	DMR/MARC	W2PUT	443.11250	+	CC1	MiARC
	FM	W2PUT	146.50000	147.50000	88.5 Hz	
	FM	W2PUT	444.10000	+	82.5 Hz	MiARC
Milford	FM	WR8DAR	444.42500	+	118.8 Hz	MiARC
Millington	DSTAR	KC8KGZ	444.65000	+	100.0 Hz	MiARC
Mio	FM	WT8G	145.35000	-		MiARC
Moline	DMR/BM	K8SN	442.26250	+	CC1	MiARC
	FM	N8JPR	223.92000	-	94.8 Hz	
	FM	K8SN	442.17500	+	103.5 Hz	MiARC

MICHIGAN   195

Location	Mode	Call sign	Output	Input	Access	Coordinator
Monroe, Downtown						
	FM	W8YZ	145.31000	-	100.0 Hz	MiARC
	FM	W8OTC	446.07500	-		MiARC
Moorestown	FM	KA8ABM	146.96000	-	103.5 Hz	MiARC
Morley	FM	K8SN	442.07500	+	103.5 Hz	
Mount Clemens	FM	WA8MAC	147.20000	+	100.0 Hz	MiARC
	FUSION	K8UO	444.77500	+	123.0 Hz	MiARC
Mount Pleasant	FM	KC8RTU	442.82500	+	100.0 Hz	MiARC
Mt Clemens	DMR/MARC	K9DPD	443.95000	+	CC1	
Mt. Clemens	FM	KC8UMP	443.62500	+	151.4 Hz	MiARC
	FM	W8FSM	927.25000	902.25000	131.8 Hz	MiARC
Munith	DMR/MARC	N8URW	443.89000	+	CC0	
Muskegon	DMR/MARC	K8WNJ	444.95000	+	CC1	MiARC
	DSTAR	WD8MKG	145.36000	-		MiARC
	DSTAR	WD8MKG	444.01250	+		MiARC
	FM	K8COP	52.80000	-	94.8 Hz	MiARC
	FM	W8IRA	145.33000	-	94.8 Hz	MiARC
	FM	K8WNJ	146.82000	-	94.8 Hz	MiARC
	FM	W8ZHO	146.94000	-	94.8 Hz	MiARC
	FM	KE8LZ	147.38000	+		MiARC
	FM	N8KQQ	224.70000	-	94.8 Hz	MiARC
	FM	N8UKF	442.30000	+		MiARC
	FM	N8KQQ	442.95000	+	94.8 Hz	MiARC
	FM	W8ZHO	444.55000	+	94.8 Hz	MiARC
New Hudson	FM	N8BK	442.77500	+	107.2 Hz	MiARC
Newberry	FM	W8NBY	146.61000	-	114.8 Hz	UPARRA
Niles	FM	KC8BRS	147.18000	+	94.8 Hz	MiARC
	FM	WB9YPA	224.50000	-	94.8 Hz	
	FM	WB9WYR	444.12500	+	94.8 Hz	MiARC
North Branch	FM	KG8ID	443.45000	+	100.0 Hz	MiARC
Northville	FM	WR8DAR	443.10000	+	82.5 Hz	MiARC
Norton Shores	FM	N8UKF	443.20000	+		MiARC
Novi	DMR/MARC	KC8LTS	442.21250	+	CC1	MiARC
	FM	N8OVI	444.80000	+	110.9 Hz	MiARC
Oak Park	FM	W8HP	146.64000	-	100.0 Hz	MiARC
Ovid	FM	N8TSK	51.92000	-	203.5 Hz	MiARC
	FM	N8TSK	444.00000	+	225.7 Hz	MiARC
	FM	KD8AGP	445.50000	-	94.8 Hz	MiARC
Owosso	DSTAR	W8SHI	145.24000	-		MiARC
	DSTAR	W8SHI	444.30000	+		MiARC
	FM	N8DVH	147.02000	+	100.0 Hz	MiARC
	FM	N8DVH	442.40000	+	100.0 Hz	MiARC
Oxford	FM	KA8CSH	443.00000	+		MiARC
Paw Paw	DSTAR	W8VY	145.34000	-		
	DSTAR	W8VY	444.07500	+		MiARC
	FM	W8GDS	147.20000	+	94.8 Hz	MiARC
Pellston	FM	WA8EFE	444.95000	+	103.5 Hz	MiARC
Phoenix	DMR/MARC	W8YY	444.75000	+	CC1	UPARRA
Pickford	FM	W8EUP	146.64000	-	107.2 Hz	UPARRA
Pinckney	DMR/BM	W2GLD	442.67500	+	CC8	MiARC
Pine Stump Junction						
	FM	W8NBY	147.09000	+		UPARRA
Pleasant Lake	FM	KC8LMI	224.18000	-	88.5 Hz	MiARC
	FM	KC8LMI	446.80000	-	136.5 Hz	
Pleasant Lk.	DMR/MARC	KC8LMI	442.51250	+	CC1	
Pontiac	DMR/MARC	W8OAK	444.32500	+	CC1	
	FM	W8OAK	146.90000	-	100.0 Hz	MiARC
	FM	WN8G	443.82500	+		MiARC
	FM	KB9WIS	927.45000	902.45000		MiARC
Port Huron	FM	AA8K	146.72000	-	186.2 Hz	MiARC
	FM	N8YTV	443.70000	+	100.0 Hz	MiARC
	FM	KE8JOT	444.90000	+	131.8 Hz	

196   MICHIGAN

Location	Mode	Call sign	Output	Input	Access	Coordinator
Portage	DMR/BM	KM8CC	443.40000	+	CC1	
Portland	FM	N8ZMT	145.13000	-	94.8 Hz	MiARC
Potterville	FM	W8BCI	145.39000	-	100.0 Hz	MiARC
	FM	N8JI	442.02500	+	173.8 Hz	MiARC
Quanicassee	FM	N8BBR	145.31000	-	131.8 Hz	MiARC
Republic	FM	K8LOD	146.82000	-	100.0 Hz	UPARRA
Riverview	FM	KC8LTS	927.48750	902.48750	131.8 Hz	MiARC
Rockford	FM	W8AGT	927.68750	902.68750	131.8 Hz	MiARC
Rogers City	FM	WB8TQZ	147.02000	+	103.5 Hz	MiARC
Romeo	P25	K8FBI	442.07500	+	123.0 Hz	MiARC
Romulus	FM	W8TX	442.27500	+	107.2 Hz	MiARC
Roscommon	FM	N8QOP	52.64000	-		MiARC
	FM	N8QOP	145.45000	-	141.3 Hz	MiARC
	FM	N8QOP	443.10000	+	146.2 Hz	MiARC
Rose City	FM	W8DMI	145.11000	-	131.8 Hz	MiARC
Roseville	FM	N8EDV	147.22000	+	100.0 Hz	MiARC
	FM	N8EDV	224.46000	-	100.0 Hz	MiARC
Rust	FM	NM8RC	146.82000	-		
Saginaw, Rosine Tower						
	FM	K8DAC	147.24000	+	103.5 Hz	MiARC
Saint Johns	FM	W8CLI	443.52500	+	100.0 Hz	MiARC
Saint Joseph	FM	KB8VIM	146.72000	-	131.8 Hz	MiARC
Sandusky	FM	W8AX	146.86000	-		MiARC
Saranac	FM	N8MRC	144.35000	147.85000	67.0 Hz	
Saranac, Saranac Bus Garage						
	FM	WA8RRA	444.72500	+	94.8 Hz	MiARC
Saugatuck	FM	AC8GN	146.96000	-	94.8 Hz	MiARC
	FM	AC8GN	442.70000	+		MiARC
Sault Sainte Marie						
	DMR/MARC	KE8FJW	444.13750	+	CC1	UPARRA
Sault St Marie	FM	W8EUP	147.21000	+	107.2 Hz	UPARRA
Sherman	FM	W8QPO	147.10000	+	100.0 Hz	MiARC
Sister Lakes	FM	W8MAI	146.82000	-	88.5 Hz	MiARC
South Lyon	FM	N8SL	147.04000	+	110.9 Hz	MiARC
Southfield	FM	W8HD	52.68000	-		MiARC
Southgate	DMR/MARC	KC8LTS	443.32500	+	CC1	MiARC
Sparta	FM	W8USA	145.23000	-	94.8 Hz	MiARC
St Johns	DSTAR	KD8IEI	145.44000	-		MiARC
	DSTAR	KD8IEI	442.93750	+		MiARC
St. Joseph	DSTAR	W8MAI	442.27500	+		MiARC
Sterling	FM	K8WBR	147.06000	+	103.5 Hz	MiARC
Sterling Heights	FM	N8LC	147.08000	+	100.0 Hz	MiARC
	FM	N8LC	442.92500	+		MiARC
	FM	KD8EYF	927.28750	902.28750		
Strongs	FM	W8ARS	147.33000	+	107.2 Hz	UPARRA
Stutsmanville	DMR/BM	W8FSM	442.08750	+	CC1	
	DSTAR	W8CCE	443.37500	+		MiARC
	FM	W8GQN	146.68000	-	110.9 Hz	MiARC
	FM	WB8DEL	224.56000	-	100.0 Hz	MiARC
	FM	N8DNX	442.37500	+	107.2 Hz	MiARC
Sumnerville	DSTAR	KE8GVB	145.14000	-		MiARC
	DSTAR	KE8GVB	442.82500	+		MiARC
Tawas	FM	W8ICC	146.64000	-	103.5 Hz	
Tba	DMR/MARC	N8OBU	444.71250	5444.71250	CC2	
Tecumseh	FM	W8MSU	442.90000	+		
Temperance	FM	K8RPT	444.55000	+	100.0 Hz	MiARC
Test Id	DMR/MARC	N8OBU	444.45000	5444.45000	CC2	
Thompsonville, Crystal Mountai						
	FM	W8BNZ	147.04000	+	114.8 Hz	MiARC
Traverse City	DSTAR	KD8OXV	443.31250	+		
	FM	W8IRA	145.15000	-	114.8 Hz	MiARC
	FM	W8TVC	145.27000	-	114.8 Hz	MiARC

MICHIGAN 197

Location	Mode	Call sign	Output	Input	Access	Coordinator
Traverse City	FM	W8TCM	146.86000	-	114.8 Hz	
	FM	W8QPO	147.10000	+	100.0 Hz	
	FM	W8TCM	442.50000	+	114.8 Hz	MiARC
	FM	W8QPO	442.90000	+	114.8 Hz	MiARC
	FM	KJ4KFJ	443.00000	+	114.8 Hz	MiARC
	FM	N8CN	444.40000	+		
Trenary	FM	W8FYZ	147.03000	+	100.0 Hz	UPARRA
Trenton	FM	WY8DOT	147.24000	+	100.0 Hz	MiARC
Troy, PNC Bank Building (Forme						
	FM	N8KD	147.14000	+	100.0 Hz	MiARC
Ubly	FM	KC8KOD	442.32500	+	103.5 Hz	MiARC
Utica	FM	K8UO	147.18000	+	100.0 Hz	MiARC
Vanderbilt	FM	W8IRA	145.29000	-	103.5 Hz	MiARC
Walkerville	FM	NW8J	145.43000	-	94.8 Hz	MiARC
Wallace	FM	KS8O	444.65000	+	DCS 125	UPARRA
Warren	DMR/BM	K8FBI	442.60000	+	CC1	MiARC
	DMR/BM	K8FBI	444.48750	+	CC1	
	DMR/MARC	KA8WYN	442.03750	+	CC3	
	DSTAR	WA8BRO	442.03750	+		MiARC
Waterford	DMR/MARC	N8QQS	442.83750	+	CC1	MiARC
Waterloo	DMR/MARC	KB8POO	147.31000	+	CC1	
Watrousville	FM	N8UT	147.32000	+	110.9 Hz	MiARC
	FM	N8UT	442.50000	+	91.5 Hz	MiARC
West Bloomfield	FM	WB8ARC	442.50000	+	107.2 Hz	
West Branch	DMR/MARC	W8FSM	443.95000	+	CC1	
	FM	W8YUC	145.41000	-	91.5 Hz	MiARC
West Branch, Pointer Hill Park						
	FM	K8OAR	146.94000	-	103.5 Hz	MiARC
	FM	K8OAR	444.97500	+	103.5 Hz	MiARC
West Branch, WBMI Tower						
	FM	KD8NCN	444.22500	+	107.2 Hz	MiARC
West Olive	DMR/MARC	K8OEC	443.57500	+	CC1	MiARC
Westland	DSTAR	W8DTW	145.17000	-		MiARC
	DSTAR	W8DTW	444.72500	+		MiARC
	DSTAR	W8DTW	1284.40000	-		MiARC
	DSTAR	W8DTW	1298.40000			MiARC
	FM	K8WX	443.15000	+	107.2 Hz	MiARC
	FM	N8ISK	443.27500	+	107.2 Hz	MiARC
Wetmore	FM	KC8BAN	145.41000	-	100.0 Hz	UPARRA
Wetmore, Michigan, Holiday Gas						
	FM	WB8Q	446.10000	-	100.0 Hz	
White Cloud, Maike Fire Lookou						
	FM	KB8IFE	145.45000	-		MiARC
	FM	KB8IFE	444.97500	+		MiARC
White Lake	DMR/BM	N8JY	444.65000	+	CC10	
	FM	N8BIT	145.49000	-	67.0 Hz	MiARC
White Pine	FM	AA8YF	147.30000	+	100.0 Hz	UPARRA
Whitehall	DMR/MARC	K8COP	443.25000	+	CC1	MiARC
	FM	K8COP	146.68000	-	94.8 Hz	MiARC
Winona	FM	W8UXG	146.73000	-		UPARRA
Wixom	FM	AC8IL	145.25000	-	100.0 Hz	
Yale	FM	N8ERV	443.30000	+	100.0 Hz	MiARC
Ypsilanti, St. Joseph Mercy Ho						
	FM	W8FSA	146.92000	-	100.0 Hz	MiARC
MINNESOTA						
Aitkin	FM	KC0QXC	146.80500	+	127.3 Hz	Minnesota RC
	FM	N0BZZ	147.36000	+	203.5 Hz	Minnesota RC
Albert Lea	FM	WA0RAX	443.52500	+	100.0 Hz	Minnesota RC
Alexandria	FM	W0ALX	146.79000	-	146.2 Hz	Minnesota RC
	FM	W0ALX	442.02500	+	146.2 Hz	Minnesota RC
Arden Hills	FM	KA0PQW	223.94000	-	100.0 Hz	Minnesota RC

198   MINNESOTA

Location	Mode	Call sign	Output	Input	Access	Coordinator
Arden Hills	FM	WI9WIN	442.07500	+	110.9 Hz	Minnesota RC
Askov	FM	W0MDT	146.95500	-	146.2 Hz	
Aurora	FM	N0BZZ	147.24000	+	156.7 Hz	Minnesota RC
Austin	FM	W0AZR	145.47000	-	100.0 Hz	Minnesota RC
	FM	W0AZR	146.73000	-	100.0 Hz	Minnesota RC
	FM	N0RZO	443.50000	+		Minnesota RC
Avon	FM	K0STC	147.10500	+	85.4 Hz	Minnesota RC
	FM	K0VSC	443.50000	+		
	FM	KG0CV	443.65000	+	85.4 Hz	Minnesota RC
Barnesville	FM	KC0SD	147.06000	+	123.0 Hz	Minnesota RC
Becker, MN, Excel Energy Power						
	FM	KD0YLG	443.47500	+	DCS 172	Minnesota RC
Becker, Power Plant Stack						
	FM	KD0YLG	147.34500	+	DCS 172	Minnesota RC
Bemidji	FM	KB0MM	145.45000	-		Minnesota RC
	FM	W0BJI	146.73000	-		Minnesota RC
	FM	WA0IUJ	147.18000	147.68000	82.5 Hz	Minnesota RC
	FM	W0BJI	444.02500	+	71.9 Hz	Minnesota RC
	FM	NI0K	444.95000	+	123.0 Hz	Minnesota RC
Bertha, Water Tower						
	FM	N0WN	147.12000	+	123.0 Hz	Minnesota RC
	FM	N0WN	444.75000	+		Minnesota RC
Big Falls	FM	N0NKC	146.91000	-	103.5 Hz	Minnesota RC
Big Lake	FM	K0SCA	145.49000	-	146.2 Hz	Minnesota RC
	FM	N0JDH	443.60000	+	114.8 Hz	Minnesota RC
Blaine	FM	W0YFZ	146.67000	-	114.8 Hz	Minnesota RC
Bloomington	DMR/BM	WA0CQG	442.15000	+	CC1	
	DMR/BM	N0NKI	443.10000	+	CC1	Minnesota RC
	DSTAR	WT0O	442.90000	+		
	FM	KD0CL	147.09000	+	100.0 Hz	Minnesota RC
	FM	N0BVE	444.32500	+	131.8 Hz	Minnesota RC
Bloomington, City Hall						
	FM	KD0CL	443.17500	+	DCS 047	Minnesota RC
Blue Earth	FM	N0PBA	147.00000	+	136.5 Hz	
	FM	KE0RTF	443.02500	+	114.8 Hz	
Brainerd	FM	W0UJ	53.11000	52.11000	123.0 Hz	Minnesota RC
	FM	W0UJ	146.70000	-	141.3 Hz	Minnesota RC
Brooklyn Park	DMR/MARC	KA0KMJ	445.42500	-	CC1	
Buck Hill-Burnsville						
	DMR/MARC	N0AGI	443.12500	+	CC1	Minnesota RC
Buffalo	FM	N0FWG	444.37500	+	156.7 Hz	Minnesota RC
Burnsvile	FM	W0BU	224.54000	-	100.0 Hz	Minnesota RC
Burnsville	FM	W0BU	53.37000	52.37000	100.0 Hz	Minnesota RC
	FM	W0BU	147.21000	+	100.0 Hz	Minnesota RC
	FUSION	W0BU	444.30000	+	114.8 Hz	Minnesota RC
Carlton	FM	KC0RTX	146.79000	-	103.5 Hz	Minnesota RC
Carver	FM	KB0FXK	443.92500	+		
Carver, Carver Water Tower						
	FM	WB0RMK	147.16500	+		Minnesota RC
Centervill	DMR/MARC	K0GOI	443.67500	+	CC11	
Centerville	DMR/BM	K0GOI	443.62500	+	CC11	Minnesota RC
Chaska	DSTAR	KD0JOS	53.27000	52.27000	114.8 Hz	Minnesota RC
	DSTAR	KD0JOS	147.27000	+		Minnesota RC
	DSTAR	KD0JOT	1282.50000	1262.50000		
	DSTAR	KD0JOS	1283.50000	1263.50000		Minnesota RC
	FM	N0BVE	53.45000	52.45000		Minnesota RC
	FM	N0BVE	145.23000	-	114.8 Hz	
	FM	KD0JOS	442.12500	+		Minnesota RC
Clara City	DMR/BM	K0WPD	443.35000	+	CC3	
Cloquet	FM	WA0GWI	146.67000	-		Minnesota RC
	FM	KB0YHX	443.10000	+		
Cohasset	FM	KB0CIM	146.98500	-	118.8 Hz	Minnesota RC

MINNESOTA 199

Location	Mode	Call sign	Output	Input	Access	Coordinator
Cohasset	FM	KB0CIM	444.15000	+	114.8 Hz	Minnesota RC
Cold Spring	DMR/BM	W0SAV	442.30000	+	CC3	Minnesota RC
Coleraine	FM	KB0QYC	147.16500	+	114.8 Hz	Minnesota RC
Collegeville	FM	W0SV	147.01500	+	100.0 Hz	Minnesota RC
	FM	W0SV	442.22500	+		
Cologne	FM	N0KP	444.60000	+		Minnesota RC
Columbia Heights	FM	K0FCC	224.50000	-	114.8 Hz	Minnesota RC
	FM	K0FCC	224.66000	-	114.8 Hz	Minnesota RC
	FM	K0FCC	444.75000	+	114.8 Hz	Minnesota RC
Cook	FM	N0BZZ	147.36000	+	162.2 Hz	Minnesota RC
Coon Rapids	P25	KD0ORH	444.40000	+		Minnesota RC
Cottage Grove	FM	W0CGM	147.18000	+	74.4 Hz	Minnesota RC
Crookston	FM	KB0BSJ	147.12000	+	123.0 Hz	
Crosby	FM	W0UJ	147.22500	+	141.3 Hz	Minnesota RC
	FM	W0UJ	444.92500	+		Minnesota RC
Crosslake	FM	W0UJ	147.03000	+	141.3 Hz	Minnesota RC
Crown	FM	N0GEF	145.21000	-	114.8 Hz	Minnesota RC
Dalton	FM	KB0JPT	224.08000	-	225.7 Hz	Minnesota RC
Darwin	FM	W0CRC	146.68500	-	146.2 Hz	Minnesota RC
Dayton	DMR/MARC	KA0KMJ	442.85000	+	CC1	
	FM	W0MDT	443.25000	+	DCS 023	Minnesota RC
Deer Creek, Water Tower						
	FM	N0WN	146.92500	-		Minnesota RC
Deer River	FM	KD0JFI	146.62500	-	103.5 Hz	Minnesota RC
Detroit Lakes	FM	W0EMZ	147.19500	+		Minnesota RC
Duluth	DMR/BM	N0NKI	443.30000	+	CC1	Minnesota RC
	DMR/BM	K0OE	444.90000	+	CC1	
	DMR/BM	K0OE	902.10000	947.10000	CC1	
	DSTAR	N0EO	147.37500	+		Minnesota RC
	DSTAR	N0EO	442.20000	+		Minnesota RC
	FM	KB0QYC	53.13000	52.13000	103.5 Hz	Minnesota RC
	FM	N0EO	145.31000	-	110.9 Hz	Minnesota RC
	FM	KC0HXC	145.41000	-	100.0 Hz	Minnesota RC
	FM	KC0RTX	145.45000	-	103.5 Hz	Minnesota RC
	FM	W0GKP	146.94000	-	103.5 Hz	Minnesota RC
	FM	KA0TMW	147.18000	+	103.5 Hz	Minnesota RC
	FM	KB0QYC	442.80000	+		Minnesota RC
	FM	W0GKP	444.10000	+		Minnesota RC
	FM	KC0RTX	444.20000	+	103.5 Hz	Minnesota RC
	FM	N0EO	444.30000	+	103.5 Hz	Minnesota RC
	FM	KB0QYC	927.48750	902.48750	103.5 Hz	Minnesota RC
	FM	KB0QYC	927.60000	902.60000	114.8 Hz	Minnesota RC
Duxbury	FM	KE0ACL	146.91000	-	146.2 Hz	Minnesota RC
East Grand Forks	FM	WA0VFY	147.39000	+		
Eden Prairie	DMR/MARC	N0VZC	442.47500	+	CC11	
	FM	K0EPR	146.88000	-		Minnesota RC
Eden Prairie, MN Hwy 5 And Del						
	FM	W5RTQ	444.35000	+	114.8 Hz	Minnesota RC
Edina	FM	WC0HC	145.43000	-	127.3 Hz	Minnesota RC
	FM	WC0HC	444.20000	+	127.3 Hz	Minnesota RC
	FM	KG0BP	444.85000	+	114.8 Hz	Minnesota RC
Elk River	FM	K0CJD	146.97000	-		Minnesota RC
	FM	K0SCA	147.28500	+	131.8 Hz	Minnesota RC
Ellendale	DMR/BM	KD0TGF	442.02500	+	CC1	Minnesota RC
	FM	KA0PQW	224.64000	-	110.9 Hz	
	FM	KA0PQW	442.92500	+	114.8 Hz	Minnesota RC
Ely	FM	N0OIW	146.64000	-	151.4 Hz	Minnesota RC
Ely, Water Tower	FM	K0VRC	147.19500	+	151.4 Hz	Minnesota RC
Emmaville, Camp Wilderness Boy						
	FM	K0NLC	147.39000	+		Minnesota RC
Fairmont	FM	K0SXR	146.64000	-		Minnesota RC
	FM	N0PBA	444.35000	+	136.5 Hz	Minnesota RC

200 MINNESOTA

Location	Mode	Call sign	Output	Input	Access	Coordinator
Falcon Heights	FM	W0YC	53.15000	147.15000		Minnesota RC
Falcon Heights, U Of M, St. Pa						
	FM	W0YC	147.15000	+	114.8 Hz	Minnesota RC
Falcon Heights, UMSP						
	FM	W0YC	444.42500	+	114.8 Hz	Minnesota RC
Faribault	DMR/MARC	KD0YRF	442.17500	+	CC1	
	DMR/MARC	N0PQK	444.57500	+	CC1	Minnesota RC
	DSTAR	KD0ZSA	444.62500	+		Minnesota RC
	FM	N0ZR	145.19000	-	100.0 Hz	Minnesota RC
	FM	KD0ZSA	146.79000	-	100.0 Hz	Minnesota RC
	FM	KB0IOA	444.70000	+		Minnesota RC
Fergus Falls	FM	K0QIK	146.64000	-		Minnesota RC
	FM	K0QIK	147.28500	+	91.5 Hz	Minnesota RC
	FM	K0QIK	444.20000	+	151.4 Hz	Minnesota RC
Finland	FM	N0BZZ	145.41000	-	114.8 Hz	Minnesota RC
Fisher	FM	KC0SD	146.70000	-		Minnesota RC
Foley	FM	KD0NRL	147.07500	+	85.4 Hz	
Foreston	FM	N0GOI	146.74500	-	107.2 Hz	Minnesota RC
	FM	N0GOI	443.67500	+	114.8 Hz	Minnesota RC
Fulda	FM	W0DRK	147.36000	+	141.3 Hz	
	FM	W0DRK	444.25000	+	141.3 Hz	
Gaylord	FM	KC0QNA	146.80500	-	141.3 Hz	Minnesota RC
Gem Lake	FM	K0LAV	224.10000	-		Minnesota RC
	FM	K0LAV	444.95000	+	114.8 Hz	Minnesota RC
	FM	K0LAV	919.10000	894.10000		Minnesota RC
Giese	FM	KB0QYC	146.86500	-	146.2 Hz	Minnesota RC
Gilbert	FM	KB0QYC	443.50000	+	141.3 Hz	Minnesota RC
Gilbert, Water Tower						
	FM	NT0B	147.15000	+		
Glenville	FM	NX0P	146.68500	-	100.0 Hz	Minnesota RC
	FM	WA0RAX	146.88000	-	100.0 Hz	Minnesota RC
	FM	WA0YCT	444.97500	+	100.0 Hz	
Golden Valley	DSTAR	KD0JOV	145.15000	-		
	FM	W0PZT	146.82000	-	127.3 Hz	Minnesota RC
	FM	WC0HC	444.17500	+	127.3 Hz	
Grand Marais	FM	W0BBN	444.25000	+	151.4 Hz	Minnesota RC
Grand Marais, Gunflint Lake						
	FM	W0BBN	146.73000	-	151.4 Hz	Minnesota RC
Grand Marais, Maple Hill						
	FM	W0BBN	146.89500	-	151.4 Hz	Minnesota RC
Grand Portage	FM	W0BBN	146.65500	-	151.4 Hz	Minnesota RC
Grand Rapids	FM	KB0CIM	53.29000	52.29000	146.2 Hz	Minnesota RC
	FM	K0GPZ	444.55000	+	123.0 Hz	Minnesota RC
Grand Rapids, Coleraine						
	FM	K0GPZ	146.88000	-		Minnesota RC
Green Isle	FM	KC0QNA	443.82500	+	141.3 Hz	
Ham Lake	DSTAR	W0ANA	145.40500	-		Minnesota RC
	DSTAR	W0ANA	443.77500	+		Minnesota RC
	DSTAR	W0ANA	1287.00000	1267.00000		Minnesota RC
	FM	W0YC	224.94000	-	100.0 Hz	
	FM	K9EQ	444.02500	+	DCS 026	Minnesota RC
Hampton	FM	K0JTA	147.36000	+	136.5 Hz	Minnesota RC
Hastings	FM	W0CGM	146.98500	-		Minnesota RC
Hibbing, Water Tower						
	FM	N0AGX	147.12000	+		
Hinckley	FM	KB0QYC	444.57500	+	146.2 Hz	
Hugo	FM	N0SBU	443.05000	+	118.8 Hz	Minnesota RC
Hutchinson	FM	KB0WJP	147.37500	+	146.2 Hz	Minnesota RC
International Falls	FM	K0HKZ	146.97000	-		Minnesota RC
Isabella	FM	KB0QYC	147.30000	+	114.8 Hz	Minnesota RC
Isanti	DSTAR	KE0KKN	442.27500	+		Minnesota RC
	FM	N0JOL	146.64000	-	146.2 Hz	Minnesota RC

MINNESOTA 201

Location	Mode	Call sign	Output	Input	Access	Coordinator
Isle, Mille Lacs Lake						
	FM	W0REA	146.61000	-	141.3 Hz	Minnesota RC
Janesville	DMR/BM	KD0TGF	442.50000	+	CC1	
Karlstad	DMR/BM	KB0ISW	443.97500	+	CC6	Minnesota RC
	FM	KA0NWV	145.47000	-	123.0 Hz	
	FM	KB0ISW	146.65500	-	127.3 Hz	Minnesota RC
Knife River	FM	KC0RTX	147.13500	+	103.5 Hz	
La Crescent	FM	WR9ARC	146.97000	-	131.8 Hz	Minnesota RC
Le Center , County Courthouse						
	FM	KC0LSR	444.22500	+	136.5 Hz	
Le Sueur	FM	WB0ERN	146.61000	-	136.5 Hz	Minnesota RC
Lengby	FM	W0BJI	147.27000	+		Minnesota RC
Litchfield	DMR/BM	KC0CAP	443.70000	+	CC3	
	DMR/MARC	KC0CAP	443.80000	+	CC3	Minnesota RC
	FM	AE0GD	146.62500	-	146.2 Hz	Minnesota RC
	FM	K0MCR	147.30000	+	146.2 Hz	Minnesota RC
Little Falls	DSTAR	W0REA	444.00000	+		Minnesota RC
	FM	N0RND	443.07500	+	146.2 Hz	Minnesota RC
	FM	KA0JSW	443.12500	+	123.0 Hz	Minnesota RC
Little Falls, Little Falls AT&						
	FM	W0REA	147.13500	+	123.0 Hz	Minnesota RC
Littlefork	FM	KA0WRT	444.90000	+	103.5 Hz	Minnesota RC
Long Prairie	FM	KC0TAF	146.65500	-		Minnesota RC
Madison	FM	NY0I	444.90000	+		Minnesota RC
Mahnomen	FM	W0BJI	444.50000	+		Minnesota RC
Mahtowa	FM	KB0TNB	53.17000	52.17000	103.5 Hz	Minnesota RC
	FM	W0GKP	147.00000	-	103.5 Hz	Minnesota RC
Mankato	FM	W0WCL	147.04500	+	136.5 Hz	Minnesota RC
	FM	W0WCL	147.24000	+	136.5 Hz	Minnesota RC
	FM	WA2OFZ	442.82500	+	136.5 Hz	Minnesota RC
	FM	W0WCL	444.67500	+	100.0 Hz	Minnesota RC
	P25	K2KLN	443.65000	+	114.8 Hz	Minnesota RC
Mankato , Good Counsel Hill						
	FM	K0JCR	442.52500	+	136.5 Hz	
Maple Grove	FM	K0LTC	443.55000	+	114.8 Hz	Minnesota RC
Maple Plain	FM	K0LTC	147.00000	+	114.8 Hz	Minnesota RC
Maplewood	FM	W0MR	147.12000	+		Minnesota RC
	FM	K0AGF	442.45000	+		Minnesota RC
	FM	KC0MQW	442.60000	+	156.7 Hz	Minnesota RC
	FM	W0MR	444.82500	+		Minnesota RC
	FM	W0MR	1285.00000	1265.00000		
Marcell	FM	K0GPZ	147.07500	+		Minnesota RC
Marshall	DMR/BM	N0NKI	443.70000	+	CC1	
	FUSION	W0WX	146.95500	-		
Marshall, Avera Marshall Regio						
	FM	W0WX	146.95500	-	141.3 Hz	Minnesota RC
Mendota	FM	N0BVE	444.32500	+		
Milaca	FM	KD0JOU	145.35000	-	141.3 Hz	Minnesota RC
Minneapolis	DMR/BM	NH7CY	442.42500	+	CC1	Minnesota RC
	DMR/MARC	N0BVE	442.65000	+	CC2	Minnesota RC
	DSTAR	KD0JOU	145.11000	-		
	DSTAR	W1AFV	442.95000	+		
	DSTAR	KD0JOU	444.87500	+		Minnesota RC
	DSTAR	KD0JOU	1283.30000	1263.30000		Minnesota RC
	FM	K0MSP	145.37000	-	107.2 Hz	Minnesota RC
	FM	KD0JOU	147.03000	+	114.8 Hz	Minnesota RC
	FM	WB0ZKB	147.27000	+	114.8 Hz	Minnesota RC
	FM	KD0WIL	443.57500	+	114.8 Hz	Minnesota RC
	FM	KB0FJB	443.80000	+	114.8 Hz	Minnesota RC
	FM	N0BVE	444.65000	+	114.8 Hz	Minnesota RC
Minneapolis, U Of Minn						
	FM	KA0KMJ	444.42500	+	114.8 Hz	Minnesota RC

202   MINNESOTA

Location	Mode	Call sign	Output	Input	Access	Coordinator
Minnetonka	DMR/MARC	N0BVE	442.67500	+	CC2	
	FM	N0BVE	145.45000	-		Minnesota RC
	FM	KA0KMJ	443.00000	+	100.0 Hz	Minnesota RC
Montevideo	FM	NY0I	147.12000	+	146.2 Hz	Minnesota RC
Moorhead	FM	W0ILO	145.35000	-	123.0 Hz	Minnesota RC
	FM	W0JPJ	442.50000	+	DCS 065	Minnesota RC
	FM	W0ILO	444.87500	+	123.0 Hz	Minnesota RC
Moorhead, Clay County Social S						
	FM	W0JPJ	145.15000	-		Minnesota RC
Morris	FM	NG0W	444.40000	+	103.5 Hz	Minnesota RC
Mounds View, Medtronic						
	FM	W0MDT	444.52500	+	DCS 025	Minnesota RC
Mounds View, Medtronic CRDM Ca						
	FM	K9EQ	444.07500	+	100.0 Hz	Minnesota RC
MSP-Airport	DMR/MARC	N0BVE	444.92500	+	CC11	
New Brighton	FM	K0FCC	145.29000	-	114.8 Hz	Minnesota RC
	FM	N0MNB	1250.00000	427.20000		Minnesota RC
	FM	N0MNB	1253.25000	1233.25000		Minnesota RC
New Richland	FM	N0RPJ	145.33000	-	100.0 Hz	Minnesota RC
Nisswa, WJJY Tower						
	FM	W0UJ	443.92500	+	110.9 Hz	Minnesota RC
North Branch	FM	K0GOI	147.31500	+	91.5 Hz	Minnesota RC
Northfield	FM	N0OTL	146.65500	-	136.5 Hz	Minnesota RC
Northwest Angle	FM	N0MHO	147.21000	+	123.0 Hz	Minnesota RC
Oakdale	FM	WD0HWT	146.85000	-		
Ogilvie	FM	KD0CI	147.24000	+	146.2 Hz	Minnesota RC
Ortonville	FM	NY0I	444.50000	+	146.2 Hz	Minnesota RC
Outing	FM	WR0G	145.43000	-	127.3 Hz	Minnesota RC
Owatonna	FM	K0HNY	145.49000	-	100.0 Hz	
	FM	WB0VAJ	147.10500	+	100.0 Hz	Minnesota RC
	FM	WB0VAK	444.45000	+	100.0 Hz	Minnesota RC
Park Rapids	FM	K0GUV	147.30000	+		Minnesota RC
Paynesville	FM	KD0YLG	224.80000	-		Minnesota RC
	FM	N0ANC	444.62500	+		Minnesota RC
Paynesville MN	FM	WD0DEH	145.27000	-	DCS 172	Minnesota RC
Pequot Lakes, Maple Hill						
	FM	W0REA	147.09000	+	123.0 Hz	Minnesota RC
Perham	FM	K0QIK	147.15000	+		
Pinewood	FM	KC0FTV	442.22500	447.32500	118.8 Hz	Minnesota RC
Pipestone	FM	W0DRK	147.07500	+	141.3 Hz	
Plymouth	FM	WC0HC	146.70000	-	127.3 Hz	Minnesota RC
	FM	N0FWG	444.37500	+	114.8 Hz	Minnesota RC
	FM	W0PZT	444.50000	+	127.3 Hz	Minnesota RC
Princeton	DMR/BM	W9YZI	442.50000	+	CC3	
	FM	KD0YEQ	146.77500	-	146.2 Hz	Minnesota RC
	FM	K0SCA	444.70000	+	146.2 Hz	Minnesota RC
Prior Lake	DMR/MARC	N0AGI	443.07500	+	CC1	Minnesota RC
Proctor	FM	N0BZZ	147.33000	+	151.4 Hz	Minnesota RC
Ramsey	DSTAR	KE0MVE	442.52500	+		Minnesota RC
	DSTAR	KE0MVE	1287.10000	1267.10000		Minnesota RC
	FM	K0MSP	444.97500	+	114.8 Hz	Minnesota RC
Red Wing	DMR/MARC	N0BVE	442.47500	+	CC2	
	FM	AA0RW	147.30000	+	136.5 Hz	Minnesota RC
	FM	AA0RW	442.25000	+	136.5 Hz	Minnesota RC
Redwood Falls	FM	KB0CGJ	146.86500	-	141.3 Hz	Minnesota RC
Richfield	FM	W0RRC	145.39000	-	DCS 047	Minnesota RC
	FM	W0RRC	444.47500	+	118.8 Hz	Minnesota RC
Robbinsdale	FM	K0YTH	444.77500	+	114.8 Hz	Minnesota RC
Rochester	DMR/MARC	KD0YRF	443.97500	+	CC2	Minnesota RC
	FM	W0MXW	146.82000	-	100.0 Hz	Minnesota RC
	FM	W0EAS	147.25500	+	100.0 Hz	Minnesota RC
	FM	KD0EBO	147.85500	146.95500	100.0 Hz	

MINNESOTA   203

Location	Mode	Call sign	Output	Input	Access	Coordinator
Rochester, Mayo Clinic						
	DSTAR	W0MXW	443.85000	+		Minnesota RC
	FM	W0MXW	146.62500	-	100.0 Hz	Minnesota RC
Roosevelt	FM	N0MHO	147.00000	-	123.0 Hz	
Rosemount	FM	W0EIB	224.94000	-		Minnesota RC
Roseville	DMR/MARC	N0NMZ	443.52500	+	CC11	Minnesota RC
Rush City	FM	K0ECM	145.33000	-	146.2 Hz	Minnesota RC
Sabin, Water Tower						
	FM	WB0BIN	146.89500	-	100.0 Hz	
Saint Paul	DMR/MARC	K0GOI	442.02500	+	CC11	
Saint Cloud	DMR/MARC	KC0ARX	442.22500	+	CC3	
Sauk Centre	FM	W0ALX	147.25500	+		Minnesota RC
Sauk Rapids, Water Tower On To						
	FM	W0SV	146.94000	-	100.0 Hz	Minnesota RC
Sebeka, Water Tower						
	FM	N0WN	147.33000	+		Minnesota RC
Shakopee	DMR/BM	KC0NPA	444.72500	+	CC4	
Silver Lake	FM	KB0WJP	443.40000	+	146.2 Hz	
Slayton, DOT On HWY 59						
	FM	W0DRK	146.79000	-	141.3 Hz	Minnesota RC
Spring Lake Park	DMR/MARC	N8AGJ	443.87500	+	CC11	
Spring Valley	FM	N0ZOD	147.01500	+	110.9 Hz	Minnesota RC
St Cloud	DMR/MARC	W0SAV	442.22500	+	CC3	
St. Bonifacius	DMR/MARC	N0VZC	444.72500	+	CC2	
St. Cloud	DMR/BM	KC0ARX	442.32500	+	CC3	Minnesota RC
	DSTAR	KD0YLG	443.85000	+		Minnesota RC
	FM	N0OYQ	146.83500	-	85.4 Hz	Minnesota RC
	FM	K0VSC	443.45000	+	123.0 Hz	Minnesota RC
St. Cloud, St. Cloud Hospital						
	FM	N0ANC	145.19000	-	146.2 Hz	Minnesota RC
	FM	N0ANC	442.15000	+		Minnesota RC
St. Louis Park	FM	W0EF	146.76000	-	114.8 Hz	Minnesota RC
	FM	W0EF	444.10000	+	114.8 Hz	Minnesota RC
St. Paul	DSTAR	W8WRR	444.32500	+		
	FM	K0GOI	53.47000	52.47000		Minnesota RC
	FM	K0GOI	145.17000	-	100.0 Hz	Minnesota RC
	FM	KE0NA	223.90000	-	100.0 Hz	Minnesota RC
	FM	N0GOI	444.05000	+		Minnesota RC
	FM	WD0HWT	444.80000	+	114.8 Hz	Minnesota RC
St. Paul, Univ. Of St. Thomas						
	FM	K0AGF	145.31000	-	114.8 Hz	Minnesota RC
St. Peter	FM	WQ0A	147.13500	+	100.0 Hz	Minnesota RC
	FM	N0KP	224.52000	+		Minnesota RC
	FM	WQ0A	444.15000	+		Minnesota RC
Stewartville	FM	N0ZQB	444.25000	+	136.5 Hz	Minnesota RC
Stillwater	FM	W0JH	147.06000	+	DCS 026	Minnesota RC
Tamarack	FM	N0BZZ	443.20000	+	114.8 Hz	Minnesota RC
Thief River Falls	DMR/BM	WB0WTI	444.80000	+	CC6	Minnesota RC
	FM	WB0WTI	146.85000	-	123.0 Hz	Minnesota RC
Tofte, Lutsen	FM	W0BBN	146.86500	-	151.4 Hz	Minnesota RC
Tracy	FM	W0DRK	147.15000	+	141.3 Hz	
Two Harbors	FM	KB0TNB	53.02000	52.02000	103.5 Hz	Minnesota RC
	FM	WB0DGK	147.27000	+	103.5 Hz	
Tyler	FM	WB6AMY	145.11000	-	146.2 Hz	Minnesota RC
	FM	W0ZZY	145.35000	-	156.7 Hz	Minnesota RC
Ulen	FM	W0QQK	146.68500	-		Minnesota RC
Virginia	FM	KB0QYC	53.15000	52.15000	103.5 Hz	Minnesota RC
Wabasha	FM	WA0UNB	146.74500	-	136.5 Hz	Minnesota RC
	FM	WA0UNB	421.25000	439.25000		Minnesota RC
Wabasso	FM	KB0CGJ	444.52500	+	141.3 Hz	Minnesota RC
Walker	FM	NA0RC	146.79000	-	103.5 Hz	Minnesota RC
Wanda	DSTAR	KD0IAI	444.02500	+		Minnesota RC

204   MINNESOTA

Location	Mode	Call sign	Output	Input	Access	Coordinator
Wannaska	FM	KC0IGT	147.09000	+	123.0 Hz	Minnesota RC
Warroad	DMR/BM	N0MHO	443.00000	+	CC1	
Waseca	FM	KB0UJL	146.71500	-	141.3 Hz	Minnesota RC
	FM	KB0UJL	442.30000	+	141.3 Hz	Minnesota RC
Waseca, Water Tower						
	FM	WA0CJU	146.94000	-		Minnesota RC
White Bear Lake	DMR/MARC	W0REA	147.39000	+	CC11	Minnesota RC
	DSTAR	KA0JSW	444.00000	+		Minnesota RC
	DSTAR	KA0JSW	1285.50000	1265.50000		Minnesota RC
	FM	WD0HWT	444.25000	+	100.0 Hz	Minnesota RC
Willmar	FM	W0SW	146.91000	-		Minnesota RC
	FM	W0SW	147.03000	-		Minnesota RC
	FM	KB0MNU	444.80000	+	146.2 Hz	Minnesota RC
Windom	FM	W0DRK	147.25500	+	141.3 Hz	Minnesota RC
Winona	FM	W0NE	146.64000	-	100.0 Hz	Minnesota RC
	FM	W0NE	146.83500	-	131.8 Hz	
	FM	N0PDD	147.28500	+	100.0 Hz	Minnesota RC
	FUSION	W0NE	442.15000	+		
Woodbury	DMR/MARC	K0GOI	442.82500	+	CC11	
Woodbury, Woodwinds Hospital						
	FM	N0GOI	442.82500	+		Minnesota RC
Worthington	FM	K0QBI	146.67000	-	141.3 Hz	Minnesota RC
	FM	W0DRK	444.85000	+	141.3 Hz	
Wyoming	DMR/BM	N0DZQ	444.23125	+	CC3	Minnesota RC
	FM	N0VOW	146.89500	-	82.5 Hz	
Zimmerman	FM	KE0ATF	444.45000	+	DCS 023	Minnesota RC

MISSISSIPPI

Location	Mode	Call sign	Output	Input	Access	Coordinator
Abbeville	FM	WB5VYH	145.47000	-	107.2 Hz	SERA
Aberdeen	FM	WB5TZN	147.27000	+	210.7 Hz	
	FM	WB5TZN	444.45000	+	210.7 Hz	SERA
Ackerman, ETV Tower						
	FM	NO5N	147.12000	+	136.5 Hz	SERA
Amory	DMR/BM	AD5T	443.20000	+	CC1	SERA
	FM	KB5DWX	146.94000	-	192.8 Hz	SERA
Bay Springs	FM	W5NRU	145.49000	-	DCS 023	SERA
Bay St Louis	FM	WO5V	444.75000	+	179.9 Hz	SERA
Biloxi, Cableone Tower						
	FM	W5SGL	146.73000	-	136.5 Hz	SERA
Booneville	FM	WX5F	147.15000	+	110.9 Hz	SERA
	FM	WX5F	441.85000	+	118.8 Hz	
	FM	KG5UYK	442.97500	+	203.5 Hz	
Boonville	DSTAR	W5NEM	146.83500	-		
Brandon	FM	K5RKN	147.34500	+	100.0 Hz	SERA
Brookhaven, Water Tower						
	FM	W5WQ	146.85000	-	103.5 Hz	SERA
Clinton	DSTAR	W5PFR	441.81250	+		SERA
Cockrum	DMR/BM	KB5DMT	441.92500	+	CC1	
Coldwater	DMR/MARC	W5GWD	444.92500	+	CC1	
Collins	FM	W5NRU	146.95500	-	179.9 Hz	
	FM	W5NRU	146.98500	-	DCS 023	SERA
Columbia	FM	N5LJC	147.28500	+	123.0 Hz	SERA
	FM	N5LJC	444.80000	+		
Columbus	DSTAR	KC5ULN	444.92500	+		SERA
	FM	KC5ULN	146.62500	-	136.5 Hz	
	FM	KC5ULN	147.00000	+	136.5 Hz	
Coonwood	FM	N5EYM	146.97000	-	110.9 Hz	
Corinth	DMR/MARC	K5WHB	147.28500	+	CC1	
	DMR/MARC	K5WHB	444.60000	+	CC1	
	FM	WF5D	145.39000	-	100.0 Hz	
	FM	W5AWP	146.92500	-	107.2 Hz	
	FM	KC5CO	147.34500	+	123.0 Hz	

MISSISSIPPI 205

Location	Mode	Call sign	Output	Input	Access	Coordinator
Corinth	FM	W5AWP	441.80000	+	107.2 Hz	SERA
	FM	KJ5CO	443.90000	+	123.0 Hz	SERA
	FM	KE5HYT	444.90000	+	100.0 Hz	SERA
Corinth, 1 Mile East Of LakeHi						
	DMR/BM	K5WHB	444.98750	+	CC1	
Corinth, LakeHill Motors						
	FM	K5WHB	53.07000	52.07000	203.5 Hz	
Corinth, Pine Mountain						
	FM	KB5YNM	147.00000	-	203.5 Hz	
Diamondhead, Near Diamondhead						
	FM	KA5EPR	444.45000	+	136.5 Hz	
Edinburg	FM	W5PPB	145.33000	-	77.0 Hz	SERA
Ellisville	FM	W5NRU	145.23000	-	DCS 023	SERA
	FM	W5NRU	224.88000	-	179.9 Hz	
	FM	W5NRU	442.25000	+	136.5 Hz	SERA
	FM	W5NRU	443.60000	+	179.9 Hz	SERA
	FM	W5NRU	443.65000	+	77.0 Hz	SERA
	FUSION	W5NRU	442.30000	+	179.9 Hz	SERA
Ellisville, South, YSF Auto/Au						
	FM	W5NRU	442.90000	+	179.9 Hz	SERA
Forest	FM	KF5SEB	145.24000	-	100.0 Hz	SERA
Fulton	FM	WX5P	145.45000	-	192.8 Hz	SERA
Gloster	FM	KX5E	145.43000	-	136.5 Hz	SERA
	FM	N5ZNS	443.82500	+		SERA
Grenada	FM	W5LV	146.70000	-		SERA
	FM	AD5IT	444.70000	+	107.2 Hz	SERA
Gulfport	FM	WD5BJT	444.15000	+	77.0 Hz	SERA
Guntown	FM	WJ5D	145.15000	-	156.7 Hz	
Hattiesburg	DMR/BM	W5NRU	442.70000	+	CC1	SERA
	FM	KD5MIS	147.31500	+	136.5 Hz	SERA
	FUSION	K5IJX	146.67000	-	136.5 Hz	
Hattiesburg, Hburg-Laurel Airp						
	FM	K5IJX	444.77500	+	136.5 Hz	SERA
Hattiesburg, Oak Grove High Sc						
	FUSION	W5CJR	146.77500	-	136.5 Hz	
Hattiesburg, Richburg Hill						
	FM	N5LRQ	442.72500	+	167.9 Hz	SERA
Hattiesburg, Wesley Medical Ce						
	FM	W5CJR	147.36000	+		SERA
Heidelberg	FM	W5NRU	147.57000		179.9 Hz	
	FM	KC5RC	444.30000	+	100.0 Hz	SERA
Hernando	FM	N5PYQ	145.37000	-	107.2 Hz	
	FM	W5GWD	146.91000	-	107.2 Hz	
	FM	N5PYQ	444.92500	+	107.2 Hz	SERA
Holly Springs	FM	KD5VMV	147.22500	+	107.2 Hz	
Horn Lake	DSTAR	W5AV	144.96000	+		SERA
	DSTAR	W5AV	145.55000	-		
	FM	N5NBG	145.27000	-	107.2 Hz	SERA
	FM	N5NBG	444.65000	+		SERA
Horn Lake, MS	DMR/BM	W5AV	442.01250	+	CC1	
Houston	FM	KD5YBU	146.89500	-	141.3 Hz	SERA
Indianola	FM	AB5DU	444.85000	+	136.5 Hz	SERA
Iuka	FM	W5TCR	146.85000	-	141.3 Hz	SERA
Jackson	FM	W5PFC	146.76000	-	77.0 Hz	SERA
	FM	KA5SBK	146.94000	-	100.0 Hz	
	FM	NC5Y	444.70000	+	77.0 Hz	SERA
Jackson, St. Dominic's Hospita						
	FM	W5PFC	146.88000	-	77.0 Hz	
Kosciusko	FM	KB5ZEA	146.85000	-		SERA
Laurel	FM	KC5PIA	53.45000	52.45000	136.5 Hz	
	FM	W5NRU	146.61000	-	136.5 Hz	SERA
	FM	WV5D	147.03000	-	136.5 Hz	SERA

206   MISSISSIPPI

Location	Mode	Call sign	Output	Input	Access	Coordinator
Laurel	FM	W5FSJ	147.06000	+	146.2 Hz	SERA
	FM	W5NRU	147.13500	+	DCS 023	SERA
	FM	WV5D	442.37500	+	136.5 Hz	SERA
	FM	KC5PIA	444.97500	+	136.5 Hz	SERA
Leakesville	FM	KE5WGF	147.00000	+	136.5 Hz	SERA
	FM	KE5WGF	444.22500	+	136.5 Hz	SERA
Long Beach	FM	K5XXV	145.33000	-	136.5 Hz	SERA
Love Station	DMR/BM	KB5DMT	431.92500	446.92500	CC1	
Lucedale	DMR/BM	KD4VVZ	444.20000	+	CC1	
Madison	FM	K5XU	146.64000	-	77.0 Hz	SERA
McComb	FM	W5WQ	444.87500	+	100.0 Hz	SERA
Mccomb, SW MS Reg Med Ctr						
	FM	W5WQ	146.94000	-	103.5 Hz	SERA
McHenry	DSTAR	KI4TMJ	145.17000	-		SERA
	DSTAR	KI4TMJ	444.47500	+		
	DSTAR	KI4TMJ	1250.00000			SERA
	DSTAR	KI4TMJ	1284.00000	1264.00000		SERA
	FM	KA5VFU	147.37500	+	136.5 Hz	SERA
McVille	FM	KG5EVY	444.20000	+	103.5 Hz	
Meridian	FM	W5LRG	145.41000	-		
	FM	KC5ZZH	146.65500	-		
	FM	W5FQ	146.70000	-	100.0 Hz	
	FM	NO5C	146.97000	-	100.0 Hz	
	FM	W5LRG	444.50000	+	107.2 Hz	SERA
Monticello, 379 Firetower Road						
	FM	N5JHK	147.01500	+		SERA
Moselle, Hattiesburg-Laurel Re						
	FM	K5IJX	145.37000	-	136.5 Hz	
Mount Zion	FM	K5YVY	146.74500	-		
Natchez	FM	K5OCM	146.91000	-	91.5 Hz	SERA
	FM	K5SVC	147.36000	+	100.0 Hz	
New Albany, Hospital						
	FM	K5YVY	146.74500	-	88.5 Hz	SERA
New Albany, Union County Fairg						
	FM	NA5MS	146.67000	-	131.8 Hz	
Ocean Springs	FM	KB5CSQ	444.25000	+	77.0 Hz	SERA
Olive Branch, Lewisburg Water						
	FM	W5OBM	444.70000	+	107.2 Hz	SERA
Olive Branch, Water Tower Hwy						
	FM	W5OBM	147.25500	+	79.7 Hz	SERA
Oxford	DMR/MARC	KC5KLW	145.47000	290.34000	CC1	
	DMR/MARC	KC5KLW	444.98750	894.97450	CC1	
	FM	W5LAF	147.33000	+	107.2 Hz	SERA
	FM	W5LAF	444.35000	+	107.2 Hz	SERA
Pascagoula	FM	KC5LCW	443.45000	+	123.0 Hz	SERA
Pelahatchie	FM	W5PPB	145.39000	-	77.0 Hz	SERA
Perkinston	FM	K5GVR	147.16500	+	136.5 Hz	SERA
	FM	K5GVR	442.47500	+	136.5 Hz	SERA
Petal	FM	W5CJR	145.19000	-	136.5 Hz	SERA
	FM	N5YH	443.15000	+	136.5 Hz	
Petal, PHS Football Pressbox						
	FM	N5QXX	444.82500	+	114.8 Hz	
Philadelphia	FM	N5EPP	147.33000	+		SERA
	FM	WB5YGI	444.95000	+		SERA
Picayune	FM	KE5LT	443.72500	+	179.9 Hz	SERA
Pontotoc, Silo	FM	AF5FM	444.50000	+	131.8 Hz	
Poplarville	FM	K5PRC	145.15000	-	136.5 Hz	
	FM	W5PMS	145.21000	-	136.5 Hz	SERA
Poplarville, Hillsdale						
	FM	W5NRU	145.41000	-	136.5 Hz	SERA
Port Gibson	FM	AF5OQ	146.62500	-	141.3 Hz	SERA
Quitman	FM	KF5MWE	147.39000	+	100.0 Hz	SERA

MISSISSIPPI 207

Location	Mode	Call sign	Output	Input	Access	Coordinator
Raymond	DSTAR	W5DRA	147.38000	+	77.0 Hz	SERA
	FUSION	W5PFR	444.00000	+	77.0 Hz	SERA
Ridgeland	FM	N5WDG	443.70000	+	77.0 Hz	SERA
Robinsonville, Horseshoe Casin						
	FM	W5GWD	443.30000	+	107.2 Hz	
Saltillo	DMR/BM	KI5CMA	444.42500	+	CC1	
	DMR/BM	KI5CMA	444.87500	+	CC1	
Sharon	FM	W5PPB	145.45000	-	77.0 Hz	SERA
Shiloh	FM	KF5SEB	147.04500	+	100.0 Hz	SERA
Slidell, Stennis Space Center						
	FM	N5GJB	147.21000	+	136.5 Hz	SERA
Soso	FM	WV5D	444.27500	+	136.5 Hz	SERA
Southaven	FM	W5GWD	145.35000	-	107.2 Hz	
Starkville	DMR/BM	AD5HM	440.55000	+	CC1	
	FM	K5DY	146.73000	-	210.7 Hz	SERA
	FM	W5YD	146.80500	-		
	FM	K5DY	444.75000	+	136.5 Hz	SERA
Sumner	FM	W5JWW	147.09000	+		SERA
Sumrall	FM	K5PN	443.35000	+	136.5 Hz	SERA
Taylorsville	FM	W5NRU	224.48000	-	179.9 Hz	SERA
Tupelo	DMR/BM	AB5OR	443.50000	+	CC1	
	DMR/BM	AB5OR	444.82500	894.65000	CC1	
	DMR/BM	N5VGK	444.95000	+	CC1	
	DSTAR	KE5LUX	146.64000	-		
	FM	N5VGK	145.49000	-	141.3 Hz	
	FM	W5NEM	147.07500	+	103.5 Hz	
	FM	N5VGK	444.82500	+	141.3 Hz	
Tupelo, Tupelo Airport						
	FM	K5TUP	147.24000	+	100.0 Hz	
Union	FM	K5SZN	147.24000	+	103.5 Hz	SERA
Vancleave	FM	W5WA	145.11000	-	123.0 Hz	SERA
Vicksburg	FM	W5WAF	145.41000	-	100.0 Hz	SERA
	FM	K5ZRO	147.27000	+	100.0 Hz	SERA
	FM	K5ZRO	444.85000	+	100.0 Hz	SERA
Water Valley	DMR/BM	KD5NDU	147.27000	+	CC1	SERA
Waynesboro	DMR/BM	N5IDX	450.00000	+	CC1	
	FM	KB5AAB	147.10500	+	136.5 Hz	SERA
West Point	FM	KD5RZQ	147.18000	+		SERA
	FM	N5WXD	443.45000	+		SERA
Wiggins	FM	N5UDK	145.27000	-	136.5 Hz	SERA
	FM	N5UDK	443.30000	+	167.9 Hz	SERA
Yazoo City	FM	KE5YES	147.22500	+	77.0 Hz	
MISSOURI						
Arcadia	FM	KA0CUU	146.95500	-	100.0 Hz	
Arnold	DMR/BM	KD0BQS	442.45000	+	CC2	MRC
Ashland	FM	KB0IRV	52.89000	51.19000	127.3 Hz	MRC
	FM	KD0SAF	444.17500	+	107.2 Hz	MRC
Ava	FM	N0RFI	146.62500	-	110.9 Hz	MRC
	FM	K0DCA	442.75000	+		MRC
Barnett, Golden Beach/Gravois						
	FM	AA0IY	442.92500	+	127.3 Hz	MRC
Barnhart	FM	K0AMC	50.50000	50.40000		
	FM	K0AMC	443.72500	+	192.8 Hz	MRC
Belle	DMR/BM	N0NOE	442.60000	+	CC2	MRC
Bethany	FM	N2OYJ	443.07500	+	100.0 Hz	
Blackwell	FM	K0GDI	53.41000	51.71000	100.0 Hz	MRC
Bloomfield, MSHP Tower						
	FM	KM0HP	147.33000	+	100.0 Hz	MRC
Blue Springs	FM	KB0VBN	147.01500	+	151.4 Hz	MRC
	FM	KB0VBN	444.95000	+	107.2 Hz	MRC
Bolivar	DMR/BM	WB0LVR	145.29000	-	CC5	MRC

208   MISSOURI

Location	Mode	Call sign	Output	Input	Access	Coordinator
Bolivar	DMR/BM	WB0LVR	443.67500	+	CC5	MRC
	FM	WB0LVR	147.06000	+	DCS 165	MRC
Bonne Terre, Bonne Terre Airpo						
	FM	W0EMM	442.32500	+	151.4 Hz	MRC
Boonville	FM	W0BRC	147.36000	+		MRC
	FM	KA0GFC	442.70000	+	77.0 Hz	MRC
	FM	KA0GFC	444.70000	+	77.0 Hz	MRC
Bourbon	FM	AA0GB	146.98500	-	141.3 Hz	
Branson	DMR/BM	KC0M	147.15000	+	CC5	MRC
	DMR/BM	K0NXA	444.45000	+	CC5	MRC
	FM	KJ6TQ	147.10500	+	136.5 Hz	MRC
	FM	KC0M	147.19500	+	162.2 Hz	MRC
	FM	KJ6TQ	224.10000	-	162.2 Hz	MRC
	FM	KJ6TQ	443.55000	+	162.2 Hz	MRC
Branson, Water Tower - Near Ti						
	P25	K0NXA	146.65500	-	91.5 Hz	MRC
Bridgeton, SSM De Paul Hospita						
	FM	W0KE	146.73000	-	141.3 Hz	MRC
	FM	W0KE	443.45000	+		MRC
Brinktown	FM	N0GYE	146.89500	-		MRC
Brookfield	FM	W0CIT	147.34500	+		MRC
Bunker	FM	KD0IM	147.27000	+	156.7 Hz	
Butler	FM	KD0PVP	147.22500	+	91.5 Hz	MRC
Calm	FM	N0IQM	442.52500	+		MRC
Calwood	FM	KS0B	444.95000	+	127.3 Hz	MRC
Cape Girardeau	FM	KE0UWK	444.45000	+	100.0 Hz	
Cape Girardeau, River Radio Tr						
	FM	W0QMF	146.68500	-	100.0 Hz	MRC
	FM	W0RMS	444.20000	+		MRC
Carrollton, East Water Tower						
	FM	N0SAX	146.65500	-	94.8 Hz	MRC
Carthage	FM	W0LF	442.32500	+	103.5 Hz	MRC
Centralia	FM	KM0R	443.02500	+	77.0 Hz	MRC
Cherryville	FM	N0BJM	146.64000	-	110.9 Hz	MRC
Chillicothe	DMR/BM	KB0YAS	444.40000	+	CC7	MRC
	FM	K0MPT	147.22500	+		MRC
Clayton	FM	W0SRC	146.94000	-	141.3 Hz	MRC
	FM	W0SRC	442.10000	+	141.3 Hz	MRC
Clever	FM	K0NXA	224.28000	-	162.2 Hz	MRC
	FM	K0NXA	442.42500	+	162.2 Hz	MRC
Columbia	DMR/BM	K0SI	444.42500	+	CC3	MRC
	FM	K0SI	146.76000	-	127.3 Hz	MRC
	FM	WX0BC	442.32500	+		MRC
Columbia, Near I-70						
	FM	WX0BC	146.61000	-	127.3 Hz	MRC
Concordia	FM	KE0PKD	147.10500	+	156.7 Hz	MRC
Conway	DMR/BM	K0NXA	145.47000	-	CC12	MRC
	FM	K0LH	146.70000	-	88.5 Hz	MRC
Crane, Water Tower						
	FM	K0NXA	442.15000	+	162.2 Hz	MRC
	FM	KB0NHX	927.11250	902.11250	162.2 Hz	MRC
Crystal City, Buck Nob						
	FM	KD0RIS	146.77500	-	100.0 Hz	MRC
Cuba	FM	KD0JOX	147.34500	+	110.9 Hz	MRC
Da Juh	FM	W0IN	145.19000	-	91.5 Hz	MRC
De Soto	FM	K0MGU	442.85000	+		MRC
Deepwater	DMR/BM	KM0HP	443.30000	+	CC4	
Deepwater, MSHP Tower						
	DMR/BM	W0DR	443.30000	+	CC4	MRC
Defiance	DMR/BM	N0RVC	443.52500	+	CC2	MRC
Des Peres, West County Mall						
	FM	W0SRC	146.91000	-	141.3 Hz	MRC

MISSOURI   209

Location	Mode	Call sign	Output	Input	Access	Coordinator
Dexter	FM	N0GK	147.00000	-	100.0 Hz	MRC
	FM	N0DAN	443.90000	+	100.0 Hz	MRC
Dexter, Hospital	FM	N0DAN	147.15000	+	100.0 Hz	MRC
Dixon	FM	W0GS	146.79000	-	88.5 Hz	MRC
El Dorado Springs, Cedar Sprin						
	FM	W0BRN	146.67000	-		MRC
Eldon	FM	KC0KWL	53.05000	51.35000	127.3 Hz	MRC
	FM	AA0NC	146.62500	-	131.8 Hz	MRC
	FM	N0QVO	147.27000	+	123.0 Hz	MRC
	FM	N0GYE	224.58000	-		MRC
Eldridge	FM	K0LH	146.70000	-	88.5 Hz	MRC
Eminence	FM	KN0D	145.31000	-	100.0 Hz	MRC
Eolia	FM	KA0EJQ	145.19000	-		MRC
Excelsior Springs	DMR/BM	K0BSJ	145.19000	-	CC4	MRC
	DMR/BM	K0AMJ	443.32500	+	CC4	
	FM	W0MRM	53.29000	51.59000	146.2 Hz	
	FM	K0ESM	147.37500	+	156.7 Hz	MRC
	FM	K0ESM	444.65000	+	156.7 Hz	MRC
Farmington	FM	K0EOR	147.03000	+	100.0 Hz	MRC
Festus, Hwy CC	FM	W0KLX	443.95000	+	71.9 Hz	MRC
Fordland	FM	N0NWS	145.49000	-	136.5 Hz	MRC
Fulton	DMR/BM	WB8SQS	442.95000	+	CC1	MRC
	DMR/BM	WB8SQS	444.95000	+	CC1	
	FM	KC0MV	147.31500	-	127.3 Hz	
Gainesville	FM	WB0JJJ	147.39000	+	110.9 Hz	MRC
	FM	WB0JJJ	444.35000	+	110.9 Hz	MRC
Gladstone, Gladstone Water Tan						
	FM	W0MB	145.43000	-		MRC
Granby, MSHP Tower						
	FM	KM0HP	145.39000	-	91.5 Hz	MRC
Grant City	FM	W0BYU	147.06000	+	94.8 Hz	MRC
Greenville	FM	KD0MRV	145.23000	-	123.0 Hz	MRC
Hamilton	FM	WD0BBR	146.74500	-	141.3 Hz	MRC
Hannibal	FM	W0KEM	146.62500	-	103.5 Hz	MRC
Hannibal, Grape Street Hill						
	FM	W0KEM	146.88000	-	103.5 Hz	MRC
Harrisonville, 3 Miles NW						
	FM	W0JD	443.70000	+		MRC
Harvester	FM	W0ECA	145.49000	-	141.3 Hz	MRC
Hayti, Pemiscot-Dunklin Co-op						
	FM	KB0UFL	146.98500	-	100.0 Hz	MRC
Higginsville	FM	KC0HJG	442.95000	+	94.8 Hz	
High Ridge	FM	K0AMC	146.92500	-	192.8 Hz	MRC
	FM	K0AMC	444.55000	+	192.8 Hz	
	FM	K0AMC	444.75000	+	192.8 Hz	MRC
	FM	K0AMC	444.85000	+		MRC
Highlandville	FM	WA6JGM	145.23000	-	162.2 Hz	MRC
Hillsboro	FM	KB0TLL	147.07500	+	141.3 Hz	MRC
Holden	FM	W0AU	53.55000	51.85000		MRC
	FM	W0AU	224.88000	-		MRC
	FM	KM0HP	444.52500	+	107.2 Hz	
Holliday	FM	KM0HP	145.29000	-	127.3 Hz	MRC
Holts Summit	DMR/MARC	KB4VSP	145.00000	290.00000	CC3	
	DMR/MARC	KB4VSP	146.00000	292.00000	CC3	
	DMR/MARC	KB4VSP	147.42000		CC3	
	DMR/MARC	KB4VSP	449.87500	-	CC3	
	FM	KD0SAF	443.80000	+		MRC
	FM	KB4VSP	444.87500	+	127.3 Hz	MRC
Hoover	FM	KA0FKL	442.07500	+	151.4 Hz	MRC
Houston	FM	KB0MPO	147.13500	+	100.0 Hz	MRC
Iconium	DMR/BM	WB0YRG	444.47500	+	CC4	
Imperial	FM	KB0TLL	147.10500	+	141.3 Hz	MRC

210   MISSOURI

Location	Mode	Call sign	Output	Input	Access	Coordinator
Independence, Bank Of America						
	FM	W0TOJ	147.09000	+		MRC
Independence, I-435 And 23rd S						
	FM	K0GQ	442.40000	+	151.4 Hz	MRC
Independence, IFD Fire Station						
	FM	W0SHQ	146.73000	-		MRC
Independence, IFD Station #1						
	FM	K0EJC	145.31000	-		MRC
Jane, Walmart	FM	K5QBX	147.25500	+	162.2 Hz	ARC
Jefferson City	DMR/BM	KB4VSP	146.86500	-	CC3	MRC
	FM	K0ETY	147.00000	-	127.3 Hz	MRC
	FM	KB4VSP	443.17500	+	127.3 Hz	MRC
Jefferson City, Capitol						
	FM	K0ETY	442.15000	+	127.3 Hz	MRC
Jefferson City, MO						
	DMR/BM	KM0HP	443.00000	+	CC3	
Joplin	DMR/BM	W0IN	444.50000	+	CC5	
	FM	N0NWS	145.35000	-	91.5 Hz	MRC
	FM	WB0IYC	147.00000	+		MRC
	FM	W0IN	147.21000	+		MRC
	FM	WB0UPB	443.47500	+		MRC
	FM	N0ARM	444.62500	+	91.5 Hz	
Joplin, Mercy Hospital						
	FM	W0IN	145.19000	-	91.5 Hz	MRC
Kansas City	DMR/BM	WB0YRG	224.24000	-	CC4	
	DMR/BM	W0NQX	420.20000	433.20000	CC4	
	DMR/BM	WB0YRG	442.55000	+	CC1	MRC
	DMR/BM	WB0YRG	444.46250	+	CC4	MRC
	DMR/BM	WB0YRG	444.97500	+	CC4	MRC
	DMR/BM	WB0YRG	927.01250	902.01250	CC1	MRC
	DMR/BM	WB0YRG	927.11250	902.11250	CC1	MRC
	DMR/BM	WB0YRG	927.48750	902.48750	CC1	
	DMR/MARC	W0WJB	443.45000	+	CC1	
	FM	K0HAM	53.13000	51.43000	88.5 Hz	MRC
	FM	K0GXL	53.19000	51.49000		MRC
	FM	W0TE	146.79000	-	107.2 Hz	MRC
	FM	W0WJB	146.97000	-		MRC
	FM	WA0SMG	147.27000	+		MRC
	FM	N0EQW	443.05000	+		MRC
	FM	W0WJB	443.17500	+		
	FM	WV0T	443.35000	+		MRC
	FM	WA0NQA	443.77500	+	110.9 Hz	MRC
	FM	N0NKX	444.12500	+	123.0 Hz	MRC
	FM	W0CW	1285.05000	1265.05000		MRC
Kansas City, Booth						
	FM	WB0KIA	224.20000	-		
Kansas City, City Hall						
	FM	W0OEM	443.25000	+	131.8 Hz	MRC
Kansas City, NWS Central Regio						
	FM	WA0QFJ	147.33000	+		MRC
Kansas City, Plaza/Midtown						
	FM	N0WW	443.27500	+	151.4 Hz	MRC
Kansas City, Top Of KCMO City						
	FM	K0HAM	146.82000	-	151.4 Hz	MRC
Kansas City, VA Hospital						
	FM	KC0VA	443.50000	+	151.4 Hz	
Kearney	FM	N0TIX	147.04500	+		MRC
	FM	KB0EQV	443.90000	+	127.3 Hz	MRC
Kennet	FM	KC0LAT	147.19500	+		MRC
Kennett	FM	KC0LAT	444.57500	+	107.2 Hz	MRC
Kimberling City	FM	K0EI	147.34500	+	162.2 Hz	MRC
	FM	K0EI	444.30000	+		MRC

MISSOURI 211

Location	Mode	Call sign	Output	Input	Access	Coordinator
Kingston	FM	KC0GP	444.67500	+	107.2 Hz	MRC
Kingston, 3.5mi West - State R						
	FM	W0BYU	443.37500	+	192.8 Hz	MRC
Kirksville, KTVO Studio Link T						
	FM	W0CBL	145.13000	-		MRC
Kirkwood, Near Kirkwood Train						
	FM	K0ATT	147.15000	+	141.3 Hz	MRC
Laurie	FM	KA0RFO	146.95500	-	192.8 Hz	MRC
Lawson	FM	KZ0G	443.82500	+	151.4 Hz	MRC
Lebanon	DMR/BM	N0GW	443.50000	+	CC12	
Lee's Summit	FM	K0HAM	147.31500	+		MRC
	FM	K0MRR	444.30000	+	131.8 Hz	MRC
	FM	N0AAP	444.35000	+	110.9 Hz	MRC
Lee's Summit, Missouri State H						
	FM	KC0SKY	146.70000	-	107.2 Hz	MRC
	FM	KM0HP	444.77500	+	151.4 Hz	MRC
Lees Summit	DMR/BM	K0MGS	443.60000	+	CC4	MRC
Liberty	FM	N0ELK	145.11000	-		MRC
Linn, MSHP Tower						
	FM	KM0HP	145.39000	-	127.3 Hz	MRC
Macomb	FM	N0NWS	146.74500	-	136.5 Hz	MRC
Macon	FM	N0PR	146.80500	-	156.7 Hz	MRC
Madison	FM	N0SYL	146.98500	-	110.9 Hz	MRC
Marceline	FM	KD0ETV	443.15000	+	110.9 Hz	
Marshall	FM	WB0WMM	147.16500	+	127.3 Hz	MRC
Marshall Junction, MSHP Tower						
	FM	KM0HP	442.17500	+	127.3 Hz	MRC
Marshfield	FM	K0NI	146.86500	-	156.7 Hz	MRC
Maryland Heights	FM	WD0FCH	426.00000	440.00000		MRC
Maryville	DMR/BM	N0GGU	444.47500	+	CC7	MRC
	FM	W0BYU	146.68500	-		MRC
Mexico, KWWR Backup Tower						
	FM	AA0RC	147.25500	+	127.3 Hz	MRC
	FM	AA0RC	443.42500	+		MRC
Mexico, SSM Audrain						
	FM	AA0RC	444.82500	+	127.3 Hz	MRC
Milan	FM	AC0OK	147.18000	+	103.5 Hz	MRC
Moberly	FM	K0MOB	147.09000	+	127.3 Hz	MRC
	FM	K0MOB	443.97500	+		
Monett	FM	W0OAR	146.97000	-	162.2 Hz	MRC
	FM	K0SQS	147.30000	+		MRC
	FM	K0SQS	444.65000	+		MRC
Monroe City	FM	KA0EJQ	146.70000	-		MRC
	FM	KA0EJQ	444.20000	+	141.3 Hz	MRC
Mosby	FM	K0MRR	444.30000	+	131.8 Hz	MRC
Mountain Grove	FM	KG0LF	147.28500	+		MRC
Neosho	DMR/BM	KC0NQE	444.52500	+	CC4	
	FM	KC0FDQ	146.80500	-	127.3 Hz	MRC
Nevada	DMR/BM	W0HL	443.97500	+	CC5	
	FM	W0HL	145.45000	-	91.5 Hz	MRC
	FM	WB0NYD	147.13500	+		
	FM	K0CB	444.00000	+		MRC
	FM	W0HL	444.22500	+	91.5 Hz	MRC
New Madrid	FM	KB0UFL	146.92500	-		MRC
Nixa, Water Tower						
	FM	K0NXA	145.27000	-	162.2 Hz	MRC
Norwood	FM	K0DCA	147.16500	+	162.2 Hz	MRC
O'Fallon	FM	KA0EJQ	444.20000	+		MRC
	FM	W0ECA	444.47500	+	141.3 Hz	MRC
O'Fallon, Water Tower						
	FM	WB0HSI	146.67000	-		MRC
Oak Grove	FM	KB0THQ	444.27500	+	123.0 Hz	MRC

212   MISSOURI

Location	Mode	Call sign	Output	Input	Access	Coordinator
Odin	FM	KC0ROS	147.09000	+	162.2 Hz	MRC
Olivette, Lindbergh And Olive						
	FM	W0SRC	146.85000	-	141.3 Hz	MRC
	FM	W0SRC	224.52000	-	141.3 Hz	MRC
	FM	W0SRC	443.07500	+		MRC
Ongo	FM	K0DCA	145.15000	-	162.2 Hz	MRC
	FM	K0DCA	443.92500	+		MRC
Osage Beach	FM	KB8KGU	442.20000	+	100.0 Hz	MRC
	FM	N0QVO	444.50000	+	127.3 Hz	MRC
Osborn	FM	WD0SKY	145.15000	-	107.2 Hz	MRC
	FM	N0SWP	442.67500	+	127.3 Hz	
Overland Park	DMR/BM	K0XM	927.23750	902.23750	CC1	
Owensville	DMR/BM	N0NOE	443.60000	+	CC2	
Ozark	DMR/BM	K0NXA	146.77500	-	CC5	MRC
Pacific, Nike Missile Base Con						
	FM	KD0ZEA	146.45000	+		
	FM	WA0FYA	224.94000	-	141.3 Hz	MRC
	FM	KD0ZEA	442.30000	+		MRC
Paris	FM	N0SYL	146.83500	-		MRC
Peculiar	DMR/BM	WB0YRG	444.02500	+	CC1	MRC
	FM	WB0YRG	444.32500	+	151.4 Hz	MRC
Piedmont	FM	K0WCR	147.37500	+	100.0 Hz	MRC
Pineville	FM	WB6ARF	147.07500	+	162.2 Hz	MRC
Platte City	FM	W0USI	444.15000	+	88.5 Hz	MRC
Plattsburg	DMR/BM	KC0QLU	146.89500	-	CC1	MRC
Polk, KRBK-TV Tower						
	FM	N0NWS	147.18000	+	136.5 Hz	MRC
Poplar Bluff	FM	AB0JW	444.92500	+	179.9 Hz	MRC
Poplar Bluff, MSHP Station, No						
	FM	KM0HP	146.91000	-	100.0 Hz	MRC
Poplar Bluff, R-1 School Maint						
	FM	AB0JW	145.35000	-	100.0 Hz	MRC
Potosi	FM	N0WNC	147.19500	+	141.3 Hz	MRC
	FM	N0WNC	444.40000	+	100.0 Hz	MRC
Prescott	FM	N0KBC	146.85000	-	114.8 Hz	MRC
Raymore	DMR/MARC	K0DAN	443.47500	+	CC1	MRC
	FM	WB0YRG	146.86500	-	151.4 Hz	
	FM	N0HV	147.12000	+	151.4 Hz	MRC
Raytown, Water Tower						
	FM	K0GQ	145.17000	-	151.4 Hz	MRC
Renick	FM	KM0HP	145.29000	-	127.3 Hz	MRC
Republic	FM	K0NXA	53.27000	51.57000	162.2 Hz	MRC
	FM	K0EAR	146.82000	-	162.2 Hz	MRC
	FM	W6OQS	444.60000	+	77.0 Hz	MRC
Richmond	FM	AC0JR	442.02500	+		MRC
Riverside	DMR/MARC	W0OES	448.45000	-	CC5	
	DMR/MARC	W0OES	448.70000	-	CC4	
Rockport	DMR/BM	WB0OKX	444.77500	+	CC7	
Rolla	DMR/BM	W0EEE	442.67500	+	CC12	
	FM	W0GS	147.21000	+	88.5 Hz	MRC
Rolla, Phelps Health Medical O						
	DSTAR	W0CMD	444.00000	+		MRC
Rolla, TJ Residence Hall						
	DMR/BM	W0EEE	443.82500	+	CC12	
	FM	W0EEE	145.45000	-	110.9 Hz	MRC
Saint Charles	DMR/BM	WB0HSI	444.65000	+	CC2	MRC
Saint Joseph	FM	WA0HBX	443.95000	+	100.0 Hz	
Saint Joseph, KQ2 TV Tower						
	FM	W0NH	146.85000	-	100.0 Hz	MRC
Salem	FM	K0GGM	146.97000	-	110.9 Hz	MRC
Scopus, KMHM Tower						
	FM	W0QMF	146.82000	-		MRC

MISSOURI   213

Location	Mode	Call sign	Output	Input	Access	Coordinator
Sedalia	FM	WA0SDO	147.03000	-	179.9 Hz	MRC
	FM	WB0LRX	224.44000	-	107.2 Hz	MRC
Seymour	FM	K0DCA	145.37000	-		MRC
Seymour, Highway 60 And Route						
	FM	W9IQ	145.19000	-	123.0 Hz	MRC
Shell Knob, Fire Lookout Tower						
	FM	AC0JK	145.21000	-	162.2 Hz	MRC
Sikeston	DMR/BM	KE0MQF	147.21000	+	CC1	MRC
	FM	KB0ZAW	147.07500	+	100.0 Hz	MRC
Smithville	DMR/MARC	WB0YRG	146.64000	-	CC4	
	FM	N0NKX	224.46000	-	151.4 Hz	MRC
Springfield	DMR/BM	K0NXA	146.68500	-	CC5	MRC
	DMR/BM	K0NXA	443.40000	+	CC5	MRC
	DMR/BM	KG0PE	444.67500	+	CC5	MRC
	DMR/MARC	W0PM	442.37500	+	CC1	
	FM	KA0FKF	145.43000	-	107.2 Hz	MRC
	FM	K0NXA	147.01500	+	162.2 Hz	MRC
	FM	WX0OEM	443.57500	+	162.2 Hz	MRC
	FM	W0EBE	444.40000	+	162.2 Hz	MRC
	FM	W0YKE	444.72500	+	136.5 Hz	MRC
	FM	K0NXA	927.01250	902.01250	162.2 Hz	MRC
Springfield, Cox South Hospita						
	FM	KC0DBU	145.33000	-	156.7 Hz	MRC
Springfield, Hammons Tower						
	FM	W0EBE	146.91000	-	162.2 Hz	MRC
Springfield, Rayfield Communic						
	FM	W0PM	147.22500	+	162.2 Hz	
Springfield, Troop D MSHP HQ						
	FM	W0EBE	146.64000	-	162.2 Hz	MRC
St Charles	DMR/BM	N0KQG	443.25000	+	CC2	MRC
St Charles, Water Tower - Lind						
	FM	KO0A	145.33000	-		MRC
St Genevieve	FM	K0QOD	146.62500	-	100.0 Hz	MRC
St Louis	FM	N0ARS	442.37500	+	141.3 Hz	
St. Louis	FM	WB0QXW	145.21000	-	123.0 Hz	MRC
	FM	N0FLC	147.22500	+		MRC
	FM	N0FLC	443.15000	+		MRC
	FM	WB0QXW	444.15000	+	146.2 Hz	MRC
St. Louis, Bates And Virginia						
	FM	KC0TPS	146.61000	-		MRC
St. Louis, Dorchester Apartmen						
	FM	W0SRC	146.97000	-	141.3 Hz	MRC
St. Louis, Forest Park						
	FM	K0KYZ	145.17000	-		
	FM	K0GFM	442.82500	+	127.3 Hz	MRC
St. Louis, I-270 And McDonnell						
	FM	W0MA	147.06000	+	141.3 Hz	MRC
	FM	W0MA	442.87500	+	141.3 Hz	MRC
St. Louis, KDNL ABC 30 TV Towe						
	FM	W9AIU	443.32500	+	141.3 Hz	MRC
St. Louis, KDNL Link Tower						
	FM	W9AIU	146.76000	-	141.3 Hz	MRC
St. Louis, Mercy Hospital						
	FM	W0SLW	147.39000	+	100.0 Hz	MRC
St. Louis, Pine Street And Tuc						
	FM	N9ES	443.77500	+	141.3 Hz	MRC
St. Louis, Red Cross - Lindber						
	FM	W0MDG	147.36000	+		MRC
	FM	K0GOB	224.98000	-		MRC
	FM	K0MDG	442.57500	+		MRC
	FM	K0GOB	442.70000	+		MRC
	FM	W0MDG	442.97500	+	141.3 Hz	

214   MISSOURI

Location	Mode	Call sign	Output	Input	Access	Coordinator
St. Louis, SLU Hospital						
	FM	WD0EFP	443.47500	+	77.0 Hz	MRC
St. Louis, Tilles Park						
	FM	N0ARS	145.35000	-	123.0 Hz	MRC
St. Paul	FM	N0EEA	224.66000	-		MRC
Stockton Lake	FM	K0NXA	444.97500	+	162.2 Hz	MRC
Stover	FM	KB0QWQ	147.39000	+	127.3 Hz	MRC
	FM	KB0QWQ	444.92500	+	127.3 Hz	MRC
Sullivan	DMR/BM	N0NOE	444.60000	+	CC2	MRC
	FM	K0CSM	145.15000	-		MRC
Sullivan, West Sullivan Water						
	FM	KC0DBS	146.80500	-	110.9 Hz	MRC
Sunrise Beach	FM	N0ZS	146.73000	-		MRC
Thayer	FM	W0WZR	146.80500	-	110.9 Hz	MRC
Trenton, Courthouse						
	FM	KB0RPJ	146.95500	-	156.7 Hz	MRC
Turney	DMR/BM	N0MIJ	440.00000	+	CC7	
Van Buren	FM	N0IBV	146.86500	-	100.0 Hz	MRC
Viburnum	FM	KD0KIB	147.30000	+	100.0 Hz	MRC
	FM	KD0KIB	442.05000	+	100.0 Hz	MRC
Walnut Grove	FM	AK0C	147.33000	+	162.2 Hz	MRC
Warrensburg	FM	W0AU	146.88000	-	107.2 Hz	MRC
	FM	W0APR	146.94000	-	107.2 Hz	
	FM	W0AU	443.20000	+	107.2 Hz	MRC
Warrenton	FM	KA0CWU	147.04500	+	141.3 Hz	MRC
Warrenton, MSHP Tower West Of						
	FM	WA0EMA	147.33000	+	123.0 Hz	MRC
Warsaw	DMR/MARC	KF0KR	442.77500	+	CC3	
	FM	WB0EM	146.92500	-	107.2 Hz	MRC
	FM	KD0CNC	147.30000	+		MRC
Warsaw, MSHP Tower						
	FM	KM0HP	147.07500	+	127.3 Hz	MRC
Washburn	FM	KE0CZQ	444.85000	+	123.0 Hz	MRC
Washington	DMR/BM	KE0WXR	443.90000	+	CC2	
	DMR/BM	K0FDG	444.10000	+	CC2	
	FM	WA0FYA	147.24000	+	141.3 Hz	MRC
	FM	WA0FYA	444.35000	+	141.3 Hz	MRC
Washington, Indian Prairie Sch						
	FM	WA0FYA	147.18000	+	141.3 Hz	
West Plains	FM	KD0AIZ	145.25000	-	110.9 Hz	MRC
	FM	W0HCA	146.94000	-	110.9 Hz	MRC
Willow Springs	DMR/BM	K0NXA	146.67000	-	CC12	
Windsor	DMR/BM	N0TLE	443.87500	+	CC3	MRC
	FM	K0UG	147.19500	+	107.2 Hz	MRC
Wright City	P25	AD0JA	146.47000	147.47000	173.8 Hz	MRC

MONTANA

Location	Mode	Call sign	Output	Input	Access	Coordinator
Anaconda	FM	KB7IQN	53.03000	52.03000	131.8 Hz	
	FM	KB7IQO	147.02000	+		
	FM	K0PP	147.08000	+	107.2 Hz	
	FM	KA7NBR	446.80000	-		
Belgrade	FM	WB7USV	448.85000	-	100.0 Hz	
Big Sky	DMR/BM	KL7JGS	447.00000	-	CC1	MRCC
	FM	W7LR	146.82000	-	82.5 Hz	MRCC
Big Timber	FM	NU7Q	146.64000	-	100.0 Hz	MRCC
Bigfork	FM	KA5LXG	146.62000	-	100.0 Hz	
	FM	KA5LXG	442.07500	+	88.5 Hz	
Billings	DMR/BM	KC7NP	448.25000	-	CC1	
	DMR/BM	N7YHE	448.65000	-	CC1	
	DSTAR	K7EFA	449.00000	-		MRCC
	FM	WR7MT	147.08000	+		
	FM	N7YHE	147.10000	+	100.0 Hz	MRCC

MONTANA 215

Location	Mode	Call sign	Output	Input	Access	Coordinator
Billings	FM	K7EFA	147.24000	+		
	FM	K7EFA	147.30000	+	100.0 Hz	
	FM	W7JDX	449.25000	-	100.0 Hz	
	FM	N7VR	449.75000	-		
Billings, Red Lodge Mountain						
	FM	K7EFA	147.36000	+	100.0 Hz	
Blacktail Mountain	FM	K7LYY	147.18000	+	100.0 Hz	
Boulder, Depot Hill						
	FM	KC7MRQ	146.70000	-	162.2 Hz	MRCC
Bozeman	DMR/BM	KL7JGS	444.10000	+	CC1	
	DMR/BM	WA7U	447.95000	-	CC1	
	DMR/BM	KL7JGS	448.35000	-	CC1	
	FM	KD7TQM	145.01000			
	FM	WR7MT	147.18000	+	100.0 Hz	
	FM	KB7KB	448.35000	-	100.0 Hz	
Bozeman, Bridger Bowl						
	FM	W7YB	146.88000	-	100.0 Hz	
Bozeman, MSU	FM	KI7XF	447.70000	-	77.0 Hz	
Butte	FM	W7ROE	146.94000	-	100.0 Hz	
	FM	W7VNE	147.02000	+	107.2 Hz	
Carlton	DMR/BM	DMR	422.02500	429.02500	CC1	
Colstrip, Little Wolf Mountain						
	FM	KC7KCF	146.90000	-		
Dillon	FM	N7AFS	146.76000	-	107.2 Hz	
Dixon	FM	K7KTR	444.55000	+		
East Glacier, Mt Baldy						
	FM	K7HR	146.70000	-	103.5 Hz	
Eureka	FM	KC7CUE	147.34000	+	100.0 Hz	
Eureka, Pinkham Mountain						
	FM	KC7CUE	147.34000	+		
	FM	WR7DW	444.25000	+		
Eureka, Sams Hill	FM	KC7CUE	145.43000	-	100.0 Hz	MRCC
Eureka, Virginia Hill						
	FM	WR7DW	145.39000	-	100.0 Hz	MRCC
	FM	WR7DW	443.80000	+	100.0 Hz	MRCC
Fairfield	FM	W7ECA	147.26000	+	100.0 Hz	MRCC
Forsyth	FM	KC7BOB	147.20000	+	100.0 Hz	MRCC
Glasgow	FM	WX7GGW	146.84000	-	100.0 Hz	
Glendive	FM	W7DXQ	146.76000	-	100.0 Hz	
Great Falls	DMR/BM	WR7HLN	449.20000	-	CC2	MRCC
	FM	AA7GS	146.68000	-	100.0 Hz	
	FM	AE7HW	147.14000	+	100.0 Hz	MRCC
	FM	AA7GS	147.24000	+		MRCC
	FM	W7GMC	147.36000	+	100.0 Hz	MRCC
Great Falls, 1708 6 St NW						
	FM	AG7AD	145.23000	-	123.0 Hz	
Great Falls, Gore Hill						
	FM	W7ECA	147.30000	+		
Great Falls, Highwood Baldy Mo						
	FM	W7ECA	146.74000	-	100.0 Hz	MRCC
Great Falls, Porphyry Peak						
	FM	W7ECA	147.12000	+		
Greycliff	FM	WR7MT	147.28000	+	100.0 Hz	MRCC
Hamilton	DMR/MARC	KG6MQE	422.02500	429.02500	CC11	
	FM	W7FTX	146.72000	-	203.5 Hz	MRCC
Havre	FM	KC7NV	146.91000	-	100.0 Hz	
Havre, Havre Air Base						
	FM	W7HAV	146.98000	-	100.0 Hz	
Helena	DMR/BM	WR7HLN	147.10000	+	CC1	MRCC
Helena, Hogback Mountain						
	FM	W7MRI	145.45000	-	100.0 Hz	

216 MONTANA

Location	Mode	Call sign	Output	Input	Access	Coordinator
Helena, Mount Belmont						
	FM	N7RB	147.22000	+	100.0 Hz	
Kalispell	FM	N7LT	448.45000	-	67.0 Hz	
Kalispell, Blacktail Mtn						
	FM	N7LT	147.36000	+	100.0 Hz	
Lakeside, Blacktail Mountain						
	FM	K7LYY	146.76000	-	100.0 Hz	
Lewistown	FM	K7VH	442.00000	+	100.0 Hz	
Lewistown, Judith Peak.						
	FM	K7VH	146.96000	-	100.0 Hz	
Libby	FM	KB7SQE	53.15000	52.15000		
	FM	KB7SQE	444.35000	+		MRCC
	FM	KB7SQE	444.82500	+	100.0 Hz	
Libby, Blue Mountain						
	FM	W3YAK	444.22500	+	88.5 Hz	
Libby, King Mountain						
	FM	K7LBY	146.84000	-	100.0 Hz	MRCC
Libby, Meadow Peak						
	FM	AG7FF	145.31000	-	100.0 Hz	MRCC
Lookout Pass	DMR/BM	WR7HLN	444.20000	+	CC2	MRCC
Miles City	FM	K7HWK	146.92000	-		
	FM	K7RNS	444.95000	+		
Missoula	DMR/BM	W1KGK	147.12000	+	CC1	
Missoula, Point Six						
	FM	W7PX	146.90000	-	88.5 Hz	
	FM	W7PX	147.04000	+		
	FM	W7PX	444.80000	+	88.5 Hz	
Missoula, University Mountain						
	FM	NZ7S	147.00000	+		
Noxon, Green Mountain						
	FM	KD7OCP	145.33000	-	123.0 Hz	
Olney, Werner Peak						
	FM	WR7DW	444.65000	+	100.0 Hz	
Plains	FM	K7KTR	147.14000	+	103.5 Hz	
Polson	FM	W7CMA	145.35000	-	100.0 Hz	
Pompeys Pillar	FM	KF7FW	147.18000	+	123.0 Hz	MRCC
Red Lodge	FM	N7YHE	147.20000	+	100.0 Hz	
Red Lodge, Grizzly Peak Mtn						
	FM	WB7RIS	147.00000	+	100.0 Hz	MRCC
Red Lodge, Palisades Peak, Rad						
	FM	KE7FEL	449.90000	-		
Ronan	FM	KC7MRQ	146.70000	-	162.2 Hz	
Round Butte	FM	K7KTR	147.14000	+	103.5 Hz	
Roundup, Bull Mountain						
	FM	K7EFA	145.41000	-	100.0 Hz	
Saltese, Lookout Pass						
	FM	K7HPT	147.02000	+		
Scobey	FM	N0PL	443.50000	+		
Shepherd	FM	KD0CST	146.40000	+	118.8 Hz	
Sidney	FM	W7DXQ	147.38000	+		
	FM	W7DXQ	444.50000	+		
St Ignatius	FM	KD7YAC	145.43000	-		
St. Mary, Hudson Bay Divide						
	FM	K7JAQ	146.82000	-	100.0 Hz	
	FM	K7JAQ	441.20000	+		
Stevensville	DMR/BM	AE7OD	449.42500	-	CC1	
	DMR/BM	KD7HP	449.42500	-	CC1	
	FM	W7FTX	447.50000	-	203.5 Hz	
	FM	KD7HP	927.75000	902.75000	114.8 Hz	
Superior, Thompson Peek						
	FM	W7PX	146.96000	-	88.5 Hz	

MONTANA 217

Location	Mode	Call sign	Output	Input	Access	Coordinator
Thompson Falls, Clarks Peak So						
	FM	W1KGK	146.68000	-	100.0 Hz	
Three Forks	FM	KL7JGS	224.72000	-		
	FM	KB7KB	448.35000	-		
Three Forks, Round Spring						
	FM	WR7MT	147.38000	+	100.0 Hz	MRCC
Toston	DMR/BM	WR7HLN	449.30000	-	CC1	MRCC
West Glacier, West Entrance To						
	FM	W7YP	447.50000	-		
West Yellowstone	DMR/BM	KL7JGS	449.90000	-	CC1	
Whitefish, Big Mountain						
	FM	KO8N	145.27000	-	100.0 Hz	
Whitefish, Sandy Hill						
	FM	K7LYY	147.38000	+		
Zortman	FM	N7ARA	146.79000	-		
	FM	W7ECA	147.26000	+	100.0 Hz	
NEBRASKA						
Ainsworth	FM	WM0L	147.36000	+	225.7 Hz	
Albion	FM	KB0TLX	147.37500	+		
Alliance	FM	N0NEB	146.71500	-	103.5 Hz	
	FM	N2VHZ	147.01500	+		
Alma	FM	KA0RCZ	145.20500	-		WB0CMC
Ashland, KUON-TV Tower						
	FM	K0ASH	145.31000	-		WB0CMC
Aurora	FM	W0CUO	147.18000	+	123.0 Hz	
Axtell, NTV Studio	FM	KA0RCZ	444.62500	+		WB0CMC
Beatrice	FM	KC0SWG	145.22000	-	131.8 Hz	WB0CMC
	FM	KC0MLT	145.34000	-	100.0 Hz	WB0CMC
Beatrice, Homestead Monument						
	FM	K0ORU	146.79000	-		WB0CMC
Bellevue	DMR/BM	KB0ZZT	443.65000	+	CC2	WB0CMC
	DMR/MARC	KB0ZZT	442.27500	+	CC1	WB0CMC
	FM	WB0QQK	145.11500	-	179.9 Hz	
	FM	WB0EMU	147.06000	+	131.8 Hz	
	FM	W0WYV	147.39000	+	131.8 Hz	
	FM	WB0QQK	443.35000	+	179.9 Hz	
	FM	W0JJK	443.82500	+	179.9 Hz	
	FM	WB0QQK	444.87500	+	179.9 Hz	
Broken Bow	FM	KR0A	147.06000	+		
Brownville, Indian Cave State						
	FM	K0TIK	444.22500	+		WB0CMC
Burwell	FM	W0EJL	147.09000	+		WB0CMC
Campbell, Mid-Rivers E911 Disp						
	FM	W0WWV	444.47500	+	136.5 Hz	WB0CMC
Chadron	FM	W0FLO	145.62000		100.0 Hz	
Chadron, Nebraska National For						
	FM	W0FLO	147.36000	+	123.0 Hz	WB0CMC
Columbus	DSTAR	WA0COL	146.95500	-		WB0CMC
	DSTAR	WA0COL	442.17500	+		
	FM	WA0COL	146.64000	-		
	FM	N0RHM	146.77500	-	131.8 Hz	
	FM	WA0COL	442.05000	+	167.9 Hz	
Edison	FM	KD0AN	146.74500	-		WB0CMC
Elba	FM	W0CUO	147.24000	+	123.0 Hz	
Fairbury	FM	WB0RMO	147.12000	+		
Fremont	DMR/MARC	KA0NCR	444.85000	+	CC1	
	FM	W0UVQ	444.17500	+		
Fremont, Stanton Tower (high-r						
	FM	KD0EFC	146.67000	-	100.0 Hz	WB0CMC
Grand Island	FM	W0CUO	53.35000	51.65000		
	FM	W0CUO	146.94000	-	123.0 Hz	

218   NEBRASKA

Location	Mode	Call sign	Output	Input	Access	Coordinator
Grand Island	FM	NI0P	443.95000	+	123.0 Hz	WB0CMC
	FUSION	KD0ENX	147.34500	+	123.0 Hz	WB0CMC
Grand Island, Heartland Event						
	FM	KD0ENX	444.75000	+		WB0CMC
Grand Island, Nebraska State F						
	FM	W0CUO	145.41500	-	123.0 Hz	WB0CMC
Gretna	FM	W0MAO	444.90000	+		WB0CMC
Harrisburg	FM	N0NEB	147.00000	-	103.5 Hz	
Hastings, Hastings Dog-Walk Pa						
	FM	W0WWV	443.20000	+	82.5 Hz	WB0CMC
Hastings, KCNT Tower						
	FM	W0WWV	145.13000	-	123.0 Hz	WB0CMC
Heartwell, KGIN TV Tower						
	FM	W0WWV	146.82000	-	123.0 Hz	
Humboldt	DMR/MARC	NV8Q	443.57500	+	CC1	
Jackson, KFHC Radio Tower						
	FM	N0DCA	146.79000	-	110.9 Hz	WB0CMC
Jefferson	FM	K2WMA	444.95000	+	141.3 Hz	
Julian	FM	W0MAO	444.62500	+		
Kearney	FM	W0KY	146.62500	-	123.0 Hz	WB0CMC
	FM	KA0RCZ	147.31500	+	123.0 Hz	
	FM	KA0DBK	147.39000	+	123.0 Hz	
	FM	KA0RCZ	444.85000	+	74.4 Hz	WB0CMC
Kearney, County Courthouse						
	DSTAR	KD0PBW	147.03000	+		
Kearney, Kearney Water Tower						
	FM	KC0WZL	147.00000	-	123.0 Hz	
Lexington	FM	N0VL	146.85000	-	123.0 Hz	WB0CMC
Lincoln	DMR/MARC	WB0QQK	442.42500	+	CC1	WB0CMC
	FM	N0FER	145.19000	-		
	FM	N0UNL	145.32500	-		
	FM	K0KKV	146.76000	-		WB0CMC
	FM	K0LNE	146.85000	-		WB0CMC
	FM	N0FER	147.04500	+		
	FM	K0SIL	147.19500	+		
	FM	N0FER	147.24000	+		
	FM	W0MAO	147.33000	+		WB0CMC
	FM	N0FER	224.98000	-		
	FM	K0KKV	442.70000	+	146.2 Hz	
	FM	N0FER	443.00000	+		
	FM	N0FER	443.50000	+	162.2 Hz	
	FM	W0DMS	444.10000	+		
	FUSION	WB0KBK	146.62500	-		WB0CMC
Lincoln, Capitol	DSTAR	W0MAO	145.25000	-		
	DSTAR	W0MAO	442.15000	+		
Loomis	FM	K0PCA	146.89500	-	123.0 Hz	WB0CMC
McCook	FM	N7UVW	444.50000	+	151.4 Hz	
	FM	N7UVW	444.80000	+	162.2 Hz	
McCook, 180' Tower S. Of McCoo						
	FM	K0TAJ	147.27000	+		
Mitchell	FM	WD0BQM	445.10000	-	100.0 Hz	
Murray	FM	KA0IJY	147.21000	+	103.5 Hz	
	FM	KA0IJY	442.57500	+	100.0 Hz	
Nebraska City	FM	K0TIK	146.70000	-		
	FM	K0TIK	442.10000	+		
Nehawka	FM	K0LNE	146.85000	-		
Neligh	FM	KB0TRU	146.83500	-		WB0CMC
Nelson	FM	W0WWV	147.21000	+	123.0 Hz	WB0CMC
Norfolk	FM	W0OFK	442.25000	+		WB0CMC
	FM	W0OFK	444.25000	+	88.5 Hz	
North Platte	FM	K0KDC	146.83500	-		
	FM	N0UGO	146.94000	-	123.0 Hz	WB0CMC

NEBRASKA 219

Location	Mode	Call sign	Output	Input	Access	Coordinator
North Platte	FM	N0IQ	147.33000	+	123.0 Hz	
	FM	N0IQ	444.40000	+	123.0 Hz	
O'Neill	DMR/BM	KB0GRP	146.61000	-	CC1	
	FM	KB0GRP	444.87500	+	146.2 Hz	
Oakhurst	FM	W2GSA	145.04500		67.0 Hz	
Oceanview	DSTAR	NJ2CM B	440.09375	+		
Ogallala	FM	N0UGO	146.76000	-	123.0 Hz	
Omaha	DMR/BM	K0OQL	442.00000	+	CC1	
	DMR/BM	KW1RKY	442.32500	+	CC1	
	DMR/BM	KW1RKY	443.97500	+	CC1	
	DMR/MARC	K0BOY	442.65000	+	CC1	WB0CMC
	DMR/MARC	W0AAI	442.82500	+	CC1	WB0CMC
	DMR/MARC	KI0PY	444.97500	+	CC1	WB0CMC
	DSTAR	KD0CGR	145.17500	-		
	DSTAR	KD0CGR	442.12500	+		WB0CMC
	FM	K0BOY	145.45000	-	131.8 Hz	WB0CMC
	FM	WB0CMC	147.00000	+		
	FM	WA0WTL	147.30000	+		WB0CMC
	FM	WB0YLA	224.76000	-	146.2 Hz	WB0CMC
	FM	KA0IJY	442.47500	+	100.0 Hz	WB0CMC
	FM	KC0YUR	442.95000	+	146.2 Hz	
	FM	KB0SMX	443.45000	+	100.0 Hz	
	FM	KF6SWL	443.72500	+	100.0 Hz	
	FM	KG0S	443.92500	+	103.5 Hz	
	FM	WB0WXS	444.05000	+		WB0CMC
	FUSION	K0OQL	444.42500	+		WB0CMC
Omaha, Benson High School						
	FM	WB0CMC	421.25000	434.05000		WB0CMC
Omaha, CHI Immanuel Hospital						
	FM	N0YMJ	145.37000	-		WB0CMC
Omaha, Douglas County Communic						
	FM	K0BOY	147.36000	+		
Omaha, KETV Tower (Crown Point						
	FM	K0USA	146.94000	-		WB0CMC
Omaha, KPTM	FM	WB0CMC	444.95000	+		
Omaha, KPTM/KXVO TV Tower						
	FM	W0EQU	443.77500	+		WB0CMC
Omaha, Methodist Women's Hosp						
	DMR/BM	KC0YUR	443.97500	+	CC1	WB0CMC
Omaha, WOWT TV Tower 36 + Farn						
	FM	K0SWI	442.22500	+	136.5 Hz	WB0CMC
Papillion	FM	WB0EMU	145.23500	-	131.8 Hz	
	FM	WB0EMU	146.71500	-		
Papillion, Sarpy County Courth						
	FM	WB0EMU	442.72500	+	131.8 Hz	WB0CMC
Pilger	FM	KG0S	444.12500	+	131.8 Hz	
Scottsbluff	FM	WD0BQM	145.47500	-		
	FM	AG0N	146.44500		88.5 Hz	
	FM	N0NEB	147.07500	+		
	FM	W0KAV	444.82500	+		
Scottsbluff, Stage Hill						
	FM	WB7GR	444.12500	+	114.8 Hz	
Scribner	FM	KE0IBU	147.10500	+		WB0CMC
South Sioux City	FM	K9NHP	147.47000			
South Sioux City, KBWF TV Towe						
	FM	N0DCA	443.75000	+	110.9 Hz	WB0CMC
South Sioux City, Law Enforcem						
	FM	W0MEG	145.35500	-		WB0CMC
Spencer	FM	KC0HMN	147.33000	+	131.8 Hz	WB0CMC
St Libory, Microwave Tower						
	FM	NI0P	444.92500	+	100.0 Hz	WB0CMC
Thurman/Tabor	DMR/MARC	WA0RJR	442.80000	+	CC1	WB0CMC

220 NEBRASKA

Location	Mode	Call sign	Output	Input	Access	Coordinator
Valley, City Water Tower						
	FM	KD0PGV	145.26500	-		WB0CMC
Wahoo	DMR/MARC	KB0ZZT	443.60000	+	CC1	WB0CMC
Wayne	FM	N0ZQR	147.03000	+		
Weeping Water	DMR/MARC	KC0HYI	442.77500	+	CC1	WB0CMC
Wilber	FM	KD0VKC	146.98500	-	100.0 Hz	WB0CMC
Winslow	FM	KD0PGV	444.37500	+	100.0 Hz	WB0CMC
Winslow, NE	FM	WB0IEN	147.16500	+		WB0CMC
Wood River	DMR/BM	KE0HZX	444.70000	+	CC1	
	DSTAR	NI0P	444.70000	+	167.9 Hz	WB0CMC
York	FM	WA0HOU	147.27000	+		WB0CMC
	FM	WA0HOU	444.20000	+		
Yorktown Heights	FM	K2HR	146.94000	-	127.3 Hz	

NEVADA

Location	Mode	Call sign	Output	Input	Access	Coordinator
Amargosa Valley	FM	NV7AV	146.76000	-	123.0 Hz	
Battle Mountain, Mount Lewis						
	FM	WA6TLW	52.52500	51.52500		CARCON
	FM	WA6TLW	146.79000	-		
	FM	W7LKO	443.90000	+		
Battle Mountain, Mt Lewis						
	FM	W7LKO	146.91000	-	100.0 Hz	
	FM	WA6TLW	444.85000	+	94.8 Hz	
Beatty, Sawtooth Mtn						
	FM	WB6TNP	449.22500	-	141.3 Hz	SNRC
	FM	K6DLP	449.86000	-		
Boulder City, Opal Mountain						
	FM	WB6TNP	448.80000	-	141.3 Hz	SNRC
	FM	WB6TNP	927.31250	902.31250	DCS 606	SNRC
Buffalo, New York	DMR/MARC	N4NJJ	445.97500	+	CC2	SNRC
Bullhead City, Spirit Mountain						
	DSTAR	K7RLW	446.22500	-		SNRC
	FM	N7SKO	145.27000	-	131.8 Hz	SNRC
	FM	WR7NV	224.98000	-	100.0 Hz	SNRC
	FM	K7GIL	446.37500	-		SNRC
	FM	WR7RED	448.20000	-		SNRC
	FM	WB6TNP	448.70000	-	141.3 Hz	SNRC
	FM	N6JFO	449.30000	-	131.8 Hz	SNRC
	FM	KI7D	449.32500	-		SNRC
	FM	WB6TNP	927.88750	902.88750	DCS 606	SNRC
Carlin	FM	W7LKO	146.85000	-	100.0 Hz	
	FM	W7LKO	444.20000	+	100.0 Hz	
Carlin, Mary's Mountain						
	FM	KO7G	443.85000	+		
Carlin, Swales Mountain						
	FM	WB7BTS	449.35000	-	186.2 Hz	
Carson City	DMR/MARC	W7TA	442.05000	+	CC1	
	FM	WA7DG	146.82000	-	123.0 Hz	
	FM	K5BLS	442.12500	+	156.7 Hz	CARCON
	FM	WA6JQV	444.32500	+	127.3 Hz	
	FM	WA6JQV	927.50000	902.50000	127.3 Hz	
Carson City, McClellan						
	FM	KB7MF	145.24000	-		CARCON
Carson City, McClellan Peak						
	FM	K6LNK	443.32500	+		CARCON
	FM	WA6JQV	444.55000	+	127.3 Hz	
Carson City, McClellan Pk.						
	FM	K5BLS	145.41000	-	156.7 Hz	
Carson City, Nevada DEM Buildi						
	FM	W7DEM	442.90000	+	156.7 Hz	CARCON
Cold Springs, QTH						
	FM	KE7DZZ	448.62500	-	100.0 Hz	

NEVADA 221

Location	Mode	Call sign	Output	Input	Access	Coordinator
Elko	DMR/BM	KB7SJZ	442.40000	+	CC1	
	DMR/BM	KB7SJZ	442.47500	+	CC1	
	FM	WB7BTS	146.94000	-	100.0 Hz	
	FM	KE7LKO	147.33000	+	100.0 Hz	
	FM	K9VX	147.39000	+		
	FM	KI6V	440.32500	+		
	FM	W7LKO	442.05000	+	100.0 Hz	
	FM	W7LKO	444.80000	+		CARCON
	FM	W7LKO	444.95000	+	100.0 Hz	
	FM	W7LKO	449.75000	-		
Elko, Adobe	FM	KE7LKO	53.01000	52.01000	100.0 Hz	CARCON
Elko, Adobe Summit						
	FM	W7LKO	444.70000	+		
Elko, Grindstone Mountain						
	FM	W7LKO	53.25000	52.25000	100.0 Hz	CARCON
Elko, Lamoille Summit						
	FM	W7LKO	147.21000	+	100.0 Hz	
Elko, SnoBowl	FM	KI6V	443.37500	+	100.0 Hz	
Elko, Swales Mountain						
	FM	KE7LKO	29.68000	-	100.0 Hz	CARCON
Ely, Kimberly Mountain						
	FM	N7ELY	147.18000	+	114.8 Hz	CARCON
Ely, Squaw Peak	FM	N7ELY	146.88000	-	114.8 Hz	CARCON
	FM	N7ELY	442.10000	+		
Fallon	FM	K5BLS	442.12500	441.52500		
	FM	WA6KDW	444.37500	+	100.0 Hz	
Fallon, Eagle Ridge						
	FM	K5BLS	442.22500	+	156.7 Hz	CARCON
Fallon, Fairview Peak						
	FM	KE6QK	145.35000	-	123.0 Hz	CARCON
Fallon, Lahontan Park						
	FM	KE6QK	146.97000	-	103.5 Hz	
Fallon, NV	FM	K5BLS	147.34500	+		
Fernley	DMR/MARC	W7TA	441.95000	+	CC1	CARCON
	FM	W7JA	443.25000	+	192.8 Hz	
	FM	K7UI	443.90000	+	103.5 Hz	
	FM	WA6JQV	444.45000	+		
	FM	WA6TLW	444.75000	+		
	FM	N7TR	444.90000	+	123.0 Hz	
	FM	N7RMK	446.87500	-	DCS 516	
Fernley, Eagle Ridge						
	FM	N7PLQ	147.36000	+	123.0 Hz	
	FM	N7PLQ	443.50000	+	100.0 Hz	
	FM	WA6TLW	444.70000	+	94.8 Hz	
Gardnerville	DMR/BM	KD7FPK	442.15000	+	CC4	CARCON
	DMR/BM	KD7FPK	443.97500	+	CC4	CARCON
Genoa	FM	NH7M	443.77500	+		CARCON
	FM	WA6JQV	444.00000	+		
Gerlach	FM	K1C	147.03000	+	100.0 Hz	
Gerlach, Fox Mountain						
	FM	K7UI	441.70000	+	146.2 Hz	
Gerlach, Granite Ridge						
	FM	KD7YIM	145.23000	-	123.0 Hz	CARCON
Glenbrook	FM	N3KD	146.70000	-		
Glendale, Beacon Hill						
	FM	N7SGV	145.30000	-	DCS 244	SNRC
Goldfield	FM	WB7WTS	223.70000	-		
Goldfield, Montezuma Peak						
	FM	WB7WTS	146.64000	-		CARCON
	FM	WA6TLW	444.85000	+	94.8 Hz	
Hawthorne, Corey Peak						
	FM	WA6TLW	146.79000	-		CARCON

222   NEVADA

Location	Mode	Call sign	Output	Input	Access	Coordinator
Hawthorne, Corey Peak						
	FM	K6LNK	440.72500	+		CARCON
	FM	WA6BXP	1284.97500	1264.97500		
Hawthorne, Montgomery Pass						
	FM	KE6VVB	146.67000	-		
	FM	KE6VVB	444.20000	+		
Henderson	FM	KH7R	447.37500	-		SNRC
	FM	KE7OPJ	447.65000	-	123.0 Hz	
	FM	N8HC	447.72500	-	114.8 Hz	
Henderson, Black Mountain						
	FM	NK2V	145.39000	-	100.0 Hz	SNRC
	FM	N7OK	147.09000	+	100.0 Hz	SNRC
	FM	NX7R	1293.62500	1273.62500	114.8 Hz	SNRC
Henderson, CSN Henderson						
	P25	N4NJJ	420.85000	+		
Henderson, Downtown						
	FM	K7RSW	448.87500	-	114.8 Hz	SNRC
Henderson, Seven Hills						
	FM	W7HEN	447.92500	-	156.7 Hz	
Henderson, Sun City Anthem						
	FM	WA7SCA	445.68000	-	123.0 Hz	SNRC
Henderson, Sun City Anthem, In						
	FM	WA7SCA	145.28000	-	123.0 Hz	SNRC
Incline	FM	K6LNK	441.55000	+		CARCON
Incline Village	DMR/MARC	W7TA	443.95000	+	CC1	CARCON
	FM	N7VXB	441.20000	+	100.0 Hz	
Incline Village, Relay Peak, M						
	FM	W7TA	147.15000	+	123.0 Hz	CARCON
Jackpot	FM	W7GK	147.27000	+	100.0 Hz	
Jarbridge, Deer Mountain						
	FM	KA7CVV	147.16000	+		
Las Vegas	DMR/BM	KB6XN	146.79000	-	CC15	SNRC
	DMR/BM	N5VAE	147.36000	+	CC1	
	DMR/BM	N5VAE	441.95000	+	CC1	
	DMR/BM	KB6XN	445.75000	-	CC2	
	DMR/BM	KB6XN	445.80000	-	CC15	
	DMR/BM	KB6XN	445.82500	-	CC15	
	DMR/BM	W6OLI	446.03750	-	CC1	
	DMR/BM	N7ARR	446.45000	-	CC1	SNRC
	DMR/BM	KG7SS	447.35000	-	CC15	
	DMR/BM	K7IZA	447.80000	-	CC1	
	DMR/BM	KB6XN	449.97500	-	CC15	SNRC
	DMR/BM	KB6XN	927.91250	902.91250	CC15	
	DMR/MARC	KB6XN	146.73000	-	CC2	
	DMR/MARC	KB6XN	146.97000	-	CC15	SNRC
	DMR/MARC	KB6XN	147.43500	146.43500	CC2	
	DMR/MARC	WB6EGR	445.01250	-	CC15	SNRC
	DMR/MARC	KB6XN	445.70000	+	CC2	SNRC
	DMR/MARC	KB6XN	445.72500	-	CC2	
	DMR/MARC	KB6XN	445.85000	-	CC15	
	DMR/MARC	KG7SS	447.27500	-	CC15	
	DMR/MARC	KG7SS	447.30000	-	CC15	
	DMR/MARC	KF6FM	448.27500	-	CC1	
	DMR/MARC	KG6DTL	448.32500	-	CC15	
	DMR/MARC	WB9STH	448.45000	-	CC15	SNRC
	DMR/MARC	WB9STH	448.97500	-	CC15	
	DMR/MARC	KB6XN	449.32500	-	CC15	
	DMR/MARC	KB6XN	927.93750	902.93750	CC15	
	DSTAR	N7ARR	145.17500	-		SNRC
	DSTAR	W7AES	147.97500	-		SNRC
	DSTAR	N7ARR	446.80000	-		SNRC
	DSTAR	W7AES	449.57500	-		SNRC

NEVADA 223

Location	Mode	Call sign	Output	Input	Access	Coordinator
Las Vegas	DSTAR	W7AES	1282.39000	1262.39000		SNRC
	FM	KE6DV	145.16000	-		SNRC
	FM	W7EB	223.58000	-		SNRC
	FM	WR7NV	224.98000	-	100.0 Hz	
	FM	W7HEN	446.47500	-	156.7 Hz	
	FM	KG7OKC	446.57500	-	199.5 Hz	SNRC
	FM	KF6QYX	448.05000	-	123.0 Hz	
	FM	NO7BS	448.47500	-	100.0 Hz	SNRC
	FM	WA7HXO	449.15000	-	136.5 Hz	
	FM	WB7RAA	449.35000	-		SNRC
	FM	KK7AV	449.47500	-	114.8 Hz	
	FM	WA7GIC	449.52500	-		SNRC
	FM	KE7CCH	449.60000	-	107.2 Hz	SNRC
	FM	WB6TNP	927.03750	902.03750		
	FM	K7RRC	927.71250	902.71250	127.3 Hz	SNRC
	P25	KG7SS	927.05000	902.05000	DCS 411	
Las Vegas, Angel Peak						
	FM	N9CZV	53.19000	52.19000	110.9 Hz	SNRC
	FM	N7ARR	145.37000	-	123.0 Hz	SNRC
	FM	N7SGV	147.30000	+	127.3 Hz	SNRC
	FM	WB6TNP	224.50000	-	131.8 Hz	SNRC
	FM	N7OK	447.47500	-	110.9 Hz	SNRC
	FM	WB6TNP	448.57500	-	141.3 Hz	SNRC
	FM	WR7WHT	449.02500	-	156.7 Hz	SNRC
	FM	KI7D	449.20000	-	114.8 Hz	SNRC
	FM	N7TND	449.50000	-	146.2 Hz	SNRC
	FM	N6JFO	449.80000	-	131.8 Hz	SNRC
	FM	KI7D	449.85000	-		SNRC
	FM	WB6TNP	927.25000	902.25000	114.8 Hz	SNRC
Las Vegas, Angel Pk						
	FM	N7SGV	147.18000	+	DCS 244	SNRC
Las Vegas, Apex	FM	KG7SS	927.05000	902.05000	DCS 411	SNRC
Las Vegas, Apex Mountain						
	FM	KC7TMC	147.06000	+	100.0 Hz	SNRC
	FM	N7YOR	447.62500	-	114.8 Hz	SNRC
	FM	KD5MSS	447.85000	-	127.3 Hz	SNRC
	FM	KC7TMC	449.87500	-	127.3 Hz	SNRC
Las Vegas, Blue Diamond Hill						
	FM	W7HTL	446.20000	-	77.0 Hz	SNRC
	FM	W7HTL	446.22500	-	85.4 Hz	SNRC
	FM	K6JSI	447.95000	-	100.0 Hz	SNRC
	FM	W7HTL	447.97500	-	82.5 Hz	SNRC
Las Vegas, Blue Diamond Mounta						
	FM	N7ARR	447.00000	-	123.0 Hz	SNRC
Las Vegas, CSN Henderson						
	FM	N4NJJ	445.37500	-	107.2 Hz	SNRC
Las Vegas, Fitzgeralds Hotel						
	FM	WN9ANF	448.07500	-	127.3 Hz	SNRC
Las Vegas, Frenchman Mountain						
	FM	N3TOY	927.11250	902.11250	DCS 432	SNRC
	FM	N7OK	927.28750	902.28750	151.4 Hz	SNRC
Las Vegas, Hi Potosi Mountain						
	FM	WA7HXO	146.88000	-	100.0 Hz	SNRC
	FM	WA6TLW	447.17500	-		SNRC
	FM	N6DD	449.00000	-	131.8 Hz	SNRC
	FM	WA7HXO	449.17500	-	136.5 Hz	SNRC
Las Vegas, Lo Potosi Mountain						
	FM	W7EB	224.48000	-	110.9 Hz	SNRC
	FM	WB6ORK	447.90000	-	156.7 Hz	SNRC
	FM	WR7BLU	448.52500	-	156.7 Hz	SNRC
	FM	KG6ALU	448.82500	-	146.2 Hz	SNRC
	FM	WB6TNP	449.25000	-	141.3 Hz	SNRC

224   NEVADA

Location	Mode	Call sign	Output	Input	Access	Coordinator
Las Vegas, Lo Potosi Mountain						
	FM	W7OQF	449.40000	-	136.5 Hz	SNRC
	FM	WB9STH	449.42500	-		SNRC
	FM	WB6TNP	927.03750	902.03750	141.3 Hz	SNRC
	FM	WB9STH	927.42500	902.42500		SNRC
	FM	KB6XN	1290.00000	1270.00000	100.0 Hz	SNRC
Las Vegas, Low Potosi						
	FM	N6LXX	447.22500	-	110.9 Hz	
Las Vegas, MGM Grand						
	FM	N7RMB	145.30000	-	100.0 Hz	SNRC
Las Vegas, NV	FM	W7JCA	445.87500	-	100.0 Hz	SNRC
Las Vegas, Plaza Hotel						
	FM	K6JSI	145.25000	-	100.0 Hz	SNRC
Las Vegas, QTH	FM	N7ARR	147.00000	-	123.0 Hz	SNRC
	FM	W0JAY	147.10500	+		SNRC
	FM	KE7KD	448.30000	-	118.8 Hz	SNRC
	FM	N8DBM	448.67500	-	77.0 Hz	SNRC
Las Vegas, Red Mountain						
	FM	WA7LAT	449.10000	-	136.5 Hz	SNRC
Las Vegas, Southern Hills Hosp						
	FM	K7UGE	448.50000	-	100.0 Hz	SNRC
Las Vegas, Strip	FM	N7ARR	446.70000	-	118.8 Hz	SNRC
Las Vegas, Summerlin						
	FM	WB6TNP	927.67500	902.67500	82.5 Hz	SNRC
Las Vegas, Suncoast Casino						
	FM	WB6TNP	449.22500	-	141.3 Hz	SNRC
Las Vegas, Sunrise Hospital						
	FM	KC7DB	147.27000	+		SNRC
Las Vegas, Sunrise Mountain						
	FM	W7AOR	449.72500	-	97.4 Hz	SNRC
	FM	N7OK	927.18750	902.18750	151.4 Hz	SNRC
Las Vegas, Westgate Hotel And						
	FM	K7UGE	146.94000	-	100.0 Hz	SNRC
Laughlin	FM	KC6ZTB	147.30000	+	156.7 Hz	
Laughlin, Spirit Mountain						
	FM	K7MPR	146.82000	-	123.0 Hz	SNRC
Lincoln, Highland Peak						
	FM	WA7HXO	145.22000	-	100.0 Hz	SNRC
Logandale, Beacon Hill						
	FM	W7MVR	147.39000	+		SNRC
Lovelock, Tia's Hill						
	FM	KE7INV	145.31000	-		CARCON
Lovelock, Toulon Peak						
	FM	W7TA	146.92500	-	123.0 Hz	CARCON
Mesquite	DMR/BM	N7OKD	446.03750	-	CC15	
	FM	WA7HXO	448.02000	-	136.5 Hz	SNRC
	FM	N7ARR	449.82500	-	123.0 Hz	SNRC
Middlegate, Fairview Peak						
	FM	NV7CC	444.50000	+	123.0 Hz	
Minden	FM	W7DI	147.27000	+		
	FM	W7DI	443.75000	+	123.0 Hz	
Minden, Leviathan Peak						
	FM	NV7CV	147.33000	+	123.0 Hz	
Mission Hills Hend						
	DMR/BM	NX7R	447.62500	-	CC15	
Moapa Valley, Overton						
	DMR/BM	AA4Z	447.05000	-	CC15	
Mountains Edge LV						
	DMR/BM	NX7R	420.62500	+	CC15	
New Washoe City	FM	NH7M	53.00000	103.50000	103.5 Hz	CARCON
	FM	NH7M	145.41000	-	97.4 Hz	CARCON
	FM	NH7M	440.37500	+	97.4 Hz	

NEVADA 225

Location	Mode	Call sign	Output	Input	Access	Coordinator
New Washoe City, McClellan Pea						
	FM	W7RHC	145.31000	-	110.9 Hz	CARCON
North Las Vegas	DMR/BM	W7XM	442.85000	+	CC1	
	DMR/BM	W7XM	445.30000	885.60000	CC15	
	DMR/MARC	W1PAA	149.77000	-	CC1	
	DMR/MARC	W1PAA	444.77000	+	CC1	
	DMR/MARC	W1PAA	449.77000	-	CC15	
	FM	KE7ZHN	145.46000	-	100.0 Hz	SNRC
	FM	WA7CYC	146.67000	-	136.5 Hz	SNRC
	FM	WH6CYB	448.77500	-		SNRC
North Las Vegas , Nellis AFB						
	FM	KP4UZ	447.77500	-	114.8 Hz	SNRC
North Las Vegas, Apex Mountain						
	FM	W7HEN	449.92500	-	131.8 Hz	
Overton	DMR/BM	AA4Z	445.52500	-	CC15	
Overton, Overton	FM	KG7OUI	447.05000	-	114.8 Hz	SNRC
Pahrump	DMR/BM	WB7DRJ	449.27500	-	CC5	
	FM	W7NYE	145.13000	-	100.0 Hz	SNRC
	FM	W7NYE	145.49000	-	100.0 Hz	SNRC
	FM	W6NYK	147.03000	+	100.0 Hz	SNRC
	FM	AD7DP	147.12000	+	94.8 Hz	
	FM	ND7M	147.36000	+		
	FM	N7HYV	223.84000	-		SNRC
	FM	KF7DXU	446.32500	-	141.3 Hz	SNRC
	FM	K6JSI	447.40000	-	100.0 Hz	
	FM	WB6TNP	448.72500	-	131.8 Hz	SNRC
	FM	N7HYV	448.85000	-	127.3 Hz	
	FM	N7ARR	449.75000	-	123.0 Hz	SNRC
	FM	WB6AMT	449.85000	-	156.7 Hz	
Pahrump, Desert View Reg Med C						
	FM	W7NYE	447.70000	-	123.0 Hz	
Pequop, Pequop Summit						
	FM	WA6TLW	444.85000	+	94.8 Hz	
Pronto	FM	WO7I	146.73000	-	88.5 Hz	CARCON
	FM	WO7I	442.82500	+	141.3 Hz	CARCON
Reno	DMR/MARC	W7TA	444.92500	+	CC1	CARCON
	FM	WA6TLW	52.61000	51.61000		
	FM	KD7DTN	145.15000	-	123.0 Hz	CARCON
	FM	KA7ZAU	146.89500	-	100.0 Hz	CARCON
	FM	N7PLQ	147.06000	+	123.0 Hz	
	FM	KK7RON	223.92000	-		
	FM	K7IY	224.06000	-	123.0 Hz	
	FM	W7UIZ	224.10000	-		
	FM	WA7RPS	224.42000	-	88.5 Hz	
	FM	KH6UG	440.00000	+	110.9 Hz	
	FM	WA6CBA	440.10000	+	123.0 Hz	
	FM	KD7DTN	441.30000	+	114.8 Hz	
	FM	WA6TLW	442.17500	+	127.3 Hz	
	FM	W7NIK	442.55000	+	110.9 Hz	
	FM	N7TGB	443.17500	+	123.0 Hz	
	FM	K7VI	443.60000	+	103.5 Hz	
	FM	K7VI	443.70000	+	103.5 Hz	
	FM	KR7EK	444.42500	+		
	FM	KE7KD	444.50000	+	114.8 Hz	
	FM	N7PLQ	444.52500	+	146.2 Hz	
	FM	W7NIK	445.00000	-	123.0 Hz	
Reno, Airport	FM	W7NIK	442.37500	+		
Reno, Chimney Peak						
	FM	N7PLQ	146.55000			CARCON
Reno, Grand Sierra Resort Hote						
	FM	W7TA	147.30000	+	123.0 Hz	CARCON

226 NEVADA

Location	Mode	Call sign	Output	Input	Access	Coordinator
Reno, Lemon Valley						
	FM	KB6TDJ	52.90000	51.90000	107.2 Hz	CARCON
	FM	KB6TDJ	440.20000	+	107.2 Hz	
Reno, Lower Peavine Peak						
	FM	K6LNK	440.75000	+		
	IDAS	NT7Q	442.25000	+	141.3 Hz	CARCON
Reno, McClellen Peak						
	FM	KR7EK	444.25000	+	146.2 Hz	
Reno, Mt. Rose Relay Station						
	FM	AE7I	224.58000	-		CARCON
Reno, Northern Nevada Hosp						
	FM	KE7R	444.35000	+	123.0 Hz	
Reno, Olinghouse	FM	W7NV	51.80000	50.80000	110.9 Hz	CARCON
Reno, Ophir	FM	W7TA	147.39000	+	123.0 Hz	CARCON
Reno, Ophir Hill	FM	W7TA	146.61000	-	123.0 Hz	CARCON
	FM	W7TA	146.67000	-		
	FM	W7TA	147.00000	+		
	FM	N7PLQ	443.05000	+	123.0 Hz	
	FM	W7TA	443.07500	+	123.0 Hz	
Reno, Peavine Lookout						
	FM	W7UIZ	145.39000	-		CARCON
Reno, Peavine Peak						
	FM	W7OFT	146.73000	-	123.0 Hz	
	FM	K7AN	146.76000	-	123.0 Hz	
	FM	W7TA	147.21000	+	100.0 Hz	CARCON
	FM	AE7I	224.54000	-		CARCON
	FM	N7TUA	440.07500	+		
	FM	N7ARR	441.65000	+	123.0 Hz	
	FM	W7TA	444.12500	+	123.0 Hz	CARCON
	FM	N7PLQ	444.80000	+	123.0 Hz	
	FM	N7PLQ	444.97500	+	146.2 Hz	CARCON
Reno, Peppermill Hotel						
	FM	WA7RPS	444.02500	+		CARCON
Reno, Pond Peak	FM	K7AN	145.45000	-	123.0 Hz	CARCON
	FM	K7AN	444.40000	+	118.8 Hz	
Reno, QTH	FM	WA7RPS	440.12500	+		
	FM	KD7DTN	440.72500	+	123.0 Hz	CARCON
	FM	W9CI	444.60000	+		
Reno, Red Peak	FM	W7UNR	145.29000	-	123.0 Hz	CARCON
	FM	N7PLQ	440.02500	+	123.0 Hz	
	FM	WA7NHJ	443.12500	+	136.5 Hz	
Reno, RHC	FM	W7RHC	145.21000	-		CARCON
Reno, Slide Mountain						
	FM	N7VXB	146.94000	-	123.0 Hz	
	FM	WA7NHJ	440.15000	+	110.9 Hz	
	FM	W6KCS	442.02500	+	156.7 Hz	
	FM	WA6TLW	444.65000	+	94.8 Hz	
	FM	WA6TLW	905.90000	930.90000		
	FM	W6CYX	1282.00000	1262.00000		
Reno, VA Hosp	FM	KE7R	147.12000	+	123.0 Hz	
Reno, Virginia City						
	FM	W7TA	147.39000	+	100.0 Hz	CARCON
Reno, Virginia Peak						
	FM	WA7WOP	147.18000	+		
	FM	WA6TLW	444.85000	+	94.8 Hz	
Reno, West Of The Reno-Stead A						
	FM	K7PTT	445.50000	-	77.0 Hz	
Reno, Windy Hill	FM	W7UIZ	145.37000	-		
Schurz, Bald Mountain						
	FM	K7UI	441.90000	+	123.0 Hz	
Silver Springs	FM	KE6QK	146.97000	-	103.5 Hz	CARCON
So Lake Tahoe	FM	W6SUV	442.82500	+	88.5 Hz	CARCON

NEVADA   227

Location	Mode	Call sign	Output	Input	Access	Coordinator
South Lake Tahoe						
	DMR/BM	WA6EWV	442.47500	+	CC3	CARCON
	FM	K5BLS	442.30000	+	156.7 Hz	CARCON
Spanish Springs, Sparks, Virgi						
	FM	W7TA	147.03000	+	123.0 Hz	CARCON
Sparks	FM	N7KP	52.80000	51.80000	123.0 Hz	CARCON
	FM	KK7RON	146.86500	-		
	FM	N7VN	147.09000	+	127.3 Hz	
	FM	K1SER	147.36000	+	100.0 Hz	
	FM	KK7RON	443.40000	+	103.5 Hz	
	FM	N7KP	443.62500	+	103.5 Hz	
	FM	N7KP	443.80000	+	123.0 Hz	
	FM	KD7DTN	444.22500	+		
	FM	N7PLQ	444.80000	+	123.0 Hz	
	FM	N7KP	927.35000	902.35000	114.8 Hz	
Sparks / Fernley	DMR/MARC	W7TA	442.87500	+	CC1	
Sparks, Liberty Hill						
	FM	N7KP	145.23000	-		CARCON
Spring Creek	FM	KB7IWQ	444.20000	+		
Spring Valley	FM	WB9STH	447.42500	-	173.8 Hz	SNRC
Sun Valley	FM	KC7STW	440.30000	+	141.3 Hz	
Sun Valley, McCellan Peak						
	FM	KC7STW	145.27000	-	141.3 Hz	CARCON
Tonopah	FM	N7ARR	146.64000	-	123.0 Hz	SNRC
	FM	KB7PPG	147.12000	+	114.8 Hz	
Touplon	FM	K6ALT	445.55000	-		
Tuscarora	FM	KD7CWA	147.30000	+	100.0 Hz	
	FM	W7LKO	444.50000	+	100.0 Hz	
	FM	WA6TLW	444.65000	+	94.8 Hz	
Upper Potosi	DMR/BM	KB6CRE	449.95000	-	CC1	SNRC
Vacaville	FM	W6KCS	927.06250	902.06250	100.0 Hz	CARCON
	FM	KI6SSF	927.33750	902.33750	162.2 Hz	CARCON
VC Highlands, Comstock Memoria						
	FM	K7RC	146.86500	-	123.0 Hz	CARCON
	FM	K7RC	441.62500	+	114.8 Hz	
Virginia City	DMR/MARC	W7TA	444.82500	+	CC1	CARCON
	FM	WA7UEK	145.47000	-	123.0 Hz	CARCON
Virginia City, McClellan Peak						
	FM	W6JA	145.49000	-	123.0 Hz	
Virginia City, Ophir Hill						
	FM	KC7ARS	444.37500	+	156.7 Hz	CARCON
VoTech SE. Las Vegas						
	DMR/BM	NX7R	445.62500	-	CC15	SNRC
Walker	FM	N7TR	443.27500	+		
Warm Springs, Warm Springs Sum						
	FM	WB7WTS	146.85000	-		
Washoe Valley	DMR/BM	KM6CQ	440.27500	+	CC1	
	FM	W7RHC	440.55000	+		CCARC
Wellington, Lobdell Peak						
	FM	KD7NHC	146.88000	-	123.0 Hz	
	FM	KD7NHC	440.05000	+		
Wells	FM	W7LKO	146.96000	-	100.0 Hz	
	FM	W7LKO	444.90000	+		
Winnemucca, Winnemucca Mountai						
	FM	W7TA	146.67000	-	123.0 Hz	CARCON
Yerington	FM	WA6JQV	444.30000	+		
Yerington, Lobdell Summit						
	FM	W7DED	444.87500	+	100.0 Hz	CARCON

NEW HAMPSHIRE

Location	Mode	Call sign	Output	Input	Access	Coordinator
Acworth	DMR/BM	WX1NH	446.42500	-	CC5	
Alton	FM	K1JEK	146.86500	-		

228 NEW HAMPSHIRE

Location	Mode	Call sign	Output	Input	Access	Coordinator
Alton	FM	K1JEK	444.05000	+	88.5 Hz	
Amherst	FM	K1ZQ	224.02000	-	136.5 Hz	
Bedford	DMR/BM	W3UA	442.75000	+	CC7	
	FM	N1QC	146.68500	-		
Berlin, Cates Hill	FM	W1COS	145.68500	+	100.0 Hz	
Berlin, Mount Washington						
	FM	WA1PBJ	448.22500	-	88.5 Hz	NESMC
Bethlehem, Mt Agassiz						
	FM	N1PCE	442.95000	+		
Bow	DMR/MARC	K1OX	145.17000	-	CC8	
	FM	KB1QV	53.15000	52.15000	71.9 Hz	
Bow, Quimby Mountain						
	FM	KA1SU	447.32500	-	88.5 Hz	
Brattleboro	DMR/MARC	WR1VT	444.40000	+	CC1	NESMC
Brattleboro, Schofield Mtn						
	FM	WR1VT	146.86500	-	100.0 Hz	NESMC
Brookline	FM	N1IMO	53.41000	52.41000	88.5 Hz	
Center Barnstead	FM	K1DED	446.47500	-	88.5 Hz	NESMC
Charlestown	FM	NX1DX	146.92500	-	118.8 Hz	
Chester	DMR/MARC	K1OX	145.19000	-	CC9	
	FM	K1OX	224.20000	-		
	FM	N1IMO	224.50000	-	88.5 Hz	
	FM	K1JC	442.55000	+	88.5 Hz	
Chesterfield	FM	KK1CW	223.98000	-	100.0 Hz	
Claremont	DMR/BM	WX1NH	446.32500	-	CC7	
	FM	WX1NH	53.07000	52.07000	97.4 Hz	
	FM	KU1R	147.28500	+	103.5 Hz	
	FM	WX1NH	224.06000	-	97.4 Hz	
	FM	KU1R	443.95000	+	103.5 Hz	
Clarksville	DMR/MARC	W1COS	145.25000	-	CC0	NESMC
Clarksville, Ben Young Hill						
	FM	KB1IZU	146.71500	-	100.0 Hz	
Colebrook	FM	W1HJF	147.30000	+	110.9 Hz	
Concord	DMR/MARC	KB1CFL	145.42000	-	CC7	
Crotched Mtn	DMR/MARC	WA1ZYX	446.97500	-	CC10	
Deerfield, Saddleback Mountain						
	FM	W1SRA	147.00000	-	100.0 Hz	
Deerfield, WENH Tower						
	FM	WA1ZYX	449.45000	-	123.0 Hz	
Derry	DMR/MARC	K1QVC	145.31000	-	CC1	
	DSTAR	K1QVC	447.37500	-	85.4 Hz	
	FM	K1LVA	145.71000		100.0 Hz	
	FM	NM1D	146.74500	-	114.8 Hz	
	FM	K1CA	146.85000	-	85.4 Hz	
	FM	KC2LT	147.21000	+	107.2 Hz	
	FM	KC2LT	441.30000	+	107.2 Hz	
	FM	W1AJI	441.55000	+	127.3 Hz	
	FM	N1VQQ	447.82500	-	88.5 Hz	
	FM	K1CA	449.62500	-	85.4 Hz	
Derry, Warner Hill	DSTAR	NN1PA	447.22500	-		
Dover	DMR/BM	KC1AWV	441.47500	+	CC1	
Dublin, Snow Hill	FM	WQ2H	441.70000	+	100.0 Hz	NESMC
East Kingston	FM	KC1EWP	441.35000	+	88.5 Hz	
East Unity	DMR/BM	WX1NH	446.52500	-	CC7	
Enfield	FM	KA1UAG	444.90000	+	131.8 Hz	
Epsom, Fort Mountain						
	P25	W1ASS	443.85000	+		
Exeter	FM	K1KN	224.22000	-	67.0 Hz	
Francestown	FM	KA1BBG	147.06000	+	123.0 Hz	
Franconia	DMR/BM	K1EME	145.43000	-	CC1	NESMC
Franklin, Veterans Memorial Sk						
	DSTAR	NE1DS	145.48000	-		

NEW HAMPSHIRE   229

Location	Mode	Call sign	Output	Input	Access	Coordinator
Franklin, Veterans Memorial Sk						
	FM	W1JY	147.30000	+	88.5 Hz	NESMC
Fremont	DSTAR	N1HIT	448.87500	-		NESMC
	FM	N1HIT	145.38000	-		
Gilford	DMR/MARC	K1RE	449.42500	-	CC3	NESMC
Gilford, Gunstock Mountain						
	DSTAR	W1CNH	447.77500	-		
	FM	W1JY	146.98500	-	123.0 Hz	
	WX	K1RJZ	53.77000	52.77000	71.9 Hz	NESMC
Gilsum	FM	KA1BBG	147.36000	+	123.0 Hz	
Goffstown	DMR/MARC	NN1PA	145.20000	-	CC10	
	DMR/MARC	KM3T	444.30000	+	CC10	NESMC
	DSTAR	NE1DV	446.57500	-		
	FM	W1AKS	147.13500	+	100.0 Hz	
	FM	W1ASS	927.71250	902.71250	100.0 Hz	
Goffstown, Mt Uncanoonic						
	FM	NN1PA	444.20000	+	186.2 Hz	
Gorham, Mount Washington						
	FM	W1NH	146.65500	-	100.0 Hz	NESMC
Greenfield	FM	KA1BBG	448.52500	-	123.0 Hz	
Greenland	FM	KB1UVE	446.72500	-		
Gunstock Mtn	DMR/MARC	K1RE	145.36000	-	CC3	
	DMR/MARC	K1RJZ	447.87500	-	CC9	
Hampton Beach	DSTAR	K1HBR	145.44000	-		
	DSTAR	K1HBR	449.47500	-		
Hanover	FM	W1FN	443.55000	+	136.5 Hz	
	FM	W1ET	444.95000	+	88.5 Hz	
Hanover, Moose Mountain						
	FM	W1FN	145.33000	-	100.0 Hz	
Henniker, Pats Peak						
	FM	K1BKE	146.89500	-	100.0 Hz	NESMC
Hollis	FM	N1IMO	146.73000	-	88.5 Hz	
	FM	N1IMO	443.50000	+	88.5 Hz	
	FM	N1VQQ	444.25000	+	107.2 Hz	
Hudson	DMR/MARC	K1MOT	145.26000	-	CC5	
	DMR/MARC	NE1B	147.10500	+	CC0	
	DMR/MARC	K1MOT	447.72500	-	CC1	
	DSTAR	KB1UAP	449.97500	-		
	FM	N1VQQ	53.97000	52.97000	100.0 Hz	
	FM	KC2LT	448.27500	-	107.2 Hz	
Keene	DMR/MARC	WA1ZYX	444.65000	+	CC1	
	FM	WA1ZYX	53.73000	52.73000	141.3 Hz	
Kensington	DSTAR	KB1TIX	145.40000	144.60000		
	FM	W1WQM	145.15000	-	127.3 Hz	
	FM	W1WQM	444.40000	+	100.0 Hz	
Kensington, Waymouth Hill						
	FM	KB1VTL	443.45000	+	88.5 Hz	
Lancaster	FM	KC1SR	444.85000	+		NESMC
Lincoln, Cannon Mtn						
	FM	K1EME	224.08000	-	114.8 Hz	
Littleton	FM	K1EME	147.34500	+	114.8 Hz	
Londonderry	FM	K1DED	442.00000	+	100.0 Hz	
Madbury	FM	N1HIT	448.87500	-		NESMC
Manchester	DMR/MARC	W1RCF	145.22000	-	CC11	
	FM	N1SM	147.33000	+	141.3 Hz	
Mason	FM	K1TLV	443.75000	+		
Milford	FM	K3RQ	146.44500	147.44500	123.0 Hz	
Mont Vernon	FM	K3RQ	448.42500	-	88.5 Hz	
Moultonborough	FM	N1TZE	145.31000	-	88.5 Hz	
	FM	N1EMS	147.25500	+	156.7 Hz	
Moultonborough, Red Hill						
	FM	W1JY	147.39000	+	123.0 Hz	

230 NEW HAMPSHIRE

Location	Mode	Call sign	Output	Input	Access	Coordinator
Mount Vernon	FM	WA1HCO	447.12500	-	88.5 Hz	
Mt. Washington	DMR/MARC	W1IMD	448.97500	-	CC2	
Nashua	DMR/BM	N1DAS	446.42500	-	CC5	
	FM	WW1Y	147.04500	+	100.0 Hz	
	FM	N1KXT	444.80000	+	131.8 Hz	
	FM	N1IMO	448.82500	-	88.5 Hz	
New Boston, Chestnut Hill						
	FM	W1VTP	147.37500	+	88.5 Hz	
New Ipswich	FM	N3LEE	443.15000	+	71.9 Hz	
New London	FM	W1VN	145.25000	-	88.5 Hz	
North Conway	DMR/MARC	W1MWV	448.77500	-	CC2	
North Conway, Mt Cranmore						
	WX	W1MWV	145.45000	-	100.0 Hz	
Northwood	FM	K1JEK	146.70000	-	88.5 Hz	
Ossipee	FM	W1BST	51.64000	-	131.8 Hz	
	FM	W1BST	147.03000	+	88.5 Hz	
	FM	W1WU	224.60000	-	123.0 Hz	
Pack Monadnock	FM	KA1OKQ	443.35000	+	110.9 Hz	
	FM	W1XOJ	448.00000	-	203.5 Hz	
	FM	N1IMO	449.37500	-	88.5 Hz	
Pelham	DMR/BM	W1STT	146.50000	145.00000	CC1	
Pembroke	FM	KA1OKQ	147.22500	+	100.0 Hz	
	FM	N1KXT	443.65000	+	131.8 Hz	
Pembroke, QTH	FM	W1ALE	146.94000	-	114.8 Hz	NESMC
Peterborough	FM	N1IMO	147.19500	+	88.5 Hz	
Peterborough, Temple Mountain						
	FM	WA1ZYX	447.42500	-	141.3 Hz	
Pittsfield	FM	N1IMO	146.79000	-	88.5 Hz	
	FM	N1AKE	224.54000	-	103.5 Hz	
Portsmouth	FM	KB1ZDR	441.95000	+		
Rindge	FM	WA1UNN	146.77500	-	123.0 Hz	NESMC
	FM	WA1HOG	223.92000	-	100.0 Hz	
Rochester	DMR/MARC	K1LTM	145.24000	-	CC3	
Salem	DSTAR	K1HRO	444.35000	+		NESMC
	DSTAR	K1HRO	1293.00000	1273.00000		
	FM	NY1Z	147.16500	+	136.5 Hz	
	FM	NY1Z	449.77500	-		
Salem, HRO	DSTAR	K1HRO	145.32000	-		
Sanbornton	DMR/MARC	K1JC	145.18000	-	CC6	
Sanbornton, Steele Hill Resort						
	DSTAR	W1VN	449.67500	-		NESMC
	FM	W1JY	146.67000	-	123.0 Hz	
Seabrook	FM	WA1NH	146.61000	-	141.3 Hz	NESMC
Somersworth	DMR/MARC	W1WNS	145.18000	-	CC5	
Stratford	FM	KC1FZQ	443.95000	+	91.5 Hz	
Sunapee	FM	K1JY	442.35000	+	88.5 Hz	
Surry	FM	KB1HPK	224.98000	-	123.0 Hz	NESMC
Unity	FM	KA1BBG	147.18000	+	123.0 Hz	
Wakefield	DMR/MARC	K1LTM	145.28000	-	CC7	
Walpole	FM	WA1ZYX	443.80000	+	141.3 Hz	
Walpole, WEKW Tower						
	FM	K1PH	147.03000	+		
West Lebanon	FM	KA1UAG	443.50000	+	131.8 Hz	NESMC
West Ossipee	DMR/MARC	K1LTM	147.07500	+	CC6	
Westmoreland	FM	K1TQY	146.80500	-	100.0 Hz	
Whitefield	DMR/BM	N1PCE	442.30000	+	CC1	
	FM	N1PCE	145.37000	-	114.8 Hz	NESMC
	FM	N1PCE	449.82500	-	82.5 Hz	NESMC
Windham	DSTAR	KC1EGN	146.64000	145.14000		

ADVANCED SPECIALTIES INC.

New Jersey's Communications Store

YAESU
The radio
Authorized Dealer

ASTATIC • OPEK • COMET • MALDOL • MFJ
UNIDEN • ANLI • YAESU • WHISTLER • PROCOMM

Yaesu FTM-7250D
Digital
Dual Band

Yaesu FT-3DR
2m/70cm
Dual-Band
Digital Handheld
Transceiver

FT-857D
HF, 6M, 2M, 70 CM

VX-6R
Tri Band
Submersible HT

New Uniden SDS100
True IQ
Phase II
Digital Scanner

FT-2980R
80W 2M Mobile

Closed Sunday & Monday
**Orders/Quotes
(201)-VHF-2067**
114 Essex Street ■ Lodi, NJ 07644

Visit us at:
www.advancedspecialties.net

AMATEUR RADIO – SCANNERS – BOOKS – ANTENNAS
MOUNTS – FILTERS
ACCESSORIES AND MORE!

232   NEW JERSEY

Location	Mode	Call sign	Output	Input	Access	Coordinator
NEW JERSEY						
Absecon	FM	N2HQX	147.21000	+	123.0 Hz	ARCC
Allenwood	FM	N2MO	145.11000	-	127.3 Hz	MetroCor
Alpine	FM	W2VH	224.98000	-	107.2 Hz	MetroCor
	FM	K2FJ	442.90000	+		MetroCor
	FM	WB2ZZO	444.20000	+	136.5 Hz	
Alpine, Armstrong Tower						
	FM	W2MR	442.70000	+	97.4 Hz	MetroCor
Atlantic City	DMR/BM	AG2NJ	449.55630	-	CC1	
	DMR/MARC	K2ACY	445.23125	-	CC1	ARCC
	FM	KC2GUM	146.44500	147.44500		
	FM	AA2BP	444.35000	+	107.2 Hz	ARCC
Avon	FM	KB2MMR	442.15000	+	141.3 Hz	
Barnegat	FM	N2NF	224.28000	-		ARCC
	FM	N2AYM	927.83750	902.83750	162.2 Hz	ARCC
Bayonne	FM	W2ODV	145.43000	-	123.0 Hz	
	FM	W2CTL	446.62500	-	141.3 Hz	
Bayville	FM	K2HES	145.31000	-	141.3 Hz	
	FM	N2IXU	448.47500	-	74.4 Hz	
Beach Haven	FM	WA2NEW	448.57500	-	141.3 Hz	ARCC
Beachwood	FM	N2MDX	441.50000	+		
Belle Mead	FM	KD2ARB	442.40000	-	123.0 Hz	
Belle Meade	FM	KB2EAR	224.04000	-	151.4 Hz	
	FM	KB2EAR	927.36250	902.36250	141.3 Hz	
Bergenfield	DSTAR	K9GTM	145.42000	-		
Bloomsbury	FM	N3MSK	449.57500	-	151.4 Hz	ARCC
	FM	N3MSK	449.58000	-	151.4 Hz	
Blue Anchor	FM	KB2AYS	445.12500	-	91.5 Hz	
Boonton	DMR/BM	N2WNS	449.77500	-	CC1	
Boonton, Sheep Hill						
	FM	W2TW	147.03000	+	151.4 Hz	MetroCor
Brick	DMR/MARC	W2NJR	146.49000	147.49000	CC1	
Bricktown	FM	K2RFI	146.49000	147.49000	141.3 Hz	
Bridgewater	FM	WA2OCN	444.95000	+	141.3 Hz	ARCC
Brigantine Island	FM	K2ACY	447.57500	-	156.7 Hz	ARCC
Brown Mills	DMR/BM	K2JZO	449.02500	-	CC1	ARCC
Browns Mills	FM	K2JZO	224.86000	-	131.8 Hz	ARCC
	FM	K2JZO	449.67500	-	141.3 Hz	ARCC
Budd Lake	DMR/MARC	K2DMR	440.80000	+	CC1	
	FM	WR2M	223.86000	-	136.5 Hz	
	FM	WS2P	448.67500	-		MetroCor
Camden	FM	N3KZ	145.43000	-	79.7 Hz	
	FM	N2KDV	442.15000	+		ARCC
	FM	WB3EHB	444.30000	+	131.8 Hz	ARCC
	FM	N2HQX	448.02500	-	131.8 Hz	ARCC
Cape May	DMR/BM	N2ICV	443.05000	+	CC1	
	FM	KC2JPP	449.87500	-	146.2 Hz	ARCC
Cape May Court House						
	FM	W2CMC	147.24000	+	146.2 Hz	ARCC
	FM	W2CMC	442.00000	+	146.2 Hz	
	FM	W2CMC	443.60000	+	146.2 Hz	
	FM	NJ2DS	447.47500	-		ARCC
Carteret	FM	K2ZV	447.67500	-	136.5 Hz	MetroCor
Cedar Grove	DSTAR	W2DGL	146.44500	147.44500		
Chatsworth	FM	KC2QVT	145.47000	-	127.3 Hz	ARCC
Cherry Hill	DMR/BM	KB3MMJ	432.10000		CC1	
	FM	NJ2CH	145.37000	-	91.5 Hz	ARCC
Cherryville	FM	N3MSK	53.25000	52.25000	146.2 Hz	ARCC
	FM	WB2NQV	147.37500	+	151.4 Hz	ARCC
	FM	K2PM	224.60000	-	203.5 Hz	ARCC
	FM	W2CRA	444.85000	+	141.3 Hz	ARCC

NEW JERSEY 233

Location	Mode	Call sign	Output	Input	Access	Coordinator
Cinnaminson	FM	K2CPD	445.62500	-	127.3 Hz	ARCC
Clark	FM	NJ5R	444.70000	+	88.5 Hz	MetroCor
Cliffside Park	FM	N2OFY	445.77500	-	141.3 Hz	
	FM	KB2OOJ	447.42500	-	127.3 Hz	MetroCor
	FM	WA2YYX	447.57500	-		
Clifton, Bohn Hall MSU						
	FM	KB2N	224.36000	-	141.3 Hz	MetroCor
Collings Lakes	FM	N3YYZ	444.85000	+	91.5 Hz	
Columbia, Hainesburg						
	FM	WB2NMI	146.47500	147.47500	110.9 Hz	
Corbin City	DMR/MARC	WR3IRS	440.40000	+	CC1	ARCC
	FM	W2FLY	440.75000	+		ARCC
	FM	N3KZ	441.35000	447.35000	131.8 Hz	ARCC
Denville	FM	KD2EKH	449.37500	-	141.3 Hz	ARCC
Denville, St Clares Hospital						
	FM	KC2DEQ	442.05000	+	123.0 Hz	
Dover	FM	N4TCT	224.72000	-	82.5 Hz	ARCC
Eagleswood Township						
	FM	KA2PFL	442.75000	+	131.8 Hz	ARCC
Egg Harbor City	FM	W3BXW	53.91000	52.91000	131.8 Hz	ARCC
	FM	W3BXW	146.64000	-	131.8 Hz	ARCC
	FM	AG2NJ	147.16500	+	91.5 Hz	
Egg Harbor Township						
	FM	K2BR	448.77500	-	118.8 Hz	ARCC
	FM	KD2KVZ	448.97500	-	123.0 Hz	ARCC
Egg Harbor Township, Atlantic						
	FM	K2BR	146.74500	-	146.2 Hz	ARCC
Elizabeth	DMR/BM	KB2OOJ	448.52500	-	CC1	
	DMR/MARC	K2ZZ	449.77500	-	CC3	
	FM	W2JDS	145.41000	-	107.2 Hz	MetroCor
	FM	K2ETS	443.15000	+	141.3 Hz	ARCC
Elizabeth , Trinitas Regional						
	FM	WB2CMN	442.40000	+	141.3 Hz	
Ellisdale	FM	K2NI	224.16000	-	131.8 Hz	MetroCor
	FM	K2NI	447.53000	-	123.0 Hz	MetroCor
Fair Lawn	DSTAR	W2KBF	145.68000			
Fort Lee	FM	W2MPX	145.45000	-	141.3 Hz	
	FM	W2QAQ	224.24000	-	107.2 Hz	
	FM	K2QW	442.95000	+	141.3 Hz	
Franklin Lakes	FM	W2IP	441.30000	+	114.8 Hz	
Frenchtown	FM	K2PM	448.12500	-	151.4 Hz	ARCC
Galloway	DMR/MARC	AG2NJ	446.02500	-	CC1	
	FM	KC2TGB	444.65000	+		
Glassboro, Rowan University						
	FM	KD2LNB	440.10625	+		
Glen Gardner	DMR/BM	KD2DMU	446.50000	+	CC1	
Glen Gardner, Mt Kipp						
	FM	WB2NQV	147.01500	+		
	FM	K2PM	224.12000	-		ARCC
	FM	W2CRA	446.47500	-	141.3 Hz	
Gloucester City	FM	NJ2GC	447.77500	-		ARCC
Green Brook	DMR/MARC	W2QW	442.25000	+	CC1	MetroCor
	FM	WB2BQW	145.25000	-	100.0 Hz	MetroCor
	FM	N2NSV	444.50000	+	131.8 Hz	MetroCor
Greenbrook	FM	W2FUV	224.24000	-	141.3 Hz	
Greenwich	FM	N3KZ	443.70000	+	131.8 Hz	
Guttenberg	DMR/MARC	N2DXZ	444.83750	+	CC1	
Hackensack	FM	W2AKR	444.10000	+	141.3 Hz	
Hackettstown, Strand Theater/H						
	FM	WW2BSA	448.07500	-	141.3 Hz	ARCC
Haledon	FM	WB2CKD	442.30000	+	141.3 Hz	ARCC

234   NEW JERSEY

Location	Mode	Call sign	Output	Input	Access	Coordinator
Hamburg, Hamburg Mt						
	FM	N2BEI	446.32500	-	151.4 Hz	
Hardyston	FM	W2VER	51.72000	50.72000	136.5 Hz	ARCC
	FM	W2VER	449.08000	-	141.3 Hz	ARCC
Harrington Park, Oradell Reser						
	DSTAR	K2MCI	145.55000			
	FM	K2MCI	439.92500			
Hasbrouck Heights						
	FM	K2OMP	442.50000	+	141.3 Hz	ARCC
Hawthorne	FM	WA2CAI	444.90000	+	114.8 Hz	ARCC
Hillsborough	FM	K2NJ	147.13500	+	151.4 Hz	MetroCor
Hillsborough, Sourland Mountai						
	FM	N2BEI	449.32500	-		
Hoboken	DMR/BM	K2XDX	446.52500	-	CC8	
	FM	K2XDX	223.88000	-	250.3 Hz	
Holland Township	FM	WA2GWA	146.85000	-	151.4 Hz	ARCC
Hopatcong	DMR/MARC	K2DMR	446.77500	-	CC1	
	DMR/MARC	N2VUG	449.45000	-	CC1	
	FM	N2QJN	224.28000	-	88.5 Hz	ARCC
	FM	N2OZO	448.17850	-	141.3 Hz	
Hopatcong Borough						
	FM	K2SRT	446.77500	-		
	FM	N2OZO	448.17500	-		ARCC
Howell	DMR/MARC	KB2RF	440.30000	+	CC1	MetroCor
Hudson	P25	KC2GOW	146.88000	-		
Jackson	DMR/MARC	K2NYX	443.61250	+	CC6	
	FM	N2RDM	224.30000	-	127.3 Hz	ARCC
Jefferson	DMR/MARC	K2DMR	448.62500	-	CC1	
	FM	WR2M	53.39000	52.39000	146.2 Hz	
	FM	WR2M	440.85000	+	94.8 Hz	MetroCor
Jersey City	FM	NY4Z	440.62500	+	74.4 Hz	MetroCor
Jersey City, Christ Hospital						
	FM	N2DCS	441.20000	+		MetroCor
Kendall Park	FM	KB2EAR	444.91250	+	141.3 Hz	
Keyport	FM	KB2SEY	224.96000	-		ARCC
Lacey, Bamber Lakes Fire Co						
	FM	KC2GUM	146.44500	147.44500	131.8 Hz	
Lake Hopatcong	DSTAR	NJ2MC	145.18000	-		MetroCor
	DSTAR	NJ2MC	441.60000	+		MetroCor
	FM	KC2DEQ	51.70000	-	123.0 Hz	
	FM	WR2M	53.39000	52.39000		ARCC
	FM	WA2EPI	224.62000	-	107.2 Hz	MetroCor
Lakehurst	FM	W2DOR	443.35000	+	141.3 Hz	
	FM	W3BXW	447.22500	-	131.8 Hz	ARCC
	FM	NJ2AR	927.32500	902.32500		
Lakewood	FM	W2RAP	146.95500	-	103.5 Hz	
	FM	N2AYM	223.82000	-	162.2 Hz	ARCC
	FM	NE2E	449.37500	-	141.3 Hz	ARCC
	FM	NE2E	449.38000	-	123.0 Hz	
	FM	N2AYM	1295.00000	1275.00000	127.3 Hz	ARCC
Landing	DMR/BM	KE2GKB	438.80000	-	CC1	
Lawrenceville, Ch.52 WNJT TOWE						
	FM	N2RE	146.46000	147.46000	131.8 Hz	ARCC
Leesburg	FM	WT2Y	147.00000	+	110.9 Hz	ARCC
Lindenwold	FM	K2EOC	440.24375	+		
Little Falls	FM	W2XTV	443.05000	+		MetroCor
Little Falls, Monclair State U						
	FM	WO2X	443.45000	+		ARCC
Little Ferry	FM	W2NIW	441.85000	+	136.5 Hz	
Livingston	FM	NE2S	146.59500	147.59500	127.3 Hz	MetroCor
Mahwah	P25	KD2IBK	442.22500	+	141.3 Hz	
Manahawkin	DMR/MARC	W2NJR	445.07500	-	CC1	

NEW JERSEY 235

Location	Mode	Call sign	Output	Input	Access	Coordinator
Manahawkin	DMR/MARC	K2HR	445.42500	-	CC3	
	DMR/MARC	WA3BXW	448.07500	-	CC1	ARCC
	FM	N2OO	146.83500	-	127.3 Hz	
Manasquan	FM	N2IXU	445.67500	-	94.8 Hz	
Manchester, Manchester Water T						
	FM	WA2RES	145.17000	-	131.8 Hz	ARCC
Mantua	FM	W2FHO	449.97500	-	131.8 Hz	ARCC
Martinsville	DMR/MARC	W2NJR	147.28500	+	CC1	
	DMR/MARC	K1DO	447.07500	-	CC1	MetroCor
	DSTAR	NJ2DG	145.14000	-		
	DSTAR	NJ2DG	441.65000	+		MetroCor
	DSTAR	NJ2DG	1250.50000			
	DSTAR	NJ2DG	1284.00000	1264.00000		MetroCor
	FM	N2ZAV	224.64000	-	151.4 Hz	MetroCor
	FM	WX3K	224.88000	-	103.5 Hz	MetroCor
	FM	N3MSK	445.72500	-	136.5 Hz	
	FM	WA2OCN	448.18000	-	141.3 Hz	MetroCor
Medford	FM	K2AA	145.29000	-	91.5 Hz	ARCC
Middletown	FM	AA2OW	145.48500	-	151.4 Hz	ARCC
	FM	N2DR	448.72500	-	151.4 Hz	MetroCor
Millville	FM	W2SCR	449.62500	-	123.0 Hz	ARCC
Minotola	FM	KE2CK	146.80500	-	118.8 Hz	ARCC
	FM	KE2CK	448.92500	-	192.8 Hz	
Monroe Township	FM	KA2CAF	53.71000	52.71000		
Monroe Township, Englishtown S						
	FM	KA2CAF	224.50000	-	131.8 Hz	
Montclair-Off Air	DMR/MARC	N2DCE	447.22500	-	CC1	
Montgaue	FM	N2KMB	441.35000	+	192.8 Hz	
Montvale	FM	K2ZD	446.97500	-	141.3 Hz	MetroCor
Morristown	FM	WS2Q	145.37000	-	151.4 Hz	
	FM	WS2Q	224.94000	-	107.2 Hz	
	FM	WS2Q	443.25000	+	141.3 Hz	
Mount Arlington	FM	WB2SLJ	224.72000	-	141.3 Hz	ARCC
Mount Freedom	FM	WS2Q	146.89500	-	151.4 Hz	MetroCor
Mount Laurel, BCC						
	DSTAR	KC2QVT	445.33125	-		
Mountain View	DMR/MARC	N2WNS	439.78750	-	CC1	MetroCor
Murray Hill	DMR/MARC	W2NJR	147.25500	+	CC1	
	DMR/MARC	W2NJR	449.97500	-	CC1	
	FM	W2LI	147.25500	+	141.3 Hz	MetroCor
	FM	W2LI	449.97500	-	141.3 Hz	MetroCor
Mystic Island	FM	KA2PFL	449.47500	-	131.8 Hz	ARCC
New Brunswick	FM	NE2E	440.45000	+	123.0 Hz	
Newark	DMR/BM	N2DMJ	443.09000	+	CC1	
	DMR/BM	W2RLA	445.25000	-	CC2	
	FM	KD2HQY	145.12000	-	91.5 Hz	
	FM	W2RLA	145.35000	-	114.8 Hz	MetroCor
	FM	K2MFF	147.22500	+	141.3 Hz	MetroCor
	FM	WB2MFC	147.28500	+		MetroCor
	FM	W2KB	224.28000	-	123.0 Hz	
	FM	N2BEI	446.90000	-	141.3 Hz	MetroCor
Newton	FM	W2LV	147.21000	+	151.4 Hz	
	FM	W2LV	147.30000	+	151.4 Hz	
	FM	W2LV	147.33000	+	151.4 Hz	
	FM	W2LV	443.00000	+	103.5 Hz	
Newton, Kittatiny Mountain						
	FM	W2LV	224.50000	-	141.3 Hz	
North Caldwell	FM	W2JT	147.18000	+	156.7 Hz	
Nutley	FM	N2SMI	441.75000	+		
	FM	W2FOY	446.33000	-	151.4 Hz	
Oakhurst	FM	WW2ARC	443.00000	+	127.3 Hz	
Oakland	DMR/MARC	N2JTI	446.16250	-	CC1	

236   NEW JERSEY

Location	Mode	Call sign	Output	Input	Access	Coordinator
Oakland	DMR/MARC	KM4WUD	448.82500	-	CC0	
Ocean City	FM	WA3UNG	448.62500	-	131.8 Hz	ARCC
Ocean View	DSTAR	NJ2CM	440.09375	+		
Old Bridge	FM	WB2HKK	444.05000	+	141.3 Hz	
Paramus	DMR/MARC	KM4WUD	444.15000	+	CC0	MetroCor
	FM	W2AKR	146.79000	-	141.3 Hz	MetroCor
	FM	KA2MRK	441.95000	+	114.8 Hz	
	IDAS	KM4WUD	448.88750	-		MetroCor
Paramus, Bergen Toll Plaza						
	FM	WB2MAZ	53.49000	52.49000	136.5 Hz	
Parin	FM	W2CJA	53.67000	52.67000		MetroCor
Parlin	FM	W2CJA	147.12000	+		MetroCor
	FM	W2CJA	446.17500	+	114.8 Hz	MetroCor
Parsippany	FM	WB2JTE	440.10000	+	141.3 Hz	
Paterson	DMR/MARC	NJ2PC	440.95000	+	CC1	
Phillipsburg, St. Luke's Hospi						
	FM	W2MCC	447.92500	-	DCS 205	
Pilesgrove	FM	N2SRQ	445.03125	-		
Pine Hill	FM	K2UK	146.86500	-	131.8 Hz	ARCC
	FM	K2UK	442.35000	+	131.8 Hz	ARCC
Pitman	FM	W2MMD	147.18000	+	131.8 Hz	ARCC
	FM	W2MMD	442.10000	+	131.8 Hz	ARCC
	FM	W2MMD	1284.40000	1264.40000		ARCC
Plainfield	DMR/BM	KD2HUY	444.40000	+	CC1	
Princeton	FM	KD2ARB	223.40000	-	100.0 Hz	
	FM	N3KZ	442.85000	+	131.8 Hz	ARCC
Quinton	FM	N2KEJ	53.71000	52.71000	74.4 Hz	ARCC
Randolph	FM	W2GCM	441.50000	+	141.3 Hz	MetroCor
Rockaway	FM	WK3SS	446.62500	-		
Roselle, Firehouse						
	FM	N2PSU	445.92500	-	141.3 Hz	ARCC
Rosenhayn	FM	KC2TXB	147.25500	+	179.9 Hz	
	FM	KC2TXB	445.31875	-		
	FM	KE2CK	448.12500	-	192.8 Hz	ARCC
Roxbury, Mooney Mountain Meado						
	FM	N2XP	146.98500	-	131.8 Hz	
Saddle Brook	FM	WB2IZC	224.42000	-	88.5 Hz	MetroCor
Salem	FM	N2KEJ	224.46000	-	74.4 Hz	ARCC
Sayervlle	FM	K2MID	440.80000	+	141.3 Hz	
Sayreville	FM	K2GE	145.05000			
	FM	K2GE	146.76000	-	156.7 Hz	
	FM	K2MID	224.56000	-	141.3 Hz	
	FM	K2GE	443.20000	+	141.3 Hz	
Secaucus, Impreveduto Towers						
	FM	KC2IES	441.55000	+	88.5 Hz	
South Brunswick	FM	KA2RLM	443.40000	+	141.3 Hz	
	FM	KC2CWP	449.28000	-	151.4 Hz	
South Orange	FM	N2NSS	446.12500	-	162.2 Hz	MetroCor
South Plainfield	FM	W2LPC	445.27500	-	67.0 Hz	
South Plainfield, Police Stati						
	FM	NJ2SP	146.97000	-		
South River	FM	NE2E	224.78000	-	123.0 Hz	
	FM	NE2E	444.25000	+	123.0 Hz	
	FM	NE2E	1291.20000	1271.20000		
Springfield	FM	WA2BAT	147.50500	146.50500	123.0 Hz	MetroCor
	FM	W2FCC	224.14000	-	123.0 Hz	
Springfield NJ	DMR/BM	W2NJR	449.42500	-	CC1	
Succasunna	FM	N2XP	447.77500	-	136.5 Hz	
	FM	WT2S	447.78000	-	136.5 Hz	
Tinton Falls	FM	W2GSA	147.04500	+		MetroCor
Tinton Falls, Garden State Pkw						
	FM	K2EPD	448.92500	-	141.3 Hz	MetroCor

NEW JERSEY 237

Location	Mode	Call sign	Output	Input	Access	Coordinator
Toms River	DMR/MARC	N2IXU	441.90000	+	CC1	
	DMR/MARC	WA2JWR	445.77500	-	CC3	
	FM	W2DOR	223.92000	-		
	FM	KE2HC	224.70000	-		ARCC
	FM	WA2OTP	444.00000	+	141.3 Hz	ARCC
	FM	NJ2AR	448.62500	-	141.3 Hz	ARCC
	FM	WA2RES	449.82500	-	131.8 Hz	ARCC
Toms River, Holiday City						
	FM	WA2JWR	146.65500	-	127.3 Hz	
Toms River, The Bennett Bubble						
	FM	W2DOR	146.91000	-	127.3 Hz	
Tuckerton	FM	N2NF	146.70000	-	192.8 Hz	ARCC
Union	FM	W2FCC	446.37500	-	141.3 Hz	MetroCor
Union City	FM	KD2VN	224.20000	-	131.8 Hz	MetroCor
Union NJ	DMR/MARC	W2NJR	449.47500	-	CC1	
Vernon	FM	W2VER	146.92500	-	141.3 Hz	ARCC
	FM	W2VER	449.07500	-	141.3 Hz	ARCC
	FM	W2VER	918.07500	893.07500	141.3 Hz	ARCC
	FM	W2VER	927.33750	902.33750	141.3 Hz	ARCC
Verona	FM	W2UHF	448.87500	-	151.4 Hz	MetroCor
Villas	FM	KC2DOK	447.82500	-	162.2 Hz	ARCC
Vineland	DSTAR	K2GOD	446.62500	-		
	FM	WA2WUN	145.49000	-	179.9 Hz	ARCC
Voorhees	FM	K2EOC	146.89500	-	192.8 Hz	
Wall Twp	FM	N2CTD	146.77500	-	103.5 Hz	MetroCor
	FM	WB2ANM	444.35000	+	141.3 Hz	
Wanaque	FM	WA2SNA	146.49000	147.49000	107.2 Hz	MetroCor
	FM	WA2SNA	446.17500	-	107.2 Hz	MetroCor
Warren	DMR/MARC	WR3IRS`	440.28750	+	CC1	
	FM	K2ETS	223.96000	-	110.9 Hz	MetroCor
Warrensville	FM	K2PM	224.00000	-	151.4 Hz	
Warrenville	FM	W2QW	146.62500	-	141.3 Hz	
Washington	FM	WC2EM	223.78000	-	110.9 Hz	
	FM	W2SJT	443.85000	+	110.9 Hz	ARCC
Washington, Montana Mountain						
	FM	W2SJT	146.82000	-	110.9 Hz	ARCC
Watchung	FM	K2ETS	146.94000	-	141.3 Hz	MetroCor
Waterford Works	DMR/MARC	WR3IRS	443.30000	+	CC1	ARCC
	FM	KA2PFL	52.60000	52.80000	131.8 Hz	ARCC
	FM	W2FLY	145.21000	-	DCS 174	ARCC
	FM	WA3BXW	147.34500	+	127.3 Hz	ARCC
	FM	W2MX	224.62000	-		ARCC
	FM	KA2PFL	442.30000	+	131.8 Hz	ARCC
	FM	N3KZ	442.70000	+	131.8 Hz	ARCC
	FM	W2FLY	444.45000	+		ARCC
Wayne	FM	WA2SQQ	442.00000	+		
Wayne, St Josephs Hospital						
	FM	KD2KWT	444.80000	+	79.7 Hz	MetroCor
Wayne, St. Josephs Hospital						
	FM	KD2KWT	145.21000	-	79.7 Hz	MetroCor
West Atlantic City	FM	W2HRW	146.98500	-	146.2 Hz	ARCC
	FM	W2HRW	443.25000	+		ARCC
West Orange	DMR/BM	W2NJR	440.05000	+	CC1	
	DMR/BM	N2MH	442.60000	+	CC1	
	DMR/MARC	KC2NFB	446.22500	-	CC1	
	FM	WA2JSB	146.41500	147.41500	85.4 Hz	MetroCor
	FM	WA2JSB	447.87500	-	156.7 Hz	
West Trenton	DMR/MARC	KM4WUD	438.00000	-	CC0	
	FM	W2ZQ	146.67000	-	131.8 Hz	ARCC
	FM	NJ2EM	224.32000	-	67.0 Hz	
	FUSION	W2ZQ	442.65000	+	131.8 Hz	ARCC
West Windsor	FM	W2MER	147.10500	+		

238   NEW JERSEY

Location	Mode	Call sign	Output	Input	Access	Coordinator
Westampton	FM	KC2QVT	147.15000	+	127.3 Hz	ARCC
	FM	KC2QVT	448.32500	-	127.3 Hz	ARCC
Wildwood	FM	WA2WUN	146.67000	-		
Willingboro	FM	WB2YGO	146.92500	-	131.8 Hz	ARCC
	FM	WB2YGO	223.88000	-	118.8 Hz	ARCC
	FM	WB2YGO	442.05000	+	118.8 Hz	ARCC
Winslow	FM	K2AX	145.15000	-	91.5 Hz	ARCC
Woodbine	FM	N2CMC	146.61000	-	88.5 Hz	ARCC
Woodbury	FM	KC2DUX	147.12000	+	123.0 Hz	
Woodland Park, Rifle Camp Park						
	FM	NJ2BS	146.61000	-	141.3 Hz	
Woodridge	FM	W2RN	443.75000	+		

NEW MEXICO

Location	Mode	Call sign	Output	Input	Access	Coordinator
Alamogordo	DMR/BM	KE5MIQ	147.22000	+	CC1	NMFCC
	DMR/MARC	KD5OH	442.65000	+	CC1	
	DMR/MARC	N6CID	442.95000	+	CC3	NMFCC
	FM	KA5BYL	53.41000	52.41000		NMFCC
	FM	K5LRW	146.80000	-	127.3 Hz	
	FM	N6CID	146.86000	-	100.0 Hz	NMFCC
	FM	KC5OWL	146.90000	-	77.0 Hz	NMFCC
	FM	K5LRW	224.04000	-	100.0 Hz	
	FM	KA5BYL	224.60000	-		NMFCC
	FM	KF5LGO	444.97500	+	100.0 Hz	
Alamogordo, La Luz						
	DSTAR	W6DHS	144.96000	144.56000		
Alamogordo, Long Ridge						
	FM	WA5IHL	145.35000	-	88.5 Hz	
Albquerque	DMR/MARC	WA5IHL	442.25000	+	CC1	NMFCC
Albuquerque	DMR/BM	K5RKE	443.30000	+	CC1	
	DMR/BM	WR7HLN	443.30000	+	CC1	NMFCC
	DMR/BM	NM5SH	443.65000	+	CC1	NMFCC
	DMR/BM	AJ5Z	444.60000	+	CC1	
	DMR/MARC	KA8JMW	442.90000	+	CC7	NMFCC
	FM	KD5MHQ	146.74000	-	146.2 Hz	NMFCC
	FM	KF5ERC	146.92000	-	67.0 Hz	
	FM	K5LXP	147.32000	+	100.0 Hz	NMFCC
	FM	N5GU	147.38000	+		NMFCC
	FM	W5MHG	147.46500		88.5 Hz	
	FM	KB5UGU	223.90000	-		
	FM	KA5BIW	224.38000	-	100.0 Hz	
	FM	KH6JTM	224.58000	-	100.0 Hz	NMFCC
	FM	KE5XE	443.50000	-	123.0 Hz	
	FM	KC0QIZ	443.85000	+	103.5 Hz	NMFCC
Albuquerque NM	FM	K5URR	146.64000	-	67.0 Hz	
	FM	K5URR	146.94000	-	100.0 Hz	
Albuquerque, BCFD Fire Station						
	FM	K5BIQ	145.13000	-	100.0 Hz	NMFCC
Albuquerque, Northeast Heights						
	FM	K6LIE	224.48000	-	100.0 Hz	NMFCC
Albuquerque, Northern Sandia M						
	FM	K5FIQ	442.45000	+	67.0 Hz	
Albuquerque, Op Center						
	DSTAR	W5URR	449.45000	-		NMFCC
Albuquerque, Op Ctr						
	FM	K5FIQ	146.90000	-	67.0 Hz	NMFCC
	FM	K5FIQ	449.55000	-	71.9 Hz	NMFCC
	FM	K5FIQ	449.80000	-	67.0 Hz	NMFCC
Albuquerque, Sandia Crest						
	DSTAR	W5MPZ	443.80000	+		NMFCC
	FM	W5SCA	145.01000			
	FM	NM5ML	145.29000	-	100.0 Hz	NMFCC

NEW MEXICO 239

Location	Mode	Call sign	Output	Input	Access	Coordinator
Albuquerque, Sandia Crest						
	FM	W5CSY	145.33000	-	100.0 Hz	NMFCC
	FM	KB5GAS	442.10000	+	162.2 Hz	
	FM	W5CSY	444.00000	+	100.0 Hz	NMFCC
Albuquerque, Sandia Mnts						
	FM	WA5IHL	145.29000	-	100.0 Hz	
Albuquerque, Sandia Peak						
	FM	K5FSB	442.60000	+	100.0 Hz	NMFCC
Albuquerque, Southeast						
	DMR/BM	N5GU	444.60000	+	CC1	NMFCC
	FM	KC5ZXW	444.10000	+	100.0 Hz	NMFCC
Albuquerque, UNM Campus						
	FM	K5PRN	442.52500	+	100.0 Hz	NMFCC
Albuquerque, West Mesa						
	FM	K5BIQ	442.05000	+	100.0 Hz	NMFCC
Alto	DMR/BM	W5JXT	444.40000	+	CC1	
Angel Fire, Agua Fria Peak						
	FM	N5LEM	147.34000	+	100.0 Hz	NMFCC
Artesia	FM	KU5J	442.00000	+		NMFCC
	FM	K5CNM	442.45000	+	162.2 Hz	NMFCC
	FM	W5COW	444.97500	+	156.7 Hz	NMFCC
Aztec	DMR/MARC	N5UBJ	442.25000	+	CC1	NMFCC
	DSTAR	KF5VBE	1291.10000	1271.10000		NMFCC
	FM	NM5SJ	146.74000	-	100.0 Hz	NMFCC
	FM	KB5ITS	447.45000	-	107.2 Hz	NMFCC
Aztec, Tank Mtn	FM	KB5ITS	146.88000	-	100.0 Hz	NMFCC
Belen	FM	KC5OUR	146.70000	-	100.0 Hz	
Bloomfield	FM	K5WXI	146.92000	-	100.0 Hz	NMFCC
	FM	KB5ITS	448.65000	-	127.3 Hz	NMFCC
Bloomfield, Harris Mesa						
	DSTAR	KF5VBD	1292.30000	1272.30000		NMFCC
	FM	NM5ML	147.28000	+	67.0 Hz	NMFCC
Caballo	DMR/MARC	NM5C	448.50000	-	CC1	NMFCC
Caballo (Truth Or Consequences						
	FM	NM5EM	145.13000	-	141.3 Hz	
	FM	K7EAR	145.47000	-	141.3 Hz	
Capitan	FM	KC5QVN	146.61000	-	100.0 Hz	NMFCC
Caprock, Caudill Ranch						
	FM	NM5EM	145.25000	-	141.3 Hz	
Carlsbad	FM	KD6WJG	53.08000	52.08000	127.3 Hz	NMFCC
	FM	KG5BOM	145.39000	-	100.0 Hz	
	FM	KD6WJG	146.76000	-	127.3 Hz	NMFCC
	FM	N5CNM	146.88000	-	88.5 Hz	NMFCC
	FM	K5CNM	147.28000	+	123.0 Hz	NMFCC
	FM	N5MJ	224.46000	-	127.3 Hz	NMFCC
	FM	N5MJ	444.45000	+	127.3 Hz	NMFCC
Chama, Overlook Mtn						
	FM	W5SF	147.08000	+	162.2 Hz	NMFCC
Cimarron	FM	N5GDR	145.21000	-	110.9 Hz	NMFCC
Clines Corners, Tapia Mesa						
	FM	K5FIQ	147.06000	+	67.0 Hz	NMFCC
Cloudcroft, Benson Ridge						
	FM	K5BEN	145.23000	-	123.0 Hz	NMFCC
	FM	NM5EM	145.37000	-	156.7 Hz	NMFCC
Cloudcroft, James Ridge						
	FM	KE5MIQ	147.34000	+	151.4 Hz	NMFCC
Clovis	DMR/MARC	K5NEC	442.32500	+	CC1	
	FM	KA5B	147.24000	+	67.0 Hz	NMFCC
	FM	WS5D	147.32000	+	71.9 Hz	NMFCC
	FM	NM5ML	442.52500	+	67.0 Hz	NMFCC
	FM	KA5B	443.45000	+	131.8 Hz	NMFCC
	FM	WS5D	444.45000	+	88.5 Hz	

240   NEW MEXICO

Location	Mode	Call sign	Output	Input	Access	Coordinator
Clovis, Claud	FM	NM5EM	145.37000	-	141.3 Hz	NMFCC
Columbus	FM	W5DAR	145.43000	-	88.5 Hz	NMFCC
Conchas Dam, Mesa Rica						
	FM	NM5ML	147.36000	+	100.0 Hz	NMFCC
Corona, Gallinas Peak						
	FM	NM5EM	145.51500	-	141.3 Hz	
	FM	NM5ML	147.28000	+	100.0 Hz	NMFCC
Corrales	DMR/BM	N5QD	443.20000	+	CC1	
Cuba	FM	NM5SC	443.10000	+		
Cuba, Eureka Mesa						
	FM	NM5ML	147.24000	+	67.0 Hz	NMFCC
Datil, Davenport Lookout						
	FM	NM5ML	147.04000	+	100.0 Hz	NMFCC
Datil, Luera Peak	FM	NM5ML	147.14000	+	100.0 Hz	NMFCC
Datil, Madre Mtn.	FM	NM5EM	147.32000	+	141.3 Hz	NMFCC
Deming	FM	W5JX	147.08000	+		NMFCC
	FM	NM2J	147.12000	+	88.5 Hz	NMFCC
	FM	WA6RT	449.47500	-	77.0 Hz	NMFCC
	FM	N5WSB	449.85000	-	100.0 Hz	NMFCC
Deming, Little Florida Mtn						
	FM	W5DAR	146.82000	-	88.5 Hz	NMFCC
	FM	NM5ML	147.02000	+	100.0 Hz	NMFCC
	FM	N5IA	147.04000	+		NMFCC
	FM	K7EAR	147.06000	+	141.3 Hz	NMFCC
Des Moines, Sierra Grande						
	FM	NM5EM	147.17500	+	141.3 Hz	NMFCC
Dixon, Cerro Abajo						
	FM	KD5PX	147.18000	+	100.0 Hz	NMFCC
Dona Ana	FM	KC5SJQ	224.34000	-		NMFCC
Dulce	FM	NM5SJ	145.43000	-	136.5 Hz	NMFCC
Eagle Nest, Iron Mountain						
	FM	NM5ML	444.35000	+	100.0 Hz	NMFCC
Eagle Nest, Touch Me Not						
	FM	NM5EM	147.04000	+	141.3 Hz	NMFCC
Farmington	DMR/MARC	N5UBJ	442.32500	+	CC1	NMFCC
	DSTAR	KF5VBE	145.11500	-		
	DSTAR	KF5VBF	444.15000	+		
	FM	KB5ITS	53.01000	52.01000	131.8 Hz	NMFCC
	FM	KB5ITS	146.76000	-	100.0 Hz	NMFCC
	FM	K5WXI	146.85000	-	100.0 Hz	NMFCC
	FM	KB5ITS	147.00000	+	100.0 Hz	NMFCC
	FM	K5WY	449.00000	-	100.0 Hz	NMFCC
Farmington, Farmington Bluffs						
	DSTAR	KF5VBF	1291.70000	1271.70000		NMFCC
Fort Sumner	FM	KB5ZFA	147.14000	+	100.0 Hz	NMFCC
Gallup	FM	KC5WDV	449.75000	-	100.0 Hz	NMFCC
Gallup, Deza Bluff	FM	KC5WDV	448.20000	-		
Gallup, Deza Bluffs						
	FM	NM5ML	147.22000	+	67.0 Hz	NMFCC
	FM	KC5WDV	448.20000	-	100.0 Hz	NMFCC
Gallup, Gibson Peak						
	FM	KC5WDV	147.26000	+	100.0 Hz	NMFCC
Glenwood	FM	WY5G	448.77500	-	103.5 Hz	
Glenwood, Brushy Mountain						
	FM	WY5G	448.77500	-	103.5 Hz	NMFCC
Grants	DMR/BM	WR7HLN	443.35000	+	CC1	NMFCC
	DMR/BM	KE5FYL	444.65000	+	CC1	NMFCC
	FM	KE5FYL	147.18000	+		NMFCC
Grants, 515 West High Street						
	FM	K5EMO	444.97500	+	67.0 Hz	NMFCC
Grants, La Mosca	FM	NM5ML	444.80000	+	67.0 Hz	NMFCC

NEW MEXICO   241

Location	Mode	Call sign	Output	Input	Access	Coordinator
Grants, La Mosca Peak						
	FM	KE5FYL	444.95000	+	100.0 Hz	NMFCC
Grants, Microwave Ridge						
	FM	NM5ML	146.66000	-	100.0 Hz	NMFCC
Grants, Mt Taylor	FM	K5URR	146.94000	-	100.0 Hz	
	FM	NM5EM	146.98000	-	141.3 Hz	
Grants, MW Ridge						
	FM	K5URR	146.64000	-	67.0 Hz	NMFCC
High Rolls	FM	W5AKU	442.80000	-	100.0 Hz	NMFCC
High Rolls, Cloudcroft/ Alamog						
	FM	KF5MQH	147.00000	+	100.0 Hz	
Hobbs	FM	N5LEA	444.15000	+	162.2 Hz	NMFCC
Hobbs, Professional Communicat						
	FM	AH2AZ	444.27500	+	162.2 Hz	
Hope	FM	K5CNM	147.38000	+	123.0 Hz	NMFCC
Hurley, Murray Tank						
	FM	WD5EZC	147.06000	+		NMFCC
Jal	FM	N5SVI	147.10000	+		NMFCC
	FM	N5LEA	444.25000	+	100.0 Hz	
Johns Place	FM	KD5MHQ	146.74000	+		
Kenton	FM	WB5NJU	147.39000	+	88.5 Hz	
La Cueva	FM	N9PGQ	146.84000	-	107.2 Hz	NMFCC
	FM	N9PGQ	442.12500	+	107.2 Hz	NMFCC
La Luz	DMR/BM	W6DHS	440.60000	+	CC1	
Las Cruces	DSTAR	W5GB	146.84000	-		NMFCC
	FM	N5BL	146.64000	-	100.0 Hz	NMFCC
	FM	KA5ECS	146.94000	-		NMFCC
	FM	NM5ML	147.18000	+	100.0 Hz	NMFCC
	FM	W7DXX	147.38000	+	100.0 Hz	NMFCC
	FM	N5IAC	223.94000	-		NMFCC
	FM	N5IAC	447.50000	-		NMFCC
	FM	KC5EVR	449.57500	-	100.0 Hz	
	FM	KC5IEC	449.80000	-	100.0 Hz	NMFCC
	FM	WA8FBN	449.90000	-		NMFCC
Las Cruces, Twin Peaks						
	FM	N5BL	448.20000	-	100.0 Hz	NMFCC
Las Vegas, Elk Mountain						
	FM	W5SF	147.30000	+	162.2 Hz	NMFCC
Las Vegas, Mesa Apache						
	FM	WA5IHL	444.37500	+	100.0 Hz	NMFCC
Little Florida	FM	WB5QHS	448.72500	-	100.0 Hz	
	FM	W5DAR	449.47500	-	77.0 Hz	NMFCC
	FM	W5CF	449.85000	-	127.3 Hz	
Logan, Ute Lake	FM	K5DST	147.34000	+	131.8 Hz	NMFCC
Lordsburg	DMR/MARC	N5IA	440.82500	+	CC1	NMFCC
Lordsburg, Jacks Peak						
	FM	NM5EM	145.14500	-	141.3 Hz	
	FM	N5IA	145.17000	-	100.0 Hz	NMFCC
	FM	WB5QHS	145.25000	-	88.5 Hz	NMFCC
	FM	N5IA	449.00000	-	100.0 Hz	
Los Alamos	DMR/BM	KA5BIW	442.00000	+	CC1	NMFCC
	DMR/MARC	NM5BB	442.22500	+	CC7	NMFCC
	DSTAR	NM5WR	442.42500	+		
	FM	KC2HSO	144.26500		203.5 Hz	
	FM	KA5BIW	224.04000	-	100.0 Hz	NMFCC
	FM	WD9CMS	444.77500	+		NMFCC
	FM	WD9CMS	927.90000	902.90000		NMFCC
Los Alamos, Barranca Mesa						
	FM	W5PDO	146.88000	-		NMFCC
Los Alamos, Pajarito Mountain						
	DSTAR	NM5EC	145.19000	-		
	FM	W5SF	145.19000	-	162.2 Hz	NMFCC

242   NEW MEXICO

Location	Mode	Call sign	Output	Input	Access	Coordinator
Los Alamos, Pajarito Mountain						
	FM	KB5RX	223.94000	-		NMFCC
Los Lunas	FM	WA5TSV	444.12500	+	100.0 Hz	NMFCC
Loving	FM	K5CNM	147.36000	+	123.0 Hz	NMFCC
Lybrook	FM	NM5SJ	145.49000	-	100.0 Hz	NMFCC
Maljamar	FM	NM5ML	147.14000	+	67.0 Hz	NMFCC
	FM	N5LEA	444.35000	+	162.2 Hz	
McIntosh, McIntosh Post Office						
	FM	KB5VPZ	447.27500	-	71.9 Hz	
Melrose	FM	NM5ML	147.28000	+	67.0 Hz	NMFCC
Midway (Portales)	FM	KE5RUE	147.00000	+	67.0 Hz	
Mountainair, Capilla Peak						
	FM	W5NES	444.07500	+	100.0 Hz	NMFCC
Navajo Dam, Navajo Lake						
	FM	KB5ITS	147.36000	+	100.0 Hz	NMFCC
Nebo, Cedar Hill	FM	NM5SJ	147.06000	+	100.0 Hz	NMFCC
Ora Vista	FM	W5TWY	147.30000	+	100.0 Hz	NMFCC
Organ	FM	KC5SJQ	1293.90000	1273.90000	127.3 Hz	NMFCC
Organ, San Augustin Peak						
	FM	N5IAC	146.78000	-	100.0 Hz	NMFCC
Pecos, Elk Mountain						
	FM	NM5ML	147.26000	+	67.0 Hz	NMFCC
Pinos Altos	FM	NM5ML	145.11500	-	67.0 Hz	NMFCC
	FM	WB5QHS	448.30000	-	100.0 Hz	
Placitas, La Madera						
	FM	NM5SC	147.08000	+	100.0 Hz	NMFCC
Placitas, Sandia Peak						
	FM	NM5SV	443.40000	+	100.0 Hz	NMFCC
Portales	DMR/MARC	W5OMU	443.75000	+	CC3	
	FM	KE5RUE	146.82000	-	67.0 Hz	NMFCC
Portales, Eastern New Mexico U						
	DSTAR	KE5RUE	443.80000	+		NMFCC
Portales, ENMU Greyhound Stadi						
	FM	W5OMU	147.00000	+	67.0 Hz	
Queen, Queen Fire Station						
	FM	K5CNM	147.30000	+	123.0 Hz	NMFCC
Quemado, Fox Mtn						
	FM	WB7EGF	448.05000	-	103.5 Hz	
Raton, Raton Pass						
	DSTAR	KD0RDI	446.77500	-		NMFCC
Raton, Sierra Grande						
	FM	NM5ML	147.28000	+	100.0 Hz	NMFCC
Red River, Valle Vidal						
	FM	KF5PFO	145.25000	-	123.0 Hz	NMFCC
Redlake	FM	KE5RUE	146.84000	-	67.0 Hz	NMFCC
	FM	KE5RUE	442.25000	+	67.0 Hz	NMFCC
Reserve, Frisco	FM	NM5EM	147.34000	+	141.3 Hz	NMFCC
Reserve, Frisco Divide						
	FM	NM5ML	147.36000	+	67.0 Hz	NMFCC
Rio Rancho	FM	NM5HD	145.37000	-	162.2 Hz	
	FM	WA5OLD	442.35000	+	100.0 Hz	NMFCC
	FM	KC5IPK	442.75000	+	162.2 Hz	NMFCC
	FM	NM5F	443.70000	+	100.0 Hz	NMFCC
Rio Rancho, Intel Corp						
	FM	K5CPU	444.70000	+	100.0 Hz	NMFCC
Rio Rancho, Rainbow						
	FM	NM5RR	147.10000	+	100.0 Hz	NMFCC
	FM	NM5RR	443.00000	+	100.0 Hz	NMFCC
Roswell	DMR/BM	W5JXT	444.95000	+	CC1	NMFCC
	DMR/MARC	KJ5UFO	444.30000	+	CC1	
	FM	W5GNB	52.94000	-	100.0 Hz	
	FM	NM5ML	147.26000	+	100.0 Hz	NMFCC

NEW MEXICO 243

Location	Mode	Call sign	Output	Input	Access	Coordinator
Roswell	FM	W5GNB	147.32000	+	146.2 Hz	NMFCC
	FM	W5GNB	444.95000	+	179.9 Hz	
Roswell, Capitan Peak						
	FM	WA5IHL	146.66000	-	67.0 Hz	NMFCC
Roswell, Comanche Hill						
	DSTAR	W5ZU	444.42500	+		NMFCC
	FM	W5GNB	52.94000	-	100.0 Hz	NMFCC
	FM	N5IMJ	444.00000	+	100.0 Hz	NMFCC
	FM	W5GNB	444.55000	+	162.2 Hz	NMFCC
Roswell, Nmmi	FM	N5MMI	146.64000	-	100.0 Hz	NMFCC
Ruidoso	FM	N5SN	443.60000	+	85.4 Hz	
Ruidoso, Alto Crest						
	FM	KR5NM	146.92000	-	100.0 Hz	NMFCC
Ruidoso, Buck Mountain						
	FM	NM5ML	444.37500	+	67.0 Hz	NMFCC
Ruidoso, Buck Mtn						
	FM	K5RIC	146.98000	-	100.0 Hz	NMFCC
	FM	K5RIC	443.92500	+	100.0 Hz	NMFCC
San Luis, Cabezon						
	FM	K5YEJ	443.00000	+	100.0 Hz	NMFCC
San Ysidro, Pajarito Peak						
	FM	NM5SC	443.10000	+	100.0 Hz	NMFCC
Sandia Park	FM	W5AOX	444.15000	+	100.0 Hz	NMFCC
Santa Fe	DMR/BM	W6EZY	432.55000		CC1	
	DMR/BM	WR7HLN	443.15000	+	CC1	NMFCC
	DMR/BM	WR7HLN	444.40000	+	CC1	
	DSTAR	W5SF	145.21000	-		
	DSTAR	W5SF	444.57500	+		
	FM	KF5SGT	447.77500	-	100.0 Hz	
	FM	K9GAJ	449.27500	-	146.2 Hz	
Santa Fe, St Vincent Hospital						
	FM	W5SF	147.20000	+	162.2 Hz	NMFCC
Santa Fe, Tesuque Peak						
	FM	W5SF	146.82000	-	162.2 Hz	NMFCC
	FM	NM5EM	147.02000	+	141.3 Hz	NMFCC
	FM	KB5ZQE	442.82500	+	131.8 Hz	NMFCC
Santa Rosa, Moon Ranch						
	FM	NM5EM	147.04000	+	141.3 Hz	NMFCC
Silver City	FM	WY5G	448.80000	-	100.0 Hz	NMFCC
	FM	WA7ACA	448.87500	-	100.0 Hz	NMFCC
Silver City, Black Peak						
	FM	K5GAR	146.98000	-	103.5 Hz	NMFCC
Socorro, M Mountain						
	DSTAR	W5AQA	444.50000	+		
	FM	W5AQA	146.68000	-	100.0 Hz	NMFCC
Socorro, Socorro Peak						
	FM	NM5EM	145.17500	-	141.3 Hz	NMFCC
	FM	NM5ML	147.24000	+	100.0 Hz	NMFCC
Socorro, West Peak						
	FM	KC5ORO	442.12500	+	123.0 Hz	NMFCC
T 0r C, Caballo Pk						
	FM	WB5QHS	448.97500	-	114.8 Hz	
T Or C, Caballo Peak						
	FM	NM5ML	147.26000	+	100.0 Hz	NMFCC
T Or C, Caballo Pk						
	FM	N5BL	146.76000	-	100.0 Hz	NMFCC
	FM	WB5QHS	448.17500	-	100.0 Hz	
	FM	WB5QHS	448.97500	-	114.8 Hz	NMFCC
Tank Mountain	FM	KB5ITS	146.88000	-	100.0 Hz	
Taos, Picuris Peak						
	FM	KF5PFO	51.50000	-	100.0 Hz	
	FM	KF5PFO	147.12000	+	67.0 Hz	NMFCC

244 NEW MEXICO

Location	Mode	Call sign	Output	Input	Access	Coordinator
Taos, Picuris Peak						
	FM	KF5PFO	224.40000	-	225.7 Hz	NMFCC
Taos, Ski Valley	FM	NM5ML	147.14000	+	67.0 Hz	
Taos, Wheeler Peak						
	FM	N5TSV	444.97500	+	123.0 Hz	NMFCC
Tijeras, Cedro Peak						
	FM	K5BIQ	145.15000	-	100.0 Hz	NMFCC
	FM	NM5ML	147.34000	+	67.0 Hz	NMFCC
Tijeras, Raven Road						
	FM	K5CQH	146.72000	-	100.0 Hz	NMFCC
Tres Piedras, San Antonio Moun						
	FM	NM5ML	147.22000	+	100.0 Hz	NMFCC
Tucumcari	FM	WA5EMA	146.88000	-		NMFCC
	FM	WA5EMA	224.98000	-		NMFCC
	FM	WA5EMA	443.75000	+		NMFCC
Tucumcari, Tucumcari Mountain						
	FM	NM5ML	147.22000	+	100.0 Hz	NMFCC
Tularosa	FM	W5TYW	443.90000	+	100.0 Hz	NMFCC
Wagon Mound	FM	NM5ML	147.20000	+	67.0 Hz	
Wagon Mound, Turkey Mountain						
	FM	NM5ML	147.20000	+	67.0 Hz	
Weed, Weed Lookout						
	FM	KE5MIQ	146.96000	-	151.4 Hz	NMFCC
Window Rock	FM	KD7LEN	442.00000	+	100.0 Hz	
Zuni	FM	KD5SAR	145.43000	-	162.2 Hz	NMFCC

LIMARC
Long Island Mobile Amateur Radio Club

Serving The Amateur Radio Community Since 1965

Our 55th Year

The **Long Island Mobile Amateur Radio Club** is an ARRL Affiliated Special Service Club serving the Amateur Radio community since 1965. LIMARC, one of the largest Amateur Radio clubs in the USA, is a nonprofit organization, dedicated to the advancement of Amateur Radio, public service and assistance to fellow amateurs.

LIMARC operates eight club repeaters, all using a 136.5 PL. (except DMR)

Repeater Frequencies: Our two 2m repeaters are linked.
W2VL 146.850 [-] (Glen Oaks)
ECHOLINK W2VL-R, Node 487981

W2KPQ 147.375 [+] (Selden)
ECHOLINK W2KPQ-R, Node 503075

IRLP via Reflector Node 9126

WA2LQO 146.745 [-] Fusion (Plainview)

W2VL 1288.00 [-] (Glen Oaks)

W2KPQ 449.125 [-] IRLP Node 4969,
ECHOLINK W2KPQ-L, Node 500940

W2KPQ 224.820 [-] (Glen Oaks)

W2KPQ DMR 449.375 [-] CC1 (Plainview)

W2KPQ DMR 449.3625 [-] CC1 (Selden)

Packet Node W2KPQ and BBS W2KPQ-4 on 145.07MHz

Repeater Trustees: W2VL, W2QZ; **W2KPQ,** WB2WAK

Special Events Callsign WV2LI: Trustee N2GA

Some of LIMARC's regular activities are
General Meetings: 2nd Wednesday (except July and August) at Levittown Hall, Hicksville, NY @ 8:00 PM
VE Tests: 2nd Saturday in odd numbered months at Levittown Hall – check our web-site for additional information
License Classes and **Field Day**
Co- SPONSOR OF THE SCHOOL CLUB ROUNDUP

2020 Events
Hamfests-February, June, and October
Special Event Stations K2CAM, (Cradle of Aviation Museum)
May - Lindbergh Flight, July - Apollo 11

Weekly Nets
Technical Net: Sunday @ 8:00 PM
Club Info Net: Monday @ 8:30 PM followed by the **Swap & Shop Net**

Other Regularly Scheduled Nets
Computer Net: 3rd and 4th Wednesday at 8:30 PM
Astronomy Net: 1st and 3rd Tuesday at 8:30 PM
Scanner Net: 1st and 3rd Thursday at 8:30 PM
Nostalgia/Trivia Net: @ 8:30 PM on the fifth Wednesday of those months where one occurs.
Note: All Nets are linked between the 146.850 and 147.375 Repeaters

---

For more information on current LIMARC events:
Access LIMARC on the World Wide Web:
http://www.limarc.org or e-mail us at: **limarc@limarc.org**
Write: **LIMARC, P.O. Box 392, Levittown, NY 11756**
Phone 516-450-5153

246   NEW YORK

Location	Mode	Call sign	Output	Input	Access	Coordinator
NEW YORK						
Albany	DMR/MARC	KM4WUD	444.00000	+	CC0	UNYREPCO
	DMR/MARC	WB2ERS	444.75000	+	CC1	MetroCor
	FM	W2GBO	53.41000	52.41000	100.0 Hz	UNYREPCO
	FM	KA2QYE	147.37500	+	100.0 Hz	UNYREPCO
	FM	K2AD	444.00000	+	100.0 Hz	UNYREPCO
	FM	KB2SIY	444.70000	+	94.8 Hz	UNYREPCO
Albany, Blue Hill	FM	WA2MMX	927.21250	902.21250	114.8 Hz	UNYREPCO
Albany, Helderberg Mtn						
	FM	K2CT	145.19000	-	103.5 Hz	UNYREPCO
Albany, St. Peter's Hospital						
	FM	K2ALB	146.64000	-	100.0 Hz	
Albion	FM	WA2DQL	145.27000	-	141.3 Hz	WNYSORC
	FM	K2SRV	442.87500	+		UNYREPCO
Alfred	FM	K2BVD	146.95500	-	127.3 Hz	UNYREPCO
Alma, Alma Hill	FM	KA2AJH	147.21000	+	123.0 Hz	WNYSORC
	FM	KA2AJH	444.10000	+	107.2 Hz	
Amherst - North	DMR/BM	N2CID	442.18750	+	CC1	WNYSORC
Arcade	FM	KC2PES	442.27500	+	107.2 Hz	WNYSORC
Arcadia, Brantling Ski						
	DSTAR	WA2EMO	444.75000	+		UNYREPCO
	FM	WA2EMO	146.68500	-	71.9 Hz	UNYREPCO
Arkport	FM	KC2FSW	147.04500	+		UNYREPCO
Arkwright	FM	K2XZ	146.67000	-	88.5 Hz	WNYSORC
Armonk, Kensico Reservoir						
	FM	W2TWY	224.30000	-	114.8 Hz	MetroCor
Athens	DMR/MARC	KM4WUD	442.57500	+	CC0	
	DSTAR	K2MCI	145.55000			
Auburn	FM	K2INH	53.05000	52.05000	71.9 Hz	UNYREPCO
	FM	W2QYT	145.23000		103.5 Hz	UNYREPCO
	FM	W2QYT	147.00000	+	71.9 Hz	UNYREPCO
	FM	K2RSY	147.27000	+	71.9 Hz	UNYREPCO
Austerlitz	DMR/MARC	N2JTI	445.18750	-	CC1	
	FM	KQ2H	442.75000	+		
	FM	N2ACF	445.12500	-	114.8 Hz	UNYREPCO
Austerlitz, Austerlitz Mountai						
	FM	WA2ZPX	442.85000	+	156.7 Hz	UNYREPCO
Ava	FM	W2OFQ	146.88000	-	71.9 Hz	
Averill Park	FM	W1GRM	449.35000	-	71.9 Hz	
Avon	FM	WR2AHL	146.94000	-	162.2 Hz	
Bald Mountain	FM	KT2D	927.88750	902.88750	100.0 Hz	
Baldwinsville	FM	WA2DAD	444.90000	+	131.8 Hz	UNYREPCO
Barkersville, Lake Nancy						
	FM	K2DLL	147.24000	+	91.5 Hz	
Batavia	FM	W2SO	147.28500	+	141.3 Hz	WNYSORC
Batavia, NY State School For T						
	FM	W2SO	444.27500	+		
Bath	FM	N2AAR	145.19000	-		UNYREPCO
	FM	N2HLT	146.80500	-	151.4 Hz	UNYREPCO
Bayside	DMR/MARC	K2JRC	438.58750	-	CC3	MetroCor
Beacon	DMR/MARC	K2HR	145.39625	-	CC5	
	DMR/MARC	NY4Z	441.45000	+	CC7	
	DMR/MARC	K2HR	443.15625	+	CC9	
Beacon, Mt Beacon						
	FM	K2ROB	53.31000	52.31000	114.8 Hz	UNYREPCO
	FM	AE2AN	146.97000	-	100.0 Hz	
	FM	W2GIO	223.92000	-	100.0 Hz	UNYREPCO
	FM	KC2OUR	443.55000	+	156.7 Hz	UNYREPCO
	FM	WA2GZW	449.57500	-	100.0 Hz	UNYREPCO
	FM	KC2VTJ	927.48750	902.48750	136.5 Hz	
Bear Mountain	FM	N2ACF	447.80000	-	114.8 Hz	

NEW YORK   247

Location	Mode	Call sign	Output	Input	Access	Coordinator
Bemus Point	FM	WA2LPB	145.29000	-	127.3 Hz	WNYSORC
Bethany	FM	WA1W	146.36000	+	123.0 Hz	
Bethpage	FM	K2ATT	449.30000	-	156.7 Hz	MetroCor
Binghamton	FM	N2YOW	444.30000	+		
	FM	AA2EQ	444.55000	+		
Binghamton Township						
	FM	W2EWM	145.47000	-		UNYREPCO
Binghamton, Airport						
	FM	WA2QEL	146.86500	-	146.2 Hz	UNYREPCO
Binghamton, Ingraham Hill						
	FM	K2TDV	146.73000	-	100.0 Hz	UNYREPCO
Binghamton, Trim Street						
	FM	K2VQ	147.07500	+		
Binghamton/Endicott						
	FM	AC2YS	442.80000	+	100.0 Hz	
Blue Mountain Lake						
	FM	W2CJS	146.86500	-	162.2 Hz	UNYREPCO
Bohemia	FM	N2HBA	444.60000	+	136.5 Hz	MetroCor
Boonville	FM	WD2ADX	146.65500	-		UNYREPCO
Boston	DMR/BM	W2BRW	442.15000	+	CC1	
	FM	WB2JQK	29.68000	-	107.2 Hz	
	FM	N2ZDU	224.82000	-	107.2 Hz	
Branchport	FM	N2LSJ	442.60000	+	110.9 Hz	UNYREPCO
Brewster	FM	WA2ZPX	147.39000	+	151.4 Hz	UNYREPCO
Briarcliff Manor	DMR/MARC	K2HR	443.60000	+	CC1	
Bristol	FM	W2IMT	224.68000	-	110.9 Hz	UNYREPCO
Bristol, Worden Hill						
	FM	WR2AHL	145.11000	-	110.9 Hz	UNYREPCO
Brockport	FM	N2HJD	147.22500	+	110.9 Hz	UNYREPCO
Bronx	DMR/MARC	N2NSA	440.36750	+	CC1	
	DMR/MARC	N2NSA	443.30000	+	CC1	MetroCor
	DMR/MARC	N2NSA	443.35000	+	CC1	MetroCor
	DMR/MARC	KC2IVF	445.42500	-	CC2	
	FM	WB2KVO	224.58000	-		MetroCor
	FM	KB2NGU	440.20000	+	88.5 Hz	MetroCor
	FM	W2MGF	441.20000	+	DCS 225	
	FM	N2YN	442.75000	+	173.8 Hz	
	FM	N2HBA	447.62500	-	136.5 Hz	MetroCor
Bronx , New York	FM	K2CSX	145.67000		DCS 226	
Brookhaven	FM	W2OFD	444.85000	+	123.0 Hz	
Brooklyn	DMR/BM	KB2RNI	146.61000	-	CC1	
	DMR/BM	KB2RNI	441.15000	+	CC1	
	DMR/MARC	WB2ZEX	438.27500	-	CC1	
	DMR/MARC	NY4Z	442.09375	+	CC1	
	DMR/MARC	KC2BVP	449.77500	-	CC0	
	FM	WB2HWW	29.66000	-	114.8 Hz	SLVRC
	FM	KB2RQE	145.31000	-	100.0 Hz	UNYREPCO
	FM	KC2RA	146.43000	147.43000	136.5 Hz	MetroCor
	FM	NB2A	146.74500	-	136.5 Hz	MetroCor
	FM	KB2NGU	147.30000	+	146.2 Hz	MetroCor
	FM	W2SN	224.60000	-	100.0 Hz	MetroCor
	FM	NN2N	224.74000	-	123.0 Hz	MetroCor
	FM	KB2NGU	439.50000	-		
	FM	N2ROW	441.10000	+	136.5 Hz	MetroCor
	FM	N2UOL	446.17500	-	136.5 Hz	MetroCor
	FM	WA2JNF	446.67500	-	114.8 Hz	MetroCor
	FM	KB2NGU	448.37500	-	162.2 Hz	
	FM	KB2PRV	448.97500	-	136.5 Hz	MetroCor
	FM	K2MAK	449.77500	-		MetroCor
	FM	N2HBA	927.88750	902.88750	151.4 Hz	MetroCor
Brooklyn Heights	FM	W2CMA	145.23000	-	114.8 Hz	UNYREPCO

248 NEW YORK

Location	Mode	Call sign	Output	Input	Access	Coordinator
Brooklyn, Downtown						
	DSTAR	WG2MSK	445.47500	-		MetroCor
	FM	K2RMX	446.82500	-	141.3 Hz	MetroCor
Brunswick, Bald Mountain						
	FM	W2GBO	146.94000	-		
	FM	WA2MMX	927.21250	902.21250	114.8 Hz	
Bufallo	FM	W2ERD	927.22500	902.22500	88.5 Hz	
Buffalo	DMR/BM	W2BRW	444.92500	+	CC1	
	DMR/BM	W1FPW	445.97500	-	CC1	
	FM	AB2UK	29.68000	-	107.2 Hz	SLVRC
	FM	WB2ECR	146.86500	+	151.4 Hz	WNYSORC
	FM	WB2ECR	224.76000	-	107.2 Hz	
	FM	WB2ECR	443.52500	+		
	FM	WB2ECR	443.97500	+	141.3 Hz	
	FM	WA2HKS	444.00000	+		WNYSORC
	FM	N2LYJ	927.32500	902.32500	88.5 Hz	WNYSORC
	FM	WA2WWK	927.47500	902.47500	131.8 Hz	
Buffalo, UB Kimball Tower						
	FM	WA2HKS	444.06250	+		
Burlington	FM	W2EES	146.71500	-	167.9 Hz	UNYREPCO
Cairo	FM	KB2DYB	146.74500	-	210.7 Hz	UNYREPCO
	FM	N2SQW	147.09000	+		UNYREPCO
Camillus	FM	N2PYK	146.62500	-	103.5 Hz	
Canadice, Bald Hill						
	FM	W2XRX	145.29000	-	110.9 Hz	UNYREPCO
	FM	WR2AHL	444.95000	+	110.9 Hz	
Canandaigua	FM	K2BWK	146.82000	-	110.9 Hz	UNYREPCO
Candor, Candor Hill						
	FM	K2OQ	147.30000	+	91.5 Hz	UNYREPCO
Capital District	FM	KC2ARE	146.55500	-		
Carle Place	FM	N2YXZ	445.97500	-	118.8 Hz	MetroCor
Carmel	DMR/MARC	KC2CWT	438.36250	430.76250	CC1	
	FM	KC2CWT	224.02000	-	136.5 Hz	UNYREPCO
Carmel, Mt. Ninham						
	DSTAR	K2PUT	445.87500	-		UNYREPCO
	FM	K2PUT	145.13000	-	136.5 Hz	UNYREPCO
Carmel-HYTERA	DMR/MARC	KC2CWT	446.18750	445.68750	CC1	
Cazenovia	FM	N2LZI	147.07500	+	97.4 Hz	UNYREPCO
Cazenovia, Route 92						
	FM	W2CM	147.21000	+	103.5 Hz	UNYREPCO
Cedarhurst	FM	N2XPM	445.52500	-	131.8 Hz	
Central Islip	FM	WB2ROL	147.03000	+	136.5 Hz	ASMA
	FM	WB2GLW	927.85000	902.85000	146.2 Hz	
Cherry Valley	FM	NC2C	145.35000	-	167.9 Hz	
	FM	WA2IJE	224.98000	-		UNYREPCO
Chili	FM	WR2AHL	146.76000	-	110.9 Hz	
Churchville	FM	KB2CHM	443.10000	+	110.9 Hz	
Clarence	FM	AG2AA	147.36000	+	107.2 Hz	
Clay	FM	KD2CDY	146.64000	-	131.8 Hz	UNYREPCO
	FM	WA2DAD	444.25000	+	131.8 Hz	UNYREPCO
Clyde	DMR/BM	KA2NDW	443.07500	+	CC3	UNYREPCO
	FM	KA2NDW	53.47000	52.47000	82.5 Hz	UNYREPCO
	FM	KA2NDW	145.47000	-	82.5 Hz	UNYREPCO
	FM	KA2NDW	224.47000	-	82.5 Hz	UNYREPCO
	FM	KA2NDW	449.07500	-	82.5 Hz	
Cobleskill	FM	WA2ZWM	146.61000	-	123.0 Hz	UNYREPCO
Cohoes	DMR/BM	WA2CW	442.40000	+	CC1	
Colden	FM	W2IVB	53.57000	52.57000	88.5 Hz	WNYSORC
	FM	W2IVB	145.31000	-	88.5 Hz	WNYSORC
	FM	W2IVB	442.10000	+	88.5 Hz	WNYSORC
	FM	WB2JPQ	444.10000	+	88.5 Hz	

NEW YORK 249

Location	Mode	Call sign	Output	Input	Access	Coordinator
Colden, WIVB Tower						
	FM	WB2ELW	147.09000	+	107.2 Hz	WNYSORC
Colesville	FM	WA2QEL	146.82000	-	146.2 Hz	UNYREPCO
Cooperstown	FM	NC2C	146.64000	-		UNYREPCO
Corinth	FM	K2DLL	448.22500	-	91.5 Hz	
Corinth, Spruce Mountian						
	FM	K2DLL	147.00000	+	91.5 Hz	UNYREPCO
Corning	FM	N2IED	147.01500	+	123.0 Hz	
Corona, Hall Of Science						
	FM	KC2PXT	145.27000	-	136.5 Hz	MetroCor
Cortland	FM	KB2LUV	145.49000	-	71.9 Hz	UNYREPCO
	FM	KB2FAF	147.03000	+	71.9 Hz	UNYREPCO
	FM	K2IWR	147.18000	+	71.9 Hz	UNYREPCO
	FM	KB2FAF	147.22500	+	71.9 Hz	UNYREPCO
	FM	KB2FAF	442.85000	+	71.9 Hz	UNYREPCO
Coxsackie	DMR/BM	N2LEN	445.02500	-	CC1	
	FM	N2LEN	147.09000	+		
Cragsmoor	DMR/BM	N2LEN	443.40000	+	CC1	
	FM	WB2BQW	29.69000	-	100.0 Hz	UNYREPCO
	FM	WB2BQW	53.33000	52.33000	100.0 Hz	UNYREPCO
Cranberry Lake	DMR/MARC	K2WW	444.75000	+	CC1	
Dansville	DMR/BM	KC2REY	445.28750	-	CC1	
Deer Park	DMR/BM	N2GQ	440.50000	+	CC1	
Deerfield	FM	W2JIT	146.76000	-		
Deerfield, Smith Hill						
	FM	WA2CAV	224.66000	-		UNYREPCO
Defreestville	FM	K2CWW	444.30000	+	100.0 Hz	
Delevan	DMR/MARC	N2CID	444.20000	+	CC1	
	FM	WB2JPQ	51.62000	50.62000	88.5 Hz	WNYSORC
	FM	K2XZ	145.39000	-		WNYSORC
	FM	K2XZ	444.17500	+	88.5 Hz	
Dexter	FM	AC2GE	147.03000	+	151.4 Hz	
	FM	KD2CPX	443.15000	+	151.4 Hz	
Dix Hills	DMR/MARC	W2RGM	147.07500	+	CC1	MetroCor
	DMR/MARC	W2RGM	448.52500	-	CC1	MetroCor
	FM	W2RGM	53.85000	52.85000	114.8 Hz	ASMA
	FM	W2RGM	224.56000	-	136.5 Hz	
Dobbs Ferry	FM	K2UTB	442.85000	+	118.8 Hz	MetroCor
Dundee	FM	KD2NOL	146.98500	-	82.5 Hz	
East Bay Park	FM	K2RRA	146.88000	-		
East Elmhurst	DMR/BM	N2YGI	439.65000	-	CC13	
East Fishkill	FM	N2SPF	53.61000	52.61000	100.0 Hz	UNYREPCO
East Galway	FM	K2DLL	147.36000	+		
East Hampton	FM	W2HLI	224.60000	-		MetroCor
East Meadow	DMR/MARC	KC2NFB	444.40000	+	CC1	
	FM	WB2CYN	147.13500	+	136.5 Hz	
	FM	K2CX	443.32500	+	141.3 Hz	MetroCor
	FM	KB2BWV	443.52500	+	114.8 Hz	MetroCor
	FM	AA2UC	443.80000	+	141.3 Hz	MetroCor
	FM	NC2PD	444.88750	+	179.9 Hz	MetroCor
East Rockaway	FM	WA2YUD	224.54000	-	136.5 Hz	MetroCor
Eastern Long Island						
	DMR/BM	K1IMD	433.63750	+	CC1	
Eden	FM	WB2JPQ	146.83500	-	88.5 Hz	WNYSORC
	FM	WB2JPQ	444.20000	+	88.5 Hz	
Ellenville	FM	KQ2H	53.73000	52.73000	146.2 Hz	UNYREPCO
Ellenville, Sam's Point						
	FM	KQ2H	147.07500	+	94.8 Hz	
Elmira	DMR/BM	N2NUO	444.60000	+	CC1	
	DMR/MARC	KC2EQ	443.00000	+	CC3	
	FM	W2ZJ	146.70000	-	100.0 Hz	UNYREPCO
	FM	N3AQ	147.36000	+		UNYREPCO

250   NEW YORK

Location	Mode	Call sign	Output	Input	Access	Coordinator
Elmira	FM	NR2P	223.98000	-	100.0 Hz	UNYREPCO
	FM	KA3EVQ	444.20000	+	100.0 Hz	UNYREPCO
	FM	N3AQ	444.90000	+		
Endicott	FM	N2YR	145.39000	-	123.0 Hz	UNYREPCO
	FM	WA2VCS	147.25500	-	100.0 Hz	UNYREPCO
Esopus	FM	N2ACF	445.82500	-	114.8 Hz	
	FM	N2ACF	447.80000	-	114.8 Hz	
Fairport	DSTAR	KB2VZS	444.80000	+		UNYREPCO
Falconer	DSTAR	KD2LWX	444.97500	+	127.3 Hz	WNYSORC
Farmingville	FM	WB2BQW	145.25000	-	100.0 Hz	ASMA
	FM	K2SPD	145.31000	-	118.8 Hz	ASMA
	FM	WA2UMD	146.71500	-	136.5 Hz	MetroCor
	FM	WA2LIR	224.86000	-	107.2 Hz	ASMA
	FM	K2LI	445.72500	-	91.5 Hz	MetroCor
	FM	K2SPD	446.51250	-	118.8 Hz	UNYREPCO
Farmingville, NY	FM	WR2UHF	444.70000	+	114.8 Hz	MetroCor
Farmingville, Telescope Hill						
	FM	WA2DCI	446.32500	-	127.3 Hz	
Finchville	FM	WA2VDX	146.76000	-	100.0 Hz	UNYREPCO
	FM	WA2ZPX	147.39000	+	156.7 Hz	UNYREPCO
	FM	WA2VDX	449.52500	-	123.0 Hz	UNYREPCO
Fine	FM	WA2NAN	147.13500	+	151.4 Hz	SLVRC
Fine, Adirondack Foothills						
	FM	WA2NAN	442.02500	+		
Fleischmanns	FM	WA2SEI	53.47000	52.47000	107.2 Hz	UNYREPCO
Flushing	FM	K2HAM	147.09000	+	114.8 Hz	MetroCor
	FM	KB2HRA	444.95000	+	114.8 Hz	MetroCor
Flushing, Queens College						
	FM	WB2HWW	53.47000	52.47000		
	FM	WB2HWW	440.70000	+	114.8 Hz	MetroCor
Fredonia	DMR/BM	KD2MNA	147.19500	+	CC1	WNYSORC
	FM	W2SB	146.62500	-	127.3 Hz	WNYSORC
Fresh Meadows	DMR/MARC	W2KTU	449.72500	-	CC1	
Frewsburg	FM	W2DRZ	146.79000	-	127.3 Hz	WNYSORC
Fulton	FM	K2QQY	146.85000	-	123.0 Hz	UNYREPCO
	FM	WN8Z	444.35000	+	103.5 Hz	UNYREPCO
Ga	DMR/MARC	MMDVM	443.50000	+	CC1	
Gainesville	FM	WB2JPQ	145.49000	-	88.5 Hz	WNYSORC
Geneva	FM	W2ONT	147.09000	+	110.9 Hz	UNYREPCO
Glen	FM	K2JJI	146.97000	-		
	FM	N2MNT	147.19500	+	156.7 Hz	UNYREPCO
Glen Oaks	DMR/MARC	WB2WAK	438.51250	-	CC1	MetroCor
	DMR/MARC	K2JRC	438.61250	-	CC3	
	FM	K2CJP	145.41000	-	114.8 Hz	MetroCor
	FM	W2VL	146.85000	-	136.5 Hz	MetroCor
	FM	W2KPQ	224.82000	-	136.5 Hz	MetroCor
	FM	KB2EKX	440.75000	+	162.2 Hz	MetroCor
	FM	K2CJP	445.37500	-	114.8 Hz	MetroCor
	FM	WB2WAK	447.02500	-	DCS 516	
	FM	WB2VTJ	448.77500	-	114.8 Hz	MetroCor
Glen Spey	FM	N2ACF	446.12500	-	114.8 Hz	
Glenmont	FM	K2QY	147.12000	+	100.0 Hz	
Glenville	FM	W2IR	146.79000	-	100.0 Hz	UNYREPCO
Gloversville	DMR/BM	KW2Y	445.07500	-	CC1	
	FM	W8NUD	53.13000	51.13000		
	FM	K2JJI	224.70000	-		UNYREPCO
	FM	K1YMI	443.70000	+	100.0 Hz	UNYREPCO
Gloversville, Gloversville Hig						
	FM	K2JJI	146.70000	-	100.0 Hz	
Goshen, Orange County Emergenc						
	FM	KC2OUR	449.67500	-	162.2 Hz	UNYREPCO
Grafton	FM	K2CBA	145.31000	-		UNYREPCO

NEW YORK   251

Location	Mode	Call sign	Output	Input	Access	Coordinator
Grafton	FM	K2REN	147.18000	+	100.0 Hz	UNYREPCO
	FM	WB2HZT	147.33000	+	146.2 Hz	UNYREPCO
Greenport	FM	W2AMC	440.05000	+	107.2 Hz	MetroCor
Greenwich	DMR/MARC	N2LEN	447.27500	-	CC1	
Groveland	FM	AA2GV	147.03000	+	110.9 Hz	UNYREPCO
Half Moon	FM	W2GBO	448.87500	-	203.5 Hz	
Hampton Bays	FM	WA2UEG	147.19500	+	136.5 Hz	
Harriman	DMR/MARC	N2JTI	443.80000	+	CC1	
	FM	W2AEE	53.17000	52.17000	136.5 Hz	UNYREPCO
	FM	W2AEE	223.80000	-	107.2 Hz	UNYREPCO
	FM	N2ACF	439.87500	+	118.8 Hz	
Harriman, Arden Mt						
	FM	N2JTI	147.10500	+	114.8 Hz	UNYREPCO
Hauppague	FM	WB2ROL	442.85000	+	151.4 Hz	
Hauppauge	DSTAR	WD2NY	444.23750	+		
	FM	WA2LQO	145.33000	-	136.5 Hz	ASMA
	FM	W2LRC	145.43000	-	136.5 Hz	ASMA
Haverstraw	FM	W2LGB	447.87500	-	114.8 Hz	UNYREPCO
Hemlock	DSTAR	K2BWK	147.37500	+		UNYREPCO
	DSTAR	K2BWK	443.50000	+		UNYREPCO
Herkimer	FM	N2ZWO	147.09000	+		UNYREPCO
Highbridge	FM	K2CSX	438.10000		91.5 Hz	
Highland	DMR/MARC	KC2OBW	440.31250	+	CC11	
Highland, Illinois Mountain						
	FM	N2OXV	147.04500	+	100.0 Hz	UNYREPCO
Highland, WRWD	FM	KC2OBW	927.65000	902.65000	136.5 Hz	
Holtsville	FM	WB2MOT	146.94000	-	136.5 Hz	ASMA
	FM	AG2I	442.05000	+	114.8 Hz	ASMA
	FM	W2SBL	449.17500	-	114.8 Hz	MetroCor
Hoosick Falls, Hoosick Falls						
	FM	K2FCR	146.65500		100.0 Hz	
Hornell	FM	KD2WA	51.66000	-	110.9 Hz	UNYREPCO
Hudson	FM	N2LEN	147.15000	+	114.8 Hz	
	FM	K2RVW	147.21000	+	110.9 Hz	UNYREPCO
Hunter	FM	WB2UYR	145.15000	-		UNYREPCO
Hunter, Colonels Chair Summit						
	DMR/BM	N2LEN	448.27500	-	CC1	
Huntington	FM	WR2ABA	147.21000	+	136.5 Hz	
	FM	KC2QHN	448.22500	-		
	FM	WR2ABA	448.67500	-	114.8 Hz	MetroCor
Huntington, Huntington Hospita						
	FM	KF2GV	448.22500	-	69.3 Hz	MetroCor
Ilion	FM	N2ZWO	145.11000	-	167.9 Hz	UNYREPCO
	FM	N3SQ	146.80500	-		UNYREPCO
Inlet	DMR/MARC	N2LBT	147.43750	144.93750	CC1	
Islip	FM	K2IRG	147.34500	+	100.0 Hz	ASMA
Islip Terrace	FM	WA2UMD	447.77500	-	114.8 Hz	MetroCor
Ithaca	DSTAR	AF2A	449.02500	-		
	FM	K2ZG	146.89500	-	107.2 Hz	UNYREPCO
Ithaca, Connecticut Hill						
	FM	AF2A	146.97000	-	103.5 Hz	UNYREPCO
Ithaca, Cornell Univ						
	FM	W2CXM	146.61000	-	103.5 Hz	UNYREPCO
Ithaca, Hungerford Hill						
	FM	AF2A	146.94000	-	103.5 Hz	UNYREPCO
Jamestown	FM	KS2D	145.33000	-	127.3 Hz	WNYSORC
	FM	K2LUC	146.94000	-	127.3 Hz	WNYSORC
Jasper	FM	KC2JLQ	147.33000	+	110.9 Hz	UNYREPCO
Jeffereson Valley	DMR/MARC	NY4Z	448.92500	-	CC1	
Jewett	FM	W1EQX	145.45000	-		SLVRC
Johnsburg	FM	W2CDY	146.46000		114.8 Hz	
Kenmore	FM	K2LED	147.00000	+	107.2 Hz	

252　NEW YORK

Location	Mode	Call sign	Output	Input	Access	Coordinator
Kew Gardens	FM	NB2A	927.28750	902.58750	DCS 271	MetroCor
Kew Gardens, Ny	FM	N2XBA	224.46000	-	141.3 Hz	
Kingston	DMR/BM	N2MCI	448.62500	-	CC1	
	FM	WA2MJM	146.80500	-	103.5 Hz	
	FM	WA2MJM	147.25500	+	103.5 Hz	UNYREPCO
	FM	WA2MJM	448.62500	-	77.0 Hz	UNYREPCO
Kirkland	FM	K1DCC	147.24000	+	71.9 Hz	
	FM	KA2FWN	443.85000	+		
Knapp Creek	FM	W3VG	146.85000	-	127.3 Hz	WNYSORC
Krumville	FM	KC2BYY	146.74500	-	123.0 Hz	UNYREPCO
Lackawanna	FM	WB2JPQ	444.15000	+	88.5 Hz	
Lake George	DMR/MARC	W2WCR	443.25000	+	CC1	
	FM	W2WCR	443.45000	+	100.0 Hz	
	FM	N2ACF	444.45000	+	114.8 Hz	
Lake George, Prospect Mountain						
	FM	W2WCR	224.78000	-		UNYREPCO
Lake Peekskill	FM	W2NYW	146.67000	-	156.7 Hz	UNYREPCO
	FM	KB2CQE	449.92500	-	179.9 Hz	UNYREPCO
Lake Placid	DMR/MARC	N2NGK	446.67500	-	CC1	
	DMR/MARC	N2NGK	446.97500	-	CC1	
	DMR/MARC	N2NGK	449.67500	-	CC1	
	FM	N2NGK	147.30000	+	100.0 Hz	VIRCC
Lake Placid, Blue Mountain						
	FM	N2JKG	442.75000	+	123.0 Hz	VIRCC
Lancaster	DMR/BM	W2BRW	444.08750	+	CC1	
	FM	W2SO	53.17000	52.17000	107.2 Hz	WNYSORC
	FM	W2SO	147.25500	+	107.2 Hz	WNYSORC
	FM	W2SO	224.64000	+		
	FM	W2SO	443.85000	+		
Liberty	FM	KC2AXO	147.13500	+	94.8 Hz	UNYREPCO
	FM	N2ACF	441.95000	+	114.8 Hz	UNYREPCO
Limestone	FM	W3VG	53.31000	52.31000	127.3 Hz	WNYSORC
Liverpool	DMR/BM	WB2WGH	443.80000	+	CC1	
Lockport	DMR/MARC	W2OM	444.62500	+	CC1	
	DSTAR	K2MJ	144.46500			
Lockport, Niagara County EOC						
	FM	W2RUI	146.82000	-	107.2 Hz	WNYSORC
Long Island	FM	N2HBA	927.96250	902.96250	151.4 Hz	UNYREPCO
Long Lake	FM	KD2BAD	146.64000	-	162.2 Hz	UNYREPCO
	FM	KD2BAD	443.85000	+	162.2 Hz	UNYREPCO
Lowville	FM	W2RHM	146.95500	-	156.7 Hz	UNYREPCO
Lyon Mountain	FM	W2UXC	147.28500	+	123.0 Hz	VIRCC
	FM	WA2LRE	224.02000	-	123.0 Hz	VIRCC
Macedon	FM	W1YX	147.18000	+		
	FM	N2HJD	442.92500	+	110.9 Hz	
	FM	KA1CNF	444.77500	+		UNYREPCO
Mahattan, Lincoln Center						
	FM	N2BEI	147.19500	+		
Mahopac	DMR/MARC	NY4Z	145.39000	-	CC1	UNYREPCO
	DMR/MARC	K2HR	146.91000	-	CC5	
	DMR/MARC	NY4Z	446.27500	-	CC10	
	FM	K2HR	29.66000	-	74.4 Hz	SLVRC
	FM	K2HR	224.00000	-	79.7 Hz	UNYREPCO
	FM	NY4Z	224.70000	-	79.7 Hz	UNYREPCO
Malone	FM	WB2RYB	53.15000	52.15000	123.0 Hz	SLVRC
	FM	NG2C	147.09000	+		SLVRC
	FM	WB2RYB	147.22500	+	100.0 Hz	
	FM	NG2C	444.75000	+		SLVRC
Malverne	DMR/MARC	WB2WAK	446.42500	-	CC1	MetroCor
	FUSION	WB2WAK	447.02500	-		
Manhattan	DMR/MARC	KD2LS	433.40000	+	CC1	
	DMR/MARC	K2HR	440.60000	+	CC1	

NEW YORK 253

Location	Mode	Call sign	Output	Input	Access	Coordinator
Manhattan	DMR/MARC	NY4Z	442.05000	+	CC7	MetroCor
	DSTAR	K2DIG	445.27500	-		MetroCor
	FM	KQ2H	29.62000	-	146.2 Hz	SLVRC
	FM	K2HR	145.29000	-	94.8 Hz	MetroCor
	FM	WR2MSN	145.57000	146.57000	192.8 Hz	MetroCor
	FM	WB2ZSE	147.00000	-	136.5 Hz	MetroCor
	FM	W2ABC	147.27000	+	141.3 Hz	
	FM	WR2MSN	223.76000	-	192.8 Hz	MetroCor
	FM	KB2TM	223.90000	-	141.3 Hz	MetroCor
	FM	WA2HDE	224.66000	-	127.3 Hz	MetroCor
	FM	KQ2H	224.80000	-	141.3 Hz	
	FM	WR2MSN	440.42500	+	156.7 Hz	MetroCor
	FM	NY4Z	440.60000	+	141.3 Hz	MetroCor
	FM	NE2E	441.45000	+	123.0 Hz	MetroCor
	FM	KB2RQE	442.45000	+	179.9 Hz	MetroCor
	FM	N2YN	443.65000	+	77.0 Hz	
	FM	WA2CBS	445.07500	-	114.8 Hz	MetroCor
	FM	K2NYR	445.22500	-	74.4 Hz	MetroCor
	FM	KF2GV	446.92500	-	69.3 Hz	MetroCor
	FM	WB2ZTH	447.17500	-	141.3 Hz	MetroCor
	FM	KE2EJ	447.20000	-	100.0 Hz	MetroCor
	FM	KC2IMB	448.43000	-	107.2 Hz	MetroCor
	FM	N2JDW	449.02500	-	123.0 Hz	MetroCor
	FM	WB2ZSE	449.80000	-	114.8 Hz	MetroCor
Manhattan , Bowling Green						
	FM	K2IRT	441.70000	+	100.0 Hz	
Manhattan NYC	DMR/MARC	K2JRC	438.56250	-	CC2	MetroCor
Manhattan West Side						
	DMR/MARC	K2JRC	443.70000	+	CC3	
Manhattan, Empire State Buildi						
	DMR/MARC	K2MAK	448.27500	-	CC3	MetroCor
	FM	KQ2H	449.22500	-	82.5 Hz	MetroCor
Manhattan, Lincoln Center - Up						
	FM	N2BEI	449.32500	-	136.5 Hz	MetroCor
Manhattan, Rockefeller Center						
	FM	WA2ZLB	147.36000	+	107.2 Hz	MetroCor
	FM	WA2ZLB	223.94000	-	107.2 Hz	MetroCor
	FM	WA2ZLB	447.82500	-	107.2 Hz	MetroCor
Manhattan, Sheraton Hotel						
	FM	WB2SEB	449.62500	-	179.9 Hz	MetroCor
Manhattan, Times Square						
	P25	KC2LEB	440.55000	+	141.3 Hz	MetroCor
Manhattan, Washington Heights						
	FM	N2JDW	147.15000	+	136.5 Hz	MetroCor
Manorville	FM	N2NFI	145.37000	-	136.5 Hz	UNYREPCO
Martindale, Forest Lake						
	FM	K2RVW	224.28000	-		
	FM	K2RVW	449.92500	-	110.9 Hz	
Mattituck	DMR/BM	K1IMD	449.67500	-	CC1	
	DMR/MARC	K1IMD	449.62500	-	CC1	
Mayville	FM	WB2EDV	444.45000	+		
Mayville, Emergency Operations						
	DSTAR	KD2LYO	146.76000	-	127.3 Hz	
Melville	FM	WB2CIK	53.11000	51.31000	107.2 Hz	MetroCor
	FM	KB2AKH	147.28500	+	97.4 Hz	MetroCor
	FM	WB2CIK	442.95000	+	114.8 Hz	MetroCor
Middle Grove	FM	WB2BGI	145.43000	-	156.7 Hz	UNYREPCO
Middle Island	FM	W2OQI	146.82000	-	136.5 Hz	MetroCor
Middletown	FM	WR2MSN	224.54000	-	156.7 Hz	UNYREPCO
Middletown, Scotchtown Ave						
	DSTAR	K9CEO	145.60000			
Millbrook	FM	N2EYH	146.89500	-	100.0 Hz	

254   NEW YORK

Location	Mode	Call sign	Output	Input	Access	Coordinator
Mineola	FM	W2EJ	146.64000	-	100.0 Hz	MetroCor
	FM	KC2DVQ	443.25000	+	123.0 Hz	MetroCor
Mineville	FM	WA2LRE	53.35000	52.35000	123.0 Hz	VIRCC
	FM	WA2LRE	147.25500	+	123.0 Hz	VIRCC
Mohawk	DMR/BM	KA2FWN	449.42500	-	CC1	
Montauk	DMR/MARC	WZ2Y	443.20000	+	CC1	
	FM	WZ2Y	145.27000	-	136.5 Hz	MetroCor
Monticello, Sackett Lake						
	DSTAR	K2ASS	147.55000			
Morrisville	FM	WA2DTN	444.60000	+	162.2 Hz	UNYREPCO
Mount Vernon	FM	K2UQT	145.49500	-		MetroCor
Mt. Beacon	DMR/MARC	K2ATY	441.01875	+	CC10	
Munnsville	DMR/BM	N2ADK	443.00000	+	CC1	
	DMR/BM	N2ADK	448.00000	-	CC1	UNYREPCO
Nanuet	FM	KC2EHA	147.53500	145.53500	123.0 Hz	UNYREPCO
	FM	WR2I	443.35000	+	114.8 Hz	UNYREPCO
Napanoch	FM	KC2FBI	147.53500	145.53500	100.0 Hz	
Naples	FM	NO2W	146.92000	-		UNYREPCO
Naples-Gannet Hill						
	FM	W2ONT	442.20000	+	110.9 Hz	UNYREPCO
Naples-Gannett Hill						
	FM	W2ONT	145.45000	-	110.9 Hz	UNYREPCO
Nassau	DMR/MARC	WW2FD	145.41000	-	CC1	
New Baltimore	DMR/BM	N2LEN	449.02500	-	CC1	
New Hartford	DMR/BM	N2USB	146.73000	-	CC12	
	DMR/BM	N2USB	147.01500	+	CC1	UNYREPCO
	DMR/BM	N2USB	441.26000	5441.26000	CC1	
New Oregon	FM	WB2JPQ	444.37500	+	88.5 Hz	
New Rochelle	DMR/BM	N2YGI	447.32500	-	CC13	
	DMR/MARC	NY1FD	145.49500	-	CC1	
	DMR/MARC	KC2TOM	147.04000	+	CC1	
	DMR/MARC	N2YGI	439.58750	-	CC1	
	DMR/MARC	NY4Z	446.28125	-	CC1	
	FM	N2YGI	147.58500	145.08500	88.5 Hz	
	FM	NY4Z	446.72500	-	192.8 Hz	
New Scotland	FM	K2CWW	145.33000	-		UNYREPCO
New Windsor	FM	KD2ANX	146.48500	-	88.5 Hz	
New York	DMR/MARC	K2ZZ	448.27500	-	CC3	
New York City	DMR/MARC	N2NSA	443.88750	+	CC1	MetroCor
New York, New York						
	FM	N2MCC	448.42500	-		
Newark	DMR/BM	N2MKT	443.25000	+	CC1	UNYREPCO
Newark, Water Tower						
	FM	WA2AAZ	146.74500	-	71.9 Hz	UNYREPCO
	FM	KA2NDW	927.21250	902.21250	82.5 Hz	
Newark, Water Tower Hill						
	FM	WA2AAZ	224.90000	-	110.9 Hz	UNYREPCO
Newburgh	FM	N2HEP	449.47500	-	71.9 Hz	UNYREPCO
Newburgh, Cronomer Hill						
	FM	N2HEP	146.43000	147.43000	71.9 Hz	UNYREPCO
North Babylon	FM	KB2UR	224.12000	-	131.8 Hz	MetroCor
North Babylon, Babylon Town Ha						
	FM	KB2UR	147.25500	+		
	FM	KB2UR	446.77500	-	110.9 Hz	
North Chatham	FM	W2JWR	449.12500	-	100.0 Hz	
	FM	W2JWR	927.12500	902.12500	100.0 Hz	
North Chili	FM	W2XRX	444.82500	+		
	P25	KD2AWT	146.70000	-		
North Creek	DMR/BM	W2WCR	442.25000	+	CC1	
North Creek, Gore Mountain						
	FM	W2WCR	147.13500	+	123.0 Hz	UNYREPCO
North Hebron	DMR/MARC	N2ZTC	442.30000	+	CC7	

NEW YORK   255

Location	Mode	Call sign	Output	Input	Access	Coordinator
North Lindenhurst	DSTAR	W2TOB	440.25000	+		MetroCor
	FM	W2GSB	146.68500	-	110.9 Hz	
North Tonawanda	DMR/BM	W2BRW	442.48750	+	CC1	
	FM	W2SEX	146.95500	-	151.4 Hz	WNYSORC
	FM	N2WUT	443.60000	+	67.0 Hz	
Norway	FM	N2ZWO	147.04500	+	167.9 Hz	UNYREPCO
Norwich	FM	W2RME	146.68500	-	110.9 Hz	UNYREPCO
Nyack	FM	N2ACF	29.64000	-	114.8 Hz	SLVRC
Nyack, Tappan Zee Bridge						
	FM	WR2I	449.42500	-	114.8 Hz	UNYREPCO
Oceanside	DMR/MARC	N2ION	447.92500	447.42500	CC7	MetroCor
Ogdensburg	DMR/MARC	KE2EMS	442.85000	442.35000	CC4	SLVRC
Ogdensburg.NY	DMR/BM	KC2KVE	146.79000	147.79000	CC1	
Olean	FM	K2XZ	444.85000	+	88.5 Hz	WNYSORC
Oneida	DMR/BM	KA2FWN	444.80000	+	CC1	
	FM	W2MO	145.17000	-		UNYREPCO
	FM	W2MO	443.65000	+	103.5 Hz	
Oneonta	DMR/MARC	N2ZNH	448.17500	-	CC1	
	FM	W2SEU	146.85000	-	167.9 Hz	
Orange County RACES						
	FM	N2TMT	147.28500	+	118.8 Hz	
Orangeburg	DMR/MARC	N2JTI	444.00000	+	CC1	UNYREPCO
Orangetown	FM	N2ACF	53.37000	52.37000	114.8 Hz	UNYREPCO
	FM	WB2RRA	147.16500	+	114.8 Hz	UNYREPCO
	FM	WA2MLG	224.38000	-	114.8 Hz	UNYREPCO
	FM	N2ACF	443.85000	+	114.8 Hz	UNYREPCO
	FM	N2ACF	927.85000	902.05000	114.8 Hz	UNYREPCO
Oriskany	FM	KD2MCI	146.23500	+	71.9 Hz	
Oriskany, Oneida County Sherif						
	FM	KD2MCI	146.83500	-	71.9 Hz	UNYREPCO
Oswegatchie	DMR/BM	WA2NAN	442.17500	+	CC1	SLVRC
Oswego	FM	W2OSC	147.15000	+	103.5 Hz	UNYREPCO
Otisville	DSTAR	KC2YYF	145.63000			
Otisville, Goshen Turnpike						
	DSTAR	K9RRD	448.57500	-		UNYREPCO
Otisville, Graham Hill						
	FM	KC2OUR	448.32500	-	123.0 Hz	
Otisville, Graham Mountain						
	FM	KC2OUR	448.32500	-	123.0 Hz	
Owego	FM	W2VDX	146.76000	-		
	FM	K2OQ	147.39000	+	91.5 Hz	UNYREPCO
Parishville	DMR/MARC	K2WW	147.07500	+	CC1	
	FM	W2LCA	444.85000	+	151.4 Hz	SLVRC
Patchogue	DMR/BM	WD2NY	147.03000	+	CC1	
Patterson	FM	K2CQS	224.88000	-		UNYREPCO
Pearl River	FM	NJ2BS	146.83500	-	151.4 Hz	UNYREPCO
Perinton, Baker Hill						
	FM	KB2VZS	146.71500	-	110.9 Hz	
Perrysburg	FM	KC2DKP	444.90000	+	107.2 Hz	WNYSORC
Peru	DMR/MARC	NV2M	442.28750	+	CC1	VIRCC
	FM	NV2M	224.02000	-	123.0 Hz	
Peru, Terry Mountain						
	FM	WA2LRE	145.49000	-	123.0 Hz	VIRCC
Pine Bush	FM	AA2XX	443.70000	+	141.3 Hz	
Pine Hill	FM	KQ2H	444.05000	+		
Pittsford	DMR/BM	W2RDK	443.00000	+	CC1	
Plainview	DMR/MARC	W2KPQ	449.37500	-	CC1	
	DMR/MARC	K2LIE	449.41250	-	CC5	
	FM	WA2LQO	146.74500	-	136.5 Hz	MetroCor
	FM	WB2WAK	146.80500	-	136.5 Hz	MetroCor
	FM	WA2UZE	147.33000	+	136.5 Hz	MetroCor
	FM	KC2AOY	441.40000	+	151.4 Hz	MetroCor

256 NEW YORK

Location	Mode	Call sign	Output	Input	Access	Coordinator
Plainview	FM	WB2WAK	446.47500	-	136.5 Hz	MetroCor
	FM	KE2EJ	447.20000	-	136.5 Hz	MetroCor
	FM	N2FLF	447.35000	-	114.8 Hz	MetroCor
	FM	W2KPQ	449.12500	-	136.5 Hz	MetroCor
Plattsburgh	DMR/BM	KD2MAJ	145.07000	147.44500	CC1	
	FM	WA2LRE	53.59000	52.59000	123.0 Hz	VIRCC
	FM	W2UXC	147.15000	+	123.0 Hz	VIRCC
	FM	NV2M	447.57500	-	123.0 Hz	VIRCC
	FM	WA2LRE	448.07500	-	123.0 Hz	
Poland	FM	N2CNY	145.21000	-	167.9 Hz	
Pomfret, Concord Drive						
	FM	W2SB	444.35000	+	88.5 Hz	WNYSORC
Pomona	FM	N2ACF	145.17000	-	114.8 Hz	
	FM	N2ACF	146.46000	147.46000	77.0 Hz	UNYREPCO
	FM	N2ACF	223.82000	-	114.8 Hz	UNYREPCO
	FM	N2ACF	444.45000	+	114.8 Hz	UNYREPCO
Pompey	FM	WW2N	145.27000	-	103.5 Hz	UNYREPCO
Pompey Hill	FM	KD2AYD	144.39000			
Pompey, Pompey Hill						
	FM	W2CNY	146.77500	-	151.4 Hz	UNYREPCO
	FM	W2CM	146.91000	-	103.5 Hz	UNYREPCO
Port Jefferson	FM	W2RC	449.52500	-	114.8 Hz	MetroCor
Port Jefferson, St. Charles Ho						
	FM	W2RC	145.15000	-	136.5 Hz	UNYREPCO
Port Jervis	FM	N2ACF	449.12500	-	114.8 Hz	
Portable	DMR/MARC	K1IMD	449.97500	-	CC1	MetroCor
Potsdam	DMR/BM	K2CC	443.35000	+	CC1	SLVRC
Poughquag	DMR/MARC	N2ZWN	447.07500	-	CC3	
Putnam Valley	FM	N2CBH	448.72500	-	107.2 Hz	UNYREPCO
Queens	DMR/MARC	N2YGI	439.56250	-	CC1	
	DMR/MARC	KC2CQR	449.72500	-	CC1	MetroCor
	FM	KB2NYC	224.18000	-	114.8 Hz	
	FM	NB2A	445.17500	-	141.3 Hz	MetroCor
	FM	KB2NYC	446.38750	-	114.8 Hz	
	FM	KC2LAI	446.81250	-	71.9 Hz	MetroCor
Queens Village	FM	WB2QBP	442.65000	+	141.3 Hz	MetroCor
Queensbury, NY	DSTAR	W2CDY	145.55000			
Rand Hill	DMR/MARC	WA2LRE	145.07000	147.44500	CC1	
Remsen	FM	KB2AUJ	145.33000	-	71.9 Hz	UNYREPCO
Ripley	FM	K2OAD	145.47000	-	127.3 Hz	WNYSORC
Riverhead	DMR/MARC	N2NFI	442.30000	+	CC1	UNYREPCO
Rochester	DMR/BM	W2RIT	442.07500	+	CC1	
	DMR/BM	KD2FRD	443.60000	+	CC1	
	FM	N2HJD	29.68000	-	123.0 Hz	SLVRC
	FM	N2HJD	53.33000	52.33000	123.0 Hz	
	FM	N2HJD	146.92500	-	110.9 Hz	UNYREPCO
	FM	N2HJD	224.58000	-	110.9 Hz	
	FM	N2HJD	442.80000	+	110.9 Hz	UNYREPCO
	FM	W2RFC	444.40000	+	110.9 Hz	UNYREPCO
	FM	WB2KAO	444.85000	+	110.9 Hz	WNYSORC
	FM	WR2AHL	444.95000	+		
	FM	W2JLD	446.02500	-	114.8 Hz	
Rochester , Cobbs Hill						
	FM	N2MPE	444.45000	+		
Rochester, Cobbs Hill						
	FM	N2MPE	146.61000	-	110.9 Hz	
Rochester, Highland Hospital						
	FM	WR2ROC	146.79000	-	110.9 Hz	
Rochester, Monroe County EOC B						
	FM	K2RRA	146.88000	-	110.9 Hz	
Rochester, RIT	FM	K2GXT	147.07500	+	110.9 Hz	
	FM	K2GXT	442.07500	+	110.9 Hz	

NEW YORK 257

Location	Mode	Call sign	Output	Input	Access	Coordinator
Rochester, Senica Towers						
	FM	N2HJD	444.70000	+	110.9 Hz	UNYREPCO
Rochester, URMC	FM	WR2ROC	147.31500	+	110.9 Hz	UNYREPCO
Rockaway Park	DMR/BM	KE4DYI	438.40000	-	CC0	
Rocky Point	FM	N2FXE	146.59500	+	136.5 Hz	MetroCor
	FM	N2FXE	443.90000	+	123.0 Hz	
Rosedale	FM	K2EAR	145.35000	-	114.8 Hz	MetroCor
Royalton	FM	KD2WA	29.66000	-	107.2 Hz	SLVRC
	FM	KD2WA	443.45000	+	107.2 Hz	
	FM	KD2WA	927.45000	902.45000		
Rush	DMR/MARC	N2CHP	444.22500	+	CC1	
Rush/East Avon, Watts Electron						
	FM	N2YCK	29.64000	-	110.9 Hz	UNYREPCO
	FM	N2YCK	53.37000	52.37000	110.9 Hz	UNYREPCO
	FM	N2YCK	145.35000	-	110.9 Hz	UNYREPCO
	FM	N2YCK	224.02000	-	110.9 Hz	UNYREPCO
	FM	N2YCK	443.75000	+	110.9 Hz	UNYREPCO
	FM	N2YCK	927.85000	902.85000	110.9 Hz	UNYREPCO
Russell, Kimball Hill						
	FM	KA2JXI	146.92500	-	151.4 Hz	SLVRC
Rye Brook	FM	KB2GTE	444.65000	+	114.8 Hz	MetroCor
S. Bristol , Bristol Mountain						
	FM	W2SIX	53.63000	52.63000	110.9 Hz	
S. Bristol, Bristol Mountain S						
	FM	WR2AHL	444.55000	+	110.9 Hz	
S. Bristol, Mees Observatory						
	FM	WR2ROC	146.65500	-	110.9 Hz	
Sag Harbor	DMR/MARC	K1IMD	448.67500	-	CC1	
	FM	K2GLP	449.98000	-	94.8 Hz	ASMA
Sams Point, New York						
	FM	KC2OUR	53.55000	52.55000	156.7 Hz	
Saranac Lake	FM	W2TLR	145.31000	-	127.3 Hz	SLVRC
	FM	W2WIZ	147.03000	+		VIRCC
Schenectady	DMR/BM	N2LEN	147.30000	+	CC1	
	FM	K2AE	444.20000	+		
Schenectady, Crawford Hill						
	FM	K2AE	147.06000	+		
Schenevus	FM	KC2AWM	223.96000	-	100.0 Hz	UNYREPCO
Selden	DMR/BM	WD5TAR	444.70000	+	CC1	
	DMR/MARC	WA2VNV	448.82500	-	CC1	
	DMR/MARC	W2KPQ	449.36250	-	CC1	
	FM	WA2VNV	146.76000	-	136.5 Hz	
	FM	W2KPQ	147.37500	+	136.5 Hz	MetroCor
	FM	WA2UMD	447.52500	-		MetroCor
	FM	WA2UMD	447.80000	-	114.8 Hz	MetroCor
Setauket	FM	K2YBW	147.04500	+	136.5 Hz	WNYSORC
Sherburne	FM	KD2HKB	443.05000	+	179.9 Hz	
Sherman	FM	WB2EDV	53.61000	52.61000	127.3 Hz	WNYSORC
	FM	WB2EDV	442.75000	+		
Shirley	DMR/BM	KC2WCB	443.52500	+	CC1	
Sidney	FM	AC2KP	146.95500	-		
Sloatsburg	FM	N2ACF	444.85000	+	114.8 Hz	UNYREPCO
Smithtown	FM	W2LRC	224.62000	-		MetroCor
South Blooming Grove						
	FM	N2OKB	449.62500	-	136.5 Hz	
South Bristol	FM	NR2M	224.46000	-	110.9 Hz	UNYREPCO
Southampton	FM	WA2UEG	147.19500	+	136.5 Hz	UNYREPCO
Southold	FM	W2OQI	448.32500	-	107.2 Hz	MetroCor
Speculator	FM	KA2VHF	147.16500	+		
Stamford	FM	K2NK	53.27000	52.27000	107.2 Hz	UNYREPCO
	FM	KQ2H	449.22500	-	82.5 Hz	
Stanley	DMR/BM	W2ACC	444.30000	+	CC3	

258 NEW YORK

Location	Mode	Call sign	Output	Input	Access	Coordinator
Stanley	DMR/BM	KD2HVC	444.30000	+	CC3	
	FM	W2ACC	224.26000	-	110.9 Hz	
Staten Island	DMR/BM	KB2EA	447.72500	-	CC1	MetroCor
	DMR/MARC	KC2RQR	442.30000	+	CC3	
	FM	WA2IAF	146.88000	-	141.3 Hz	MetroCor
	FM	KA2PBT	445.82500	-	156.7 Hz	MetroCor
	FM	N2BBO	445.87500	-	136.5 Hz	
	FM	WA2IAF	447.37500	-	141.3 Hz	MetroCor
	FM	N2IXU	448.47500	-	94.8 Hz	MetroCor
	FM	KC2GOW	927.43750	902.43750	100.0 Hz	MetroCor
Staten Island, Grymes Hill						
	FM	KC2GOW	224.10000	-	141.3 Hz	MetroCor
Staten Island, North Shore						
	FM	N2EHN	445.57500	-	141.3 Hz	MetroCor
Staten Island, Todt Hill						
	FM	KC2GOW	53.83000	52.83000	136.5 Hz	MetroCor
	FM	KC2GOW	147.31500	+		MetroCor
	FM	KC2GOW	445.12500	-		MetroCor
	FM	KC2GOW	927.70000	902.00000	100.0 Hz	MetroCor
Stockton	FM	K2HE	146.88000	-	127.3 Hz	
Syosset	FM	WB2CYN	447.97500	-	136.5 Hz	MetroCor
	FM	N2HBA	448.02500	-	136.5 Hz	MetroCor
Syracuse	DMR/MARC	W2CM	443.30000	+	CC1	UNYREPCO
	FM	KD2SL	53.67000	52.67000	103.5 Hz	UNYREPCO
	FM	KC2VER	145.31000	-		
	FM	KD2SL	146.67000	-	103.5 Hz	UNYREPCO
	FM	WA2AUL	443.10000	+	103.5 Hz	UNYREPCO
Syracuse, Museum Of Science An						
	FM	K2MST	443.15000	+	71.9 Hz	
Syracuse, Onondaga Community C						
	FM	K2OCR	147.30000	+	67.0 Hz	
Syracuse, Sentinal Heights						
	FM	KD2SL	444.00000	+	103.5 Hz	
Syracuse, Sentinel Heights						
	FM	KD2SL	145.15000	-	123.0 Hz	
Thiells	DMR/MARC	KD2EQY	441.80000	+	CC1	
	DMR/MARC	KD2EQY	449.18750	-	CC1	
	FM	W2LGB	449.77500	-		
Thiells, Rosman Center						
	FM	W2LGB	441.58750	+		
	FM	W2LGB	442.18750	+		
	FM	W2LGB	449.18750	-	114.8 Hz	UNYREPCO
Todt Hill, Staten Island						
	FM	W2RJR	223.84000	-	141.3 Hz	
Troy	FM	N2TY	145.17000	-	127.3 Hz	UNYREPCO
	FM	K2REN	146.76000	-	103.5 Hz	UNYREPCO
	FM	W2SZ	146.82000	-		
	FM	W2SZ	224.42000	-		UNYREPCO
	FM	KB2HPW	224.64000	-		UNYREPCO
	FM	W2SZ	443.00000	+		
	FM	N2TY	447.07500	-	127.3 Hz	UNYREPCO
	FM	W2GBO	448.42500	-		
Tupper Lake	DMR/MARC	W2TUP	446.02500	-	CC1	
	FM	W2TUP	147.33000	+	100.0 Hz	UNYREPCO
Tupper Lake, Big Tupper Ski Ar						
	FM	W2TUP	444.70000	+	110.9 Hz	UNYREPCO
Tuxedo Park	DSTAR	K9GOD	145.67000			
Upton	FM	K2BNL	442.40000	+	114.8 Hz	MetroCor
Utica	DMR/BM	KA2FWN	447.57500	-	CC1	
	FM	KC2WLK	147.99500	144.49500	100.0 Hz	
Valhalla	DSTAR	W2ECA	448.18750	-		
	FM	WB2ZII	147.06000	+	114.8 Hz	MetroCor

NEW YORK   259

Location	Mode	Call sign	Output	Input	Access	Coordinator
Valhalla	FM	WB2ZII	224.40000	-	114.8 Hz	MetroCor
	FM	K2XD	440.65000	+	114.8 Hz	MetroCor
	FM	WB2ZII	447.47500	-	114.8 Hz	MetroCor
Valhalla, Grasslands Tower						
	FM	WB2ZII	927.98750	902.98750	114.8 Hz	MetroCor
Valley Stream	FM	WB2IIQ	444.65000	+	103.5 Hz	MetroCor
	FM	N2ZEI	448.62500	-	136.5 Hz	MetroCor
Verona	DMR/BM	K1DCC	442.25000	+	CC1	
	FM	K1DCC	146.94000	-	71.9 Hz	
Vestal	FM	AA2EQ	224.48000	-	88.5 Hz	UNYREPCO
	FM	KD2HNW	443.40000	+		
	FM	KD2HNW	446.02500	-		
Victor, Baker Hill	FM	N2HJD	145.41000	-	110.9 Hz	UNYREPCO
	FM	N2HJD	442.90000	+	110.9 Hz	
Voorheesville	DSTAR	WA2UMX	443.30000	+		
Walden	FM	KC2OUR	146.62500	147.92500	127.3 Hz	UNYREPCO
Walton	FM	K2NK	29.66000	-	107.2 Hz	SLVRC
Warrensburg	DMR/MARC	N2LEN	442.05000	+	CC1	
Warwick	FM	N2ACF	448.22500	-	114.8 Hz	
Warwick, Mt Peter	FM	N2IXA	147.63000	-	107.2 Hz	
Washingtonville	FM	N2ACF	443.80000	+	114.8 Hz	
	FM	KQ2H	445.90000	-	82.5 Hz	
Washingtonville, Schunnemunk M						
	FM	WB2BQW	145.25000	-	100.0 Hz	UNYREPCO
Waterloo	FM	W2ACC	145.13000	-	110.9 Hz	UNYREPCO
	FM	W2ACC	442.22500	447.32500	82.5 Hz	UNYREPCO
Watertown	FM	WB2OOY	146.70000	-	151.4 Hz	
	FM	KA2QJO	147.25500	+	151.4 Hz	
Waterville	DMR/BM	KA2FWN	449.32500	-	CC1	
Watkins Glen	FM	KA2IFE	147.16500	+		UNYREPCO
	FM	WR2M	224.96000	-	88.5 Hz	UNYREPCO
Waverly	DMR/BM	N2NUO	444.65000	+	CC1	
Wellsville, Madison Hill						
	FM	WB2MOD	444.47500	+		
West Islip	FM	W2GSB	223.86000	-	110.9 Hz	MetroCor
	FM	W2GSB	440.85000	+	110.9 Hz	MetroCor
	FM	W2YMM	927.31250	902.31250	DCS 606	
West Point	FM	W2KGY	145.27000	-		
West Shokan, Ashokan Reservoir						
	FM	N2NCP	51.76000	-	103.5 Hz	UNYREPCO
Westbury	DMR/MARC	KD2IVF	438.46000	-	CC1	
	DSTAR	NC2EC	146.67000	-		
	DSTAR	NC2EC	448.57500	-		
Wethersfield	FM	K2ISO	145.17000	-	110.9 Hz	WNYSORC
	FM	K2XZ	146.64000	-		WNYSORC
	FM	N2FQN	147.10500	+	141.3 Hz	WNYSORC
	FM	KC2QNX	147.31500	+	141.3 Hz	WNYSORC
Whiteface, Whiteface Mountain						
	FM	N2JKG	447.77500	-	123.0 Hz	VIRCC
Whitestone, Whitestone Bridge						
	FM	W2BAT	444.90000	+	225.7 Hz	MetroCor
Wilmington, Whiteface Mountain						
	FM	N2JKG	145.11000	-	123.0 Hz	VIRCC
Woodstock	FM	N2WCY	53.11000	52.11000	77.0 Hz	UNYREPCO
Wurtsboro	FM	KQ2H	447.52500	-	82.5 Hz	UNYREPCO
	FM	N2ACF	449.87500	-	114.8 Hz	UNYREPCO
Wurtsburo, Catskill Mountains						
	FM	KQ2H	29.62000	-	146.2 Hz	SLVRC
Yaphank	FM	KA2RGI	53.79000	52.79000	156.7 Hz	ASMA
	FM	W2DQ	145.21000	-	136.5 Hz	ASMA
	FM	W2DQ	446.62500	-	110.9 Hz	ASMA
Yonkers	FM	W2YRC	146.86500	-	110.9 Hz	MetroCor

260   NEW YORK

Location	Mode	Call sign	Output	Input	Access	Coordinator
Yonkers	FM	K2JQB	146.91000	-	114.8 Hz	MetroCor
	FM	N2PAL	224.08000	-	114.8 Hz	
	FM	W2YRC	224.94000	-	88.5 Hz	MetroCor
	FM	W2YRC	440.15000	+	88.5 Hz	MetroCor
	FM	N2QNB	445.42500	-	136.5 Hz	MetroCor
York Town Hts	DMR/BM	KB2LFH	441.56250	+	CC3	
Yorktown	DMR/MARC	K2HR	440.62500	+	CC7	
	FM	WB2IXR	147.01500	+	114.8 Hz	MetroCor
Yorktown Heights	DMR/MARC	NY4Z	443.15000	+	CC1	
	FM	WA2TOW	146.94000	-	162.2 Hz	MetroCor
	FM	AF2C	443.15000	+	88.5 Hz	MetroCor
Yorktown Hts	DMR/MARC	K2HPS	446.56250	-	CC3	

NORTH CAROLINA

Location	Mode	Call sign	Output	Input	Access	Coordinator
Aberdeen, Aberdeen Fire Depart						
	FM	KW1B	147.30000	+	100.0 Hz	SERA
Ahoskie	DMR/MARC	WB4YNF	444.33750	+	CC1	
	FM	WB4YNF	145.13000	-	131.8 Hz	SERA
	FM	KG4GEJ	146.91000	-	131.8 Hz	
	FM	WB4YNF	224.12000	-	131.8 Hz	
	FM	WB4YNF	444.20000	+	131.8 Hz	
Albemarle	DMR/BM	K4DVA	440.56250	+	CC1	
	DSTAR	K4DVA	144.92000	147.42000		
	DSTAR	K4DVA	440.68750	+		
	FM	K4OGB	146.98500	-	77.0 Hz	SERA
	FM	N4HRS	444.90000	+	110.9 Hz	
Albemarle, Morrow Mountain Sta						
	FM	K4DVA	443.52500	+	77.0 Hz	
Alexander	FM	KG4LGY	53.19000	52.19000	100.0 Hz	
Andrews	FM	K4AIH	224.88000	-		
	FM	K4AIH	443.65000	+	151.4 Hz	
Andrews, Joanna Bald						
	FM	WD4NWV	442.60000	+	151.4 Hz	SERA
Andrews, Joanna Mtn						
	FM	K4AIH	147.04500	-	151.4 Hz	
Angier	FM	KA0GMY	147.01500	+	110.9 Hz	SERA
Angier 900MHZ	DMR/MARC	NC4RA	927.62500	902.62500	CC1	
Apex	DMR/MARC	KI4EMS	147.07500	-	CC1	
Archers Lodge, Spring Hill Tow						
	FM	K4JDR	444.00000	+	100.0 Hz	SERA
Asheville	DMR/BM	K4HCU	440.62500	+	CC1	
	DMR/MARC	WA4TOG	442.55000	+	CC1	
	FM	KI4DNY	224.60000	-	94.8 Hz	
	FM	KE4MU	442.15000	+	94.8 Hz	
	FM	AC4JK	442.42500	+	DCS 053	
Asheville, Mount Mitchell						
	FM	N2GE	145.19000	-		SERA
	FM	N2GE	224.54000	-		
Asheville, Oteen / Azalea						
	FM	W4DCD	444.56250	+		SERA
Asheville, Spivey Mountain						
	FM	W4MOE	146.91000	-	91.5 Hz	
	FM	K4HCU	442.65000	+	100.0 Hz	
Asheville, Stradley Mountain						
	FM	W4MOE	224.52000	-	91.5 Hz	
Auburn	DSTAR	K4ITL	442.21250	+		
	FM	AK4H	147.27000	+		
	FM	K4ITL	224.16000	-	91.5 Hz	
Auburn, WRAL Tower						
	FM	K4ITL	145.21000	-	82.5 Hz	
Bakersville	FM	W4LNZ	222.12000	-		
	FM	W4LNZ	444.87500	+		

NORTH CAROLINA   261

Location	Mode	Call sign	Output	Input	Access	Coordinator
Bakersville, Locust Knob						
	FM	KK4MAR	145.31000	-	123.0 Hz	
Banner Elk, Sugartop						
	FM	KX4CZ	442.17500	+	123.0 Hz	SERA
Bath, Hunters Bridge Water Tow						
	FM	NC4ES	442.55000	+	82.5 Hz	SERA
Bearwallow Mountain						
	DSTAR	NC4BS	442.96250	+		SERA
Beech Mtn	FM	WA4NC	444.57500	+	151.4 Hz	
Benson	FM	K4JDR	444.02500	+	100.0 Hz	
Bethel	FM	KD4EAD	147.37500	+	151.4 Hz	
Bolivia	FM	K4PPD	145.37000	-	88.5 Hz	
	FM	K4PPD	147.31500	+	118.8 Hz	
	FM	KE4TUD	444.70000	+	100.0 Hz	
Boone	DMR/MARC	WA4NC	440.75000	+	CC1	
	DMR/MARC	WA4NC	443.03750	+	CC1	SERA
Boone, Rich Mountain						
	FM	WA4J	147.36000	+	103.5 Hz	
Brasstown, Poorhouse Mtn						
	FM	K4CTE	147.31500	+	103.5 Hz	
Brevard, Rich Mountain						
	FM	AG4AZ	442.85000	+		
Broadway	FM	K4ITL	147.10500	+	82.5 Hz	
Browns Summit	FM	N2DMR	146.76000	-	156.7 Hz	SERA
Bryson City	FM	N0SU	443.40000	+	151.4 Hz	
Bunn	FM	KC4WDI	444.25000	+	88.5 Hz	
Burgaw	FM	N4JDW	442.02500	+	88.5 Hz	
Burlington	FM	K4EG	443.60000	+	123.0 Hz	SERA
Burnsville	FM	KF4LCG	146.95500	-		
	FM	KD4GER	441.92500	+		
	FM	KF4LCG	443.65000	+		
Burnsville, Phillips Knob						
	FM	KD4WAR	147.37500	+	123.0 Hz	SERA
Butner	FM	WA4IZG	146.94000	-	100.0 Hz	SERA
	FM	KC4WDI	443.20000	+	100.0 Hz	SERA
Buxton	DMR/BM	K4OBX	146.62500	-	CC1	SERA
	DMR/MARC	K4OBX	444.06250	+	CC1	SERA
	FM	K4OBX	53.01000	52.01000	131.8 Hz	SERA
	FM	K4OBX	145.15000	-	131.8 Hz	SERA
	FM	K4OBX	442.42500	+	100.0 Hz	
	FM	K4OBX	444.92500	+	131.8 Hz	
Calabash	FM	N4DBM	145.33000	-	162.2 Hz	SERA
	FM	KD4GHL	444.75000	+	118.8 Hz	SERA
Canton	FM	KI4GMA	444.85000	+	100.0 Hz	
Canton, Mount Pisgah						
	FM	N2GE	146.76000	-		SERA
	FM	N2GE	224.26000	-		
Carthage	FM	NC4ML	147.24000	+	91.5 Hz	SERA
Carthage, John Deere Dealer						
	FM	N1RIK	442.85000	+	107.2 Hz	SERA
Cary	DMR/BM	W1CKD	441.36250	+	CC1	
	DMR/BM	N1FTE	441.78750	+	CC1	
	DMR/MARC	KB4CTS	443.78750	+	CC1	
Cary, Vertical Bridge Tower						
	FM	K4JDR	444.77500	+	100.0 Hz	SERA
Castalia, Water Tower						
	FM	N4JEH	444.95000	+	107.2 Hz	
Chalybeate Springs, Kipling						
	FM	W4RLH	443.10000	+	100.0 Hz	
Chapel Hill	DSTAR	KR4RDU	442.53750	+		
	FM	W4UNC	53.45000	52.45000	107.2 Hz	SERA
	FM	K4ITL	147.13500	+	82.5 Hz	SERA

262   NORTH CAROLINA

Location	Mode	Call sign	Output	Input	Access	Coordinator
Chapel Hill	FM	W4UNC	442.15000	+	131.8 Hz	SERA
Charlotte	DMR/BM	KA4YMY	927.01250	902.01250	CC1	
	DMR/MARC	KC4YPB	440.80000	+	CC1	
	DMR/MARC	W4ZO	442.41250	+	CC1	SERA
	DMR/MARC	WG8E	443.22500	+	CC4	
	DMR/MARC	KM4BRM	443.43750	+	CC1	
	DMR/MARC	KI4WXS	443.86250	+	CC1	
	FM	W4BFB	145.29000	-	118.8 Hz	
	FM	W4BFB	146.94000	-	118.8 Hz	
	FM	W4CQ	147.06000	-		
	FM	W4BFB	224.40000	-		
	FM	KD4ADL	442.72500	+	110.9 Hz	
	FM	K4CBA	444.05000	+	136.5 Hz	
	FM	K4KAY	444.35000	+	151.4 Hz	
	FM	W4BFB	444.60000	+	118.8 Hz	
	FM	K4KAY	927.61250	902.61250	118.8 Hz	
Charlotte, Hood Road						
	DMR/BM	KI4WXS	444.02500	+	CC1	SERA
Charlotte, UNC Charlotte						
	FM	W0UNC	442.65000	+	88.5 Hz	
Cherryville	FM	N4DWP	224.96000	-		
China Grove	FM	N4UH	145.41000	-	136.5 Hz	
	FM	N4UH	443.25000	+	136.5 Hz	SERA
Chocowinity, WLGT-FM Tower						
	FM	K4BCH	147.25500	+	131.8 Hz	
Clayton	FM	N4TCP	443.67500	+		SERA
Clemmons	FM	WB9SZL	224.70000	-	100.0 Hz	SERA
Cleveland	FM	N4YR	53.25000	52.25000	100.0 Hz	SERA
	FM	W4SNA	53.31000	52.31000	100.0 Hz	
Cleveland , Young Mountain						
	FM	KU4PT	146.73000	-	94.8 Hz	
Cleveland, Young Mountain						
	FM	W4SNA	53.31000	52.31000	100.0 Hz	SERA
	FM	K4CH	443.70000	+	127.3 Hz	
Clinton	FM	W4TLP	146.79000	-	88.5 Hz	
	FM	W4TLP	224.28000	-	91.5 Hz	
Clinton, Taylors Bridge						
	FM	N4JDW	443.07500	+	100.0 Hz	
Coats, Old ATT Tower/American						
	FM	K4JDR	444.55000	+	100.0 Hz	SERA
Columbia	DMR/MARC	KX4NC	440.58750	+	CC1	
Columbia, WUND-TV Tower						
	FM	KX4NC	146.83500	-	131.8 Hz	SERA
	FM	K4OBX	442.72500	+	131.8 Hz	SERA
	FM	KX4NC	443.30000	+	131.8 Hz	SERA
Concord	FM	K4CEB	146.65500	-		
	FM	KD4ADL	147.21000	+	110.9 Hz	
	FM	N2QJI	442.52500	+	94.8 Hz	
	FM	W4ZO	444.25000	+		
	FM	KD4ADL	444.77500	+	110.9 Hz	
Concord, Cabarrus Sheriff's Of						
	FM	K4WC	443.35000	+	136.5 Hz	
Connelly Springs, South Mounta						
	FM	NA4CC	442.57500	+	82.5 Hz	SERA
Corolla	DMR/MARC	K4OBX	444.33750	+	CC1	
Creedmoor	FM	N4MEC	146.98500	-	100.0 Hz	SERA
Crowders Mt	FUSION	KA4YMZ	443.43750	+		
Cullowhee	DMR/MARC	W3WDD	444.97500	+	CC1	
Cullowhee, Western Carolina Un						
	FM	K4WCU	444.97500	+	100.0 Hz	
Dallas	FM	W4CQ	444.45000	+	82.5 Hz	
Dallas, TV Tower	FM	KA4YMZ	224.02000	-	82.5 Hz	SERA

NORTH CAROLINA 263

Location	Mode	Call sign	Output	Input	Access	Coordinator
Delco, UNC-TV Tower Delco						
	FM	AD4DN	224.50000	-	88.5 Hz	
Denton	FM	KD4LHP	442.75000	+	118.8 Hz	SERA
Dobson	FM	W4DCA	53.97000	52.97000	100.0 Hz	
	FM	N4DAJ	146.92500	-	100.0 Hz	SERA
	FUSION	W4DCA	444.52500	+	100.0 Hz	SERA
Duck	DMR/MARC	K4OBX	442.63750	+	CC1	
Dunn	FM	W4PEQ	146.70000	-	82.5 Hz	
Durham	DMR/BM	W4BAD	444.57500	+	CC1	
	DMR/MARC	W4BAD	147.36000	+	CC1	SERA
	DMR/MARC	KI4EMS	441.97500	+	CC2	
	FM	K4WCV	53.63000	52.63000	88.5 Hz	SERA
	FM	W4BAD	145.37000	-	100.0 Hz	SERA
	FM	KB4WGA	444.92500	+	94.8 Hz	
Durham, Camden Ave City-County						
	FM	K4JDR	444.92500	+	94.8 Hz	SERA
Durham, Durham VA Hospital						
	FM	WR4AGC	444.45000	+	100.0 Hz	SERA
Durham, TV Hill	FM	WR4AGC	145.45000	-	82.5 Hz	
	FM	NC4TV	421.25000	434.05000		
	FM	WR4AGC	444.10000	+	82.5 Hz	SERA
Eastover	FM	KN4ZZ	443.90000	+	100.0 Hz	
Edenton	FM	W4UUU	443.32500	+	123.0 Hz	
Elizabeth City	DMR/MARC	WA4VTX	440.56250	+	CC1	
	FM	WA4VTX	146.65500	-	131.8 Hz	SERA
Elizabethtown	FM	N4DBM	224.38000	-	91.5 Hz	SERA
Elizabethtown, NCSHP Tower						
	FM	N4DBM	146.98500	-	162.2 Hz	SERA
Elk Park	FM	KX4CZ	443.40000	+	123.0 Hz	
Elm City	FM	K2IMO	442.32500	+	88.5 Hz	
Engelhard	FM	K4OBX	146.71500	-	131.8 Hz	SERA
Englehard	DMR/MARC	WB4YNF	442.46250	+	CC1	SERA
Fayetteville	FM	W4EBM	53.81000	52.81000		
	FM	K4MN	444.40000	+	100.0 Hz	
Fayetteville, CC Child Support						
	FM	K4MN	146.91000	-	100.0 Hz	
Fayetteville, WAMC						
	FM	WA4FLR	147.33000	+	100.0 Hz	
Flat Rock, Sauratown Mountain						
	FM	KQ1E	443.05000	+	136.5 Hz	
Forest City	FM	K4OI	146.67000	-	114.8 Hz	
	FM	AI4M	442.00000	+	114.8 Hz	
Forest City, Cherry Mountain						
	FM	AI4M	443.30000	+	123.0 Hz	
Fountain	FM	N4HAJ	444.42500	+	88.5 Hz	
Franklin	DMR/MARC	W3WDD	444.20000	+	CC1	SERA
	DMR/MARC	N4DTR	444.37500	+	CC1	
	FM	W4GHZ	147.24000	+	151.4 Hz	
Franklin, Cowee Bald						
	FM	K2BHQ	145.49000	-	167.9 Hz	
Fuquay-Varina	FM	KK4OSI	443.30000	+	100.0 Hz	
Galax, VA	FM	N4VL	145.13000	-	103.5 Hz	
Garner, TV Tower East Of Ralei						
	FM	KD4PBS	442.07500	+	114.8 Hz	
Gastonia	DMR/BM	N4GAS	445.72500	-	CC1	
	DMR/MARC	KA4YMZ	443.91250	+	CC1	SERA
	FM	KQ1E	442.05000	+	DCS 152	
Gastonia, Crowder's Mtn						
	FM	K4CBA	442.05000	+	DCS 152	
Gastonia, Crowders Mountain						
	DSTAR	KK4JDH	443.86250	+		SERA
	FM	K4GNC	146.80500	-	100.0 Hz	

264   NORTH CAROLINA

Location	Mode	Call sign	Output	Input	Access	Coordinator
Gastonia, Crowders Mountain						
	FM	KC4IRA	224.62000	-	127.3 Hz	
	FM	KC4IRA	442.70000	+	100.0 Hz	
	FM	KA4YMZ	443.43750	+		SERA
	FM	KC4IRA	927.03750	902.03750	94.8 Hz	
Gastonia, Spencer Mountain						
	FM	KA4YMZ	927.01250	902.01250		
	IDAS	KA4YMY	444.70000	+		SERA
Gibsonville, Fire Station 28						
	FM	WB4IKY	145.49000	-	107.2 Hz	
Goldsboro	DMR/MARC	KB4CTS	442.36250	+	CC1	
	FM	K4CYP	146.85000	-	88.5 Hz	
	FM	K4CYP	443.00000	+	88.5 Hz	SERA
Goldsboro, Mar Mac / Busco Bea						
	FM	WA4DAN	224.46000	-	91.5 Hz	
Goldsboro, Rosewood Community						
	FM	K4JDR	145.33000	-	100.0 Hz	SERA
Graham	DMR/MARC	N2DMR	443.72500	+	CC1	SERA
Grantsboro	FM	KR4LO	444.35000	+		
Grantsboro, WMGV Tower						
	FM	KF4IXW	145.23000	-	85.4 Hz	SERA
	FM	KF4IXW	444.87500	+	85.4 Hz	
Greensboro	DMR/BM	ND4L	441.92500	+	CC1	
	DMR/MARC	N4DUB	441.86250	+	CC11	SERA
	DMR/MARC	NC4RA	442.88750	+	CC1	
	DMR/MARC	W4GG	444.22500	+	CC1	SERA
	FM	W4GSO	145.15000	-	100.0 Hz	SERA
Greenville	DMR/BM	NC4ES	444.62500	+	CC1	SERA
	DMR/MARC	WB4PMQ	444.80000	+	CC1	SERA
Greenville, Brody School Of Me						
	FM	N4HAJ	444.72500	+	91.5 Hz	
Greenville, WNCT-TV Tower						
	FM	W4GDF	147.09000	+	131.8 Hz	SERA
Grifton	FM	WA4DAN	224.84000	-	91.5 Hz	
Grifton, WNCT/WITN TV TOWER						
	FM	W4NBR	146.68500	-	88.5 Hz	
Hampstead, Topsail Fire Tower						
	FM	NC4PC	443.55000	+	100.0 Hz	SERA
Hamstead	FM	N4JDW	146.94000	-	88.5 Hz	
Haw River	FM	KD4JFN	224.62000	-	107.2 Hz	
Hayesville	DMR/MARC	K1DMR	443.03750	+	CC1	
	FM	KC4CBQ	53.27000	52.27000	186.2 Hz	SERA
	FM	KC4CBQ	444.67500	+	186.2 Hz	SERA
Hayesville, Cherry Mountain						
	FM	KG4JIA	442.50000	+	94.8 Hz	
Henderson, South Of Kerr Lake						
	FM	K4JDR	444.37500	+	100.0 Hz	SERA
Hendersonville	DMR/MARC	W4FOT	441.88750	+	CC1	
	DMR/MARC	WA4TOG	442.45000	+	CC1	
	DMR/MARC	WA4TOG	927.55000	902.55000	CC1	SERA
	DSTAR	KJ4JAL	147.25500	+		
	FM	W4FOT	53.13000	52.13000	100.0 Hz	
	FM	WB4YAO	146.64000	-	91.5 Hz	
	FM	WA4KNI	147.10500	+	91.5 Hz	SERA
	FM	WA4KNI	224.24000	-		
	FM	N4KOX	927.56250	902.56250	127.3 Hz	
Hendersonville, Bearwallow Mou						
	FM	WA4KNI	145.27000	-	91.5 Hz	
Hendersonville, Bearwallow Mtn						
	FM	WA4KNI	444.25000	+	91.5 Hz	
Hendersonville, Pinnacle Mount						
	DSTAR	KJ4JAL	442.02500	+		

NORTH CAROLINA 265

Location	Mode	Call sign	Output	Input	Access	Coordinator
Hertford	DMR/MARC	WB4YNF	442.16250	+	CC1	
	FM	WA4VTX	147.33000	+	131.8 Hz	SERA
	FM	WA4VTX	444.30000	+	131.8 Hz	SERA
Hickory	FM	WA4PXV	53.05000	52.05000	151.4 Hz	
	FM	WA4PXV	146.85000	-		SERA
Hickory, Barretts Mountain						
	FM	WA4PXV	442.37500	+	131.8 Hz	SERA
High Point	FM	KF4OVA	442.97500	+	107.2 Hz	
	FM	N2DMR	444.97500	+	107.2 Hz	SERA
High Point, High Point Regiona						
	FM	W4UA	147.16500	+	67.0 Hz	SERA
High Point, NC	FM	NC4AR	145.29000	-	88.5 Hz	
High Point, The Radio Building						
	FM	KF4OVA	444.62500	+	107.2 Hz	SERA
Hillsborough	DMR/MARC	WR4AGC	443.13750	+	CC1	
	FM	WR4AGC	147.22500	+	82.5 Hz	SERA
	FM	WR4AGC	224.26000	-		
Holly Springs, Holly Springs W						
	FM	KF4AUF	444.32500	+	100.0 Hz	SERA
Hope Mills, Water Tower						
	FM	W4KMU	146.83500	+		
Hubert	FM	NC4OC	147.00000	-	88.5 Hz	SERA
	FM	KE4FHH	443.31250	+		SERA
	FM	KE4FHH	444.67500	+	88.5 Hz	SERA
Jackson	FM	KB4CTS	444.32500	+	107.2 Hz	
Jacksonville	DMR/MARC	KE4FHH	441.83750	+	CC1	SERA
Jefferson, Phoenix Mountain						
	FM	W4YSB	147.30000	+	103.5 Hz	
	FM	W4JWO	224.22000	-	88.5 Hz	
	FM	W4JWO	443.07500	+	94.8 Hz	
Kelly	FM	WA4DAN	224.54000	-	91.5 Hz	
Kernersville	FM	KF4OVA	53.01000	52.01000	88.5 Hz	SERA
	FM	KF4OVA	146.86500	+	88.5 Hz	SERA
	FM	KF4OEV	224.24000	-	107.2 Hz	
	FM	KF4OVA	224.34000	-	88.5 Hz	SERA
Kilby Gap	FM	WA4PXV	443.57500	+	173.8 Hz	SERA
King	DMR/MARC	W4SNA	442.63750	+	CC1	
	DMR/MARC	W4SNA	442.68750	+	CC1	
	FM	K4GW	146.31500	+	100.0 Hz	
	FM	K4GW	444.12500	+	100.0 Hz	SERA
	FM	KE4QEA	444.20000	+	107.2 Hz	
King, Sauratown Mountain						
	FM	W4SNA	53.95000	52.95000	100.0 Hz	
	FM	W4NC	145.47000	-	100.0 Hz	SERA
	FM	W4SNA	146.79000	-	107.2 Hz	SERA
	FM	K4GW	147.31500	+	100.0 Hz	SERA
	FM	W4SNA	444.75000	+	100.0 Hz	SERA
King, Sauratown Mtn						
	FM	W4WAU	224.72000	-	114.8 Hz	SERA
Kings Mountain	DSTAR	W4NYR	145.08000	146.48000		SERA
	DSTAR	W4NYR	444.18750	+		
Kinston	DMR/MARC	N4DEA	440.65000	+	CC1	
	FM	W4OIX	145.47000	-	88.5 Hz	
	FM	N4HAJ	442.00000	+	88.5 Hz	
Kornegay	FM	N4HAJ	444.12500	+	91.5 Hz	
Lansing	FM	WB4ZCP	224.84000	-	103.5 Hz	
Lansing, Phoenix Mountain						
	FM	W4MLN	444.30000	+	103.5 Hz	
Laurinburg	FM	KI4RR	146.62500	-		
Leland	FM	N4JDW	145.17000	-	88.5 Hz	
Lenoir	DMR/MARC	KG4BCC	443.18750	+	CC1	
	FM	N4LNR	146.62500	-	94.8 Hz	SERA

266   NORTH CAROLINA

Location	Mode	Call sign	Output	Input	Access	Coordinator
Lenoir, Hibriten Mountain						
	FM	KG4BCC	147.33000	+	141.3 Hz	
Level Cross, WFMY Tower						
	FM	W4GG	145.25000	-	88.5 Hz	SERA
Lexington	DMR/MARC	N4TZD	441.93750	+	CC1	
	FM	N4LEX	145.31000	-	107.2 Hz	
	FM	W4PAR	146.91000	-	107.2 Hz	
	FM	K4AE	441.90000	+	127.3 Hz	SERA
	FM	KO0NTZ	442.27500	+	146.2 Hz	
	FM	W4PAR	444.50000	+	146.2 Hz	SERA
Lincolnton	FM	NC4LC	147.01500	+	141.3 Hz	
	FM	WA4YGD	442.35000	+	141.3 Hz	
Locust	FM	W4DEX	147.39000	+	77.0 Hz	SERA
	FM	W4DEX	224.48000	-		SERA
Louisburg	FM	AA4RV	146.80500	-	118.8 Hz	SERA
	FM	KD4CPV	224.22000	-		
Lumberton	FM	W4LBT	147.36000	+	82.5 Hz	
Madison, Second Chance Ranch						
	FM	N4IV	147.34500	+	103.5 Hz	SERA
Maiden	FM	KT4NC	145.17000	-	88.5 Hz	
Malmo	FM	N4ILM	147.06000	+	88.5 Hz	
Mamie	DMR/BM	W4PCN	147.06000	+	CC1	
	DMR/MARC	W4PCN	442.85000	+	CC1	
Mamie, Powells Point						
	FM	W4PCN	146.94000	-	131.8 Hz	SERA
Margarettsville	DMR/MARC	K4MJO	442.35000	+	CC1	
Marion	DMR/MARC	WD4PVE	444.85000	+	CC1	
	FM	WD4PVE	146.98500	-	118.8 Hz	
Mars Hill, Wolf Ridge						
	FM	N2GE	224.66000	-		SERA
Mars Hill, Wolf Ridge Ski Area						
	FM	KI4DNY	442.72500	+	100.0 Hz	SERA
Marshall, Duckett Top Mountain						
	FM	K4HCU	224.36000	-	79.7 Hz	SERA
McCain, McCain Prison Watertan						
	FM	N1RIK	146.80500	-	107.2 Hz	
	FM	N1RIK	442.25000	+	107.2 Hz	SERA
McCain, McCain Watertank						
	FM	N1RIK	927.13750	902.13750	131.8 Hz	SERA
Millers Creek	FM	N4GGN	146.71500	-	94.8 Hz	
Mocksville	DMR/BM	NG8M	444.80000	+	CC8	
	DMR/BM	NG8M	449.80000	-	CC8	
Monroe	FM	NC4UC	145.39000	-	94.8 Hz	
	FM	W4ZO	444.30000	+	100.0 Hz	
	FM	NC4UC	444.42500	+	94.8 Hz	
Mooresville	FM	WG8E	443.82500	+	110.9 Hz	
Moravian Falls	FM	KK4OVN	53.77000	52.77000	100.0 Hz	SERA
	FM	N1KKD	147.22500	+	162.2 Hz	
	FM	KA2NAX	224.12000	-	123.0 Hz	
	FM	KA2NAX	442.67500	+	88.5 Hz	
Morehead City	DMR/MARC	W4YMI	444.97500	+	CC1	
	FM	KF4IXW	53.09000	52.09000	162.2 Hz	
Morganton, High Peak						
	FM	KM4VIQ	147.15000	+	94.8 Hz	SERA
Morganton, High Peak Mountain						
	FM	K4OLC	145.21000	-	94.8 Hz	
Morganton, Jonas Ridge						
	FM	N4HRS	444.62500	+	110.9 Hz	
Morganton, Walker Top Mountain						
	FM	KC4QPR	146.74500	-	94.8 Hz	
Morrisville, Cisco Campus, Bui						
	FM	KC4SCO	444.07500	+	100.0 Hz	

NORTH CAROLINA 267

Location	Mode	Call sign	Output	Input	Access	Coordinator
Mount Airy	FM	KF4UY	444.82500	+	100.0 Hz	
Mount Airy, Fisher Peak						
	FM	KD4ADL	443.42500	+	110.9 Hz	
Mount Gilead, Uwharrie Nationa						
	FM	KI4DH	147.09000	+	100.0 Hz	
	FM	KI4DH	442.20000	+	100.0 Hz	
Mount Mitchell	FM	N4YR	53.63000	52.63000	100.0 Hz	
Moyock	FM	W4NV	443.02500	+		
Mt Mitchell	FM	N4YR	442.22500	+	107.2 Hz	
Murphy	DMR/MARC	N4DTR	444.10000	+	CC1	
	FM	KE4EST	444.75000	+	100.0 Hz	
Nags Head, Bodie Island						
	FM	K4OBX	145.29000	-	131.8 Hz	SERA
Nashville	DMR/MARC	KB4CTS	442.61250	+	CC1	SERA
Needmore	FM	KT4WO	444.55000	+	88.5 Hz	
New Bern	FM	WO3F	442.07500	+	100.0 Hz	
Newell	FM	WT4IX	442.12500	+	156.7 Hz	
Newport	DMR/MARC	WO3F	444.82500	+	CC1	
	FM	K4GRW	29.66000	-		
	FM	K4GRW	145.45000	-	100.0 Hz	
Newton	FM	K4CCR	147.07500	+	88.5 Hz	
North Wilkesboro	FM	N4VL	145.13000	-	103.5 Hz	
Oak Island, Oak Island Water T						
	FM	N4GM	444.60000	+		SERA
Oriental	FM	W4SLH	147.21000	+	151.4 Hz	
Oxford	FM	W4BAD	145.17000	-	100.0 Hz	
	FM	NO4EL	444.60000	+	100.0 Hz	
Pikeville	FM	KI4RK	444.47500	+		SERA
Pinehurst, Moore Regional Hosp						
	FM	N1RIK	145.27000	-	107.2 Hz	SERA
	FM	N1RIK	444.70000	+	107.2 Hz	SERA
Polkville	FM	N4DWP	444.97500	+		
Pollocksville	FM	W4EWN	146.61000	-	100.0 Hz	
Portable Repeater	DMR/MARC	KM4BRM	440.80000	+	CC1	
Purlear	FM	W4NCG	442.25000	+	127.3 Hz	
Raleigh	DMR/MARC	K4JDR	440.91250	+	CC1	
	DMR/MARC	K4HA	442.51250	+	CC1	
	DMR/MARC	K4ITL	442.51250	+	CC1	
	DMR/MARC	K4ITL	443.33750	+	CC1	SERA
	FM	K4ITL	53.03000	52.03000		
	FM	K4GWH	145.19000	-	156.7 Hz	
	FM	W4DW	146.64000	-		
	FM	KD4RAA	146.77500	-	88.5 Hz	
	FM	WB4TQD	146.88000	-	82.5 Hz	SERA
	FM	KA0GMY	147.01500	+	110.9 Hz	
	FM	KC4WDI	441.60000	+	77.0 Hz	
	FM	W4MLU	442.57500	+	79.7 Hz	
	FM	W4RNC	444.52500	+	82.5 Hz	
	FM	K4GWH	444.82500	+	146.2 Hz	
Raleigh, Downtown						
	FM	W4BAD	443.17500	+	100.0 Hz	
Raleigh, North Carolina State						
	FM	W4ATC	442.67500	+	100.0 Hz	
Raleigh, State Fairgrounds						
	FM	K4JDR	441.72500	+	100.0 Hz	SERA
Randleman	FM	K4ITL	147.25500	+	82.5 Hz	
	FM	N2DMR	147.37500	+	114.8 Hz	SERA
	FM	K4ITL	442.82500	+	82.5 Hz	
Ranlo, Spencer Mountain						
	FM	W4BFB	145.23000	-	118.8 Hz	
	FM	N4GAS	147.12000	+	100.0 Hz	SERA
	FM	N4GAS	224.86000	-		

268 NORTH CAROLINA

Location	Mode	Call sign	Output	Input	Access	Coordinator
Ranlo, Spencer Mountain						
	FM	N4GAS	444.55000	+	100.0 Hz	SERA
Reidsville, Cone Health Annie						
	FM	N4IV	146.85000	-	103.5 Hz	SERA
Research Triangle Park						
	FM	W4DW	145.13000	-	82.5 Hz	
Richfield	DMR/BM	KI4UDZ	440.56250	+	CC1	
Richlands	DMR/MARC	KK4VBH	443.97500	+	CC1	
Roanoke Rapids	FM	N4WFU	146.74500	-	131.8 Hz	SERA
Roaring Gap	FM	WA4PXV	53.05000	52.05000	151.4 Hz	
Robbinsville	FM	N4GSM	145.11000	-	151.4 Hz	
	FM	K4KVE	442.37500	+	103.5 Hz	
Robbinsville, Joanna Bald Moun						
	FM	K4KVE	53.31000	52.31000	167.9 Hz	SERA
Rockfish, Rockfish Watertank						
	FM	KG4HDV	442.10000	+	100.0 Hz	
Rockingham	FM	K4RNC	146.95500	-	88.5 Hz	SERA
	FM	KF4DBW	442.58750	+		SERA
Rocky Mount	FM	N4JEH	29.66000	-		
	FM	WR4RM	147.12000	+	131.8 Hz	
	FM	K4ITL	147.18000	+	82.5 Hz	
	FM	KR4AA	224.58000	-	91.5 Hz	
	FM	NC4ES	444.70000	+	107.2 Hz	SERA
	FM	WN4Z	444.85000	+	131.8 Hz	
Rolesville	FM	AA4RV	444.95000	+	88.5 Hz	
Rosman	FM	W4TWX	444.87500	+		
Rougemont	FM	NC4CD	443.27500	+	100.0 Hz	
Roxboro	FM	W4BAD	441.67500	+	100.0 Hz	
Salisbury	FM	KU4PT	224.76000	-		
Sanford	DMR/MARC	K4ITL	442.33750	+	CC1	SERA
	DSTAR	KE4DSU	147.50500	146.50500		
	FM	KB4HG	441.95000	+	136.5 Hz	
	P25	W0SMT	146.61000	-		
Scotland Neck	FM	NC4FM	444.07500	+	203.5 Hz	
Selma, Bailey Feed Mill Elevat						
	FM	K4JDR	444.15000	+	100.0 Hz	SERA
Shannon	FM	N4DBM	444.57500	+	77.0 Hz	SERA
Shelby	FM	W4NYR	147.34500	+		
	FM	W4NYR	224.06000	-		
	FM	N4DWP	224.46000	-		
	FM	AE6JI	444.27500	+	127.3 Hz	
Snow Camp, Cane Creek Mtn						
	DSTAR	AK4EG	145.32000	-		
	DSTAR	AK4EG	444.88750	+		
	DSTAR	AK4EG	1284.40000	1264.40000		
Sophia	DMR/MARC	K4NWJ	440.71250	+	CC1	
	FM	WR4BEG	224.14000	-		
	FM	WR4BEG	443.07500	+		
Southern Pines	DMR/MARC	KF4DBW	443.76250	+	CC1	
Sparta	FM	W4DCA	53.97000	52.97000	100.0 Hz	
	FM	W4DCA	443.95000	+	100.0 Hz	
	FM	N1RIK	927.02500	902.02500	107.2 Hz	SERA
Sparta, Air Bellows Gap						
	FM	WA4PXV	29.67000	-	151.4 Hz	
	FM	WA4PXV	442.15000	+	192.8 Hz	SERA
	FM	WA4PXV	442.60000	+	94.8 Hz	SERA
Sparta, Green Mountain						
	FM	K4ITL	145.43000	-		
Sparta, Roaring Gap						
	FM	WA4PXV	443.57500	+	173.8 Hz	
Spruce Pine	DMR/MARC	KC4TVO	443.38750	+	CC1	

NORTH CAROLINA 269

Location	Mode	Call sign	Output	Input	Access	Coordinator
Spruce Pine, Iowa Hill						
	FM	KK4MAR	147.21000	+	123.0 Hz	SERA
Spruce Pine, Woodys Knob						
	FM	KK4MAR	443.92500	+	123.0 Hz	
St Pauls, UNC-TV 31 Tower						
	FM	W4LBT	147.04500	+		SERA
Stacy	FM	KD4KTO	444.00000	+	131.8 Hz	
Stanfield	FM	W4DEX	443.20000	+	77.0 Hz	
Statesville	FM	KK4OVN	443.77500	+	118.8 Hz	SERA
	FM	N4SZF	443.85000	+		
Statesville, WSIC Tower						
	FM	W4SNC	146.68500	-	77.0 Hz	SERA
Sugar Mountain	DMR/MARC	WA4NC	442.08750	+	CC1	
Sunshine, Cherry Mountain						
	FM	KG4JIA	147.24000	+	131.8 Hz	
	FM	KG4JIA	224.64000	-	71.9 Hz	
Swansboro, Swansboro Water Tow						
	FM	KE4FHH	146.76000	-	88.5 Hz	
Swepsonville	FM	K4EG	146.67000	-		SERA
Sylva	DMR/MARC	W3WDD	440.68750	+	CC1	
	DMR/MARC	W3WDD	442.97500	+	CC1	SERA
	FM	KJ4VKD	444.15000	+		
Sylva, Kings Mountain						
	FM	KF4DTL	147.34500	+	151.4 Hz	SERA
Tarboro, Vidant Edgecombe Hosp						
	FM	NC4ES	444.50000	+	100.0 Hz	SERA
Taylorsville	FM	W4ERT	147.19500	+	94.8 Hz	
	FM	W4ERT	441.62500	+	123.0 Hz	
Thomasville	FM	W4TNC	441.80000	+	127.3 Hz	
Thomasville, Water Tank On Com						
	FM	WW4DC	443.32500	+	88.5 Hz	
Tobaccoville	FM	KK4GAF	447.40000	-	103.5 Hz	
Tryon	DSTAR	KK4LVF	442.87500	+		
	FM	KF4JVI	145.33000	-	91.5 Hz	
	FM	KJ4SPF	442.55000	+	107.2 Hz	
	FM	K4SV	442.87500	+	123.0 Hz	
Tryon, White Oak Mountain						
	FM	W4RCW	147.28500	+		
Wadesboro	FM	W4USH	147.31500	+	74.4 Hz	
Wake Forest	DMR/BM	KD2LH	444.30000	+	CC1	
Washington	DMR/BM	NC4ES	440.52500	+	CC1	
Waves	FM	K4OBX	444.32500	+	131.8 Hz	
Waxhaw	FM	K4WBT	444.52500	+	94.8 Hz	
Waynesville	DMR/MARC	K4KGB	443.85000	+	CC1	
	DMR/MARC	N4DTR	444.45000	+	CC1	SERA
	DMR/MARC	N4DTR	927.66250	902.66250	CC1	
Waynesville, Chambers Mountain						
	FM	N4DTR	147.39000	+	94.8 Hz	SERA
Waynesville, Sylva						
	FM	K4RCC	444.87500	+	131.8 Hz	SERA
Wendell, NC Viper Tower						
	FM	KD4RAA	444.87500	+	100.0 Hz	SERA
West Jefferson	DMR/MARC	W4MLN	444.73750	+	CC1	
Wilkesboro	DMR/MARC	W4FAR	440.53750	+	CC1	
	FM	W4MIS	145.04000	-	100.0 Hz	
	FM	WB4PZA	146.82000	-	94.8 Hz	SERA
Wilkesboro, Pores Knob						
	FM	W4FAR	145.37000	-	94.8 Hz	
Williamston	DMR/MARC	K4SER	442.38750	+	CC1	SERA
	FM	K4SER	53.31000	52.31000	131.8 Hz	SERA
	FM	K4SER	145.41000	-	131.8 Hz	SERA
	FM	K4SER	444.25000	+	131.8 Hz	SERA

270 NORTH CAROLINA

Location	Mode	Call sign	Output	Input	Access	Coordinator
Wilmington	FM	N4ILM	29.66000	-	88.5 Hz	
	FM	N4JDW	53.33000	52.33000	88.5 Hz	
	FM	AD4DN	53.43000	52.43000	88.5 Hz	
	FM	NI4SR	145.41000	-	67.0 Hz	
	FM	AD4DN	146.67000	-	88.5 Hz	SERA
	FM	AC4RC	147.18000	+	88.5 Hz	
	FM	WA4US	224.20000	-		
	FM	KD4MEA	224.68000	-	91.5 Hz	
	FM	WA4US	442.17500	+		
	FM	WA4US	442.20000	+		
	FM	AC4RC	442.50000	+	88.5 Hz	
	FM	KB4FXC	442.75000	+	67.0 Hz	
	FM	AD4DN	443.40000	+	88.5 Hz	
	FM	WA4US	443.85000	+		
	FM	WA4US	443.95000	+		
	FM	WA4US	444.20000	+		
	FM	WA4US	444.45000	+		
	FM	KB4FXC	444.50000	+	67.0 Hz	
	FM	WA4US	444.65000	+		
	FM	N4PLY	444.77500	+	131.8 Hz	
Wilmington, Porter's Neck						
	FM	N4ILM	146.73000	-	88.5 Hz	
Wilson	FM	WA4WAR	146.76000	-	131.8 Hz	
	FM	WA4AEC	444.90000	+	179.9 Hz	
Wingate	DMR/MARC	W4ZO	444.38750	+	CC1	
	DSTAR	W4FAN	444.86250	+		
	FM	N4HRS	443.95000	+	110.9 Hz	SERA
Winnabow	FM	N4ILM	146.82000	-	88.5 Hz	SERA
	FM	K4ITL	147.34500	+	88.5 Hz	
	FM	N4ILM	444.85000	+	88.5 Hz	
Winston-Salem	FM	N4YR	224.60000	-	107.2 Hz	
	FM	KD4MMP	444.72500	+	107.2 Hz	
Winston-Salem, NC Baptist Hosp						
	FM	W4NC	146.64000	-	100.0 Hz	SERA
	FM	W4NC	444.27500	+	100.0 Hz	SERA
Wolf Ridge, Ski Area						
	FM	K4MFD	147.18000	+	94.8 Hz	
Yadkinville	FM	KM4MHZ	146.61000	-	85.4 Hz	SERA
	FM	N4YSB	147.01500	-	100.0 Hz	
	FM	KD4KMK	442.02500	+	100.0 Hz	
	FM	N4AAD	442.80000	+	100.0 Hz	SERA
Youngsville	FM	WB4TQD	145.39000	-	82.5 Hz	
	FM	N4TAB	442.17500	+	82.5 Hz	
	FM	WB4IUY	442.30000	+	100.0 Hz	
Zebulon	FM	WB4IUY	224.80000	-	88.5 Hz	
NORTH DAKOTA						
Bismarck	FM	W0ZRT	146.85000	-	107.2 Hz	
	FM	KC0AHL	444.65000	+	107.2 Hz	
Bowman, Twin Buttes						
	FM	KB0DYA	145.31000	-		
Carrington, Big Chief						
	FM	K0BND	147.58500		100.0 Hz	
Cavalier	FM	N0CAV	147.15000	+		
Cleveland	FM	W0FX	53.01000	52.01000		
Fargo	FM	K0EED	145.49000	-	82.5 Hz	
	FM	W0RRW	146.97000	-		
Fargo, Multiband Tower						
	FM	K0EED	145.49000	-	82.5 Hz	
Fargo, NDSU	FM	W0HSC	147.09000	+		
	FM	KD0SWQ	444.00000	+		
Forbes	FM	KC5ZCH	145.33000	-	136.5 Hz	

NORTH DAKOTA 271

Location	Mode	Call sign	Output	Input	Access	Coordinator
Glen Ullin	FM	KD7RDD	147.30000	+	162.2 Hz	
Grand Forks	FM	N0LAC	147.33000	+	123.0 Hz	
Grandin	FM	W0ILO	146.76000	-	123.0 Hz	
Hannover	FM	W0ZRT	145.43000	-		
Harlow	FM	KF0HR	147.01500	+	123.0 Hz	
Jamestown	FM	WB0TWN	444.92500	+		NDFC
Killdeer	FM	K0ND	146.64000	-		
Lisbon	FM	N0BQY	147.00000	-		
Maddock	FM	KF0HR	147.24000	+	141.3 Hz	
	FM	K0AJW	449.80000	-	67.0 Hz	
Mandan	FM	W0ZRT	146.94000	-		
Mandan, Old Red Trail						
	FM	N0FAZ	444.20000	+	103.5 Hz	
Minot	FM	K0AJW	146.97000	-	67.0 Hz	
	FM	W0ND	444.50000	+	67.0 Hz	
	FM	K0AJW	444.80000	+		
Rugby	FM	WB0ATB	147.06000	+		
Sentinal Butte	FM	K0ND	146.73000	-		NDFC
St. Anthony	FM	W0ZRT	146.85000	-	123.0 Hz	NDFC
Stanley	FM	K0WSN	146.61000	-	100.0 Hz	
	FM	K0PHH	146.79000	-		
	FM	K0PHH	443.25000	+		
Wahpeton, College						
	FM	W0END	147.37500	+		
	FM	W0END	443.80000	+		
Williston	FM	AB0JX	147.21000	+	100.0 Hz	
	FM	K0WSN	443.85000	+		

The ARRL Operating Manual
Your Guide to On-Air Operating Activities

Talking to friends. Serving your community. Chasing DX. Adding a new band. Trying a new mode. Working a contest. Taking your radio on the road or out in the field. These are just a few of the exciting and fun on-air operating activities to try once you have earned your license and mastered the basics.

The ARRL Operating Manual is as a comprehensive guide to amateur radio operating. This edition has been updated by experienced hams who are active on the air. They are happy to share what they have learned so that you can get involved and on the air too.

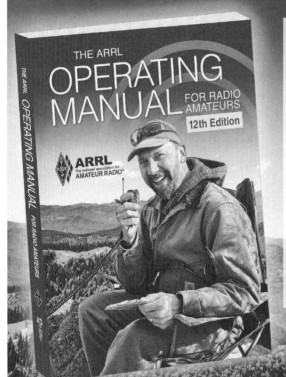

Get On the Air!

- Learn basic operating procedures and how to set up your station.
- Become an active ham, contact radio amateurs around the world.
- Join a local radio club and enjoy amateur radio group activities.
- Volunteer your radio skills to serve your community.
- Participate in contests, and earn operating awards.

Includes new material on FT8 and other *WSJT-X* digital modes.

The ARRL Operating Manual
ARRL Item No. 1205
Only $24.95

ARRL The national association for AMATEUR RADIO

www.arrl.org/shop
Toll-Free US 888-277-5289,
or elsewhere +1-860-594-0355

OHIO 273

Location	Mode	Call sign	Output	Input	Access	Coordinator
OHIO						
21550 Ball Ave	DMR/MARC	N8CXU	449.85000	+	CC1	
Akron	DMR/BM	W8UPD	443.11250	+	CC1	OARC
	DMR/BM	WB8AVD	444.51250	+	CC7	
	DMR/MARC	KD8ASA	440.50000	+	CC1	
	FM	N8XPK	53.17000	52.17000	107.2 Hz	OARC
	FM	W8UPD	145.17000	-		OARC
	FM	WB8HHP	146.95500	-	110.9 Hz	OARC
	FM	N8XPK	1292.20000	1272.20000		
Akron, Spring Hill Apartments						
	FM	N8XPK	147.13500	+	110.9 Hz	OARC
Alliance	FM	W8LKY	145.37000	-	110.9 Hz	OARC
	FM	W8LKY	442.35000	+	131.8 Hz	
Amherst	DMR/MARC	N1TVI	444.33750	+	CC1	
	FM	WD8OCS	146.62500	-	110.9 Hz	OARC
Andover	DMR/BM	W8VFD	444.96250	+	CC2	
	DMR/MARC	N8OHU	444.25000	+	CC1	
Archbold	FM	N8RLD	145.41000	-	107.2 Hz	OARC
Ashland	DMR/BM	N8IHI	444.03750	+	CC7	
Ashley	FM	KC8BVF	444.62500	+		OARC
Athens	FM	W8UKE	145.15000	-		OARC
	FM	KC8AAV	147.15000	+		OARC
Attica	FM	N0CZV	443.67500	+	131.8 Hz	OARC
Atwater	DMR/BM	W8FAA	435.00000		CC7	
	DMR/BM	W8FAA	443.97500	+	CC7	
	FM	WA8LCA	224.14000	-		
Avon	FM	N8SIW	443.70000	+		
Bainbridge	DMR/MARC	KD8SPV	443.58750	+	CC7	
	FM	W8BAP	146.92500	-	74.4 Hz	OARC
	FM	WR8ANN	147.06000	+	110.9 Hz	OARC
	FM	KD8GRN	443.62500	+		
Barberton	DMR/BM	KC8ZKI	442.13750	+	CC1	
	FM	WB8OVQ	147.09000	+	110.9 Hz	OARC
Barnesville	FM	WB8WJT	146.64000	-	100.0 Hz	OARC
Bascom	FM	KB8EOC	145.15000	-		
	FM	N8VWZ	146.68500	-		OARC
	FM	KB8EOC	442.30000	+		
Batavia, WOBO Tower						
	FM	K8YOJ	224.00000	-	123.0 Hz	OARC
Beavercreek	FM	N8NPT	53.73000	52.73000		OARC
	FM	N8DCP	224.58000	-		OARC
Bellbrook	DMR/BM	N8NQH	444.87500	+	CC13	OARC
	FM	W8GCA	146.91000	-		OARC
	FM	W8DGN	147.04500	+	118.8 Hz	OARC
	FM	W8DGN	443.67500	+		
Bellefontaine, Campbell Hill						
	FM	W8FTV	147.00000	+	100.0 Hz	OARC
	FM	W8FTV	443.82500	+	186.2 Hz	OARC
Bellevue	FM	NF8E	442.62500	+	110.9 Hz	OARC
Belpre	FM	N8NBL	146.97000	-	91.5 Hz	OARC
	FM	KI8JK	147.31500	+	103.5 Hz	
Berlin Hts	FM	WB8LLY	146.80500	-	110.9 Hz	OARC
Bethel, Ohio	FM	W8GFT	444.27500	+	141.3 Hz	
Birmingham	FM	K8KXA	443.52500	+		OARC
Blanchester	FM	KB8CWC	145.25000	-	162.2 Hz	OARC
	FM	KB8CWC	442.02500	+		OARC
Bloomfield	DMR/BM	KD8MST	442.05000	+	CC2	
Boardman	FM	KD8ODF	444.52500	+	110.9 Hz	
Bowling Green	DMR/BM	WD8LEI	443.91250	+	CC1	
	FM	KD8BTI	146.79000	-	103.5 Hz	OARC
	FM	K8TIH	147.18000	+	67.0 Hz	OARC

274 OHIO

Location	Mode	Call sign	Output	Input	Access	Coordinator
Bowling Green	FM	K8TIH	442.12500	+	67.0 Hz	OARC
	FM	KD8BTI	443.51250	+	103.5 Hz	OARC
	FM	K8TIH	444.47500	+	67.0 Hz	OARC
Brecksville	FM	K8IIU	442.65000	+	131.8 Hz	OARC
Brunswick	FM	N8OVW	53.19000	52.19000		OARC
	FM	K8SCI	145.29000	-	110.9 Hz	OARC
	FM	N8OVW	443.02500	+	131.8 Hz	
Bryan	FM	KA8OFE	146.82000	-		OARC
Bucyrus	DMR/BM	KD8NCL	443.87500	+	CC1	OARC
Cable	FM	WB8UCD	224.86000	-	100.0 Hz	OARC
Cadiz	FM	WB8FPN	146.65500	-	114.8 Hz	OARC
Caldwell	FM	NC8OH	147.28500	+		OARC
Cambridge	FM	W8VP	146.85000	-	91.5 Hz	OARC
	FM	KB8ZMI	147.00000	+	91.5 Hz	OARC
	FM	KB8ZMI	444.37500	+	91.5 Hz	OARC
Canal Winchester	FM	KA8ZNY	224.38000	-		
Canfield	FM	KD8DWV	145.27000	-	110.9 Hz	OARC
	FM	KC8WY	442.75000	+	131.8 Hz	OARC
Canton	DMR/BM	KG8DQ	444.78750	+	CC7	
	FM	KB8MIB	53.57000	52.57000	110.9 Hz	OARC
	FM	W8AL	146.79000	-	141.3 Hz	OARC
	FM	W8TUY	443.85000	+		OARC
Canton, Mercy Medical Center						
	FM	N8ATZ	147.12000	+	110.9 Hz	OARC
Carrollton	FM	K8VPJ	51.86000	50.86000		
	FM	K8VPJ	442.40000	+		OARC
	FM	N8RQU	442.58750	+		OARC
	FM	K8VPJ	442.62500	+		OARC
Centerville	DMR/MARC	WB8SCT	147.13500	+	CC1	
	FM	WB8ART	145.43000	-	88.5 Hz	OARC
	FM	KC8QGP	444.60000	+		OARC
Chagrin Falls	FM	KF8YK	444.22500	+	131.8 Hz	
Chardon	FM	W8DES	146.94000	-	110.9 Hz	OARC
	FM	KF8YK	927.56250	902.56250	131.8 Hz	OARC
	P25	KF8YK	444.81250	+		OARC
Chesapeake	FM	W8SOE	146.71500	-	103.5 Hz	OARC
Chesterland	FM	K9IC	224.40000	-	141.3 Hz	
	FM	K9IC	444.60000	+	131.8 Hz	OARC
Chillicothe	DMR/MARC	KD8SPV	444.35000	+	CC7	OARC
	DMR/MARC	KD8SPV	444.42500	+	CC7	
Chillicothe, Scioto Trail Stat						
	DSTAR	W8BAP	147.01500	+		OARC
	FM	KA8WWI	53.23000	52.23000		OARC
	FM	W8BAP	146.85000	-	74.4 Hz	OARC
Cincinnati	DMR/MARC	WB8FXJ	442.17500	+	CC1	
	DMR/MARC	WB8CRS	443.40000	+	CC1	
	DMR/MARC	K8BIG	443.90000	+	CC1	
	DSTAR	K8BIG	145.35000	-		OARC
	FM	KD8TE	53.19000	52.19000	114.8 Hz	OARC
	FM	N8SIM	145.31000	-	123.0 Hz	OARC
	FM	K8YOJ	145.37000	-		OARC
	FM	K8SCH	146.67000	-	123.0 Hz	
	FM	WR8CRA	146.76000	-	162.2 Hz	OARC
	FM	K8YOJ	146.85000	-	123.0 Hz	OARC
	FM	K8SCH	146.92500	-	123.0 Hz	OARC
	FM	K8BIG	147.09000	+	123.0 Hz	OARC
	FM	W8VND	147.24000	+	123.0 Hz	OARC
	FM	K8YOJ	224.06000	-	100.0 Hz	OARC
	FM	W8ESS	224.62000	-	110.9 Hz	OARC
	FM	N8JRX	442.20000	+	192.8 Hz	
	FM	W8NWS	443.70000	+		OARC
	FM	KB8SBN	444.07500	+	110.9 Hz	OARC

OHIO 275

Location	Mode	Call sign	Output	Input	Access	Coordinator
Cincinnati	FM	W8ESS	444.22500	+	110.9 Hz	OARC
	FM	N8TVU	444.30000	+	118.8 Hz	OARC
	FM	KB8BWE	444.75000	+		OARC
	FM	K8CF	444.92500	+		OARC
	FM	W8ESS	927.55000	902.55000	110.9 Hz	OARC
	P25	K8BIG	147.15000	+		OARC
	P25	K8BIG	444.00000	+	123.0 Hz	
Cincinnati, Star Tower						
	P25	K8BIG	927.02500	902.02500	123.0 Hz	
Circleville	FM	KD8HIJ	147.18000	+	74.4 Hz	OARC
	FM	KD8HIJ	442.70000	+		OARC
Clarksfield	DMR/MARC	KD8DRG	442.17500	+	CC7	
Cleveland	DMR/MARC	KD8TWG	444.95000	+	CC1	
	DSTAR	WB8THD	145.35000	-		OARC
	DSTAR	WB8THD	442.32500	+		OARC
	FM	NA8SA	147.19500	+	110.9 Hz	OARC
	FM	WX8CLE	442.12500	+	82.5 Hz	
	FM	N8OND	444.27500	+	131.8 Hz	OARC
Clyde	FM	NF8E	145.35000	-	110.9 Hz	OARC
Coldwater	FM	W8MCA	145.25000	-	107.2 Hz	OARC
Colerain Township						
	FM	K8CR	443.57500	+	123.0 Hz	OARC
Columbus	ATV	WR8ATV	444.97500	+		OARC
	DMR/BM	KD8USF	444.85000	+	CC7	
	DMR/MARC	W8TRB	145.37000	-	CC1	
	DMR/MARC	N8RQJ	441.50000	+	CC1	
	DMR/MARC	W8TRB	443.15000	+	CC0	OARC
	DMR/MARC	W5BSG	443.15000	+	CC1	
	DMR/MARC	N8RQJ	443.42500	+	CC1	OARC
	DMR/MARC	W8TRB	927.01250	902.01250	CC1	
	DMR/MARC	W8TRB	927.52500	902.52500	CC1	
	DSTAR	W8DIG	145.39000	-		OARC
	DSTAR	W8CMH	145.49000	-		OARC
	DSTAR	W8DIG	442.65000	+		OARC
	DSTAR	W8CMH	444.00000	+		OARC
	DSTAR	W8DIG	1285.00000	1265.00000		OARC
	FM	WC8OH	145.11000	-	100.0 Hz	OARC
	FM	WB8MMR	145.23000	-		OARC
	FM	WA8PYR	145.27000	-	82.5 Hz	
	FM	N8PVC	145.43000	-	123.0 Hz	OARC
	FM	W8ZPF	146.67000	-	131.8 Hz	OARC
	FM	W8AIC	146.76000	-	123.0 Hz	OARC
	FM	WB8LAP	146.80500	-		OARC
	FM	W8RRJ	146.97000	-	123.0 Hz	OARC
	FM	K8DDG	147.06000	+	94.8 Hz	OARC
	FM	W8CQK	147.15000	+		OARC
	FM	N8OIF	147.21000	+		OARC
	FM	K8DRE	147.24000	+		OARC
	FM	WR8ATV	427.25000	439.25000		OARC
	FM	K8NIO	442.80000	+	151.4 Hz	
	FM	WB8YOJ	443.57500	+		OARC
	FM	K8MK	443.81250	+		OARC
	FM	N8YMT	444.10000	+		OARC
	FM	WA8PYR	444.17500	+	82.5 Hz	
	FM	W8AIC	444.20000	+	151.4 Hz	OARC
	FM	WB8INY	444.27500	+	94.8 Hz	OARC
	FM	WB8YOJ	444.30000	+		OARC
	FM	N8ADL	444.90000	+		OARC
	FM	KA8ZNY	927.48750	902.48750	131.8 Hz	OARC
	FM	WR8ATV	1250.00000	1280.00000		OARC
	FM	WR8ATV	1258.00000	1280.00000		OARC
	P25	W8DIG	443.65000	+		OARC

276   OHIO

Location	Mode	Call sign	Output	Input	Access	Coordinator
Columbus	P25	W8DIG	927.03750	902.03750	131.8 Hz	
Columbus Grove	DMR/BM	K8TEK	444.30000	+	CC1	
Columbus, Ohio State Universit						
	FM	W8LT	442.60000	+	114.8 Hz	OARC
Conneaut	FM	W8BHZ	147.39000	+	DCS 047	OARC
Constitution, On The WTAP Towe						
	FM	N8MDG	444.92500	+	146.2 Hz	
Constitution, WTAP Tower						
	FM	W8TAP	146.74500	-	114.8 Hz	OARC
Copley	FM	WD8BIW	147.24000	+		
Cortland	FM	N8GZE	147.10500	+	114.8 Hz	OARC
	FM	WA8ILI	443.87500	+		OARC
Coshocton	FM	KE8BDF	145.23000	-	71.9 Hz	OARC
Cridersville	FM	W8EQ	444.92500	+		OARC
Cuyahoga Falls	FM	W8VPV	147.27000	+	110.9 Hz	OARC
	FM	W8VPV	444.85000	+	110.9 Hz	OARC
Dayton	DMR/MARC	WA8PLZ	147.36000	+	CC1	OARC
	DMR/MARC	W8AK	442.87500	+	CC1	
	DMR/MARC	WB8SCT	444.60000	+	CC1	OARC
	DSTAR	W8RTL	1283.50000	1263.50000		OARC
	FM	WC8OH	145.11000	-	67.0 Hz	OARC
	FM	K8MCA	146.64000	-	123.0 Hz	OARC
	FM	WA8PLZ	146.82000	-	77.0 Hz	OARC
	FM	W8BI	146.94000	-	123.0 Hz	OARC
	FM	WF8M	146.98500	-	123.0 Hz	OARC
	FM	W8BI	223.94000	-		OARC
	FM	WC8OH	224.16000	-		OARC
	FM	KB8CSL	224.72000	-		
	FM	W8BI	442.10000	+	123.0 Hz	OARC
	FM	WB8VSU	442.30000	+	123.0 Hz	OARC
	FM	W6CDR	442.75000	+	94.8 Hz	OARC
	FM	KD8GRN	443.22500	+		
	FM	KB8CSL	443.60000	+		OARC
	FM	WF8M	443.77500	+	131.8 Hz	OARC
	FM	W8KSE	444.25000	+	123.0 Hz	OARC
	FM	N8YFM	927.48750	902.48750		
Defiance	FM	K8VON	147.09000	+	107.2 Hz	OARC
	FM	K8VON	442.57500	+	107.2 Hz	OARC
Defiance, OH	FM	KT8EMA	444.73750	+		
Delaware	DMR/BM	KC8BPE	442.13750	+	CC1	
	FM	W8SMK	145.17000	-		OARC
	FM	N8DCA	145.19000	-		OARC
	FM	KA8IWB	145.29000	-	123.0 Hz	
	FUSION	KE8O	443.55000	+	151.4 Hz	OARC
Delphos	FM	W8YEK	147.12000	+		OARC
	FM	KB8UDX	443.15000	+	107.2 Hz	OARC
Delta	FM	K8LI	147.28500	+	103.5 Hz	OARC
	FM	K8LI	444.45000	+	103.5 Hz	OARC
Deshler	FM	KC8QYH	444.93750	+	103.5 Hz	
Doylestown	DMR/BM	W8WKY	442.27500	+	CC1	OARC
	FM	W8WKY	147.39000	+	114.8 Hz	OARC
East Liverpool	FM	K8BLP	146.70000	-	162.2 Hz	OARC
	FM	K8BLP	442.17500	+		OARC
East Palestine	FM	W8GMM	146.77500	-		OARC
Eaton	FM	K8YR	145.47000	-	100.0 Hz	OARC
	FM	W8VFR	444.02500	+	74.4 Hz	OARC
	FM	W8VFR	444.93750	+	97.4 Hz	OARC
Eaton, Preble County EMA Offic						
	FM	K8YR	442.90000	+	100.0 Hz	
Elyria	DMR/BM	NW8S	444.31250	+	CC1	
	DSTAR	WA8DIG	443.43750	+		OARC
	FM	W8HF	145.23000	-	110.9 Hz	OARC

OHIO 277

Location	Mode	Call sign	Output	Input	Access	Coordinator
Elyria	FM	K8KRG	146.70000	-	110.9 Hz	OARC
	FM	WD8CHL	224.04000	-	141.3 Hz	
	FM	KA8VDW	443.98750	+	162.2 Hz	OARC
	FM	KB8O	444.31250	+	131.8 Hz	OARC
	FM	K8KRG	444.80000	+	131.8 Hz	OARC
Elyria, American Red Cross Cha						
	FM	KC8BED	147.15000	+	110.9 Hz	OARC
Elyria, Lorain County Communit						
	FM	KC8BED	444.17500	+	131.8 Hz	
Englewood	FM	KB8ZR	443.50000	+	103.5 Hz	OARC
Euclid	DMR/MARC	N8CXU	444.85000	+	CC1	
Fairborn	FM	KI6SZ	51.66000	50.66000		OARC
	FM	K8FBN	145.41000	-	118.8 Hz	OARC
	FM	K8FBN	442.37500	+	118.8 Hz	OARC
	FM	N8QBS	442.57500	+	127.3 Hz	OARC
	FM	N8QBS	442.82500	+		OARC
	FM	KI6SZ	444.31250	+		OARC
Fairfield	FM	WC8VOA	145.39000	-		OARC
	FM	W8WRK	146.70000	-	123.0 Hz	OARC
	FM	W8KJ	443.57500	+		
Fairfield, Jungle Jim's						
	FM	KD8WDU	443.95000	+	123.0 Hz	
Fairlawn	FM	N8NOQ	443.75000	+	131.8 Hz	OARC
Fairport Hrbr	FM	N8JCV	224.08000	-	141.3 Hz	OARC
Fayette	FM	KB8GOM	442.07500	+	103.5 Hz	OARC
Findlay	DMR/BM	W8FT	443.70000	+	CC1	
	FM	W8FT	147.15000	+	88.5 Hz	OARC
	FM	N8PC	442.87500	+	100.0 Hz	OARC
	FM	W8JES	443.42500	+	91.5 Hz	
Fort Recovery	FM	KB8SCR	442.67500	+	107.2 Hz	OARC
	FM	KB8SCR	444.81250	+	107.2 Hz	OARC
Franklin	FM	WB8ZVL	145.29000	-	118.8 Hz	OARC
	FM	WE8N	442.42500	+	77.0 Hz	OARC
	FM	WB8ZVL	443.15000	+	118.8 Hz	OARC
Fremont	DMR/BM	N8SCA	442.06250	+	CC7	
	DMR/BM	N8FIS	443.00000	+	CC7	OARC
	FM	W8NCK	145.25000	-	186.2 Hz	OARC
	FM	N8SCA	145.49000	-	107.2 Hz	OARC
	FM	KC8EPF	443.45000	+		OARC
Fresno	FM	KB9JSC	443.53750	+	162.2 Hz	OARC
Ft. Recovery	FM	KB8SCR	223.96000	-	107.2 Hz	
	FM	KB8SCR	442.67500	+	107.2 Hz	OARC
Gahanna	FM	KB8SXJ	442.50000	+	82.5 Hz	OARC
Galena	FM	W8RUT	224.66000	-		
Galion	FM	W8BAE	146.85000	-	71.9 Hz	OARC
Gallipolis	DMR/MARC	KD8SPV	444.11250	+	CC7	
	FM	KC8ZAB	147.06000	+	74.4 Hz	OARC
Gallipolis, Mound Hill						
	FM	KC8ZAB	442.27500	+	74.4 Hz	OARC
Gallipolis, Warehime Hill						
	FM	K8ATG	146.73000	-	91.5 Hz	
Georgetown	FM	N1DJS	146.73000	-	162.2 Hz	OARC
Germantown	FM	WG8ARS	443.18750	+	123.0 Hz	
Gibsonburg	FM	K8KXA	444.76250	+	131.8 Hz	
Goshen	FM	K8DV	443.45000	+	123.0 Hz	OARC
Greenfield, WVNU Radio Tower						
	FM	K8HO	146.68500	-		OARC
	FM	K8HO	444.77500	+		OARC
Greenhills, 740' Tower						
	FM	WB8CRS	146.88000	-	123.0 Hz	OARC
Greenville	DMR/BM	W8FLH	441.37500	+	CC1	
	FM	W8QIY	146.79000	-	94.8 Hz	OARC

278   OHIO

Location	Mode	Call sign	Output	Input	Access	Coordinator
Greenville	FM	N8OBE	444.17500	+		OARC
Hamilton	DMR/MARC	KD8VLU	442.96250	+	CC1	
	DSTAR	W8RNL	145.15000	-		OARC
	DSTAR	W8RNL	442.62500	+		OARC
	DSTAR	KD8TUZ	442.65000	+		OARC
	DSTAR	W8RNL	1293.00000	1273.00000		OARC
	FM	W8CCI	146.97000	-		
	FM	W8WRK	147.00000	+	123.0 Hz	OARC
	FM	W8CCI	147.33000	+	118.8 Hz	OARC
	FM	WB8TCB	224.54000	-		
	FM	W8WRK	443.33750	+		OARC
	FM	KD8EYB	444.11250	+	118.8 Hz	OARC
	FM	K8KY	444.65000	+	123.0 Hz	OARC
Hamilton, North C Street Hill						
	FM	W8AJT	224.88000	-		
Hannibal	FM	WB8CSW	147.24000	+		OARC
Hanoverton, Guilford Lake Stat						
	FM	NN8B	444.91250	+		
Hayesville	FM	N8IHI	147.10500	+	71.9 Hz	OARC
Hayesville, Ashland/Mansfield						
	FM	KD8BIW	224.58000	-	110.9 Hz	OARC
Highland Hills	FM	WB8APD	444.95000	+		OARC
Hillsboro	DMR/BM	W8CTC	443.07500	+	CC1	
	DMR/MARC	KD8RWK	443.07500	+	CC1	
	FM	K8HO	147.21000	+	100.0 Hz	OARC
Hillsboro, Rocky Fork State Pa						
	FM	K8HO	444.67500	+		OARC
Hinckley Township						
	FM	W8WGD	443.42500	+	131.8 Hz	OARC
Holland	FM	K8TTE	444.21250	+	103.5 Hz	
	FM	W8AK	444.27500	+	107.2 Hz	OARC
Hubbard	FM	W8IZC	443.10000	+		OARC
Huber Heights	DMR/BM	K1CCN	443.26250	+	CC7	
	FM	NO8I	224.30000	-	123.0 Hz	OARC
	FM	W8AK	442.92500	+	123.0 Hz	OARC
	FM	NO8I	442.95000	+	118.8 Hz	
Huber Heights, W8BI Clubhouse						
	DSTAR	W8HEQ	145.27000	-		
	DSTAR	W8HEQ	444.08750	+		
Hudson	DMR/BM	W8HD	444.08750	+	CC1	
	FM	K8KSW	145.25000	-		OARC
	FM	KD8FL	443.47500	+		
Ironton	FM	W8SOE	444.62500	+	103.5 Hz	OARC
Jacksonville	FM	KC8QDQ	147.22500	+		OARC
Jersey	FM	W8RRJ	52.70000	52.90000	123.0 Hz	
	FM	KA8ZNY	927.58750	902.58750		
Johnstown	DMR/BM	AC8GI	145.47000	-	CC7	
	DMR/BM	KD8USF	145.47000	-	CC7	
	DMR/BM	AC8GI	443.28750	+	CC7	
	DMR/BM	KD8USF	443.28750	+	CC7	
Keene	FM	W8CCA	147.04500	+		OARC
Kent	DMR/BM	N8ZPS	442.02500	+	CC1	OARC
	DMR/BM	K8GI	443.06250	+	CC10	
	DMR/BM	N8BHU	444.30000	+	CC7	OARC
	FM	N8ZPS	145.11000	-		
	FM	K8IV	146.89500	-	118.8 Hz	
	FM	N8BHU	224.02000	-	141.3 Hz	OARC
Kenton	FM	W8VMV	146.62500	-	85.4 Hz	OARC
Kettering	DMR/MARC	W8AK	443.75000	+	CC1	OARC
	FM	W8GUC	444.66250	+	123.0 Hz	OARC
Kingsville	FM	N8XUA	443.65000	+	103.5 Hz	OARC

OHIO 279

Location	Mode	Call sign	Output	Input	Access	Coordinator
Lafayette Township						
	FM	W8EOC	444.92500	+	131.8 Hz	OARC
Lakewood	FM	WR8ABC	146.88000	-	110.9 Hz	OARC
	FM	WR8ABC	224.90000	-	141.3 Hz	OARC
	FM	WR8ABC	444.70000	+		OARC
Lancaster	DMR/MARC	KC8MLN	147.37500	+	CC8	
	DMR/MARC	KC8MLN	442.97500	+	CC8	
	FM	K8QIK	53.09000	52.09000	71.9 Hz	
	FM	K8QIK	146.70000	-	94.8 Hz	OARC
	FM	K8QIK	147.03000	+	71.9 Hz	OARC
	FM	K8QIK	443.87500	+	71.9 Hz	OARC
Lebanon	DSTAR	KE8AOQ	443.15000	+		
	FM	WC8EMA	444.18750	+	88.5 Hz	OARC
Lexington	FM	WD8Q	443.22500	+	146.2 Hz	OARC
Lexington, Mid-Ohio Sports Car						
	FM	N8YOA	147.01500	+		
Lima	FM	WB8ULC	146.67000	-		OARC
	FM	W8AOH	146.74500	-	100.0 Hz	OARC
	FM	W8EQ	146.94000	-		OARC
	FM	K8TCF	147.03000	+		OARC
	FM	N8GCH	444.07500	+		OARC
Lima, Husky Refinery Complex						
	FM	N8GCH	145.17000	-		OARC
Lima, WLIO Tower						
	DSTAR	KT8APR	443.62500	+		OARC
	FM	KT8APR	53.63000	52.63000	107.2 Hz	OARC
	FM	KT8APR	145.37000	-	107.2 Hz	OARC
Lisbon	DMR/BM	KD8XB	443.93750	+	CC10	
	FM	KC8PHW	421.25000	434.05000		OARC
	FM	KD8XB	442.52500	+	162.2 Hz	OARC
Lisbon, MARCS Tower						
	FM	K8GQB	146.80500	-	162.2 Hz	OARC
Litchfield	FM	W8EOC	51.66000	-	107.2 Hz	OARC
Lithopolis	FM	KB8WQ	145.21000	-	71.9 Hz	
Logan	DMR/BM	KD8ORN	442.35000	+	CC1	
	FM	K8LGN	443.12500	+		OARC
London	FM	KE8RV	147.28500	+	82.5 Hz	OARC
Lorain	FM	WA8CAE	443.60000	+	131.8 Hz	OARC
	FM	WD8CHL	444.12500	+	131.8 Hz	OARC
Loudonville	FM	W3YXS	146.74500	-	71.9 Hz	
Loveland	FM	WD8KPU	442.77500	+		OARC
	FM	WB8BFS	443.80000	+		OARC
	FM	WU8S	444.52500	+	100.0 Hz	OARC
Macedonia	FM	WR8ABC	444.40000	+	131.8 Hz	OARC
Malvern	FM	K8VPJ	147.07500	+	131.8 Hz	OARC
Mansfield	DMR/BM	KD8EVR	444.70000	+	CC7	
	DMR/BM	W8WE	444.70000	+	CC7	OARC
	FM	W8NDB	145.33000	-	71.9 Hz	OARC
	FM	K8HF	147.36000	+	71.9 Hz	OARC
	FM	KA8VDW	443.07500	+	151.4 Hz	OARC
Mantua	DMR/MARC	KD8DRG	443.55000	+	CC7	
Marietta	DMR/BM	N8OJ	145.33000	-	CC7	
	DMR/BM	N8OJ	442.47500	+	CC7	OARC
	DMR/BM	N8OJ	442.72500	+	CC7	OARC
	DMR/BM	N8OJ	442.90000	+	CC7	
	DMR/BM	N8OJ	444.10000	+	CC1	
	DMR/MARC	W8JL	442.90000	+	CC7	
	DMR/MARC	N8OJ	443.05000	+	CC1	
	FM	KI8JK	146.80500	-		OARC
	FM	W8HH	146.88000	-	91.5 Hz	OARC
	FM	W8JTW	443.05000	+	186.2 Hz	OARC
	FM	W8HH	443.40000	+	91.5 Hz	OARC

280 OHIO

Location	Mode	Call sign	Output	Input	Access	Coordinator
Marion	DMR/BM	KC8BPE	442.18750	+	CC1	
	DMR/MARC	KC8BPE	442.01250	+	CC1	
	FM	WW8MRN	146.89500	-	250.3 Hz	OARC
	FM	WW8MRN	147.30000	+	250.3 Hz	OARC
	FM	KC8BPE	444.98750	+	110.9 Hz	
	FM	WW8MRN	447.01250	-	123.0 Hz	
	FM	W8MRN	449.73750	-	123.0 Hz	
Marysville	DMR/BM	K8JWL	443.45000	+	CC7	OARC
	FM	N8IG	145.35000	-	127.3 Hz	OARC
	FM	N8YRF	147.39000	+		OARC
Mason	DMR/BM	KB8ML	444.95000	+	CC1	
	FM	W8BRQ	145.13000	-		OARC
	FM	W8ESS	442.27500	+	110.9 Hz	OARC
	FM	WB8WFG	444.15000	+		OARC
	FM	W8SAI	444.95000	+	131.8 Hz	OARC
Massillon	FM	WA8GXM	53.05000	52.05000	136.5 Hz	OARC
	FM	W8NP	147.18000	+	110.9 Hz	OARC
	FM	W8NP	442.85000	+	131.8 Hz	OARC
	FM	WA8GXM	443.67500	+	DCS 023	OARC
Maumee	FM	W8TER	444.70000	+	103.5 Hz	OARC
Mayfield Heights	FM	N8QBB	51.62000	-	107.2 Hz	OARC
McArthur	FM	W8VCO	147.10500	+	88.5 Hz	OARC
McConnelsvill	FM	WB8VQV	147.19500	+		OARC
McDermott	FM	KB8RBT	443.32500	+		OARC
Medina	DMR/MARC	N8OND	443.32500	+	CC1	
	FM	W8EOC	147.03000	+	141.3 Hz	OARC
	FM	W8HN	147.28500	+	110.9 Hz	
	FM	W8EOC	224.86000	-		OARC
	FM	N8OND	444.27500	+		
	FM	N8OND	927.68750	902.68750	131.8 Hz	OARC
Mentor	DMR/MARC	N8NOD	443.47500	+	CC1	
	FM	N9AGC	145.21000	-	110.9 Hz	OARC
	FM	N8BC	147.16500	+	110.9 Hz	OARC
	FM	WB8PHI	147.25500	+		OARC
	FM	N9AGC	444.47500	+	131.8 Hz	OARC
Mentor, Tri-Point Hospital						
	FM	N8BC	224.50000	-	141.3 Hz	OARC
	FM	N8BC	444.65000	+	131.8 Hz	OARC
Mentor, Tripoint Hospital						
	FM	N8BC	147.21000	+	110.9 Hz	OARC
Miamisburg	FM	W6CDR	146.77500	-	77.0 Hz	OARC
	FM	W8NCI	147.01500	+	77.0 Hz	OARC
	FM	W8DYY	147.19500	+		OARC
	FM	NV8E	442.45000	+	123.0 Hz	OARC
	FM	W8DYY	443.00000	+	88.5 Hz	OARC
Miamisburg, Sycamore Hospital						
	FM	W8DYY	145.33000	-		OARC
Middle Point	FM	W8FY	146.85000	-		OARC
Middlefield	FM	KC8IBR	442.25000	+		OARC
Middletown	DMR/MARC	W8BLV	444.36250	+	CC1	
	FM	W8BLV	146.61000	-	77.0 Hz	OARC
	FM	N8COZ	146.71500	-		OARC
	FM	W8JEU	147.31500	+	77.0 Hz	OARC
	FM	W8MUM	443.53750	+	118.8 Hz	
	FM	AG8Y	444.47500	+	100.0 Hz	OARC
	FM	W8BLV	444.82500	+	77.0 Hz	OARC
Miller City	FM	NO8C	443.56250	+	107.2 Hz	OARC
Millersburg	FM	KD8QGQ	146.67000	-	71.9 Hz	OARC
	FM	KD8CJ	444.87500	+	131.8 Hz	OARC
Minerva	FM	KD8XB	442.95000	+	162.2 Hz	OARC
Minford	DMR/MARC	KD8SPV	443.11250	+	CC7	
	FM	KF8YO	224.92000	-	114.8 Hz	

OHIO   281

Location	Mode	Call sign	Output	Input	Access	Coordinator
Mogadore	FM	KC8MXW	444.48750	+		OARC
Monroe	FM	WA8MU	442.55000	+	118.8 Hz	OARC
Montville	FM	N8XUA	145.33000	-		OARC
	FM	N8XUA	443.45000	+		OARC
Montville Township (						
	DMR/BM	AL7OP	442.47500	+	CC7	OARC
Mount Gilead	DMR/BM	WN7C	444.91250	+	CC7	OARC
Mount Vernon	DMR/BM	KD8EVR	442.10000	+	CC7	OARC
Mt Eaton	FM	KB8PXM	53.33000	52.33000	100.0 Hz	OARC
Mt Gilead	DMR/BM	W8NL	444.86250	+	CC1	OARC
	FM	WY8G	224.94000	-	71.9 Hz	OARC
Mt Gilead, US 42	FM	W8NL	146.77500	-	107.2 Hz	OARC
Munroe Falls	FM	WB8CXO	147.33000	+	110.9 Hz	
Napoleon	FM	K8TII	147.22500	+		OARC
	FM	K8TII	147.31500	+		OARC
Nashport	DMR/MARC	K8QG	145.25000	-	CC1	
New Concord	DMR/BM	KD8USF	442.05000	+	CC2	
New Lexington	DMR/BM	KD8MST	442.07500	+	CC2	
	DMR/BM	KD8USF	442.07500	+	CC2	
New Springfield	DMR/BM	KD8XB	444.88750	+	CC10	
	FM	K8TKA	147.31500	+	156.7 Hz	
New Springfield, Glacier Hills						
	FM	K8TKA	9999.99999	10000.99999	156.7 Hz	
New Springfld	FM	KF8YF	443.52500	+	162.2 Hz	OARC
Newark	FM	W8WRP	146.88000	-	141.3 Hz	OARC
	FM	WD8RVK	442.05000	+		OARC
	FM	W8AJL	443.60000	+	71.9 Hz	
	FM	W8WRP	444.50000	+	141.3 Hz	OARC
Newbury	FM	K8SGX	52.68000	52.88000	107.2 Hz	OARC
	FM	W8LYD	146.85000	-	110.9 Hz	OARC
	FM	WB8QGR	444.62500	+	131.8 Hz	OARC
	FM	K8SGX	444.97500	+	DCS 251	OARC
Newton Falls	FM	N8VPR	147.22500	+		OARC
North Ridgeville	FM	K8IC	444.50000	+		OARC
North Royalton	FM	K8KRG	145.15000	-	110.9 Hz	OARC
	FM	K8SCI	224.76000	-		OARC
	FM	K8SCI	443.15000	+	131.8 Hz	OARC
	FM	WA8CEW	443.90000	+		OARC
	FM	K8YSE	444.07500	+	131.8 Hz	OARC
Northwood	FM	KB8YVY	442.42500	+	103.5 Hz	OARC
Norton	DMR/BM	KE8LDH	442.51250	+	CC1	OARC
	DMR/BM	KD8DRG	443.85000	+	CC7	
	DMR/MARC	KD8DRG	444.00000	+	CC7	
	FM	WB8UTW	53.15000	52.15000		OARC
	FM	WA8DBW	146.68000	-	110.9 Hz	OARC
	FM	WB8UTW	224.06000	-		OARC
	FM	WD8KNL	444.00000	+		OARC
Norwalk	FM	KA8VDW	442.67500	+	162.2 Hz	OARC
Oak Harbor	FM	K8VXH	147.07500	+	100.0 Hz	OARC
	FM	K8VXH	442.25000	+	100.0 Hz	
	FM	WB8JLT	443.85000	+	186.2 Hz	OARC
Oregon	FM	N8UAS	443.30000	+	103.5 Hz	OARC
	FM	W8MTU	444.92500	+	103.5 Hz	
Orrville	FM	KD8SQ	146.71500	-		OARC
Orwell	FM	KF8YF	146.65500	-		OARC
	FM	KF8YF	444.25000	+		OARC
Ottawa	FM	W8ZRZ	443.88750	+	107.2 Hz	OARC
Ottawa, Putnam County Health D						
	FM	W8MCB	146.71500	-	107.2 Hz	OARC
Owensville	FM	W8MRC	147.34500	+	123.0 Hz	
Oxford	FM	W8CCI	145.21000	-	118.8 Hz	OARC
Painesville	DMR/MARC	N1TVI	443.33750	+	CC1	

282   OHIO

Location	Mode	Call sign	Output	Input	Access	Coordinator
Parma	FM	WA8Q	145.31000	-	110.9 Hz	OARC
	FM	KB8WLW	145.41000	-	110.9 Hz	OARC
	FM	K8ZFR	146.82000	-	110.9 Hz	OARC
	FM	KB8WLW	224.48000	-	141.3 Hz	OARC
	FM	W8DRZ	442.05000	+	131.8 Hz	OARC
	FM	KB8WLW	442.22500	+	131.8 Hz	OARC
	FM	KB8WLW	442.45000	+		OARC
	FM	K8ZFR	443.82500	+	131.8 Hz	
	FM	WR8SS	444.77500	+	131.8 Hz	OARC
	FM	W8CJB	444.90000	+	131.8 Hz	OARC
	FM	KB8WLW	927.61250	902.61250	131.8 Hz	
Parma, WJW Tower						
	FM	W8DRZ	444.05000	+	131.8 Hz	OARC
Pataskala	FM	W8NBA	147.33000	+	123.0 Hz	OARC
Paulding	FM	KE8FJX	147.13500	+	141.3 Hz	OARC
	FM	KE8FJX	442.27500	+	141.3 Hz	OARC
Peebles	FM	KJ8I	145.17000	-		OARC
	FM	KJ8I	442.67500	+		OARC
	FM	WD8LSN	444.02500	+		OARC
Perrysburg	FM	W8ODR	444.51250	+		OARC
Philo	FM	W8ZZV	146.61000	-	74.4 Hz	OARC
Piqua	FM	N8OWV	442.12500	+	123.0 Hz	OARC
	FM	W8AK	444.72500	+	123.0 Hz	OARC
Piqua, Water Tower						
	FM	W8SWS	147.21000	+	67.0 Hz	
Polk	FM	N8SIW	145.13000	-	110.9 Hz	OARC
	FM	WA8TJC	442.15000	+	88.5 Hz	OARC
	FM	N8SIW	442.97500	+	131.8 Hz	OARC
	FM	KA8VDW	443.62500	+	162.2 Hz	OARC
Pomeroy	FM	KC8LOE	146.86500	-	88.5 Hz	OARC
	FM	K8ATG	443.70000	+	91.5 Hz	OARC
Portable	DMR/MARC	W8JL	442.60000	+	CC7	OARC
Portsmouth	FM	KC8BBU	147.36000	+	136.5 Hz	OARC
	FM	N8QA	444.60000	+		
Ravenna	FM	KB8ZHP	145.39000	-		OARC
	FM	KB8ZHP	442.87500	+	100.0 Hz	OARC
	FM	K8IV	444.57500	+		OARC
Ray	DMR/MARC	KD8SPV	146.89500	-	CC7	OARC
	DMR/MARC	KD8SPV	442.22500	+	CC7	OARC
	FM	W8YUL	146.79000	-		OARC
Republic	FM	KC8RCI	147.25500	+	107.2 Hz	OARC
	FM	KC8RCI	444.43750	+	107.2 Hz	OARC
Reynoldsburg	FM	W8FEH	146.91000	-	71.9 Hz	OARC
	FM	W8LAD	443.95000	+		OARC
Richfield	DMR/BM	KA8OAD	444.51250	+	CC7	OARC
	FM	N8CPI	442.55000	+	131.8 Hz	OARC
	FM	KA8JOY	443.92500	+	131.8 Hz	OARC
Rittman	DMR/BM	WW8TF	442.37500	+	CC1	OARC
	DMR/BM	KE8LDG	442.73750	+	CC1	OARC
	DMR/BM	N8OFP	443.73750	+	CC7	
Roseville	FM	N8ROA	442.17500	+	91.5 Hz	OARC
Salem	FM	KB8MFV	29.68000	-		
	FM	KB8MFV	53.03000	52.03000	88.5 Hz	OARC
	FM	KB8MFV	146.86500	-	110.9 Hz	OARC
	FM	KA8OEB	147.25500	+	156.7 Hz	OARC
	FM	KB8MFV	147.28500	+	88.5 Hz	
	FM	KB8MFV	442.10000	+	88.5 Hz	OARC
	FM	KA8OEB	444.67500	+	156.7 Hz	OARC
	FM	K8BTP	444.96250	+		
Sandusky	DMR/BM	KD8AVO	444.73750	+	CC7	
	FM	W8LBZ	444.37500	+	110.9 Hz	OARC
Sciotoville	FM	W8KKC	145.45000	-	114.8 Hz	

OHIO   283

Location	Mode	Call sign	Output	Input	Access	Coordinator
Seaman	FM	KC8FBG	444.51250	+		OARC
Shaker Heights	FM	KD8LDE	224.38000	-	131.8 Hz	
	FM	K8ZFR	444.75000	+	131.8 Hz	OARC
Shaker Heights, Shaker Towers						
	FM	KD8LDE	443.80000	+		OARC
Shelby	FM	W8DZN	147.16500	+	88.5 Hz	
	FM	W8DZN	442.52500	+	88.5 Hz	OARC
Sidney	DMR/MARC	N8YFM	443.90000	+	CC1	OARC
	FM	W8AK	147.34500	+	107.2 Hz	OARC
	FM	W8JSG	442.47500	+		OARC
	FM	KE8BCY	443.20000	+		OARC
	FM	N6JSX	444.88750	+	107.2 Hz	OARC
Solon	FM	KD8ZNQ	147.50500			
South Charleston	FM	KB8GJG	53.39000	52.39000		OARC
South Euclid	FM	N8APU	146.79000	-	88.5 Hz	OARC
South Webster	FM	N8QA	145.39000	-		OARC
South Zanesville	DMR/BM	KD8USF	443.10000	+	CC2	
	DMR/BM	KD8USF	444.07500	+	CC2	
	DMR/BM	KD8MST	444.40000	+	CC2	
	DMR/BM	KD8MST	444.95000	+	CC2	
	FM	KD8MST	442.24000	+	DCS 143	
Springboro	DMR/MARC	W8CYE	147.13500	+	CC1	
	FM	W8CYE	145.49000	-	77.0 Hz	OARC
	FM	K8DZ	223.82000	-	77.0 Hz	OARC
	FM	N8RXL	224.22000	-	100.0 Hz	OARC
Springfield	DMR/MARC	KC8NYH	442.51250	+	CC1	
	DMR/MARC	KC8NYH	443.41250	+	CC1	OARC
	FM	W8OG	145.31000	-	82.5 Hz	
	FM	K8IRA	145.45000	-	123.0 Hz	OARC
	FM	W8OG	146.73000	-	77.0 Hz	OARC
	FM	KA8HMJ	147.22500	+	100.0 Hz	OARC
	FM	KA8HMJ	444.41250	+		OARC
Springfield, Ohio Masonic Home						
	FM	W8BUZ	443.30000	+	146.2 Hz	OARC
St Clairsville	FM	W8GBH	145.21000	-		OARC
St Marys	FM	W8GZ	146.80500	-	107.2 Hz	OARC
	FM	K8QYL	147.33000	+	107.2 Hz	OARC
St. Paris	FM	WB8UCD	224.60000	-	100.0 Hz	OARC
Steubenville, WTOV-TV Tower						
	FM	WD8IIJ	147.06000	+	114.8 Hz	OARC
Stone Creek	FM	W8ZX	444.82500	+	71.9 Hz	
Stonecreek	FM	W8ZX	146.73000	-	71.9 Hz	OARC
Stoutsville	DMR/MARC	KD8SPV	442.20000	+	CC7	
	FM	KD8GRN	443.06250	+		OARC
Stow	FM	AF1K	442.42500	+	131.8 Hz	OARC
Sugarcreek	FM	W8ZX	146.92500	-	71.9 Hz	OARC
SWANTON, Ohio	FM	N3VGX	442.36250	+	103.5 Hz	
Sylvania	FM	KC8GWH	443.77500	+	103.5 Hz	OARC
Thompson	FM	KB8FKM	224.96000	-	141.3 Hz	OARC
	FM	WB5OD	443.70000	+		
	FM	KF8YK	444.56250	+	131.8 Hz	OARC
Thornville	DMR/BM	AC8GI	145.25000	-	CC7	OARC
	DMR/BM	AC8GI	442.45000	+	CC7	OARC
	DMR/BM	AC8GI	444.85000	+	CC7	OARC
	FM	KB8ZMI	146.83500	-	91.5 Hz	OARC
Tiffin	DMR/BM	W3BWW	444.82500	+	CC1	OARC
	FM	K8EMR	443.80000	+	107.2 Hz	OARC
Tiffin, Kiwanis Manor						
	FM	W8ID	145.45000	-	107.2 Hz	OARC
Tipp City	FM	N8RVS	444.53750	+		OARC
Toledo	DMR/BM	WA8SRV	443.75000	+	CC1	
	DMR/BM	WB8WEA	444.16250	+	CC1	

284   OHIO

Location	Mode	Call sign	Output	Input	Access	Coordinator
Toledo	DMR/MARC	N8EFJ	444.85000	+	CC1	OARC
	DMR/MARC	KD8KCF	446.12500	-	CC1	
	DSTAR	W8HHF	442.75000	+		OARC
	DSTAR	KD8QOF	444.25000	+		
	FM	K8ALB	146.61000	-	103.5 Hz	OARC
	FM	WJ8E	147.34500	+	103.5 Hz	OARC
	FM	W8RZM	147.37500	+	103.5 Hz	OARC
	FM	WJ8E	442.95000	+	103.5 Hz	OARC
	FM	N8LPQ	443.97500	+		OARC
	FM	N8LPQ	444.95000	+	103.5 Hz	OARC
	FM	W8HHF	927.02500	902.02500	131.8 Hz	OARC
	FM	KD8KCF	927.91250	902.91250	131.8 Hz	
	FM	WJ8E	1285.00000	1265.00000	103.5 Hz	OARC
	FM	WJ8E	1287.00000	1267.00000		
	FUSION	W8HHF	146.83500	-		OARC
Toledo, University Of Toledo						
	FM	W8HHF	224.14000	-	103.5 Hz	OARC
Toledo, University Of Toledo C						
	FM	W8HHF	147.27000	+	103.5 Hz	OARC
	FM	W8HHF	442.85000	+	103.5 Hz	OARC
Troy	DMR/BM	WI8DX	446.60000	-	CC1	
	FM	W8FW	145.23000	-	100.0 Hz	OARC
	FM	KD8KID	147.24000	+		OARC
	FM	N8OWV	223.98000	-		OARC
	FM	WD8CMD	442.97500	+		OARC
	FM	KB8MUV	443.63750	+		OARC
Uhrichsville	FM	K8CQA	443.50000	+		OARC
Uniontown	FM	WD8BIW	53.25000	52.25000		OARC
	FM	WD8BIW	145.45000	-	110.9 Hz	OARC
	FM	WB8OVQ	442.00000	+	131.8 Hz	OARC
Upper Sandusky	FM	KE8PX	147.21000	+	107.2 Hz	OARC
Urbana	FM	K8VOR	146.95500	-	100.0 Hz	OARC
	FM	K7GUN	147.22500	+		
	FM	WB8UCD	147.37500	+	100.0 Hz	OARC
	FM	K7GUN	224.98000	-	100.0 Hz	OARC
	FM	WB8UCD	443.17500	+	123.0 Hz	OARC
	FM	K7DN	443.35000	+		OARC
Van Wert	FM	W8FY	146.70000	-		OARC
	FM	W8FY	434.00000	923.30000		OARC
	FM	W8FY	444.85000	+	136.5 Hz	OARC
	FM	N8IHP	446.02500	447.02500	156.7 Hz	
Vermilion	FM	W8DRZ	443.05000	+	131.8 Hz	OARC
Vermillion	FM	KA8VDW	53.29000	52.29000	107.2 Hz	OARC
Vienna	FM	N8NVI	147.04500	+		OARC
Wadsworth	FM	KD8DRG	145.49000	-	110.9 Hz	OARC
	FM	WB8UTW	927.66250	902.66250		OARC
Wapakoneta	FM	KD8CQL	442.15000	+	107.2 Hz	OARC
Warren	FM	N8DOD	146.83500	-	131.8 Hz	OARC
	FM	N8DOD	224.16000	-	131.8 Hz	OARC
	FM	KA9YTS	442.82500	+	131.8 Hz	OARC
	FM	W8VTD	443.00000	+		OARC
	FM	N8DOD	443.72500	+	131.8 Hz	OARC
	FM	WA8ILI	444.83750	+		OARC
Warrensville Heights						
	DMR/MARC	N8NOD	442.08750	+	CC1	OARC
Washington Court House						
	FM	N8EMZ	147.27000	+		OARC
	FM	N8QLA	442.07500	+	77.0 Hz	OARC
	FM	N8EMZ	444.61250	+		OARC
Waterville	DMR/BM	N8XLJ	443.77500	+	CC1	
Wauseon	FM	KB8MDF	53.41000	52.41000		OARC
	FM	K8BXQ	147.19500	+	103.5 Hz	OARC

OHIO 285

Location	Mode	Call sign	Output	Input	Access	Coordinator
Waynesburg	FM	KC8ONY	442.20000	+		OARC
Wellington	FM	K8TV	444.66250	+		OARC
Wellston	DMR/BM	N8OJ	147.37500	+	CC7	OARC
	DMR/BM	N8OJ	443.25000	+	CC1	
West Carrollton	FM	K8ZQ	444.50000	+		OARC
West Chester	DMR/BM	W8SDR	443.73750	+	CC1	
	DMR/MARC	W8SDR	444.97500	+	CC1	
	FM	WC8RA	442.32500	+	123.0 Hz	OARC
	FM	W8VVL	442.70000	+	123.0 Hz	OARC
West Chester, VOA Museum						
	FM	WC8VOA	443.65000	+		OARC
West Lafayette	FM	WX8OH	443.32500	+	71.9 Hz	OARC
West Milton	FM	N8EIO	444.56250	+		OARC
West Salem	FM	KE8X	53.27000	52.27000	107.2 Hz	OARC
	FM	KE8X	443.30000	+	131.8 Hz	OARC
Wickliffe	FM	WA8PKB	444.15000	+		OARC
	FM	WA8PKB	444.72500	+		OARC
Willow Wood	FM	W8SOE	146.61000	-	103.5 Hz	OARC
Wilmington	FM	WB8ZZR	147.12000	+	123.0 Hz	OARC
	FM	WB8ZZR	442.15000	+		OARC
	FM	K8IO	443.37500	+	123.0 Hz	OARC
	FM	WB8ZZR	444.57500	+	141.3 Hz	OARC
Wintersville, Blessed Sacramen						
	FM	WD8IIJ	443.77500	+		OARC
Woodsfield	FM	WD8RED	147.27000	+		
Wooster	DMR/BM	W8WOO	443.17500	+	CC1	OARC
	FM	W8WOO	147.21000	+		OARC
	FM	WB8VPG	147.34500	+	110.9 Hz	OARC
	FM	K8WAY	443.40000	+		OARC
	FM	KD8EU	444.25000	+	131.8 Hz	OARC
Wrightsville	FM	KF8RC	147.18000	+	118.8 Hz	OARC
Xenia	FM	W8XRN	147.16500	+	123.0 Hz	OARC
	FM	W8XRN	443.10000	+	123.0 Hz	OARC
Youngstown	DMR/BM	W8IZC	443.25000	+	CC1	OARC
	FM	W8QLY	146.74500	-	110.9 Hz	OARC
	FM	KB8N	147.00000	+		OARC
Youngstown, WHOT Tower						
	FM	W8IZC	146.91000	-	110.9 Hz	OARC
Zanesville	DMR/BM	KD8MST	444.10000	+	CC2	OARC
	DMR/BM	KD8USF	444.10000	+	CC2	
	FM	KB8ZMI	147.07500	+	91.5 Hz	OARC
	FM	KJ8N	224.94000	-		OARC
	FM	KB8ZMI	442.25000	+	91.5 Hz	OARC
	FM	W8TJT	444.95000	+		OARC
OKLAHOMA						
Ada	FM	KE5GLC	145.27000	-	141.3 Hz	ORSI
	FM	WB5NBA	147.28500	+	114.8 Hz	ORSI
Alfalfa, Perry Boradcasting To						
	FM	K5GSM	145.11000	-	DCS 712	ORSI
Altus	DMR/BM	WX5ASA	146.89500	-	CC1	
	DSTAR	AJ5Q	442.22500	447.32500		ORSI
	FM	WX5ASA	147.28500	+	100.0 Hz	ORSI
	FM	WX5ASA	442.05000	+	100.0 Hz	ORSI
Altus, Navajo Mountain						
	FM	WX5ASA	146.79000	-	100.0 Hz	ORSI
Altus, OK, Navajo Mountain						
	DSTAR	AJ5Q	432.20000			
Alva, Hardtner	FM	W5ALZ	444.90000	+	103.5 Hz	ORSI
Anadarko	FM	WX5LAW	147.27000	+		ORSI
Antlers	FM	KI5KC	145.49000	-		ORSI
	FM	KD5DAR	444.20000	+	88.5 Hz	ORSI

286   OKLAHOMA

Location	Mode	Call sign	Output	Input	Access	Coordinator
Antlers	FM	KI5KC	444.92500	+	114.8 Hz	ORSI
Ardmore	FM	W5BLW	146.97000	-		ORSI
	FM	KB5LLI	147.07500	+	123.0 Hz	ORSI
	FM	W5CVE	433.50000		DCS 023	
Bartlesville	DMR/BM	W5RAB	442.18750	+	CC1	ORSI
	DMR/BM	W5RAB	444.47500	+	CC2	
	FM	W5NS	146.65500	-	88.5 Hz	ORSI
	FM	W5NS	146.76000	-	88.5 Hz	ORSI
	FM	W5RAB	224.26000	-	88.5 Hz	ORSI
	FM	KD5IMA	443.12500	+	88.5 Hz	ORSI
	FM	W5IAS	444.42500	+	88.5 Hz	ORSI
	FM	W5RAB	927.65000	902.65000	DCS 074	ORSI
	IDAS	KD5IMA	145.15000	-	88.5 Hz	ORSI
	IDAS	KF5FWE	442.27500	+	100.0 Hz	ORSI
	IDAS	KF5FWE	444.25500	+	100.0 Hz	
Bethany	FM	N5USR	224.96000	-	103.5 Hz	ORSI
	FM	N5USR	442.85000	+	103.5 Hz	ORSI
	FM	WA5CZN	444.05000	+	192.8 Hz	ORSI
Big Cabin	DMR/BM	W5RAB	443.82500	+	CC1	
Big Cabin, Big Cabin						
	FM	W5RAB	223.94000	-	88.5 Hz	
Big Cedar	FM	N5JMG	147.37500	+	123.0 Hz	ORSI
Bixby, Leonard Mountain						
	FM	W5RAB	224.18000	-	88.5 Hz	
Blackwell	FM	KD5MTT	145.31000	-		ORSI
Blackwell, Water Tower						
	FM	KD5MTT	444.95000	+		ORSI
Blanchard	DSTAR	KF5ZLE	442.97500	+		ORSI
	FM	W0DXA	442.00000	+	141.3 Hz	ORSI
	FM	W5LHG	444.62500	+	127.3 Hz	ORSI
Bridge Creek	FM	KS5B	145.47000	-	141.3 Hz	ORSI
	FM	W5PAA	146.85000	-	141.3 Hz	ORSI
	FM	W5PAA	444.85000	+	141.3 Hz	ORSI
Broken Arrow	DMR/BM	WX5OU	443.50000	+	CC1	
Broken Arrow, Tiger Hill						
	FM	W5DRZ	146.91000	-	88.5 Hz	ORSI
	FM	W5DRZ	444.00000	+		ORSI
Broken Bow	FM	K7RM	53.03000	51.33000	100.0 Hz	ORSI
Broken Bow, Carter Mountain						
	FM	K5DYW	147.13500	+	67.0 Hz	ORSI
Buffalo	FM	W5HFZ	52.81000	51.11000	131.8 Hz	ORSI
	FM	W5HFZ	145.13000	-	131.8 Hz	ORSI
	FM	W5GPR	147.12000	+	203.5 Hz	ORSI
	FM	W5HFZ	442.07500	+	131.8 Hz	ORSI
Calumet	DMR/BM	AE5DN	444.32500	+	CC1	
	FM	WA5FLT	146.61000	-		ORSI
Cement	FM	KW5FAA	444.40000	+	141.3 Hz	ORSI
	WX	WX5LAW	444.45000	+	123.0 Hz	ORSI
Chandler	FM	WB5BCR	147.36000	+	192.8 Hz	ORSI
Chickasha	FM	W5KS	145.23000	-	141.3 Hz	ORSI
Choctaw	FM	K5CAR	147.09000	+	141.3 Hz	ORSI
Chouteau	FM	K5LEE	145.13000	-		ORSI
Claremore	FUSION	WX5RC	444.35000	+		
Claremore, Rogers State Univer						
	FM	WX5RC	444.35000	+		
Clarita	FM	KF5IUL	442.02500	+	114.8 Hz	ORSI
Clayton	FM	KM5VK	146.73000	-	114.8 Hz	ORSI
	FM	N5AVV	444.10000	+		
Cleora	DMR/BM	W5BIV	442.62500	+	CC1	
Clinton	FM	K5ODN	147.30000	+	127.3 Hz	ORSI
Clinton, At 200 Ft On Wright R						
	FM	KE5RRK	145.19000	-	88.5 Hz	ORSI

OKLAHOMA 287

Location	Mode	Call sign	Output	Input	Access	Coordinator
Coalgate, Water Tower At Airpo						
	FM	KF5IUL	145.25000	-	123.0 Hz	ORSI
Coleman	FM	WG5B	147.16500	+	131.8 Hz	ORSI
Cyril	FM	KB5LLI	147.04500	+	123.0 Hz	ORSI
	WX	KB5LLI	442.27500	+	123.0 Hz	ORSI
Cyrill	FM	KB5LLI	147.00000	+		ORSI
Daisy	FM	W5CUQ	145.21000	-	100.0 Hz	ORSI
Daisy, Bald Mountain						
	FM	W5CUQ	145.21000	-	114.8 Hz	ORSI
Davis	FM	KN6UG	146.86500	-	192.8 Hz	ORSI
Davis, Arbuckle Mountains						
	FM	WG5B	145.23000	-	179.9 Hz	
	FM	WG5B	147.15000	+	131.8 Hz	ORSI
	FM	WG5B	443.07500	+	107.2 Hz	ORSI
Del City	FM	W5DEL	146.70000	-	103.5 Hz	ORSI
	FM	WN5J	443.10000	+	100.0 Hz	ORSI
	FM	W5DEL	443.30000	+	103.5 Hz	ORSI
Del City, Arvest Bank						
	FM	W5MWC	442.95000	+		
Depew	DMR/BM	WD5ETD	444.37500	+	CC1	
Duncan	FM	WD5IYF	146.73000	-	118.8 Hz	ORSI
	FM	WD5IYF	444.82500	+	118.8 Hz	ORSI
Durant	FM	K5CGE	147.25500	+	114.8 Hz	ORSI
	FM	K5KIE	147.39000	+	118.8 Hz	ORSI
	WX	K5BQG	146.98500	-	118.8 Hz	ORSI
Edmond	DMR/BM	W5RLW	442.32500	+	CC1	
	DMR/BM	KP4DJT	443.02500	+	CC7	
	DMR/BM	W5RLW	443.05000	+	CC1	ORSI
	FM	K5CPT	145.21000	-	131.8 Hz	ORSI
	FM	K5SBH	145.27000	-	100.0 Hz	ORSI
	FM	K5EOK	147.13500	+	79.7 Hz	ORSI
	FM	KD5SKS	442.20000	+	123.0 Hz	ORSI
	FM	K5CPT	442.22500	447.32500	131.8 Hz	ORSI
	FM	KC5GEP	443.15000	+	79.7 Hz	ORSI
	FM	K5EOK	443.42500	+	88.5 Hz	ORSI
	FM	KD5SKS	444.42500	+		ORSI
El Reno	FM	K5OL	442.25000	+	141.3 Hz	ORSI
	FM	W5ELR	444.25000	+		ORSI
Elk City	DMR/MARC	W5GY	147.22500	+	CC1	ORSI
	DSTAR	K5ELK	146.68500	-		ORSI
	DSTAR	K5ELK	443.02500	+		ORSI
	FM	WX5BSA	444.52500	+	88.5 Hz	ORSI
Elmore City	WX	KB5LLI	146.74500	-	141.3 Hz	ORSI
Enid	FM	W5HTK	145.29000	-	141.3 Hz	ORSI
	FM	W5HTK	147.15000	+	88.5 Hz	ORSI
	FM	N5LWT	147.37500	+		ORSI
	FM	WA5QYE	444.40000	+	141.3 Hz	ORSI
	FM	N5LWT	444.82500	+		ORSI
Enterprise	FM	N5JMG	147.27000	+	141.3 Hz	ORSI
	FM	KA5HET	442.10000	+	123.0 Hz	ORSI
	FM	W5CUQ	444.62500	+	88.5 Hz	ORSI
Enterprise, Blue Mtn						
	FM	W5CUQ	147.10500	+	114.8 Hz	ORSI
Fairland	FM	W0GMM	147.28500	+	110.9 Hz	ORSI
Fairview	FM	WK5V	147.07500	+		ORSI
Forgan	FM	N5AKN	147.39000	+		ORSI
Fort Gibson	DMR/BM	WA5VMS	442.12500	+	CC1	ORSI
	FM	WA5VMS	927.65000	902.65000		ORSI
Fort Smith, Cavanal Mountain						
	FM	W5ANR	146.64000	-	88.5 Hz	ORSI
	FM	W5ANR	444.50000	+	88.5 Hz	ORSI
Ft. Gibson	NXDN	WA5VMS	444.80000	+		ORSI

288   OKLAHOMA

Location	Mode	Call sign	Output	Input	Access	Coordinator
Goltry	FM	W5ALZ	443.20000	+	103.5 Hz	
Grandfield	FM	KB5LLI	147.25500	+	192.8 Hz	ORSI
	FM	WX5LAW	442.20000	+	123.0 Hz	ORSI
Granfield	FM	KD5BAK	147.07500	+	97.4 Hz	ORSI
Granite	DMR/MARC	K5XTL	442.07500	+	CC1	ORSI
	DSTAR	WX5ASA	147.34500	+		ORSI
	FM	KE5HRS	224.92000	-	151.4 Hz	ORSI
	FM	WX5ASA	432.17500			
Granite, Walsh Mountain						
	DSTAR	WX5ASA	432.17500			
	FM	K5XTL	146.98500	-	156.7 Hz	ORSI
Guthrie	FM	W5KSU	147.10500	+		ORSI
	FM	W5IAS	443.92500	+	88.5 Hz	ORSI
Guymon	FM	N5DFQ	147.15000	+	88.5 Hz	ORSI
	FM	N5DFQ	444.97500	+	88.5 Hz	ORSI
Hartshorne	DMR/MARC	N5AVV	147.30000	+	CC1	ORSI
Headrick	FM	WX5LAW	443.30000	+	123.0 Hz	ORSI
Hichita	WX	W5IAS	444.60000	+	88.5 Hz	ORSI
Hugo	FM	KB5JTR	146.61000	-	114.8 Hz	ORSI
Keetonville	DMR/BM	W5RAB	442.35000	+	CC1	ORSI
Ketchum	FM	W5RAB	444.87500	+	88.5 Hz	ORSI
Kingfisher	FM	W5GLD	146.64000	-	100.0 Hz	ORSI
	FM	W5GLD	444.97500	+	100.0 Hz	ORSI
Kingston	FM	N4SME	443.45000	+	127.3 Hz	ORSI
	FM	N5BCW	444.97500	+		
Lawton	DMR/BM	KB5SKY	145.25000	454.75000	CC4	
	DMR/BM	N4RDB	442.00000	+	CC1	
	DMR/BM	N5PLV	443.60000	+	CC4	ORSI
	DSTAR	KG5ACV B	442.65000	+		
	FM	KC5AVY	145.17000	-		ORSI
	FM	K5VHF	145.43000	-		ORSI
	FM	N5PYD	146.80500	-		ORSI
	FM	W5KS	146.91000	-		ORSI
	FM	WX5LAW	147.18000	+	123.0 Hz	ORSI
	FM	W5KS	147.36000	+	173.8 Hz	ORSI
	FM	N4RDB	442.45000	+	123.0 Hz	ORSI
	FM	WX5LAW	442.52500	+	123.0 Hz	ORSI
	FM	W5KS	443.85000	+		ORSI
	FM	KD5IAE	444.70000	+	141.3 Hz	ORSI
	FM	K5VHF	444.90000	+	118.8 Hz	ORSI
Leonard	DMR/BM	W5RAB	443.65000	+	CC1	ORSI
Liberty, Jerimah Mountain						
	ATV	W5JWT	145.11000	-	114.8 Hz	ORSI
Lindsay	FM	N5RAK	444.87500	+	131.8 Hz	ORSI
Lone Grove	FM	W5BLW	146.79000	-		ORSI
Mangum, Navajo Mt.						
	FM	KF5CRF	442.10000	+	123.0 Hz	ORSI
Mannford	DMR/BM	W5IAS	444.45000	+	CC1	ORSI
	FM	W5IAS	147.04000	+	88.5 Hz	ORSI
	FM	W5IAS	442.80000	+	88.5 Hz	
Marietta	FM	N5KEY	146.83500	-	173.8 Hz	ORSI
Marlow	FM	K5UM	146.95500	-		ORSI
McAlester	FM	W5CUQ	145.37000	-	71.9 Hz	ORSI
	FM	W5CUQ	444.62500	+	88.5 Hz	ORSI
Medford, Salt Plains Lake						
	FM	KB0HH	147.30000	+	103.5 Hz	ORSI
Medicine Park	FM	WX5LAW	444.07500	+	123.0 Hz	ORSI
Medicine Park, Big Rock						
	WX	AF5Q	224.50000	-	123.0 Hz	
Miami	FM	KG5FJI	442.65000	+		ORSI
Midwest City	FM	W5MWC	444.00000	+	151.4 Hz	ORSI
Mooreland	FM	K5GUD	145.39000	-	88.5 Hz	ORSI

OKLAHOMA 289

Location	Mode	Call sign	Output	Input	Access	Coordinator
Mooreland	FM	K5GUD	444.27500	+		ORSI
Mounds	DMR/BM	W5RAB	444.85000	+	CC1	
	FM	N5DRW	147.12000	+	88.5 Hz	ORSI
	FM	WB5NJU	147.39000	+	88.5 Hz	ORSI
	FM	W5IAS	444.60000	+	88.5 Hz	ORSI
Muskogee	FM	KK5I	146.74500	-	88.5 Hz	ORSI
	FM	KK5I	146.85000	-		
	FM	WA5VMS	147.33000	+	88.5 Hz	ORSI
	FM	KK5I	224.34000	-	88.5 Hz	ORSI
	FM	W5IAS	443.10000	+	88.5 Hz	ORSI
Nashoba, Cripple Creek Mountai						
	FM	KM5VK	145.29000	-	162.2 Hz	ORSI
	FM	KM5VK	145.33000	-	162.2 Hz	ORSI
	FM	KM5VK	224.56000	-	114.8 Hz	ORSI
	FM	KM5VK	442.90000	+	114.8 Hz	ORSI
Newalla	FM	K5UV	145.15000	-	114.8 Hz	ORSI
Newcastle	FM	KX5MOT	444.67500	+	192.8 Hz	ORSI
Norman	DMR/BM	N5MS	443.82500	+	CC1	ORSI
	FM	K9KK	224.44000	-		ORSI
	FM	WA5LKS	442.12500	+	107.2 Hz	ORSI
	FM	N5KUK	444.35000	+	141.3 Hz	ORSI
Norman, County Tower						
	FM	W5NOR	147.06000	+	141.3 Hz	ORSI
	FM	W5NOR	443.70000	+	141.3 Hz	ORSI
Norman, National Weather Cente						
	DSTAR	W5TC	444.75000	+		ORSI
Norman, OU Physical Sciences C						
	FM	W5OU	146.88000	-		ORSI
Nowata	FM	N5ZZX	145.37000	-		ORSI
Oklahoma City	DMR/BM	N5KNU	442.50000	+	CC1	ORSI
	DMR/BM	KB5KWV	443.17500	+	CC1	ORSI
	DMR/BM	W5GDL	443.22500	+	CC1	ORSI
	FM	WN5J	145.33000	-		ORSI
	FM	KK5FM	145.37000	-	141.3 Hz	ORSI
	FM	KD5AHH	145.49000	-	131.8 Hz	ORSI
	FM	W5PAA	146.76000	-	141.3 Hz	ORSI
	FM	W5RLW	146.79000	-	100.0 Hz	
	FM	W5MEL	146.82000	-	151.4 Hz	ORSI
	FM	W5PAA	146.98500	-	141.3 Hz	ORSI
	FM	W5MEL	147.21000	+	141.3 Hz	ORSI
	FM	W5PAA	224.10000	-		ORSI
	FM	NZ5W	224.30000	-	123.0 Hz	ORSI
	FM	WN5J	224.88000	-		ORSI
	FM	W5PAA	442.70000	+	141.3 Hz	ORSI
	FM	KK5FM	442.77500	+	141.3 Hz	ORSI
	FM	WX5OKC	444.25000	+	141.3 Hz	ORSI
	FM	W5MEL	444.30000	+	141.3 Hz	ORSI
	FM	W5PAA	927.91250	902.91250	123.0 Hz	
	FM	WN5J	1283.10000	1263.10000		ORSI
	WX	WX5OKC	145.41000	-	141.3 Hz	ORSI
Oklahoma City, KFOR Tower						
	FM	WD5AII	147.03000	+	167.9 Hz	ORSI
	FM	W5MWC	444.20000	+	167.9 Hz	ORSI
Oklahoma City, KFOR-TV Tower						
	FM	W5PAA	146.67000	-	151.4 Hz	ORSI
	FM	W5PAA	444.10000	+	141.3 Hz	ORSI
Oklahoma City, Sandbridge Buil						
	FM	W5GDL	444.21250	+	141.3 Hz	ORSI
Oklahoma City, Sandridge Build						
	FM	W5GDL	224.92000	-	141.3 Hz	ORSI
Owasso, Keetonville						
	FM	WD5ETD	224.48000	-	88.5 Hz	

290   OKLAHOMA

Location	Mode	Call sign	Output	Input	Access	Coordinator
Pawhuska	FM	KC5KLM	147.27000	+	88.5 Hz	ORSI
	FM	W5RAB	444.97500	+	167.9 Hz	ORSI
Pecola	FM	KB5SWA	444.02500	+		ORSI
Perry	FM	KL7MA	442.92500	+	141.3 Hz	ORSI
	WX	KF5RDI	146.86500	-	141.3 Hz	ORSI
Pittsburgh, Tiger Mtn						
	FM	AD5MC	146.95500	-	151.4 Hz	ORSI
Ponca City	DMR/BM	K5BOX	444.75000	+	CC1	
	FM	K5BOX	53.77000	52.07000		ORSI
	FM	W5HZZ	146.97000	-	88.5 Hz	ORSI
	FM	W5BE	442.67500	+	88.5 Hz	ORSI
	FM	AF5VB	443.50000	+	88.5 Hz	ORSI
	FM	W5RAB	444.60000	+	167.9 Hz	ORSI
	FM	W5HZZ	444.70000	+		ORSI
Ponca City, McCord Water Tower						
	FM	N5PC	146.73000	-	88.5 Hz	
Pond Creek	FM	KW5FAA	442.30000	+	141.3 Hz	ORSI
Porum	FM	W4NFD	53.05000	52.05000	114.8 Hz	
	FM	W4NFD	444.52500	+	107.2 Hz	
Poteau	WX	K6CKS	444.55000	+	136.5 Hz	ORSI
Prague	FUSION	N0LOZ	447.00000	-	100.0 Hz	
Preston	FM	W5KO	145.33000	-	88.5 Hz	ORSI
	FM	WX5OKM	147.22500	+	88.5 Hz	ORSI
	FM	KD5FMU	444.17500	+	88.5 Hz	ORSI
Pryor	FM	WX5MC	444.67500	+	88.5 Hz	ORSI
	WX	WX5MC	147.06000	+	88.5 Hz	ORSI
Rose	FM	KC5DBH	146.98500	-	110.9 Hz	
Sapulpa, Creek County Rural Wa						
	FUSION	KC5RBH	145.43000	-	88.5 Hz	ORSI
Seiling	FM	W5OKT	444.92500	+	103.5 Hz	ORSI
Selling	FM	W5OKT	444.92500	+	103.5 Hz	
Seminole	FM	WJ5F	147.01500	+		ORSI
	FM	WJ5F	147.19500	+	141.3 Hz	ORSI
Seward	FM	KA5LSU	145.17000	-	100.0 Hz	ORSI
	FM	KA5LSU	442.07500	+		ORSI
	FM	KK5FM	444.77500	+	141.3 Hz	ORSI
Sharon	DSTAR	K5GUD	147.31500	+		ORSI
Shawnee	FM	W5SXA	145.39000	-	131.8 Hz	ORSI
Shawnee, Shawnee Twin Lakes						
	FM	KG5BGO	145.19000	-	131.8 Hz	
Skiatook, 3-10 Y Tower						
	FM	WA5LVT	444.72500	+		
Stillwater	FM	K5SRC	145.35000	-	107.2 Hz	ORSI
	FM	K5FVL	147.25500	+	107.2 Hz	ORSI
	FM	K5FVL	442.60000	+	103.5 Hz	ORSI
	FM	K5FVL	444.47500	+	100.0 Hz	ORSI
	WX	K5FVL	444.52500	+	88.5 Hz	ORSI
	WX	K5FVL	444.90000	+	141.3 Hz	ORSI
Stuart	FM	W5CUQ	146.89500	-	114.8 Hz	ORSI
	FM	W5CUQ	443.72500	+	127.3 Hz	
Tahlequah	FM	N5ROX	442.70000	+		
	WX	N5ROX	147.24000	+	88.5 Hz	ORSI
	WX	N5ROX	442.22500	447.32500	88.5 Hz	ORSI
Talihina, Sycamore Mountain						
	WX	N5CST	145.41000	-	88.5 Hz	ORSI
Tecumseh	FM	KD5WAV	146.62500	-	131.8 Hz	ORSI
The Village	FM	KB5QND	443.40000	+	141.3 Hz	ORSI
Tishomingo	FM	W5JTM	145.31000	0.60000	141.3 Hz	
	FM	W5JTB	145.31000	-	141.3 Hz	
Tulsa	DMR/MARC	WD5ETD	441.97500	+	CC1	
	FM	W5IAS	145.11000	-	88.5 Hz	ORSI
	FM	WA5LVT	146.80500	-	88.5 Hz	ORSI

OKLAHOMA 291

Location	Mode	Call sign	Output	Input	Access	Coordinator
Tulsa	FM	K5JME	147.00000	+		ORSI
	FM	AE5RH	224.86000	222.86000	88.5 Hz	
	FM	W5IAS	443.85000	+	88.5 Hz	ORSI
	FM	K5LAD	444.30000	+	100.0 Hz	ORSI
Tulsa Central	DMR/MARC	WA5LVT	442.47500	+	CC2	ORSI
Tulsa, Asbury Methodist Church						
	DSTAR	KN5V	442.60000	+		
Tulsa, Channel 8 TV Studios						
	FM	W5IAS	443.75000	+	88.5 Hz	
Tulsa, CityPlex Towers						
	FM	WD5ETD	927.70000	902.70000	DCS 074	ORSI
	WX	WA5LVT	146.88000	-	88.5 Hz	ORSI
Tulsa, Lookout Mountain						
	FM	W5IAS	443.00000	+		ORSI
Tulsa, Reservoir Hill						
	FM	WT5EOC	146.83500	-	103.5 Hz	ORSI
Tulsa, Sun Building						
	FM	WA5LVT	146.94000	-	88.5 Hz	ORSI
	FM	WA5LVT	444.95000	+	88.5 Hz	
Tuttle	DMR/BM	N5PTV	444.10000	+	CC1	
	FM	WA7WNM	224.68000	-		ORSI
	FM	K8TOR	443.90000	+		ORSI
Velma	FM	KC5JCO	444.95000	+		ORSI
Vici	FM	N5WO	444.95000	+	88.5 Hz	ORSI
Vinita	DSTAR	NO5RA	147.16500	+		ORSI
	FM	KC5VVT	444.37500	+	88.5 Hz	ORSI
	FUSION	N5BYS	443.07500	+	88.5 Hz	ORSI
Watonga	FM	K5GSM	442.37500	+	DCS 223	ORSI
Waurika	FM	W5KS	145.29000	-	123.0 Hz	ORSI
Waynoka	FM	W5ALZ	443.45000	+	103.5 Hz	ORSI
Weatherford	FM	KB5TOO	147.07500	+		
Wewoka, Tate Mountain						
	FM	WJ5F	147.01500	+	141.3 Hz	ORSI
Wilburton	FM	KL7JW	146.62500	-	88.5 Hz	ORSI
Winchester	FM	N5LW	147.15000	+	88.5 Hz	ORSI
Woodward	FM	W5GPR	146.73000	-	203.5 Hz	ORSI
	FM	K5GSM	147.00000	+	103.5 Hz	ORSI
	FM	W5ALZ	442.05000	+	103.5 Hz	ORSI
	WX	K5GUD	444.87500	+	103.5 Hz	ORSI
Woodward, Airport						
	FM	W5GPR	146.73000	-	203.5 Hz	ORSI
OREGON						
Agness	FM	WA7JAW	147.04000	+	88.5 Hz	ORRC
Albany	FM	KD6VLR	444.97500	+	100.0 Hz	ORRC
Aloha	FM	NM7B	443.35000	+	156.7 Hz	
Aloha, Cooper Mountain						
	FM	N7QQU	147.36000	+	107.2 Hz	ORRC
	FM	KA7OSM	442.52500	+	107.2 Hz	
Aloha, Oregon	FM	K7RPT	442.35000	+	100.0 Hz	
Alsea, Prairie Mountain						
	FM	W6PRN	444.35000	+	173.8 Hz	
Amity	FM	K7RPT	444.12500	+	100.0 Hz	
Ashland	FM	W9PCI	146.62000	-	100.0 Hz	ORRC
	FM	WX7MFR	440.70000	+	162.2 Hz	ORRC
Ashland, Mount Ashland						
	FM	WX7MFR	147.26000	+	123.0 Hz	ORRC
	FM	AB7BS	442.30000	+	DCS 125	ORRC
Ashland, Soda Mountain						
	FM	WA6RHK	147.16000	+	136.5 Hz	ORRC
Ashland, Soda Mtn						
	FM	W7PRA	146.70000	-	136.5 Hz	

292   OREGON

Location	Mode	Call sign	Output	Input	Access	Coordinator
Astoria	FM	N7HQR	440.82500	+	118.8 Hz	ORRC
	FM	W7BU	444.85000	+	118.8 Hz	
Astoria, Nicolai Mountain						
	FM	W7BU	146.76000	-	118.8 Hz	ORRC
	FM	WA6TTR	444.50000	+	118.8 Hz	ORRC
Astoria, Wickiup Mountain						
	FM	KF7TCG	146.66000	-	118.8 Hz	
	FM	K7RPT	146.72000	-	114.8 Hz	
Astoria, Wickiup Mtn						
	FM	W7BU	442.50000	+	118.8 Hz	ORRC
Athena	DSTAR	K7LW	444.90000	+		
Athena, Weston Mountain						
	FM	W7NEO	147.04000	+		
Athena, Weston Mtn						
	FM	W7NEO	441.70000	+	131.8 Hz	
Baker City, Beaver Mountain						
	FM	W7NYW	145.27000	-	110.9 Hz	
Banks, Chrysler Rd						
	FM	KC7UQB	440.87500	+	103.5 Hz	ORRC
Beavercreek, Highland Butte						
	FM	AH6LE	146.92000	-	107.2 Hz	
Beaverton	FM	WB7CRT	444.75000	+	123.0 Hz	ORRC
Beaverton, St Vincent Hosp						
	FM	W7PSV	444.85000	+	123.0 Hz	ORRC
Bend	FM	W7DCO	442.55000	+	162.2 Hz	
	FM	KB7LNR	443.65000	+	162.2 Hz	ORRC
Bend, Long Butte	FM	W7JVO	146.94000	-	162.2 Hz	
	FM	KB7LNR	147.36000	+		ORRC
Bend, Mt Bachelor						
	FM	W7DCO	145.45000	-	103.5 Hz	ORRC
Blue River, Indian Ridge						
	FM	W7EUG	145.37000	-	100.0 Hz	ORRC
Blue River, Mount Hagan						
	FM	W7EXH	145.11000	-	100.0 Hz	ORRC
	FM	W7PRA	147.34000	+	DCS 023	
	FM	K7SLA	443.10000	+	100.0 Hz	ORRC
Blue River, Mt Hagan						
	FM	AB7BS	444.80000	+	DCS 125	
Boardman, Coalfire Plant						
	FM	AI7HO	147.16000	+		
Boring, Fire Station						
	FM	KD7WGZ	441.77500	+	107.2 Hz	
Brookings, Bosley Butte						
	FM	N7UBQ	147.25000	+	88.5 Hz	ORRC
Brookings, Fire Station						
	FM	W7BKG	146.84000	-	88.5 Hz	ORRC
Brookings, Harbor	FM	KA7GNK	444.97500	+	100.0 Hz	ORRC
Brookings, Harbor Hill						
	FM	W7BKG	146.96000	-	88.5 Hz	ORRC
Brownsville	FM	W7WZA	147.30000	+	77.0 Hz	ORRC
Brownsville, Scott Mountain						
	FM	W7NK	147.26000	+	DCS 026	ORRC
Brownsville, Washburn Heights						
	FM	K7VFO	224.00000	-	100.0 Hz	ORRC
	FM	KB7KUB	442.60000	+	DCS 023	ORRC
Burns, Radar Hill	FM	KF7HPT	146.76000	-	103.5 Hz	
Burns, Sharps Ridge						
	FM	W7JVO	147.30000	+	162.2 Hz	ORRC
Burns, Steens Mountain						
	FM	W7PRA	145.11000	-	136.5 Hz	
Buxton	DMR/MARC	KB7APU	444.02500	+	CC1	

OREGON 293

Location	Mode	Call sign	Output	Input	Access	Coordinator
Camas Valley, Kenyon Mountain						
	FM	W7PRA	146.92000	-	136.5 Hz	ORRC
Canby	FM	K7CFD	224.76000	-	131.8 Hz	ORRC
	FM	WB7QAZ	442.90000	+	123.0 Hz	ORRC
	FM	K7CFD	444.45000	+	131.8 Hz	
Cannon Beach, Arch Cape						
	FM	W7BU	146.74000	-	118.8 Hz	ORRC
Canyon City, Eagle Peak						
	FM	N7LZM	146.64000	-		
Cape Meares	FM	KA7AHV	147.16000	+	118.8 Hz	ORRC
	FM	N7IS	442.97500	+	100.0 Hz	ORRC
Cascade Locks, Mt Defiance						
	FM	KF7LN	927.16250	902.16250	151.4 Hz	ORRC
Cave Junction, $8 Mountain						
	FM	WB6YQP	145.49000	-	136.5 Hz	
Cedar Mill	FM	K7RPT	147.38000	+	100.0 Hz	ORRC
	FM	N7PRM	444.80000	+	107.2 Hz	
Central Point	FM	KL7VK	147.38000	+	131.8 Hz	ORRC
	FM	WA6RHK	440.82500	+	136.5 Hz	ORRC
	FM	W9PCI	444.10000	+	100.0 Hz	ORRC
Central Point, Johns Peak						
	FM	KB7SKB	147.10000	+	136.5 Hz	
Chemult	FM	WA7TYD	444.92500	+	100.0 Hz	ORRC
Chemult, Walker Mountain						
	FM	WA7TYD	145.47000	-	162.2 Hz	ORRC
Chiloquin, Train Mtn						
	FM	K7LNK	444.95000	+	136.5 Hz	ORRC
Clackamas	FM	KJ7IY	146.80000	-	107.2 Hz	ORRC
	FM	KB7WUK	1291.00000	1271.00000	107.2 Hz	ORRC
Clackamas, Mount Scott						
	FM	KR7IS	29.68000	-	162.2 Hz	ORRC
Clackamas, Mt Scott						
	FM	W7LT	147.18000	+		
	FM	WB7QIW	147.28000	+	167.9 Hz	ORRC
	FM	KC7MZM	440.30000	+	167.9 Hz	
	FM	KJ7IY	443.15000	+	107.2 Hz	
	FM	WB7QIW	443.47500	+	167.9 Hz	ORRC
Coburg, Buck Mountain						
	FM	W7EXH	224.70000	-	100.0 Hz	ORRC
Coburg, Coburg Ridge						
	FM	K7QT	441.12500	+	141.3 Hz	ORRC
Colton	DMR/MARC	KB7APU	442.75000	+	CC1	
	FM	WB7DZG	442.92500	+	107.2 Hz	ORRC
Colton, Goat Mountain						
	FM	W7OTV	146.96000	-	127.3 Hz	ORRC
	FM	WB6EGS	441.40000	+	88.5 Hz	
	FM	N7PIR	443.70000	+	103.5 Hz	ORRC
Coos Bay	FM	W7OC	147.10000	+	110.9 Hz	ORRC
	FM	W7OC	441.72500	+	131.8 Hz	ORRC
Coos Bay, Blossom Hill						
	FM	W7OC	147.28000	+	146.2 Hz	ORRC
	FM	WA7JAW	440.80000	+	103.5 Hz	ORRC
Coos Bay, Noah Butte						
	FM	K7TVL	146.88000	-	136.5 Hz	ORRC
Coquille, Beaver Hill						
	FM	K7CCH	146.61000	-	110.9 Hz	ORRC
Corbett	DMR/MARC	N7LF	444.15000	+	CC1	
	FM	K7NE	145.41000	-	100.0 Hz	
Corvallis, Elmers	FM	W7OSU	443.05000	+	100.0 Hz	
Corvallis, Good Sam Hospital						
	FM	N8GFO	442.30000	+	162.2 Hz	ORRC

294   OREGON

Location	Mode	Call sign	Output	Input	Access	Coordinator
Corvallis, Mary's Peak						
	FM	K7LNK	440.42500	+	DCS 125	ORRC
Corvallis, Marys Peak						
	FM	WA7TUV	146.82000	-	100.0 Hz	
Corvallis, Vineyard Mountain						
	FM	W7OSU	147.16000	+	100.0 Hz	ORRC
Cottage Grove	FM	WB7LCS	440.87500	+		
	FM	K7SLA	443.02500	+	156.7 Hz	ORRC
Cottage Grove, Bear Mtn						
	FM	AB7BS	145.39000	-	146.2 Hz	ORRC
	FM	K7LNK	444.55000	+	100.0 Hz	
Cottage Grove, Fairview Peak						
	FM	W7SLA	145.23000	-	110.9 Hz	ORRC
Cottage Grove, Harness Mountai						
	FM	W7ZQE	146.66000	-	100.0 Hz	
Creswell	FM	N7NPA	440.10000	+	110.9 Hz	ORRC
Culver	DMR/BM	KC7DMF	444.42500	+	CC1	
Deer Island	FM	N7EI	224.38000	-	114.8 Hz	ORRC
Deer Island, Meissner Lookout						
	FM	N7EI	146.88000	-	114.8 Hz	ORRC
Diamond Lake, Cinnamon Butte						
	FM	K7RBG	147.22000	+	67.0 Hz	
Dufur, Tygh Ridge	FM	WC7EC	147.26000	+	82.5 Hz	ORRC
Elgin, Spout Springs Ski Resor						
	FM	WF7S	146.80000	-	123.0 Hz	ORRC
Estacada	FM	KD7DEG	440.85000	+	107.2 Hz	ORRC
Eugene	DMR/BM	N7GWA	442.42500	+	CC1	
	DMR/MARC	N7MAQ	442.96250	+	CC1	
	FM	W7EXH	443.27500	+	100.0 Hz	ORRC
Eugene, Blanton Heights						
	FM	W7ARD	145.45000	-	123.0 Hz	ORRC
	FM	K7TBL	146.88000	-	100.0 Hz	ORRC
	FM	W7NK	147.26000	+	100.0 Hz	ORRC
	FM	W7NK	147.36000	+	123.0 Hz	ORRC
	FM	W7DTV	442.80000	+	77.0 Hz	
	FM	W7NK	442.90000	+	110.9 Hz	ORRC
Eugene, Buck Mountain						
	FM	W7NK	145.17000	-	100.0 Hz	ORRC
Eugene, Coburg Ridge						
	FM	W7CQZ	147.08000	+	100.0 Hz	ORRC
	FM	K7UND	441.32500	+	100.0 Hz	
	FM	K7THO	442.12500	+	DCS 125	ORRC
Falls City, Laurel Mountain						
	FM	W7SRA	147.02000	+	186.2 Hz	ORRC
Florence, Glenada Hill						
	FM	W7FLO	441.10000	+		ORRC
Florence, Herman Peak						
	FM	W7FLO	146.80000	-	100.0 Hz	ORRC
	FM	W7FLO	442.57500	+	DCS 125	ORRC
Forest Grove	FM	K7RPT	442.35000	+	100.0 Hz	
Forest Grove, South Saddle Mou						
	FM	K7RPT	147.32000	+	100.0 Hz	ORRC
	FM	K7RPT	442.32500	+	100.0 Hz	ORRC
Fossil, Snowboard Ridge						
	FM	KI7DEL	146.68000	-	162.2 Hz	ORRC
Frenchglen, Steens Mountain						
	FM	AB7BS	145.13000	-	136.5 Hz	
Gales Creek	FM	KJ7IY	927.11250	902.11250	107.2 Hz	ORRC
Gates	FM	WA7ABU	145.19000	-	100.0 Hz	ORRC
Gearhart	FM	W7BU	146.80000	-	118.8 Hz	
Glendale, Mount Reuben						
	FM	AB7BS	146.86000	-	123.0 Hz	ORRC

OREGON   295

Location	Mode	Call sign	Output	Input	Access	Coordinator
Glide	FM	WB7RKR	444.87500	+	127.3 Hz	
Glide, Lane Mountain						
	FM	WA7BWT	147.24000	+	136.5 Hz	ORRC
Glide, Scott Mountain						
	FM	WA7BWT	145.43000	-	88.5 Hz	
	FM	WB7RKR	224.10000	-	114.8 Hz	ORRC
	FM	WB7RKR	444.62500	+	91.5 Hz	ORRC
Gold Beach, Grizzly Mountain						
	FM	K7SEG	146.74000	-	88.5 Hz	ORRC
Government Camp						
	DMR/MARC	KB7APU	442.98750	+	CC1	
	FM	N7PIR	443.87500	+	103.5 Hz	ORRC
Government Camp, Mount Hood						
	FM	KB7APU	224.78000	-	136.5 Hz	ORRC
	FM	KB7APU	444.10000	+	DCS 432	ORRC
Government Camp, Timberline Lo						
	FM	WB7DZG	52.97000	51.27000	107.2 Hz	ORRC
Grants Pass	DSTAR	KE7LKX	440.46250	+		
	FM	WA6OTP	147.22000	+		ORRC
Grants Pass, Bluey Mountain						
	FM	WB6QYP	147.14000	+	162.2 Hz	
Grants Pass, Gilbert Peak						
	FM	K7LIX	146.64000	-		
	FM	K7LIX	147.30000	+	136.5 Hz	
	FM	AB7BS	444.55000	+	173.8 Hz	ORRC
	FM	W6PRN	444.67500	+	100.0 Hz	
Grants Pass, Onion Mountain						
	FM	AB7BS	440.55000	+	173.8 Hz	ORRC
Gresham	DMR/MARC	N7LF	443.10000	+	CC1	
	FM	N7DOD	446.27500	-	167.9 Hz	
Gresham, Walters Hill						
	FM	KE7AWR	441.62500	+	146.2 Hz	ORRC
Hampton	FM	K7SQ	223.98000	-	103.5 Hz	
Hampton, Glass Butte						
	FM	W7JVO	147.20000	+	162.2 Hz	ORRC
Harrisburg, Buck Mountain						
	FM	W7PRA	444.82500	+	DCS 023	ORRC
Hebo, Mount Hebo						
	FM	N7QFT	147.26000	+	162.2 Hz	ORRC
Hebo, Mt Hebo	DSTAR	W7GC	147.39000	+		
	FM	W7LI	147.22000	+	100.0 Hz	ORRC
	FM	W7GC	440.90000	+	118.8 Hz	ORRC
	FM	W7LI	441.25000	+	118.8 Hz	
	FM	WB7QIW	443.07500	+	167.9 Hz	ORRC
Heppner , Black Mountain						
	FM	KC7SOY	145.23000	-	67.0 Hz	
Heppner, Black Mountain						
	FM	KC7SOY	146.78000	-	67.0 Hz	
Hermiston	FM	WB7ILL	443.80000	+	123.0 Hz	
Hillsboro	FM	KE7AWR	441.22500	+	100.0 Hz	ORRC
Hillsboro, Intel's Jones Farm						
	FM	K7CPU	444.97500	+	107.2 Hz	ORRC
Hillsboro, Synopsys						
	DSTAR	N7QQU	440.55000	+		ORRC
Hood River, Columbia River Gor						
	FM	AF7YV	440.60000	+	100.0 Hz	
Hood River, Middle Mountain						
	FM	KA7HRC	444.90000	+	100.0 Hz	
Hood River, Mount Defiance						
	FM	WA7ROB	145.15000	-	94.8 Hz	ORRC
	FM	KF7LN	147.10000	+	100.0 Hz	ORRC

296 OREGON

Location	Mode	Call sign	Output	Input	Access	Coordinator
Horton, Prairie Mountain						
	FM	W7NK	443.50000	+	103.5 Hz	ORRC
Horton, Prairie Mtn						
	FM	W7EUG	146.68000	-	100.0 Hz	
Huntington, Lime Hill						
	FM	K7OJI	147.12000	+	100.0 Hz	
	FM	K7OJI	444.15000	+	100.0 Hz	
Jackson	FM	W7UIV	440.95000	+		
Jacksonville	FM	W9PCI	145.33000	-	100.0 Hz	ORRC
	FM	W9PCI	444.20000	+	100.0 Hz	ORRC
	FM	W9PCI	444.30000	+	100.0 Hz	ORRC
John Day, Airport	FM	W7JVO	145.24000	-	162.2 Hz	
John Day, Fall Mountain						
	FM	N7LZM	147.22000	+	123.0 Hz	
Jordan Valley	FM	KC7GLR	444.80000	+		ORRC
Joseph	FM	KB7DZR	147.00000	+	103.5 Hz	ORRC
Junction City, Prairie Mountai						
	FM	AB7BS	443.57500	+	DCS 125	ORRC
Junction City, Prairie Mtn						
	FM	W7PRA	145.13000	-	DCS 023	ORRC
Junction City, Prairie Peak						
	FM	W7PXL	146.72000	-	100.0 Hz	
Keizer	FM	KD7PFG	53.23000	51.53000	100.0 Hz	ORRC
	FM	KD7PFG	440.80000	+	156.7 Hz	ORRC
Keno	FM	KE7CSD	146.86000	-		ORRC
	FM	KD7TNG	442.52500	+	118.8 Hz	ORRC
Keno, Chase Mountain						
	FM	WA6RHK	440.67500	+	173.8 Hz	ORRC
Keno, Hamaker Mountain						
	FM	N6MRX	147.18000	+	100.0 Hz	
	FM	K6PRN	443.20000	+	100.0 Hz	
Klamath Falls, Chase Mountain						
	FM	W7PRA	147.20000	+	136.5 Hz	ORRC
Klamath Falls, Hamaker Mountai						
	FM	W7VW	146.85000	-	118.8 Hz	
Klamath Falls, Hogback Mountai						
	FM	W7VW	146.61000	-	118.8 Hz	
Klamath Falls, Hogsback Mtn						
	FM	W7VW	443.90000	+	118.8 Hz	
La Grande	FM	K7RPT	147.38000	+	100.0 Hz	
	FUSION	W7GRA	146.98000	-		
La Grande, Grande Ronde Hospit						
	FM	W7GRA	146.98000	-	100.0 Hz	ORRC
La Grande, Mount Fanny						
	FM	W7NYW	145.15000	-	110.9 Hz	
	FM	K7RPT	147.26000	+	103.5 Hz	ORRC
La Grande, Mount Harris						
	FM	KF7GOR	444.92500	+	146.2 Hz	
La Pine	FM	WA7TYD	145.49000	-	77.0 Hz	ORRC
Lake Oswego	FM	WA7LO	444.30000	+	82.5 Hz	
Lakeside, Roman Nose Mtn						
	FM	W7PRA	147.14000	+	136.5 Hz	ORRC
Lakeview, Drake Peak						
	FM	KE7QP	145.31000	-	173.8 Hz	ORRC
Lakeview, Grizzly Peak						
	FM	KE7QP	147.34000	+	173.8 Hz	
Langlois, Stone Butte						
	FM	KA7GNK	145.21000	-	88.5 Hz	ORRC
Laurelwood, Bald Peak						
	FM	K7AUO	443.65000	+	100.0 Hz	ORRC
	FM	K7AUO	1291.50000	1271.50000		ORRC
Lime, Lime Hill	FM	W7NYW	145.17000	-	110.9 Hz	

OREGON 297

Location	Mode	Call sign	Output	Input	Access	Coordinator
Lincoln City	FM	W7VTW	147.04000	+	100.0 Hz	ORRC
Longview	DMR/MARC	N3EG	441.70000	+	CC1	
Lyons , McCully Mtn						
	FM	W1ARK	147.06000	+	100.0 Hz	ORRC
Lyons, McCulley Mountain						
	FM	K7RTL	440.60000	+	100.0 Hz	ORRC
Madras, Eagle Butte						
	FM	K7RPT	442.22500	+	114.8 Hz	
Mapleton, Roman Nose Mountain						
	FM	AB7BS	147.14000	+	136.5 Hz	ORRC
McMinnville, Eola Hill						
	FM	W7YAM	441.80000	+	114.8 Hz	ORRC
McMinnville, Eola Hills						
	FM	W7RXJ	146.64000	-	100.0 Hz	ORRC
Medford	DSTAR	KG7FOJ	145.24000	-		ORRC
	DSTAR	KG7FOJ	444.65000	+		
	DSTAR	KG7FOJ	1293.12500	1313.12500		
	FM	KG7FOJ	1248.50000	1228.50000		ORRC
Medford, Baldy	DSTAR	KE7MVI	443.77500	+		ORRC
	FM	K7JAX	146.84000	-	123.0 Hz	ORRC
Medford, Downtown						
	FM	K7RPT	147.02000	+	100.0 Hz	ORRC
Medford, Rogue Valley Manor						
	FM	K7RVM	147.00000	+	123.0 Hz	ORRC
	FM	K7RVM	444.45000	+	100.0 Hz	ORRC
Mill City	DMR/MARC	K7MTW	442.97500	+	CC1	
	DMR/MARC	N7MAQ	443.00000	+	CC1	
Millican, Pine Mountain						
	FM	W7JVO	146.70000	-	162.2 Hz	
	FM	W7JVO	444.75000	+	156.7 Hz	ORRC
Milton-Freewater, Pikes Peak						
	FM	KD7DDQ	147.28000	+	103.5 Hz	
Mitchell, Stephenson Mountain						
	FM	W7JVO	147.18000	+	162.2 Hz	ORRC
Molalla	DMR/BM	WB7AWL	440.27500	+	CC1	
Molalla, Molalla Fire District						
	FM	W7DTV	440.70000	+	77.0 Hz	ORRC
Monmouth	FM	KE7AAJ	927.70000	902.70000	162.2 Hz	ORRC
Monroe, Prairie Peak						
	FM	W7ARD	53.03000	51.33000	100.0 Hz	ORRC
Myrtle Creek, Boomer Hill						
	FM	KC7UAV	147.12000	+	100.0 Hz	ORRC
Myrtle Creek, Sheep Hill						
	FM	WA6KHG	444.15000	+		
Myrtle Point	FM	W7OC	444.17500	+	146.2 Hz	ORRC
Myrtle Point, Bennett Butte						
	FM	W7OC	145.19000	-	146.2 Hz	
	FM	KD7IOP	444.52500	+	123.0 Hz	ORRC
Newberg	DMR/MARC	N7MAQ	444.48750	+	CC1	
	FM	KR7IS	52.83000	51.13000	107.2 Hz	ORRC
	FM	KR7IS	224.06000	-	107.2 Hz	ORRC
	FM	K7RPT	444.12500	+	100.0 Hz	
Newberg, Bald Peak						
	FM	K0INK	442.55000	+	114.8 Hz	ORRC
Newberg, Chehalem Mountain						
	FM	KB7PSM	145.11000	-	103.5 Hz	ORRC
	FM	K7WWG	146.90000	-	127.3 Hz	ORRC
	FM	KB7PSM	442.15000	+		ORRC
Newberg, Chehalem Ridge						
	FM	K7RPT	443.75000	+	100.0 Hz	ORRC
Newberg, Parrett Mountain						
	FM	AH6LE	442.67500	+	100.0 Hz	ORRC

298   OREGON

Location	Mode	Call sign	Output	Input	Access	Coordinator
Newport, Cape Foulweather						
	FM	W7VTW	145.37000	-	167.9 Hz	
Newport, Courthouse						
	FM	W7VTW	145.47000	-	167.9 Hz	ORRC
Newport, Otter Crest						
	FM	W7PRA	145.39000	-	DCS 023	ORRC
	FM	W7VTW	147.30000	+	156.7 Hz	ORRC
	FM	W7GC	444.75000	+	118.8 Hz	ORRC
North Bend	FM	WA7JAW	444.97500	+		ORRC
North Plains	FM	KE7DC	145.45000	-	136.5 Hz	ORRC
	FM	KE7DC	442.40000	+	136.5 Hz	ORRC
Oakridge, Wolf Mountain						
	FM	W7ARD	53.07000	51.37000	100.0 Hz	ORRC
	FM	N7EXH	146.98000	-	100.0 Hz	ORRC
	FM	W7PRA	441.67500	+	100.0 Hz	ORRC
Ontario	FM	K7RHB	443.15000	+	100.0 Hz	
Ontario, Malheir	FM	W7NYW	145.17000	-	110.9 Hz	
Ontario, Malheur Butte						
	FM	K7OJI	147.10000	+	100.0 Hz	ORRC
Oregon City	DMR/BM	WB7AWL	145.25000	-	CC1	
	FM	K7RTL	145.21000	-	110.9 Hz	ORRC
Oregon City, Boynton Standpipe						
	FM	W7ZRS	442.07500	+	103.5 Hz	ORRC
Pendleton, Cabbage Hill						
	FM	W7ZKH	146.88000	-	103.5 Hz	
	FM	N7NKT	224.56000	-		
	FM	W7URG	444.97500	+	136.5 Hz	
	FM	N7ERT	927.50000	902.50000		
Pendleton, Mission						
	FM	W7NEO	444.82500	+		
Pleasant Valley	FM	KJ7IY	145.27000	-		
Port Orford	FM	K7POH	147.20000	+	118.8 Hz	ORRC
Port Orford, Cape Blanco						
	FM	KD7IOP	440.72500	+	114.8 Hz	ORRC
Portland	DMR/MARC	N7LF	440.00000	+	CC1	
	DMR/MARC	KB7APU	440.62500	+	CC1	
	DMR/MARC	KA7AGH	443.27500	+	CC7	ORRC
	DMR/MARC	WA7HAA	444.83750	+	CC7	
	FM	WB2QHS	426.25000	910.25000		ORRC
	FM	W7PMC	443.22500	+	107.2 Hz	ORRC
	FM	KB7OYI	443.85000	+	107.2 Hz	ORRC
	FM	KG7KOU	1249.75000	1229.75000		ORRC
	FM	W7AMQ	1257.00000	426.20000		ORRC
Portland, Council Crest						
	FM	K7LJ	145.23000	-		ORRC
	FM	WB2QHS	440.67500	+	136.5 Hz	ORRC
	FM	K7LJ	442.65000	+	100.0 Hz	ORRC
	FM	K7NE	443.30000	+	100.0 Hz	
	FM	W7PGE	444.67500	+		ORRC
Portland, Emanuel Hospital						
	FM	K7LHS	440.82500	+	110.9 Hz	ORRC
Portland, Garden Home						
	FM	K7GDS	443.55000	+	100.0 Hz	
Portland, Healy Heights						
	FM	N7NLL	53.09000	51.39000	107.2 Hz	
Portland, KGW Tower						
	FM	N7EXH	146.98000	-	DCS 023	ORRC
	FM	AB7BS	440.50000	+	DCS 125	
	FM	K7QDX	927.12500	902.12500	103.5 Hz	ORRC
Portland, KOIN Tower						
	FM	W7RAT	440.40000	+	123.0 Hz	ORRC

OREGON   299

Location	Mode	Call sign	Output	Input	Access	Coordinator
Portland, KPDX Tower						
	FM	W7PM	442.25000	+	100.0 Hz	ORRC
Portland, Mt Tabor						
	FM	K7LJ	145.39000	-	100.0 Hz	ORRC
Portland, OHSU	FM	K7LTA	442.70000	+	100.0 Hz	ORRC
	FM	K0HSU	444.00000	+	100.0 Hz	
Portland, Skyline	FM	N7EXH	145.31000	-	123.0 Hz	ORRC
	FM	W7AC	147.14000	+	107.2 Hz	ORRC
	FM	N7PIR	440.45000	+	103.5 Hz	ORRC
	FM	W7EXH	441.35000	+	100.0 Hz	
	FM	K6PRN	443.20000	+	173.8 Hz	
	FM	W7DTV	443.62500	+	77.0 Hz	ORRC
Portland, Stonehenge Tower						
	FM	KE7AWR	146.70000	-	100.0 Hz	ORRC
	FM	KB7OYI	440.35000	+	127.3 Hz	
	FM	WA7BND	443.05000	+		ORRC
Portland, Sylvan KOIN TX Tower						
	FM	K7RPT	442.22500	+	100.0 Hz	
Portland, Sylvan TV Tower						
	FM	K7RPT	147.04000	+	100.0 Hz	ORRC
Portland, VA Hospital						
	FM	KE7FBE	145.25000	-	136.5 Hz	
Portland, West Hills						
	FM	N7NLL	442.02500	+	100.0 Hz	
Prineville, Grizzly Mountain						
	FM	N7CCO	147.38000	+	162.2 Hz	ORRC
	FM	N7PIR	444.17500	+	103.5 Hz	ORRC
Prineville, Round Mtn						
	FM	N7CCO	145.21000	-	141.3 Hz	ORRC
Rainier, Rainier Hill						
	FM	NU7D	224.66000	-	114.8 Hz	ORRC
Redmond, Cinder Butte						
	FM	K7RPT	147.04000	+	114.8 Hz	ORRC
Redmond, Cline Butte						
	FM	W7PRA	145.13000	-	DCS 026	ORRC
Redmond, Oregon						
	DMR/BM	KC7DMF	444.45000	+	CC1	ORRC
Reedsport, Roman Nose Mtn						
	FM	AB7BS	443.55000	+	DCS 125	ORRC
Reedsport, Winchester Hill						
	FM	W7OC	147.18000	+	146.2 Hz	ORRC
	FM	W7OC	444.40000	+		
Rockaway Beach	FM	W7GC	442.75000	+	118.8 Hz	ORRC
Rogue River	FM	N7AGX	53.17000	51.47000		ORRC
Rogue River, Elk Mountain						
	FM	WA6HWW	145.27000	-	136.5 Hz	
Rogue River, Fielder Mountain						
	FM	AB7BS	440.85000	+	94.8 Hz	ORRC
Roseburg	FM	WB6MFV	441.87500	+	88.5 Hz	ORRC
Roseburg, Chilcoot Mountain						
	FM	W7PRA	145.21000	-	136.5 Hz	ORRC
Roseburg, Lane Mountain						
	FM	KC7TLY	146.90000	-	100.0 Hz	
Roseburg, Scott Mountain						
	FM	W7PRA	146.70000	-	136.5 Hz	ORRC
	FM	AB7BS	441.85000	+	DCS 125	ORRC
Ruch	FM	KL7VK	146.72000	-	131.8 Hz	ORRC
Saginaw	FM	W7NK	444.85000	+		ORRC
Saginaw, Bear Mountain						
	FM	N7EXH	146.76000	-	123.0 Hz	ORRC
Salem	DMR/MARC	N7MAQ	442.88750	+	CC1	
	FM	WA7ABU	145.29000	-		ORRC

300   OREGON

Location	Mode	Call sign	Output	Input	Access	Coordinator
Salem	FM	WB7RKR	224.24000	-	100.0 Hz	ORRC
	FM	KE7DLA	224.60000	-	100.0 Hz	ORRC
	FM	AC7RF	440.27500	+	107.2 Hz	ORRC
	FM	KB7PPM	440.72500	+		ORRC
	FM	AB7F	441.37500	+	123.0 Hz	ORRC
	FM	KE7DLA	442.50000	+	100.0 Hz	ORRC
	FM	KC7CFS	443.45000	+	123.0 Hz	ORRC
Salem, Bald Hill (Eagle Crest)						
	FM	N7PIR	440.07500	+	103.5 Hz	ORRC
	FM	W7NK	441.75000	+	123.0 Hz	
Salem, CCC	FM	AD7ET	443.17500	+	88.5 Hz	ORRC
Salem, Eagle Crest						
	FM	W7SRA	145.33000	-	186.2 Hz	ORRC
	FM	AB7BS	442.45000	+	DCS 125	
Salem, Popcorn Hill						
	FM	K7MRR	145.35000	-	186.2 Hz	ORRC
	FM	K7UN	441.70000	+	186.2 Hz	ORRC
Salem, Prospect Hill						
	FM	W7PRA	145.49000	-	136.5 Hz	ORRC
	FM	W7SRA	146.86000	-	186.2 Hz	ORRC
	FM	W7DTV	441.17500	+	77.0 Hz	
	FM	W7SRA	441.27500	+	100.0 Hz	ORRC
	FM	W7SRA	443.72500	+	186.2 Hz	ORRC
Salem, Salem Hills						
	FM	W6WHD	224.92000	-	136.5 Hz	ORRC
	FM	W6WHD	444.10000	+	136.5 Hz	ORRC
Salem, Silver Creek Falls						
	FM	WA7ABU	444.95000	+	100.0 Hz	ORRC
Sandy	FM	KJ7IY	53.35000	51.65000	107.2 Hz	ORRC
	FM	KJ7IY	145.43000	-	107.2 Hz	ORRC
	FM	KJ7IY	442.87500	+	107.2 Hz	ORRC
Scio, Rodgers Mountain						
	FM	KA7ENW	146.61000	-	167.9 Hz	
	FM	KA7ENW	442.85000	+	167.9 Hz	ORRC
Seaside	FM	WA7PIX	147.00000	+	118.8 Hz	
Seaside, Reservoir						
	FM	WA7VE	145.49000	-	118.8 Hz	ORRC
Sheridan	FM	AC7ZQ	224.56000	-	100.0 Hz	ORRC
Sherwood	DSTAR	WB7DZG	146.62000	-		ORRC
	DSTAR	WB7DZG	444.31250	+		ORRC
	DSTAR	WB7DZG	1292.00000	1272.00000		ORRC
	FM	KR7IS	145.47000	-	107.2 Hz	ORRC
	FM	KJ7IY	442.27500	+	107.2 Hz	
	FM	KR7IS	443.42500	+	107.2 Hz	ORRC
Silverton, Silver Creek Falls						
	FM	WA7ABU	52.99000	51.29000	100.0 Hz	ORRC
Silverton, Silver Falls						
	FM	W7SAA	147.34000	+	77.0 Hz	
	FM	KE7DLA	224.16000	-	100.0 Hz	ORRC
	FM	W7SAA	444.25000	+	100.0 Hz	ORRC
Sisters	FM	W7EXH	147.34000	+	100.0 Hz	ORRC
Sisters, Fivemile Butte						
	FM	W7DUX	146.90000	-	123.0 Hz	ORRC
Sisters, Hoodoo Butte						
	FM	W7PRA	441.62500	+	100.0 Hz	ORRC
Springfield, Willamette Height						
	FM	WA7FQD	146.74000	-	100.0 Hz	ORRC
Summer Lake, Dead Indian Mount						
	FM	KE7QP	146.80000	-	173.8 Hz	ORRC
Sunriver, Spring River Butte						
	FM	WA7TYD	146.64000	-	123.0 Hz	ORRC

OREGON 301

Location	Mode	Call sign	Output	Input	Access	Coordinator
Sweet Home, Marks Ridge						
	FM	KG7BZ	52.91000	51.21000	100.0 Hz	
	FM	K7ENW	147.20000	+	167.9 Hz	ORRC
	FM	KG7BZ	443.12500	+	77.0 Hz	ORRC
Terrebonne, Crooked River Ranc						
	FM	W7JVO	147.06000	+	162.2 Hz	ORRC
The Dalles	FM	KC7LDD	146.74000	-	100.0 Hz	ORRC
The Dalles, Stacker Butte						
	FM	KF7LN	444.70000	+	100.0 Hz	ORRC
Tigard	FM	KK7TJ	442.57500	+	107.2 Hz	ORRC
Tigard, Bull Mountain						
	FM	KF7TTS	440.17500	+	110.9 Hz	ORRC
Tigard, QTH	FM	K7ICY	440.10000	+	162.2 Hz	
Tillamook, Triangulation Point						
	FM	W7EM	440.52500	+	77.0 Hz	ORRC
Timber	FM	KR7IS	52.85000	51.15000	107.2 Hz	ORRC
	FM	KJ7IY	145.27000	-	107.2 Hz	ORRC
Timber, Hoffman Hill						
	FM	KJ7IY	441.82500	+	107.2 Hz	ORRC
Timberline	FM	K7RPT	147.12000	+	100.0 Hz	ORRC
Timberline, Magic Mile Chairli						
	FM	K7RPT	444.22500	+	100.0 Hz	ORRC
Ukiah, Carney Butte						
	FM	W7URG	444.95000	+	136.5 Hz	ORRC
Vale, Lime Hill	FM	K7OJI	444.15000	+	100.0 Hz	
Vernonia, Corey Hill						
	FM	W7VER	145.25000	-	114.8 Hz	
Waldport, Table Mountain						
	FM	W7VTW	147.00000	+	136.5 Hz	ORRC
Walton, Walker Point						
	FM	W7EXH	145.31000	-	100.0 Hz	ORRC
	FM	K7LNK	444.92500	+	DCS 125	
Warren	FM	W6WHD	224.92000	-	107.2 Hz	
Warren, McNulty Water Tower						
	FM	N7EI	146.68000	-	114.8 Hz	ORRC
	FM	N7EI	444.62500	+	107.2 Hz	ORRC
West Linn	FUSION	WA7BND	443.05000	+		
Wilamina	FM	W7GRT	442.82500	+	123.0 Hz	
Wilhoit	FM	WA7ABU	145.29000	-		
Willamina	FM	W7GRT	146.66000	-	136.5 Hz	
Wolf Creek	FM	N7EXH	146.84000	-	74.4 Hz	ORRC
Wolf Creek, King Mountain						
	FM	W7PRA	146.94000	-	DCS 023	ORRC
	FM	WB6YQP	147.34000	+	136.5 Hz	
	FM	W6PRN	443.52500	+	173.8 Hz	
	FM	AB7BS	444.50000	+	DCS 125	ORRC
Woodburn	DMR/MARC	N7MAQ	441.32500	+	CC1	
Yoncalla, Harness Mtn						
	FM	AB7BS	444.90000	+	DCS 125	

PENNSYLVANIA

Location	Mode	Call sign	Output	Input	Access	Coordinator
Abington Township, Abington Ho						
	FM	K3DN	441.15000	+	88.5 Hz	
Abington, Abington Memorial Ho						
	FM	WA3DSP	223.76000	-	131.8 Hz	ARCC
Acme	FM	N3QZU	51.78000	-		
	FM	W3SDR	146.67000	-	131.8 Hz	
	FM	W3NBN	421.25000	439.25000		
Albion	FM	WA3WYZ	53.55000	52.55000	186.2 Hz	
	FM	WA3USH	223.94000	-		
Allentown	DMR/BM	KC2LHJ	441.40000	+	CC1	
	DMR/MARC	N3RPV	441.60000	+	CC0	

302 PENNSYLVANIA

Location	Mode	Call sign	Output	Input	Access	Coordinator
Allentown	DMR/MARC	N3RPV	448.71250	-	CC0	
	DSTAR	W3OI	147.16500	+		
	DSTAR	W3OI	445.02500	-		
	FM	W3OI	147.13500	+	DCS 315	ARCC
	FM	WA3VHL	147.22500	+	151.4 Hz	ARCC
	FM	KA3NRJ	224.08000	-	203.5 Hz	
	FM	N3HES	443.50000	+	156.7 Hz	ARCC
	FM	N3KZ	444.15000	+	131.8 Hz	ARCC
	FM	N3MFT	448.77500	-	131.8 Hz	ARCC
Allentown, Lehigh Mountain						
	FM	W3OI	146.94000	-	71.9 Hz	ARCC
Allentown, Scholl Woodlands Pr						
	FM	N3XG	443.35000	+	100.0 Hz	
Allentown, South Mountain						
	FM	K4MTP	224.40000	-		
	FM	KA3NRJ	444.10000	+	151.4 Hz	ARCC
Almont	FM	K3MFI	53.23000	52.23000	146.2 Hz	ARCC
Altoona, Wopsy Ridge						
	FM	W3QZF	146.61000	-	123.0 Hz	WPRC
	FM	W3QW	146.82000	-	123.0 Hz	WPRC
Apollo	FM	N1RS	29.68000		131.8 Hz	
	FM	N1RS	51.90000	-	141.3 Hz	
	FM	N1RS	146.97000	-	131.8 Hz	
	FM	N1RS	224.30000	-	131.8 Hz	
	FM	KB3UEM	224.64000	-	131.8 Hz	
Arnot	DMR/MARC	N3FE	444.85000	+	CC1	
Bangor	DMR/MARC	N2DCE	445.21875	-	CC1	
	FM	WA3MDP	147.04500	+	131.8 Hz	
	FM	N3TXG	447.22500	-	131.8 Hz	ARCC
Barnesville	DMR/BM	KB2MXV	440.75000	+	CC5	
Barnesville, Mountain Valley G						
	FM	W3TWA	449.77500	-	131.8 Hz	ARCC
Bear Creek	FM	N3SQ	147.76500	-	127.3 Hz	
Beaver	FM	KA3IRT	442.45000	+	100.0 Hz	
Beaver Falls	FM	W3SGJ	145.31000	-	131.8 Hz	
Beaver, Brighton Township Wate						
	FM	WW3AAA	146.85000	-	131.8 Hz	
	FM	N3TN	224.88000	-	131.8 Hz	
	FM	N3TN	444.25000	+	131.8 Hz	
Beaver, Heritage Valley						
	FM	N3TN	147.13500	+	131.8 Hz	
Bedford	FM	K3NQT	145.49000	-	123.0 Hz	
	FM	K3NQT	224.48000	-	123.0 Hz	WPRC
Bedford, Kinton Knob						
	FM	K3NQT	444.20000	+	123.0 Hz	
Bellefonte	DMR/MARC	K3ARL	147.10500	+	CC1	
Bensalem	DMR/MARC	WB0YLE	445.05250	-	CC1	
	FM	W3BXW	444.20000	+	131.8 Hz	ARCC
Bentleyville	FM	WA3QYV	224.58000	-		
Benton, Red Rock	FM	N3VTH	147.00000	+	77.0 Hz	
	FM	N3KZ	441.80000	+	131.8 Hz	
Berwick	FM	WC3H	53.59000	52.59000	77.0 Hz	
	FM	KB3BJO	147.22500	+	85.4 Hz	ARCC
Bethel, Booths Corner Farmers						
	FM	W3KG	224.22000	-	88.5 Hz	ARCC
Bethlehem	DMR/MARC	N3RPV	145.23000	-	CC1	
	DMR/MARC	K3IHI	444.37500	-	CC1	
	FM	K3LPR	146.77500	-	136.5 Hz	ARCC
	FM	KC2ABV	445.17500	-	100.0 Hz	
Biglerville	FM	W3KGN	443.10000	+	103.5 Hz	ARCC
Biglerville, Big Flat So Mt						
	FM	W3BD	443.05000	+	151.4 Hz	ARCC

PENNSYLVANIA 303

Location	Mode	Call sign	Output	Input	Access	Coordinator
Birdsboro	DMR/BM	N3GAR	442.85000	+	CC1	
Blairsville	FM	W3BMD	146.91000	-	131.8 Hz	
	FM	W3BMD	444.97500	+	110.9 Hz	WPRC
Blakeslee	DMR/MARC	KG3I	447.27500	-	CC0	ARCC
Bloomsburg	FM	WB3DUC	53.13000	52.13000	131.8 Hz	ARCC
	FM	WC3H	145.13000	-	77.0 Hz	ARCC
	FM	WB3DUC	147.12000	+	131.8 Hz	ARCC
	FM	WB3DUC	447.02500	-	91.5 Hz	
Blossburg, Bloss Mountain						
	FM	N3TJJ	442.40000	+		
Blue Knob, Ski Resort						
	FM	KB3KWD	147.15000	+	167.9 Hz	
Booth's Corner	FM	W3KG	223.94000	-	88.5 Hz	ARCC
Boothwyn	FM	W3KG	446.77500	-	88.5 Hz	
Brackinridge	FM	WA3WOM	145.37000	-	131.8 Hz	
Bradford	FM	KD3OH	147.24000	+	173.8 Hz	
Brentwood, Brentwood Fire Depa						
	FM	AB3PJ	145.33000	-	131.8 Hz	WPRC
Brentwood, Brentwood Fire Hall						
	FM	KW3LO	443.60000	+	131.8 Hz	
Bridgeville	FM	N3WX	51.94000	-	131.8 Hz	WPRC
	FM	KS3R	145.13000	-		
Briggsville	FM	WC3H	145.13000	-	77.0 Hz	
Bryn Mawr	FM	WB3JOE	224.42000	-	131.8 Hz	ARCC
Buck	FM	AK3E	224.78000	-	151.4 Hz	
Bucktown	FM	N3KZ	147.27000	+	77.0 Hz	
	FM	W3EOC	446.17500	-	100.0 Hz	ARCC
Burgettstown	FM	K3PSP	147.39000	+	131.8 Hz	
Butler, Alameda Park						
	FM	K3PSG	147.30000	+	131.8 Hz	
Butler, Center Township Water						
	FM	W3UDX	147.36000	+	131.8 Hz	
Butler, Sunnyview Long Term Ca						
	FM	K3PSG	443.90000	+	131.8 Hz	WPRC
California	FM	KA3FLU	145.11000	-		
Camelback Mountain						
	FM	N3KZ	53.79000	52.79000	131.8 Hz	ARCC
Canonsburg	FM	N3FB	443.65000	+	131.8 Hz	
Carbondale	DMR/BM	KC3LEE	145.49000	145.43000	CC8	
	DMR/BM	K1NRA	440.05625	+	CC8	ARCC
Carbondale, Salem Mountain						
	FM	K1NRA	145.49000	-	85.4 Hz	
Carnegie	FM	W3KWH	147.03000	+	123.0 Hz	
	FM	W3KWH	426.25000	439.25000		
Carnegie, Settlers Cabin Park						
	FM	W3KWH	444.45000	+	103.5 Hz	
Carrick	FM	W3PGH	29.64000	-	131.8 Hz	
	FM	W3PGH	52.64000	51.64000	131.8 Hz	
	FM	W3PGH	146.61000	-	131.8 Hz	
	FM	W3PGH	444.95000	+	131.8 Hz	
Carroltown, NEXT TO WATER TOWE						
	FM	N3LAD	146.77500	-	123.0 Hz	WPRC
Castle Shannon	FM	K3CSF	440.60000	+		
Catasauqua	DMR/MARC	KC2ABV	441.10000	+	CC1	
Catawissa, Catawissa Mountain						
	FM	K4MTP	53.05000	52.05000	131.8 Hz	
Centerville	FM	N3NQT	146.79000	-	123.0 Hz	
Central City	FM	WR3AJL	146.62500	-	123.0 Hz	
Central City, Statler Hill						
	FM	KE3UC	443.57500	+	123.0 Hz	
Chalfont	FM	W3DBZ	223.90000	-	107.2 Hz	ARCC
Chambersburg	DMR/BM	WR3IRS	441.90000	+	CC1	ARCC

304 PENNSYLVANIA

Location	Mode	Call sign	Output	Input	Access	Coordinator
Chambersburg	DMR/MARC	KA3LAO	146.41500	147.41500	CC1	WPRC
	FM	N3KZ	443.70000	+	131.8 Hz	
Charlestown	DMR/MARC	WR3IRS	440.20000	+	CC1	ARCC
Cherry Hill	FM	N8XUA	146.76000	-	186.2 Hz	
	FM	WA3USH	444.92500	+		
Cherry Valley	FM	KC8PHW	442.32500	+		
Chester	DMR/MARC	KM3W	443.92500	+	CC1	
	FM	KM3W	444.70000	+	131.8 Hz	ARCC
Chester, Crozer Chester Medica						
	FM	KM3W	224.70000	-	131.8 Hz	ARCC
Clarion	FM	KE3EI	442.65000	+		
	FM	N3HZX	444.32500	+	110.9 Hz	
Clearfield	FM	K3EDD	444.62500	+	173.8 Hz	
Clearfield, Rockton Mountain						
	FM	N5NWC	52.90000	-	173.8 Hz	
	FM	W3CPA	146.46000	147.46000	173.8 Hz	
Clifford	FM	WR2M	224.86000	-		
	FM	WR2M	445.47500	-		
Clinton	FM	K3KEM	147.21000	+	100.0 Hz	
	FM	K3KEM	443.00000	+		
	FM	K3KEM	444.85000	+	131.8 Hz	
Coatesville	FM	W3EOC	441.95000	+	100.0 Hz	ARCC
Cochranville	FM	WB3LGG	449.67500	-	94.8 Hz	
Connellsville	FM	WB3JNP	145.17000	-	131.8 Hz	
	FM	W3NAV	146.89500	-	131.8 Hz	
	FM	N3LGY	444.82500	+	151.4 Hz	
Coopersburg	FM	W3LR	443.59000	+	151.4 Hz	ARCC
	FM	W3LR	449.27500	-	151.4 Hz	ARCC
Coraopolis	FM	KA3IRT	444.15000	+		
Cornwall	DMR/BM	W3AD	449.02500	-	CC1	ARCC
	DMR/MARC	KA0JQO	147.10500	+	CC1	
	FM	K3LV	147.31500	+	82.5 Hz	ARCC
	FM	N3TPL	224.82000	-	114.8 Hz	ARCC
	FM	KA3CNT	224.84000	-	131.8 Hz	ARCC
	FM	W3BXW	442.15000	+	131.8 Hz	ARCC
	FM	N3FYI	446.47500	-	114.8 Hz	
Cornwall Furnace	DMR/MARC	K3LV	447.92500	-	CC1	
Cornwall Mountain						
	FM	N3TUQ	927.58750	902.58750	114.8 Hz	ARCC
Cornwall, Cornwall Mtn						
	FM	W3AD	145.39000	-	118.8 Hz	ARCC
Corry	FM	W3YXE	147.09000	+	186.2 Hz	
	FM	KE3PD	224.06000	-	186.2 Hz	
	FM	W3CCB	443.50000	+		
	FM	W3YXE	444.80000	+		
Coudersport	FM	N3PC	146.68500	-	173.8 Hz	WPRC
	FM	KB3EAR	443.30000	+		
Cowansville	FM	N1RS	146.50500	147.50500	131.8 Hz	WPRC
	FM	KA3HUK	224.18000	-		
	FM	N1RS	444.30000	+	131.8 Hz	
Dallas	FM	W3LR	449.27500	-	151.4 Hz	
Darby, Mercy Fitzgerald Hospit						
	FM	W3UER	147.36000	+	131.8 Hz	ARCC
Delano	FM	W3SC	53.31000	52.31000		
	FM	W3SC	145.37000	-	123.0 Hz	ARCC
Derry	FM	W3CRC	145.15000	-	131.8 Hz	
	FM	N1RS	146.49000	147.49000	131.8 Hz	
	FM	KE3PO	442.27500	+	131.8 Hz	
Dingmans Ferry	FM	AA2HA	145.33000	-	141.3 Hz	ARCC
Dover	FM	WB3EPJ	442.70000	+	74.4 Hz	ARCC
Downington	DMR/MARC	N3JCS	448.05000	-	CC1	
Doylestown	FM	WA3EPA	145.35000	-	131.8 Hz	ARCC

PENNSYLVANIA 305

Location	Mode	Call sign	Output	Input	Access	Coordinator
Drexel Hill	DSTAR	N3AEC B	440.04380	+		
	DSTAR	N3AEC A	1299.70000			
Du Bois, Rockton Mountain						
	FM	N3QC	147.31500	+	173.8 Hz	WPRC
Dunmore	FM	KC3MN	147.00000	+	77.0 Hz	
Eagle	DMR/MARC	WR3IRS	441.30000	+	CC1	
Eagles Peak	FM	N3KZ	442.40000	+	131.8 Hz	ARCC
Eagleville	DSTAR	AA3E A	1255.57500	+		
	FM	AA3E	146.83500	-	88.5 Hz	ARCC
	FM	K3CX	449.92500	-	100.0 Hz	ARCC
Earlville	FM	N3KZ	53.87000	52.87000	131.8 Hz	ARCC
	FM	K3ZMC	147.21000	+	131.8 Hz	
	FM	K3ZMC	443.55000	+	131.8 Hz	
East Monongahela						
	FM	N3OVP	223.90000	-		
	FM	W3CSL	442.42500	+		
	FM	W3CDU	443.35000	+		
East Stroudsburg	DSTAR	KB3TEM C	146.55000	-		
Easton	FM	N3LWY	51.82000	-	88.5 Hz	ARCC
	FM	KB3AJF	224.74000	-	100.0 Hz	ARCC
	FM	N2ZAV	1294.00000	1274.00000		ARCC
Easton, Braden Airport						
	FM	KB3VPK	145.27000	-	151.4 Hz	
Eau Claire	FM	W3ZIC	145.19000	-	186.2 Hz	
Edinboro, Franklin Township EO						
	FM	KB3PSL	146.98500	-	186.2 Hz	
Effort	DMR/MARC	K4MTP	445.37500	-	CC1	ARCC
Elizabethville, Berry Mountain						
	FM	KB3VDL	147.24000	+	123.0 Hz	
Elk Mountain	FM	N3HPY	447.37500	-	131.8 Hz	ARCC
Ellendale	FM	KB3NIA	147.07500	+	123.0 Hz	ARCC
Ellwood City	FM	N3ZJM	443.62500	+		
Elverson	DMR/MARC	W3SJS	145.22500	+	CC1	
Emporium	FM	N3FYD	146.80500	-		
	FM	N3SNN	147.18000	+	173.8 Hz	
Ephrata	DMR/MARC	K3TUF	443.60000	+	CC1	
	FM	W3XP	145.45000	-	100.0 Hz	ARCC
	FM	W3XP	444.65000	+	131.8 Hz	ARCC
Erie	FM	W3GV	146.61000	-	186.2 Hz	
	FM	KB5ELV	442.15000	+	186.2 Hz	
	FM	N3APP	443.37500	+	186.2 Hz	
Erie, Presque Isle State Park						
	FM	N3APP	147.27000	+	186.2 Hz	WPRC
Evans City	FM	KA3HUK	224.98000	-		
	FM	N3XCD	442.67500	+		
	FM	KB3LSM	443.70000	+	131.8 Hz	WPRC
Fairfield	DMR/MARC	KA3LAO	146.47500	147.47500	CC1	ARCC
Fairless Hills	FM	W3BXW	147.30000	+	131.8 Hz	ARCC
Fairview Township						
	DMR/MARC	N2JEH	444.05625	+	CC1	ARCC
Fairview Village	FM	N3CVJ	224.20000	-	88.5 Hz	ARCC
Falls Creek, KDUJ Airport						
	FM	N3QC	443.85000	+	173.8 Hz	WPRC
Feasterville	FM	N3SP	223.80000	-	131.8 Hz	ARCC
	FM	WB3BLG	224.98000	-		
Finleyville	FM	N3OVP	51.98000	-	141.3 Hz	WPRC
Forkston	DMR/BM	WR3IRS	440.10000	+	CC1	
	FM	N3KZ	442.00000	+	131.8 Hz	ARCC
Franklin	FM	W3ZIC	145.23000	-	186.2 Hz	
	FM	N3QCR	224.74000	-	186.2 Hz	
Freeland, Water Tower						
	FM	W3HZL	449.42500	-	103.5 Hz	

306   PENNSYLVANIA

Location	Mode	Call sign	Output	Input	Access	Coordinator
Galeton	FM	KB3EAR	147.34500	+	131.8 Hz	
Gatchellville	DMR/BM	N3CNJ	145.23000	-	CC1	
	DMR/BM	N3CNJ	447.72500	-	CC1	
Georgeville	FM	KB3CNS	442.85000	+		
Gettysburg	FM	W3KGN	145.35000	-	103.5 Hz	ARCC
Girard, West Erie Co Emergency						
	FM	WE3OPS	146.95500	-	186.2 Hz	
Glen Mills	FM	W3LW	224.98000	-	94.8 Hz	ARCC
	FM	W3LW	1295.80000	1280.80000	94.8 Hz	ARCC
Grand Valley	FM	W3GFD	443.05000	+		WPRC
Grantville	FM	AA3RG	448.22500	-	192.8 Hz	ARCC
Grantville, Blue Mtn						
	FM	K3LV	147.16500	+	82.5 Hz	
Green Lane	FM	AA3E	449.12500	-	88.5 Hz	ARCC
Greensburg	DMR/BM	WC3PS	145.44000	-	CC1	
	DSTAR	WC3PS A	1250.00000			
	DSTAR	WC3PS	1286.00000	1266.00000		
	FM	W3LWW	147.18000	+	131.8 Hz	
Greensburg, D.M.V. Center						
	FM	N3RSJ	444.80000	+	123.0 Hz	
Greentown	FM	WA2AHF	444.65000	+	114.8 Hz	ARCC
Greenville	DMR/BM	KE3JP	444.37500	+	CC8	WPRC
	DSTAR	K3WRB	145.43000	-		WPRC
Greenville, Western Union Towe						
	FM	KE3JP	146.44500	147.44500	186.2 Hz	WPRC
	FM	KE3JP	443.42500	+	186.2 Hz	WPRC
Grove City, Grove City Medical						
	FM	W3LIF	146.68500	-	"	
Hanover, Pigeon Hills						
	FM	W3MUM	147.33000	+	123.0 Hz	ARCC
Harrisburg	DMR/MARC	W3ND	147.37500	+	CC1	ARCC
	FM	KA3TKW	145.21000	-	123.0 Hz	ARCC
	FM	W3UU	146.76000	-	100.0 Hz	ARCC
	FM	KA3RKW	224.18000	-	123.0 Hz	ARCC
	FM	W3ND	448.07500	-	123.0 Hz	ARCC
Harrisburg, Blue Mountain						
	FM	W3ND	145.29000	-	123.0 Hz	ARCC
	FM	W3ND	145.47000	-	123.0 Hz	ARCC
	FM	W3ND	444.45000	+	123.0 Hz	ARCC
Harrisburg, Ellendale Forge						
	FM	W3ND	145.11000	-	131.8 Hz	ARCC
Hatfield	FM	WA3RYQ	147.33000	+		ARCC
Hazelwood	FM	WA3PBD	145.47000	-	71.9 Hz	
	FM	KA3IDK	444.05000	+		
	FM	WA3PBD	444.10000	+		
	FM	WA3PBD	923.25000	910.25000		
Hazleton	FM	WC3H	53.59000	52.59000	77.0 Hz	ARCC
	FM	W3OHX	146.67000	-	103.5 Hz	ARCC
	FM	W3RC	224.60000	-	77.0 Hz	ARCC
	FM	W3RC	441.90000	+	114.8 Hz	ARCC
	FM	W3RC	927.32500	902.32500	DCS 311	ARCC
Hegins	DMR/BM	KB3VXB	147.39000	+	CC7	
	DMR/BM	W3TWA	442.29000	+	CC7	
Hermitage	FM	K3AWS	147.15000	+		
Hilltown	FM	W3HJ	145.33000	-	131.8 Hz	ARCC
	FM	W3HJ	147.39000	+	100.0 Hz	ARCC
	FM	W3CCX	224.58000	-		ARCC
	FM	W3HJ	442.90000	+	123.0 Hz	
	FM	K3DN	443.95000	+	131.8 Hz	ARCC
	FM	K3BUX	927.31250	902.31250	DCS 131	ARCC
Hollidaysburg	DMR/BM	NU3T	442.10000	+	CC1	
Holtwood	DMR/BM	KX3B	146.74500	-	CC1	

PENNSYLVANIA 307

Location	Mode	Call sign	Output	Input	Access	Coordinator
Holtwood	DMR/BM	KX3B	448.75000	-	CC1	
Homestead	FM	WA3PBD	29.62000	-	118.8 Hz	
	FM	WA3PBD	51.74000	-	100.0 Hz	
	FM	WA3PBD	146.73000	-	100.0 Hz	
	FM	WA3PBD	223.94000	-	118.8 Hz	
	FM	KA3IDK	224.14000	-		
Honey Brook	FM	K3CX	53.33000	52.33000	131.8 Hz	
Honeybrook	DMR/BM	N3TJJ	449.27500	-	CC1	
Hopwood	FM	W3PIE	443.75000	+	131.8 Hz	
Horsham	FM	K3JJO	147.16500	+	162.2 Hz	ARCC
	FM	WA3TSW	444.55000	+	100.0 Hz	ARCC
Hunlock Creek	FM	N3CSE	146.80500	-	82.5 Hz	ARCC
Huntersville, Long Ridge						
	FM	KB3DXU	145.45000	-	167.9 Hz	ARCC
Huntingdon	FM	W3WIV	145.31000	-		
	FM	WB3CJB	146.70000	-		
	FM	WO3T	442.60000	+	123.0 Hz	
Independence	FM	KA3IRT	145.45000	-		
Indiana	FM	W3BMD	29.66000	-	131.8 Hz	WPRC
Industry	FM	N3CYR	146.41500	147.41500		
Irwin	FM	W3OC	147.12000	+	131.8 Hz	WPRC
	FM	W3OC	442.25000	+	131.8 Hz	WPRC
Jeannette/Hempfield						
	FM	N3RSJ	444.22500	+	123.0 Hz	
Jennerstown, Laurel Mountain						
	FM	NJ3T	444.47500	+	123.0 Hz	
Jennersville, Route 1 At Route						
	FM	N3SLC	145.25000	-		
Jim Thorpe	FM	W3HA	147.25500	+	162.2 Hz	ARCC
Johnstown	DMR/BM	N3YFO	145.39000	-	CC1	WPRC
	DMR/BM	N3YFO	442.82500	+	CC1	WPRC
	DMR/BM	KE3UC	444.80000	+	CC1	WPRC
	FM	N3FQQ	51.80000	-		
	FM	N3LZX	147.37500	+		
	FM	W3IW	224.26000	-		
	FM	K3WS	224.68000	-	123.0 Hz	
Johnstown, Lural Mountain						
	FM	WA3WGN	146.94000	-	123.0 Hz	
Joliett, Keffers Fire Tower						
	FM	K4MTP	53.05000	52.05000	131.8 Hz	ARCC
	FM	K4MTP	223.82000	-		
Kane	FM	WB3IGM	53.47000	52.47000	173.8 Hz	
	FM	WB3IGM	146.73000	-	173.8 Hz	WPRC
Kilbuck	FM	WA3RSP	53.29000	52.29000		
Kittanning	FM	K3TTK	145.41000	-	173.8 Hz	
Kittanning, Armstrong Co EOC						
	FM	K3QY	443.97500	+		WPRC
Kylertown	DSTAR	W3PHB C	146.25500	+		
Lake Harmony	FM	N3KZ	442.10000	+	131.8 Hz	ARCC
Lake Wallenpaupack						
	FM	WA2ZPX	442.35000	+		ARCC
Lancaster	DMR/BM	W3PC	448.57500	-	CC1	
	FM	KA3CNT	449.32500	-		ARCC
Lancaster, Gap Hill						
	FM	N3EDM	449.52500	-	114.8 Hz	
Lancaster, Lancaster General H						
	FM	W3RRR	147.01500	+	118.8 Hz	ARCC
	FM	W3RRR	449.57500	-	114.8 Hz	ARCC
Lancaster, Pennsylvania						
	FM	N3FYI	224.32000	-	114.8 Hz	
	FM	N3FYI	446.47500	-	114.8 Hz	ARCC
Landisburg	DMR/MARC	N3TWT	146.68500	-	CC1	ARCC

308 PENNSYLVANIA

Location	Mode	Call sign	Output	Input	Access	Coordinator
Laporte	FM	N3XXH	145.31000	-	167.9 Hz	ARCC
	FM	W3NOD	446.92500	-	82.5 Hz	ARCC
	FM	W3NOD	449.92500	-	82.5 Hz	ARCC
Lawrenceville, Childrens Hospi						
	FM	KF2CHP	443.40000	+	131.8 Hz	WPRC
Lebanon	FM	W3WAN	146.88000	-	74.4 Hz	
	FM	KE3RG	147.24000	+	82.5 Hz	ARCC
	P25	K3LV	447.67500	-	82.5 Hz	ARCC
Leechburg	FM	K3QY	147.33000	+	173.8 Hz	WPRC
Lehigh Valley-Easton, High Ato						
	FM	N3LWY	146.65500	-	136.5 Hz	ARCC
Lewisburg	FM	K3FLT	146.62500	-	110.9 Hz	ARCC
Lewistown	FM	K3DNA	146.91000	-	123.0 Hz	WPRC
Lewistown, Strodes Mills, Gran						
	FM	W3DBB	145.19000	-	123.0 Hz	
Liberty	FM	WA2JOC	145.29000	-		ARCC
Lima	DSTAR	W3AEC B	440.05630	+		
	DSTAR	W3AEC A	1299.90000			
Lima, Fair Acres	DSTAR	W3AEC	440.05625	+		ARCC
	DSTAR	W3AEC	1255.55000	1275.55000		ARCC
Lititz	FM	KA3CNT	224.44000	-	131.8 Hz	ARCC
	FM	KA3CNT	449.22500	-		ARCC
Little Offset	FM	KA2QEP	448.52500	-	131.8 Hz	ARCC
Lock Haven	DMR/BM	W3LHU	146.47500	147.47500	CC1	
Lock Haven, Swissdale						
	FM	K3KR	147.36000	+		WPRC
Loganton, Riansares Mountain						
	FM	N3SSL	53.77000	52.77000	173.8 Hz	
	FM	N3XXH	224.30000	-	85.4 Hz	
Long Branch	FM	W3RON	443.12500	+		
Long Pond	FM	KB3WW	146.44500	147.44500	131.8 Hz	ARCC
	FM	KB3WW	224.34000	-	131.8 Hz	ARCC
	FM	K4MTP	224.92000	-	127.3 Hz	ARCC
	FM	N3BUB	448.27500	-	131.8 Hz	ARCC
	FM	N3VAE	448.47500	-	123.0 Hz	ARCC
Long Pond, Near The Pocono Int						
	FM	KG3I	446.57500	-	151.4 Hz	ARCC
Lykens	FM	KB3VDL	449.82500	-	123.0 Hz	
Manheim	FM	K3IR	145.23000	-	118.8 Hz	ARCC
	FM	K3IR	449.97500	-	114.8 Hz	ARCC
Mansfield	DMR/BM	N3FE	444.60000	+	CC1	
	DMR/MARC	N3FE	441.61250	+	CC1	
	DMR/MARC	N3FE	441.71250	+	CC1	
Mars	FM	K3RS	224.94000	-		WPRC
Meadowbrook	FM	WA3UTI	146.71500	-	131.8 Hz	ARCC
	FM	WA3UTI	443.15000	+	131.8 Hz	ARCC
Meadville	FM	W3MIE	147.21000	+	186.2 Hz	
	FM	W3MIE	444.07500	+	186.2 Hz	
Meadville, Crawford County Fai						
	FM	W3MIE	145.13000	-	186.2 Hz	
Mechanicsburg	FM	N3TWT	443.30000	+	67.0 Hz	ARCC
Mechanicsburg, Three Square Ho						
	FM	N3TWT	146.46000	147.46000	67.0 Hz	ARCC
Media	FM	W3AWA	145.23000	-	131.8 Hz	ARCC
	FM	W3AEC	146.94000	-	131.8 Hz	ARCC
Media, Fair Acres Geriatric Ce						
	FM	W3KG	146.94000	-	131.8 Hz	
Mehoopany	FM	WA3PYI	53.35000	52.35000	131.8 Hz	ARCC
Mehoopany, Forkston Mountain						
	FM	N3KZ	147.21000	+	77.0 Hz	ARCC
Meyersdale	DMR/BM	W3DCW	444.37500	+	CC1	
	FM	KK3L	224.52000	-		

PENNSYLVANIA 309

Location	Mode	Call sign	Output	Input	Access	Coordinator
Meyersdale	FM	N3KZ	442.20000	+	131.8 Hz	
Meyersdale, Hays Mill Fire Tow						
	DMR/BM	W3DCW	441.50000	+	CC1	
Meyersdale, Hays Mills Fire To						
	FM	KQ3M	145.27000	-	123.0 Hz	
Meyersdale, Mount Davis						
	DMR/BM	WA3P	443.72500	+	CC1	
Middleburg	FM	K3SNY	146.82000	-	100.0 Hz	
Mildred	FM	N3VTH	146.88000	-	77.0 Hz	
Milton	FM	K3FLT	146.98500	-	110.9 Hz	ARCC
Monongahela	FM	KA3BFI	147.22500	+		
Monroeville	FM	WA3PBD	444.00000	+	131.8 Hz	
Montana Mt	DMR/MARC	WB3EHB	444.29375	+	CC1	
Montoursville	DMR/BM	KB3HLL	145.49000	-	CC1	ARCC
	DMR/BM	KC3FOW	447.06000	-	CC2	
	FM	KB3DXU	443.50000	+	167.9 Hz	ARCC
Montrose	FM	K3SQO	147.52000	+		
Montrose Borough						
	FM	K3SQO	147.24000	+	107.2 Hz	ARCC
Montrose, Montrose Fairgrounds						
	FM	N3KZ	147.37500	+	77.0 Hz	ARCC
Morrisville	DMR/BM	W2FUV	145.25000	-	CC1	
	DMR/BM	W2FUV	927.65000	902.65000	CC1	
	FM	WR3B	447.47500	-	103.5 Hz	ARCC
	FM	WB0YLE	927.65000	902.65000	141.3 Hz	ARCC
Mount Holly Springs						
	FM	KG3S	444.30000	+	151.4 Hz	
Mount Lebanon	FM	N3SH	146.95500	-	131.8 Hz	WPRC
	FM	N3SH	442.55000	+	131.8 Hz	
Mount Pleasant	FM	KA3JSD	51.96000	-	141.3 Hz	
	FM	KA3JSD	147.01500	+	127.3 Hz	WPRC
	FM	KA3JSD	444.87500	+		
Mt Holly	FM	N3TWT	145.43000	-	67.0 Hz	ARCC
Murrysville	FM	W3GKE	443.50000	+		
Nazareth	FM	W3OK	51.76000	-	151.4 Hz	ARCC
	FM	W3OK	145.11000	-		ARCC
	FM	W3OK	146.70000	-	151.4 Hz	ARCC
	FM	KB3KKZ	443.45000	+	127.3 Hz	ARCC
	FM	W3OK	444.90000	+	151.4 Hz	ARCC
New Bethlehem	FM	N3TNA	442.72500	+		
	FM	N3TNA	444.42500	+	186.2 Hz	WPRC
New Castle	DMR/BM	KB3YBB	444.02500	+	CC8	
	DSTAR	KB3YBB	443.07500	+		
	FM	KC3BDF	146.62500	-	131.8 Hz	WPRC
	FM	N3ETV	147.19500	+	131.8 Hz	
	FM	N3ETV	444.72500	+		
New Cumberland	DMR/MARC	KB3TWW	146.44500	145.44500	CC1	ARCC
	DMR/MARC	KB3TWW	444.05625	+	CC1	
New Galilee	FM	KE3ED	51.72000	-		
New Germany	DMR/BM	KC3DVR	440.85000	+	CC1	
	FM	KC3DES	145.21000	-	123.0 Hz	
New Kensington	DMR/MARC	W3PRL	146.46000	147.46000	CC1	
	FM	N1RS	146.49000	147.49000	131.8 Hz	
	FM	K3MJW	146.64000	-	131.8 Hz	WPRC
	FM	N1RS	442.80000	-	141.3 Hz	
	FM	K3MJW	444.52500	+	131.8 Hz	
New London	FM	KB3DRX	448.97500	-	107.2 Hz	ARCC
New Ringgold	FM	W3TWA	147.27000	+	131.8 Hz	
Newfoundland	FM	KB3WW	147.19500	+	131.8 Hz	ARCC
Newmanstown	FM	N3SWH	147.28500	+	131.8 Hz	ARCC
Newport	FM	W3ND	444.55000	+	123.0 Hz	
Newtown Square	DMR/BM	WR3IRS	441.20000	+	CC1	ARCC

310 PENNSYLVANIA

Location	Mode	Call sign	Output	Input	Access	Coordinator
Newtown Square	FM	WB3JOE	147.06000	+	131.8 Hz	
	FM	W3DI	147.19500	+	100.0 Hz	ARCC
	FM	W3DI	442.25000	+	131.8 Hz	ARCC
	FM	WA3NNA	442.60000	+	131.8 Hz	ARCC
Newville	FM	N3TWT	146.46000	147.46000	67.0 Hz	
Norristown	FM	N3CDP	223.86000	-	131.8 Hz	ARCC
	FM	N3CB	448.67500	-	131.8 Hz	ARCC
Norrisville	FM	W3MIE	147.03000	+	186.2 Hz	
North Hills	FM	W3EXW	444.40000	+	88.5 Hz	
North Washington, North Washin						
	FM	K3PSG	442.90000	+	131.8 Hz	
Oakland	FM	WA3YOA	443.55000	+		
Oakland, University Of Pittsbu						
	FM	W3YJ	443.45000	+	100.0 Hz	
Oil City	DMR/BM	W3ZIC	444.12500	+	CC1	WPRC
	DMR/BM	W3ZIC	445.12500	+	CC1	
Orson, Mount Ararat						
	FM	K4MTP	53.05000	52.05000	131.8 Hz	ARCC
Oxford	FM	W3EOC	448.87500	-	100.0 Hz	ARCC
Palmerton	FM	N3DVF	224.26000	-	94.8 Hz	ARCC
Palmerton, Blue Mountain						
	FM	W3EPE	449.37500	-	131.8 Hz	ARCC
Paoli	FM	WB3JOE	145.13000	-	131.8 Hz	
Paoli, Hospital	FM	WB3JOE	445.67500			ARCC
Parkesburg	FM	W3GMS	146.98500	-	100.0 Hz	ARCC
	FM	KJ6AL	442.00000	+	94.8 Hz	ARCC
Pennfield Manor	FM	N3RZL	442.35000	+		
Perkasie	DMR/MARC	K3BUX	445.57500	-	CC1	
	FM	W3AI	145.31000	-	131.8 Hz	ARCC
Philadelphia	DMR/MARC	WR3IRS	440.65000	+	CC1	ARCC
	DMR/MARC	AB3LI	445.97500	-	CC1	
	DMR/MARC	AB3LI	446.05000	-	CC1	
	DSTAR	K3PDR B	445.18130	-		
	FM	KD3WT	145.41000	-	127.3 Hz	ARCC
	FM	WB3EHB	224.06000	-	131.8 Hz	ARCC
	FM	K3TU	224.80000	-	131.8 Hz	ARCC
	FM	K3CX	440.15000	+	100.0 Hz	ARCC
	FM	K3TU	442.80000	+	131.8 Hz	ARCC
	FM	KD3WT	446.87500	-	131.8 Hz	ARCC
	FM	W3PHL	923.25000	910.25000		ARCC
Philadelphia, Oxford Circle						
	FM	WB0CPR	449.77500	-	141.3 Hz	ARCC
Philadelphia, Roxborough Tower						
	FM	W3SBE	442.55000	+	91.5 Hz	ARCC
Philadelphia, University Of Pe						
	FM	WM3PEN	146.68500	-	146.2 Hz	ARCC
Philipsburg	DSTAR	KC3BMB	444.75000	+		
Philipsburg, Rattlesnake Mount						
	FM	W3PHB	146.43000	147.43000	173.8 Hz	WPRC
Pine Grove	FM	AA3RG	145.17000	-	110.9 Hz	ARCC
	FM	AA3RG	146.64000	-	82.5 Hz	ARCC
Pittsburgh	FM	KB3CNN	444.10000	+	71.9 Hz	
	FM	W3VC	444.65000	+	131.8 Hz	
Pittsburgh, Carnegie Mellon Un						
	FM	W3VC	145.19000	-	131.8 Hz	
Pittsburgh, WQED TV-13 Tower						
	DSTAR	W3EXW	146.82000	-		
	FM	W3EXW	146.88000	-	88.5 Hz	
Pittston	DMR/BM	KB3TEM	446.50000	-	CC1	
Pittston, Suscon	FM	W3RC	147.03000	+	77.0 Hz	ARCC
	FM	N3FCK	443.60000	+	100.0 Hz	
Pleasant Mount	FM	K2KQZ	53.07000	52.07000	136.5 Hz	ARCC

PENNSYLVANIA 311

Location	Mode	Call sign	Output	Input	Access	Coordinator
Pleasant Township						
	FM	W3YZR	146.97000	-	186.2 Hz	
Pleasantville	FM	W3ZIC	147.12000	+	186.2 Hz	WPRC
Plumstead	FM	N3EXA	449.72500	-	136.5 Hz	ARCC
Plumsteadville	FM	KB3AJF	447.97500	-	131.8 Hz	ARCC
Pocopson	DSTAR	W3EOC	445.06875	-		
	DSTAR	W3EOC B	445.06880	-		
	DSTAR	W3EOC A	1299.40000			
	FM	W3EOC	1255.50000	1275.50000		ARCC
Port Royal, Tuscarora Mountain						
	FM	K3TAR	147.04500	+	146.2 Hz	ARCC
Pottstown	FM	K3ZMC	224.02000	-	131.8 Hz	ARCC
	FM	KI3I	442.75000	+	141.3 Hz	ARCC
Pottsville	DMR/BM	W3TWA	442.30000	+	CC5	
	DMR/MARC	K3VNN	442.30000	+	CC1	
	FM	WX3N	443.00000	+	77.0 Hz	ARCC
Pottsville, Delano	FM	W3SC	444.95000	+	123.0 Hz	
Prospect	FM	N3HWW	224.24000	-		
Punxsutawney	FM	N5NWC	146.71500	-	173.8 Hz	
	FM	N3GPM	442.47500	+		
	FM	N5NWC	443.47500	+		
Quakertown	FM	WA3IPP	146.88000	-	131.8 Hz	ARCC
	FM	WA3KEY	443.20000	+	114.8 Hz	ARCC
Quarryville	FM	N3EIO	448.17500	-	94.8 Hz	ARCC
Ransom	FM	N3EVW	53.43000	52.43000	136.5 Hz	ARCC
Rawlinsville	DMR/BM	N3TPL	448.62500	-	CC1	ARCC
	FM	WA3WPA	224.32000	-	131.8 Hz	ARCC
Reading	DMR/BM	K3TI	444.35000	+	CC1	ARCC
	DMR/BM	KB3WLV	445.17500	-	CC1	
	FM	KA3KDL	29.64000	-		
	FM	KA3KDL	146.62500	-		
	FM	W3BN	146.91000	-	131.8 Hz	ARCC
	FM	K3SJH	446.92500	-	156.7 Hz	
	FM	W3MEL	448.72500	-	146.2 Hz	ARCC
	FM	K3CX	449.62500	-	100.0 Hz	ARCC
Reading , Mt Penn						
	FM	KB3OUC	443.97500	+	114.8 Hz	
Reading, Mount Penn						
	FM	K3TI	224.64000	-	114.8 Hz	ARCC
	FUSION	K3TI	145.49000	-	114.8 Hz	ARCC
Reading, Mt Penn	FM	K3TI	145.15000	-		ARCC
Reading, Mt Penn Fire Tower						
	FM	K2SEH	147.18000	+	110.9 Hz	ARCC
Red Lion	DMR/BM	N3CNJ	448.52500	-	CC1	
	FM	W3ZGD	146.86500	-	123.0 Hz	ARCC
	FM	KA3CNT	224.84000	-	131.8 Hz	ARCC
	FM	N3NRN	449.42500	-	123.0 Hz	ARCC
Reesers Summit	FM	W3ND	446.42500	-	123.0 Hz	ARCC
Richboro	FM	WA3DSP	53.03000	52.03000	131.8 Hz	
	FM	N3TS	146.79000	+	131.8 Hz	ARCC
Richfield, Shade Mountain						
	FM	K3SNY	146.82000	-	100.0 Hz	
Ridgway	FM	N3NIA	147.00000	+	173.8 Hz	WPRC
	FM	N3NWL	147.28500	+		
	FM	N3RZL	442.20000	+		
	FM	N3NIA	443.80000	+		
Rochester Mills	FM	KB3CNS	224.90000	-		
Rockton	DMR/BM	N3QC	442.95000	+	CC1	WPRC
	FM	KE3DR	147.39000	+	173.8 Hz	
Ross Township	FM	N3TXG	446.22500	-	131.8 Hz	
Rossiter	FM	KE3DR	146.65500	-		
	FM	N3FXN	444.57500	+		

312 PENNSYLVANIA

Location	Mode	Call sign	Output	Input	Access	Coordinator
Roxborough	FM	W3QV	147.03000	+	91.5 Hz	ARCC
	FM	W3WAN	441.70000	+	74.4 Hz	ARCC
	FM	W3QV	444.80000	+	186.2 Hz	ARCC
Saint Clair	FM	W3SC	147.34500	+	123.0 Hz	ARCC
Sandy Lake	FM	WA3NSM	443.17500	+	186.2 Hz	WPRC
Scandia, Kinzua Reservoir						
	FM	W3GFD	444.47500	+		WPRC
Scenery Hill	FM	K3PSP	147.28500	+	131.8 Hz	
Schaefferstown	FM	N3JOZ	448.92500	-	146.2 Hz	ARCC
Schahola	FM	K3TSA	145.35000	-	100.0 Hz	ARCC
Schnecksville	DSTAR	W3EPE B	445.22500	-		
Schnecksville, Lehigh Career &						
	FM	K3TE	443.10000	+	131.8 Hz	
Schwenksville	FM	AA3RE	51.94000	-	88.5 Hz	ARCC
Scranton	DMR/MARC	N3MBK	441.35000	+	CC2	
	FM	KC3MN	146.83500	-		
	FM	K3CSG	146.94000	-	127.3 Hz	ARCC
	FM	WB3FEQ	147.28500	+	136.5 Hz	
	FM	KC3MN	224.56000	-	136.5 Hz	
	FM	N3EVW	448.82500	-	136.5 Hz	ARCC
Scranton, Top Of Morgan Highwa						
	FM	N3FCK	442.55000	+	100.0 Hz	ARCC
Seeley	FM	N3YCT	146.79000	-	179.9 Hz	ARCC
Selinsgrove, Selinsgrove Cente						
	FM	K3SNY	147.18000	+	100.0 Hz	ARCC
Sellersville	FM	W3AI	444.75000	+	103.5 Hz	ARCC
Seneca	FM	W3ZIC	145.25000	-	123.0 Hz	
Seven Springs	DMR/MARC	W3WGX	146.83500	-	CC1	WPRC
	FM	KB9WCX	443.92500	+	123.0 Hz	
Sharon, Keel Ridge						
	FM	W3LIF	145.35000	-	186.2 Hz	
Sharpsville	FM	KB3GRF	444.37500	+	186.2 Hz	
Sheffield	FM	N3KZ	442.70000	+	186.2 Hz	
Shickshinny, PA, State Game La						
	FM	N3FCK	444.50000	+	100.0 Hz	ARCC
Shohola Falls	FM	K3TSA	146.71500	-	82.5 Hz	ARCC
Shohola, Walker Lake						
	FM	WB1FXX	444.32500	+	107.2 Hz	
Sigel	FM	N3JGT	147.10500	+	173.8 Hz	
	FM	N3GPM	443.27500	+	110.9 Hz	
Slippery Rock	FM	KA3HUK	224.84000	-		
Smethport	FM	NJ3K	147.30000	+	173.8 Hz	WPRC
Somerset	FM	K3SMT	147.19500	+	123.0 Hz	
	FM	N3VFG	443.95000	+	88.5 Hz	
Souderton	FM	N3ZA	145.19000	-		ARCC
Southampton	FM	W3SK	146.79000	-		
	FUSION	W3SK	448.22500	-	131.8 Hz	ARCC
Spring Grove	FM	N3TVL	53.03000	52.03000	123.0 Hz	
	FM	KB3SNM	446.30000	-	88.5 Hz	
Springboro	FM	KF8YF	146.46000	147.46000	100.0 Hz	
Springtown	FM	W3BXW	442.95000	+	131.8 Hz	ARCC
State College	FM	K3CR	145.45000	-	146.2 Hz	
	FM	W3GA	146.76000	-	146.2 Hz	WPRC
	FM	K3CR	443.65000	+	146.2 Hz	WPRC
State College, Black Moshannon						
	FM	N3EB	444.70000	+	114.8 Hz	WPRC
Steelton	FM	N3NJB	147.30000	+	100.0 Hz	ARCC
Stowe	FM	KB3OZC	145.25000	-	100.0 Hz	ARCC
Strattanville	FM	N3HZV	444.22500	+		
Stroudsburg	P25	N3GRQ	448.22500	-	131.8 Hz	
	P25	N3GRQ	927.22500	902.22500	131.8 Hz	
Sugar Grove	FM	W3GFD	145.11000	-	186.2 Hz	WPRC

PENNSYLVANIA 313

Location	Mode	Call sign	Output	Input	Access	Coordinator
Summerdale	FM	N3KZ	442.20000	+	131.8 Hz	ARCC
Summerdale PA	DMR/MARC	W3BXW	442.45000	+	CC1	ARCC
Summit Station, Blue Mountain						
	FM	N3TJJ	224.14000	-	114.8 Hz	
Sunbury, Areffe Knob						
	FM	KC3FIT	146.65500	-	107.2 Hz	
Susquehanna	FM	WB2BQW	145.25000	-	100.0 Hz	ARCC
Tamaqua, High Rise						
	FM	W3TWA	146.49000	147.49000	131.8 Hz	
Tannersville	FM	W3WAN	145.23000	-	77.0 Hz	ARCC
	FM	WX3OES	146.86500	-	100.0 Hz	ARCC
Tannersville, Camelback Mounta						
	FM	K4MTP	53.05000	52.05000	131.8 Hz	
Texter Mtn	FM	N3SWH	449.07500	-	131.8 Hz	ARCC
Thompson	FM	N3DUG	147.04500	+	141.3 Hz	
Thorndale	FM	AA3VI	224.36000	-	123.0 Hz	ARCC
	FM	AA3VI	447.07500	-	123.0 Hz	ARCC
Tioga, Jackson Summit						
	FM	KB3EAR	146.62500	-	131.8 Hz	ARCC
Titusville	FM	WB3KFO	51.82000	-	82.5 Hz	WPRC
Towanda	FM	K3ABC	147.28500	+	82.5 Hz	ARCC
	FM	WA3GGS	444.25000	+	151.4 Hz	ARCC
Towanda, Kellogg Mountain						
	FM	N3XXH	224.24000	-	85.4 Hz	
	FM	N3XXH	441.85000	+	85.4 Hz	
Trout Run, Shrivers Ridge						
	FM	KB3DXU	145.15000	-	167.9 Hz	ARCC
Troy	FM	KB3DOL	444.05000	+	100.0 Hz	ARCC
Tunkhannock, Forkston						
	FM	N3FCK	441.15000	+	100.0 Hz	
Tuscarora Mtn.	FM	KI3D	147.04500	+	146.2 Hz	
Ulysses	FM	KB3HJC	145.43000	-	127.3 Hz	WPRC
Union City	DMR/BM	WA3UC	441.90000	+	CC1	
	FM	WA3UC	146.70000	-	186.2 Hz	
Union Dale, Elk Mountain						
	FM	N3DUG	448.77500	-	141.3 Hz	
Union Dale, Elk Mountain Ski R						
	FM	N3KZ	145.43000	-	77.0 Hz	ARCC
Uniondale, Elk Mountain						
	FM	N3DUG	146.74500	-	141.3 Hz	ARCC
	FM	N3SQ	449.40000	-	127.3 Hz	
Uniontown	FM	W3PIE	147.04500	+	131.8 Hz	
	FM	W3PIE	147.25500	+		
	FM	WA3UVV	442.02500	+	131.8 Hz	WPRC
Unityville, Bloody Run Creek M						
	FM	K3PPK	442.35000	+		
University Park, Penn State Un						
	DSTAR	K3CR	145.37000	-		
Upper Darby, Delaware Co. Memo						
	DSTAR	N3AEC	440.04375	+		ARCC
	DSTAR	N3AEC	1255.52500	1275.52500		ARCC
Upper Potsgrove	FM	W3PS	445.82500	-	156.7 Hz	ARCC
Upper Strasburg	FM	W3ACH	147.12000	+	100.0 Hz	
Utica, Western Union Tower						
	FM	KE3JP	442.60000	+		WPRC
Valley Forge	FM	W3PHL	53.41000	52.41000	131.8 Hz	ARCC
	FM	W3PHL	146.76000	-	131.8 Hz	ARCC
	FM	W3PHL	224.94000	-	131.8 Hz	ARCC
	FM	W3PHL	919.20000	894.20000	131.8 Hz	ARCC
	FM	W3PHL	1292.00000	1272.00000	131.8 Hz	ARCC
Valley Forge, Valley Forge Mou						
	FM	W3PHL	443.90000	+		ARCC

314   PENNSYLVANIA

Location	Mode	Call sign	Output	Input	Access	Coordinator
Vowinckel	FM	N3GPM	51.70000	-	186.2 Hz	WPRC
Vowinkel	FM	N3UOH	147.07500	+	110.9 Hz	WPRC
Warminster	FM	K3MFI	53.37000	52.37000	131.8 Hz	ARCC
Warminster, Township Building						
	FM	K3DN	147.09000	+	131.8 Hz	ARCC
Warren	DMR/MARC	W3KKC	145.15000	-	CC1	WPRC
	FM	KB3ORS	145.27000	-	173.8 Hz	WPRC
	FM	KB3KOP	146.76000	-	186.2 Hz	
	FM	N3MWD	443.90000	+		WPRC
Warren, Hearts Content						
	FM	W3GFD	147.01500	+		
Warren, Kinzua Dam						
	DSTAR	KC3CYM	442.22500	+		
Warrington	FM	WA3ZID	147.00000	+	131.8 Hz	ARCC
Washington	FM	W3CYO	145.49000	-		
	FM	K3PSP	146.79000	-		
	FM	W3PLP	147.34500	+	131.8 Hz	
	FM	W3CYO	224.40000	-		
	FM	W3PLP	442.12500	+	131.8 Hz	
	FM	W3CYO	443.30000	+		
Waterford	FM	KE3JP	145.43000	-		
	FM	W3GV	146.82000	-	186.2 Hz	
	FM	KF8YF	443.95000	+	100.0 Hz	
Waterville, Ramsey						
	FM	KB3DXU	145.35000	-	167.9 Hz	ARCC
Waymart	FM	WB3KGD	146.65500	-	146.2 Hz	ARCC
Waynesburg	FM	N3GC	146.43000	147.43000	131.8 Hz	WPRC
	FM	K3PSP	147.31500	+	131.8 Hz	
Wellsboro	DMR/BM	NR3K	443.85000	+	CC1	
	DMR/BM	K3LSY	446.50000	-	CC1	
	FM	NR3K	145.27000	-	127.3 Hz	ARCC
	FM	NR3K	147.06000	+	127.3 Hz	ARCC
Wellsboro, Dutch Hill						
	FM	NR3K	147.06000	+	127.3 Hz	
West Alexander	FM	K3PSP	145.25000	-	131.8 Hz	
West Chester	FM	W3EOC	446.52500	-	100.0 Hz	ARCC
West Hazleton	FM	KB3LVC	927.40000	902.40000	100.0 Hz	
West Mifflin	FM	KA3IDK	444.50000	+		
	FM	KA3IDK	444.52500	+		
	FM	KA3IDK	444.55000	+		
	FM	KA3IDK	1285.00000	1265.00000		
West Newton	FM	N3OVP	442.70000	+		
Wexford, Large Tower Northeast						
	FM	K3SAL	147.24000	+		
Whitnyville	DMR/MARC	N3FE	444.10000	+	CC1	
Widnoon	FM	K3QY	145.41000	-	173.8 Hz	WPRC
Wilkes Barre, Bunker Hill						
	FM	K3YTL	145.45000	-	82.5 Hz	ARCC
Wilkes-Barre	DMR/BM	WR3IRS	441.75000	+	CC1	
	FM	WB3FKQ	146.61000	-	82.5 Hz	ARCC
	FM	N3KZ	442.20000	+	131.8 Hz	ARCC
	FM	WB3FKQ	927.81250	902.81250	82.5 Hz	
Wilkes-Barre, Bear Creek						
	FM	N3FCK	146.46000	147.46000	100.0 Hz	ARCC
Wilkes-Barre, Penobscot Knob						
	FM	N3DAP	224.42000	-	94.8 Hz	ARCC
Williamsport	DMR/BM	KB3HLL	147.09000	+	CC1	ARCC
	DMR/BM	KB3HLL	443.05000	+	CC1	
	FM	N3XXH	224.28000	-	85.4 Hz	ARCC
	FM	N3PFC	1285.00000	1265.00000		ARCC
Williamsport, Armstrong Mounta						
	FM	K4MTP	53.05000	52.05000	131.8 Hz	

PENNSYLVANIA 315

Location	Mode	Call sign	Output	Input	Access	Coordinator
Williamsport, Bald Eagle Mount						
	FM	KB3AWQ	51.52000	-	173.8 Hz	
	FM	KB3DXU	145.33000	-	167.9 Hz	ARCC
	FM	W3AVK	146.73000	-	151.4 Hz	ARCC
	FM	WX3N	443.20000	+	77.0 Hz	ARCC
	FM	KB3AWQ	444.90000	+	173.8 Hz	ARCC
Williamsport, Williamsport Are						
	FM	W3AHS	147.30000	+	151.4 Hz	ARCC
Willimsport	DMR/BM	W3AVK	444.00000	+	CC1	
Wind Gap	FM	N3MSK	53.29000	52.29000		ARCC
	FM	KA3HJW	443.70000	+	151.4 Hz	ARCC
	FM	W3BXW	447.57500	-	131.8 Hz	ARCC
	FM	KC2IRV	449.87500	-	131.8 Hz	
Windber	FM	N3LZV	443.15000	+	123.0 Hz	
Wooddale	FM	N3JNZ	448.37500	-	91.5 Hz	ARCC
Wyncote	FM	N3FSC	224.38000	-	107.2 Hz	ARCC
Wyndmoor	DSTAR	K3PDR	146.61000	-		
	DSTAR	K3PDR	445.18125	-		
	FM	K3PDR	447.47500	-		
Wyndmoor. PA	DMR/BM	K3PDR	447.62500	-	CC1	
Yeadon	FM	N3UXQ	449.97500	-	131.8 Hz	ARCC
Yeadon,pa	DMR/BM	N3TPE	440.35000	+	CC15	
York	FM	W3HZU	53.97000	52.97000	127.3 Hz	ARCC
	FM	N3KZ	442.05000	+	131.8 Hz	ARCC
	FM	W3SBA	444.25000	+	146.2 Hz	
York, Rocky Ridge County Park						
	FM	W3HZU	146.97000	-	123.0 Hz	ARCC
Youngsville	FM	W3GFD	442.07500	+	186.2 Hz	WPRC
PUERTO RICO						
Adjuntas	FM	KP4ST	147.35000	+	127.3 Hz	
Adjuntas, Cerro Cerca Del Ciel						
	FM	KP4NET	448.60000	-	123.0 Hz	WIRCI
Aguada	FM	WP4S	147.03000	+		
Aguada, Atalaya	FM	WP3OF	449.02500	-		
Aguas Buenas	FM	WP4OCD	145.23000	-	77.0 Hz	
	FM	KP4RF	145.45000	-		
	FM	WP4JLH	146.73000	-	123.0 Hz	
	FM	KP4CK	146.85000	-	127.3 Hz	
	FM	WP4YF	449.17500	-	127.3 Hz	WIRCI
Aguas Buenas, Cerro Marquesa						
	FM	KP3BR	447.22500	-	77.0 Hz	WIRCI
Aguas Buenas, La Mesa						
	FM	KP4RF	447.60000	-		WIRCI
Barranquitas	DMR/MARC	NP4UG	449.42500	-	CC1	
	DMR/MARC	KP3JD	449.57500	-	CC1	
	FM	KP4RF	145.41000	-	127.3 Hz	
	FM	KP4LP	145.49000	-	82.5 Hz	
	FM	KP3AJ	147.33000	+	151.4 Hz	
	FM	KP4LP	224.08000	-		
	FM	KP4LP	447.25000	-		
	FM	WP4YF	447.65000	-	136.5 Hz	
	FM	NP3EF	448.77500	-	67.0 Hz	WIRCI
Bayamon	FM	WP3BM	51.62000	50.62000	136.5 Hz	
	FM	KP4ZZ	440.17500	+	141.3 Hz	
	FM	KP3ZZ	447.12500	-	94.8 Hz	
BayamÃ³n	FM	WP4O	448.07500	-		WIRCI
Bayamon	FM	WP4F	448.22500	-		WIRCI
	FM	NP3A	448.68750	51.31250	100.0 Hz	WIRCI
	FM	WP4XI	449.60000	-	100.0 Hz	
	FM	WP4KMB	449.62500	-	250.3 Hz	WIRCI
	FM	WP4KMB	449.65000	-		

316　PUERTO RICO

Location	Mode	Call sign	Output	Input	Access	Coordinator
Bayamon	FM	KP4XC	449.67500	-		
	FM	WP4MXY	449.90000	-		
Beatriz	FM	KP4LDR	449.22500	-		WIRCI
Caguas	FM	KP4NB	224.10000	-		
	FM	NP4H	224.98000	-		
	FM	WP4FMX	447.90000	-		
	FM	KP3AB	927.10000	902.10000		
Camuy, Quebrada						
	FM	KP4EYT	449.30000	-	123.0 Hz	WIRCI
Canovanas	FM	WP4JP	147.17000	+		
	FM	KP4LO	147.23000	+		
	FM	KP4IN	224.18000	-		
	FM	NP3FV	449.37500	-		
	FM	WP3OF	449.91250	-		WIRCI
Carolina	FM	WP4BVS	224.06000	-		
	FM	NP4VG	224.16000	-		
	FM	WP4N	224.22000	-		
	FM	KP4PR	448.27500	-	127.3 Hz	
Cayey	DMR/BM	WP4MXB	447.20000	-	CC1	WIRCI
	FM	KP4ID	147.09000	+	127.3 Hz	
	FM	KP4GE	223.94000	-		
	FM	KP3AB	223.98000	-		
	FM	NP4H	224.86000	-		
	FM	WP4MXB	447.37500	-		WIRCI
	FM	KP4MSR	447.55000	-	127.3 Hz	
	FM	KP4LST	448.57500	-	225.7 Hz	WIRCI
	FM	KP4LDR	449.22500	-	110.9 Hz	
	FM	KP4ZZ	449.30000	-	127.3 Hz	
	FM	KP3AB	449.97500	-		
Cayey, Puerto Rico						
	DSTAR	WP4MXB	1293.00000	1273.00000		
Ceiba	FM	WP4DE	448.55000	-		
Cerro Viviana San Lorenzo						
	FM	KP4MCR	147.19000	+		
Ciales	FM	KP3AB	145.35000	-	123.0 Hz	
Corozal	FM	KP3AV	29.66000	-		
	FM	WP3XH	146.61000	-	151.4 Hz	
	FM	KP4DH	146.67000	-	94.8 Hz	
	FM	NP4CB	224.30000	-		
	FM	KP4DH	224.46000	-		
	FM	WP4AIX	447.50000	-		
	FM	KP3I	447.70000	-	151.4 Hz	
	FM	KP4AOB	447.95000	-		
	FM	WP4Q	448.47500	-	88.5 Hz	WIRCI
Fajardo	FM	KP3AB	145.19000	-	100.0 Hz	
	FM	WP3CB	145.25000	-	88.5 Hz	
	FM	NP3CB	145.47000	-		
	FM	NP3H	146.69000	-		
	FM	NP3H	224.24000	-		
	FM	KP4MTG	448.05000	-	127.3 Hz	WIRCI
	FM	NP3H	448.25000	-		
	FM	KP4GX	448.72500	-	151.4 Hz	
	FM	WP4MQZ	449.50000	5.00000		WIRCI
Guayama	FM	KP4JMT	448.68750	51.31250		WIRCI
Guaynabo	DMR/BM	KP4AG	447.75000	-	CC1	WIRCI
	FM	KP4CE	147.27000	+		
	FM	KP4KA	224.58000	-		
	FM	KP4XK	449.82500	-		
Gurabo	FM	KP4FAK	145.31000	-		
	FM	WP4LXE	146.79000	-		
	FM	WP4SE	146.87000	-		
	FM	WP4B	146.95000	-	127.3 Hz	

PUERTO RICO 317

Location	Mode	Call sign	Output	Input	Access	Coordinator
Gurabo	FM	WP4KAG	449.00000	-		
Humacao	FM	WP4GUL	224.52000	-		
	FM	KP4BW	449.63750	-		WIRCI
Isabela	DSTAR	WP4QYG	447.50000	-		WIRCI
	FM	WP4MMR	447.40000	-	100.0 Hz	
	FM	KP4MSR	447.55000	-	127.3 Hz	
Jayuya	FM	KP4ILO	146.75000	-		
	FM	NP3H	146.99000	-		
	FM	WP4IFU	147.15000	+	100.0 Hz	
	FM	KP4IN	147.25000	+		
	FM	KP4IN	224.04000	-		
	FM	KP4PK	224.42000	-		
	FM	WP4AZT	447.05000	-	127.3 Hz	
	FM	KP4PK	447.17500	-		
	FM	WP4CBC	447.72500	-		
	FM	KP3AJ	447.80000	-	136.5 Hz	
	FM	WP4CRG	448.30000	-	100.0 Hz	
	FM	KP4SA	448.37500	-	123.0 Hz	WIRCI
	FM	WP4AZT	448.50000	-	127.3 Hz	
	FM	KP4IS	448.52500	-		
	FM	KP4CAR	449.80000	-		WIRCI
	FM	WP4MJP	449.85000	-		
Jayuya, Cerro Puntas						
	FM	KP4CAR	147.21000	+		
Juana Diaz	FM	KP4LRL	449.87500	-	110.9 Hz	
Juncos	FM	KP4IJ	224.38000	-		
	FM	KP4GX	447.47500	-		
Lajas, PR	FM	WP4FD	440.20000	+		WIRCI
Lares	FM	KP4ARN	146.79000	-	79.7 Hz	
Las Piedras	FM	KP4EGM	448.12500	-	123.0 Hz	WIRCI
Luquillo	FM	NP4ZB	51.72000	50.72000		
	FM	NP3EF	147.35000	+		
	FM	WP4NPX	447.02500	-	136.5 Hz	
	FM	WP3HY	447.90000	-	100.0 Hz	
Manati	DMR/BM	KP3IV	448.95000	-	CC1	
	FM	KP4PG	224.90000	-		
Maricao	DMR/BM	KP4IP	448.67500	-	CC1	WIRCI
	FM	KP3AB	146.77000	-	123.0 Hz	WIRCI
	FM	KP4BKY	146.87000	-		
	FM	KP4FRA	146.93000	-	136.5 Hz	
	FM	WP4CPV	147.07000	+	146.2 Hz	
	FM	WP4CPV	147.23000	+	146.2 Hz	
	FM	KP3AB	224.92000	-		
	FM	KP4UK	447.55000	-		
	FM	KP4NIN	449.77500	-	123.0 Hz	WIRCI
Maricao, Monte Del Estado						
	FM	KP4SE	147.29000	+	127.3 Hz	
Maricao, Monte Del Estado, Pue						
	FM	KP4SE	448.97500	443.87500	136.5 Hz	
Mayaguez	FM	KP4AIC	146.69000	-	88.5 Hz	
	FM	WP4MPR	447.92500	-	107.2 Hz	
Mayaguez, Monte Del Estado						
	FM	KP4IP	147.13000	+	100.0 Hz	WIRCI
	FM	KP4IP	448.65000	-	100.0 Hz	WIRCI
Moca	FM	KP4KJI	145.17000	-	114.8 Hz	
	FM	NP4MM	447.63750	5.00000	88.5 Hz	
Moca, Cordillera Jaicoa Moca						
	FM	KP4JED	447.63750	-	88.5 Hz	
Morovis	FM	WP4KY	145.76000		100.0 Hz	
Naguabo	FM	KP4BW	440.92500	+	141.3 Hz	WIRCI
Naranjito	FM	WP4FHR	147.01000	+		
	FM	WP4NPC	147.23000	+		

318 PUERTO RICO

Location	Mode	Call sign	Output	Input	Access	Coordinator
Naranjito	FM	KP4FO	224.50000	-		
	FM	KP4AMV	448.80000	-	136.5 Hz	WIRCI
	FM	WP4FUI	449.92500	-	179.9 Hz	
Orocovis	DMR/BM	KP3AV	448.90000	-	CC1	WIRCI
	FM	KP4OG	145.39000	-	94.8 Hz	
	FM	KP4DEU	146.71000	-	88.5 Hz	
	FM	KP4BCQ	147.39000	+		
	FM	NP4TX	447.32500	-	136.5 Hz	
	FM	KP4FRE	447.52500	-	136.5 Hz	
Patillas	FM	KP4KGZ	447.15000	-	107.2 Hz	
Ponce	FM	WP4JLQ	147.39000	+		
	FM	KP4ALA	447.77500	-		
	FM	KP4EGY	448.17500	-		WIRCI
	FM	WP4NQR	448.82500	-		WIRCI
	FM	KP4GBF	449.10000	-		
	FM	KP4ASD	449.15000	-	110.9 Hz	WIRCI
	IDAS	NP3WP	449.55000	-	127.3 Hz	
Quebradillas	FM	WP3OF	146.87000	-		
Rio Grande	FM	KP4FRA	146.93000	-		
	FM	NP3H	146.99000	-		
	FM	NP4SB	147.03000	+	151.4 Hz	
	FM	NP3EF	147.31000	+	88.5 Hz	
	FM	NP3CB	224.70000	-		
	FM	KP4SQ	447.37500	-		
	FM	KB9EZX	448.15000	-	100.0 Hz	WIRCI
	FM	WP4FWN	449.27500	-		
	FM	NP3EF	449.75000	-	100.0 Hz	
Rio Piedras	FM	WP4MQQ	449.52500	-		
Roncador Utuado	FM	WP3OF	447.57500	-		WIRCI
Salinas	DMR/BM	WP3JM	448.75000	-	CC1	WIRCI
	FM	WP4NVY	448.02500	-	173.8 Hz	
	FM	WP4NWR	448.85000	-	100.0 Hz	WIRCI
	FM	WP4NZE	449.32500	-	123.0 Hz	
San German	FM	WP4GAV	448.32500	-	100.0 Hz	
San Juan	FM	KP4FAK	147.39000	+		
	FM	WP4BFC	224.66000	-		
	FM	KP4ZZ	440.12500	+	127.3 Hz	
	FM	WP4CIE	447.10000	-	100.0 Hz	
	FM	WP4GZO	447.35000	-		
	FM	KP4GA	447.85000	-	127.3 Hz	WIRCI
	FM	WP4X	448.42500	-	127.3 Hz	
	FM	NP3A	449.17500	-		
	FM	WP4POX	449.58750	-		WIRCI
	FM	KP4AIC	449.72500	-		
	FM	N1TKK	449.95000	-		
San Juan, Cerro Collores						
	FM	KP4MCR	146.89000	-	DCS 114	WIRCI
San Lorenzo	FM	WP4LTR	147.03000	+		
	FM	KP4YS	447.55000	-	100.0 Hz	
	FM	WP4MHS	448.45000	-		
San Patricio	FM	WP4POX	449.95000	-	123.0 Hz	WIRCI
San Sebastian	FM	WP4HVS	147.29000	+	123.0 Hz	
	FM	WP4HVS	449.45000	-		
San Sebastian , Colinas Verdes						
	FM	NP4PC	449.73750	-	88.5 Hz	
Toa Alta	DMR/MARC	NP4UG	449.47500	-	CC2	
	FM	WP4BCK	224.34000	-		
	FM	KP4KC	449.35000	-	123.0 Hz	
Toa Baja	FM	WP3ZQ	145.27000	-		
	FM	WP3TM	447.27500	-	100.0 Hz	
Utuado	FM	KP4CD	145.43000	-		
	FM	NP4PS	449.40000	-	67.0 Hz	

PUERTO RICO 319

Location	Mode	Call sign	Output	Input	Access	Coordinator
Utuado, Roncador	FM	WP3R	147.37000	+	77.0 Hz	
Vieques	FM	NP3MR	448.62500	-		
Villalba	FM	KP3AB	145.19000	-	136.5 Hz	
	FM	WP4IZI	146.41000	-		
	FM	NP4WI	146.97000	-	127.3 Hz	
	FM	WP4AZT	147.05000	+	127.3 Hz	
	FM	KP3AB	224.02000	-		
	FM	KP4FRA	448.40000	-	136.5 Hz	
Villlalba	FM	KP3AB	448.10000	-	123.0 Hz	WIRCI
Yabucoa	FM	WP4BV	145.33000	-		
	FM	KP4MCR	447.62500	-		
	FM	KP4DDF	447.67500	-	100.0 Hz	
	FM	KH2RU	447.82500	-		
	FM	WP4BV	448.35000	-		
Yauco	FM	NP4QH	147.07000	+		
	FM	KP3AB	147.35000	+		
Yauco, Monte Alto De La Bander						
	FM	KP4ILO	145.47000	-		WIRCI
	FM	KP4KJI	448.20000	-	100.0 Hz	WIRCI

RHODE ISLAND

Location	Mode	Call sign	Output	Input	Access	Coordinator
Bristol	DMR/MARC	K1CW	145.33000	-	CC2	
	FM	K1CW	443.15000	+	94.8 Hz	
Coventry	FM	N1JBC	147.16500	+	67.0 Hz	
	FM	KA1ABI	223.90000	-		
Cranston	DSTAR	KB1TIA	145.40000	144.60000		
	FM	K1CR	146.70000	-		
	FM	W1PHR	147.10500	+	103.5 Hz	
	FM	N1NTP	147.28000	+	67.0 Hz	
Cumberland	DMR/BM	W1DMR	146.62500	-	CC1	
	FM	NB1RI	145.17000	-	67.0 Hz	
	FM	KR1RI	146.94000	-	67.0 Hz	
	FM	W1DMR	927.67500	902.07500	DCS 245	NESMC
	NXDN	KB1ISZ	146.44000	144.94000		
East Providence	FM	W1AQ	147.33000	+	173.8 Hz	
Exeter	FM	NB1RI	146.98500	-	67.0 Hz	
Foster	FM	NB1RI	146.91000	-	67.0 Hz	NESMC
	FM	KB1TUG	444.40000	+	67.0 Hz	
Greenville	FM	N1MIX	53.87000	52.87000	85.4 Hz	
	FM	N1MIX	146.85000	-	DCS 411	
	FM	N1MIX	448.07500	-	DCS 411	
	FM	N1MIX	927.82500	902.82500		
Johnston	FM	KB1TOT	146.83500	-		
	FM	W1OP	223.98000	-		
Kingston	DMR/MARC	KA1REO	446.52500	-	CC1	
Lincoln	FM	NB1RI	146.46000	144.96000	67.0 Hz	
	FM	N1BS	224.62000	-	67.0 Hz	NESMC
	FM	K1RSR	447.77500	-	67.0 Hz	
Newport	FM	WC1R	146.88000	-	100.0 Hz	
	FM	KC2GDF	448.32500	-	67.0 Hz	
North Providence	DMR/MARC	N1JBC	145.37000	-	CC2	
	FM	NB1RI	224.56000	-		
	FM	N1JBC	224.92000	-		
	FM	N1JBC	449.22500	-	141.3 Hz	
	FM	KA1EZH	927.76250	902.76250	67.0 Hz	
Portsmouth	FM	NB1RI	53.17000	52.17000	67.0 Hz	
	FM	W1AAD	145.30000	-		NESMC
	FM	W1SYE	145.45000	-	100.0 Hz	NESMC
	FM	NB1RI	147.07500	+	67.0 Hz	
Providence	FM	N1BS	29.64000	-	67.0 Hz	NESMC
	FM	N1BS	145.35000	-	67.0 Hz	NESMC
	FM	N1RWX	147.12000	+	67.0 Hz	

320 RHODE ISLAND

Location	Mode	Call sign	Output	Input	Access	Coordinator
Providence	FM	N1RWX	444.20000	+	88.5 Hz	
	FM	N1MIX	446.77500	-	146.2 Hz	
	FM	K1CR	448.92500	-	88.5 Hz	
	FM	N1RWX	927.51250	902.51250	127.3 Hz	
	NXDN	K2BUI	449.97500	-		
Saunderstown	FM	K1NQG	146.71500	-	67.0 Hz	
Scituate	FM	KB1NZZ	53.55000	52.55000	203.5 Hz	
	FM	KR1RI	146.76000	-	67.0 Hz	
	FM	K1KYI	223.76000	-		
Warwick	DMR/BM	KA1MXL	145.53000		CC1	
	DMR/BM	KA1MXL	444.83750	+	CC1	
	DMR/MARC	KA1MXL	433.10000	432.10000	CC8	
	DMR/MARC	KA1MXL	434.20000	+	CC4	
	FM	KA1MXL	53.58000	52.58000	141.3 Hz	
	FM	KA1LMX	223.92000	-		
	FM	WA1ABC	447.02500	-	67.0 Hz	
	FM	WA1ABC	927.75000	902.05000	67.0 Hz	
West Greenwich	FM	KA1RCI	449.32500	-	127.3 Hz	
	FM	W1WNS	449.93750	-	DCS 244	
West Greenwich, RI						
	DMR/BM	K1EWG	448.82500	-	CC2	
West Warwick	DSTAR	W1HDN	147.04500	+		
	FM	W1RI	145.13000	-	77.0 Hz	
	FM	K1WYC	224.30000	-	100.0 Hz	
	FM	KA1SOO	224.76000	-		
West Warwick, Court House						
	FM	KC1DGM	446.72500	-	100.0 Hz	
Westerly	FM	N1LMA	147.24000	+	100.0 Hz	
	FM	W1WRI	147.31500	+	110.9 Hz	
	FM	NB1RI	147.39000	+	67.0 Hz	
	FM	N1LMA	224.98000	-	136.5 Hz	
	FM	N1LMA	449.67500	-	127.3 Hz	

SOUTH CAROLINA

Location	Mode	Call sign	Output	Input	Access	Coordinator
Aiken	DMR/MARC	N2ZZ	443.46250	+	CC1	SERA - SC
	FM	W4ZKM	145.45000	-	123.0 Hz	
	FM	N4ADM	147.28500	+	100.0 Hz	
	FM	WR4SC	441.52500	+	91.5 Hz	SERA - SC
Aiken, Aiken Regional Medical						
	FM	N2ZZ	145.35000	-	156.7 Hz	
Aiken, Old Aiken Hospital						
	DSTAR	KR4AIK	145.16000	-		
	DSTAR	KR4AIK	443.41250	+		
Anderson	DMR/MARC	N4LRD	444.53750	+	CC1	
Anderson, Anderson Memorial Ho						
	FM	N4AW	146.97000	-		SERA - SC
Anderson, Anmed Hospital						
	FM	KB4JDH	442.82500	+	127.3 Hz	SERA - SC
Aynor	FM	WR4SC	146.71500	-	162.2 Hz	
	FM	NE4SC	147.09000	+	123.0 Hz	SERA - SC
	FM	WR4SC	441.67500	+	162.2 Hz	SERA - SC
Bamberg	FM	WB4TGK	145.33000	-	156.7 Hz	
Barnwell	DMR/MARC	WR4SC	440.68750	+	CC1	SERA - SC
	FM	KK4BQ	147.03000	+	156.7 Hz	SERA - SC
	FM	WR4SC	442.00000	+	91.5 Hz	
	FM	KK4BQ	449.25000	-	156.7 Hz	
Beach Island	DMR/MARC	WR4SC	444.28750	+	CC1	SERA - SC
Beaufort	DMR/MARC	WR4SC	441.98750	+	CC1	
	FM	W4BFT	145.13000	-	88.5 Hz	SERA - SC
Beaufort, Beaufort Memorial Ho						
	FM	W4BFT	443.85000	+	123.0 Hz	SERA - SC
Beaufort, WJWJ	FM	W4BFT	146.65500	-		SERA - SC

SOUTH CAROLINA   321

Location	Mode	Call sign	Output	Input	Access	Coordinator
Beech Island	FM	WR4SC	147.34500	+	91.5 Hz	
	FM	WR4SC	443.12500	+	91.5 Hz	
Bennettsville	FM	KG4HIE	443.00000	+	123.0 Hz	
Blacksburg, Whitacker Mtn						
	FM	KF4SCG	442.62500	+	107.2 Hz	
Blacksburg, Whitaker Mountain						
	FM	W4NYR	146.88000	-		
	FM	W4NYR	444.32500	+		
Bluffton	FM	W4IAR	442.67500	+	100.0 Hz	
Caesar's Head Mountain						
	FM	WR4SC	145.13000	-	123.0 Hz	
	FM	K4ECG	443.12500	+	123.0 Hz	
Calhoun Falls	FM	KI4CCZ	444.00000	+	118.8 Hz	SERA - SC
	FM	KI4CCZ	444.57500	+	103.5 Hz	SERA - SC
Charleston	DMR/BM	KD4TXX	443.45000	+	CC1	
	DMR/MARC	WR4SC	443.03750	+	CC1	SERA - SC
	DSTAR	KR4CHS	145.12000	-		
	DSTAR	W4HRS	145.16000	-		
	DSTAR	W4HRS	444.11250	+		
	FM	WR4SC	146.76000	-	123.0 Hz	
	FM	WR4SC	147.10500	+	123.0 Hz	
	FM	WR4SC	441.57500	+	123.0 Hz	
	FM	WR4SC	441.72500	+	123.0 Hz	
	FM	W4HRS	444.77500	+	123.0 Hz	
Charleston, Adams Run						
	FM	W4ANK	147.34500	+	123.0 Hz	
Charleston, Rutledge Tower (MU						
	FM	W4HRS	145.45000	-	123.0 Hz	
	FM	W4HRS	444.82500	+	123.0 Hz	
Charleston, Tree Top						
	FM	K4IUG	443.37500	+	100.0 Hz	SERA - SC
Charleston, USS Yorktown						
	FM	WA4USN	146.79000	-	123.0 Hz	
Cheraw	FM	W4APE	145.49000	-	123.0 Hz	SERA - SC
	FM	K4CCC	147.13500	+	123.0 Hz	
	FM	KG4HIE	444.37500	+	91.5 Hz	
Chester, Near Chester County H						
	FM	W4CHR	145.31000	-	167.9 Hz	SERA - SC
Clemson	DMR/MARC	K4BAN	442.23750	+	CC1	
	DSTAR	WD4EOG	444.22500	+		
Clemson, Clemson University						
	FM	WD4EOG	145.45000	-		SERA - SC
	FM	WD4EOG	444.62500	+	156.7 Hz	SERA - SC
Clemson, Kite Hill - WSBF-FM						
	FM	WD4EOG	147.38250	+	123.0 Hz	
	FM	WD4EOG	444.38750	+		
Clover	FM	KC4KPJ	443.72500	+		SERA - SC
Columbia	DMR/MARC	WR4SC	440.61250	+	CC1	SERA - SC
	DMR/MARC	WR4SC	442.51250	+	CC1	
	DSTAR	KJ4FCS	145.38000	-		SERA - SC
	DSTAR	KJ4FCS	442.77500	+		
	DSTAR	W4CAE	443.20000	+		
	DSTAR	W4CAE	1284.22500	1264.22500		
	FM	KJ4BWK	145.40000	144.60000		
	FM	WR4SC	146.71500	-	91.5 Hz	
	FM	W4CAE	147.33000	+	156.7 Hz	SERA - SC
	FM	K4HI	147.36000	+	100.0 Hz	SERA - SC
	FM	WR4SC	441.72500	+	91.5 Hz	
	FM	N7GZT	442.20000	+		
	FM	N5CWH	444.20000	+		
	FM	K9OH	444.87500	+	91.5 Hz	SERA - SC
	FM	N5CWH	927.63750	902.63750	DCS 132	

322 SOUTH CAROLINA

Location	Mode	Call sign	Output	Input	Access	Coordinator
Conway	DMR/BM	NE4SC	443.66250	+	CC1	
	DSTAR	NE4SC	144.98000	147.48000		
	DSTAR	NE4SC	442.78750	+		
	FM	W4GS	145.11000	-	85.4 Hz	
	FM	AA2UC	443.80000	+	141.3 Hz	
	FM	W4GS	444.67500	+	85.4 Hz	
Darlington	FM	KB4RRC	147.25500	+	162.2 Hz	SERA - SC
	FM	KJ4OEF	444.60000	+	162.2 Hz	
Darlington, Florence Darlingto						
	FM	KB4RRC	444.80000	+	91.5 Hz	SERA - SC
Dillon	DMR/MARC	WR4SC	443.16250	+	CC1	SERA - SC
	FM	KJ4OEF	444.95000	+	162.2 Hz	
Dillon, WPDE News 15 Tower						
	DSTAR	W4PDE	145.34000	-		
	DSTAR	W4PDE	443.88750	+		
	FM	W4PDE	146.74500	-	82.5 Hz	SERA - SC
	FM	NE4SC	224.04000	-	123.0 Hz	SERA - SC
Dorchester	FM	W4HNK	147.18000	+	123.0 Hz	
	FM	W4HNK	443.80000	+	123.0 Hz	
Edgefield	FM	WR4EC	146.85000	-	91.5 Hz	SERA - SC
Elgin	FUSION	KI4IVP	147.24000	+	103.5 Hz	SERA - SC
Florence	DMR/BM	WX4ARC	442.28750	+	CC1	
	DMR/MARC	WR4SC	442.16250	+	CC1	SERA - SC
	FM	WR4SC	146.68500	-	91.5 Hz	SERA - SC
	FM	W4APE	147.19500	+	123.0 Hz	SERA - SC
	FM	WR4SC	441.57500	+	91.5 Hz	SERA - SC
	FM	W4APE	442.05000	+	123.0 Hz	SERA - SC
Florence , MUSC Florence						
	FM	W4ULH	444.00000	+		
Florence, City County Complex						
	FM	W4GEY	146.97000	-	167.9 Hz	SERA - SC
Florence, ETV Tower						
	FM	W4ULH	146.85000	-		SERA - SC
Florence, McLeod Regional Medi						
	FM	KJ4OEF	444.85000	+	162.2 Hz	
Fork	FM	K4KNJ	146.52000		100.0 Hz	
Fort Jackson	FM	W4CAE	146.77500	-	156.7 Hz	
Fort Mill	FM	KT4TF	145.11000	-	110.9 Hz	
	FM	KT4TF	224.80000	-	110.9 Hz	
Gaffney	FM	KG4JIA	224.50000	-	123.0 Hz	
	FM	N6WOX	443.62500	+	162.2 Hz	
Gaffney, Draytonville Mtn						
	FM	KE4MDP	145.43000	-	162.2 Hz	
Georgetown	DMR/MARC	WR4SC	441.81250	+	CC1	
Georgetown, Memorial Hospital						
	FM	W4HRS	146.70000	-	123.0 Hz	
Greeleyville	FM	KG4AQH	145.23000	-	123.0 Hz	SERA - SC
	FM	KG4AQH	444.75000	+	123.0 Hz	SERA - SC
Green Pond	DSTAR	KJ4LNJ	145.48000	-		SERA - SC
	DSTAR	KJ4LNJ	444.07500	+		SERA - SC
Greenville	DMR/MARC	WR4SC	443.11250	+	CC1	SERA - SC
	FM	WR4SC	145.37000	-	123.0 Hz	
	FM	W4IQQ	146.94000	-	107.2 Hz	
	FM	W4ILY	224.20000	-		
	FM	WR4SC	441.67500	+	91.5 Hz	
	FM	WA4MWC	1250.00000	1280.00000		
Greenville, Caesars Head Mount						
	ATV	N4VDE	421.25000	434.05000		SERA - SC
Greenville, Ceasars Head						
	FM	W4NYK	146.61000	-		SERA - SC
Greenville, Paris Mountain						
	FM	W4NYK	146.82000	-		

SOUTH CAROLINA   323

Location	Mode	Call sign	Output	Input	Access	Coordinator
Greenville, Paris Mountain						
	FM	KB4PQA	442.25000	+		
Greenwood	DMR/MARC	WR4SC	443.83750	+	CC1	
	FM	W4GWD	147.16500	+	107.2 Hz	SERA - SC
	FM	WR4SC	441.62500	+	91.5 Hz	
	FM	W4GWD	443.90000	+	107.2 Hz	
Greenwood, Self Regional						
	DSTAR	W4GWM	145.42000	-		
Greenwood, Tower						
	FM	KK4SM	444.82500	+	162.2 Hz	SERA - SC
Greer	DMR/MARC	KO4MZ	145.43000	-	CC1	
Hardeeville	FM	KK4ONS	147.06000	+	123.0 Hz	
Hilton Head Island						
	DMR/BM	N4MSE	442.57500	+	CC1	
	FM	W4IAR	145.31000	-	100.0 Hz	
	FM	W4IAR	147.24000	+	100.0 Hz	
	FM	W4IAR	444.35000	+	123.0 Hz	
Honea Path	FM	KJ4VLT	443.77500	+	127.3 Hz	
Indianland, Water Tank						
	FM	KT4TF	443.47500	+	110.9 Hz	
Inman, Greenville	FM	W4IQQ	443.35000	+		
Iva	FM	AI4JE	443.25000	+	123.0 Hz	
Jackson , SRS ?C? Road Near Ce						
	FM	W4ZKM	145.45000	-	123.0 Hz	
Jedburg	FM	W4ANK	147.27000	+	123.0 Hz	
Knightsville	FM	WA4USN	146.94000	-	123.0 Hz	
	FM	WA4USN	441.45000	+	123.0 Hz	
Ladson	FM	N2OBS	146.86500	-	123.0 Hz	
Lake City	DMR/MARC	WR4SC	440.63750	+	CC1	
Landrum, NEAR Lake Lanier At A						
	FM	N2PNE	444.92500	+		SERA - SC
Laurens	FM	KD4HLH	146.86500	-	107.2 Hz	
Leesville	DSTAR	AK2H	146.65500	-		
	DSTAR	AK2H	443.50000	+		
	FM	N5CWH	53.27000	52.27000	162.2 Hz	
	FM	W4RRC	147.25500	+	123.0 Hz	SERA - SC
	FM	N5CWH	224.56000	-	162.2 Hz	
	FM	N5CWH	443.32500	+	162.2 Hz	
Lewis, Near Lewis Fire Departm						
	FM	KW4BET	442.47500	+	162.2 Hz	SERA - SC
Lexington	FM	KA4FEC	147.39000	+	156.7 Hz	
Lexington, South Carolina						
	P25	N4SCG	147.00000	+		
Liberty	FM	N4VDE	443.97500	+	103.5 Hz	SERA - SC
Little Mountain	DMR/MARC	WR4SC	443.53750	+	CC1	SERA - SC
	DSTAR	KJ4MKV	145.24000	+		
	DSTAR	KJ4MKV	443.32500	+		
	FM	KJ4MKV	53.21000	52.21000	88.5 Hz	
	FM	K4AVU	147.21000	+	156.7 Hz	
	FM	N4UHF	444.65000	+		
Longs	FM	N2CUE	440.80000	+	179.9 Hz	
Loris	DMR/MARC	K2PJ	147.28500	+	CC1	
Lucknow	FM	W4APE	146.92500	-	123.0 Hz	SERA - SC
Lugoff	FM	KI4RAX	146.82000	-	91.5 Hz	
	FM	KI4RAX	441.80000	+	91.5 Hz	
Manning	FM	KM4ABW	145.15000	-	91.5 Hz	SERA - SC
Marietta, Caesars Head Mountai						
	FM	K9OH	145.47000	-	91.5 Hz	SERA - SC
Marion	FM	KO4L	147.00000	-	91.5 Hz	SERA - SC
Moncks Corner	FM	W4BRK	146.61000	-	123.0 Hz	SERA - SC
	FM	WD4NUN	147.15000	+	91.5 Hz	SERA - SC

324 SOUTH CAROLINA

Location	Mode	Call sign	Output	Input	Access	Coordinator
Moncks Corner, EOC						
	FM	W4HRS	145.49000	-	103.5 Hz	
Mount Pleasant	FM	WA4USN	441.45000	+	123.0 Hz	
Mountain Rest, Long Mtn						
	FM	KJ4YLP	147.03000	-	123.0 Hz	
	FM	N4LRD	442.77500	+	127.3 Hz	
Mountain Rest, Oconee State Pa						
	FM	KJ4YLO	442.20000	+	123.0 Hz	SERA - SC
Mt. Pleasant, East Cooper Medi						
	FM	KK4ZBE	146.68500	-	162.2 Hz	SERA - SC
Mullins	FM	W4APE	145.47000	-	123.0 Hz	SERA - SC
Murrells Inlet	DMR/MARC	WR4SC	441.88750	+	CC1	SERA - SC
	FM	W4GS	146.80500	-	85.4 Hz	
Myrtle Beach	DMR/MARC	N3TX	431.40000	438.00000	CC1	
	DMR/MARC	WR4SC	441.91250	+	CC1	SERA - SC
	FM	NE4SC	53.05000	52.05000	123.0 Hz	
	FM	W4GS	147.12000	+	85.4 Hz	
	FM	NE4SC	444.90000	+	100.0 Hz	
	FM	NE4SC	444.97500	+	123.0 Hz	
Myrtle Beach, Sheraton Myrtle						
	FM	W4GS	145.29000	-	85.4 Hz	
North Augusta	FM	K4NAB	146.73000	-		
	FM	KE4RAP	444.80000	+	146.2 Hz	SERA - SC
	FM	KG4HIR	927.80000	902.80000		
North Charleston	DMR/MARC	WR4SC	442.46250	+	CC1	
North Charleston, Trident Hosp						
	FM	W4HRS	146.73000	-	123.0 Hz	
North Myrtle Beach						
	DMR/BM	NE4SC	444.08750	+	CC1	SERA - SC
Orangeburg	DMR/MARC	WR4SC	440.58750	+	CC1	SERA - SC
	DSTAR	KX4DOR	145.28000	-		SERA - SC
	FM	KJ4QLH	146.80500	-	156.7 Hz	
	FM	WR4SC	146.88000	-	123.0 Hz	
	FM	KJ4QJH	147.09000	+	156.7 Hz	SERA - SC
	FM	KO4BR	224.78000	-		
	FM	WR4SC	441.75000	+	123.0 Hz	
	FM	AD4U	444.97500	+		
Pageland	FM	W4JMY	53.37000	52.37000	100.0 Hz	SERA - SC
	FM	W4APE	146.89500	-	123.0 Hz	SERA - SC
Pickens	DMR/BM	AA2C	444.75000	+	CC1	
	DMR/MARC	WX4PG	442.31250	+	CC1	
	FM	AC4RZ	53.35000	52.35000	162.2 Hz	
	FM	WT4F	146.70000	-	107.2 Hz	
	FM	WB4LZT	927.51250	902.51250	100.0 Hz	SERA - SC
Pickens, Caesars Head Mountain						
	FM	WR4XM	224.14000	-	131.8 Hz	
	FM	WR4XM	442.40000	+	127.3 Hz	
Pickens, Glassy Mountain						
	FM	WB4YXZ	147.00000	-	151.4 Hz	
	FM	WR4XM	224.40000	-	131.8 Hz	
	FM	AC4RZ	443.45000	+	110.9 Hz	SERA - SC
Pickens, Glassy Mtn						
	FM	WX4PG	147.19500	+	141.3 Hz	
	FM	KN4SWB	444.35000	+	127.3 Hz	
Pickens, Sassafras Mtn						
	FM	N4AW	224.32000	-	131.8 Hz	SERA - SC
Pickens, Sassafrass Mountain						
	FM	N4AW	146.79000	-		SERA - SC
Richburg	FM	KC4KPJ	444.72500	+	110.9 Hz	
Rock Hill	DMR/MARC	WR4SC	440.51250	+	CC1	
	FM	K4YTZ	147.03000	-	88.5 Hz	SERA - SC
	FM	KD4EOD	147.22500	+	110.9 Hz	

SOUTH CAROLINA   325

Location	Mode	Call sign	Output	Input	Access	Coordinator
Rock Hill	FM	KB4GA	224.84000	-	123.0 Hz	
	FM	W3SPC	440.25000	+		
	FM	WR4SC	441.52500	+	162.2 Hz	
	FM	W4FTK	444.92500	+	91.5 Hz	
Russellville	FM	KK4B	147.30000	+	162.2 Hz	SERA - SC
Saint George	FM	K4ILT	147.04500	+	103.5 Hz	
Saluda, Water Tank						
	FM	W4DEW	146.91000	-	123.0 Hz	SERA - SC
Seabrook Island	FM	WA4USN	145.41000	-	123.0 Hz	
Shoals Junction	DMR/BM	WJ4X	442.60000	+	CC1	SERA - SC
Simpsonville	FM	WA4UKX	146.73000	-	100.0 Hz	
Six Mile	FM	W4TWX	441.80000	+	110.9 Hz	SERA - SC
Six Mile, Six Mile Mountain						
	FM	WA4SSJ	224.10000	-	131.8 Hz	SERA - SC
	FM	W4TWX	441.87500	+	110.9 Hz	SERA - SC
Six Mile, Six Mile Mountian						
	FM	W4TWX	145.17000	-	162.2 Hz	SERA - SC
Socastee	FM	W2SOC	443.00000	+	100.0 Hz	
South Charleston	DMR/MARC	WR4SC	442.38750	+	CC1	SERA - SC
Spartanburg	DMR/MARC	WR4SC	440.66250	+	CC1	SERA - SC
	FM	K4JLA	147.31500	+	123.0 Hz	SERA - SC
	FM	K4II	224.44000	-		SERA - SC
	FM	WR4SC	441.95000	+	162.2 Hz	SERA - SC
	FM	K4II	442.07500	+	123.0 Hz	SERA - SC
Spartanburg, Camp Croft						
	FM	WR4SC	147.09000	+	162.2 Hz	SERA - SC
Spartanburg, Spartanburg Downt						
	FM	K4CDN	444.07500	+	100.0 Hz	SERA - SC
Spartanburg, Spartanburg Regio						
	FM	KI4WVC	442.80000	+		
St George	DMR/MARC	WR4SC	440.65000	+	CC1	
St Matthews	FM	AD4U	146.67000	-	156.7 Hz	
Summerville, Summerville Med.						
	FM	W1GRE	146.98500	-	123.0 Hz	
Sumter	DMR/MARC	WR4SC	442.31250	+	CC1	
	FM	W4GL	53.77000	52.77000		
	FM	W4GL	146.64000	-	156.7 Hz	
	FM	W4GL	147.01500	+	156.7 Hz	SERA - SC
	FM	W4GL	224.12000	-		
	FM	W4VFR	224.66000	-		
	FM	WR4SC	441.62500	+	162.2 Hz	
	FM	W4GL	444.15000	+	123.0 Hz	
Sumter Downtown						
	DMR/MARC	WR4SC	441.83750	+	CC1	SERA - SC
Travelers Rest	FM	AC4RZ	442.92500	+	162.2 Hz	
Trenton, SC	FM	W4DV	145.49000	-	71.9 Hz	SERA - SC
Union	FM	K4USC	145.15000	-		SERA - SC
	FM	K4USC	146.68500	-		
	FM	K4USC	442.10000	+		
Walhalla	DMR/MARC	KN4SWB	442.53750	+	CC1	
Walhalla, Long Mountain						
	FM	K4WD	145.29000	-	162.2 Hz	SERA - SC
Walhalla, Stumphouse Mtn						
	FM	KJ4YLO	443.70000	+	91.5 Hz	
Wallace	FM	W4APE	29.60000	-	136.5 Hz	
Walterboro	FM	KG4BZN	53.31000	52.31000	123.0 Hz	
	FM	KG4BZN	145.39000	-		
	FM	KG4BZN	147.13500	+		
	FM	KG4BZN	444.55000	+	123.0 Hz	
Walterboro, Colleton County Me						
	FM	W4HRS	444.85000	+	123.0 Hz	
Wedgefield	FM	WB4BZA	145.43000	-	156.7 Hz	

326   SOUTH CAROLINA

Location	Mode	Call sign	Output	Input	Access	Coordinator
White Hall	FM	WR4SC	146.71500	-	123.0 Hz	
	FM	WA4SJS	146.91000	-	156.7 Hz	
	FM	WR4SC	441.67500	+	123.0 Hz	
York	FM	W4PSC	443.37500	+	110.9 Hz	

SOUTH DAKOTA

Location	Mode	Call sign	Output	Input	Access	Coordinator
Aberdeen	FM	WB0TPF	147.03000	+	146.2 Hz	
Beresford	FM	KA0VHV	147.24000	+	141.3 Hz	
Bowdle	FM	N0AHL	147.12000	+	146.2 Hz	
Brookings	FM	W0BXO	146.94000	-	110.9 Hz	
Bruce, Oakwood Lakes						
	FM	KC0FLK	444.25000	+	146.2 Hz	
Castlewood	FM	W0WTN	443.10000	+	146.2 Hz	SouDak
Clear Lake	FM	W0GC	444.30000	+	136.5 Hz	SouDak
Colonial Pine Hills	FM	K0LGB	444.45000	+	146.2 Hz	
Crandell, Sweetwater Lake						
	FM	N0AHL	146.79000	-	146.2 Hz	
Custer, Bear Mountain						
	FM	KC0BXH	146.85000	-	146.2 Hz	
Custer, Mt Coolidge						
	FM	K0HS	147.12000	+	146.2 Hz	SouDak
Custer, Water Tank Hill						
	FM	WN6QJN	147.09000	+	146.2 Hz	SouDak
Fairview	FM	N0JPE	146.56000		67.0 Hz	
Flandreau	FM	K0TGA	146.98500	-	146.2 Hz	
Hot Springs	FM	W0FUI	146.47500		146.2 Hz	
Hot Springs, Battle Mountain						
	FM	K0HS	146.70000	-	146.2 Hz	SouDak
Humboldt	FM	N0LCL	147.28500	+	103.5 Hz	
Huron	FM	K0OH	147.09000	+		
	FM	W0NOZ	443.85000	+		
Lead, Terry Peak	FM	KC0BXH	146.76000	-	146.2 Hz	
	FM	WB0JEK	147.03000	+	146.2 Hz	
Lead, Terry Peak Summit						
	FM	WB0JEK	443.65000	+		
Madison	FM	KB0MRG	440.17500	+	131.8 Hz	
Miller	FM	WV8CW	144.15000		91.5 Hz	
	FM	KC0WNG	440.65000	+	241.8 Hz	
Mitchell	FM	W0ZSJ	146.64000	-	146.2 Hz	
Mobridge	FM	N3NTV	145.50500			
	FM	W0YMB	147.21000	+	146.2 Hz	
Murdo	FM	AA0CT	147.30000	+	146.2 Hz	
New Underwood	FM	KA1OTT	146.97000	-		
Philip	FM	N0OMP	147.37500	+	146.2 Hz	
Pierpont	FM	W0JOZ	147.33000	+	146.2 Hz	
Pierre	FM	W0PIR	145.35000	-	146.2 Hz	
	FM	KD0S	146.54000			
	FM	KD0S	146.56500		146.2 Hz	
	FM	KD0S	146.73000	-	146.2 Hz	
Rapid City	FM	W0BLK	146.94000	-	146.2 Hz	
	FM	WA0MFZ	147.57000		146.2 Hz	
	FM	W0RE	443.85000	+		
	FM	W0RE	444.20000	+	82.5 Hz	
	FM	W0BLK	444.57500	+	146.2 Hz	
	FM	W0BLK	444.70000	+	146.2 Hz	
Rapid City, NWS Office						
	FM	K0VVY	147.50000			
Redfield	FM	WD0BIA	147.15000	+	146.2 Hz	
Reliance	FM	N0NPO	146.94000	-	146.2 Hz	
Salem	FM	W0SD	145.51000	-	146.2 Hz	
Sioux Falls	DMR/BM	WD0EXR	443.40000	+	CC1	
	DSTAR	W0ZWY	442.75000	+		

SOUTH DAKOTA   327

Location	Mode	Call sign	Output	Input	Access	Coordinator
Sioux Falls	FM	W0ZWY	146.89500	-	146.2 Hz	
	FM	W0FSD	223.86000	-		
	FM	KB0WSW	443.77500	+	146.2 Hz	
	FM	W0ZWY	444.20000	+	82.5 Hz	
	FM	KD0ZP	444.82500	+	146.2 Hz	
	FM	KD0ZP	444.90000	+	146.2 Hz	
Sisseton	FM	W0WM	146.88000	-		
Springfield, Norwegian Hill						
	FM	W0OJY	147.21000	+	146.2 Hz	
Tabor	FM	KC0TOW	147.31500	+		
Toronto	FM	KC0OVC	146.77500	-	146.2 Hz	
Turkey Ridge	FM	W0SD	444.97500	+	146.2 Hz	
Vermillion, Cement Plant						
	FM	W0OJY	147.37500	+	146.2 Hz	SouDak
Volga	FUSION	N0VEK	444.05000	+	136.5 Hz	SouDak
Watertown	FM	W0WTN	146.85000	-		
	FM	K0TY	442.00000	+	146.2 Hz	
Webster	FM	KC0MYX	442.10000	+	88.5 Hz	
	FM	N0PTW	443.92500	+		
Wessington Springs						
	FM	AA0F	147.34500	+	146.2 Hz	
Yankton	FM	W0OJY	146.85000	-	146.2 Hz	SouDak

TENNESSEE

Location	Mode	Call sign	Output	Input	Access	Coordinator
Adamsville	FM	AG4NX	145.00000			
Alcoa	DMR/BM	KD4CWB	442.47500	+	CC1	
	FM	K4BTL	442.00000	+		
Altamont	FM	KF4TNP	146.88000	-	167.9 Hz	SERA
Arrington	DMR/BM	N4ULM	441.90000	+	CC3	
Athens	FM	KG4FZR	147.06000	-	141.3 Hz	SERA
	FM	KG4FZR	442.27500	+	141.3 Hz	
Bartlett	FM	N4GMT	1284.25000	1264.25000	107.2 Hz	SERA
Bean Station	FM	W2IQ	443.45000	+	156.7 Hz	SERA
Bean Station, Clinch Mtn						
	FM	W2IQ	147.03000	+	156.7 Hz	SERA
Benton	DMR/BM	W4DMM	443.82500	+	CC1	
	DSTAR	KM4MCN	144.98000	147.48000		
Benton, Chilhowee Mountain						
	FM	WD4DES	147.18000	+	118.8 Hz	SERA
	FM	WD4DES	442.25000	+	118.8 Hz	SERA
	FM	KA4ELN	444.80000	+	123.0 Hz	SERA
Benton, Chilhowee Mountain/Osw						
	FM	KA4ELN	147.37500	+		SERA
Bethpage	DMR/BM	K4OZECHO	442.91250	+	CC1	
Bethpage, Mutton Hollow						
	FM	W4LKZ	145.39000	-	114.8 Hz	SERA
	FM	W4LKZ	443.30000	+	107.2 Hz	SERA
Blountville	FM	W4CBX	147.00000	+		
Brentwood	FM	WC4EOC	145.21000	-	173.8 Hz	
Brighton	FM	KE4ZBI	145.49000	-		
Bristol	FM	W4UD	146.67000	-		SERA
	FM	KE4CCB	146.70000	-		SERA
Bristol, Bristol Regional Hosp						
	FM	W4DOH	441.95000	+	146.2 Hz	SERA
Bristol, Holston Mountain						
	FM	KG4VBS	145.11000	-		SERA
	FM	W4CBX	224.20000	-		SERA
	FM	KE4CCB	442.20000	+	100.0 Hz	SERA
Brownsville, Brownsville-Haywo						
	FM	KI4BXI	444.52500	+	107.2 Hz	SERA
Carthage	FM	W4HTL	146.73000	-		
Caryville	DSTAR	KK4VQG	145.02000	+		

328   TENNESSEE

Location	Mode	Call sign	Output	Input	Access	Coordinator
Caryville	FM	KB4PNG	444.55000	+	77.0 Hz	SERA
	FM	W4HKL	444.67500	+	100.0 Hz	
Caryville, Caryville Mountain						
	FM	KA4OAK	224.28000	-		SERA
Centerville	FM	KI4DAD	442.55000	+	123.0 Hz	SERA
	FM	N4XW	443.70000	+	123.0 Hz	SERA
Chattanooga	DMR/BM	W4DMM	440.51250	+	CC1	
	DMR/BM	W4PL	444.15000	+	CC1	SERA
	DMR/BM	W4EDP	444.72500	+	CC1	
	DSTAR	W4RRG	444.72500	+		SERA
	FM	K4VCM	53.35000	52.35000		SERA
	FM	K4CMY	145.13000	-		SERA
	FM	K4VCM	224.78000	-		SERA
Chattanooga, Lookout Mountain						
	DSTAR	N4LMC	145.16000	-		SERA
	FM	W4AM	146.61000	-	107.2 Hz	
	FM	N4LMC	224.12000	-	146.2 Hz	
	FM	W4EDP	442.65000	+		SERA
	FM	W4AM	444.10000	+		SERA
Chattanooga, Monteagle Mountai						
	FM	NQ4Y	145.41000	-	114.8 Hz	SERA
Chattanooga, Signal Mountain						
	DMR/BM	W4YI	442.42500	+	CC1	SERA
	DSTAR	W4PL	145.29000	-		
	DSTAR	W4PL	443.15000	+		
	DSTAR	W4PL	1291.00000	1271.00000		
	FM	N4LMC	144.92000	147.42000	146.2 Hz	
	FM	W4AM	145.39000	-	107.2 Hz	SERA
	FM	K4VCM	224.42000	-		SERA
Chattanooga, WDEF Tower						
	FM	K4VCM	146.79000	-		SERA
Clarksville, Indian Mound						
	FM	AA4TA	146.92500	-	110.9 Hz	SERA
Cleveland	DMR/BM	WB4BSD	431.61000	+	CC1	
	DMR/MARC	WB4JGI	440.80000	+	CC1	
	DMR/MARC	WB4JGI	443.82500	+	CC1	
	DSTAR	KK4BXE	145.48000	-		SERA
	DSTAR	KK4BXE	440.52500	+		
	FM	KA4ELN	442.40000	+	123.0 Hz	SERA
	FM	W4OAR	442.92500	+	100.0 Hz	SERA
	FM	W4GZX	444.27500	+	114.8 Hz	SERA
	P25	KA4ELN	145.45000	147.95000		
Cleveland, CARC Clubhouse						
	FM	W4GZX	146.92500	-	114.8 Hz	SERA
	FM	KA4ELN	224.10000	-	123.0 Hz	SERA
Clinton, I 75	FM	WX4RP	442.15000	+	100.0 Hz	
Collegedale, White Oak Mountai						
	FM	KA6UHV	147.00000	+	131.8 Hz	SERA
	FM	KA6UHV	443.57500	+	131.8 Hz	
Collierville	FM	KA7UEC	443.30000	+	107.2 Hz	SERA
	FM	KA7UEC	443.62500	+	107.2 Hz	SERA
	FM	KJ4FYA	444.12500	+	107.2 Hz	SERA
Columbia	FM	W4GGM	147.12000	+	127.3 Hz	
	FM	KG4LUY	442.72500	+		
	FM	W4GGM	443.17500	+	100.0 Hz	SERA
Cookeville	DMR/BM	KK4TD	145.11000	-	CC1	
	FM	W4HPL	147.21000	+		SERA
	FM	W4EOC	444.60000	+	107.2 Hz	
Cookeville, TTU	FM	WA4UCE	145.43000	-		
Counce	FM	WV4P	444.80000	+	131.8 Hz	
Cross Plains	FM	AF4TZ	147.34500	+	114.8 Hz	
Crossville	FM	W8EYU	146.86500	-	118.8 Hz	SERA

TENNESSEE 329

Location	Mode	Call sign	Output	Input	Access	Coordinator
Crossville	FM	W4KEV	147.34500	+	118.8 Hz	SERA
	FM	W4KEV	443.87500	+	88.5 Hz	SERA
Crossville, Exit 320 I-40, Old						
	FM	W8EYU	146.89500	-	118.8 Hz	
Crossville, Fairfield Glade Re						
	FM	W8EYU	443.85000	+		
Crossville, Hinch Mountain						
	FM	W4KEV	53.93000	52.93000		
Cumberland Furnace						
	DMR/BM	N4GRW	441.81250	+	CC1	SERA
	FM	N4GRW	53.15000	52.15000	91.5 Hz	SERA
	FM	N4GRW	224.42000	-		SERA
	FM	N4GRW	927.52500	902.52500	107.2 Hz	SERA
Cumberland Gap , Kentucky Bord						
	FM	WA4ROB	442.85000	+	103.5 Hz	SERA
Dandridge	FM	W4KEV	146.89500	-	100.0 Hz	SERA
	FM	KD4TUD	444.62500	+	173.8 Hz	
Dayton	DMR/BM	KK4GGK	442.07500	+	CC1	SERA
Dayton, Evensville Mountain						
	FM	K4DPD	147.39000	+		
Decatur	FM	KG4FZR	145.15000	-	141.3 Hz	
	FM	KG4FZR	443.27500	+	100.0 Hz	
Decaturville	FM	KA4P	443.32500	+	131.8 Hz	
Dresden	DMR/BM	KA4BNI	442.15000	+	CC1	
	FM	KB4IBW	53.13000	52.13000	107.2 Hz	SERA
	FM	KB4IBW	145.15000	-	100.0 Hz	SERA
Dunlap	FM	KB4ACS	444.70000	+		
Dupont Springs, Green Top Mtn						
	FM	KD4CWB	444.00000	+	100.0 Hz	
Dupont Springs, Greentop						
	FM	KD4CWB	53.99000	52.99000	100.0 Hz	
	FM	KD4CWB	147.00000	-	100.0 Hz	SERA
	FM	KD4CWB	224.86000	-	100.0 Hz	SERA
Dyersburg	FM	K4DYR	444.47500	+	100.0 Hz	
Eaton	FM	KI4OAS	147.27000	+		SERA
	FM	KJ4HRM	442.30000	+	100.0 Hz	
Elizabethton	FM	KN4E	53.89000	52.89000	88.5 Hz	SERA
	FM	K4LNS	224.88000	-		SERA
	FM	WM4T	441.80000	+		SERA
	FM	KN4E	442.75000	+	88.5 Hz	SERA
Elizabethton, Holston Mountain						
	FM	WR4CC	145.29000	-	103.5 Hz	
	FM	K4LNS	147.27000	+	88.5 Hz	SERA
Erwin	FM	WB4IXU	53.95000	52.95000		
	FM	KC4DSY	147.16500	+		SERA
	FM	WB4IXU	224.78000	-		
Estill Springs	DMR/MARC	AJ4YS	442.96250	+	CC1	
Etowah, Starr Mtn	FM	KG4FZR	441.80000	+	141.3 Hz	
Everglades City	DMR/BM	WD2E	444.75000	+	CC1	
Fairview	DMR/BM	N4ULM	442.32500	+	CC2	
Falling Water, Signal Mountain						
	FM	K4VCM	443.12500	+	103.5 Hz	SERA
Fayetteville	FM	W4BV	147.03000	+	114.8 Hz	SERA
Franklin	FM	WA4BGK	53.05000	52.05000	114.8 Hz	
Franklin, Cool Springs Mall						
	FM	W4SQE	146.79000	-		
Gallatin	DMR/MARC	K4OZE	442.91250	+	CC1	
	FM	W4CAT	444.45000	+	107.2 Hz	SERA
Gallatin, Mockingbird Hill						
	FM	W4LKZ	147.27000	+	114.8 Hz	
Gallatin, Music Mountain						
	FM	W4LKZ	146.88000	-		

330   TENNESSEE

Location	Mode	Call sign	Output	Input	Access	Coordinator
Gallatin, Music Mountain						
	FM	W4CAT	147.30000	+		SERA
Gatlinburg	DMR/BM	KD4CWB	440.00000	+	CC1	
Gatlinburg, Glades Rd						
	FM	W4UO	147.42000			
Gatlinburg, Ski Mountain						
	FM	W4KEV	147.19500	+	100.0 Hz	SERA
	FM	W4KEV	444.90000	+	100.0 Hz	SERA
Georgetown	FM	WE4MB	442.02500	+	100.0 Hz	SERA
Germantown, Water Tower						
	FM	W4BS	146.62500	-	107.2 Hz	
Gleason	FM	KA4BNI	145.47000	-	100.0 Hz	
Gray	DMR/BM	WM4T	146.88000	-	CC7	
	FM	WM4T	145.25000	-		SERA
	FM	W4YSF	442.50000	+		SERA
Greeneville	DMR/BM	K4MFD	444.03750	+	CC1	SERA
	FM	W4BWW	53.01000	52.01000	100.0 Hz	
	FM	W4WC	53.01000	52.01000	100.0 Hz	SERA
	FM	N4FV	147.06000	+	88.5 Hz	SERA
	FM	KI4OTR	441.85000	+	100.0 Hz	SERA
	FM	N4FV	441.97500	+	100.0 Hz	SERA
	FM	N4CAG	443.15000	+	100.0 Hz	
	FM	K4GNR	443.27500	+	100.0 Hz	
Greeneville, Bald Mtn						
	FM	W4KEV	145.41000	-	127.3 Hz	SERA
Greeneville, Camp Creek Bald M						
	FM	W4WC	145.39000	-	88.5 Hz	SERA
	FM	W4WC	443.20000	+	100.0 Hz	SERA
Greeneville, Round Knob Mounta						
	FM	K4ETN	444.75000	+		SERA
Greeneville, Viking Mountain						
	FM	K4MFD	29.64000	-	118.8 Hz	
	FM	K4MFD	145.15000	-	118.8 Hz	SERA
	FM	K4MFD	224.44000	-	118.8 Hz	SERA
	FM	K4MFD	444.20000	+	118.8 Hz	SERA
	FM	K4MFD	927.05000	902.05000	118.8 Hz	
Greentop	DMR/BM	KD4CWB	443.50000	+	CC1	
Henderson, City Fire Station #						
	FM	KU4RT	147.10500	+	156.7 Hz	SERA
Hendersonville	FM	AK4GS	442.42500	+	107.2 Hz	SERA
	FM	W4LKZ	444.00000	+	107.2 Hz	
Hixson	FM	N4YH	442.90000	+	156.7 Hz	SERA
	FM	WJ9J	444.45000	+		
	FM	N4YH	444.87500	+	156.7 Hz	SERA
Hollow Rock	FM	KM4TFZ	146.71500	-		
Howardville	FM	W4CLM	146.91000	-		
Huntingdon	FM	KA4ZGK	146.83500	-	123.0 Hz	SERA
Jackson	DMR/BM	KA4BNI	443.71250	+	CC1	
	DSTAR	NT4MC	145.08000	146.48000		
	FM	KF4SC	145.31000	-	107.2 Hz	
	FM	WF4Q	147.21000	+	107.2 Hz	
	FM	KA4BNI	224.24000	-	131.8 Hz	SERA
	FM	NE4MA	444.45000	+	123.0 Hz	SERA
	FM	KA4BNI	444.87500	+	131.8 Hz	SERA
Jamestown, Round Mtn						
	FM	KC4MJN	443.62500	+		
Jamestown, Round Mtn Rd						
	FM	KI4KIL	147.09000	+		
Jasper	FM	KD4ATW	443.10000	+	88.5 Hz	SERA
Joelton	FM	KC4PRD	146.98500	-		
	FM	KF4TNP	442.95000	+	167.9 Hz	SERA
Johnson City	FM	AE4BT	442.87500	+	100.0 Hz	SERA

TENNESSEE 331

Location	Mode	Call sign	Output	Input	Access	Coordinator
Johnson City	FM	W4YSF	443.95000	+	118.8 Hz	SERA
	FM	K4LNS	444.10000	+	103.5 Hz	SERA
	FM	W4YSF	444.28750	+		SERA
Johnson City, ETSU						
	FM	W4DOH	442.77500	+		
Johnson City, Franklin Woods C						
	FM	W4ABR	146.79000	-	131.8 Hz	SERA
Johnson City, Tennessee						
	FM	W4BUC	443.25000	+		SERA
Jonesborough	FM	K4ETN	443.10000	+		SERA
Kingsport	FM	W4TRC	146.97000	-		SERA
Kingsport, Bays Mountain						
	FM	K4DWQ	443.17500	+	100.0 Hz	
	FM	W4TRC	443.32500	+		SERA
	FM	W4YSF	443.56250	+		SERA
	FM	W4TRC	927.02500	902.02500	123.0 Hz	
Knoxville	DMR/BM	W4KEV	440.55000	+	CC1	
	DMR/BM	K1LNX	443.56250	+	CC1	
	DMR/BM	KB4REC	444.17500	+	CC1	SERA
	DMR/MARC	K1LNX	443.66250	+	CC1	
	DSTAR	K4HXD	144.94000	147.44000		SERA
	FM	KB4REC	53.47000	52.47000	100.0 Hz	SERA
	FM	N4OQJ	224.38000	-	100.0 Hz	SERA
	FM	AA4UT	443.00000	+		SERA
	FM	W4KEV	444.50000	+	100.0 Hz	SERA
Knoxville, Bays Mtn - WJBZ Tow						
	FM	WB4GBI	145.43000	-	118.8 Hz	SERA
Knoxville, Beaver Ridge						
	FM	KB4REC	444.52500	+	123.0 Hz	
Knoxville, Chilhowee Mountain						
	FM	W4BBB	53.77000	52.77000	100.0 Hz	
Knoxville, Cross Mountain						
	FM	WB4GBI	145.47000	-	118.8 Hz	SERA
Knoxville, McKinney Ridge						
	FM	WB4GBI	147.07500	+		SERA
Knoxville, Sharp's Ridge						
	FM	W4BBB	145.21000	-	100.0 Hz	
	FM	WA4FLH	443.25000	+	88.5 Hz	SERA
Knoxville, Sharps Ridge						
	DSTAR	W4KEV	145.10000	-		
	FM	W4KEV	29.68000	-	127.3 Hz	
	FM	W4KEV	145.37000	-	100.0 Hz	SERA
	FM	W4KEV	146.88000	-		SERA
	FM	W4KEV	224.20000	-	100.0 Hz	
	FM	W4KEV	927.61250	902.61250	146.2 Hz	SERA
	P25	W4KEV	441.83750	+		SERA
Knoxville, View Park Hill						
	FM	WB4GBI	145.17000	-	118.8 Hz	SERA
Knoxville, WIMZ Tower						
	FM	W4KEV	53.25000	52.25000		
	FM	W4KEV	145.23000	-	103.5 Hz	SERA
Knoxville, WJXB Tower						
	FM	W4KEV	442.50000	+	100.0 Hz	SERA
Kodak	DMR/MARC	KD4CWB	921.50000	-	CC1	
La Vergne	FM	W4CAT	145.23000	-	114.8 Hz	SERA
LaFollette	DMR/BM	WB4CDK	444.85000	+	CC1	SERA
LaFollette, Cross Mountain						
	FM	KA4OAK	147.36000	+	100.0 Hz	SERA
LaFollette, Demory Community						
	FM	KA4OAK	145.13000	-	100.0 Hz	
Lawrenceburg	FM	KG4LUY	146.65500	-	100.0 Hz	
	FM	WD4RAT	147.39000	+		

332 TENNESSEE

Location	Mode	Call sign	Output	Input	Access	Coordinator
Lebanon	FM	W4EAO	442.12500	+	100.0 Hz	
Lebanon, Sparta Pike						
	FM	KM4GHM	444.95000	+		
Lebanon, WJFB-TV Tower						
	FM	WC4AR	147.10500	+	156.7 Hz	SERA
Lenoir City	FM	KF4DKW	444.25000	+	127.3 Hz	SERA
	FM	W4WVJ	444.60000	+		
Lenoir City, Greenback Communi						
	FM	W4WVJ	443.05000	+	100.0 Hz	SERA
Lewisburg	FM	N4MRS	146.62500	-	107.2 Hz	SERA
	FM	KF4TNP	442.10000	+	107.2 Hz	SERA
Linden, Top Of Hill West Of Li						
	FM	KJ4TVS	443.50000	+	100.0 Hz	
Lobelville	FM	WA4VVX	145.43000	-	114.8 Hz	
	FM	WA4VVX	442.85000	+	107.2 Hz	SERA
Loiusville	FM	W4KEV	927.03750	902.03750	100.0 Hz	SERA
Lookout Mountain	DMR/BM	W4EDP	442.72500	+	CC1	SERA
	DMR/BM	N4LMC	444.71250	+	CC1	SERA
Louisville, Red Hill						
	FM	N4ABV	443.37500	+	100.0 Hz	
Lyles	FM	KI4DAD	443.85000	+	123.0 Hz	
Lynchburg	FM	KF4TNP	145.45000	-	127.3 Hz	SERA
Madison	DMR/MARC	KJ4RVN	145.44000	-	CC1	
Madison, Candelabra TV Towers						
	FM	K4SNG	444.62500	+	107.2 Hz	
Madisonville	FM	K4JFT	443.60000	+	100.0 Hz	SERA
Manchester	DMR/MARC	KF4TNP	441.88750	+	CC1	
	DMR/MARC	KF4TNP	442.86250	+	CC1	
	FM	K4EGC	146.70000	-	114.8 Hz	SERA
	FM	KF4TNP	443.22500	+	127.3 Hz	
	FM	KF4TNP	444.07500	+	127.3 Hz	SERA
Martin	DMR/BM	KA4BNI	444.71250	+	CC1	
Martin, U T Martin	FM	W4UTM	146.62500	-	100.0 Hz	SERA
Maryville	DMR/MARC	KD4CWB	443.07500	+	CC1	SERA
	FM	KK4DKW	145.27000	-	127.3 Hz	SERA
	FM	W1BEW	441.82500	+		SERA
	FM	N4ABV	442.62500	+	100.0 Hz	SERA
	FM	KK4DKW	444.77500	+	94.8 Hz	SERA
	FM	KE4FGW	444.82500	+	100.0 Hz	SERA
Maryville, Reservoir Hill						
	FM	W4OLB	146.65500	-	100.0 Hz	SERA
McEwen, Water Tower						
	FM	NO4Q	147.22500	+	114.8 Hz	
McMinnville, Harrison Ferry Mo						
	FM	WD4MWQ	146.97000	-	151.4 Hz	
Medina	FM	WF4Q	146.77500	-	107.2 Hz	SERA
	FM	WT4WA	146.97000	-	107.2 Hz	
Memphis	DMR/BM	KK4BWF	441.88750	+	CC1	
	DMR/BM	W4LET	443.01250	+	CC1	SERA
	DSTAR	W4LET	145.06000	146.46000		SERA
	DSTAR	W4LET	443.98750	+		SERA
	FM	W4GMM	147.09000	+	107.2 Hz	SERA
	FM	W4BS	147.36000	+	107.2 Hz	SERA
	FM	W4BS	443.20000	+	107.2 Hz	SERA
Memphis, East Memphis Hilton H						
	FM	WB4KOG	146.85000	-	107.2 Hz	SERA
Memphis, First Tennessee Bank						
	FM	WB4KOG	145.45000	-	107.2 Hz	
	FM	WB4KOG	444.77500	+	107.2 Hz	SERA
Memphis, Hilton East Memphis						
	DSTAR	WB4KOG	144.94000	147.44000		SERA
	DSTAR	WB4KOG	442.03750	+		SERA

TENNESSEE 333

Location	Mode	Call sign	Output	Input	Access	Coordinator
Memphis, Hilton East Memphis						
	FM	WB4KOG	53.01000	52.01000	107.2 Hz	SERA
Memphis, Hilton Hotel						
	FM	WB4KOG	224.78000	-	107.2 Hz	SERA
Memphis, Methodist North Hospi						
	FM	W4BS	224.42000	-		SERA
Memphis, SCO	FM	WB4KOG	146.88000	-		SERA
Memphis, WKNO Tower						
	FM	W4EM	927.61250	902.61250	146.2 Hz	
Memphis, WPTY Tower						
	FM	W4BS	146.82000	-	107.2 Hz	SERA
Menomonie	DMR/MARC	K4USD	442.17500	+	CC1	SERA
Milan	FM	KA4BNI	442.67500	+	131.8 Hz	SERA
Mooresburg	FM	KE4KQI	927.61250	902.61250	114.8 Hz	SERA
Mooresburg, Short Mtn						
	FM	KE4KQI	147.13500	+	114.8 Hz	SERA
Morristown	DMR/BM	N4FNB	442.38750	+	CC1	
	DSTAR	W4LDG	144.92000	+		SERA
	DSTAR	W4LDG	444.47500	+		SERA
	FM	KQ4E	145.45000	-	141.3 Hz	SERA
	FM	WB4OAH	147.22500	+	141.3 Hz	
	FM	KQ4E	147.39000	+		SERA
	FM	AK4EZ	442.95000	+		SERA
	FM	KG4GVX	444.60000	+	100.0 Hz	
	FM	KQ4E	444.97500	+		
Morristown, I-81 I-40 Split						
	FM	KM4UIP	443.42500	+		SERA
Moss	DMR/BM	W4LSX	147.31500	+	CC1	
Mountain City	FM	W4MCT	145.47000	-	103.5 Hz	SERA
	FM	K4DHT	146.61000	-	103.5 Hz	SERA
	FM	K4DHT	224.28000	-	103.5 Hz	SERA
	FM	K4DHT	441.60000	+	151.4 Hz	SERA
Mountain City, Forge Mountain						
	FM	K4DHT	443.92500	+	103.5 Hz	SERA
Mountain City, Stone Mountain						
	FM	K4DHT	53.33000	52.33000	103.5 Hz	SERA
Murfreesboro	DMR/MARC	N2YCX	146.61000	-	CC4	
	FM	KU4B	145.17000	-		SERA
Nashville	DMR/BM	NE4MA	441.92500	+	CC1	
	DMR/MARC	W4DER	440.52500	+	CC1	
	DMR/MARC	AK4GS	443.08750	+	CC4	
	DMR/MARC	WA4BGK	444.58750	+	CC1	
	DSTAR	K4CPO	147.18000	+		
	FM	WA4BGK	53.01000	52.01000	107.2 Hz	
	FM	WA4PCD	146.76000	-		
	FM	W4CAT	146.95500	-	114.8 Hz	
	FM	WR3S	147.37000	+		
	FM	K1FB	224.18000	-	67.0 Hz	SERA
	FM	WA4RCW	442.75000	+	100.0 Hz	SERA
	FM	NE4MA	442.80000	+	107.2 Hz	SERA
	FM	WA4BGK	444.52500	+	107.2 Hz	
Nashville, East Nashville						
	FM	AF4TZ	146.64000	-	114.8 Hz	SERA
Nashville, Lipscomb University						
	FM	K4LRC	443.25000	+		
Nashville, Sullivan's Ridge						
	FM	AF4TZ	146.67000	-	114.8 Hz	SERA
	FM	AF4TZ	147.01500	+	114.8 Hz	SERA
Nashville, Vanderbilt Universi						
	FM	AA4VU	443.80000	+	123.0 Hz	
Nashville, Wessex Towers						
	FM	AF4TZ	444.15000	+	107.2 Hz	SERA

334 TENNESSEE

Location	Mode	Call sign	Output	Input	Access	Coordinator
New Market	FM	W4BWW	53.05000	52.05000	100.0 Hz	SERA
Newport	DMR/BM	KG4LHC	443.75000	+	CC1	SERA
	FM	KG4LHC	147.09000	+	203.5 Hz	SERA
	FM	KG4LHC	927.70000	902.70000	203.5 Hz	SERA
Newport / Dandridge						
	FM	KM4ULP	443.42500	+		
Newport, English Mountain						
	FM	WB4GBI	146.73000	-	118.8 Hz	SERA
Newport, Halls Top						
	FM	N2UGA	442.92500	+	103.5 Hz	SERA
Nolensville	FM	WD4JYD	145.35000	-		
Nunnely	FM	KG4UHH	444.07500	+	100.0 Hz	SERA
Oak Ridge	DMR/MARC	KD4CWB	443.43750	+	CC1	
	FM	W4KEV	441.92500	+	100.0 Hz	SERA
Oak Ridge, Water Tower						
	FM	W4SKH	146.97000	-	88.5 Hz	SERA
Oakfield, Jackson, TN Area						
	FM	WA4BJY	147.39000	+	162.2 Hz	SERA
Oakland, Water Tower						
	FM	WB4KOG	146.94000	-		SERA
Oliver Springs, Windrock Mount						
	FM	WB4GBI	147.15000	+	118.8 Hz	SERA
Oneida, Signal Mountain						
	FM	KB4PNG	145.35000	-	77.0 Hz	SERA
Paris	FM	N4ZKR	147.33000	+	131.8 Hz	SERA
	FM	KJ4ISZ	147.36000	+		
Parsons	FM	W4LSR	147.19500	-	94.8 Hz	SERA
Petros, Frozen Head Mountain						
	FM	KJ4SI	147.25500	+		SERA
Pikeville	FM	KF4JPU	147.28500	+		
Portland, Music Mountain						
	FM	W4LKZ	147.24000	+	114.8 Hz	
	FM	W4LKZ	444.35000	+	107.2 Hz	
Pulaski	FM	KF4TNP	146.80500	-	114.8 Hz	SERA
Reelfoot Lake Area						
	FM	W4NWT	147.04500	+	107.2 Hz	
	FM	W4NWT	442.50000	+	114.8 Hz	
Ripley	FM	KE4NTL	145.23000	-	100.0 Hz	SERA
Rockwood, Mt Roosevelt						
	DSTAR	KE4RX	441.81250	+		SERA
	DSTAR	KE4RX	1298.50000			SERA
	FM	KE4RX	443.97500	+	110.9 Hz	SERA
Rockwood, Mt. Roosevelt						
	FM	KE4RX	147.01500	+	110.9 Hz	SERA
	FM	K4APY	147.12000	+	82.5 Hz	SERA
Rogersville	DMR/BM	KN4EHX	440.56250	+	CC1	
Savannah	FM	KA4ESF	146.70000	-	123.0 Hz	
	FM	WV4P	442.80000	+	131.8 Hz	SERA
Savannah, Courthouse						
	FM	K4SDS	146.43500			
Selmer	FM	WB4MMI	146.80500	-	107.2 Hz	
Sevier	DMR/BM	KD4CWB	145.04000	146.44000	CC1	SERA
Sevierville	DMR/BM	WB4GBI	440.57500	+	CC1	SERA
Sevierville, Bluff Mountain						
	FM	WB4GBI	146.94000	-	118.8 Hz	SERA
	FM	WB4GBI	444.30000	+	118.8 Hz	SERA
	FM	WB4GBI	927.06250	902.06250	151.4 Hz	SERA
Sevierville, Bluff Mountain (G						
	FM	WB4GBI	146.85000	-	118.8 Hz	SERA
Seymour	DMR/MARC	KD4CWB	443.80000	+	CC1	SERA
Shelbyville	FM	KI4NJJ	147.06000	+	127.3 Hz	
	FM	KK4LFI	442.70000	+	100.0 Hz	

TENNESSEE 335

Location	Mode	Call sign	Output	Input	Access	Coordinator
Shelbyville	FM	WA4AWI	443.35000	+	127.3 Hz	
Signal Mountain	FM	KB4ACS	444.70000	+		
Signal Mtn	FM	KG4OVQ	442.15000	+		SERA
Sneedville	FM	KE4KQI	147.24000	+	114.8 Hz	
	FM	KE4KQI	442.45000	+	100.0 Hz	SERA
	FM	KE4WX	443.62500	+	114.8 Hz	SERA
Sparta	FM	KD4WX	146.68500	-	114.8 Hz	
	FM	K4TAX/R	442.97500	+	100.0 Hz	
	FM	KD4WX	444.37500	+	123.0 Hz	
Spring Hill	FM	N5AAA	442.65000	+	127.3 Hz	SERA
Springfield	DMR/BM	AD4RM	145.19000	-	CC1	
	DMR/BM	AC4AM	145.19000	-	CC1	
	DMR/BM	AD4RM	443.05000	+	CC1	
	FM	N8ITF	29.68000	-	88.5 Hz	SERA
	FM	N8ITF	145.19000	-		
	FM	N8ITF	443.05000	+	88.5 Hz	
Stanton, Water Tower On Neblet						
	FM	KI4BXI	146.65500	-	156.7 Hz	SERA
Sulphur Springs	FM	K4DWQ	147.12000	+		SERA
	FM	K4DWQ	442.05000	+	100.0 Hz	SERA
Sweetwater	FM	W4YJ	145.25000	-	100.0 Hz	SERA
	FM	W4YJ	444.12500	+		
Tellico Plains	FM	W3FCC	443.35000	+	141.3 Hz	
Tellico Plains, Waucheesi Moun						
	FM	K4EZK	146.82000	-	141.3 Hz	
Tellico Village	FM	WB4BSC	442.10000	+	100.0 Hz	SERA
Top Of The World Estates, Chil						
	FM	WB4GBI	146.62500	-	118.8 Hz	SERA
Trenton	FM	KN4KP	146.86500	-		
Unicoi	FM	WB4IXU	443.02500	+	210.7 Hz	
Union City	DMR/BM	W4NWT	147.01500	+	CC5	
	DMR/BM	K4JTM	443.50000	+	CC1	
	FM	WA4YGM	146.70000	-	100.0 Hz	
Union City, Tn	DMR/BM	K4JTM	442.35000	+	CC1	
Walden	DSTAR	W4PL A	1291.00000	1271.00000		
Walland	FM	AC4JF	145.33000	-	100.0 Hz	SERA
Walland, Chilhowee Mountain						
	FM	WB4GBI	53.15000	52.15000	118.8 Hz	SERA
Walland, Chilhowee Mtn						
	FM	W4BBB	147.30000	+	100.0 Hz	SERA
	FM	N4OQJ	224.22000	-		
	FM	W4BBB	444.57500	+	100.0 Hz	
Watertown	FM	W4LYR	146.83500	-	100.0 Hz	
Waynesboro	FM	KF4TNP	443.10000	+	167.9 Hz	SERA
White Bluff	FM	KG4HDZ	147.37500	+	146.2 Hz	SERA
White House	FM	W4LKZ	443.40000	+	107.2 Hz	SERA
Williston	FM	WB4KOG	147.18000	+	107.2 Hz	SERA
	FM	WB4KOG	444.40000	+	107.2 Hz	SERA
Winchester	DMR/MARC	KF4TNP	444.95000	+	CC1	SERA
Winchester, Keith Spring Mtn						
	FM	W4UOT	146.82000	-	114.8 Hz	
Woodbury, Short Mountain						
	FM	W4YXA	146.91000	-		SERA
	FM	WR3S	147.36000	+	114.8 Hz	
Woodbury, Short Mt						
	FM	WB4LHO	145.49000	-		SERA
Woodlawn	DMR/BM	K4VL	444.32500	+	CC1	

TEXAS

Location	Mode	Call sign	Output	Input	Access	Coordinator
Abilene	DMR/BM	W5SLG	442.40000	+	CC4	
	DMR/BM	KF5JJK	444.17500	+	CC14	
	DMR/MARC	KF5JJK	443.66250	+	CC14	

336   TEXAS

Location	Mode	Call sign	Output	Input	Access	Coordinator
Abilene	DMR/MARC	KF5JJK	443.96250	+	CC14	
	FM	KC5PPI	145.35000	-	110.9 Hz	TVHFS
	FM	KI5ZS	145.49000	-	88.5 Hz	TVHFS
	FM	KC5OLO	146.76000	-	146.2 Hz	TVHFS
	FM	AI5TX	443.50000	+		TVHFS
	FM	KD5YCY	444.00000	+	167.9 Hz	TVHFS
	FM	WX5TX	444.17500	+	100.0 Hz	TVHFS
	FM	KB5GAR	444.75000	+	88.5 Hz	TVHFS
	FM	KE5OGP	444.87500	+	114.8 Hz	TVHFS
	FM	N5TEQ	927.03750	902.03750	141.3 Hz	TVHFS
Abilene, Cedar Gap Mountain						
	FM	KC5OLO	146.96000	-	146.2 Hz	TVHFS
Abilene, Near Lake Kirby						
	FM	K5CCG	444.42500	+	146.2 Hz	TVHFS
Addison	FM	W5MGM	146.44000		100.0 Hz	
Adkins	FM	KK5LA	444.77500	+	123.0 Hz	TVHFS
Albany	FM	N5TEQ	444.90000	+	114.8 Hz	TVHFS
Aledo, Lake Weatherford						
	FM	KA5HND	223.90000	-	110.9 Hz	TVHFS
	FM	KA5HND	443.20000	+	110.9 Hz	TVHFS
Alice	DMR/BM	KF5UPC	441.97500	+	CC1	
Allen	DSTAR	K5PRK	441.57500	+		TVHFS
	DSTAR	K5PRK	1295.00000	1315.00000		TVHFS
	FM	K5PRK	147.18000	+	107.2 Hz	TVHFS
	FM	N5LTN	441.50000	+	110.9 Hz	TVHFS
	FM	AA5CT	442.55000	+		TVHFS
	FM	W5AIM	442.87500	+	162.2 Hz	
	FM	K5PRK	444.25000	+	79.7 Hz	TVHFS
Allen, Prestige Water Tower						
	FM	N5UIG	441.30000	+	179.9 Hz	TVHFS
Alpine	FM	AD5BB	145.23000	-	146.2 Hz	TVHFS
	FM	K5FD	146.72000	-	146.2 Hz	TVHFS
	FM	WX5II	443.92500	+		TVHFS
	FM	K5FD	446.15000	-	146.2 Hz	
Alvarado	FM	K5AEC	147.22000	+	110.9 Hz	TVHFS
Alvin	DMR/MARC	KA5AXV	444.75000	+	CC7	TVHFS
	DSTAR	K5PLD	146.49000	147.49000		TVHFS
	FM	KA5QDG	145.11000	-	123.0 Hz	TVHFS
	FM	KA5QDG	223.96000	-	123.0 Hz	TVHFS
	FM	KA5QDG	442.20000	+	103.5 Hz	TVHFS
Amarillo	FM	N5ZLU	146.74000	-	88.5 Hz	TVHFS
	FM	N5LTZ	146.92000	-		TVHFS
	FM	W5WX	147.34000	+		TVHFS
	FM	WR9B	441.37500	+	88.5 Hz	
	FM	N5LUL	441.65000	+	127.3 Hz	TVHFS
	FM	N5LTZ	443.50000	+		TVHFS
	FM	N5LTZ	444.20000	+	88.5 Hz	TVHFS
	FM	W5JTC	444.30000	+		
	FUSION	W5WX	444.47500	+		TVHFS
Amarillo, KVII Tower (ABC 7)						
	FM	W5WX	146.94000	-		TVHFS
Anahuac	FM	KK5XQ	145.33000	-	123.0 Hz	TVHFS
	FM	KB5FLX	442.10000	+	103.5 Hz	TVHFS
Angleton	FM	N9QXT	147.18000	+	141.3 Hz	TVHFS
	FM	KE5WFD	442.30000	+	127.3 Hz	TVHFS
Anna	DMR/MARC	WS5W	440.52500	+	CC1	
	DMR/MARC	WO5J	445.72500	-	CC1	
Anthony	FM	N5ZRF	442.95000	+	67.0 Hz	TVHFS
Arlington	DMR/BM	KF5UGN	441.33750	+	CC1	
	DMR/MARC	WD5DBB	443.40000	+	CC1	TVHFS
	FM	WD5DBB	146.86000	-	110.9 Hz	TVHFS
	FM	K5SLD	147.14000	+	110.9 Hz	TVHFS

TEXAS   337

Location	Mode	Call sign	Output	Input	Access	Coordinator
Arlington	FM	K5SLD	224.80000	-	110.9 Hz	TVHFS
	FM	NR5E	441.35000	+		TVHFS
	FM	N4MSE	442.75000	+	127.3 Hz	
	FM	AI5TX	443.67500	+		TVHFS
	FM	WA5VHU	443.85000	+	110.9 Hz	TVHFS
	FM	K5SLD	444.20000	+		TVHFS
	FM	W5PSB	444.55000	+		TVHFS
Aspermont	FM	KC5ATZ	444.70000	+		
Atascocita, Atascocita Fire De						
	FM	W5SI	443.55000	+	103.5 Hz	TVHFS
Athens	FM	K5EPH	147.22000	+	136.5 Hz	TVHFS
	FM	W5ETX	442.85000	+	136.5 Hz	TVHFS
	FM	KF5WT	443.30000	+	100.0 Hz	TVHFS
	FM	AI5TX	443.70000	+		TVHFS
Atlanta	DMR/MARC	KA5AHS	440.52500	+	CC9	
	FM	K5HCM	145.25000	-	100.0 Hz	TVHFS
	FM	KA5AHS	147.22000	+	100.0 Hz	TVHFS
	FM	WA5JYZ	147.26000	+	151.4 Hz	TVHFS
Atlanta / Springdale						
	FM	N5YU	145.41000	-	100.0 Hz	
Atlanta, Caver Ranch						
	FM	K5HCM	146.98000	-	100.0 Hz	TVHFS
Atlanta, Christus St. Michaels						
	FM	K5HCM	443.30000	+	100.0 Hz	TVHFS
Atlanta, Springdale						
	FM	WX5FL	145.31000	-	100.0 Hz	TVHFS
Atlanta, Springdale Tower						
	FM	KB5SQL	444.55000	+	100.0 Hz	
Aubrey	DMR/MARC	K5RNB	443.45000	+	CC1	
	FM	K5RNB	145.23000	-	100.0 Hz	TVHFS
Austin	DMR/BM	KB5PRZ	442.15000	+	CC1	
	FM	WB5PCV	146.61000	-	103.5 Hz	TVHFS
	FM	W3MRC	146.90000	-		TVHFS
	FM	WB5PCV	224.94000	-		TVHFS
	FM	K5LBJ	441.67500	+	131.8 Hz	TVHFS
	FM	W5JWB	441.77500	+	131.8 Hz	TVHFS
	FM	KA9LAY	441.97500	+	97.4 Hz	TVHFS
	FM	W5TRI	442.02500	+	114.8 Hz	TVHFS
	FM	KB5HTB	442.06250	+		
	FM	K5AB	442.20000	+	100.0 Hz	TVHFS
	FM	WB5FNZ	442.50000	+	162.2 Hz	TVHFS
	FM	K5TRA	443.07500	+	123.0 Hz	TVHFS
	FM	WA5VTV	443.80000	+	131.8 Hz	
	FM	AI5TX	443.95000	+		TVHFS
	FM	WB5PCV	444.00000	+	107.2 Hz	TVHFS
	FM	W5KA	444.10000	+	103.5 Hz	TVHFS
	FM	K5TRA	444.50000	+	110.9 Hz	
	FM	W3MRC	444.60000	+		TVHFS
	FM	K5TRA	927.01250	902.01250	225.7 Hz	
	FM	K5TRA	927.11250	902.11250	DCS 432	TVHFS
	FM	KA5D	927.17500	902.17500	110.9 Hz	TVHFS
	P25	N0GSZ	146.86000	-	146.2 Hz	TVHFS
Austin, Buckman Mountain						
	FM	W5KA	146.94000	-	107.2 Hz	TVHFS
Austin, Buckman Mtn						
	FM	WA5VTV	147.36000	+	131.8 Hz	TVHFS
Austin, KVUE TV Tower						
	FM	N5MHI	1292.40000	1272.40000	107.2 Hz	TVHFS
Austin, Oak Hill	FUSION	N5OAK	147.32000	+	114.8 Hz	TVHFS
Austin, Pickle Research Center						
	FUSION	W5KA	146.88000	-	107.2 Hz	TVHFS

338 TEXAS

Location	Mode	Call sign	Output	Input	Access	Coordinator
Austin, South Austin Hospital						
	DSTAR	W5KA	440.65000	+		TVHFS
Austin, South Austin Med Ctr						
	DSTAR	W5KA	1293.10000	1273.10000		TVHFS
Austin, University Of Texas						
	FM	KA5D	441.32500	+	97.4 Hz	TVHFS
Austin, Water Tower						
	FM	AE5WW	444.85000	+	103.5 Hz	TVHFS
Austin, Westlake Hills						
	FM	N5ZUA	145.11000	-	103.5 Hz	TVHFS
Avinger, Lake-O-The Pines / Wa						
	FM	WX5FL	145.47000	-	136.5 Hz	
Azle	FM	WB5IDM	147.16000	+	110.9 Hz	TVHFS
	FM	WB5IDM	223.94000	-	110.9 Hz	TVHFS
	FM	WB5IDM	442.15000	+	110.9 Hz	TVHFS
	FM	WB5IDM	927.01250	902.01250	110.9 Hz	TVHFS
Balcones Heights	FM	W5ROS	147.34000	+		
Bandera	FM	W5MWI	145.31000	-	DCS 032	
	FM	W5MWI	443.30000	+	DCS 032	
Bangs	FM	KB5ZVV	147.00000	+	94.8 Hz	TVHFS
Bastrop	FM	NA6M	147.34000	+	100.0 Hz	TVHFS
	FM	N5FRT	441.55000	+	114.8 Hz	
	FM	WB6ARE	441.95000	+		TVHFS
	FM	WB6ARE	442.72500	+	114.8 Hz	TVHFS
	FM	WB6ARE	443.17500	+	114.8 Hz	TVHFS
	FM	KE5FKS	443.75000	+	114.8 Hz	TVHFS
	FM	WB5UGT	444.30000	+	203.5 Hz	TVHFS
	FM	K5TRA	927.02500	902.02500	DCS 532	
Bay City	FM	W5WTM	146.72000	-	146.2 Hz	TVHFS
Bayou Vista	DMR/BM	N5FOG	433.27500	+	CC1	
Baytown	FM	NE5TX	145.31000	-	167.9 Hz	TVHFS
	FM	K5BAY	146.78000	-	100.0 Hz	TVHFS
	FM	K5BAY	443.80000	+	100.0 Hz	TVHFS
	FM	KB5IAM	443.87500	+	173.8 Hz	
Beaumont	DMR/MARC	N5YX	145.45000	-	CC1	
	FM	W5RIN	146.70000	-	107.2 Hz	TVHFS
	FM	W5RIN	146.76000	-	107.2 Hz	TVHFS
	FM	W5APX	146.94000	-	100.0 Hz	TVHFS
	FM	W5XOM	147.30000	+	103.5 Hz	TVHFS
	FM	KB5OVJ	147.34000	+	131.8 Hz	TVHFS
	FM	W5RIN	444.70000	+	107.2 Hz	TVHFS
Beckville	FM	KA5HSA	444.50000	+	151.4 Hz	
Bedford	FM	N5VAV	442.82500	+	110.9 Hz	TVHFS
Bee Cave	FM	AI5TX	443.62500	+		TVHFS
	FM	K5GJ	443.92500	+		TVHFS
	FM	K5TRA	927.08750	902.08750	151.4 Hz	
Beeville	DMR/MARC	W6AUS	147.12000	+	CC1	
	DMR/MARC	W5DTW	444.70000	+	CC1	
Bellaire	FM	AK5G	441.82500	+		TVHFS
Bellville	FM	WR5AAA	146.88000	-	203.5 Hz	TVHFS
	FM	W5SFA	444.87500	+	103.5 Hz	TVHFS
Belton	DMR/MARC	N5JLP	442.67500	+	CC10	
	FM	WD5EMS	145.19000	-		TVHFS
	FM	WD4IFU	927.15000	902.15000	114.8 Hz	
Belton, Belton High School						
	FM	KG5PIV	145.35000	-		
Big Bend	FM	AD5BB	146.82000	-	146.2 Hz	TVHFS
Big Lake	FM	N5SOR	442.30000	+	162.2 Hz	TVHFS
Big Spring	DMR/BM	W5LND	443.22500	+	CC1	TVHFS
	DMR/MARC	KE5PL	443.43750	+	CC14	
	DSTAR	W5AW	147.24000	+		
	DSTAR	W5AW	440.68750	-		

TEXAS 339

Location	Mode	Call sign	Output	Input	Access	Coordinator
Big Spring	FM	W5AW	146.82000	-		TVHFS
	FM	KE5PL	443.95000	+		TVHFS
Big Spring, Tower Ridge						
	FM	W5LND	147.04000	+	88.5 Hz	TVHFS
	FM	W5LND	442.10000	+	162.2 Hz	TVHFS
Black	FM	N5LTZ	444.42500	+		TVHFS
Blanco	FM	KF5KOI	145.28000	-	77.0 Hz	TVHFS
Blanket	FM	N5AG	224.72000	-	94.8 Hz	TVHFS
Bluetown	DMR/MARC	KC5HWB	443.88750	+	CC9	
Boerne	FM	AB5UE	145.42000	-	123.0 Hz	
	FM	WB5CIT	441.65000	+	103.5 Hz	
	FM	AB5UE	443.60000	+	123.0 Hz	
	FM	W5VEO	444.75000	+	162.2 Hz	TVHFS
Boerne, Sheriff?s Office						
	FM	KB5TX	145.19000	-	88.5 Hz	
Boerne, Tower Road						
	FM	KB5TX	146.64000	-	88.5 Hz	
	FM	KB5TX	444.90000	+	88.5 Hz	TVHFS
Bonham	FM	K5FRC	443.75000	+	100.0 Hz	TVHFS
Boonsville	FM	K5RHV	443.90000	+	100.0 Hz	TVHFS
Borger	DMR/MARC	KD5ROK	146.13125	-	CC14	
	DMR/MARC	KE5CJ	441.33750	+	CC14	
	FM	W5WDR	147.00000	+	88.5 Hz	
Bowie	FM	K1RKH	145.39000	-	192.8 Hz	TVHFS
	FM	WX5ARC	147.32000	+	192.8 Hz	TVHFS
	FM	WX5ECT	441.77500	+		TVHFS
Boyd	FM	N5KOU	1294.50000		88.5 Hz	
Boyd, Courthouse	FM	K5JEJ	146.98000	-	110.9 Hz	TVHFS
Brackettville	FM	AA5KC	146.88000	-	127.3 Hz	TVHFS
	FM	WB5TZJ	443.62500	+		TVHFS
	FM	AA5KC	443.72500	+	100.0 Hz	TVHFS
Brady	DMR/MARC	KG5YMG	443.57500	+	CC1	
	FM	AA5JM	146.62000	-	114.8 Hz	TVHFS
	FM	KC5EZZ	146.90000	-	162.2 Hz	TVHFS
	FM	WO5OD	444.07500	+	110.9 Hz	
	FM	WA5HOT	444.87500	+	162.2 Hz	TVHFS
Brazos, Chestnut Mountain						
	FM	W5SUF	444.17500	+	114.8 Hz	TVHFS
Brenham, Blinn College						
	FM	W5AUM	147.26000	+	103.5 Hz	TVHFS
Brenham, Brenham High School						
	FM	KF5ZRT	441.32500	+	103.5 Hz	
Brenham, Brenham National Bank						
	FM	W5AUM	145.39000	-	103.5 Hz	TVHFS
	FM	N5MBM	441.65000	+	110.9 Hz	TVHFS
Brenham, Kruse Village						
	FM	N5MBM	441.92500	+	123.0 Hz	TVHFS
Brenham, Police Dept Communica						
	FM	W5AUM	443.25000	+	103.5 Hz	TVHFS
Bridgeport	FM	N5WEB	927.12500	902.12500	100.0 Hz	TVHFS
Brownfield	FM	WA5OEO	147.34000	+	162.2 Hz	TVHFS
Brownsville	DMR/MARC	KC5MAH	444.37500	+	CC1	TVHFS
	FM	K5MPH	145.52000			
	FM	W5RGV	147.04000	+	114.8 Hz	TVHFS
	FM	N5XWO	441.30000	+	151.4 Hz	TVHFS
	FM	N5RGV	444.17500	+	114.8 Hz	
	FM	W5RGV	444.40000	+		
Brownwood	FM	W5CBT	146.82000	-	94.8 Hz	TVHFS
	FM	K5BWD	146.94000	-	94.8 Hz	TVHFS
	FM	AI5TX	443.92500	+		TVHFS
	FM	K5BWD	444.70000	+	94.8 Hz	TVHFS

340   TEXAS

Location	Mode	Call sign	Output	Input	Access	Coordinator
Brownwood, Bangs Hill						
	FUSION	WD9ARW	443.90000	+	94.8 Hz	
Bruceville	DSTAR	W5NGU	440.62500	+		TVHFS
	FM	W5NCD	147.24000	+	97.4 Hz	TVHFS
	FM	W5NCD	444.47500	+		TVHFS
Bruni	DMR/MARC	KE5NL	146.86000	-	CC4	TVHFS
Bryan	DSTAR	W5AC	443.40000	+		TVHFS
	FM	KD5DLW	443.42500	+	127.3 Hz	TVHFS
Buffalo	FM	W5UOK	147.28000	+	146.2 Hz	TVHFS
Buffalo, Water Tower						
	FM	WD5EMS	444.25000	+		
Bullard	FM	KB9LFZ	443.20000	+	127.3 Hz	
Bulverde	FM	WA5KBQ	443.25000	+	103.5 Hz	TVHFS
	FM	W5DK	444.35000	+	151.4 Hz	TVHFS
Buna	FM	W5JAS	145.39000	-	118.8 Hz	TVHFS
	FM	W5JAS	442.42500	+	118.8 Hz	TVHFS
	P25	KB5OVJ	147.38000	+		TVHFS
Burkburnett	FM	W5DAD	146.70000	-	192.8 Hz	TVHFS
	FM	KD5PWT	444.02500	+	192.8 Hz	TVHFS
	FM	KD5INN	444.30000	+		
Burleson	FM	K5JCR	440.70000	+	136.5 Hz	TVHFS
	FM	WA5JRS	444.52500	+	167.9 Hz	
Burnet	FM	KB5YKJ	145.29000	-	114.8 Hz	TVHFS
	FM	K5HLA	147.02000	+	88.5 Hz	
Byers	FM	KF5DFD	146.82000	-	192.8 Hz	TVHFS
Caddo	FM	KB5WB	444.72500	+	110.9 Hz	TVHFS
Cameron, McClaren Hill						
	FM	KE5URD	147.02000	+	123.0 Hz	TVHFS
Canyon	FM	N5LTZ	443.65000	+	88.5 Hz	TVHFS
Canyon Lake	FM	W5ERX	444.45000	+	114.8 Hz	TVHFS
	FM	K5TRA	927.03750	902.03750	141.3 Hz	
Carlton	FM	W5GKY	145.27000	-	110.9 Hz	TVHFS
	FM	W5GKY	147.30000	+	100.0 Hz	TVHFS
Carrizo Springs, Center Of Tow						
	FM	W5EVH	443.30000	+	100.0 Hz	TVHFS
Carrollton	DMR/MARC	N5GDL	440.45000	+	CC1	
	FM	KB5A	145.21000	-	110.9 Hz	TVHFS
	FM	K5ZYZ	146.82000	-	110.9 Hz	TVHFS
	FM	K5JG	441.62500	+	100.0 Hz	
	FM	K5GWF	441.82500	+		TVHFS
	FM	K5AB	444.45000	+	110.9 Hz	TVHFS
Carrollton, Baylor Hospital						
	FM	KB5A	442.65000	+	110.9 Hz	TVHFS
	FM	N5KRG	927.17500	902.17500	110.9 Hz	TVHFS
Carthage	FM	KA5HSA	146.72000	-		TVHFS
	FM	KA5HSA	444.80000	+	151.4 Hz	TVHFS
Cason, Deaton Brothers Tower						
	FM	WA5OQR	146.88000	-	151.4 Hz	TVHFS
Castroville	FM	K5YDE	146.80000	-	162.2 Hz	TVHFS
	FM	KD5DX	147.20000	+	162.2 Hz	TVHFS
Cat Spring	FM	WB5UGT	444.07500	+	103.5 Hz	TVHFS
Cedar Creek	FM	WA5AP	444.05000	+	141.3 Hz	
Cedar Hill	FM	W5WB	147.06000	+	110.9 Hz	TVHFS
	FM	W5AHN	147.26000	+	85.4 Hz	TVHFS
	FM	AI5TX	224.10000	-		TVHFS
	FM	WB5YUV	442.40000	+	110.9 Hz	TVHFS
	FM	AI5TX	443.50000	+		TVHFS
	FM	N5UN	443.97500	+	156.7 Hz	TVHFS
	FM	W5AUY	444.50000	+		TVHFS
Cedar Park	DMR/MARC	KE5ZW	442.65000	+	CC1	
	FM	KC5WLF	146.84000	-	103.5 Hz	TVHFS
	FM	W2MN	146.98000	-	103.5 Hz	TVHFS

TEXAS   341

Location	Mode	Call sign	Output	Input	Access	Coordinator
Cedar Park	FM	W2MN	147.12000	+	103.5 Hz	TVHFS
	P25	KE5ZW	146.68000	-	123.0 Hz	TVHFS
	FUSION	KC5WLF	145.37000	-	103.5 Hz	TVHFS
Celina	FM	N5MRG	224.24000	-	103.5 Hz	
	FM	KE5UT	444.51250	+	123.0 Hz	TVHFS
	FM	N5BCW	444.97500	+	103.5 Hz	
Centerville	FM	K3WIV	145.21000	-		TVHFS
	FM	K3WIV	145.45000	-	114.8 Hz	TVHFS
	FM	N5HLC	146.78000	-	103.5 Hz	TVHFS
	FM	KD0RW	147.30000	+	114.8 Hz	TVHFS
	FM	WA5GED	441.65000	+	103.5 Hz	
	FM	K3WIV	442.77500	+		TVHFS
	FM	K3WIV	442.97500	+		TVHFS
	FM	AI5TX	443.67500	+		TVHFS
Chalk Mountain	FM	K5AB	145.47000	-	110.9 Hz	TVHFS
	FM	W5DNT	147.02000	+	162.2 Hz	TVHFS
Channelview	FM	KC5TCT	441.60000	+	203.5 Hz	TVHFS
	FM	KE5CWO	446.10000	-	141.3 Hz	
Channelview, I-10 E & BW8 E						
	FM	K5CAP	443.18000	+	179.9 Hz	
Chappell Hill	DMR/MARC	N5MBM	145.15000	-	CC13	TVHFS
	FM	N5MBM	52.35000	51.35000	DCS 131	TVHFS
	FM	N5MBM	224.90000	-	103.5 Hz	TVHFS
	FM	N5MBM	441.62500	+	103.5 Hz	TVHFS
	FM	N5MBM	927.15000	902.15000	103.5 Hz	TVHFS
Childress	DMR/MARC	KM5PM	444.72500	+	CC1	
	FM	N5OX	146.96000	-		TVHFS
	FM	N5OX	442.40000	+	100.0 Hz	
Chita	FM	W5IOU	145.35000	-	131.8 Hz	TVHFS
Choate	FM	K5ZZT	443.65000	+		TVHFS
Christine	FM	W5DK	443.77500	+	141.3 Hz	TVHFS
Clarendon	FM	KE5NCA	444.27500	+	127.3 Hz	
Clarksville	FM	KI5DX	147.34000	+	100.0 Hz	
Clear Lake Shores						
	FM	KA5QDG	145.39000	-	123.0 Hz	TVHFS
Clear Lake Shores, Water Tower						
	FM	WA5LQR	442.37500	+	103.5 Hz	
Clear Lake, Boeing Building						
	FM	K5HOU	442.75000	+	103.5 Hz	TVHFS
Clear Lake, Boeing Building 37						
	FM	K5HOU	146.86000	-	100.0 Hz	TVHFS
Cleburne	FM	KB5YBI	145.49000	-	88.5 Hz	TVHFS
	FM	W5JCR	224.76000	-	88.5 Hz	
Cleburne, Hulen Park						
	FM	KY5O	444.00000	+	136.5 Hz	TVHFS
Cleveland	DMR/MARC	W5JSC	443.85000	+	CC3	
	FM	N5AK	146.90000	-		TVHFS
	FM	N5AK	224.78000	-		TVHFS
	FM	N5AK	444.65000	+		TVHFS
Clifton	FM	W5BCR	147.18000	+	123.0 Hz	
	FM	W5BCR	444.40000	+	123.0 Hz	TVHFS
College Station	DMR/BM	KG5RKI	441.35000	+	CC1	
	DMR/BM	K5ZY	444.12500	+	CC1	
	FM	K5ZY	147.16000	+	88.5 Hz	
College Station, Skyline Commu						
	FM	W5BCS	146.68000	-	88.5 Hz	TVHFS
College Station, Texas A&M Uni						
	FM	W5AC	146.82000	-	88.5 Hz	TVHFS
Colleyville	FM	W5RV	441.90000	+	110.9 Hz	TVHFS
Colorado City	FM	K5WTC	444.85000	+	162.2 Hz	TVHFS
Columbus	FM	W5SFA	147.14000	+	103.5 Hz	TVHFS
	FM	WB5UGT	442.75000	+	141.3 Hz	TVHFS

342   TEXAS

Location	Mode	Call sign	Output	Input	Access	Coordinator
Comfort	FM	KD3VK	441.82500	+	82.5 Hz	TVHFS
Commerce	FM	WB5MQP	147.02000	+	167.9 Hz	TVHFS
	FM	W5AMC	444.52500	+	103.5 Hz	TVHFS
Conroe	DMR/MARC	N5LUY	444.75000	+	CC3	
	FM	NE5TH	147.02000	+	136.5 Hz	TVHFS
	FM	N5KWN	147.14000	+	136.5 Hz	TVHFS
	FM	KF5RDE	441.30000	+	118.8 Hz	
	FM	KE5PTZ	441.75000	+	123.0 Hz	TVHFS
	FM	WB5DGR	442.25000	+	103.5 Hz	TVHFS
	FM	W5SAM	442.52500	+	127.3 Hz	TVHFS
	FM	WB5DGR	442.90000	+	151.4 Hz	TVHFS
	FM	AG5EG	444.57500	+	103.5 Hz	TVHFS
Coppell	FM	W5CPL	441.37500	+	103.5 Hz	
Coppell, Coppell High School						
	FM	KC5BY	444.27500	+	114.8 Hz	
Copperas Cove	FM	K5CRA	147.26000	+	88.5 Hz	TVHFS
	FM	K5CRA	443.32500	+	88.5 Hz	TVHFS
Cornudas	FM	WS5B	147.32000	+	110.9 Hz	TVHFS
Corpus Christi	DMR/BM	W5QLD	441.35000	+	CC1	
	DMR/MARC	AD5CA	441.60000	+	CC1	
	FM	N5CRP	146.82000	-	107.2 Hz	TVHFS
	FM	N5CRP	146.88000	-	107.2 Hz	TVHFS
	FM	K5GGB	147.06000	+	107.2 Hz	TVHFS
Corpus Christi, USS Lexington						
	FM	W5LEX	444.85000	+	103.5 Hz	TVHFS
Corsicana	DMR/MARC	KC5HWB	442.92500	+	CC9	TVHFS
	DSTAR	K5NEM	147.34000	+		TVHFS
	FM	N5ZUA	444.77500	+	100.0 Hz	TVHFS
Cotulla	FM	WY5LL	146.79000	-	192.8 Hz	
Crockett	DMR/MARC	KG5OKB	443.90000	+	CC9	
	FM	W5DLC	145.31000	-	103.5 Hz	TVHFS
	FM	WA5EC	146.70000	-	123.0 Hz	TVHFS
	FM	WA5FCL	443.60000	+	100.0 Hz	TVHFS
	FM	WB5UGT	444.22500	+	123.0 Hz	TVHFS
Crosby	FM	KB5IJF	442.05000	+	103.5 Hz	TVHFS
	FM	AI5TX	443.70000	+		TVHFS
	FM	W5TWO	444.77500	+	103.5 Hz	TVHFS
Crosbyton	FM	WB5BRY	147.16000	+	179.9 Hz	TVHFS
	FM	KC5MVZ	442.27500	+	107.2 Hz	TVHFS
Cross Plains	FM	KA9DNO	444.65000	+	77.0 Hz	TVHFS
Cypress	DMR/MARC	N5LUY	444.47500	+	CC7	
	FM	N5LUY	442.65000	+	156.7 Hz	TVHFS
Daingerfield	FM	NG5F	145.23000	-	151.4 Hz	TVHFS
Dale, Flag Hill	FM	KE5AMB	145.43000	-	114.8 Hz	
	FM	KE5AMB	443.00000	+	114.8 Hz	TVHFS
Dallas	DMR/BM	N4MSE	442.02500	+	CC1	TVHFS
	DMR/BM	WA5YST	444.35000	+	CC1	
	DMR/MARC	W5EBQ	440.47500	+	CC1	
	DMR/MARC	W5EBQ	440.63750	+	CC1	
	DMR/MARC	W5EBQ	441.63750	+	CC1	
	DMR/MARC	N4MSE	927.05000	902.05000	CC1	
	DSTAR	W5FC	145.13000	-		TVHFS
	DSTAR	K5TIT	147.36000	+		TVHFS
	DSTAR	W5FC	440.57500	+		TVHFS
	DSTAR	K5TIT	442.00000	+		TVHFS
	DSTAR	K5TIT	1293.00000	1273.00000		TVHFS
	DSTAR	W5FC	1295.00000	1275.00000		TVHFS
	FM	W5EBQ	52.59000	51.59000	110.9 Hz	TVHFS
	FM	KA5CTN	145.19000	-	110.9 Hz	TVHFS
	FM	K5AHT	146.64000	-	118.8 Hz	TVHFS
	FM	W5FC	146.88000	-	110.9 Hz	TVHFS
	FM	W5DCR	146.96000	-	110.9 Hz	TVHFS

TEXAS 343

Location	Mode	Call sign	Output	Input	Access	Coordinator
Dallas	FM	N4MSE	224.70000	-	127.3 Hz	TVHFS
	FM	W5FC	224.88000	-	110.9 Hz	TVHFS
	FM	KG5LL	441.55000	+	110.9 Hz	TVHFS
	FM	W5DCR	442.07500	+	110.9 Hz	TVHFS
	FM	N5ZW	442.27500	+		TVHFS
	FM	W5FC	442.42500	+	110.9 Hz	TVHFS
	FM	WO5E	442.47500	+		TVHFS
	FM	N5ARC	442.50000	+	110.9 Hz	TVHFS
	FM	N5DA	443.00000	+	110.9 Hz	TVHFS
	FM	K5TIT	443.47500	+	156.7 Hz	TVHFS
	FM	AI5TX	443.95000	+		TVHFS
	FM	K5MET	444.07500	+	110.9 Hz	TVHFS
	FM	K5TAO	444.15000	+	100.0 Hz	
	FM	N4MSE	444.65000	+	186.2 Hz	
	FM	N4MSE	927.06250	902.06250	DCS 432	
Dallas, Green Building						
	FM	W5EBQ	146.70000	-	110.9 Hz	TVHFS
Davilla, Davilla VFD						
	FM	KG5DUO	147.00000	+	123.0 Hz	
Davy	FM	WD5IEH	147.32000	+	141.3 Hz	TVHFS
Decatur	FM	W5KFC	146.78000	-	131.8 Hz	TVHFS
	FM	WQ5A	442.60000	+	131.8 Hz	TVHFS
	FM	N5ERS	443.22500	+	110.9 Hz	TVHFS
	FM	KE5WBO	444.40000	+	156.7 Hz	TVHFS
Del Rio	FM	KD5HAM	146.82000	-	127.3 Hz	TVHFS
	FM	K5CXR	147.30000	+		TVHFS
	FM	WB5TZJ	443.50000	+		TVHFS
Del Valle	FM	KB5HTB	145.17000	-	88.5 Hz	TVHFS
Denison	FM	W5DWH	145.33000	-	100.0 Hz	TVHFS
	FM	KD5HQF	441.35000	+	100.0 Hz	
Denton	DMR/MARC	W5NGU	146.92000	-	CC1	TVHFS
	DMR/MARC	N5LS	440.66250	+	CC1	TVHFS
	DMR/MARC	N5LS	927.66250	902.66250	CC1	TVHFS
	FM	W5NGU	441.32500	+	88.5 Hz	TVHFS
	FM	N5LS	927.41250	902.41250	DCS 432	
Denton, EOC	DSTAR	W5NGU	147.45000	146.45000		TVHFS
	DSTAR	W5NGU	442.92500	-		TVHFS
	FM	W5FKN	145.17000	-	110.9 Hz	TVHFS
	FM	W5NGU	927.61250	902.61250		TVHFS
Denton, Stark Hall	FM	AF5RS	224.92000	-	110.9 Hz	
Devers	FM	N5FJX	146.98000	-	103.5 Hz	TVHFS
	FM	KA5QDG	224.92000	-	123.0 Hz	TVHFS
	FM	N6LXX	444.85000	+	151.4 Hz	TVHFS
Devine	FM	WB5LJZ	146.61000	-		TVHFS
	FM	WB5LJZ	146.88000	-	141.3 Hz	TVHFS
Dickens	FM	WX5LBB	444.32500	+	162.2 Hz	TVHFS
Dimmitt	FM	KW5KW	444.70000	+	110.9 Hz	
Donna	FM	KC5YFP	146.74000	-	114.8 Hz	TVHFS
Doss	FM	W5RP	147.16000	+	162.2 Hz	TVHFS
	FM	W5RP	442.30000	+	162.2 Hz	TVHFS
Double Mountain	FM	AI5TX	443.70000	+		TVHFS
Doucette	FM	WD5TYL	147.22000	+	100.0 Hz	TVHFS
Dripping Springs	DMR/BM	N5OAK	441.30000	+	CC1	TVHFS
	FM	W5MIX	146.74000	-	67.0 Hz	TVHFS
Dumas	DMR/MARC	KD5ROK	443.01250	+	CC14	
	FM	N5LTZ	444.35000	+	88.5 Hz	TVHFS
Duncanville	FM	KA5KEH	441.35000	+	114.8 Hz	TVHFS
Eastland	FM	KB5WB	442.72500	+	114.8 Hz	TVHFS
Eastland, Lake Leon						
	FM	KB5WB	444.80000	+	156.7 Hz	TVHFS
Eddy	FM	W5BEC	147.14000	+	123.0 Hz	TVHFS
Eden	FM	AI5TX	443.97500	+		TVHFS

344   TEXAS

Location	Mode	Call sign	Output	Input	Access	Coordinator
Edgewood	FM	W5EEY	444.20000	+	136.5 Hz	TVHFS
Edom	FM	W5ETX	146.62000	-	136.5 Hz	TVHFS
El Paso	DMR/MARC	N5RWZ	442.22500	5442.22500	CC1	
	DMR/MARC	N5RWZ	442.32500	5442.32500	CC1	
	DMR/MARC	AE5RJ	443.60000	+	CC1	
	DMR/MARC	W5WIN	444.50000	+	CC1	
	DSTAR	KG5ZPX	440.65000	+		
	DSTAR	N6TOC	444.60000	+		
	FM	K5WPH	53.55000	52.55000		TVHFS
	FM	K5WPH	145.33000	-	67.0 Hz	TVHFS
	FM	WX5ELP	145.41000	-	88.5 Hz	TVHFS
	FM	K5ELP	146.70000	-	114.8 Hz	TVHFS
	FM	N6TOC	146.94000	-	123.0 Hz	
	FM	KJ5EO	147.06000	+		TVHFS
	FM	KJ5EO	147.10000	+		TVHFS
	FM	K5ELP	147.20000	+	67.0 Hz	TVHFS
	FM	K5WPH	147.24000	+	162.2 Hz	TVHFS
	FM	KD6CUB	147.28000	+	67.0 Hz	TVHFS
	FM	K5KKO	147.32000	+	162.2 Hz	TVHFS
	FM	KE5OIB	147.36000	+		TVHFS
	FM	KD6CUB	224.82000	-	100.0 Hz	TVHFS
	FM	N5ZFF	441.70000	+	100.0 Hz	TVHFS
	FM	WB5LJO	442.10000	+	123.0 Hz	TVHFS
	FM	N5FAZ	442.12500	+	103.5 Hz	TVHFS
	FM	K5WPH	442.25000	+	100.0 Hz	TVHFS
	FM	WB5LJO	442.55000	+	100.0 Hz	TVHFS
	FM	KJ5EO	442.60000	+		TVHFS
	FM	K5ELP	442.82500	+	100.0 Hz	TVHFS
	FM	KJ5EO	443.00000	+		TVHFS
	FM	N6TOC	443.37500	+	100.0 Hz	TVHFS
	FM	K5WPH	443.40000	+	100.0 Hz	TVHFS
	FM	W5DPD	443.65000	+		TVHFS
	FM	WB5LJO	443.70000	+		TVHFS
	FM	KA5CDJ	443.92500	+		TVHFS
	FM	K5ELP	444.20000	+	100.0 Hz	TVHFS
El Paso, Montana Vista						
	FM	W5HFN	444.32500	+	103.5 Hz	TVHFS
El Paso, North Mount Franklin						
	FM	K5ELP	146.88000	-	88.5 Hz	TVHFS
El Paso, North Park						
	FM	NM5ML	147.14000	+	67.0 Hz	TVHFS
Eldorado	DMR/MARC	WM5L	443.72500	+	CC14	
	DSTAR	N5QHO	443.82500	+		
Elgin	FM	KC5WXT	442.80000	+	114.8 Hz	TVHFS
Elmendorf	FM	W5ROS	146.86000	-	123.0 Hz	TVHFS
Elmo	FM	W5EEY	421.25000	439.25000		TVHFS
Emory	FM	W5ENT	146.92000	-	88.5 Hz	TVHFS
	FM	W5ENT	443.62500	+	151.4 Hz	TVHFS
Ennis	FM	KB0BWG	444.82500	+	131.8 Hz	TVHFS
Euless	FM	W5EUL	442.90000	+	110.9 Hz	TVHFS
Eustace	FM	W5IB	444.47500	+	136.5 Hz	TVHFS
Everman	FM	AB5XD	224.86000	-	110.9 Hz	TVHFS
	FM	AB5XD	441.52500	+	110.9 Hz	TVHFS
Fabens	FM	W5PDC	147.98000	-	162.2 Hz	TVHFS
	FM	W5PDC	442.45000	+	203.5 Hz	TVHFS
Fairfield	FM	WB5YJL	145.11000	-	146.2 Hz	TVHFS
Fairview	DMR/MARC	N5ITU	441.33750	+	CC1	TVHFS
Farmersville	DMR/BM	N5SN	441.60000	+	CC1	
	DMR/MARC	WA5DKW	440.01250	+	CC1	TVHFS
	DSTAR	WA5DKW	440.61250	+		
	FM	N5MRG	147.08000	+	110.9 Hz	
	FM	N5MRG	224.64000	-	100.0 Hz	

TEXAS 345

Location	Mode	Call sign	Output	Input	Access	Coordinator
Farmersville	FM	N5MRG	443.27500	+	100.0 Hz	
	FM	W5GDC	927.07500	902.07500	DCS 432	TVHFS
Florence	FM	K5AB	442.90000	+	100.0 Hz	TVHFS
Floresville	FM	WB5LOP	441.85000	+	179.9 Hz	TVHFS
Flower Mound	DMR/MARC	KM4NNO	448.05000	-	CC4	
Floydada	FM	WA5OEO	444.77500	+	162.2 Hz	TVHFS
Fort Davis, McDonald Observato						
	FM	K5FD	146.62000	-	146.2 Hz	TVHFS
Fort Davis, Mount McElroy						
	FM	KD5CCY	442.40000	+	146.2 Hz	TVHFS
	FM	N5HYD	443.95000	+		TVHFS
Fort Davis, Prude Ranch						
	FM	W5TSP	442.80000	+	110.9 Hz	
Fort Stockton	DSTAR	KG5OXR	147.60000	-		
	FM	KB5GLA	145.37000	-	88.5 Hz	TVHFS
	FM	N5SOR	146.68000	-	88.5 Hz	TVHFS
	FM	AD5BB	146.92000	-	146.2 Hz	TVHFS
	FM	KF5AEJ	147.24000	+	88.5 Hz	TVHFS
	FM	N5SOR	444.80000	+	162.2 Hz	TVHFS
Fort Worth	DMR/MARC	K5FTW	440.53750	+	CC1	
	DMR/MARC	N5GMJ	441.97500	+	CC1	TVHFS
	DMR/MARC	KG5EEL	442.00000	+	CC1	
	DMR/MARC	KB5ASY	444.03750	+	CC1	
	DSTAR	KB5DRP	440.55000	+		TVHFS
	FM	W5DFW	29.66000	-	192.8 Hz	
	FM	K5FTW	145.11000	-	110.9 Hz	TVHFS
	FM	K5FTW	146.68000	-	110.9 Hz	TVHFS
	FM	K5FTW	146.76000	-		TVHFS
	FM	W5SH	146.84000	-	110.9 Hz	TVHFS
	FM	K5FTW	146.94000	-	110.9 Hz	TVHFS
	FM	K5COW	147.28000	+	110.9 Hz	TVHFS
	FM	K5MOT	147.32000	+	110.9 Hz	TVHFS
	FM	KF5LOG	147.54000		110.9 Hz	
	FM	W0BOD	224.68000	-	103.5 Hz	TVHFS
	FM	N5UN	224.78000	-	110.9 Hz	TVHFS
	FM	K5FTW	224.94000	-	110.9 Hz	TVHFS
	FM	N5UN	423.97500	+	156.7 Hz	TVHFS
	FM	W7YC	441.30000	+		TVHFS
	FM	W5FA	441.37500	+		
	FM	W5FWS	441.60000	+	162.2 Hz	
	FM	KA5GFH	442.12500	+	156.7 Hz	TVHFS
	FM	K5HIT	442.22500	447.32500	110.9 Hz	TVHFS
	FM	N5UA	442.97500	+	110.9 Hz	TVHFS
	FM	N5PMB	443.15000	+	110.9 Hz	TVHFS
	FM	N4MSE	443.45000	+	127.3 Hz	TVHFS
	FM	K5SXK	443.92500	+		TVHFS
	FM	N5UN	443.97500	+	131.8 Hz	TVHFS
	FM	K5FTW	444.10000	+		TVHFS
	FM	K5MOT	444.30000	+	110.9 Hz	TVHFS
	FM	W5FA	444.90000	+	110.9 Hz	TVHFS
Fort Worth, North Benbrook						
	FM	K5COW	224.42000	-	110.9 Hz	TVHFS
	FM	K5COW	442.20000	+	110.9 Hz	TVHFS
Fort Worth, North Fort Worth						
	FM	WX5ATX	444.77500	+	123.0 Hz	
Franklin	FM	W5KVN	146.96000	-	146.2 Hz	TVHFS
Fredericksburg	FM	AI5TX	443.70000	+		TVHFS
Fredericksburg, Hill Country M						
	FM	W5FBG	146.76000	-	162.2 Hz	TVHFS
Fredericksburg, N.E Gillespie						
	FM	W5FBG	145.47000	-	162.2 Hz	TVHFS
Freeport	DMR/MARC	KA5VZM	444.90000	+	CC1	TVHFS

346   TEXAS

Location	Mode	Call sign	Output	Input	Access	Coordinator
Freeport	FM	KA5VZM	147.38000	+	141.3 Hz	TVHFS
Freestone	FM	AK5G	441.82500	+	123.0 Hz	TVHFS
Fresno	DMR/MARC	K3JMC	440.68750	+	CC3	TVHFS
Friendswood	FM	KD5GR	147.12000	+		TVHFS
	FM	KD5GR	444.87500	+		TVHFS
Gail	FM	KK5MV	443.75000	+	162.2 Hz	TVHFS
Gainesville	FM	K5AGG	145.29000	-	100.0 Hz	TVHFS
	FM	WB5FHI	147.34000	+	100.0 Hz	TVHFS
	FM	WB5FHI	442.77500	+	100.0 Hz	TVHFS
	FM	K5AGG	443.12500	+	100.0 Hz	TVHFS
Galveston	DMR/BM	KC5FOG	443.27500	+	CC1	TVHFS
	FM	WB5BMB	146.68000	-	103.5 Hz	TVHFS
	FM	WB5BMB	147.04000	+		TVHFS
	FM	AI5TX	443.95000	+		TVHFS
Galveston, Moody Gardens						
	FM	KA5QDG	147.30000	+	123.0 Hz	TVHFS
	FM	KA5QDG	442.15000	+	123.0 Hz	TVHFS
Garden City	FM	KD5CCY	442.90000	+	91.5 Hz	TVHFS
Gardendale	DMR/MARC	KD4LXC	444.83750	+	CC14	
	FM	WD5MOT	442.15000	+		TVHFS
	FM	N5LTZ	444.40000	+	88.5 Hz	TVHFS
	FM	WR5FM	444.52500	+	146.2 Hz	TVHFS
	FM	WR5FM	927.06250	902.06250	203.5 Hz	TVHFS
Garland	DMR/BM	AB5U	441.95000	+	CC1	TVHFS
	FM	K5QHD	146.66000	-	110.9 Hz	TVHFS
	FM	K5QBM	147.24000	+	110.9 Hz	TVHFS
	FM	K5QHD	442.70000	+	110.9 Hz	TVHFS
Gatesville	FM	W5AMK	146.96000	-	123.0 Hz	TVHFS
Geneva	FM	K5TBR	146.74000	-	118.8 Hz	TVHFS
George West	FM	KD5FVZ	443.67500	+	156.7 Hz	TVHFS
Georgetown	DMR/MARC	NA6M	444.52500	+	CC1	TVHFS
	FM	N5TT	146.64000	-	162.2 Hz	TVHFS
	FM	NA6M	147.08000	+	100.0 Hz	TVHFS
	FM	N5KF	441.57500	+	100.0 Hz	TVHFS
	FM	K5TRA	927.06250	902.06250	203.5 Hz	TVHFS
Giddings	FM	NE5DX	147.22000	+	114.8 Hz	TVHFS
	FM	KE5DX	442.57500	+	114.8 Hz	TVHFS
Gilmer	FM	W5BWC	146.90000	-	107.2 Hz	TVHFS
Goldthwaite	FM	N5QBU	146.79000	-	94.8 Hz	
	FM	K5AB	147.10000	+	100.0 Hz	TVHFS
	FM	K5AB	442.60000	+	100.0 Hz	
Goliad	FM	WB5MCT	146.74000	-	103.5 Hz	TVHFS
Gonzales	FM	KB5RSV	147.26000	+	103.5 Hz	TVHFS
	FM	WD5IEH	443.12500	+	141.3 Hz	TVHFS
Graham	FM	N5SMX	147.00000	+	110.9 Hz	TVHFS
	FM	K7KAB	147.34000	+		TVHFS
Granbury	DMR/MARC	NA5AA	443.90000	+	CC1	TVHFS
	DSTAR	N1DRP	440.65000	+		
	DSTAR	W5HCT	441.35000	+		
	FM	KE5WEA	145.43000	-	162.2 Hz	
	FM	KE5WEA	146.74000	-	162.2 Hz	TVHFS
	FM	WD5GIC	147.08000	+	88.5 Hz	TVHFS
	FM	KE5WEA	147.24000	+	162.2 Hz	TVHFS
	FM	WD5GIC	224.34000	-	88.5 Hz	TVHFS
	FM	WD5GIC	442.02500	+	88.5 Hz	TVHFS
	FM	AI5TX	443.62500	+		TVHFS
Grape Creek	FM	K7PTZ	145.60000			
Grapevine	DMR/MARC	N5EOC	440.50000	+	CC1	TVHFS
	FM	N5EOC	145.40000	-		TVHFS
	FM	N5EOC	443.87500	+	110.9 Hz	TVHFS
	FM	N5ERS	444.85000	+	110.9 Hz	TVHFS
Greenville	DMR/BM	W5NNI	446.62500	+	CC1	

TEXAS 347

Location	Mode	Call sign	Output	Input	Access	Coordinator
Greenville	DMR/MARC	K5VOM	441.97500	+	CC1	
	FM	K5VOM	224.90000	-	100.0 Hz	
	FM	K5VOM	441.80000	+	100.0 Hz	
	FM	N5SN	443.90000	+	71.9 Hz	TVHFS
Greenville, Hunt Regional Medi						
	FM	K5GVL	146.78000	-	114.8 Hz	TVHFS
Greenville, Majors Field Airpo						
	DMR/BM	W5NNI	444.62500	+	CC1	
	FM	W5NNI	147.16000	+	100.0 Hz	TVHFS
Gun Barrel City	FM	K5CCL	146.90000	-	136.5 Hz	TVHFS
	FM	K5CCL	444.05000	+		
Gunter	DMR/BM	N5GI	444.37500	+	CC1	
Halletsville	FM	KD5RCH	147.08000	+	173.8 Hz	TVHFS
	FM	KC5RXW	444.75000	+	127.3 Hz	TVHFS
Hamilton	FM	K5AB	146.92000	-	100.0 Hz	TVHFS
	FM	AB5BX	147.20000	+	88.5 Hz	TVHFS
Harlingen	DMR/BM	N5SLI	443.87500	+	CC1	
	FM	K5VCG	145.39000	-	114.8 Hz	TVHFS
	FM	W5RGV	146.80000	-	114.8 Hz	TVHFS
	FM	W5RGV	147.10000	+	114.8 Hz	TVHFS
	FM	W5RGV	147.14000	+	114.8 Hz	TVHFS
	FM	W5STX	147.20000	+	114.8 Hz	
	FM	AK5Z	443.60000	+	114.8 Hz	TVHFS
	FM	K5RAV	444.50000	+	114.8 Hz	
Haslet	FM	W5BYT	145.25000	-	100.0 Hz	TVHFS
Hawkins	FM	W5ETX	147.24000	+	136.5 Hz	TVHFS
Heath	FM	KK5PP	441.37500	+	141.3 Hz	TVHFS
Hempstead	FM	K5FLM	441.35000	+	110.9 Hz	
Henderson	FM	W5ETX	146.92000	-	136.5 Hz	
	FM	NU5G	442.30000	+	DCS 047	TVHFS
Henderson, REA 410' Tower						
	FM	N5RCA	146.78000	-	131.8 Hz	TVHFS
Henderson, Rusk County Court H						
	FM	N5RCA	145.25000	-		
Henderson, UT Health-Henderson						
	FM	W5ETX	147.04000	+	136.5 Hz	TVHFS
Henrietta	FM	KF5DFD	146.68000	-	192.8 Hz	TVHFS
	FM	KA5WLR	146.80000	-	192.8 Hz	TVHFS
	FM	KF5DFD	146.86000	-	192.8 Hz	TVHFS
	FM	KF5DFD	444.72500	+	192.8 Hz	TVHFS
	FM	KA5WLR	444.85000	+	192.8 Hz	TVHFS
Hideaway, Golf Course						
	FM	W5ETX	145.33000	-	136.5 Hz	
Hill City	FM	KA5PQK	442.45000	+		
Hillsboro	FM	WB5YFX	443.27500	+		TVHFS
Hockley	DMR/MARC	KC5DAQ	446.00000	-	CC1	
Hondo	FM	KD5DX	145.29000	-	162.2 Hz	TVHFS
	FM	KD5DX	443.35000	+	141.3 Hz	TVHFS
Horseshoe Bay	DMR/BM	N5JFP	145.43000	-	CC1	TVHFS
	DMR/BM	N5JFP	440.60000	+	CC1	TVHFS
Houston	DMR/BM	KA5PLE	441.17500	+	CC1	
	DMR/BM	KD5HKQ	443.37500	+	CC1	TVHFS
	DMR/BM	K5WH	444.45000	+	CC3	TVHFS
	DMR/MARC	KD5DFB	441.77500	+	CC7	TVHFS
	DMR/MARC	KB5TFE	443.75000	+	CC1	TVHFS
	DMR/MARC	KB5PBM	444.92500	+	CC1	
	DSTAR	KG5FAE	440.00000	+		
	FM	WB5ITT	29.65000	-	100.0 Hz	
	FM	KA5QDG	145.17000	-	123.0 Hz	TVHFS
	FM	KD5OQS	145.23000	-	107.2 Hz	TVHFS
	FM	W5INP	145.34000	-	88.5 Hz	
	FM	KD5HKQ	145.45000	-	103.5 Hz	TVHFS

348 TEXAS

Location	Mode	Call sign	Output	Input	Access	Coordinator
Houston	FM	KA5AKG	146.66000	-	141.3 Hz	TVHFS
	FM	WA5TWT	146.70000	-	103.5 Hz	TVHFS
	FM	K5WH	146.76000	-	103.5 Hz	TVHFS
	FM	K5GZR	146.82000	-	103.5 Hz	TVHFS
	FM	WR5AAA	146.88000	-	146.2 Hz	
	FM	WB5UGT	146.92000	-	103.5 Hz	TVHFS
	FM	W5JUC	146.96000	-	103.5 Hz	TVHFS
	FM	WD5X	147.00000	+	103.5 Hz	TVHFS
	FM	W5ATP	147.08000	+	103.5 Hz	TVHFS
	FM	WA5QXE	147.32000	+	100.0 Hz	TVHFS
	FM	K5DX	147.36000	+	100.0 Hz	TVHFS
	FM	KG5FYV	147.53000			
	FM	WD5X	224.10000	-	103.5 Hz	TVHFS
	FM	WB5ITT	224.50000	-		
	FM	WD5X	224.80000	-	103.5 Hz	TVHFS
	FM	KB5ELT	441.30000	+	100.0 Hz	TVHFS
	FM	KD5OQS	442.07500	+	88.5 Hz	TVHFS
	FM	WA5F	442.50000	+	123.0 Hz	TVHFS
	FM	KC5AWF	443.07500	+	88.5 Hz	TVHFS
	FM	AD5OU	443.52500	+	136.5 Hz	TVHFS
	FM	N5TZ	443.65000	+		TVHFS
	FM	KB5TFE	443.72500	+	110.9 Hz	TVHFS
	FM	WB5UGT	443.82500	+	103.5 Hz	TVHFS
	FM	WD5X	444.22500	+	103.5 Hz	TVHFS
	FM	WD5X	444.25000	+		TVHFS
	FM	W5TMR	444.25000	+		TVHFS
	FM	W5NC	444.37500	+	103.5 Hz	TVHFS
	FUSION	WB5TUF	444.00000	+	103.5 Hz	TVHFS
Houston, Astrodome						
	FM	KA5QDG	147.06000	+	123.0 Hz	TVHFS
Houston, BP Building						
	FM	W5BSA	145.19000	-	123.0 Hz	TVHFS
Houston, Ellington Field						
	FM	KA5QDG	444.80000	+	141.3 Hz	
Houston, Galleria Area						
	FM	K5ILS	224.96000	-	88.5 Hz	TVHFS
Houston, Houston NW Medical Ce						
	FM	KD0RW	147.30000	+	151.4 Hz	TVHFS
Houston, Houston Transtar						
	FM	N5TRS	145.37000	-	123.0 Hz	TVHFS
Houston, JSC Building 1						
	DMR/BM	W5RRR	146.64000	-	CC1	TVHFS
Houston, Missouri City Antenna						
	FM	KG5EEO	146.94000	-	167.9 Hz	TVHFS
Houston, San Felipe Building						
	FM	W5VOM	1292.10000	1272.10000		TVHFS
Houston, The Galleria						
	FM	N5ZUA	444.30000	+	123.0 Hz	TVHFS
Houston, TMC	DSTAR	W5HDR	440.60000	+		TVHFS
	DSTAR	W5HDR	1293.20000	1273.20000		TVHFS
Houston, Transtar	DSTAR	W5HDR	147.10000	+		TVHFS
Howe	DMR/MARC	KC5HWB	440.20000	+	CC9	
	FM	KD5HQF	224.16000	-	100.0 Hz	
	FM	KD5HQF	442.45000	+	100.0 Hz	
Huntsville	DMR/MARC	KB5ZEQ	440.67500	+	CC1	
	FM	WA5AM	146.64000	-	131.8 Hz	TVHFS
	FM	W5HVL	146.86000	-	131.8 Hz	TVHFS
	FM	WD5CFJ	442.15000	+	103.5 Hz	TVHFS
	FM	W5SAM	442.85000	+	127.3 Hz	TVHFS
	FM	AI5TX	443.97500	+		TVHFS
Hurst	DMR/MARC	W5DMR	443.26250	+	CC1	TVHFS
	FM	W5HRC	147.10000	+	110.9 Hz	TVHFS

TEXAS 349

Location	Mode	Call sign	Output	Input	Access	Coordinator
Hurst, Hurst North Water Tower						
	FM	W5HRC	442.85000	+	110.9 Hz	TVHFS
Idalou	FM	KC5MVZ	223.90000	-	123.0 Hz	TVHFS
	FM	N5TYI	443.00000	+	67.0 Hz	TVHFS
	FM	KC5MVZ	443.27500	+	107.2 Hz	TVHFS
Independence	FM	N5MBM	145.25000	-	103.5 Hz	
Independence, Rocky Hill Fire						
	FM	N5MBM	441.85000	+	123.0 Hz	TVHFS
Industry	FM	KF5KXL	441.67500	+	103.5 Hz	TVHFS
Ingleside	DSTAR	W5ICC	444.90000	+	151.4 Hz	
	FM	NZ5J	147.22000	+	173.8 Hz	TVHFS
Ingram	FM	KD5HNM	147.08000	+	151.4 Hz	
	FM	AI5TX	443.92500	+		TVHFS
Iola	FM	K5ZY	145.29000	-		
Iowa Park	FM	N5JRF	145.49000	-	192.8 Hz	TVHFS
Iraan	FM	AI5TX	443.95000	+		TVHFS
Irving	FM	N2DFW	145.45000	-	110.9 Hz	TVHFS
	FM	WA5CKF	146.72000	-	110.9 Hz	TVHFS
	FM	WA5CKF	224.40000	-	110.9 Hz	TVHFS
	FM	WA5CKF	442.67500	+	110.9 Hz	TVHFS
	FM	AL7HH	444.80000	+	110.9 Hz	TVHFS
Italy, Downtown Water Tower						
	FM	WD5DDH	442.52500	+	88.5 Hz	TVHFS
Ivanhoe	FM	K5FRC	145.47000	-	100.0 Hz	
Jacksonville	FM	KR5Q	145.43000	-	136.5 Hz	TVHFS
	FM	K5JVL	146.80000	-	136.5 Hz	TVHFS
	FM	K5JVL	444.52500	+	136.5 Hz	TVHFS
Jamaica Beach	FM	KA5QDG	444.80000	+	103.5 Hz	TVHFS
Jasper	FM	W5JAS	147.00000	-	118.8 Hz	TVHFS
	FM	K5PFE	224.86000	-	118.8 Hz	TVHFS
	FM	W5JAS	442.20000	+	192.8 Hz	TVHFS
	FM	W5JAS	444.55000	+	118.8 Hz	TVHFS
Jewett	FM	KC5SWI	145.23000	-	146.2 Hz	TVHFS
Jollyville	FM	KA9LAY	145.21000	-	97.4 Hz	TVHFS
Jourdanton	FM	N5XO	147.24000	+	82.5 Hz	TVHFS
Junction	FM	AI5TX	443.65000	+		TVHFS
Kamay	FM	N5AAJ	444.67500	+	192.8 Hz	TVHFS
Karnes City	FM	WA5S	224.46000	-	192.8 Hz	
	FM	WA5S	442.77500	+	192.8 Hz	
Katy	FM	KF5KHM	145.34000	-		
	FM	KT5TX	147.20000	+	141.3 Hz	TVHFS
	FM	W5EMR	441.97500	+	123.0 Hz	TVHFS
	FM	WD8RZA	442.35000	+	131.8 Hz	TVHFS
	FM	N5TM	927.06250	902.06250	203.5 Hz	
Kaufman	FM	N5RSE	146.98000	-		
Keller	DMR/BM	W5DFW	424.97500	+	CC1	
	FM	KA5HND	147.20000	+	88.5 Hz	TVHFS
	FM	W5DFW	443.17500	+	100.0 Hz	TFCA
	FM	N5EOC	444.70000	+	110.9 Hz	TVHFS
Kenney	FM	WR5DC	146.64000	-	141.3 Hz	
Kent	FM	KE5PL	443.92500	+		TVHFS
Kerrville	FM	W3XO	146.79000	-	162.2 Hz	TVHFS
	FM	K5ZZT	443.62500	+		TVHFS
Kerrville, State Hospital						
	FM	K5KSH	441.31250	+	162.2 Hz	
Kilgore	DMR/BM	N5VGQ	147.30000	+	CC1	TVHFS
	DMR/MARC	N5YEY	443.95000	+	CC1	
	FM	WX5FL	145.45000	-	136.5 Hz	TVHFS
Killeen	FM	KK5AN	147.04000	+	173.8 Hz	TVHFS
Kingsville	FM	KD5QWJ	146.62000	-	107.2 Hz	TVHFS
	FM	W5KCA	444.22500	+	107.2 Hz	

350 TEXAS

Location	Mode	Call sign	Output	Input	Access	Coordinator
Kingsville, Dick Kleberg Park						
	FM	W5KCA	146.68000	-	107.2 Hz	TVHFS
Kingwood	FM	W5SI	145.43000	-		TVHFS
	FM	W5SI	147.28000	+	103.5 Hz	TVHFS
	FM	W5SI	444.82500	+		TVHFS
Klein	DMR/MARC	K5MAP	440.30000	+	CC3	
Kyle	FM	KE5LOT	147.05000	+	114.8 Hz	
La Feria	FM	W5RGV	146.70000	-	114.8 Hz	TVHFS
La Grange	FM	N5FRT	441.55000	+	114.8 Hz	TVHFS
	FM	AI5TX	443.70000	+		TVHFS
	FM	WB5UGT	444.72500	+	141.3 Hz	TVHFS
La Grange, Hostyn						
	FM	N5FRT	147.38000	+		TVHFS
	FM	N5FRT	441.30000	+	114.8 Hz	
	FM	K5TRA	927.16250	902.16250	151.4 Hz	
La Marque	FM	KA5QDG	146.90000	-	123.0 Hz	TVHFS
La Porte	DMR/MARC	N5LUY	442.85000	+	CC7	TVHFS
Lago Vista	FM	KC5WLF	224.84000	-	103.5 Hz	
	FM	K5TRA	927.12500	902.12500	103.5 Hz	TVHFS
Lake Fork	DMR/MARC	KC5HWB	440.70000	+	CC9	
Lakeway	FM	N5TXR	147.30000	+	131.8 Hz	TVHFS
	FM	WB5PCV	444.40000	+	103.5 Hz	TVHFS
Lamesa	FM	N5BNX	145.15000	-	100.0 Hz	TVHFS
	FM	N5BNX	146.86000	-	100.0 Hz	TVHFS
	FM	KD5CCY	442.70000	+	91.5 Hz	TVHFS
	FM	KE5PL	443.50000	+		TVHFS
	FM	K5WTC	444.75000	+	162.2 Hz	TVHFS
	FM	N5SVF	444.95000	+	100.0 Hz	TVHFS
Lampasas	FM	N5ZXJ	145.49000	-	123.0 Hz	
	FM	K5AB	147.06000	+	100.0 Hz	
	FM	KB5SXV	147.22000	+	88.5 Hz	TVHFS
	FM	KE5ZW	443.65000	+		TVHFS
	FM	WD5EMS	927.07500	902.07500	218.1 Hz	
Laredo	FM	W5EVH	145.15000	-	100.0 Hz	TVHFS
	FM	W5LRD	146.62000	-	100.0 Hz	
	FM	W5EVH	147.12000	+	100.0 Hz	TVHFS
	FM	W5EVH	444.00000	+	100.0 Hz	TVHFS
Laredo, Airport	FM	W5EVH	146.94000	-	100.0 Hz	TVHFS
League City	FM	WR5GC	145.41000	-	131.8 Hz	TVHFS
	FM	WR5GC	442.22500	447.32500	131.8 Hz	TVHFS
Leander, Cedar Park Hospital						
	FM	KE5RS	441.60000	+	100.0 Hz	TVHFS
Leonard	FM	KW5DX	145.41000	-	114.8 Hz	TVHFS
Levelland	FM	WB5BRY	146.78000	-	179.9 Hz	TVHFS
	FM	WB5EMR	146.88000	-	103.5 Hz	TVHFS
	FM	N5SOU	147.12000	+	162.2 Hz	TVHFS
	FM	KC5TAF	443.15000	+	136.5 Hz	TVHFS
	FM	WA5OEO	444.37500	+	162.2 Hz	TVHFS
Lewisville	DMR/MARC	N3JI	440.40000	+	CC1	TVHFS
Lindale	FM	W5NFL	145.60000		146.2 Hz	
Little Elm	DMR/BM	WB8GRS	441.38750	+	CC1	
	DMR/MARC	WB8GRS	441.37500	+	CC1	
	FM	WB8GRS	145.76000		123.0 Hz	
	FM	WB8GRS	224.98000	-	123.0 Hz	TVHFS
Littlefield	FM	WB5BRY	146.64000	-	179.9 Hz	TVHFS
	FM	WA5OEO	444.85000	+	162.2 Hz	TVHFS
Live Oak	DMR/BM	K5NKK	444.02500	+	CC1	TVHFS
	FM	KE5HBB	145.37000	-	114.8 Hz	
	FM	NH7TR	146.72000	-	114.8 Hz	
	FM	KE5HBB	441.90000	+	103.5 Hz	TVHFS
Livingston	FM	WB5HZM	147.04000	+	136.5 Hz	TVHFS
	FM	WB5HZM	147.16000	+	103.5 Hz	TVHFS

TEXAS 351

Location	Mode	Call sign	Output	Input	Access	Coordinator
Livingston	FM	WB5UGT	443.12500	+	103.5 Hz	TVHFS
Llano	FM	AI5TX	443.50000	+		TVHFS
Lockhart	FM	AD5JT	145.15000	-	136.5 Hz	TVHFS
Longfellow	FM	N5BPJ	147.34000	+	88.5 Hz	TVHFS
	FM	WX5II	443.97500	+		TVHFS
Longview	DMR/BM	K5TKR	441.30000	+	CC1	
	FM	K5TKR	147.16000	+	136.5 Hz	
	FM	K5JG	441.62500	+	100.0 Hz	TVHFS
	FM	K5JG	443.42500	+		TFCA
	FM	KD5UVB	444.72500	+	136.5 Hz	TVHFS
Longview, East Mountain						
	FM	K5LET	147.34000	+	136.5 Hz	TVHFS
Longview, East Mtn						
	FM	K5LET	146.64000	-	136.5 Hz	TVHFS
Longview, Lake Cherokee						
	FM	KB5MAR	145.30000	-	146.2 Hz	
Los Fresnos	DMR/BM	KG5EUU	444.75000	+	CC2	TVHFS
	DMR/BM	N5CEY	444.77500	+	CC1	
Louise	DMR/MARC	N5TZV	442.27500	+	CC3	
Lubbock	DMR/BM	WB5BRY	444.00000	+	CC1	TVHFS
	DMR/BM	N5ZTL	444.30000	+	CC1	
	DMR/BM	WA5TBB	444.80000	+	CC1	
	DMR/MARC	KA3IDN	444.68750	+	CC14	
	DSTAR	K5LIB	145.43000	-		TVHFS
	DSTAR	KB5KYJ	443.05000	+		
	FM	KC5CZX	146.56500		123.0 Hz	
	FM	K5LIB	146.84000	-	88.5 Hz	TVHFS
	FM	WB5BRY	146.94000	-	179.9 Hz	TVHFS
	FM	WB5BRY	147.00000	-	179.9 Hz	TVHFS
	FM	WA5OEO	147.20000	+	162.2 Hz	TVHFS
	FM	N5ZTL	147.30000	+	88.5 Hz	TVHFS
	FM	W5WAT	441.67500	+	97.4 Hz	TVHFS
	FM	W5WAT	441.97500	+	97.4 Hz	TVHFS
	FM	W5WAT	442.17500	+	97.4 Hz	TVHFS
	FM	N5UQF	442.35000	+	162.2 Hz	TVHFS
	FM	K5WAT	442.47500	+	97.4 Hz	TVHFS
	FM	K5LIB	443.07500	+	88.5 Hz	TVHFS
	FM	AI5TX	443.92500	+		TVHFS
	FM	WR5FM	444.02500	+	146.2 Hz	TVHFS
	FM	K5TTU	444.10000	+	146.2 Hz	TVHFS
	FM	KZ5JOE	444.45000	+	114.8 Hz	TVHFS
	FM	WB5BRY	444.50000	+	118.8 Hz	TVHFS
	FM	KC5OBX	444.62500	+	118.8 Hz	TVHFS
	FM	N5UQF	444.87500	+	162.2 Hz	TVHFS
	FM	WA5OEO	444.97500	+	162.2 Hz	TVHFS
	FM	KA3IDN	927.01250	902.01250	225.7 Hz	TVHFS
Lubbock, NTS Communications Bu						
	FM	WB5BRY	224.64000	-		
Lucas	DMR/BM	K5LFD	442.21250	+	CC2	
Lufkin	DMR/BM	KE5CJE	440.72500	+	CC1	
	DMR/MARC	WD5EFY	440.00000	+	CC9	
	FM	KD5TD	53.71000	52.71000	100.0 Hz	TVHFS
	FM	KD5TD	145.37000	-	100.0 Hz	TVHFS
	FM	W5IRP	146.94000	-	141.3 Hz	TVHFS
	FM	K5RKJ	147.26000	+	141.3 Hz	TVHFS
	FM	KB5LS	147.36000	+	107.2 Hz	TVHFS
	FM	WB5UGT	444.42500	+	203.5 Hz	TVHFS
	FM	KB5LS	444.57500	+	107.2 Hz	TVHFS
	FM	KD5TD	444.90000	+	107.2 Hz	TVHFS
Lufkin, Courthouse						
	FM	W5IRP	444.97500	+	107.2 Hz	TVHFS
Mabank	DMR/BM	K5CCL	146.84000	-	CC1	

352 TEXAS

Location	Mode	Call sign	Output	Input	Access	Coordinator
Madisonville	FM	W5ZYX	441.35000	+		
Magnolia	DMR/MARC	W5JSC	144.52500	+	CC1	
	FM	W5JON	442.95000	+	123.0 Hz	TVHFS
	FM	KB5FLX	443.02500	+	103.5 Hz	TVHFS
	FM	KD0RW	444.67500	+	192.8 Hz	TVHFS
Manor	FM	KI4MS	442.42500	+	100.0 Hz	TVHFS
Mansfield	FM	WA5JRS	448.77500	-	167.9 Hz	TVHFS
Marathon	FM	AI5TX	443.50000	+		TVHFS
Marble Falls	FM	N5KUQ	442.85000	+	103.5 Hz	TVHFS
Marble Falls, Hidden Falls Adv						
	FM	N5KUQ	145.39000	-	103.5 Hz	TVHFS
Marietta, Cusetta Mountain						
	FM	WX5FL	146.84000	-	100.0 Hz	
Marietta, Cussetta Mountain						
	FM	WX5FL	52.37000	51.37000	136.5 Hz	
Marietta, Cussetta Mtn.						
	FM	WX5FL	145.19000	-	151.4 Hz	TVHFS
Markham	FM	WA5SNL	444.70000	+	146.2 Hz	TVHFS
Marshall	FM	KB5MAR	146.86000	-	146.2 Hz	TVHFS
	FM	K5HR	223.94000	-		TVHFS
	FM	KB5MAR	444.15000	+	146.2 Hz	TVHFS
Maydelle	FM	KB5VQG	147.02000	+		TVHFS
Mayflower	FM	W5JAS	147.12000	+	203.5 Hz	TVHFS
McAllen	DMR/MARC	KC5MOL	443.82800	449.82800	CC1	
	DSTAR	ND5N	147.60000	-		
	FM	N5SIM	145.23000	-	114.8 Hz	TVHFS
	FM	W5RGV	146.76000	-	114.8 Hz	TVHFS
	FM	W5RGV	444.60000	+	114.8 Hz	TVHFS
Mccamey	FM	KK5MV	443.47500	+		TVHFS
Mccamey, King Mountain						
	FM	N5SOR	444.70000	+	162.2 Hz	TVHFS
McKinney	DMR/BM	N5GI	145.35000	-	CC1	TVHFS
	DMR/BM	N5GI	442.57500	+	CC1	TVHFS
	DMR/MARC	W9DXM	441.32500	+	CC1	
	FM	W5MRC	146.74000	-	110.9 Hz	TVHFS
	FM	KF5TU	442.35000	+	100.0 Hz	TVHFS
	FM	N4MSE	927.16250	902.16250	DCS 432	TVHFS
McKinney, Office Building						
	FM	KG5IAN	440.02500	+	110.9 Hz	
Melissa	FM	W5MRC	443.20000	+	100.0 Hz	TVHFS
Mercedes	FM	KR4ZAN	441.60000	+	114.8 Hz	TVHFS
Mesquite	DSTAR	NT5RN	145.15000	-		TVHFS
	DSTAR	NT5RN	443.02500	+		TVHFS
	FM	AK5DX	52.75000	51.75000	110.9 Hz	TVHFS
	FM	WJ5J	145.31000	-	110.9 Hz	TVHFS
	FM	AK5DX	147.04000	+	136.5 Hz	TVHFS
	FM	AK5DX	440.30000	+		TVHFS
	FM	AK5DX	442.62500	+	110.9 Hz	TVHFS
	FM	WJ5J	444.42500	+	156.7 Hz	TVHFS
Mexia	FM	W5NFL	145.39000	-	146.2 Hz	TVHFS
Miami	FM	KA5KQH	145.11000	-	88.5 Hz	TVHFS
	FM	N5LTZ	444.85000	+	88.5 Hz	TVHFS
Midland	DMR/MARC	KE5PL	147.22000	+	CC7	
	DMR/MARC	KE5PL	444.36250	+	CC14	
	DSTAR	W5EOC	146.70000	-		
	DSTAR	WT5ARC	441.55000	+		
	FM	W5QGG	146.76000	-	88.5 Hz	TVHFS
	FM	N5XXO	146.90000	-	88.5 Hz	TVHFS
	FM	K5MSO	147.22000	+	88.5 Hz	TVHFS
	FM	W5QGG	147.30000	+	88.5 Hz	TVHFS
	FM	N5MXE	442.02500	+	162.2 Hz	TVHFS
	FM	W5LNX	442.20000	+	162.2 Hz	TVHFS

TEXAS 353

Location	Mode	Call sign	Output	Input	Access	Coordinator
Midland	FM	K5PSA	442.97500	+	162.2 Hz	TVHFS
	FM	N5XXO	443.27500	+	162.2 Hz	TVHFS
	FM	W5WRL	443.30000	+	146.2 Hz	TVHFS
	FM	KD5CCY	443.40000	+		TVHFS
	FM	KE5PL	443.57500	+		TVHFS
	FM	KE5PL	443.65000	+		TVHFS
	FM	N5SOR	443.72500	+		TVHFS
	FM	KK5MV	443.80000	+	162.2 Hz	TVHFS
	FM	W5QGG	444.20000	+	162.2 Hz	TVHFS
	FM	KB5MBK	444.60000	+	146.2 Hz	TVHFS
	FM	W5UA	444.77500	+	88.5 Hz	TVHFS
	FM	KE5PL	927.05000	902.05000	DCS 411	TVHFS
Midlothian	DMR/MARC	WD5DDH	441.65000	+	CC9	TVHFS
	DMR/MARC	KC5HWB	441.65000	+	CC9	
Minden	FM	WB5WIA	145.25000	-	123.0 Hz	TVHFS
Mineral Wells	FM	W5ABF	146.64000	-	85.4 Hz	TVHFS
	FM	W5PPC	146.86000	-	156.7 Hz	
	FM	WB5TTS	442.70000	+	85.4 Hz	TVHFS
Mission	DMR/MARC	K5HYT	444.95000	+	CC1	
Missouri City	FM	KD5HKQ	145.25000	-	156.7 Hz	
Mobile Repeater	DMR/BM	N5GI	441.32500	+	CC1	
Mont Belvieu	FM	KK5XQ	441.80000	+	103.5 Hz	TVHFS
Montgomery	DMR/BM	WA5EOC	445.37500	-	CC7	
Moody	FM	W5ZDN	145.15000	-	123.0 Hz	TVHFS
	FM	AA5RT	442.30000	+	123.0 Hz	TVHFS
	FM	AI5TX	443.92500	+		TVHFS
Moulton	FM	KC5RXW	444.47500	+	127.3 Hz	TVHFS
Mound Creek	FM	WD5IEH	441.92500	+		TVHFS
Mount Pleasant, Purley Tower E						
	FM	W5KNO	444.95000	+	151.4 Hz	TVHFS
Mount Vernon	FM	WA5YVL	147.32000	+	151.4 Hz	TVHFS
Murphy	FM	AA5BS	441.70000	+		TVHFS
Mustang Ridge, HWY 130/21 Inte						
	FM	AC5PS	441.56250	+		TVHFS
Nacogdoches	FM	W5NAC	147.32000	+	141.3 Hz	TVHFS
	FM	KE5EXX	441.35000	+	141.3 Hz	TVHFS
	FM	KE5EXX	444.00000	+		TVHFS
	FM	W5NAC	444.05000	+	141.3 Hz	TVHFS
Nassau Bay	FM	NB5F	145.15000	-		TVHFS
	FM	NB5F	442.47500	+		TVHFS
Navasota	FM	W5JSC	146.74000	-	156.7 Hz	
	FM	KG5JRA	441.87500	+	110.9 Hz	TVHFS
Nederland	DMR/BM	W5SSV	440.72500	+	CC7	TVHFS
	DMR/MARC	KG5HFL	444.92500	+	CC3	
Nevada	FM	KA5HND	444.92500	+	110.9 Hz	TVHFS
New Boston	FM	KE5ZHF	147.20000	+	100.0 Hz	TVHFS
New Braunfels	FM	WB5LVI	147.00000	-	103.5 Hz	TVHFS
	FM	WB5LVI	147.22000	+	103.5 Hz	TVHFS
	FM	W5DK	443.50000	+	141.3 Hz	TVHFS
	FM	WB5LVI	443.85000	+	103.5 Hz	TVHFS
New Ulm, Frelsburg						
	FM	N5CNB	927.13750	902.13750	131.8 Hz	
New Waverly	FM	W5SAM	147.18000	+	136.5 Hz	TVHFS
	FM	NA5SA	442.27500	+		TVHFS
	FM	N5ZUA	442.72500	+	103.5 Hz	TVHFS
No Limits	DMR/MARC	KC5HWB	440.51250	+	CC9	
Nocona	FM	N5VAV	147.36000	+	123.0 Hz	TVHFS
Nolanville	FM	KX5DX	147.44000		100.0 Hz	
	FM	KX5DX	446.05000	+	100.0 Hz	
North Richland Hills						
	FM	AB5L	52.31000	51.31000		
	FM	K5NRH	441.75000	+	100.0 Hz	TVHFS

354 TEXAS

Location	Mode	Call sign	Output	Input	Access	Coordinator
North Richland Hills, Water To						
	FM	K5NRH	145.37000	-	110.9 Hz	TVHFS
Notrees	DMR/MARC	K5MSO	443.88750	+	CC14	
	DSTAR	W5EOC	442.35000	+		
	FM	N5XXO	147.02000	+	88.5 Hz	TVHFS
	FM	AI5TX	443.70000	+		TVHFS
	FM	N5XXO	444.67500	+	162.2 Hz	TVHFS
Oak Ridge North	FM	KW5O	441.32500	+		TVHFS
Octavia	DMR/MARC	N5LLH	443.07500	+	CC1	
Odem	FM	W5JYJ	443.50000	+		TVHFS
Odessa	DMR/BM	KG5YJT	440.72500	+	CC14	
	DMR/MARC	N5RGH	443.65000	+	CC1	
	DMR/MARC	KA3IDN	444.23750	+	CC14	
	FM	W5CDM	145.41000	-	88.5 Hz	TVHFS
	FM	KD5CCY	146.74000	-	91.5 Hz	TVHFS
	FM	KF5ZPM	441.70000	+	162.2 Hz	
	FM	N5MI	441.90000	+	173.8 Hz	TVHFS
	FM	KD5CCY	442.30000	+	91.5 Hz	TVHFS
	FM	WT5ARC	443.10000	+	162.2 Hz	TVHFS
	FM	KE5PL	443.62500	+		TVHFS
	FM	W5CDM	444.10000	+	162.2 Hz	TVHFS
	FM	KA3IDN	444.23500	+	71.9 Hz	TVHFS
	FM	WT5ARC	444.42500	+	162.2 Hz	TVHFS
	FM	K5PSA	444.97500	+	179.9 Hz	TVHFS
	FM	KA3IDN	927.07500	902.07500	82.5 Hz	TVHFS
Olmito	FM	KC5WBG	147.18000	+	114.8 Hz	TVHFS
Olney	FM	W5IPN	147.24000	+		
Orange	FM	AA5P	147.06000	+	103.5 Hz	TVHFS
	FM	W5ND	147.18000	+	103.5 Hz	TVHFS
Ovalo	FM	KD5YCY	444.97500	+	103.5 Hz	TVHFS
Ozona	FM	K5SPE	147.12000	+		
	FM	KE5PL	443.62500	+		TVHFS
Palestine	FM	KR5Q	145.49000	-	136.5 Hz	TVHFS
	FM	KR5Q	146.74000	-	136.5 Hz	TVHFS
	FM	W5DLC	147.08000	+	103.5 Hz	TVHFS
	FM	W5DLC	147.14000	+	103.5 Hz	TVHFS
	FM	KR5Q	442.37500	+	136.5 Hz	TVHFS
Palestine, NALCOM Tower						
	FM	K5PAL	444.60000	+	103.5 Hz	TVHFS
Palo Pinto	DMR/MARC	W5PPC	145.19000	-	CC1	
	DMR/MARC	W5PPC	442.87500	+	CC1	
Pampa	FM	W5TSV	146.90000	-		TVHFS
	FM	N5LTZ	444.40000	+	88.5 Hz	TVHFS
Pandale	FM	N5UFV	443.92500	+		TVHFS
Paris	DMR/MARC	KC5HWB	440.58750	+	CC9	
	DSTAR	WN5ROC	147.06000	444.26000		
	DSTAR	K5PTR	442.12500	+		
	FM	KI5DX	145.13000	-	100.0 Hz	
	FM	N5JEP	145.39000	-	114.8 Hz	TVHFS
	FM	WB5RDD	146.76000	-	203.5 Hz	TVHFS
	FM	KC5OOS	147.04000	+	100.0 Hz	TVHFS
	FM	KA5RLK	147.34000	+	110.9 Hz	TVHFS
	FM	KI5DX	444.47500	+	100.0 Hz	TVHFS
	FM	WB5RDD	444.50000	+		TVHFS
Pasadena	DMR/BM	N5KIE	440.68750	+	CC1	
	FM	KD5HKQ	145.25000	-	167.9 Hz	
	FM	W5PAS	145.29000	-	103.5 Hz	TVHFS
	FM	KD5QCZ	224.48000	-	156.7 Hz	
	FM	WB5ZMY	443.45000	+	114.8 Hz	TVHFS
	FM	KD5QCZ	443.77500	+		
	FM	W5PAS	444.27500	+	103.5 Hz	TVHFS

TEXAS 355

Location	Mode	Call sign	Output	Input	Access	Coordinator
Pasadena, El Jardin						
	FM	W5PAS	145.27000	-	123.0 Hz	
Payne Springs	FM	K5CCL	426.25000	1255.05000		TVHFS
Pearland	DMR/MARC	K3JMC	441.92500	+	CC8	
	DMR/MARC	K5PLD	443.05000	+	CC3	TVHFS
	FM	K5PLD	147.16000	+	167.9 Hz	TVHFS
	FM	K5PLD	147.22000	+	167.9 Hz	TVHFS
	FM	N5KJN	441.92500	+		TVHFS
	FM	N5KJN	443.40000	+	141.3 Hz	TVHFS
Peaster	FM	KB5WB	442.45000	+	156.7 Hz	TVHFS
Penwell	FM	N5SOR	443.67500	+		TVHFS
	FM	WR5FM	444.57500	+	146.2 Hz	TVHFS
	FM	WR5FM	927.11250	902.11250	DCS 432	TVHFS
Perryton	FM	K5IS	146.64000	-	88.5 Hz	TVHFS
Pflugerville	FM	KC5CFU	441.82500	+	114.8 Hz	TVHFS
Pine Springs	FM	N5SOR	444.05000	+		TVHFS
Pipe Creek	FM	WD5FWP	147.28000	+	156.7 Hz	TVHFS
Pipe Creek, Red Cross						
	FM	N5XO	147.12000	+	82.5 Hz	TVHFS
Plainview	FM	W5WV	146.72000	-	88.5 Hz	TVHFS
	FM	N5RNY	147.10000	+	88.5 Hz	TVHFS
	FM	AI5TX	443.95000	+		TVHFS
Plano	FM	K5CG	441.31250	+	110.9 Hz	
	FM	K5BSA	442.80000	+	110.9 Hz	TVHFS
	FM	AI5TX	443.65000	+		TVHFS
	FM	W5SUF	444.17500	+	110.9 Hz	TVHFS
Pleasanton	FM	KD5ZR	145.43000	-	162.2 Hz	TVHFS
	FM	W5ROS	147.34000	+	123.0 Hz	TVHFS
	FM	KE6LGE	441.50000	+	100.0 Hz	
	FM	NU5P	443.97500	+		TVHFS
Plum Grove	FM	WB5UGT	444.17500	+	103.5 Hz	TVHFS
Port Aransas	FM	KG5BZ	147.04000	+	107.2 Hz	TVHFS
	FM	KG5BZ	927.07500	902.07500	218.1 Hz	TVHFS
Port Aransas, The Dunes Condom						
	FM	KG5BZ	145.29000	-	110.9 Hz	TVHFS
Port Arthur	FM	KD5QDO	146.86000	-	103.5 Hz	TVHFS
Port Lavaca	DMR/BM	W5KTC	442.67500	+	CC1	TVHFS
	FM	W5KTC	147.02000	+	103.5 Hz	TVHFS
Port Neches	FM	KC5YSM	444.80000	+	118.8 Hz	
Port O'Connnor	DMR/MARC	W5VOM	443.72500	+	CC9	
Post	FM	WB5BRY	147.06000	+	179.9 Hz	TVHFS
Potosi	FM	KD5YCY	443.10000	+	88.5 Hz	TVHFS
Presidio, Cibolo Creek Ranch						
	FM	K5FD	147.12000	+	146.2 Hz	TVHFS
Purves	FM	KD5HNM	147.34000	+	107.2 Hz	
Quanah	FM	KY7D	146.64000	-		TVHFS
Quinlan, Lake Tawakoni						
	FM	K5VOM	224.96000	-		
Quitman	FM	WX5FL	147.10000	+	136.5 Hz	TVHFS
Ranger	FM	N5RMA	147.06000	+	131.8 Hz	TVHFS
	FM	AI5TX	443.67500	+		TVHFS
	FM	K6DBR	444.95000	+	88.5 Hz	TVHFS
Rankin	FM	KE5PL	443.92500	+		TVHFS
Raymondville	FM	W5RGV	146.90000	-	114.8 Hz	
	FM	W5RGV	444.90000	+	114.8 Hz	TVHFS
Refugio	FM	AD5TD	147.18000	+	136.5 Hz	TVHFS
	FM	AD5TD	443.87500	+	107.2 Hz	TVHFS
Rice	FM	W5TSM	145.47000	-	110.9 Hz	TVHFS
Richardson	DMR/MARC	K5RWK	440.37500	+	CC1	
	FM	N5CXX	441.87500	+	131.8 Hz	TVHFS
	FM	NT5NT	443.32500	+	110.9 Hz	TVHFS
	FM	WX5O	444.02500	+	110.9 Hz	TVHFS

356   TEXAS

Location	Mode	Call sign	Output	Input	Access	Coordinator
Richardson	FM	N5UA	444.67500	+	110.9 Hz	TVHFS
Richardson, Palisades Central						
	FM	K5RWK	147.12000	+	110.9 Hz	TVHFS
	FM	K5RWK	443.37500	+	110.9 Hz	TVHFS
	FM	K5RWK	444.72500	+	110.9 Hz	TVHFS
Richardson, University Of Texa						
	FM	K5UTD	223.82000	-	110.9 Hz	
Richardson, UTD Campus						
	FM	K5UTD	145.43000	-	110.9 Hz	TVHFS
Richland Hills	DMR/MARC	KC5HWB	440.08750	+	CC9	
Richmond	DMR/MARC	W5VOM	443.75000	+	CC9	
	FM	KD5HAL	145.49000	-	123.0 Hz	TVHFS
	FM	WB4KTH	438.50000	+	100.0 Hz	
	FM	KD5HAL	444.52500	+	123.0 Hz	TVHFS
Rio Medina	FM	W5TSE	443.00000	+		TVHFS
Robert Lee	FM	KC5EZZ	147.34000	+	88.5 Hz	TVHFS
Rockdale	FM	KG5DUO	146.76000	-	123.0 Hz	
	FM	AF5C	147.28000	+	162.2 Hz	TVHFS
Rockport	FM	KM5WW	147.26000	+	103.5 Hz	TVHFS
Rockwall	FM	KK5PP	441.52500	+	141.3 Hz	TVHFS
	FM	NF2W	441.73750	+	107.2 Hz	
	FM	K5GCW	443.55000	+	162.2 Hz	TVHFS
Rosanky	FM	N5FRT	145.40000	-	114.8 Hz	
Rose Hill	FM	K5SOH	53.27000	52.27000	123.0 Hz	TVHFS
	FM	K5IHK	146.72000	-	123.0 Hz	TVHFS
	FM	K5SOH	223.84000	-	123.0 Hz	TVHFS
	FM	K5IHK	443.10000	+	123.0 Hz	TVHFS
	FM	KC5PCB	927.20000	902.20000	123.0 Hz	TVHFS
Rosston	DMR/MARC	W5NGU	440.68750	+	CC1	
	DSTAR	KE5YAP	440.71250	+		
	DSTAR	KE5YAP	1293.20000	1273.20000		
	FM	WD5U	145.49000	-	85.4 Hz	TVHFS
	FM	KE5GDB	224.20000	-	110.9 Hz	TVHFS
	FM	N6LXX	443.73750	+	141.3 Hz	
	FM	W5FKN	927.05000	902.05000	110.9 Hz	
Round Rock	FM	WD5EMS	145.33000	-	162.2 Hz	TVHFS
	FM	KM5MQ	441.70000	+	110.9 Hz	TVHFS
	FM	AI5TX	443.67500	+		TVHFS
Round Rock, St Davids Surgical						
	FM	WD5EMS	29.64000	-	110.9 Hz	
	FM	WD5EMS	52.95000	51.95000	100.0 Hz	
	FM	WD5EMS	224.36000	-	100.0 Hz	TVHFS
	FM	WD5EMS	927.05000	902.05000	110.9 Hz	TVHFS
	FUSION	WD5EMS	444.87500	+	100.0 Hz	TVHFS
Rowlett	FM	AB5U	147.39000	+	85.4 Hz	
Rowlett, Kirby Road Water Tank						
	FM	AB5U	441.32500	+	162.2 Hz	TVHFS
Royse City	DMR/BM	K5VOM	441.77500	+	CC1	
Runaway Bay	FM	K5JEJ	444.82500	+	110.9 Hz	TVHFS
Rusk	FM	W5ETX	146.92000	-	136.5 Hz	TVHFS
	FM	KA5AEP	147.04000	+	110.9 Hz	TVHFS
Sachse	FM	N5LOC	145.25000	-	141.3 Hz	TVHFS
Saginaw	DMR/BM	W5BYT	444.32500	+	CC1	TVHFS
	DMR/MARC	N5GMJ	440.67500	+	CC1	
	FM	K5SAG	441.37500	+	100.0 Hz	TVHFS
Saint Hedwig	FM	WA5FSR	444.07500	+	123.0 Hz	TVHFS
San Angelo	DMR/MARC	KD5TKR	441.30000	+	CC1	
	DMR/MARC	KC5HWB	443.75000	+	CC1	
	DMR/MARC	KG5CNG	444.55000	+	CC1	
	DMR/MARC	KG5CNG	444.65000	+	CC1	
	DMR/MARC	KB5GLC	444.93750	+	CC1	
	FM	N5SVK	53.63000	52.63000	88.5 Hz	TVHFS

TEXAS 357

Location	Mode	Call sign	Output	Input	Access	Coordinator
San Angelo	FM	W5QX	145.27000	-	88.5 Hz	TVHFS
	FM	K5CMW	146.88000	-	88.5 Hz	TVHFS
	FM	KC5EZZ	146.94000	-	103.5 Hz	TVHFS
	FM	N5DE	147.06000	+		TVHFS
	FM	N5SVK	147.30000	+	88.5 Hz	TVHFS
	FM	KC5EZZ	441.75000	+	162.2 Hz	TVHFS
	FM	W5RP	442.25000	+	162.2 Hz	TVHFS
	FM	AI5TX	443.70000	+		TVHFS
	FM	KC5EZZ	444.22500	+	162.2 Hz	TVHFS
San Angelo, Corp Of Engineers						
	FM	N5RV	444.35000	+	162.2 Hz	TVHFS
San Antonio	DMR/BM	AA5RO	147.32000	+	CC1	TVHFS
	DMR/BM	N5AMD	441.76250	+	CC1	TVHFS
	DSTAR	K5VPW	145.35000	-		TVHFS
	DSTAR	WA5UNH	440.70000	+		TVHFS
	DSTAR	NV5TX	442.10000	+		TVHFS
	DSTAR	WD5STR	1293.30000	1273.30000		TVHFS
	FM	WA5KBQ	53.17000	52.17000	88.5 Hz	TVHFS
	FM	KB5BSU	53.21000	52.21000	141.3 Hz	TVHFS
	FM	W5DK	145.17000	-	141.3 Hz	TVHFS
	FM	K5NNN	145.21000	-	162.2 Hz	TVHFS
	FM	KD5GSS	145.47000	-	110.9 Hz	TVHFS
	FM	W5STA	146.66000	-	110.9 Hz	TVHFS
	FM	WS5DRC	146.78000	-	162.2 Hz	TVHFS
	FM	WA5FSR	146.82000	-	179.9 Hz	TVHFS
	FM	KF5FGL	146.84000	-	82.5 Hz	TVHFS
	FM	WB5LJZ	146.88000	-	141.3 Hz	TVHFS
	FM	WB5FWI	146.94000	-	179.9 Hz	TVHFS
	FM	WB5FNZ	146.96000	-	162.2 Hz	TVHFS
	FM	W5RRA	147.02000	+	88.5 Hz	TVHFS
	FM	KK5LA	147.04000	+	123.0 Hz	TVHFS
	FM	N5CSC	147.08000	+	162.2 Hz	TVHFS
	FM	K5EOC	147.18000	+	103.5 Hz	TVHFS
	FM	WD5FWP	147.28000	+	162.2 Hz	TVHFS
	FM	W5XW	147.30000	+	107.2 Hz	TVHFS
	FM	WA5UNH	147.36000	+	179.9 Hz	TVHFS
	FM	AA5RO	147.38000	+	162.2 Hz	TVHFS
	FM	N5WSU	147.53000		100.0 Hz	
	FM	K5VPW	147.56000			
	FM	K5VPW	147.59000		114.8 Hz	
	FM	WA5UNH	224.38000	-	179.9 Hz	
	FM	K5VPW	431.05000		136.5 Hz	
	FM	KD5GSS	442.12500	+	127.3 Hz	TVHFS
	FM	WS5DRC	443.20000	+		TVHFS
	FM	KD5GAT	443.40000	+	88.5 Hz	TVHFS
	FM	WB5FNZ	443.47500	+	162.2 Hz	TVHFS
	FM	AI5TX	443.67500	+		TVHFS
	FM	WX5II	443.70000	+		TVHFS
	FM	WX5II	443.72500	+		TVHFS
	FM	WX5II	443.95000	+		TVHFS
	FM	WB5FWI	444.10000	+	179.9 Hz	TVHFS
	FM	WA5UNH	444.12500	+	179.9 Hz	TVHFS
	FM	KG5FEC	444.66250	+		TVHFS
	FM	WA5KBQ	444.95000	+	103.5 Hz	TVHFS
San Antonio, Evans And Bulverd						
	FM	N5MRM	145.39000	-	82.5 Hz	
San Antonio, I 281 Stone Oak A						
	FM	N8IQT	442.75000	+		TVHFS
San Antonio, Near The Airport						
	FM	N5YBG	441.37500	+		TVHFS
San Antonio, Near The Medical						
	FM	WA5LNL	223.82000	-	141.3 Hz	TVHFS

358 TEXAS

Location	Mode	Call sign	Output	Input	Access	Coordinator
San Antonio, Red Cross						
	FM	N5XO	443.02500	+	82.5 Hz	
San Antonio, Red Cross Center						
	FM	K5TRA	927.07500	902.07500	218.1 Hz	
San Antonio, Texas						
	DMR/BM	K5VPW	441.62500	+	CC1	TVHFS
San Antonio, University Hospit						
	FM	AA5RO	927.05000	902.05000	110.9 Hz	TVHFS
San Felipe	FM	W0FCM	442.02500	+	123.0 Hz	
San Marcos	FM	K5MPS	147.52000		233.6 Hz	
	FM	AI5TX	443.65000	+	114.8 Hz	
	FM	K5MPS	446.15000	-		
San Marcos , Devils Backbone						
	FM	W5DK	146.92000	-	131.8 Hz	TVHFS
San Marcos, Texas State Univer						
	FM	KG5PVG	442.70000	+	123.0 Hz	
Santa Anna	FM	KE5NYB	147.12000	+	94.8 Hz	TVHFS
Santa Fe	FM	N5NWK	443.47500	+		TVHFS
Santa Maria	FM	W5RGV	444.27500	+	114.8 Hz	TVHFS
SE Houston	DMR/MARC	W5ICF	441.36000	+	CC2	
Seabrook	FM	KD5QCZ	53.03000	52.03000	156.7 Hz	TVHFS
Seabrook, City Hall						
	FM	W5SFD	147.26000	+	162.2 Hz	TVHFS
	FM	KE5VJH	443.25000	+	127.3 Hz	TVHFS
Seguin	FM	WA5GC	146.76000	-	141.3 Hz	TVHFS
Seminole	FM	N5SOR	145.45000	-	88.5 Hz	TVHFS
	FM	N5SOR	146.78000	-	88.5 Hz	TVHFS
Sheffield	FM	N5SOR	443.50000	+		TVHFS
Shepp	FM	NZ5V	145.23000	-	88.5 Hz	TVHFS
Sherman	FM	W5RVT	147.00000	+	100.0 Hz	TVHFS
	FM	W5COP	147.28000	+	107.2 Hz	TVHFS
	FM	W5RVT	444.75000	+	100.0 Hz	TVHFS
Shiner	FM	KC5QLT	146.68000	-		TVHFS
	FM	W5CTX	147.12000	+	141.3 Hz	TVHFS
	FM	WD5IEH	443.77500	+	141.3 Hz	TVHFS
	FM	WA5PA	444.27500	+	141.3 Hz	TVHFS
Sinton	FM	W5CRP	147.08000	+	107.2 Hz	TVHFS
Slidell	DSTAR	W5FKN	442.92500	+		TVHFS
Smithville	FUSION	KE5FKS	145.35000	-	114.8 Hz	TVHFS
Smyer	FM	KB5MBK	442.07500	+	146.2 Hz	TVHFS
Snyder	FM	K5SNY	146.92000	-	67.0 Hz	TVHFS
	FM	AI5TX	443.62500	+		TVHFS
Socorro	DSTAR	W5WIN	147.01000	+		
Sonora	FM	N5SOR	443.97500	+		TVHFS
South Houston	DMR/MARC	W5ICF	441.92500	+	CC2	
South Padre Island						
	FM	W5RGV	147.12000	+	114.8 Hz	TVHFS
	FM	KE5KLY	147.24000	+	114.8 Hz	TVHFS
	FM	W5RGV	444.87500	+	114.8 Hz	
Southlake	DMR/BM	KI5BLU	446.31250	-	CC1	
	FM	N1OZ	442.17500	+	110.9 Hz	TVHFS
Speaks	FM	K5SOI	442.52500	+	103.5 Hz	TVHFS
Spearman	FM	N5DFQ	147.04000	+	88.5 Hz	TVHFS
	FM	N5DFQ	442.00000	+	88.5 Hz	TVHFS
	FM	KC5WBK	443.20000	+	88.5 Hz	
Splendora	FM	W5OMR	441.70000	+	DCS 023	TVHFS
Spring	DMR/MARC	N5LUY	440.65000	+	CC3	
	DSTAR	KB2WF	147.58000			
	FM	WB5UGT	442.70000	+	103.5 Hz	TVHFS
	FM	K5JLK	442.80000	+	146.2 Hz	
	FM	KA2EEU	444.35000	+	103.5 Hz	TVHFS
Stagecoach	FM	W5NC	146.66000	-	100.0 Hz	TVHFS

TEXAS 359

Location	Mode	Call sign	Output	Input	Access	Coordinator
Stephenville	FM	K5DDL	145.29000	-	110.9 Hz	TVHFS
	FM	K5IIY	147.36000	+	110.9 Hz	TVHFS
	FM	KD5HNM	444.77500	+	88.5 Hz	TVHFS
Sterling City	FM	N5FTL	146.64000	-	88.5 Hz	TVHFS
	FM	WR5FM	441.57500	+	156.7 Hz	TVHFS
	FM	AI5TX	443.67500	+		TVHFS
Stinnett	DMR/MARC	KE5CJ	443.43750	+	CC14	
	FM	W5WDR	444.80000	+		
Sugar Land	FM	KD5HKQ	147.24000	+	127.3 Hz	TVHFS
	FM	KC5EVE	443.00000	+		TVHFS
Sulphur Springs	FM	WX5FL	145.11000	-	100.0 Hz	
	FM	K5SST	146.68000	-	151.4 Hz	TVHFS
	FM	K5SST	444.82500	+	151.4 Hz	TVHFS
	FM	WX5FL	444.90000	+	DCS 152	TVHFS
Sundown	FM	KD5SHB	444.72500	+		TVHFS
Sweet Home	FM	KF5KOI	442.05000	+	77.0 Hz	TVHFS
	FM	WB5UGT	443.82500	+	203.5 Hz	TVHFS
Sweetwater	DMR/MARC	AI5TX	443.66250	+	CC14	
	FM	KC5NOX	145.25000	-	162.2 Hz	TVHFS
	FM	KE4QFH	147.08000	+	162.2 Hz	TVHFS
	FM	AI5TX	443.65000	+		TVHFS
	FM	KC5NOX	927.11250	902.11250	DCS 432	TVHFS
Sweetwater, 9 Mile Hill						
	FM	KE5YF	444.77500	+	162.2 Hz	TVHFS
Sweetwater, 9 Mile Hill (South						
	FM	W5NCA	146.68000	-	162.2 Hz	
Tabor	FM	KD5DLW	443.52500	+	127.3 Hz	TVHFS
Taft	FM	K5YZZ	444.80000	+	107.2 Hz	TVHFS
Talco	FM	N5REL	442.20000	+	151.4 Hz	TVHFS
Tarzan	FM	K5MSO	52.65000	51.65000	123.0 Hz	TVHFS
Taylor	FM	N5TT	145.45000	-	162.2 Hz	TVHFS
	FM	N3ERC	145.47000	-	114.8 Hz	TVHFS
Temple	DMR/BM	WB5TTY	444.02500	+	CC10	
	DSTAR	K5CTX	147.34000	+		
	DSTAR	K5CTX	440.52500	+		TVHFS
	DSTAR	K5CTX	1292.10000	1272.10000		TVHFS
	FM	KG5HFI	53.23000	52.23000	162.2 Hz	
	FM	W5LM	146.82000	-	123.0 Hz	
Terrell	FM	K5RCP	441.67500	+	110.9 Hz	TVHFS
Texarkana	DMR/MARC	N5RGA	444.42500	+	CC9	
	DSTAR	KD5RCA	440.10000	+		
Texarkana, Barkman Creek						
	FM	WX5FL	145.39000	-	100.0 Hz	TVHFS
	FM	WX5FL	444.42500	+	100.0 Hz	
Texarkana, Christus St Michael						
	FM	KD5RCA	145.45000	-	100.0 Hz	
Texarkana, KTAL Tower						
	FM	KD5RCA	146.62000	-	100.0 Hz	TVHFS
Texas City	DMR/MARC	W5ZMV	440.62500	+	CC5	TVHFS
	FM	WR5TC	147.14000	+	167.9 Hz	TVHFS
	FM	K5BS	442.02500	+	103.5 Hz	TVHFS
The Colony	FM	K5LRK	147.38000	+	110.9 Hz	TVHFS
	FM	K5LRK	224.00000	-	110.9 Hz	
	FM	K5LRK	443.30000	+		
The Woodlands	DMR/MARC	KB5FLX	442.67500	+	CC7	TVHFS
	FM	N5HOU	146.80000	-	103.5 Hz	
The Woodlands, Woodlands Fire						
	FM	W5WFD	444.10000	+	136.5 Hz	TVHFS
Timpson	FM	KK5XM	145.15000	-	107.2 Hz	TVHFS
	FM	KK5XM	444.67500	+	107.2 Hz	TVHFS
Tom Bean	FM	N5MRG	52.73000	51.73000		
	FM	N5MRG	223.84000	-		

360   TEXAS

Location	Mode	Call sign	Output	Input	Access	Coordinator
Tom Bean	FM	N5MRG	441.65000	+	100.0 Hz	TVHFS
Tomball	DMR/MARC	W5ZMV	147.38000	+	CC5	TVHFS
	DMR/MARC	KB5FLX	441.60000	+	CC7	TVHFS
	DMR/MARC	W5ZMV	444.62500	+	CC5	
Trinity	FM	N5ESP	145.33000	-	103.5 Hz	TVHFS
Troy	DMR/BM	N5SIM	441.65000	+	CC10	
	DMR/MARC	N5SIM	147.39000	+	CC10	
	DMR/MARC	KE5KLY	440.70000	+	CC10	
	FM	WD5EMS	442.70000	+		TVHFS
	FM	WD5EMS	927.03750	902.03750	141.3 Hz	
Tuleta	FM	K5DJS	145.71500			
Tulia	FM	WU5Y	147.36000	+	88.5 Hz	TVHFS
Tuxedo	FM	KD5YCY	447.25000	-		
Tyler	DMR/MARC	KE5FGC	443.57500	+	CC1	
	DSTAR	W5ETX	147.12000	147.92000		TVHFS
	DSTAR	W5ETX	444.85000	+		TVHFS
	FM	W5WVH	145.37000	-	136.5 Hz	TVHFS
	FM	K5TYR	146.96000	-	136.5 Hz	TVHFS
	FM	K5TYR	147.00000	-	136.5 Hz	TVHFS
	FM	W5ETX	224.20000	-	136.5 Hz	TVHFS
	FM	W5MCT	443.10000	+	136.5 Hz	
	FM	K5TYR	444.40000	+	136.5 Hz	TVHFS
Tyler, UT Health Tyler Hospita						
	FM	W5ETX	145.21000	144.41000	136.5 Hz	TVHFS
Utopia	FM	W5FN	53.15000	52.15000	127.3 Hz	
	FM	W5FN	147.10000	+	127.3 Hz	TVHFS
Uvalde	FM	W5FN	146.76000	-	127.3 Hz	TVHFS
	FM	N5RUI	146.90000	-	100.0 Hz	TVHFS
	FM	KN5S	147.24000	+	77.0 Hz	TVHFS
	FM	W5LBD	147.26000	+	100.0 Hz	TVHFS
	FM	K5DRT	443.65000	+		TVHFS
	FM	AB5JK	444.60000	+	162.2 Hz	TVHFS
Van Alstyne	FM	W5VAL	443.80000	+	103.5 Hz	TVHFS
	FM	WB4GHY	444.12500	+		TVHFS
Vanderpool	FM	N4MUJ	145.40000	144.60000		
Venus	DMR/MARC	KN5TX	441.72500	+	CC1	
	FM	WA5FWC	145.39000	-	167.9 Hz	
	FM	WA5FWC	1292.98000	1272.98000		TVHFS
Vernon	FM	NC5Z	147.02000	+		TVHFS
	FM	N5LEZ	147.16000	+	192.8 Hz	TVHFS
	FM	WB5AFY	224.42000	-	192.8 Hz	
	FM	NC5Z	444.15000	+	192.8 Hz	TVHFS
Victoria	DMR/MARC	K5COD	443.76000	+	CC1	
	FM	W5DSC	145.13000	-	103.5 Hz	TVHFS
	FM	W5DSC	145.19000	-	103.5 Hz	TVHFS
	FM	W5DK	147.16000	+	141.3 Hz	TVHFS
	FM	WD5IEH	443.22500	+	141.3 Hz	TVHFS
	FM	WD5IEH	443.97500	+	103.5 Hz	TVHFS
	FM	K5SOI	444.65000	+	103.5 Hz	TVHFS
	FM	WB5MCT	444.67500	+	162.2 Hz	TVHFS
Victoria, Citizens Hospital						
	FM	K5VCT	146.70000	-	127.3 Hz	TVHFS
	FM	W5DSC	443.80000	+	103.5 Hz	TVHFS
Vidor	FM	KD5UNK	224.20000	-		
Violet	DMR/MARC	KC5HWB	444.60000	+	CC9	
Waco	DMR/BM	AE5CA	441.30000	+	CC10	
	DSTAR	W5ZDN	146.98000	-		TVHFS
	FM	K5AB	146.66000	-	123.0 Hz	TVHFS
	FM	WA5BU	147.16000	+	123.0 Hz	TVHFS
	FM	W5ZDN	421.25000	439.25000		TVHFS
	FM	WA5BU	442.45000	+		TVHFS
	FM	K5AB	442.80000	+	123.0 Hz	TVHFS

TEXAS 361

Location	Mode	Call sign	Output	Input	Access	Coordinator
Waco	FM	W5ZDN	442.87500	+	123.0 Hz	TVHFS
	FM	KC5QIH	443.55000	+	123.0 Hz	TVHFS
	FM	AA5RT	444.15000	+	123.0 Hz	TVHFS
Walburg	DSTAR	KE5RCS	145.13000	-		TVHFS
	DSTAR	KE5RCS	440.57500	+		TVHFS
	FM	K5AB	443.30000	+	88.5 Hz	TVHFS
	FM	KE5RCS	1293.20000	1273.20000		TVHFS
Waller, Monaville	FM	KF5GXZ	444.90000	+	100.0 Hz	TVHFS
Walnut Springs	FM	WC5WC	442.57500	+		
Watauga	FM	K0BRN	443.42500	+	100.0 Hz	TFCA
	FM	W7YC	444.57500	+	110.9 Hz	TVHFS
Waxahachie	FM	WD5DDH	927.03750	902.03750	110.9 Hz	
Wayside	FM	N5LTZ	444.57500	+	88.5 Hz	TVHFS
Weatherford	DMR/MARC	KN5TX	440.35000	+	CC1	
	FM	WB5IDM	147.04000	+	110.9 Hz	TVHFS
	FM	KA5PQK	442.45000	+	156.7 Hz	TVHFS
	FM	AI5TX	443.70000	+		TVHFS
	FM	W0BOE	444.27500	+	103.5 Hz	
Webster	FM	KA5QDG	146.74000	-	107.2 Hz	TVHFS
Wellington, DOT Tower						
	FM	N5OLP	147.24000	+		TVHFS
Weslaco	FM	KC5WBG	146.72000	-	114.8 Hz	TVHFS
	FM	WA5S	224.62000	-		TVHFS
	FM	KC5WBG	444.20000	+	114.8 Hz	TVHFS
West Columbia	DMR/MARC	KN5D	443.66250	+	CC1	TVHFS
	FM	AI5TX	443.62500	+		TVHFS
West Tawakoni, Lake Tawakoni						
	FM	K5VOM	146.80000	-	141.3 Hz	
Wharton	FM	W5DUQ	145.33000	-	167.9 Hz	TVHFS
	FM	W5DUQ	444.12500	+	167.9 Hz	TVHFS
Whitney	FM	NZ5T	146.62000	-	123.0 Hz	TVHFS
	FM	W5WK	442.20000	+	103.5 Hz	TVHFS
Wichita Falls	FM	WX5TWS	145.15000	-	192.8 Hz	TVHFS
	FM	KD5INN	146.62000	-	156.7 Hz	TVHFS
	FM	K5WFT	146.66000	-	192.8 Hz	TVHFS
	FM	KD5INN	146.88000	-	192.8 Hz	TVHFS
	FM	W5US	146.94000	-	192.8 Hz	TVHFS
	FM	W5GPO	147.06000	+	156.7 Hz	TVHFS
	FM	N5LEZ	147.12000	+	192.8 Hz	TVHFS
	FM	W5DAD	442.52500	+	192.8 Hz	TVHFS
	FM	WG5K	442.80000	+	192.8 Hz	TVHFS
	FM	N5WF	444.00000	+	192.8 Hz	TVHFS
	FM	WB5ALR	444.20000	+	118.8 Hz	TVHFS
	FM	K5WFT	444.32500	+	192.8 Hz	TVHFS
	FM	WX5TWS	444.52500	+	192.8 Hz	TVHFS
	FM	K5HRO	444.77500	+	173.8 Hz	TVHFS
	FM	N5AAJ	444.95000	+	192.8 Hz	TVHFS
	FM	N5LEZ	444.97500	+	192.8 Hz	TVHFS
Wichita Falls, American Red Cr						
	FM	N5WF	147.14000	+	192.8 Hz	TVHFS
Willis	FM	WA5EOC	224.24000	-	103.5 Hz	TVHFS
	FM	W5JSC	442.62500	+	88.5 Hz	TVHFS
Willis, Lake Conroe						
	FM	KB5ASW	441.67500	+	114.8 Hz	TVHFS
Wills Point	FM	K5RKW	145.27000	-	136.5 Hz	TVHFS
	FM	W5DLP	147.28000	+	136.5 Hz	
	FM	W5ETX	443.25000	+	136.5 Hz	TVHFS
Wimberley	FM	WA5AP	147.10000	+	141.3 Hz	TVHFS
	FM	WA5PAX	444.15000	+	114.8 Hz	TVHFS
Zapata	FM	KJ5HW	147.02000	+	114.8 Hz	TVHFS

362   U.S. VIRGIN ISLANDS

Location	Mode	Call sign	Output	Input	Access	Coordinator
U.S. VIRGIN ISLANDS						
Haymarket	DMR/MARC	W4YP	444.16250	+	CC1	
St Croix	FM	NP2OW	146.91000	-	100.0 Hz	
	FM	NP2VI	147.11000	+	100.0 Hz	
St Croix, Mount Welcome						
	FM	NP2VI	147.25000	+	100.0 Hz	
St John	FM	KP2I	146.97000	-		
St John, Bordeaux Mountain						
	FM	KP2SJ	146.63000	-	100.0 Hz	
St Thomas	DMR/MARC	KP2O	448.60000	-	CC7	
	FM	KP2O	146.81000	-	100.0 Hz	
	FM	NP2GO	146.95000	-	67.0 Hz	
	FM	K9VV	448.17500	-		WIRCI
UTAH						
Alpine, Silver Lake Flat						
	FM	K7UVA	448.22500	-	100.0 Hz	
American Fork, KD7BBC QTH						
	DSTAR	NT3ST	447.97500	-		UVHFS
American Fork, Utah Lake						
	FM	KA7EGC	449.17500	-	131.8 Hz	UVHFS
Antelope Island, Frary Peak						
	FM	K7DAV	147.04000	+		
	FM	K7DAV	447.20000	-	127.3 Hz	UVHFS
Bountiful	DMR/MARC	N2DME	449.25000	-	CC1	
	FM	W7CWK	449.35000	-	123.0 Hz	UVHFS
Bountiful, Golf Course						
	FM	K7DAV	449.92500	-	100.0 Hz	UVHFS
Bountiful, QTH	FM	KD7RTO	447.42500	+	192.8 Hz	UVHFS
Brigham City	FM	K7UB	145.29000	-	123.0 Hz	UVHFS
Brighton	FM	W7SP	146.62000	-		UVHFS
Brighton, Scott's Peak						
	FM	K7JL	449.52500	-	131.8 Hz	UVHFS
Brighton, Scotts Peak						
	FM	K7JL	145.27000	-	100.0 Hz	UVHFS
Bryce Canyon	FM	W6DZL	145.35000	-	123.0 Hz	
Butterfield Peak	FM	WB7TSQ	145.45000	454.55000	100.0 Hz	
Castle Dale, Cedar Mountain						
	FM	K7SDC	147.14000	+	88.5 Hz	UVHFS
	FM	K7YI	447.12500	-	100.0 Hz	UVHFS
	FM	K7SDC	448.55000	-	88.5 Hz	UVHFS
Castle Dale, Horn Mountain						
	FM	K7SDC	147.06000	+	88.5 Hz	UVHFS
	FM	K7YI	447.62500	-	100.0 Hz	UVHFS
	FM	WX7Y	447.70000	-	123.0 Hz	UVHFS
Castledale	FM	K7JL	147.12000	+	100.0 Hz	UVHFS
Cedar City	DSTAR	WR7AAA	145.15000	-		
	DSTAR	WR7AAA	1299.25000	1279.25000		UVHFS
	FM	WV7H	145.47000	-		UVHFS
	FM	WA7GTU	447.57500	+		UVHFS
	FM	N7DZP	448.10000	-		UVHFS
	FM	WA7GTU	449.90000	-	100.0 Hz	UVHFS
Cedar City, Blowhard Mountain						
	FM	WV7H	146.80000	-	100.0 Hz	UVHFS
	FM	KB6BOB	448.65000	-		UVHFS
Cedar City, Hospital						
	FM	N7AKK	147.06000	+		UVHFS
Cedar City, Iron Mountain						
	FM	K7JH	146.76000	-	123.0 Hz	UVHFS
	FM	N7KM	146.98000	-	100.0 Hz	UVHFS
	FM	K7JH	448.80000	-	100.0 Hz	UVHFS

UTAH 363

Location	Mode	Call sign	Output	Input	Access	Coordinator
Cedar City, Iron Mountain						
	FM	WA7GTU	449.50000	-	100.0 Hz	UVHFS
Cedar Fort, Internet Mountain						
	FM	KO7R	449.70000	-	127.3 Hz	UVHFS
Clearfield	FM	KR7K	447.15000	-	114.8 Hz	UVHFS
Clearfield, Civic Center						
	FM	NJ7J	449.95000	-	123.0 Hz	UVHFS
Coalville, Lewis Peak						
	FM	K7HEN	147.24000	+	136.5 Hz	UVHFS
	FM	WA7GIE	147.36000	+	100.0 Hz	UVHFS
	FM	WA7GIE	448.65000	-		UVHFS
	FM	WB7TSQ	448.90000	-		UVHFS
	FM	WA7GIE	449.55000	-	100.0 Hz	UVHFS
Corinne, Promontory Point						
	FM	N7TOP	146.92000	-	123.0 Hz	UVHFS
Delta	FM	KB7WQD	147.38000	+	203.5 Hz	UVHFS
Delta, Notch Peak	FM	KB7WQD	147.38000	+	203.5 Hz	UVHFS
Draper, Fire Station						
	FM	KG7EGM	447.10000	-	100.0 Hz	
Draper, Lake Mountain						
	FM	WX7Y	147.08000	+	77.0 Hz	UVHFS
Draper, Point Of The Mountain						
	FM	N7IMF	224.90000	-	156.7 Hz	UVHFS
Duchense	FM	N7PQD	147.26000	+		UVHFS
Eagle Mountain	FM	WD7N	449.25000	-	110.9 Hz	UVHFS
East Of Holden	FM	WB7REL	449.30000	-	88.5 Hz	UVHFS
East Salina, Salina Canyon						
	FM	WB7REL	146.72000	-	100.0 Hz	UVHFS
Enterprise	FM	NR7K	146.74000	-	100.0 Hz	UVHFS
Enterprise, Black Hills						
	FM	KD7YK	449.72500	-		UVHFS
Ephraim, Horseshoe						
	FM	W7DHH	146.66000	-	100.0 Hz	UVHFS
	FM	N7IMF	224.64000	-		UVHFS
Eureka, Eureka Peak						
	FM	KC7KRY	447.87500	-		
Farmington, Shepard Peak						
	FM	K7DAV	53.01000	52.01000	141.3 Hz	UVHFS
Garrison	FM	N7ELY	442.10000	+	114.8 Hz	
Glendale, Spencer Bench						
	FM	WB7REL	146.72000	-	100.0 Hz	UVHFS
Goshen, West Mountain						
	FM	WA7UAH	147.02000	+	100.0 Hz	UVHFS
	FM	K7UCS	147.34000	+	100.0 Hz	UVHFS
	FM	WA7FFM	448.95000	-		UVHFS
Gunnison, N7RVS QTH						
	FM	N7RVS	447.35000	-	88.5 Hz	UVHFS
Hanksville, Ellen Peak						
	FM	K7SDC	147.08000	+	136.5 Hz	UVHFS
Herriman, High School						
	FM	N7HRC	449.25000	-	118.8 Hz	UVHFS
Highland, KX7VC QTH						
	FM	KX7VC	447.72500	-	167.9 Hz	
Holden, Beesting Peak						
	FM	N7GGN	147.10000	+	100.0 Hz	UVHFS
Howell, Blue Springs Hill						
	FM	K7UB	145.43000	-	123.0 Hz	UVHFS
	FM	K7UB	448.30000	-	123.0 Hz	UVHFS
Huntington, Skyline Dr						
	FM	WX7Y	223.92000	-	88.5 Hz	UVHFS
Huntsville	FM	W7DBA	145.21000	-	123.0 Hz	UVHFS
	FM	W7DBA	448.02500	-	123.0 Hz	UVHFS

364   UTAH

Location	Mode	Call sign	Output	Input	Access	Coordinator
Huntsville, Herd Mountain						
	FM	N7JSQ	146.68000	-	123.0 Hz	UVHFS
Hurricane	FM	KG7FOT	449.55000	-	100.0 Hz	
Hurricane, Hurricane Mesa						
	FM	K5JCA	449.27500	-		UVHFS
Indianola, Indianola Peak						
	FM	WB7REL	146.72000	-	100.0 Hz	UVHFS
Kanab, TV Site	FM	W7NRC	146.88000	-	123.0 Hz	UVHFS
Kearns	FM	K7LNP	447.57500	-	114.8 Hz	UVHFS
Kearns, Bacchus Hill						
	WX	K2NWS	447.52500	-	107.2 Hz	UVHFS
Laketown, Bear Lake						
	FM	K7OGM	147.02000	+	100.0 Hz	UVHFS
	FM	K7OGM	448.45000	-	131.8 Hz	UVHFS
	FM	K7OGM	448.97500	-		UVHFS
	FM	K7OGM	449.70000	-	100.0 Hz	UVHFS
Layton	DMR/BM	AH2S	447.05000	-	CC1	UVHFS
Layton, Francis Peak						
	FM	K7MLA	146.96000	-	100.0 Hz	UVHFS
Layton, Kaysville Peak						
	FM	AI7J	449.60000	-	136.5 Hz	UVHFS
Layton, Shepherd Peak						
	FM	K7DAV	449.87500	-	DCS 051	UVHFS
Lehi	DMR/BM	KC7WST	447.05000	-	CC1	UVHFS
	FM	KG7QWU	448.87500	-		
Lehi, City Offices	FM	KI7USB	448.92500	-	100.0 Hz	
Levan, Levan Peak						
	FM	K7JL	145.27000	-	103.5 Hz	UVHFS
Logan	DMR/BM	NU7TS	447.00000	-	CC1	UVHFS
	DMR/BM	WA7KMF	449.25000	-	CC1	
	FM	N7RRZ	147.24000	+	79.7 Hz	UVHFS
	FM	W7BOZ	449.30000	-	103.5 Hz	UVHFS
	FM	N7RRZ	449.32500	-	156.7 Hz	UVHFS
Logan, Mt Logan	DSTAR	AC7O	1299.75000	1279.75000		
	FM	WA7KMF	146.72000	-	103.5 Hz	UVHFS
	FM	AC7O	449.62500	-	103.5 Hz	UVHFS
Logan, Mt Pisgah	FM	AC7II	449.65000	-	100.0 Hz	UVHFS
Logan, WA7KMF QTH						
	DSTAR	AC7O	145.15000	-		UVHFS
	DSTAR	AC7O	447.97500	-		
	FM	AC7O	147.20000	+	103.5 Hz	UVHFS
Magna, Kessler Peak						
	FM	W7YDO	448.85000	-		
Manti	FM	N7YFZ	449.75000	-	131.8 Hz	UVHFS
Manti, Barton Peak						
	FM	KD7YE	448.27500	-	107.2 Hz	UVHFS
Mantua, Murrays Hill						
	FM	WA7KMF	449.80000	-	103.5 Hz	UVHFS
Mapleton	FM	N6EZO	146.80000	-	100.0 Hz	UVHFS
Mendon	FM	N7PKI	448.92500	-	100.0 Hz	UVHFS
Mexican Hat	FM	KD7HLL	449.90000	-	123.0 Hz	UVHFS
Midway	FM	N7ZOI	449.95000	-		UVHFS
Midway, Wilson Peak						
	FM	N7ZOI	147.20000	+	88.5 Hz	UVHFS
Milford, Frisco Peak						
	FM	WR7AAA	146.94000	-		
	FM	K7JL	448.67500	-		UVHFS
Moab	FM	K7QEQ	146.90000	-	88.5 Hz	UVHFS
Moab, Bald Mesa	FM	K7QEQ	146.76000	-		
	FM	K7QEQ	449.10000	-	107.2 Hz	UVHFS
Monroe, Monroe Peak						
	FM	WA7HSW	146.64000	-	100.0 Hz	UVHFS

UTAH **365**

Location	Mode	Call sign	Output	Input	Access	Coordinator
Monroe, Monroe Peak						
	FM	WB7REL	146.86000	-	100.0 Hz	UVHFS
	FM	W7DHH	447.45000	-	114.8 Hz	UVHFS
Monticello, Abajo Peak						
	FM	K7SDC	146.61000	-	88.5 Hz	UVHFS
	FM	K7QEQ	447.40000	-		UVHFS
Monticello, Abajo Pk						
	FM	K7QEQ	223.94000	-		UVHFS
Morgan, TV Site.	FM	KB7ZCL	147.10000	+	123.0 Hz	UVHFS
Murray	FM	NM7P	145.35000	-	100.0 Hz	UVHFS
	FM	N7HIW	448.12500	-	100.0 Hz	UVHFS
Murray, Intermountain Med Ctr						
	DSTAR	KO7SLC	145.15000	-		
	DSTAR	KO7SLC	447.95000	-		
	DSTAR	KO7SLC	1298.75000	1278.75000		
Murray, Intermountain Medical						
	FM	KE7LMG	447.25000	-	100.0 Hz	UVHFS
Myton, Flat Top	FM	W7BYU	145.49000	-	136.5 Hz	UVHFS
North Logan	DMR/MARC	AC7JT	449.25000	-	CC1	
Ogden	FM	K7HEN	145.49000	-	123.0 Hz	UVHFS
	FM	WB7TSQ	147.38000	+		UVHFS
	FM	KC7NAT	449.07500	-	123.0 Hz	UVHFS
Ogden, Downtown						
	FM	WB6CDN	224.50000	-	167.9 Hz	UVHFS
Ogden, Foothills	FM	WB7TSQ	145.41000	-	123.0 Hz	UVHFS
Ogden, Little Mountain						
	FM	W7SU	146.82000	-	123.0 Hz	UVHFS
	FM	W7SU	448.57500	-	100.0 Hz	UVHFS
Ogden, Mt Ogden	FM	W7SU	146.90000	-	DCS 122	UVHFS
	FM	W7SU	448.60000	-	123.0 Hz	UVHFS
Ogden, Powder Mountain						
	FM	KC7SUM	145.47000	-	123.0 Hz	UVHFS
	FM	N7TOP	224.00000	-	167.9 Hz	UVHFS
	FM	N7TOP	447.77500	-	123.0 Hz	UVHFS
Ogden, Promontory Point						
	DMR/BM	NU7TS	447.35000	-	CC1	UVHFS
Ogden, Promontory Pt						
	FM	K7JL	145.49000	-	123.0 Hz	UVHFS
	FM	AC7O	147.26000	+	103.5 Hz	UVHFS
Ogden, Sheriff's Office						
	DSTAR	KE7EGG	447.95000	-		UVHFS
	DSTAR	KE7EGG	1298.75000	1278.75000		UVHFS
Ogden, Weber State University						
	FM	KD7FDH	145.25000	-	123.0 Hz	UVHFS
Orem	FM	N7IMF	224.88000	-	156.7 Hz	UVHFS
Orem, Lake Mountain						
	FM	K7UVA	146.78000	-	100.0 Hz	UVHFS
	FM	K7UVA	224.56000	-	100.0 Hz	UVHFS
	FM	K7UVA	448.20000	-	100.0 Hz	UVHFS
Orem, Point Of The Mountain						
	FM	KD7RBR	447.60000	-	162.2 Hz	UVHFS
Orem, QTH AC7DM						
	FM	N7BSA	145.47000	-	100.0 Hz	UVHFS
Page (AZ), Navajo Mountain						
	FM	W7WAC	146.96000	-	100.0 Hz	UVHFS
	FM	NA7DB	448.75000	-	100.0 Hz	UVHFS
	FM	W7CWI	449.92500	-	100.0 Hz	UVHFS
Page, Navajo Mountain						
	FM	WA7VHF	448.60000	-	100.0 Hz	UVHFS
Panguitch, Mount Dutton						
	FM	N7NKK	147.16000	+	100.0 Hz	UVHFS

366 UTAH

Location	Mode	Call sign	Output	Input	Access	Coordinator
Paradise, AC7II QTH						
	DSTAR	KF7VJO	447.95000	-		UVHFS
Park City	FM	NZ6Z	145.23000	-	100.0 Hz	UVHFS
	FM	NZ6Z	447.50000	-		UVHFS
Park City, Murdock Peak						
	FM	KB7HAF	448.47500	-		UVHFS
Payson	FM	NV7V	447.00000	-	100.0 Hz	UVHFS
	FM	KB7M	448.02500	-	100.0 Hz	UVHFS
Pleasant Grove	FM	KB7YOT	223.88000	-	156.7 Hz	UVHFS
	FM	N7UEO	449.32500	-	114.8 Hz	UVHFS
	FM	K8BKT	449.77500	-		UVHFS
Price, Wood Hill	FM	W7CEU	145.43000	-	88.5 Hz	UVHFS
	FM	K7GX	147.20000	+	88.5 Hz	UVHFS
	FM	W7CEU	224.50000	-	88.5 Hz	UVHFS
	FM	W7CEU	448.30000	-	88.5 Hz	UVHFS
Provo	FM	N7IMF	224.64000	-	156.7 Hz	UVHFS
Provo, BYU	FM	N7BYU	145.33000	-	100.0 Hz	UVHFS
	FM	N7BYU	449.07500	-	167.9 Hz	UVHFS
Provo, Lake Mountain						
	FM	K7UCS	145.23000	-	131.8 Hz	UVHFS
	FM	W7SP	146.76000	-		UVHFS
	FM	K7UCS	147.28000	+	141.3 Hz	UVHFS
	FM	W7WJC	224.42000	-	156.7 Hz	UVHFS
	FM	WA7GIE	449.47500	-	100.0 Hz	UVHFS
	FM	K7UCS	449.67500	-	173.8 Hz	UVHFS
	FM	K7UCS	449.97500	-	131.8 Hz	UVHFS
Provo, Sundance	FM	K7UVA	145.25000	-	100.0 Hz	UVHFS
Provo, UVRMC	FM	WA7FFM	449.85000	-	146.2 Hz	UVHFS
Red Spur	FM	WA7KMF	145.31000	-	103.5 Hz	UVHFS
Richfield, Monroe Peak						
	FM	WA7VHF	146.84000	-	100.0 Hz	UVHFS
Riverside	DSTAR	N7RDS	447.92500	-		UVHFS
	DSTAR	WA7KMF	449.57500	-		UVHFS
	FM	K7UB	147.22000	+	123.0 Hz	UVHFS
Riverton	FM	WB7TSQ	145.37000	-		
	FM	WB7TSQ	448.42500	-	100.0 Hz	
Roosevelt	FM	W7BYU	146.92000	-	136.5 Hz	UVHFS
Saint George	DMR/BM	W7CRC	447.35000	-	CC3	
Salina, Beesting Peak						
	FM	WB7REL	449.30000	-		
Salt Lake City	DMR/BM	N6DVZ	447.93750	-	CC1	UVHFS
	FM	W0HU	447.02500	-	100.0 Hz	UVHFS
	FM	W7SAR	447.17500	-		UVHFS
	FM	K7MRS	447.25000	-	100.0 Hz	UVHFS
	FM	KA7OEI	449.75000	-	151.4 Hz	UVHFS
	FM	W7XDX	927.58750	902.58750	DCS 432	UVHFS
Salt Lake City, Capitol						
	FM	AA7JR	145.21000	-		UVHFS
	FM	W7DES	448.00000	-	100.0 Hz	UVHFS
Salt Lake City, Ensign Peak						
	FM	KC7IIB	146.70000	-	100.0 Hz	UVHFS
	FM	WA7SNS	147.16000	+	127.3 Hz	UVHFS
	FM	KD7IMS	448.17500	-	203.5 Hz	UVHFS
	FM	KC7IIB	448.45000	-	100.0 Hz	UVHFS
	FM	K7XRD	448.52500	-		UVHFS
	FM	WA7VHF	449.27500	-	88.5 Hz	UVHFS
	FM	K7JL	449.40000	-	100.0 Hz	UVHFS
	FM	KD0J	449.90000	-	100.0 Hz	UVHFS
Salt Lake City, Farnsworth Pea						
	DSTAR	KF6RAL	145.12500	-		UVHFS
	DSTAR	KF6RAL	448.07500	-		UVHFS
	DSTAR	KF6RAL	1287.00000	1267.00000		UVHFS

UTAH 367

Location	Mode	Call sign	Output	Input	Access	Coordinator
Salt Lake City, Farnsworth Pea						
	FM	KI7DX	53.15000	52.15000	146.2 Hz	UVHFS
	FM	W7SP	146.62000	-		UVHFS
	FM	WA7VHF	146.94000	-	88.5 Hz	UVHFS
	FM	K7JL	147.12000	+	100.0 Hz	UVHFS
	FM	KI7DX	448.15000	-	127.3 Hz	UVHFS
	FM	K7JL	449.15000	-	100.0 Hz	UVHFS
	FM	K7JL	449.50000	-	100.0 Hz	UVHFS
Salt Lake City, IHC						
	FM	W7IHC	448.55000	-	100.0 Hz	UVHFS
Salt Lake City, Intermountain						
	FM	W7IHC	448.40000	-		UVHFS
Salt Lake City, Jordan Vly Hos						
	FM	N7PCE	146.84000	-		UVHFS
	FM	KD0J	224.78000	-	100.0 Hz	UVHFS
Salt Lake City, LDS HQ						
	FM	WD7SL	448.42500	-	100.0 Hz	UVHFS
Salt Lake City, Meridian Peak						
	FM	WB6CDN	224.82000	-	167.9 Hz	UVHFS
Salt Lake City, Nelson Peak						
	FM	WA7GIE	433.60000	-	100.0 Hz	UVHFS
	FM	WA7GIE	448.72500	-		UVHFS
	FM	WA7GIE	449.00000	-		UVHFS
	FM	WA7GIE	449.42500	-	100.0 Hz	UVHFS
Salt Lake City, SLC						
	FM	K7CSW	448.05000	-	100.0 Hz	UVHFS
Salt Lake City, SLCC						
	FM	KD0J	146.88000	-	88.5 Hz	UVHFS
Salt Lake City, U Of U Hospita						
	FM	KD7NX	146.74000	-	114.8 Hz	UVHFS
	FM	KD7NX	448.10000	-	114.8 Hz	UVHFS
Salt Lake, Carrigan Ridge						
	FM	KE7GHK	145.41000	-	DCS 125	UVHFS
Sandy	FM	KA7EGC	224.64000	-	156.7 Hz	
	FM	W7ROY	448.37500	-	100.0 Hz	UVHFS
Saratoga Springs	DMR/BM	KE7NHU	447.42500	-	CC1	
	FM	AC7DU	447.67500	-	100.0 Hz	UVHFS
Scofield, Boardinghouse Rdge						
	FM	WX7Y	224.98000	-	88.5 Hz	UVHFS
Scofield, Boardinghouse Ridge						
	FM	K7SDC	147.08000	+	88.5 Hz	UVHFS
Scofield, Ford Ridge						
	FM	K7SDC	145.31000	-	88.5 Hz	UVHFS
Snowbird, Hidden Peak						
	FM	K7JL	147.18000	+	100.0 Hz	UVHFS
South Salt Lake City, Fire Sta						
	FM	KF7YXL	447.70000	-	100.0 Hz	UVHFS
Springville	FM	N7KYY	447.32500	-	114.8 Hz	UVHFS
	FM	WD7N	447.47500	-	186.2 Hz	UVHFS
St George	FM	N7ARR	145.37000	-		UVHFS
	FM	KA7STK	146.70000	-		UVHFS
	FM	NR7K	449.32500	-		UVHFS
	FM	KA7STK	449.42500	-		
St George, Scrub Peak						
	FM	WB6TNP	448.72500	-		UVHFS
St George, Seegmiller Peak						
	FM	NR7K	146.91000	-	100.0 Hz	UVHFS
St George, Utah Hill						
	FM	NR7K	146.82000	-	100.0 Hz	UVHFS
	FM	K7OET	449.35000	-		
	FM	W7AOR	449.75000	-	123.0 Hz	UVHFS

368   UTAH

Location	Mode	Call sign	Output	Input	Access	Coordinator
St George, Webb Hill						
	FM	W7DRC	146.64000	-	100.0 Hz	UVHFS
St. George	DMR/BM	K6IB	447.80000	-	CC1	
St. George, Webb Hill						
	FM	K7OET	927.55000	902.55000		
Starling, Sterling	FM	WB7REL	145.29000	-	131.8 Hz	UVHFS
Sterling	DSTAR	KB7BSK	145.15000	-		
	DSTAR	KB7BSK	447.95000	-		
	FM	WB7REL	447.85000	-	131.8 Hz	UVHFS
Sterling, Salina Canyon						
	FM	WB7REL	449.25000	-	131.8 Hz	UVHFS
Sunny Side, Bruin Peak						
	FM	WX7Y	145.17500	-		
Sunnyside, Bruin Point						
	FM	K7SDC	147.32000	+	88.5 Hz	UVHFS
	FM	K7SDC	449.05000	-	88.5 Hz	UVHFS
Thiokol	FM	K7UB	448.30000	-	123.0 Hz	UVHFS
Tooele, Black Mountain						
	FM	K7HK	145.35000	-		UVHFS
	FM	W7EO	146.98000	-		UVHFS
Tooele, Home QTH						
	FM	N6RBV	447.36250	-		
Tooele, South Mountain						
	FM	W7EO	147.30000	+	100.0 Hz	UVHFS
	FM	W7EO	449.35000	-	100.0 Hz	UVHFS
Toquerville, Toquerville Hill						
	FM	W7DRC	145.45000	-	100.0 Hz	UVHFS
Torquerville, Hurricane Mesa						
	DSTAR	KF7YIX	145.15000	-		UVHFS
	DSTAR	KF7YIX	447.95000	-		UVHFS
	DSTAR	KF7YIX	1299.25000	1279.25000		UVHFS
Tremonton	DMR/BM	NU7TS	447.12500	-	CC1	UVHFS
Vernal	FM	W7BYU	449.70000	-	136.5 Hz	UVHFS
Vernal, Blue Mountain						
	FM	W7BAR	147.10000	+	136.5 Hz	UVHFS
Vernal, Grizzly Ridge						
	FM	K7HEN	145.49000	-	136.5 Hz	UVHFS
	FM	W7BAR	147.04000	+	136.5 Hz	UVHFS
Vernal, Tabby Mountain						
	FM	KG7DSO	147.34000	+	136.5 Hz	UVHFS
Vernal, Uintah County EOC						
	FM	W7BAR	449.90000	-	136.5 Hz	
Vernon, Black Crook Peak						
	FM	W7EO	145.39000	-	100.0 Hz	UVHFS
Vernon, Vernon Hills						
	FM	W7EO	449.95000	-	100.0 Hz	UVHFS
Wellsville	DSTAR	NU7TS	449.57500	-		
	FM	AF7FH	927.51250	902.51250	103.5 Hz	UVHFS
Wendover, Wendover Peak						
	FM	W7EO	147.20000	+	100.0 Hz	UVHFS
	FM	WA7GIE	449.55000	-	123.0 Hz	UVHFS
West Haven	FM	N7TOP	448.77500	-	123.0 Hz	UVHFS
	FM	N7TOP	449.77500	-	123.0 Hz	UVHFS
West Jordan	DMR/BM	N6DVZ	449.25000	-	CC2	
	FM	K7LNP	447.57500	-	114.8 Hz	UVHFS
	FM	WD7P	447.75000	-	100.0 Hz	UVHFS
West Jordan, Butterfield Peak						
	FM	WA7UAH	145.45000	-	100.0 Hz	UVHFS
	FM	K7MLA	147.14000	+	127.3 Hz	UVHFS
	FM	W7XDX	927.11250	902.11250	DCS 432	UVHFS
West Jordan, Butterfield Pk						
	FM	WA7UAH	449.72500	-	151.4 Hz	UVHFS

UTAH 369

Location	Mode	Call sign	Output	Input	Access	Coordinator
West Kaysville, Sewer Plant						
	FM	K7DOU	449.70000	-	100.0 Hz	UVHFS
West Point, West Point City Bu						
	FM	W7WPC	447.07500	-	DCS 051	UVHFS
West Valley City, WVCFD Statio						
	FM	K2WVC	448.80000	448.30000	100.0 Hz	
Woodland Hills	DMR/BM	KC7WST	449.80000	-	CC1	UVHFS
	FM	WB7RPF	447.30000	-	77.0 Hz	UVHFS
Woods Cross	DMR/BM	K7BBR	447.37500	-	CC1	

VERMONT

Location	Mode	Call sign	Output	Input	Access	Coordinator
Athens	FM	K2KDA	441.65000	+	110.9 Hz	
Bellows Falls	FM	KB1NXN	447.57500	-		
Bolton	DMR/MARC	KI1P	445.07500	-	CC7	
	FM	WB1GQR	445.02500	-		VIRCC
Bolton, Ski Area	FM	WB1GQR	145.15000	-	100.0 Hz	VIRCC
Brandon	FM	AA1PR	52.49000	51.49000	131.8 Hz	
Brandon, Center Of Village						
	FM	AA1PR	446.10000	-		
Brandon, Village Park						
	FM	AA1PR	146.47500		173.8 Hz	
Brownsville, Mount Ascutney						
	FM	W1IMD	448.12500	-	110.9 Hz	VIRCC
Burke	FM	W1AAK	449.12500	-	110.9 Hz	
Burlington	DMR/MARC	KI1P	446.47500	-	CC1	
	FM	W1KOO	146.61000	-	100.0 Hz	VIRCC
	FM	W1KOO	443.15000	+		
Cabot	FM	W1BD	146.82000	-	100.0 Hz	VIRCC
	FM	K1US	449.62500	-		
Corinth	DMR/MARC	KA1UAG	443.90000	+	CC8	VIRCC
East Barre	FM	N1IOE	147.39000	+	100.0 Hz	VIRCC
East Corinth	FM	KB1FDA	147.21000	+	100.0 Hz	VIRCC
Essex Junction	FM	KB1KJS	146.79000	-	100.0 Hz	VIRCC
Essex Junction, Fleming Elemen						
	FM	W1CTE	146.85000	-		VIRCC
Hartford	DMR/BM	WX1NH	444.00000	+	CC7	
Jay	DMR/MARC	KI1P	446.37500	-	CC7	
Jay, Jay Peak Ski Resort						
	FM	K1JAY	146.74500	-	100.0 Hz	
Jericho, Mt Mansfield						
	FM	W1KOO	146.94000	-	100.0 Hz	VIRCC
Lyndon	DMR/MARC	KI1P	448.57500	-	CC5	
Manchester	DMR/MARC	KB5VP	441.35000	+	CC7	
Manchester, Equinox Mountain						
	FM	WA1ZMS	145.39000	-	100.0 Hz	VIRCC
Manchester, Mount Equinox						
	FM	K1EQX	444.05000	+	100.0 Hz	
Marlboro, Hogback Mountain						
	FM	N1HWI	147.01500	+	100.0 Hz	VIRCC
Middlebury, Middlebury College						
	FM	KA1LM	147.36000	+	100.0 Hz	VIRCC
Monkton	DMR/MARC	W1IMD	443.75000	+	CC7	
	FM	W1AAK	444.65000	+	110.9 Hz	VIRCC
Montpelier	FM	K1VIT	449.67500	-	103.5 Hz	
Mt. Ascutney	DMR/MARC	W1UWS	448.47500	-	CC5	VIRCC
Mt. Snow	DMR/MARC	KI1P	446.27500	-	CC6	
Newfane	FM	WA1KFX	147.09000	+	110.9 Hz	VIRCC
	FM	WA1KFX	444.70000	+	110.9 Hz	
Northfield	DMR/MARC	KI1P	449.47500	-	CC6	
Pico Peak	DMR/MARC	W1IMD	444.50000	+	CC1	
Poultney	DSTAR	KC2YXS	145.60000			
Proctorsville	FM	W1TAL	146.41500		123.0 Hz	

370 VERMONT

Location	Mode	Call sign	Output	Input	Access	Coordinator
Rutland	FM	W1OOR	146.49000		123.0 Hz	
	FM	WA1ZMS	449.17500	-	100.0 Hz	
Rutland, Boardman Hill						
	FM	W1GMW	147.04500	-	100.0 Hz	VIRCC
Rutland, Killington Peak						
	FM	W1AAK	146.88000	-	110.9 Hz	VIRCC
	FM	W1ABI	444.55000	+	110.9 Hz	VIRCC
Rutland, Pico Peak						
	FM	W1IMD	444.40000	+		
Saint Albans	FM	N1STA	443.40000	+	162.2 Hz	
Saint Albans, French Hill						
	FM	N1STA	145.23000	-	100.0 Hz	VIRCC
Shaftsbury	FM	K1SV	146.83500	-	100.0 Hz	VIRCC
Tunbridge	FM	K1MOQ	146.97000	-		
Underhill, Mount Mansfield						
	FM	W1IMD	447.17500	-	82.5 Hz	VIRCC
Warren, Lincoln Peak						
	FM	K1VIT	145.47000	-	100.0 Hz	VIRCC
Wells	FM	N1VT	224.96000	-		VIRCC
White River Jct	FM	N1DAS	444.00000	+		
Williamstown	DMR/MARC	W1IMD	448.87500	-	CC7	VIRCC
	FM	W1BD	146.62500	-	100.0 Hz	VIRCC
	FM	W1AAK	444.60000	+	110.9 Hz	
Windsor, Mt Ascutney						
	FM	W1UWS	146.76000	-	110.9 Hz	VIRCC
VIRGINIA						
Abingdon	FM	NM4L	147.34500	+	103.5 Hz	
Abingdon, Brummley Mtn						
	FM	KB8KSP	442.97500	+		
Accomac	FM	K4BW	147.25500	+	156.7 Hz	
	FM	K4BW	444.30000	+		
Alexandria	DMR/MARC	W4HFH	442.41250	+	CC1	T-MARC
	DMR/MARC	N3JLT	443.10000	+	CC1	T-MARC
	DSTAR	W4HFH	145.38000	-		T-MARC
	FM	W4HFH	53.13000	52.13000	107.2 Hz	
	FM	W4HFH	147.31500	+	107.2 Hz	
	FM	W4HFH	224.82000	-	107.2 Hz	T-MARC
	FM	W4HFH	444.60000	+	107.2 Hz	T-MARC
	FM	W4HFH	927.60000	902.60000	107.2 Hz	T-MARC
	FM	W4HFH	1282.60000	1262.60000	107.2 Hz	T-MARC
Alexandria, George Washington						
	FM	K4US	146.65500	-		T-MARC
Alexandria, Inova Alexandria H						
	DSTAR	W4HFH	442.06000	+		T-MARC
	DSTAR	W4HFH	1253.60000			T-MARC
	DSTAR	W4HFH	1284.60000	1264.60000		T-MARC
Amherst	FM	K4CQ	145.49000	-	136.5 Hz	SERA-VA
Antioch, Bull Run Mountain						
	FM	WA3KOK	447.77500	-	67.0 Hz	T-MARC
	FM	N3KL	1286.10000	1266.10000		T-MARC
Arlington	DMR/MARC	N3QEM	443.06250	+	CC1	T-MARC
	FM	WB4MJF	145.15000	-		T-MARC
	FM	W4WVP	145.47000	-	107.2 Hz	T-MARC
	FM	WB4MJF	224.06000	-		T-MARC
	FM	WB4MWF	224.62000	-		T-MARC
	FM	W4CIA	441.45000	+	110.9 Hz	
	FM	AB4YP	443.20000	+	114.8 Hz	T-MARC
	FM	K4AF	444.55000	+	88.5 Hz	T-MARC
	FM	W4AVA	448.62500	-	107.2 Hz	T-MARC
	FM	W4WVP	449.32500	-	151.4 Hz	T-MARC

VIRGINIA 371

Location	Mode	Call sign	Output	Input	Access	Coordinator
Arlington, National Capitol Re						
	FM	W4AVA	146.62500	-	107.2 Hz	T-MARC
Ashburn	DMR/MARC	WB6EFW	442.13750	+	CC1	T-MARC
	DMR/MARC	N3QEM	442.90000	+	CC1	T-MARC
	FM	NV4FM	53.61000	52.61000		T-MARC
	FM	KQ4CI	448.82500	-	77.0 Hz	T-MARC
Ashland	DMR/MARC	KD4RJN	443.13750	+	CC1	
Banco, Fork Mountain						
	FM	K3HOT	443.25000	+	107.2 Hz	
Basye, Great North Mountain						
	FM	K4MRA	444.60000	+	131.8 Hz	T-MARC
Bath, Warm Springs Mountain						
	FM	W4COV	146.80500	-	107.2 Hz	
Beaverdam	DMR/MARC	WA4FC	444.61250	+	CC1	
Beaverdam, Ashland Berry Farm						
	FM	KD4RJN	147.06000	+	74.4 Hz	SERA-VA
Bedford	DMR/MARC	WA1ZMS	443.80000	+	CC1	SERA-VA
	FM	K4LYL	53.01000	52.01000		
Bedford, Apple Orchard Mtn						
	FM	WA1ZMS	146.68500	-	100.0 Hz	SERA-VA
	FM	WA1ZMS	442.65000	+	100.0 Hz	SERA-VA
Bedford, Thaxton Mountain						
	FM	K4LYL	53.15000	52.15000		SERA-VA
	FM	K4LYL	147.10500	+	136.5 Hz	SERA-VA
Bent Mountain, Slings Gap						
	FM	W4KZK	442.92500	+	107.2 Hz	SERA-VA
Blacksburg, Brush Mountain						
	FM	W9KIC	146.71500	-		SERA-VA
	FM	N4NRV	444.65000	+	107.2 Hz	
Bland	FM	KD4LMZ	145.35000	-	103.5 Hz	SERA-VA
Blue Mountain	FM	KC4CK	224.16000	-		T-MARC
Bluefield, Oneida Peak						
	FM	N8FWL	224.44000	+	123.0 Hz	T-MARC
Bluemont	DMR/MARC	N3JLT	449.92500	-	CC1	T-MARC
	FM	K8GP	53.37000	52.37000		
	FM	K8GP	224.34000	-	77.0 Hz	T-MARC
Bluemont, Blue Ridge						
	FM	WA4TSC	147.30000	+	146.2 Hz	T-MARC
Bon Air, WCVE Tower						
	FM	W4RAT	146.88000	-	74.4 Hz	SERA-VA
	FM	W4RAT	442.55000	+	74.4 Hz	SERA-VA
Buckingham	FM	WR4CV	224.40000	-	110.9 Hz	
	FM	WR4CV	444.95000	+	110.9 Hz	SERA-VA
Burke	FM	WA3TOL	448.67500	-	100.0 Hz	T-MARC
Chantilly	DMR/BM	KD4RTH	443.15000	+	CC1	
Chantilly, Dulles Airport						
	FM	N4FSC	145.31000	-	77.0 Hz	T-MARC
	FM	W4DLS	145.31000	-	77.0 Hz	
	FM	K4IAD	147.33000	+	203.5 Hz	T-MARC
	FM	K4IAD	444.75000	+	100.0 Hz	T-MARC
Charlottesville	DMR/MARC	WA4FC	444.91250	+	CC1	SERA-VA
	FM	KG4HOT	224.60000	-	151.4 Hz	SERA-VA
Charlottesville, Buck's Elbow						
	FM	WA4TFZ	224.76000	-	151.4 Hz	
Charlottesville, Carter's Moun						
	FM	KF4UCI	442.07500	+	151.4 Hz	SERA-VA
Charlottesville, Carters Mount						
	FM	K4DND	146.73000	-	151.4 Hz	
Charlottesville, Martha Jeffer						
	FM	WA4TFZ	146.92500	-	151.4 Hz	SERA-VA
	FM	WA4TFZ	444.25000	+	151.4 Hz	SERA-VA

372   VIRGINIA

Location	Mode	Call sign	Output	Input	Access	Coordinator
Charlottesville, University Of						
	FM	W4UVA	443.00000	+	151.4 Hz	
Chesapeake	DMR/MARC	KK4WTI	442.58750	+	CC1	
	FM	K4AMG	145.15000	-	103.5 Hz	
Chesapeake, Bowers Hill						
	FM	W4CAR	146.61000	-	162.2 Hz	
Chesapeake, Greenbriar						
	FM	W4CAR	146.82000	-	162.2 Hz	
	FM	W4CAR	444.00000	+	162.2 Hz	
Chester	FM	KA4CBB	147.36000	+		
Chester, Carver Middle School						
	FM	KD4KWP	145.31000	-	127.3 Hz	
Chilhowie	FM	W4DWN	442.00000	+	103.5 Hz	
Christiansburg	DMR/MARC	KD4ADL	444.25000	+	CC1	
Christiansburg, Poor Mountain						
	FM	N4VL	145.41000	-	103.5 Hz	
Clifton Forge, Warm Springs Mo						
	FM	N4HRS	444.37500	+	167.9 Hz	SERA-VA
Clinchco	FM	KB4RFN	147.15000	+	88.5 Hz	SERA-VA
Columbia	FM	WA4FC	444.35000	+	74.4 Hz	
Covesville, Heard Mountain						
	FM	WA4TFZ	146.76000	-	151.4 Hz	SERA-VA
	FM	WA4TFZ	224.76000	-	151.4 Hz	SERA-VA
	FM	WA4PGI	442.25000	+	100.0 Hz	SERA-VA
Covington, Warm Springs Mounta						
	FM	KF4YLM	146.97000	-	91.5 Hz	
Crozet, Bucks Elbow Mountain						
	FM	WA4TFZ	146.89500	-	151.4 Hz	SERA-VA
Culpeper	FM	KA4DCS	53.91000	52.91000	100.0 Hz	T-MARC
Damascus, White Top Mountain						
	FM	KM4X	443.00000	+	103.5 Hz	SERA-VA
Danville, White Oak Mtn						
	FM	K4AU	146.70000	-	107.2 Hz	SERA-VA
Dismal Peak	DMR/BM	KD4BNQ	147.13500	+	CC1	
Eastville	FM	KN4GE	147.34500	+	156.7 Hz	
Elk Creek, Iron Mtn						
	FM	N4MGQ	147.24000	+	107.2 Hz	SERA-VA
Fairfax	DMR/BM	N3QEM	442.88750	+	CC13	
	DMR/BM	N3QEM	442.91250	+	CC13	
Fairfax, Fair Oaks Hospital						
	FM	K4XY	448.37500	-		T-MARC
Fairfax, Fairfax County Public						
	FM	NV4FM	146.79000	-		T-MARC
	FM	W4YHD	224.10000	-	77.0 Hz	T-MARC
Falls Church	FM	W4AVA	447.62500	-	107.2 Hz	T-MARC
Fancy Gap	DMR/BM	WX4F	444.15000	+	CC1	
	DMR/MARC	N4YR	440.66250	+	CC14	
	DMR/MARC	WX4F	443.93750	+	CC1	
	FM	KE4QQX	29.66000	-	88.5 Hz	
	FM	N4YR	53.63000	52.63000		
	FM	N4VRD	145.33000	-	77.0 Hz	
	FM	N4YR	442.22500	+		
	FM	KB4GHT	442.32500	+	100.0 Hz	
	FM	KF4OVA	442.42500	+	107.2 Hz	SERA-VA
	FM	WA4LOY	444.10000	+	136.5 Hz	SERA-VA
Fancy Gap, Pops Peak						
	FM	K4IL	442.57500	+	100.0 Hz	SERA-VA
Farmville, Leigh Mtn						
	FM	N4HRS	146.91000	-	DCS 712	
	FM	N4HRS	444.32500	+	DCS 712	
Farmville, Water Tower						
	FM	WR4CV	146.95500	-	136.5 Hz	SERA-VA

VIRGINIA 373

Location	Mode	Call sign	Output	Input	Access	Coordinator
Farmville, Water Tower						
	FM	WR4CV	443.30000	+	136.5 Hz	SERA-VA
Floyd	FM	W4FCV	147.21000	+	114.8 Hz	SERA-VA
Floyd, Floyd School Admin Offi						
	FM	W4FCV	442.90000	+	114.8 Hz	SERA-VA
Floyd, NWS Doppler Radar Tower						
	FM	KG4MAV	443.35000	+	114.8 Hz	SERA-VA
Fork Mountain	FM	WA3KOK	443.25000	+	107.2 Hz	T-MARC
Fork Union	DMR/MARC	K4JK	444.53750	+	CC1	
Franklin	FM	N4WFU	146.74500	-	131.8 Hz	
	FM	WT4FP	147.27000	+	131.8 Hz	SERA-VA
Fredericksburg	FM	W1ZFB	51.86000	50.86000	127.3 Hz	T-MARC
Fredericksburg, Chancellor Lan						
	FM	K4TS	147.01500	+	79.7 Hz	T-MARC
	FM	K4TS	443.85000	+	79.7 Hz	
Fredericksburg, Four Mile Fork						
	FM	W1ZFB	927.03750	902.03750	91.5 Hz	T-MARC
Front Royal	FM	K4QJZ	51.94000	50.94000	141.3 Hz	T-MARC
Front Royal, High Knob Mountai						
	FM	K4QJZ	145.21000	-	141.3 Hz	T-MARC
	FM	NO4N	442.72500	+	107.2 Hz	T-MARC
Gainesville	FM	W4LAM	224.46000	-		T-MARC
Gate City	FM	K4GV	441.90000	+		
	FM	KF4VTM	444.70000	+	103.5 Hz	
Gate City, Clinch Mountain						
	FM	KF4VTM	146.82000	-		SERA-VA
Gloucester Courthouse, Walter						
	FM	WN4HRT	145.21000	-	100.0 Hz	SERA-VA
	FM	W4HZL	145.37000	-	100.0 Hz	SERA-VA
Goochland	FM	WB4IKL	147.27000	+	203.5 Hz	SERA-VA
Gordonsville	FM	KF4UCI	444.40000	+	151.4 Hz	T-MARC
Gordonsville, Gibson Mountain						
	FM	W4CUL	147.12000	+	146.2 Hz	
	FM	K3VB	224.18000	-	146.2 Hz	T-MARC
	FM	W4CUL	443.80000	+	146.2 Hz	
Gore	FM	KM4OGQ	444.25000	+	100.0 Hz	T-MARC
Greenbush	DMR/BM	K9AGR	440.67500	+	CC1	
Greene, Flat Top Mountain						
	FM	KF4UCI	145.47000	-	151.4 Hz	
Grundy	FM	K4NRR	147.31500	+		
Gum Spring	FM	KB4MIC	53.07000	52.07000	203.5 Hz	
Hampton	DMR/BM	KA4VXR	147.22500	+	CC1	SERA-VA
	FM	KE4UP	145.49000	-	100.0 Hz	
	FM	W4QR	146.73000	-	100.0 Hz	SERA-VA
	FM	K4TM	146.92500	-		SERA-VA
	FM	W4QR	444.55000	+	100.0 Hz	
Hampton, City Hall						
	DSTAR	W4HPT	145.20000	-		SERA-VA
	DSTAR	W4HPT	444.21250	+		SERA-VA
	DSTAR	W4HPT	1298.50000			SERA-VA
Hampton, Sentara Careplex Hosp						
	DSTAR	K4HPT	443.50000	+	100.0 Hz	SERA-VA
Hampton, Sentera Care Plex						
	FM	WA4ZUA	145.17000	-	131.8 Hz	SERA-VA
Harrisonburg	DMR/MARC	K4JK	444.66250	+	CC1	
	FM	N4DSL	224.50000	-	131.8 Hz	T-MARC
Harrisonburg, EMU						
	FM	K4MRA	147.31500	+	131.8 Hz	T-MARC
Harrisonburg, Lairds Knob						
	FM	KC4GXI	443.15000	+	131.8 Hz	T-MARC
Harrisonburg, Little North Mou						
	FM	K4KLH	147.22500	+	131.8 Hz	T-MARC

374 VIRGINIA

Location	Mode	Call sign	Output	Input	Access	Coordinator
Harrisonburg, Massanutten Peak						
	FM	K4MRA	145.13000	-	131.8 Hz	T-MARC
Haymarket	DMR/BM	W4BRM	448.17500	-	CC6	
	DMR/MARC	W4YP	448.97500	-	CC6	T-MARC
	DSTAR	N4USI	442.41250	+		
	FM	N3KL	145.13000	-		
	FM	N3AUY	449.02500	-	156.7 Hz	T-MARC
Haymarket, Bull Run Mountain						
	DMR/BM	W4BRM	448.32500	-	CC6	T-MARC
	FM	W4BRM	53.49000	52.49000	77.0 Hz	
	FM	W4BRM	224.40000	-	77.0 Hz	T-MARC
	FM	W4BRM	448.22500	-	77.0 Hz	T-MARC
	FM	W4BRM	927.62500	902.62500	77.0 Hz	T-MARC
Haymarket, VA	DSTAR	W4BRM	145.45000	-		T-MARC
Heathsville	FM	W4NNK	147.33000	+	100.0 Hz	SERA
Herndon	DMR/MARC	N3QEM	442.43750	+	CC1	
	DMR/MARC	N3QEM	442.86250	+	CC13	
	DMR/MARC	N3QEM	927.66250	902.66250	CC1	
	FM	W4CIA	147.21000	+	110.9 Hz	T-MARC
Hillsville	DMR/BM	K4EZ	147.39000	+	CC1	
	FM	W4GHS	145.27000	-	103.5 Hz	SERA-VA
Hillsville, VA	FM	K4EZ	147.04500	+		SERA-VA
Honaker	FM	WR4RC	145.37000	-	103.5 Hz	SERA-VA
	FM	KK4EH	146.83500	-		
Honaker, Big A Mountain						
	FM	KM4HDM	146.80500	+	103.5 Hz	SERA-VA
	FM	KM4HDM	442.10000	+		SERA-VA
Hopewell	FM	KG4DCX	147.09000	-		
Independence	FM	W4TOW	443.37500	+	103.5 Hz	SERA-VA
Isle Of Wight, County Courthou						
	FM	WT4RA	147.19500	+	100.0 Hz	SERA-VA
Jarrett	FM	N4LLE	146.62500	-	100.0 Hz	SERA-VA
Jonesville	FM	AJ4G	442.57500	+	100.0 Hz	SERA-VA
Keysville	DMR/BM	K4DJQ	443.11250	5443.11250	CC1	
Kilmarnock	FM	W4NNK	146.83500	-	100.0 Hz	SERA-VA
Kilmarnock, Fire Station						
	FM	W4GSF	145.45000	-		
King George	FM	K4GVA	448.47500	-	79.7 Hz	T-MARC
King George, Dahlgren Naval Su						
	FM	N3PZZ	145.17000	-	88.5 Hz	
King George, Dalhgren Naval Su						
	FM	N3PZZ	145.33000	-	88.5 Hz	
King George, Harry Nice Bridge						
	FM	W4KGC	146.74500	-	107.2 Hz	T-MARC
Leesburg	DMR/MARC	WR3D	443.90000	+	CC1	
	FM	WA4TXE	146.70000	-	77.0 Hz	T-MARC
	FM	WA4TXE	442.10000	+	77.0 Hz	T-MARC
Lexington	DMR/BM	W4DHW	441.92500	+	CC1	SERA-VA
Lexington, Rocky Knob						
	FM	KG4HOT	224.58000	-	136.5 Hz	
Lexington, Rocky Mountain						
	FM	W4ROC	53.01000	52.01000		
	FM	W4ROC	147.33000	+		SERA-VA
	FM	W4ROC	444.15000	+		SERA-VA
	FM	WR4CV	927.46250	902.46250	151.4 Hz	SERA-VA
Linden	DMR/MARC	N8RAT	443.16250	+	CC1	T-MARC
	FM	N3UHD	444.15000	+	77.0 Hz	T-MARC
Linden, Atop Blue Mountain At						
	FM	N8RAT	224.28000	-	100.0 Hz	T-MARC
Linden, Blue Ridge Mountains						
	FM	N3UR	442.35000	+	123.0 Hz	T-MARC
Louisa	FM	KD4OUZ	442.22500	+	131.8 Hz	SERA-VA

VIRGINIA 375

Location	Mode	Call sign	Output	Input	Access	Coordinator
Lynchburg	DMR/BM	WA4RTS	146.65500	-	CC1	
	FM	KC4RBA	145.37000	-	186.2 Hz	
Lynchburg, Candlers Mountain						
	FM	N3OG	146.61000	-	136.5 Hz	SERA-VA
	FM	N3OG	442.60000	+	136.5 Hz	SERA-VA
Lynchburg, Tobacco Row Mountai						
	FM	WA4RTS	147.19500	+	136.5 Hz	
	FM	N4HRS	443.50000	+		
	FM	K4LBG	444.75000	+	136.5 Hz	SERA-VA
Lynchburg, Tobacco Row Mtn						
	FM	K4CQ	444.50000	+	136.5 Hz	SERA-VA
Madison	DMR/BM	AE4ML	147.03000	+	CC1	T-MARC
Manassas	FM	W4OVH	146.85000	-		
	FM	W4OVH	146.97000	-	100.0 Hz	T-MARC
	FM	W4OVH	224.66000	-	100.0 Hz	T-MARC
	FM	W4OVH	442.20000	+		T-MARC
	FM	K4GVT	443.50000	+	110.9 Hz	T-MARC
Manassas, NOVEC Communications						
	DSTAR	W4OVH	146.86500	-		T-MARC
	FM	W4OVH	442.51250	+		T-MARC
Marion	FM	W4GHS	145.27000	-	103.5 Hz	
	FM	KM4X	146.64000	-	103.5 Hz	
Martinsville	DMR/BM	N2TEK	441.85000	+	CC15	SERA-VA
	FM	K4MVA	147.12000	+	107.2 Hz	
	FM	N4HRS	444.87500	+	DCS 712	SERA-VA
Martinsville, Chestnut Knob						
	FM	K4MVA	147.28500	+	107.2 Hz	
	FM	KF4RMT	441.75000	+	77.0 Hz	
Maurertown	FM	N3UHD	442.47500	+	77.0 Hz	T-MARC
Max Meadows, Hamiltons Knob						
	FM	K4IJ	442.62500	+	103.5 Hz	SERA-VA
McLean	DMR/MARC	N9KET	441.33750	+	CC1	T-MARC
Middleburg	FM	KA4DCS	29.68000	-	146.2 Hz	T-MARC
Millwood	FM	K4IAD	444.75000	+	173.8 Hz	T-MARC
Montebello, Whetstone Ridge Ra						
	FM	WR4CV	145.45000	-	136.5 Hz	SERA-VA
Monterey	FM	WD4ITN	147.18000	+		
Montross	FM	W4GMF	146.89500	-	146.2 Hz	T-MARC
	FM	KJ4PGD	442.00000	+	110.9 Hz	T-MARC
Mount Jackson	FM	KB6VAA	146.71500	-	146.2 Hz	T-MARC
Mountain Lake	DMR/BM	KD4BNQ	146.91000	-	CC1	SERA-VA
New Market	FM	KQ4D	443.35000	+		
New Market, Luray Caverns						
	FM	KQ4D	146.62500	-	131.8 Hz	T-MARC
Newport News	DSTAR	W4MT	441.81250	+		
	FM	W4MT	145.23000	-	100.0 Hz	SERA-VA
	FM	W4CM	147.16500	+	100.0 Hz	SERA-VA
Newport News, Riverside Region						
	FM	WN4HRT	147.00000	-	100.0 Hz	
Norfolk	DMR/BM	K4LCT	440.51250	+	CC7	SERA-VA
	DMR/BM	W2CID	445.51250	-	CC7	
	FM	W4VB	147.37500	+	131.8 Hz	
Norfolk, Downtown						
	FM	W4VB	145.33000	-	131.8 Hz	
	FM	W4VB	224.40000	-	131.8 Hz	SERA-VA
	FM	W4VB	442.95000	+	131.8 Hz	SERA-VA
Norfolk, Fire Station 10						
	FM	W4VB	444.47500	+	74.4 Hz	
Norfolk, Harbor Front Park						
	FM	KD4FIG	224.18000	-	103.5 Hz	SERA-VA
Norfolk, Norfolk Waterfront						
	FM	KC2HTT	147.07500	+	100.0 Hz	

376   VIRGINIA

Location	Mode	Call sign	Output	Input	Access	Coordinator
Norfolk, Norfolk Waterfront						
	FM	KC2HTT	442.45000	+	100.0 Hz	
Norton , High Knob Mt						
	FM	WD4GSM	224.42000	-		SERA-VA
Norton, High Knob						
	FM	KM4OKT	444.07500	+	136.5 Hz	SERA-VA
Palmyra	FM	K4MSR	145.17000	-	151.4 Hz	
Pearisburg, Bald Knob						
	FM	KD4BNQ	441.95000	+	107.2 Hz	SERA-VA
Pearisburg, Dismal Peak						
	FM	KE4JYN	53.47000	52.47000	107.2 Hz	
	FM	KQ4Q	224.86000	-	107.2 Hz	SERA-VA
	FM	N4HRS	444.67500	+	DCS 712	SERA-VA
Pearisburg, Giles County Court						
	FM	W4NRV	147.37500	+	100.0 Hz	SERA-VA
Pembroke	DMR/BM	WA4ONG	444.95000	+	CC1	SERA-VA
Pennington Gap	FM	KG4OXG	145.49000	-	131.8 Hz	
Petersburg	DMR/MARC	WA4FC	442.68750	+	CC1	SERA-VA
	FM	KE4SCS	146.98500	-		
	FM	KK4QAK	147.39000	+	74.4 Hz	
	FM	KK4QAK	444.97500	+	74.4 Hz	
Portsmouth	FM	W4POX	53.89000	52.89000		SERA-VA
	FM	W4POX	443.80000	+		SERA-VA
Portsmouth, Maryview Hospital						
	FM	AA4AT	146.70000	-		
Portsmouth, Portsmouth Naval H						
	FM	W4POX	146.85000	-	100.0 Hz	SERA-VA
Powhatan	DMR/MARC	N4POW	443.35000	+	CC1	SERA-VA
Powhatan, Powhatan Water Tower						
	FM	N4POW	147.31500	+	74.4 Hz	SERA-VA
Prince George, South Point Bus						
	FM	KG4YJB	444.27500	+	103.5 Hz	
Pulaski, Peaks Knob						
	FM	K4XE	442.07500	+	DCS 712	SERA-VA
Pungoteague	DMR/MARC	N4TIK	147.21000	747.21000	CC1	
	FM	N4TIK	145.11000	-	156.7 Hz	
Quantico	FM	K3FBI	147.34500	+	167.9 Hz	T-MARC
	FM	K3FBI	443.55000	+		T-MARC
Radford	FM	KB4RU	147.00000	+	107.2 Hz	SERA-VA
Radford, Cloyd's Mountain						
	FM	N4NRV	147.18000	+	103.5 Hz	T-MARC
Reston	FM	N2LEE	443.00000	+	88.5 Hz	T-MARC
Richmond	DMR/BM	WA4FC	147.18000	+	CC1	
	DMR/BM	WA4FC	443.53750	+	CC1	
	DMR/MARC	W4RAT	443.58750	+	CC1	
	DMR/MARC	WA4FC	444.53750	+	CC1	
	DMR/MARC	WA4ONG	449.95000	+	CC1	
	DSTAR	W4FJ	147.25500	+		
	DSTAR	W4FJ	1284.00000	1264.00000		
	FM	WA4MAS	145.11000	-	127.3 Hz	
	FM	WA4FC	224.52000	-	74.4 Hz	
	FM	W4FJ	443.71250	+		SERA-VA
Richmond, Downtown						
	FM	WA4FC	224.52000	-	74.4 Hz	SERA-VA
	FM	WA4FC	927.05000	902.05000		SERA-VA
	FM	WA4FC	1282.00000	1262.00000	88.5 Hz	SERA-VA
Richmond, James Monroe Buildin						
	FM	KN4SKI	146.94000	-	74.4 Hz	SERA-VA
Richmond, Southside Richmond						
	P25	WA4FC	927.07500	902.07500		SERA-VA
Richmond, WTVR-TV Tower						
	FM	KG4MRA	145.43000	-	74.4 Hz	

VIRGINIA 377

Location	Mode	Call sign	Output	Input	Access	Coordinator
Richmond, WTVR-TV Tower						
	FM	W4MEV	224.42000	-	74.4 Hz	SERA-VA
Ridgeway	FM	WS4W	224.38000	-	88.5 Hz	
Ripplemead	DMR/BM	KD4BNQ	440.80000	+	CC1	
Roanoke	DMR/MARC	K4ITL	441.88750	+	CC1	
	DMR/MARC	W5CUI	444.77500	+	CC1	SERA-VA
	FM	WB8BON	53.09000	52.09000	123.0 Hz	SERA-VA
	FM	W4KDN	146.94000	-	107.2 Hz	SERA-VA
	FM	K5JCT	442.30000	+	127.3 Hz	SERA-VA
	FM	KS4BO	443.67500	+	110.9 Hz	
Roanoke, Community Hospital						
	FM	N4HRS	444.27500	+	103.5 Hz	SERA-VA
Roanoke, Long Ridge (Sugarloaf						
	DMR/BM	KD4EG	440.70000	+	CC1	SERA-VA
Roanoke, Mill Mountain						
	FM	W4KZK	442.75000	+	107.2 Hz	SERA-VA
Roanoke, Poor Mountain						
	FM	W4CA	146.98500	-	107.2 Hz	SERA-VA
	FM	K4ARO	442.60000	+	114.8 Hz	
	FM	N4HRS	444.17500	+	DCS 712	SERA-VA
Roanoke, Poor Mtn						
	FM	K1GG	146.74500	-	107.2 Hz	SERA-VA
Roanoke, Tinker Mountain						
	FM	N4HRS	444.47500	+	DCS 712	
Roanoke, Tinker Mtn						
	FM	WB8BON	444.92500	+	107.2 Hz	
Round Hill	FM	K0QBU	449.42500	-	100.0 Hz	
Salem	DMR/BM	KD4EG	445.80000	-	CC1	
Salem, 12 O Clock Knob						
	FM	KF4RGH	146.88000	-	107.2 Hz	SERA-VA
Salem, Ft Lewis Mtn						
	FM	WB8BON	444.92500	+	77.0 Hz	
Salem, Sugar Loaf Mountain						
	FM	W4KZK	444.85000	+	107.2 Hz	SERA-VA
Salem, VA Hospital						
	FM	W4KZK	443.75000	+	107.2 Hz	
Shenandoah	FM	N4PJI	146.67000	-	114.8 Hz	T-MARC
Smithfield, Isle Of Wight Cour						
	FM	WT4RA	442.82500	+	100.0 Hz	SERA-VA
South Boston	FM	W4HCH	145.35000	-		
	FM	KF4AGO	147.06000	-		
South Hill	DMR/MARC	K4MJO	444.78750	+	CC1	
	FM	KB2AHZ	147.00000	+	77.0 Hz	SERA-VA
	FM	KB2AHZ	443.52500	+	100.0 Hz	SERA-VA
South Hill, Community Memorial						
	FM	W4CMH	145.47000	-	82.5 Hz	SERA-VA
Spotsylvania	DMR/BM	AE4ML	442.40000	+	CC1	T-MARC
	DMR/BM	AE4ML	442.85000	+	CC1	T-MARC
	FM	WW4EMC	224.26000	-		T-MARC
	FM	WW4EMC	442.70000	+	114.8 Hz	T-MARC
Stafford	DSTAR	WS4VA	145.32000	-		
	DSTAR	WU5MC	442.48750	+		
	DSTAR	WS4VA	447.27500	-		
	DSTAR	WS4VA	1282.20000	1262.20000		T-MARC
	FM	WS4VA	145.27000	-	79.7 Hz	T-MARC
	FM	WS4VA	147.37500	+	79.7 Hz	T-MARC
	FM	WS4VA	444.45000	+	79.7 Hz	T-MARC
Standardsville, Snow Mountain						
	FM	KF4UCI	443.90000	+	151.4 Hz	SERA-VA
Staunton	FM	KD4WWF	146.70000	-	131.8 Hz	
Staunton, Elliot Knob						
	FM	KG4HOT	224.30000	-	131.8 Hz	

378   VIRGINIA

Location	Mode	Call sign	Output	Input	Access	Coordinator
Staunton, Elliot Knob						
	FM	KG4HOT	444.10000	+	131.8 Hz	
Staunton, Elliott Knob						
	FM	KG4HOT	147.04500	+	131.8 Hz	
Staunton, Hermitage						
	FM	WA4ZBP	146.85000	-	131.8 Hz	
Sterling	DMR/MARC	N3QEM	442.87500	+	CC1	
Suffolk, Driver (WHRO Tower)						
	FM	N4SD	146.79000	-	100.0 Hz	
Tysons Corner	DSTAR	NV4FM	145.34000	-		T-MARC
	DSTAR	NV4FM	448.03500	-		T-MARC
	DSTAR	NV4FM	1282.80000	1262.80000		T-MARC
	FM	NV4FM	146.91000	-		T-MARC
	FM	NV4FM	447.02500	-		T-MARC
Vesta	FM	NJ1K	145.11000	-	114.8 Hz	SERA-VA
Vesuvius, Whetstone Ridge						
	FM	K4DND	145.45000	-	110.9 Hz	SERA-VA
Vienna	DMR/BM	N3QEM	145.17000	-	CC13	
	DMR/BM	N3QEM	927.67500	902.67500	CC1	
	DMR/MARC	N3QEM	927.70000	902.70000	CC1	
	FM	K4HTA	146.68500	-		T-MARC
Virginia Beach	DMR/MARC	W4BSB	442.58750	+	CC1	
	FM	KE4HGP	145.20000	-		
	FM	W4KXV	146.89500	-	141.3 Hz	
	FM	WN4HRT	147.30000	+	100.0 Hz	
Virginia Beach, Fire Station #						
	FM	W4KXV	146.97000	-	141.3 Hz	
Virginia Beach, Virginia Beach						
	FM	W4KXV	444.95000	+	141.3 Hz	
Walkerton	FM	W4TTL	146.71500	-		
Wallops Island	DMR/MARC	W4WFF	444.88000	+	CC1	
Warrenton	FM	W4VA	147.16500	+	167.9 Hz	T-MARC
	FM	W4VA	442.25000	+	167.9 Hz	T-MARC
Waynesboro , Bear Den Mountain						
	FM	W4PNT	147.07500	+	131.8 Hz	
Waynesboro, Afton Mountain						
	FM	NM9S	53.41000	52.41000		
Waynesboro, Bear Den Mountain						
	FM	KC8MTV	145.29000	-	131.8 Hz	SERA-VA
	FM	KF4UCI	444.77500	+	151.4 Hz	
Williamsburg	DMR/BM	N4ARI	145.41000	-	CC1	
	DMR/BM	KE4NYV	444.27500	+	CC1	
	DMR/BM	N4ARI	444.70000	+	CC1	SERA-VA
	FM	KB4ZIN	146.76000	-	118.8 Hz	SERA-VA
	FM	KB4ZIN	147.10500	+	118.8 Hz	
	FM	KB4ZIN	444.10000	+		SERA-VA
Winchester	FM	W3IF	442.00000	+	146.2 Hz	T-MARC
	FM	N2XIF	442.05000	+	179.9 Hz	T-MARC
	FM	NM4CC	442.60000	+	141.3 Hz	T-MARC
Winchester, Great North Mounta						
	FM	W4RKC	146.82000	-	146.2 Hz	T-MARC
	FM	W4RKC	448.77500	-	146.2 Hz	T-MARC
Winchester, North Mountain						
	FM	K4USS	145.39000	-	146.2 Hz	T-MARC
	FM	KG4Y	224.90000	-	146.2 Hz	T-MARC
	FM	KG4Y	444.55000	+		T-MARC
Wintergreen	DMR/BM	WR4CV	444.43750	+	CC1	
	DSTAR	WR4CV	444.93750	+	151.4 Hz	
	FM	WR4CV	146.82000	-	136.5 Hz	
Wintergreen, Wintergreen Resor						
	FM	WR4CV	224.40000	-	136.5 Hz	SERA-VA
	FM	WR4CV	444.55000	+	136.5 Hz	SERA-VA

VIRGINIA 379

Location	Mode	Call sign	Output	Input	Access	Coordinator
Wirtz	DMR/BM	W4JWC	440.75000	+	CC1	
Woodbridge	DSTAR	WD4HRO	1293.00000	1273.00000		
	FM	W4IY	147.24000	+	107.2 Hz	T-MARC
	FM	W4IY	224.78000	-		T-MARC
	FM	W4IY	444.90000	+	127.3 Hz	T-MARC
Woodbridge, Potomac Mills						
	FM	K4IAD	444.85000	+		
Wytheville	DMR/BM	W4VSP	442.52500	+	CC1	SERA-VA
	DMR/BM	W4VSP	444.32500	+	CC1	SERA-VA
	DMR/MARC	K4EZ	443.26250	+	CC1	SERA-VA
	FM	K4EZ	146.77500	-	103.5 Hz	SERA-VA
Wytheville, Walker Mountain						
	FM	K4YW	224.56000	-	77.0 Hz	SERA-VA
Yorktown	DMR/BM	KN4KV	444.60000	+	CC1	

"The Northwest's Largest Ham Convention"

SEA-PAC '20 ◆ SEA-PAC '21

June 5 – June 7, 2020

June 4 – June 6, 2021

ARRL Northwestern Division Convention
Seaside Convention Center, Seaside Oregon

- Commercial Exhibits • Giant Flea Market • Banquet/Entertainment
- Workshops • Seminars • Prizes • VE Testing • Special Event Station

Near the Beautiful Pacific Northwest Ocean Beach

General Info—info@seapac.org	*SEA-PAC*	Exhibitor Info—exhibitors@seapac.org
Registration Info—registration@seapac.org	Post Office Box 7263	Flea Market Info—fleamarket@seapac.or
	Aloha OR 97007-0963	

SEA-PAC on the Web: www.seapac.org

WASHINGTON 381

Location	Mode	Call sign	Output	Input	Access	Coordinator
WASHINGTON						
Aberdeen	FM	W7ZA	147.16000	+	88.5 Hz	
	FM	KA7DNK	444.60000	+	100.0 Hz	
	FM	N7UJK	444.82500	+	118.8 Hz	
Airway Heights	FM	W7TSC	443.32500	+		
Alder	FM	KB7CNN	145.45000	-	110.9 Hz	
Almira	FM	W7OHI	147.00000	+	100.0 Hz	
Anacortes	FM	KG7OCP	443.35000	+	100.0 Hz	
Anacortes, 29th St Water Tank						
	FM	KG7OCP	443.35000	+		
Anacortes, Mt Erie						
	FM	W7PSE	441.72500	+	103.5 Hz	
Anatone	FM	WF7L	446.05000	-	103.5 Hz	
Arlington	FM	N7XCG	440.40000	+	123.0 Hz	
	FM	N7NFY	443.22500	+	103.5 Hz	
Ashford	FM	K7DNR	53.39000	51.69000	100.0 Hz	
	FM	K7DNR	145.25000	-	186.2 Hz	
	FM	K7DNR	442.57500	+	141.3 Hz	
	FM	K7DNR	927.52500	902.52500	114.8 Hz	
Astoria	DMR/MARC	NA7Q	147.43750	146.33750	CC1	
Auburn	FM	K7SYE	147.24000	+	123.0 Hz	
Bainbridge Island	DSTAR	W7NPC	444.56250	+		
	DSTAR	W7NPC	1290.50000	1270.50000		
	FM	W7NPC	53.43000	51.73000	100.0 Hz	
	FM	WA6PMX	224.42000	-	88.5 Hz	
	FM	K7LD	440.20000	+	103.5 Hz	
	FM	W7NPC	444.47500	+	103.5 Hz	
Baldi Mtn	FM	N7FSP	1292.30000	-	103.5 Hz	WWARA
Belfair	FM	NM7E	145.17000	-	103.5 Hz	
	FM	KE7OYB	145.45000	-	100.0 Hz	
Bellevue	DMR/MARC	N7ERP	441.28750	+	CC1	
	DMR/MARC	AE7WZ	445.00000	-	CC3	
	DSTAR	K7LWH	146.41250	147.41250		
	DSTAR	K7LWH	443.06250	+		
	DSTAR	K7LWH	1290.00000	1270.00000		
	FM	KC7IYE	441.10000	+	156.7 Hz	
Bellevue, Cougar Mountain						
	ATV	WW7ATS	1253.25000	433.95000		
	FM	KE7GFZ	441.82500	+	103.5 Hz	
	FM	K7MMI	442.32500	+	151.4 Hz	
	FM	K7PP	443.40000	+	123.0 Hz	
	FM	K7OET	444.32500	+	100.0 Hz	
	FM	W7DME	444.85000	+	103.5 Hz	ORRC
	P25	K7SLB	146.47500	147.47500	110.9 Hz	
Bellevue, Lincoln Square						
	FM	K7LWH	444.60000	+	103.5 Hz	
Bellingham	DMR/BM	NC7Q	442.25000	+	CC1	
	DMR/BM	W7BFD	442.30000	+	CC1	
	DMR/MARC	K7SKW	146.50000	+	CC1	
	FM	N7FYU	441.92500	+	103.5 Hz	
Bellingham, King Mountain						
	FM	K7SKW	147.16000	+	103.5 Hz	
	FM	N7FYU	224.86000	-	103.5 Hz	
	FM	K7SKW	443.65000	+	103.5 Hz	
	FM	N7FYU	1290.95000	1270.95000	103.5 Hz	
Bellingham, Lookout Mountain						
	DSTAR	WC7SO	440.47500	+		
	FM	NC7Q	224.16000	-	156.7 Hz	
	FM	WA7ZWG	927.48750	902.48750	114.8 Hz	
Bellingham, Lookout Mtn						
	DSTAR	WC7SO	146.45000	147.45000		

382 WASHINGTON

Location	Mode	Call sign	Output	Input	Access	Coordinator
Bellingham, Squalicum Mountain						
	FM	K7SKW	443.75000	+	103.5 Hz	
Bellingham, Sudden Valley						
	FM	WA7SV	442.75000	+	103.5 Hz	
Blaine	FM	W7BPD	927.37500	902.37500	114.8 Hz	
Blyn, Blyn Lookout						
	FM	WR7V	53.37000	51.67000	100.0 Hz	
Boistfort, Baw Faw Peak						
	FM	W7WRG	224.08000	-	103.5 Hz	
Bothell	DSTAR	KF7UUY	441.26250	+		
	FM	K7SLB	147.47500	146.47500	114.8 Hz	
	FM	WA7HJR	442.55000	+	103.5 Hz	
Bremerton	DMR/BM	KC7Z	444.07500	+	CC1	
	DMR/MARC	AF7PR	440.72500	+	CC1	
	FM	K7OET	442.25000	+	123.0 Hz	
	FM	N7MTC	443.05000	+	179.9 Hz	
	FM	K7OET	444.80000	+	100.0 Hz	
Bremerton, Gold Mountain						
	FM	WW7RA	146.62000	-	103.5 Hz	
	FM	W7UFI	224.66000	-	103.5 Hz	
	FM	W7TWA	441.50000	+	100.0 Hz	
	FM	WW7RA	442.65000	+	103.5 Hz	
Bremerton, Green Mountain						
	FM	W7PSE	441.75000	+	103.5 Hz	
Brewster, Dyer Hill						
	FM	W7GSN	146.74000	-	110.9 Hz	IACC
Buck Mtn	FM	W7WRG	224.58000	-	103.5 Hz	WWARA
Buckley	FM	WA7LBS	443.02500	+	107.2 Hz	
Buckley, Three Sisters						
	FM	N7BUW	444.67500	+	136.5 Hz	
Buckley, Three Sisters Summit						
	FM	WB7DOB	147.30000	+	88.5 Hz	
	FM	WB7DOB	223.92000	-	103.5 Hz	
Burien	DSTAR	KF7CLD	147.50000	146.50000		
	DSTAR	KF7CLD	443.42500	+		
Burlington	DMR/BM	KF7CFR	441.95000	+	CC1	
Camano Island	FM	W7PIG	147.36000	+	127.3 Hz	
	FM	W7PIG	223.88000	-	103.5 Hz	
	FM	W7PIG	441.05000	+	103.5 Hz	
Camas, Livingston Mountain						
	FM	W7AIA	147.24000	+	94.8 Hz	
	FM	W7AIA	224.36000	-	94.8 Hz	ORRC
	FM	W7AIA	443.92500	+	94.8 Hz	ORRC
Camas, Prune Hill	FM	KE7BK	444.52500	+	103.5 Hz	ORRC
Capitol Peak	FM	K7CPR	145.47000	-	100.0 Hz	WWARA
Carnation	FM	KE7GFZ	145.59000			
	FM	W7PFB	223.90000	-	88.5 Hz	
Carson, Red Mountain						
	FM	KB7APU	145.25000	-	186.2 Hz	
	FM	KB7APU	224.02000	-	136.5 Hz	ORRC
Cathlamet, KM Hill						
	FM	NM7R	147.02000	+	118.8 Hz	
Centralia	FM	K7CEM	146.86000	-	110.9 Hz	
Centralia, Cook's Hill Fire St						
	FM	K7CEM	145.49000	-	110.9 Hz	
	FM	K7CEM	442.05000	+	110.9 Hz	
Chehalis	DMR/MARC	AF7PR	440.73750	+	CC1	
	FM	WA7UHD	145.43000	-	110.9 Hz	
Chehalis, Baw Faw Peak						
	FM	K7CH	52.93000	51.23000	100.0 Hz	
	FM	KD7HTE	444.45000	+	100.0 Hz	
	FM	K7CH	927.92500	902.92500	114.8 Hz	

WASHINGTON 383

Location	Mode	Call sign	Output	Input	Access	Coordinator
Chehalis, Baw Faw Peak (Boistf						
	FM	WA7UHD	147.06000	+	110.9 Hz	
Chehalis, Crego Hill						
	FM	K7KFM	146.74000	-	110.9 Hz	
	FM	K7KFM	443.45000	+	110.9 Hz	
Chelan	FM	K7SRG	145.45000	-		IACC
Chelan, McNeal Canyon						
	FM	K7SMX	147.10000	+		
	FM	K7SMX	444.52500	+	94.8 Hz	
Cheney	DMR/BM	WA7DRE	145.15000	-	CC1	
Chewelah	FM	AK2O	223.90000	-		IACC
Chewelah, Chewelah Peak						
	FM	W7GHJ	443.72500	+		IACC
Chewelah, Stensgar Mountain						
	FM	N7BFS	145.25000	-		IACC
Chewelah, Stranger Mountain						
	FM	K1RR	147.36000	+		IACC
Chinook, Megler Mountain						
	FM	W7BU	145.45000	-	118.8 Hz	
	FM	NM7R	147.18000	+	82.5 Hz	
	FM	W7BU	440.92500	+	118.8 Hz	
	FM	NM7R	444.92500	+	82.5 Hz	
Clarkston	FM	N7SAU	444.37500	+	131.8 Hz	IACC
Clarkston, Potter Hill						
	FM	KA7FAJ	145.39000	-		
Clarkston, Stout Ranch						
	FM	N7SAU	444.37500	+	131.8 Hz	
Cle Elum	FM	W7HNH	444.92500	+	131.8 Hz	
Cle Elum, Sky Meadows						
	DSTAR	WR7KCR	444.91250	+		
	FM	WR7KCR	147.36000	+	141.3 Hz	
Colfax, Kamiak Butte						
	FM	N7ZUF	53.75000	52.05000	100.0 Hz	IACC
	FM	W7HFI	146.74000	-		IACC
College Place	FM	KH6IHB	147.14000	+	94.8 Hz	IACC
Colville	FM	K7SRG	145.45000	-		IACC
Colville Indian Reservation, K						
	FM	KF7VSX	446.50000	-	100.0 Hz	
Colville, Monumental Mountain						
	FM	K7JAR	146.62000	-	100.0 Hz	IACC
Cosmopolis	FM	WA7ARC	444.37500	+	100.0 Hz	
Cosmopolis, Cosmopolis Hill						
	FM	W7EOC	145.39000	-	118.8 Hz	
Coupeville	FM	W7AVM	146.86000	-	127.3 Hz	
Covington	FM	N7UIC	444.90000	+	103.5 Hz	
Cowiche, Cowiche Mtn						
	FM	N7YRC	442.72500	+	127.3 Hz	IACC
Darington	FM	W7UFI	443.87500	+	103.5 Hz	
Darrington	FM	W7MB	442.67500	+	127.3 Hz	
	FM	W7UFI	443.87500	+	103.5 Hz	WWARA
Darrington, Barlow Pass						
	FM	KD7VMK	442.80000	+		
Davenport, Teel Hill						
	FM	W7OHI	147.04000	+		
Duvall	FM	N6TJQ	441.85000	+	203.5 Hz	
	FM	KE7GFZ	443.25000	+	103.5 Hz	
East Wenatchee	FM	KB7MVF	443.65000	+		IACC
	FM	K7TKR	444.87500	+	151.4 Hz	IACC
East Wenatchee, Badger Mountai						
	FM	N7RHT	444.75000	+	100.0 Hz	
Eastsound, Mt. Constitution, O						
	FM	W7MBY	53.21000	51.51000	100.0 Hz	

384   WASHINGTON

Location	Mode	Call sign	Output	Input	Access	Coordinator
Eatonville	FM	W7PFR	53.41000	51.71000	100.0 Hz	
	FM	W7EAT	224.18000	-	103.5 Hz	
	FM	W7PFR	443.97500	+	103.5 Hz	
Eatonville, Pack Forest						
	FM	W7EAT	146.70000	-	103.5 Hz	
Edmonds	DSTAR	NW7DR	146.46250	147.46250		
	DSTAR	NW7DR	444.72500	+	123.0 Hz	
	FM	W7RNB	29.68000	-		
	FM	WE7SCA	440.37500	+	103.5 Hz	
	FM	WE7SCA	444.02500	+	103.5 Hz	
Eldon	FM	WB7DVN	146.45000	+		
Electron	FM	W7UDI	444.25000	+	103.5 Hz	
Ellensburg	DMR/MARC	K7RHT	440.92500	+	CC1	
	FM	WR7KCR	442.20000	+	131.8 Hz	IACC
	FM	K7RHT	444.45000	+	131.8 Hz	IACC
	FM	K7RMR	444.82500	+	100.0 Hz	
Ellensburg, Sky Meadows						
	FM	WR7KCR	442.20000	+	131.8 Hz	
Ellensburg, Table Mountain						
	FM	K7RHT	444.45000	-	131.8 Hz	
Elma, Minot Peak	FM	W7EOC	444.05000	+	118.8 Hz	
Elmer City, Keller Butte						
	FM	KF7VSX	146.50000		100.0 Hz	
Enumclaw	DMR/MARC	NF6C	441.35000	+	CC0	
	DMR/MARC	NO7RF	902.48750	927.48750	CC1	
	FM	N7OEP	440.07500	+	103.5 Hz	
	FM	N7OEP	443.17500	+	107.2 Hz	
Enumclaw, Baldi	FM	W7WRG	224.88000	-	103.5 Hz	
Enumclaw, Baldi Mountain						
	FM	N7OEP	53.33000	51.63000	100.0 Hz	
	FM	K7MMI	146.98000	-	131.8 Hz	
	FM	WB7DOB	147.14000	+	123.0 Hz	
	FM	WB7DOB	224.76000	-	103.5 Hz	
	FM	W7TWA	441.62500	+	100.0 Hz	
	FM	W7PSE	441.70000	+	103.5 Hz	
	FM	WB7DOB	442.62500	+	103.5 Hz	
	FM	K7OET	444.80000	+	141.3 Hz	
	FM	N7FSP	1292.30000	1272.30000	103.5 Hz	
Enumclaw, Grass Mountain						
	FM	W7SIX	53.87000	52.17000	100.0 Hz	
	FM	W7AAO	145.37000	-	136.5 Hz	
Ephrata	DMR/MARC	WA7DMR	147.41250	146.41250	CC1	IACC
Ephrata, Beezly Hill						
	DSTAR	W7TT	443.90000	+		
	FM	W7TT	145.31000	-	100.0 Hz	
	FM	W7DTS	444.90000	+	103.5 Hz	
Everett	DSTAR	NR7SS	440.35000	+		
	FM	W2ZT	145.39000	-	123.0 Hz	
	FM	WA7LAW	147.18000	+	103.5 Hz	
	FM	K7UID	224.06000	-	103.5 Hz	
	FM	WA7LAW	444.57500	+	103.5 Hz	
Fairchild AFB	FM	KC5GI	440.50000	+	100.0 Hz	
Federal Way	DSTAR	WA7FW	146.84000	-		
	DSTAR	WA7FW	443.85000	+		
	DSTAR	WA7FW	1249.25000			
	DSTAR	WA7FW	1290.10000	1270.10000		
	FM	WA7FW	146.76000	-	103.5 Hz	
	FM	WA7FW	147.04000	+	103.5 Hz	
	FM	WA7FW	442.92500	+		
	FM	WA7FW	442.95000	+	103.5 Hz	
Ferndale	DMR/BM	NC7Q	440.73750	+	CC1	
	FM	NC7Q	442.82500	+	156.7 Hz	

WASHINGTON 385

Location	Mode	Call sign	Output	Input	Access	Coordinator
Forks, Mount Octupus						
	FM	K7PP	147.28000	+	123.0 Hz	
Forks, Police Department						
	FM	W7FEL	145.21000	-	100.0 Hz	
Friday Harbor	DSTAR	N7JN	145.25000	-		
	FM	N7JN	146.70000	-	131.8 Hz	
Friday Harbor, Hillview Terrac						
	DSTAR	N7JN	442.46250	+		
Gold Bar	FM	W7EAR	442.17500	+	103.5 Hz	
	FM	W7ERH	1293.00000	1273.00000	103.5 Hz	
Gold Bar, Haystack Mountain						
	FM	N7NFY	443.87500	+	127.3 Hz	
Goldendale	FM	KF7LN	443.35000	+	82.5 Hz	IACC
Goldendale, Juniper Point						
	FM	WC7EC	146.82000	-	103.5 Hz	IACC
	FM	KF7LN	443.35000	+	136.5 Hz	IACC
Goldendale, Simcoe Mountains						
	FM	KC7UTD	146.92000	-	88.5 Hz	
Graham	FM	N3KPU	145.23000	-	146.2 Hz	
Graham, Baldi Mtn						
	FM	N7BUW	444.67500	+	127.3 Hz	
Graham, Graham Hill						
	DSTAR	WA7DR	442.92500	+		
Grand Coulee, Grand Coulee Dam						
	FM	KE7NRA	146.86000	-	100.0 Hz	
Granger, Cherry Hill						
	FM	KB7CSP	147.04000	+	123.0 Hz	
Granite Falls	FM	WB7VYA	146.92000	-	123.0 Hz	
Greenacres	FM	K7HRT	442.42500	+	100.0 Hz	
Greenwater, Crystal Mountain						
	FM	WB7DOB	145.41000	-	162.2 Hz	
Hartline	FM	KB7WPU	224.42000	-	103.5 Hz	
	FM	KB7WPU	444.45000	+	100.0 Hz	
Hazel Dell	FM	KC7QPD	443.80000	+	100.0 Hz	ORRC
Highline	FM	N7IO	443.37500	+	103.5 Hz	
Hobart, Rattlesnake Ridge						
	FM	KF7BJI	442.15000	+	103.5 Hz	
Hockinson	FM	K7BPR	147.08000	+	107.2 Hz	ORRC
	FM	K7BPR	444.72500	+	107.2 Hz	ORRC
Issaquah	FM	N9VW	53.83000	52.13000	123.0 Hz	
	FM	N9VW	440.25000	+	123.0 Hz	
	FM	N9VW	442.30000	+	123.0 Hz	
	P25	K7TGU	146.42500	+		
Issaquah, East Tiger Mountain						
	FM	N7NW	223.98000	-	100.0 Hz	
	FM	K7LED	224.12000	-	103.5 Hz	
Issaquah, Squak Mountain						
	FM	N7KGJ	444.52500	+	103.5 Hz	
Issaquah, Tiger Mountain						
	FM	K7PF	146.88000	-		
	FM	W7WWI	147.08000	+	110.9 Hz	
Issaquah, Tiger Mountain East						
	DSTAR	WA7HJR	444.63750	+		
	FM	K7LED	146.82000	-	103.5 Hz	
	FM	KC7RAS	147.10000	+	123.0 Hz	
	FM	K7DNR	442.60000	+	127.3 Hz	
	FM	K7KG	443.30000	+	156.7 Hz	
	FM	N6OBY	443.32500	+	103.5 Hz	
	FM	WA7HJR	444.65000	+	131.8 Hz	
	FM	KB7CNN	1292.20000	1272.20000	103.5 Hz	
Issaquah, West Tiger Mountain						
	FM	K7NWS	145.33000	-	179.9 Hz	

386   WASHINGTON

Location	Mode	Call sign	Output	Input	Access	Coordinator
Issaquah, West Tiger Mountain						
	FM	K7NWS	224.34000	-	110.9 Hz	
	FM	K7NWS	442.07500	+	110.9 Hz	
Kalama	FM	WB7DFV	442.82500	+	131.8 Hz	
Kalama, China Garden Rd						
	FM	K7CH	927.27500	902.27500	114.8 Hz	
Kendall, Sumas Mountain						
	FM	W7BPD	145.23000	-	103.5 Hz	
Kennewick	DMR/MARC	WA7DMR	146.42500	147.42500	CC0	IACC
	FM	W7UPS	145.39000	-	103.5 Hz	IACC
	FM	K7SRG	145.45000	-	156.7 Hz	IACC
	FM	W7JWC	443.77500	+	203.5 Hz	
	FM	W7UPS	443.95000	+	123.0 Hz	
	FM	W7UPS	444.05000	+	100.0 Hz	IACC
Kennewick, Horse Heaven Hills						
	FM	W7UPS	145.39000	-	103.5 Hz	IACC
Kennewick, Johnson Butte						
	FM	W7AZ	146.64000	-	100.0 Hz	
	FM	N7LZM	147.22000	+		
Kennewick, Jump Off Joe						
	FM	N7LZM	145.41000	-	100.0 Hz	
	FM	KC7WFD	147.08000	+	94.8 Hz	
Kent	FM	K7CST	147.32000	+	103.5 Hz	
Kent, Emerald Park Elementary						
	FM	K7RFH	443.35000	+	103.5 Hz	
Kingston	FM	W7KWS	442.22500	447.32500	100.0 Hz	
Kirkland	DSTAR	N7IH	146.48750	147.48750		
	DSTAR	N7IH	443.57500	+		
	DSTAR	N7IH	1290.20000	1270.20000		
	FM	K7LWH	441.07500	+	103.5 Hz	
Kirkland , Rose Hill						
	FM	K7LWH	53.17000	51.47000	100.0 Hz	
Kirkland, Rose Hill						
	FM	K7LWH	145.49000	-	103.5 Hz	
La Center	FM	K7ABL	444.92500	+	94.8 Hz	ORRC
Lacey	FM	WC7I	146.80000	-	97.4 Hz	
	FM	W6TOZ	440.55000	+	103.5 Hz	
	FM	WC7I	442.47500	+	100.0 Hz	
Lake Forest Park	FM	WA7FUS	224.22000	-	103.5 Hz	
	FM	WE7SCA	442.00000	+	141.3 Hz	
Langley, Whidbey Island						
	FM	W7AVM	147.22000	+	127.3 Hz	
Lebam, KO Peak	FM	N7XAC	224.04000	-	118.8 Hz	
	FM	N7XAC	441.67500	+	118.8 Hz	
Liberty Lake	FM	W7TRF	443.65000	+		
Liberty Lake, Agilent Bldg						
	FM	W7TRF	443.47500	+	88.5 Hz	
Lind, Lind Hill	FM	W7UPS	448.70000	-	123.0 Hz	
Long Beach, County Building						
	FM	NM7R	444.80000	+	118.8 Hz	
Longmire	DMR/BM	WW7CH	145.13000	-	CC1	
Longview	DMR/MARC	KB7APU	444.23750	+	CC9	
	DMR/MARC	N3EG	444.98750	+	CC1	
	FM	W7DG	147.26000	+	114.8 Hz	
	FM	W7DG	224.14000	-	114.8 Hz	
	FM	AB7F	440.37500	+	123.0 Hz	ORRC
	FM	W7DG	444.90000	+	114.8 Hz	
	FM	KB7ADO	927.97500	902.97500	114.8 Hz	
Longview, Columbia Heights						
	FM	N3EG	442.12500	+	114.8 Hz	
Loon Lake	FM	WB7UCI	444.67500	+	114.8 Hz	
Lost River	DMR/MARC	NO7RF	438.38000	449.38000	CC0	

WASHINGTON 387

Location	Mode	Call sign	Output	Input	Access	Coordinator
Lyman, Lyman Hill						
	FM	W7UMH	53.09000	51.39000	100.0 Hz	
	FM	N7GDE	145.19000	-	127.3 Hz	
	FM	W7MBY	223.86000	-	103.5 Hz	
	FM	N7RIG	224.78000	-	103.5 Hz	
	FM	W7UMH	442.40000	+	107.2 Hz	
	P25	WA7ZUS	444.50000	+	103.5 Hz	
Lynnwood	DMR/MARC	K7MLR	444.15000	+	CC1	WWARA
	FM	WA7DEM	146.78000	-	162.2 Hz	
	FM	N7NFY	146.80000	-	136.5 Hz	
	FM	N6CES	444.67500	+	173.8 Hz	
Mabton, Missouri Falls						
	FM	KB7CSP	443.82500	+	100.0 Hz	
Maple Valley	DSTAR	KF7NPL	442.67500	+		
	FM	KF7NPL	147.26000	+	103.5 Hz	
Marysville	FM	WA7DEM	224.38000	-	103.5 Hz	
Mazama	DMR/MARC	NO7RF	144.51000	147.51000	CC3	
	DMR/MARC	NO7RF	145.21000	147.99000	CC3	
	DMR/MARC	N07RF	147.41250	148.41250	CC3	
Medical Lake	FM	WA7RVV	443.60000	+	100.0 Hz	
Meglar Mtn	FM	W7BU	440.92500	+	118.8 Hz	WWARA
Mercer Island	FM	W7MIR	147.16000	+	146.2 Hz	
	FM	W7MIR	440.15000	+	103.5 Hz	
Methow	DMR/MARC	NO7RF	440.71250	+	CC1	
Mill Creek	FM	WR7DS	442.72500	+	DCS 172	
	FM	N7IBF	444.92500	+	100.0 Hz	
Mineral	FM	K7HW	146.68000	-	103.5 Hz	
Monroe	FM	K7MJ	224.10000	-	123.0 Hz	
Monroe, Rattlesnake Mountain						
	FM	K7SLB	443.12500	+		
Morton, Rooster Rock						
	FM	KB7WVX	444.97500	+	110.9 Hz	
Moses Lake	DMR/MARC	NO7RF	440.92500	+	CC1	
Mount Vernon, Cultus Mountain						
	FM	WE7T	53.59000	51.89000	100.0 Hz	
	FM	K7OET	444.35000	+	100.0 Hz	
	FM	NC7Q	444.62500	+	103.5 Hz	
	FM	K7OET	927.55000	902.55000	114.8 Hz	
Mountlake Terrace						
	FM	WA7DEM	443.72500	+	103.5 Hz	
Moxee City	FM	KC7WFD	147.12000	+		
Moxee City, Elephant Mountain						
	FM	W7AQ	146.84000	-	123.0 Hz	
	FM	WA7SAR	444.60000	+	123.0 Hz	
Mukilteo, Boeing	FM	W7FLY	443.92500	+	100.0 Hz	
Naches, Whites Pass						
	FM	KD7LZN	442.47500	+	210.7 Hz	
Naselle	DMR/MARC	NA7Q	444.82500	+	CC1	
Naselle, Naselle Ridge						
	FM	NM7R	440.67500	+	118.8 Hz	
Neilton	FM	WA7ARC	146.96000	-	203.5 Hz	
Neilton Peak	FM	W7EOC	444.70000	+	118.8 Hz	WWARA
Neilton, Neilton Peak						
	FM	W7ZA	146.90000	-	88.5 Hz	
	FM	W7EOC	444.70000	+	118.8 Hz	
Newcastle	DSTAR	W7RNK	441.21250	+		
Newcastle, Cougar Mt						
	FM	KF7BJI	224.44000	-	103.5 Hz	
Newport	FM	KB7TBN	444.57500	+	100.0 Hz	IACC
Newport, Cooks Mountain						
	FM	KB7TBN	444.57500	+	100.0 Hz	

388 WASHINGTON

Location	Mode	Call sign	Output	Input	Access	Coordinator
Nordland, Marrowstone Island						
	FM	AA7MI	440.72500	+	114.8 Hz	
North Bend	DMR/MARC	WA7DMR	146.50000	147.50000	CC1	
	FM	N9VW	53.85000	52.15000	100.0 Hz	
	FM	K7SLB	147.47500	146.47500	127.3 Hz	
	FM	W7EFR	442.72500	+	123.0 Hz	
North Bend, Green Mountain						
	FM	KD7VMK	146.92000	-	103.5 Hz	
North Bend, Rattlesnake Mounta						
	FM	KC7SAR	145.11000	-	127.3 Hz	
	FM	W7PSE	441.77500	+	103.5 Hz	
	FM	W7SRG	442.30000	+	123.0 Hz	
	FM	N7NFY	927.88750	902.88750	114.8 Hz	
North Bend, Rattlesnake Ridge						
	FM	N7NFY	441.65000	+	127.3 Hz	
North Point	FM	N7QDY	146.66000	-	107.2 Hz	
Ocean Park, Fire Hall						
	FM	NM7R	145.17000	-	118.8 Hz	
Ocean Shores, City Shops						
	FM	WA7OS	441.12500	+	123.0 Hz	
Okanogan, Pitcher Mountain						
	FM	WA7MV	146.72000	-	100.0 Hz	
	FM	WA7MV	443.55000	+	100.0 Hz	
Olalla	FM	K7PAG	53.23000	51.53000	103.5 Hz	
	FM	W7ZLJ	145.35000	-	103.5 Hz	
	FM	WR7HE	440.22500	+	103.5 Hz	
Olympia	DMR/MARC	KG7KPH	440.72500	-	CC1	
	DMR/MARC	WA6VYL	441.32500	441.82500	CC1	
	FM	W7PSE	145.15000	-	103.5 Hz	
	FM	N7EBB	146.78000	-	156.7 Hz	
	FM	NT7H	440.72500	+		
	FM	KC7CKO	443.07500	+	103.5 Hz	
	FM	W7USJ	443.80000	+	146.2 Hz	
Olympia, Capitol Peak						
	FM	W7SIX	53.57000	51.87000	100.0 Hz	
	FM	K7CPR	145.47000	-	100.0 Hz	
	FM	W7WRG	224.08000	-	103.5 Hz	
	FM	W7WRG	440.50000	+	110.9 Hz	
	FM	N7UJK	444.95000	+	118.8 Hz	
	FM	W7SIX	927.30000	902.30000	114.8 Hz	
Olympia, Crawford Mountain						
	FM	NT7H	441.40000	+	103.5 Hz	
Olympia, Water Tower						
	FM	NT7H	147.36000	+	103.5 Hz	
Omak	FM	K7SRG	147.20000	+		IACC
Omak, Tunk Mountain						
	FM	WA7MV	53.11000	51.41000	100.0 Hz	IACC
	FM	WA7MV	147.32000	+	100.0 Hz	
Orcas Island, Mount Constituti						
	FM	N7JN	224.48000	-	103.5 Hz	
	FM	K7SKW	442.00000	+	110.9 Hz	
Orcas Island, Mt Constitution						
	FM	K7SKW	146.74000	-	103.5 Hz	
	FM	N7JN	146.90000	-	131.8 Hz	
	FM	WA6MPG	224.54000	-	67.0 Hz	
	FM	N7JN	443.45000	+	103.5 Hz	
	FM	K7SKW	444.05000	+	103.5 Hz	
Oroville, Buckhorn Mountain						
	FM	WA7DJ	147.14000	+	103.5 Hz	
Orting	DMR/MARC	NF6C	441.32500	+	CC0	
Oso	DMR/MARC	K7MLR	443.90000	+	CC1	
Othello	FM	N7MHE	145.35000	-	100.0 Hz	IACC

WASHINGTON 389

Location	Mode	Call sign	Output	Input	Access	Coordinator
Otis Orchards	FM	NV2Z	442.92500	+	100.0 Hz	
Otis Orchards, Fox Hill						
	FM	AD7DD	147.14000	+	127.3 Hz	IACC
Packwood	FM	K7KFL	146.74000	-	131.8 Hz	
Paradise, Mt Rainier						
	FM	WW7CH	146.78000	-	103.5 Hz	
Pasco	FUSION	WF7S	146.62000	-	123.0 Hz	
Plymouth, Sillusi Butte						
	FM	KC7RWC	145.49000	-	67.0 Hz	
	FM	AI7HO	147.02000	+	103.5 Hz	
Plymouth, Silousi Butte						
	FM	AI7HO	443.75000	+	103.5 Hz	
Point Roberts	FM	KJ1U	443.30000	+	100.0 Hz	
Port Angeles	DMR/BM	WF7W	442.12500	+	CC1	
	FM	WF7W	145.13000	-	100.0 Hz	
	FM	WA7EBH	443.70000	+	103.5 Hz	
Port Angeles, Striped Peak						
	FM	W7FEL	146.76000	-	100.0 Hz	
Port Ludlow	FM	N7PL	441.57500	+	103.5 Hz	
Port Ludlow, Mats Mats						
	FM	WR7V	442.52500	+	103.5 Hz	
Port Orchard	FM	N7IG	145.39000	-	88.5 Hz	
	FM	W6AV	441.57500	+	100.0 Hz	
	FM	K7BTZ	444.10000	+	100.0 Hz	
Port Townsend	FM	N7WGR	443.82500	+	88.5 Hz	
Port Townsend, Morgan Hill						
	FM	W7JCR	145.15000	-	114.8 Hz	
Poulsbo	FM	W7LOR	441.27500	+	123.0 Hz	
	FM	WA6PMX	442.20000	+	103.5 Hz	
Prosser	FM	W7LYV	147.38000	+	123.0 Hz	IACC
	FM	WB7WHF	444.87500	+	141.3 Hz	
Pullman	DSTAR	W7YH	443.16250	+		
	FM	KC7AUI	444.30000	+	103.5 Hz	
Pullman, SEL Campus						
	FM	K7SEL	147.10000	+	103.5 Hz	IACC
Puyallup	FM	KB7CNN	444.75000	+	103.5 Hz	
Quilcene, Buck Mountain						
	FM	W7FHZ	53.29000	51.59000	100.0 Hz	
	FM	K7PP	441.20000	+		
	FM	W2ZT	442.50000	+	123.0 Hz	
Quilcene, Buck Mtn						
	FM	K7MMI	147.20000	+	131.8 Hz	
	P25	K7SCN	440.95000	+	110.9 Hz	
Quilcene, Hood Canal						
	FM	WO7O	443.42500	+	103.5 Hz	
Randle	FM	AB7F	444.87500	+	100.0 Hz	
Rathdrum	FM	K7FVA	441.65000	+	156.7 Hz	IACC
Rattlesnake Mtn	FM	N7NFY	441.65000	+	141.3 Hz	WWARA
Raymond	FM	KA7DNK	147.24000	+	103.5 Hz	
	FM	KB7IEU	442.15000	+	127.3 Hz	
Redmond	DMR/MARC	KE7SFF	444.40000	+	CC1	
	FM	KC7IYE	145.31000	-	103.5 Hz	
	FM	N6OBY	440.67500	+	103.5 Hz	
Renton	FM	KC7IGT	441.45000	+	123.0 Hz	
	FM	K7FDF	443.60000	+	103.5 Hz	
Renton, Lake Youngs						
	FM	WB7DOB	441.37500	+	173.8 Hz	
Republic	FM	N7XAY	145.19000	-		
Richland	FM	W7VPA	146.76000	-	100.0 Hz	IACC
Richland , Rattlesnake Mtn						
	FM	W7VPA	146.76000	-	100.0 Hz	

390   WASHINGTON

Location	Mode	Call sign	Output	Input	Access	Coordinator
Richland, Rattlesnake Mtn						
	FM	W7AZ	449.10000	-	100.0 Hz	
Ritzville	FM	WD7C	146.72000	-		
	FM	W7UPS	444.05000	+	100.0 Hz	
Roche Harbor	FM	W6QC	441.60000	+	131.8 Hz	
Roosevelt	FM	W7NEO	145.19000	-		
Roy	FM	KB7UXE	444.27500	+	203.5 Hz	
Saddle Mtn	FM	W7ZA	146.90000	-	88.5 Hz	WWARA
Sammamish	DMR/MARC	KK7TR	442.05000	+	CC1	WWARA
	FM	W7SRG	440.25000	+	123.0 Hz	
	FM	KG7OI	442.12500	+	123.0 Hz	
Seatac	FM	NC7G	146.66000	-	103.5 Hz	
	FM	KE7WMH	443.10000	+	103.5 Hz	
SeaTac, Station 46						
	DSTAR	KF7BFS	440.27500	+		
Seattle	DMR/BM	WW7PSR	440.77500	+	CC2	
	DMR/BM	N7IEI	440.90000	+	CC1	
	DMR/MARC	NO7RF	433.15000	449.65000	CC0	
	DMR/MARC	KF7BJI	440.92500	+	CC1	WWARA
	DMR/MARC	W7AW	440.97500	+	CC2	
	DMR/MARC	W7ACS	441.02500	+	CC2	
	DMR/MARC	K7SLB	441.32500	+	CC1	
	DMR/MARC	NO7RF	441.32500	+	CC1	
	DMR/MARC	WW7PSR	441.77500	+	CC2	
	DMR/MARC	W7WWI	442.02500	+	CC1	
	DSTAR	W7ACS	440.76250	+		
	FM	K7SLB	440.10000	+	110.9 Hz	
	FM	W7ACS	443.00000	+	156.7 Hz	
	FM	W7ACS	443.47500	+	141.3 Hz	
	FM	K7SPG	444.00000	+	103.5 Hz	
	FM	KC7LFW	444.22500	+	123.0 Hz	
	FM	W7BMW	445.82500	-	100.0 Hz	
	P25	WA7LZO	442.90000	+	103.5 Hz	
Seattle, Beacon Hill						
	FM	W7SRZ	146.90000	-	103.5 Hz	
	FM	W7SRZ	224.68000	-	100.0 Hz	
	FM	W7ACS	440.52500	+		
	FM	W7SRZ	443.55000	+	103.5 Hz	
Seattle, Capitol Hill						
	FM	WW7PSR	146.96000	-	103.5 Hz	
	FM	WA7UHF	442.87500	+		
	FM	AJ7JA	444.37500	+	88.5 Hz	
Seattle, Columbia Center						
	FM	WW7SEA	444.55000	+	141.3 Hz	
Seattle, High Point - Myrtle R						
	FM	W7AW	53.29000	51.59000	100.0 Hz	
Seattle, KOMO TV Tower						
	FM	WW7SEA	444.42500	+	141.3 Hz	
Seattle, Maple Leaf						
	FM	W7DX	147.00000	-	103.5 Hz	
Seattle, Northwest Hospital						
	FM	W7SRZ	444.82500	+	103.5 Hz	
Seattle, Queen Anne Hill						
	FM	WW7SEA	444.70000	+	103.5 Hz	
Seattle, Roosevelt Hill						
	FM	W7ACS	443.65000	+	141.3 Hz	
Seattle, Seattle Childrens Hos						
	FM	K7IDS	444.77500	+	173.8 Hz	
Seattle, Southwest Myrtle Stre						
	FM	W7AW	441.80000	+	141.3 Hz	
Seattle, West Seattle						
	FM	W7AW	147.06000	+	107.2 Hz	

WASHINGTON 391

Location	Mode	Call sign	Output	Input	Access	Coordinator
Sedro-Wooley, Lyman Hill						
	FM	KF7VUR	147.43750	146.43750		
Selah	FM	W7HAR	444.27500	+	110.9 Hz	
Selah, Yakima Canyon						
	FM	KC7VQR	147.24000	+	192.8 Hz	
Sequim	DMR/BM	K6MBY	440.75000	+	CC1	
	DMR/BM	K6MBY	444.90000	+	CC1	
	FM	N7NFY	442.80000	+	123.0 Hz	WWARA
	FM	AF7DX	444.27500	+	100.0 Hz	
Sequim, Bell Hill	FM	KO6I	442.05000	+	103.5 Hz	
Sequim, Blyn Lookout						
	FM	KC7EQO	442.10000	+	100.0 Hz	
Sequim, Blyn Mountain						
	FM	N7NFY	442.80000	+	123.0 Hz	
Sequim, Maynard Peak						
	FM	WB0CZA	441.12500	+	123.0 Hz	
	FM	W7PSE	442.42500	+	103.5 Hz	
Shelton	FM	WB7OXJ	53.09000	51.39000	110.9 Hz	
	FM	N7SK	443.25000	+	100.0 Hz	
Shelton, South Mountain						
	FM	K7CH	145.27000	-	127.3 Hz	
	FM	K7CH	440.65000	+	100.0 Hz	
	FM	K7CH	441.92500	+	100.0 Hz	
	FM	K7CH	927.25000	902.25000	114.8 Hz	
Shelton, Water Tower						
	FM	N7SK	146.72000	-	103.5 Hz	
	FM	N7SK	927.41250	902.41250	114.8 Hz	
Shoreline	FM	KC7ONX	440.30000	+	103.5 Hz	
Shoreline, CRISTA						
	FM	W7AUX	442.82500	+	103.5 Hz	
Silverdale	FM	KD7WDG	145.43000	-	88.5 Hz	
	FM	AA7SS	445.92500	+	123.0 Hz	
Skokomish, South Mountain						
	FM	K7CH	53.03000	51.33000	100.0 Hz	
Skykomish	FM	KC7SAR	145.11000	-	123.0 Hz	
Snohomish	DMR/BM	K7LKA	444.23750	+	CC3	
	DSTAR	KG7QPU	443.90000	+	151.4 Hz	
	FM	KG7QPU	441.15000	+		
Snohomish, Clearview						
	FM	WA7DEM	442.97500	+	103.5 Hz	
Snohomish, Mt Pilchuck						
	DSTAR	NR7SS	440.32500	+		
Snoqualmie	FM	N7SNO	444.92500	+	85.4 Hz	
Snoqualmie, Rattlesnake Ridge						
	DSTAR	N7SNO	442.70000	+		
South Bend, Holy Cross Mountai						
	FM	NM7R	147.34000	+	82.5 Hz	
	FM	NM7R	224.82000	-	82.5 Hz	
	FM	NM7R	442.67500	+	118.8 Hz	
Spokane	DMR/BM	KC7AAD	444.12500	+	CC1	IACC
	DMR/MARC	KC7AAD	444.13750	+	CC1	
	DMR/MARC	KC7AAD	444.15000	+	CC1	IACC
	DSTAR	WA7DRE	1293.30000	1313.30000		
	FM	N7FM	145.33000	-	88.5 Hz	IACC
	FM	N7FM	442.60000	+	100.0 Hz	
	FM	WA7DRE	443.52500	+		
	FM	KA7ENA	443.80000	+	123.0 Hz	IACC
	FM	N7FM	444.42500	+	100.0 Hz	IACC
	FM	K7SRG	444.70000	+		IACC
Spokane Valley	FM	N7FM	146.66000	-		
Spokane Valley, Arbor Crest						
	FM	WR7VHF	444.90000	+	123.0 Hz	IACC

392   WASHINGTON

Location	Mode	Call sign	Output	Input	Access	Coordinator
Spokane Valley, Liberty Lake						
	FM	K7MMA	145.17000	-	114.8 Hz	IACC
Spokane, Booth Hill						
	FM	KA7ENA	443.80000	+	123.0 Hz	
Spokane, Brownes Mountain						
	FM	W7GBU	147.30000	+	100.0 Hz	IACC
Spokane, Downtown Spokane						
	FM	WR7VHF	147.34000	+	123.0 Hz	IACC
Spokane, Five Mile						
	FM	N7BFS	147.06000	+	77.0 Hz	IACC
	FM	N7BFS	444.17500	+		IACC
Spokane, Krell Hill						
	FM	W7UPS	145.39000	-	100.0 Hz	
	FM	WR7VHF	146.88000	-	123.0 Hz	IACC
	FM	W7RGW	927.25000	902.25000	114.8 Hz	IACC
Spokane, Krell Mtn						
	FM	K7EMF	444.50000	+		
Spokane, Liberty Lake						
	FM	K7MMA	224.40000	-	114.8 Hz	IACC
Spokane, Mica Peak						
	FM	WA7UOJ	53.29000	51.59000	100.0 Hz	IACC
	FM	WA7UOJ	145.11000	-	118.8 Hz	IACC
	FM	WA7RVV	145.15000	-	114.8 Hz	IACC
	FM	WA7HWD	147.24000	+	127.3 Hz	IACC
	FM	W7OE	147.38000	+	100.0 Hz	IACC
Spokane, Mt. Spokane						
	FM	WR7VHF	444.60000	+	123.0 Hz	IACC
Spokane, Paradise Rim						
	FM	KG7SD	147.10000	+	100.0 Hz	IACC
Spokane, Sacred Heart Hospital						
	DSTAR	WA7DRE	443.12500	+		
Spokane, South Hill						
	FM	K7EKM	147.16000	+	136.5 Hz	IACC
	FM	N1NG	444.35000	+	192.8 Hz	UVHFS
Spokane, South Valley						
	FM	W7TRF	145.21000	-	100.0 Hz	IACC
Spokane, Spokane Airport						
	FM	K7MMA	444.45000	+	77.0 Hz	IACC
Spokane, Spokane International						
	FM	K7MMA	145.29000	-		
Spokane, Tower Mountain						
	FM	KA7ENA	145.37000	-	141.3 Hz	IACC
Sultan	FM	W7SKY	53.35000	51.65000	100.0 Hz	
	FM	W7SKY	444.12500	+	103.5 Hz	
Sultan, Haystack	FM	W7WRG	224.58000	-	103.5 Hz	
Sultan, Haystack Lookout						
	FM	W7UFI	224.24000	-	103.5 Hz	
Sultan, Haystack Mountain						
	FM	WC7T	441.87500	+	103.5 Hz	
	FM	W2ZT	444.97500	+	114.8 Hz	
Sumner, White River Junction						
	FM	W7PSE	443.62500	+	103.5 Hz	
Tacoma	DMR/MARC	KG7KPH	441.42500	441.92500	CC1	
	DMR/MARC	WA7DMR	927.48750	952.48750	CC1	
	FM	K7HW	53.19000	51.49000	100.0 Hz	
	FM	KB7CNN	146.64000	-	103.5 Hz	
	FM	K7HW	146.94000	-	103.5 Hz	
	FM	W7TED	147.02000	+	103.5 Hz	
	FM	W7DK	147.28000	+	103.5 Hz	
	FM	N3KPU	147.45000	146.45000		
	FM	K7HW	224.52000	-	103.5 Hz	
	FM	W7DK	440.62500	+	103.5 Hz	

WASHINGTON 393

Location	Mode	Call sign	Output	Input	Access	Coordinator
Tacoma	FM	W7TED	442.45000	+	103.5 Hz	
	FM	WW7MST	443.67500	+	103.5 Hz	
	FM	NB7N	446.35000	-	103.5 Hz	
	FM	KB7CNN	1292.40000	1272.40000	103.5 Hz	
Tacoma WA	DMR/BM	WA7DMR	440.72500	+	CC1	
Tacoma, Ch 28 Tower						
	FM	W7DK	145.21000	-	141.3 Hz	
Tacoma, Dash Point						
	FM	N7QOR	443.95000	+	131.8 Hz	
Tacoma, Madigan Army Medical C						
	FM	KE7YYD	442.75000	+	146.2 Hz	
Tenino, Crawford Mountain						
	FM	W7DK	147.38000	+	103.5 Hz	
	FM	NT7H	224.46000	-	103.5 Hz	
Tieton	FM	AB7XQ	444.22500	+	100.0 Hz	
Tiger Mtn East	FM	WA7HJR	444.65000	+	131.8 Hz	WWARA
Tonasket	FM	W7ORC	145.47000	-	173.8 Hz	IACC
Trout Lake, King Mountain						
	FM	WA7SAR	147.08000	+	123.0 Hz	
Tukwila, Station 52						
	DSTAR	KF7BFT	440.42500	+		
Tumwater	FM	N7EHP	147.12000	+	173.8 Hz	
Tumwater, Bush Mountain						
	FM	KD7HTE	927.75000	902.75000	114.8 Hz	
Twisp	DMR/MARC	NO7RF	444.85000	+	CC1	IACC
Twisp, McClure Mountain						
	FM	WA7MV	147.22000	+	100.0 Hz	
	FM	WA7MV	444.80000	+	110.9 Hz	IACC
Underwood, Underwood Mountain						
	FM	KB7DRX	147.20000	+	100.0 Hz	
University Place	FM	K7NP	53.01000	51.31000	100.0 Hz	
	FM	K7NP	145.29000	-	114.8 Hz	
	FM	K7NP	442.37500	+	103.5 Hz	
	FM	N7EHP	443.15000	+	173.8 Hz	
	FM	K7NP	927.60000	902.60000	114.8 Hz	
Vancouver	DMR/BM	K7KSN	442.10000	+	CC1	
	DMR/BM	KB7APU	442.96250	+	CC1	ORRC
	DMR/MARC	KB7APU	146.73000	-	CC1	
	DMR/MARC	AF7PR	147.15000	+	CC1	
	FM	AB7F	224.64000	-	123.0 Hz	ORRC
	FM	K7GJT	442.10000	+		
	FM	N7XMT	444.55000	+	131.8 Hz	ORRC
	FM	W7AIA	1292.50000	1272.50000	94.8 Hz	ORRC
Vancouver, Larch Mountain						
	DSTAR	K7CLL	440.01250	+		
	FM	KB7APU	53.13000	51.43000	107.2 Hz	ORRC
	FM	W7LT	146.84000	-	123.0 Hz	ORRC
	FM	KB7APU	224.72000	-	100.0 Hz	ORRC
	FM	KB7APU	440.77500	+		ORRC
	FM	AB7F	442.37500	+	123.0 Hz	ORRC
	FM	KE7FUW	443.67500	+	107.2 Hz	
	FM	W7AIA	443.90000	+	94.8 Hz	ORRC
	FM	K5TRA	927.13750	902.13750	131.8 Hz	ORRC
Vancouver, Livingston Mountain						
	FM	AB7F	145.37000	-	123.0 Hz	ORRC
Vancouver, Peace Health						
	FM	KB7APU	443.97500	+		ORRC
Vancouver, Peace Health - Memo						
	FM	W7AIA	443.82500	+		
Vashon Island	FM	W7VMI	443.50000	+	103.5 Hz	
	FM	KG7CM	443.77500	+	103.5 Hz	
Walla Walla	FM	AL1Q	146.96000	-	74.4 Hz	

394   WASHINGTON

Location	Mode	Call sign	Output	Input	Access	Coordinator
Walla Walla	FM	KH6IHB	147.14000	+	94.8 Hz	
	FM	AL1Q	443.45000	+	123.0 Hz	IACC
	FM	KL7NA	444.25000	+		
Walla Walla, Hertzer Peak						
	FM	AL1Q	146.96000	-	74.4 Hz	
	FM	AL1Q	443.45000	+	123.0 Hz	
Wauconda	FM	KH6UG	442.55000	+		
Wenatchee	FM	N7RHT	146.78000	-	156.7 Hz	
	FM	K7SRG	147.20000	+		IACC
Wenatchee, Burch Mountain						
	FM	W7TD	146.68000	-	156.7 Hz	
Wenatchee, Mission Ridge						
	FM	WR7ADX	146.90000	-	173.8 Hz	
	FM	WR7ADX	224.74000	-	179.9 Hz	
Wenatchee, Naneum Ridge						
	FM	KB7TYR	147.26000	+	156.7 Hz	
West Richland	FM	WA7BCA	442.87500	+	203.5 Hz	
White Salmon	FM	NB7M	443.17500	+	88.5 Hz	ORRC
White Salmon, Near LDS Church						
	FM	KB7DRX	147.00000	+	100.0 Hz	
White Swan, Fort Simcoe						
	FM	KA7IJU	146.72000	-	123.0 Hz	
Winthrop	DMR/MARC	NO7RF	439.39000	449.39000	CC3	
	FM	NO7RF	146.52000		100.0 Hz	
Woodinville	FM	K6RFK	147.34000	+	100.0 Hz	
	FM	WA7TZY	442.77500	+	100.0 Hz	
Woodland	FM	N7XAC	224.30000	-	103.5 Hz	
	FM	K7LJ	444.47500	+	100.0 Hz	
Woodland, Fire Station						
	FM	W7DG	147.30000	+	114.8 Hz	
Woodland, Oil Can Henrys						
	FM	W7BO	442.17500	+		
Yacolt	FM	W7RY	440.32500	+	100.0 Hz	ORRC
	FM	W7AIA	443.12500	+	94.8 Hz	ORRC
Yacolt, Yacolt Mountain						
	FM	W7AIA	52.95000	51.25000	94.8 Hz	ORRC
	FM	K7CLL	224.42000	-	94.8 Hz	
	FM	KC7NQU	441.20000	+	107.2 Hz	ORRC
Yakima	DMR/MARC	WA7SAR	147.12000	+	CC1	
	FM	W7CCY	146.94000	-	173.8 Hz	IACC
	FM	K7SRG	147.20000	+		IACC
Yakima, Ahtanum Ridge						
	FM	W7CCY	146.94000	-	173.8 Hz	
Yakima, Bethel Ridge						
	FM	W7AQ	147.30000	+	123.0 Hz	
Yakima, Darland Mountain						
	FM	WA7SAR	146.86000	-	123.0 Hz	
Yakima, Eagle Peak						
	FM	W7AQ	146.66000	-	123.0 Hz	
Yakima, Lookout Point						
	FM	N7YRC	444.75000	+	131.8 Hz	IACC
Yakima, West Rattlesnake Peak						
	FM	KC7IDX	444.55000	+		
Yakima, Yakima Valley						
	FM	WA7SAR	147.06000	+	85.4 Hz	

WASHINGTON DC

Location	Mode	Call sign	Output	Input	Access	Coordinator
Washington	DSTAR	W3AGB	1283.10000	1263.10000		T-MARC
	FM	K3MRC	145.43000	-		
	FM	N3ADV	447.17500	-	156.7 Hz	T-MARC
	FM	KC3VO	448.87500	-		T-MARC
	FM	W3HAC	449.42500	-		T-MARC

WASHINGTON DC 395

Location	Mode	Call sign	Output	Input	Access	Coordinator
Washington	FM	WA3KOK	449.97500	-	107.2 Hz	T-MARC
	P25	W3DCA	147.27000	+	100.0 Hz	T-MARC
Washington, DC	DMR/MARC	W3AGB	147.36000	+	CC1	T-MARC
Washington, Harry S. Truman Bu						
	FM	W3DOS	145.19000	-	151.4 Hz	T-MARC
Washington, Union Station						
	FM	K3WS	447.37500	-	123.0 Hz	T-MARC
Washington, Wilbur J. Cohen Bu						
	FM	K3VOA	147.04500	+	77.0 Hz	T-MARC
	FM	K3VOA	448.57500	-	77.0 Hz	T-MARC

WEST VIRGINIA

Location	Mode	Call sign	Output	Input	Access	Coordinator
Alderson	DMR/MARC	KC8AFH	442.87500	+	CC1	SERA
	FM	KE4QOX	53.23000	52.23000	123.0 Hz	
Alderson, Keeney Knob						
	FM	KD8BBO	146.76000	-	162.2 Hz	SERA
Ansted	DMR/BM	KC8OGK	444.27500	+	CC1	
Beckley	DMR/BM	K8DLT	440.51250	+	CC1	
	FM	WV8BD	145.37000	-	100.0 Hz	
	FM	N8FWL	147.36000	+		
	FM	N8FWL	443.05000	+		
Beckley, Courthouse						
	FM	W8VT	443.80000	+	141.3 Hz	SERA
Beckley, IVY Knob						
	FM	KC8AFH	444.85000	+		SERA
Belington	FM	N8SCS	53.65000	52.65000	141.3 Hz	
	FM	KC8AJH	145.23000	-	103.5 Hz	SERA
	FM	N8SCS	444.90000	+	141.3 Hz	
Berkeley Springs	FM	WA3KOK	442.45000	+	107.2 Hz	T-MARC
	FM	KK3L	443.85000	+	123.0 Hz	
	FM	K7SOB	444.75000	+	127.3 Hz	SERA
Berkeley Springs, Cacapon Moun						
	FM	KK3L	146.74500	-	123.0 Hz	T-MARC
	FM	W3VLG	224.70000	-	123.0 Hz	
Berkeley Springs/Romney						
	FM	N8RAT	444.95000	+	123.0 Hz	T-MARC
Birch River	FM	N8FMD	145.27000	-	103.5 Hz	SERA
Bluefield	DMR/BM	W8MOP	444.45000	+	CC1	SERA
	FM	W8MOP	442.45000	+	103.5 Hz	SERA
Bluefield, East River Mountain						
	FM	W8MOP	145.49000	-	103.5 Hz	SERA
Bluefield, Windmill Gap						
	FM	W8MOP	147.06000	-	103.5 Hz	SERA
Bolt	DMR/MARC	W8LG	443.22500	+	CC1	
Bolt, Ivy Knob	FM	KC8AFH	145.17000	-	100.0 Hz	
Bridgeport	FM	W8SLH	147.12000	+	118.8 Hz	
Bruceton Mills	DMR/BM	KC8TAI	444.56250	+	CC1	
Buckhannon	DMR/BM	W8LD	442.67500	+	CC1	
	FM	N8ZAR	53.11000	52.11000	103.5 Hz	SERA
	FM	K8VE	146.85000	-		
	FM	N8ZAR	147.03000	+	103.5 Hz	SERA
	FM	N8ZAR	444.25000	+	103.5 Hz	
	FM	K8VE	444.47500	+	146.2 Hz	
Buckhannon, West Virginia Wesl						
	FM	W8LD	145.41000	-	103.5 Hz	SERA
Cameron	FM	KC8FZH	146.91000	-	123.0 Hz	SERA
Charles Town	FM	WV8VRC	441.95000	+		T-MARC
	FM	N3EAQ	444.35000	+		T-MARC
Charleston	FM	W8GK	145.35000	-	91.5 Hz	
	FM	WB8CQV	444.95000	+	203.5 Hz	
Charleston, Kanawha City						
	FM	WB8YST	444.35000	+	107.2 Hz	SERA

396 WEST VIRGINIA

Location	Mode	Call sign	Output	Input	Access	Coordinator
Charleston, Middle Ridge Of Da						
	FM	W8KTM	147.18000	+	91.5 Hz	
	FM	W8KTM	444.40000	+	91.5 Hz	
Charleston, Nitro	FM	WB8YST	224.36000	-	107.2 Hz	SERA
Chestnut Knob, Man Mountain						
	FM	N8LVE	444.12500	+	107.2 Hz	
Circleville, Spruce Knob						
	FM	N8HON	147.28500	+	103.5 Hz	
Clarksburg	FM	N8FMD	146.68500	-	103.5 Hz	SERA
	FM	WV8HC	147.16500	+		
	FM	N8FMD	147.21000	+		
	FM	N8FMD	444.17500	+		SERA
Cowen	FM	KC8ECX	146.83500	-	110.9 Hz	
Craigsville	FM	KC8LRN	224.48000	-	91.5 Hz	
Crawford, Union Hill						
	FM	W8OO	145.13000	-	103.5 Hz	SERA
	FM	W8OO	147.06000	+	103.5 Hz	SERA
Danese, Man Mountain, Chestnut						
	FM	WV8B	145.31000	-	100.0 Hz	
Danville, Drawdy Mountain						
	FM	W8NAM	147.12000	+	203.5 Hz	SERA
Danville, WZAC Tower						
	FM	W8NAM	146.68500	-	203.5 Hz	SERA
Davis	FM	KC8AJH	147.13500	+	103.5 Hz	SERA
Droop, Briary Knob NNW Of Droo						
	FM	N8KUK	147.39000	+	100.0 Hz	
Dry Fork	FM	WV8ZH	29.64000	-	162.2 Hz	
Elkins	FM	WV8ZH	53.03000	52.03000	162.2 Hz	
	FM	WV8ZH	145.21000	-	162.2 Hz	
	FM	K8VE	146.74500	-	103.5 Hz	
	FM	KB8BWZ	146.77500	-		
	FM	N8RLR	444.85000	+	162.2 Hz	SERA
Elkins, Rich Mountain						
	FM	KD8JCS	442.10000	+	162.2 Hz	
Fairmont	FM	W8SP	443.87500	+	103.5 Hz	
Fairmont, Valley Falls State P						
	FM	W8SP	145.35000	-	103.5 Hz	SERA
Fayetteville, Gauley Mountain						
	FM	KC8ZQZ	146.79000	-		SERA
Flat Top	FM	WV8B	224.12000	-	100.0 Hz	
	FM	WV8B	927.52500	902.52500		
Flat Top, Huff Knob						
	FM	WV8B	146.62500	-	100.0 Hz	
Franklin	FM	KC8FPC	147.34500	+		
Frost	FM	N8RV	145.11000	-	107.2 Hz	
Gilbert	DMR/MARC	KB8PCW	442.72500	+	CC1	
Glenville	FM	WB8WV	145.29000	-	91.5 Hz	
	FM	KA8ZXP	444.32500	+		
Grafton	FM	W8SLH	147.37500	+	103.5 Hz	
	FM	W8SLH	444.75000	+	118.8 Hz	SERA
Green Bank	FM	KC8CSE	224.52000	-	123.0 Hz	
Hamlin	FM	N8IKT	443.95000	+	123.0 Hz	
Hernshaw	FM	WB8CQV	146.82000	-	203.5 Hz	
	FM	WB8CQV	444.70000	+	203.5 Hz	
Hillsboro	FM	KC8LRN	224.22000	-	91.5 Hz	
Hillsboro, Droop Mountain						
	FM	W3ATE	145.33000	-	136.5 Hz	
Hillsboro, Droop Mtn						
	FM	W3ATE	147.31500	+	136.5 Hz	SERA
Hinton	DMR/MARC	N8OCY	443.90000	+	CC1	
	FM	KC8CNL	147.25500	+	100.0 Hz	
	FM	KC8CNL	443.90000	+	100.0 Hz	

WEST VIRGINIA 397

Location	Mode	Call sign	Output	Input	Access	Coordinator
Huntington	DMR/MARC	KB8TGK	443.55000	+	CC1	
Huntington, Barker's Ridge						
	FM	KB8TGK	443.85000	+	162.2 Hz	
Huntington, Barkers Ridge						
	FM	N8HZ	146.64000	-		
Huntington, Rotary Park						
	FM	W8VA	146.76000	-	131.8 Hz	
Huntington, Veterans Medical C						
	FM	N8OLC	146.98500	-	131.8 Hz	
Iaeger	FM	N8SNW	146.65500	-	100.0 Hz	SERA
Kanawha City, Kanawha City						
	FM	WB8YST	145.43000	-	107.2 Hz	SERA
Kenna, Kenna Water Tower						
	FM	WD8JNU	443.72500	+	127.3 Hz	SERA
Kenova	FM	KC8PFI	53.87000	52.87000	91.5 Hz	
Keyser	FM	WV8BS	147.39000	+	123.0 Hz	SERA
	FM	WV8BS	444.12500	+	103.5 Hz	SERA
Lenore	FM	AI4UK	145.39000	-	100.0 Hz	
Letart, Missile Tree Hill						
	FM	KC8MNR	444.80000	+	100.0 Hz	SERA
Lewisburg	DMR/MARC	KE8DID	444.25000	+	CC1	
Lewisburg, Muddy Creek Mtn						
	FM	KF4YLM	147.22500	+	100.0 Hz	
Lick Knob, Paint Mountain						
	FM	N8FWL	145.23000	-		
Lobelia, Briery Knob						
	FM	KC8CSE	224.60000	-	123.0 Hz	
Logan	DMR/MARC	KB8PCW	444.43750	+	CC1	SERA
Logan, Ward Rock Mountain						
	FM	KA8GMX	146.97000	-	100.0 Hz	
Madison, Workman's Knob						
	FM	WV8CCC	442.55000	+	100.0 Hz	SERA
Marlinton, Sharp Knob						
	FM	N8PKP	147.09000	+	162.2 Hz	
Martinsburg	FM	W8ORS	145.15000	-	179.9 Hz	T-MARC
	FM	WB8YZV	147.25500	+	123.0 Hz	T-MARC
	FM	K1LLS	442.70000	+	DCS 023	
Martinsburg, North Mountain						
	FM	N8RAT	442.85000	+	100.0 Hz	T-MARC
Middlebourne	FM	WV8TC	147.36000	+	110.9 Hz	
Millwood, Evergreen Hill						
	FM	KD8OOF	145.49000	-	123.0 Hz	SERA
	FM	KD8OOF	443.50000	+		SERA
Moorefield	FM	N8VAA	145.19000	-	118.8 Hz	T-MARC
	FM	K7SOB	442.40000	+	127.3 Hz	
Moorefield, Branch Mountain						
	FM	KD8IFP	146.98500	-	123.0 Hz	T-MARC
	FM	KD8AZC	447.32500	-	103.5 Hz	SERA
Moorefield, Brnach Mountain						
	FM	KD8IFP	444.40000	+	103.5 Hz	SERA
Moorefield, Nathanial Wildlife						
	FM	N8RAT	442.50000	+	123.0 Hz	T-MARC
Morgantown	DMR/BM	W8CUL	440.63750	+	CC1	SERA
	DMR/BM	KD8YNY	441.92500	+	CC1	SERA
	FM	W8MWA	145.43000	-		
	FM	AA8CC	146.92500	-	103.5 Hz	
	FM	KD8BMI	147.07500	+	103.5 Hz	
	FM	W8MWA	444.70000	+		
Morgantown, Chestnut Ridge						
	FM	K8MCR	442.60000	+		
Morgantown, WVU Engineering Sc						
	FM	W8CUL	146.76000	-	103.5 Hz	

398   WEST VIRGINIA

Location	Mode	Call sign	Output	Input	Access	Coordinator
Morgantown, WVU Engineering Sc						
	FM	W8CUL	444.80000	+	103.5 Hz	
Mossy	DMR/MARC	KC8AFH	443.87500	+	CC1	SERA
Moundsville, Grand Vue State P						
	FM	KC8FZH	146.71500	-	110.9 Hz	
	FM	KC8FZH	441.95000	+	123.0 Hz	
	FM	KC8FZH	444.07500	+	123.0 Hz	SERA
Mount Hope, Garden Ground						
	FM	WV8BSA	444.02500	+	123.0 Hz	SERA
Mount Hope, Garden Grounds						
	DSTAR	WV8BSA	441.81250	+		
	FM	WV8BSA	146.70000	-	123.0 Hz	SERA
Mountain View, Horse Pen Mount						
	FM	KB8PCW	146.85000	-	100.0 Hz	SERA
Mt Zion	FM	N8LGY	145.45000	-	107.2 Hz	
Mullens	FM	KC8IT	147.03000	+		
New Martinsville	FM	KF8LL	146.98500	-		
Newell	FM	W8LPN	442.30000	+	162.2 Hz	
Parkersburg	FM	WD8CYV	147.39000	+	91.5 Hz	
Parkersburg, Sand Hill						
	FM	WC8EC	147.25500	+	131.8 Hz	SERA
	FM	WA8LLM	443.17500	+	146.2 Hz	SERA
Parsons	FM	KD8MIV	145.37000	-		SERA
	FM	KD8MIV	444.67500	+	103.5 Hz	SERA
Pennsboro	FM	WV8RAG	147.30000	+	107.2 Hz	
	FM	WB8NSL	442.85000	+	103.5 Hz	
Philippi	FM	K8VE	145.15000	-	103.5 Hz	
Pt. Pleasant, Water Tower On J						
	FM	KE8COT	442.40000	+	91.5 Hz	
Richwood	FM	WB8YST	53.71000	52.71000	107.2 Hz	
	FM	WA8YWO	53.83000	52.83000	100.0 Hz	SERA
	FM	WB8YJJ	145.19000	-	146.2 Hz	
	FM	KB8YDG	147.15000	+	100.0 Hz	
	FM	WB8YST	223.86000	-	107.2 Hz	
	FM	KC8SDN	443.37500	+	100.0 Hz	SERA
Richwood, Grasshopper Knob						
	FM	W8TFC	147.01500	+		SERA
Ripley, Ripley 911 Tower						
	FM	WD8JNU	146.67000	-	107.2 Hz	
Rock Branch, Nitro 1,000 AGL P						
	FM	AB8DY	444.50000	+	151.4 Hz	
Rockport	FM	KC8LTG	147.13500	+	123.0 Hz	SERA
Saint Joseph, German Settlemen						
	FM	KC8FZH	444.87500	+	123.0 Hz	
Salt Rock, Porter Knob						
	FM	K8SA	145.11000	-	110.9 Hz	SERA
Scott Depot	FM	WV8AR	441.82500	+	123.0 Hz	SERA
Scott Depot, Coal Mountain						
	FM	WV8AR	147.27000	+		
Shirley	FM	KB8TJH	53.31000	52.31000		
	FM	KB8TJH	145.31000	-		
	FM	KB8TJH	444.42500	+		
Skyline	FM	K7SOB	147.36000	+	127.3 Hz	SERA
Snowshoe	FM	KC8CSE	53.33000	52.33000	156.7 Hz	SERA
South Charleston	FM	WB8CQV	146.88000	-		
Spencer	FM	KA8AUW	147.10500	+	107.2 Hz	SERA
Stanaford	FM	KD8PIQ	443.40000	+	67.0 Hz	SERA
Stonewood	FM	KD8TC	443.27500	+	91.5 Hz	
Sumerco	DMR/MARC	KB8PCW	440.52500	+	CC1	
Sumerco, Buck Knob						
	FM	KD8CVI	147.34500	+	100.0 Hz	
Summersville	FM	KE4QOX	29.66000	-	88.5 Hz	SERA

WEST VIRGINIA 399

Location	Mode	Call sign	Output	Input	Access	Coordinator
Summersville	FM	N8YHK	145.47000	-	100.0 Hz	
Sutton	FM	W8COX	224.40000	-	123.0 Hz	
Terra Alta	FM	KC8KCI	147.00000	+	103.5 Hz	
Thomas	FM	K7SOB	441.90000	+	103.5 Hz	
Valley Head	FM	KC8AJH	146.67000	-	103.5 Hz	
Webster Springs	FM	KC8HFG	146.89500	-	123.0 Hz	
	FM	KC8CSE	224.66000	-	123.0 Hz	
Weirton, Weirton Medical Cente						
	FUSION	W8CWO	146.94000	-	114.8 Hz	T-MARC
Welch	FM	WV8ED	145.45000	-	100.0 Hz	
	FM	KE8BRP	147.33000	+	100.0 Hz	
	FM	N8SNY	443.72500	+		
West Union	FM	K8DCA	146.95500	-	103.5 Hz	
Weston	FM	WD8EOM	145.39000	-		
	FM	N8MIN	443.97500	+	123.0 Hz	
Wharton, Pilot Knob, Bolt Mtn.						
	FM	WV8CCC	147.19500	+	100.0 Hz	
Wheeling	FM	KA8YEZ	145.19000	-	156.7 Hz	
	FM	W8ZQ	146.76000	-	67.0 Hz	SERA
	FM	KA8YEZ	443.02500	+	156.7 Hz	
	FM	N8EKT	444.97500	+		
Wheeling, Mount Olivet						
	FM	W8MSD	444.57500	+	156.7 Hz	
Williamson, EKB Tower						
	FM	KB8QEU	145.33000	-	127.3 Hz	SERA
	FM	KB8QEU	224.14000	-	127.3 Hz	SERA
	FM	KB8QEU	443.20000	+	127.3 Hz	SERA
Winchester, VA	FM	N8RAT	443.20000	+	123.0 Hz	T-MARC

WISCONSIN

Location	Mode	Call sign	Output	Input	Access	Coordinator
Adams	DSTAR	AC9AR	442.26875	+		WAR
	FM	AC9AR	147.03000	+	123.0 Hz	WAR
Allenton	DMR/MARC	W9RCG	442.03125	+	CC9	
	FM	N9GMT	442.35000	+	123.0 Hz	WAR
Amberg	FM	WI9WIN	443.70000	+	100.0 Hz	WAR
Antigo	FM	W9SM	145.31000	-	114.8 Hz	WAR
	FM	W9SM	147.25500	+	114.8 Hz	
	FM	N9TEV	147.31500	+	114.8 Hz	WAR
Appleton	DMR/BM	N9KRG	444.05000	+	CC6	WAR
	FM	K9STN	145.15000	-	100.0 Hz	WAR
	FM	W9ZL	145.33000	-	100.0 Hz	WAR
	FM	KB9BYP	146.65500	-	100.0 Hz	WAR
	FM	W9ZL	146.76000	-	100.0 Hz	WAR
	FM	W9RIC	442.17500	+	100.0 Hz	WAR
	FM	W9ZL	443.65000	+	100.0 Hz	WAR
Appleton, Darboy	FM	WJ9K	224.50000	-	100.0 Hz	WAR
Arlington	FM	KC9HEA	443.35000	+	123.0 Hz	WAR
Baldwin	FM	WE9COM	145.25000	-	110.9 Hz	WAR
	FM	N9UPC	442.22500	447.32500	110.9 Hz	WAR
Balsam Lake	FM	N9XH	147.19500	+	110.9 Hz	WAR
	FM	N9XH	443.72500	+	110.9 Hz	WAR
Baraboo	FM	WR9ABE	146.88000	-	123.0 Hz	WAR
	FM	N9BDR	443.57500	+	123.0 Hz	WAR
	FM	WI9WIN	443.90000	+	77.0 Hz	WAR
	FM	N9GMT	444.50000	+	123.0 Hz	WAR
Baraboo, Baraboo Bluffs						
	DSTAR	WB9FDZ	145.31500	-		WAR
	FM	WB9FDZ	147.31500	+	123.0 Hz	WAR
Baraboo, City EOC						
	FM	KC9MIO	443.82500	+	123.0 Hz	WAR
Barron	FM	KD9EJA	146.71500	-	110.9 Hz	WAR
	FM	KD9EJA	443.65000	+	DCS 351	WAR

400   WISCONSIN

Location	Mode	Call sign	Output	Input	Access	Coordinator
Barronett	FM	N9PHS	147.04500	+	110.9 Hz	WAR
Bay City	DSTAR	KB9LUK	145.21000	-		
Bayfield	FM	N0BZZ	146.70000	-	103.5 Hz	WAR
	FM	N0BZZ	443.85000	+		WAR
Beetown	FM	N0WLU	146.89500	-	114.8 Hz	
Beldenville	FM	W0MDT	147.22500	+	DCS 351	
Beloit	FM	WA9JTX	147.12000	+	123.0 Hz	WAR
Black River Falls	FM	KC9GEA	145.39000	-		
	FM	WI9WIN	443.55000	+	131.8 Hz	WAR
Bloomer, Home Shop						
	FM	W9EJH	146.65500	-	110.9 Hz	WAR
Boscobel	FM	WI9WIN	444.45000	+	131.8 Hz	
Brooklyn	FM	K9QB	29.65000		123.0 Hz	
	FM	K9QB	145.17000	-	123.0 Hz	
	FM	K9QB	224.06000	-	123.0 Hz	WAR
	FM	K9QB	927.33750	902.33750	123.0 Hz	
Brussels	FM	WE9COM	146.80500	-	146.2 Hz	
	FM	W9DOR	444.00000	+	107.2 Hz	WAR
Burlington	DMR/BM	WB9COW	442.85000	+	CC9	WAR
Cambria	FM	KC9CZH	147.01500	+	123.0 Hz	WAR
Cedarburg	FM	W9CQO	146.97000	-	127.3 Hz	WAR
	FM	W9CQO	224.18000	-	127.3 Hz	WAR
	FM	K9QLP	442.10000	+	127.3 Hz	WAR
	FM	W9DHI	444.97500	+		
Chaffey	FM	KC9EMI	147.10500	+	110.9 Hz	WAR
	FM	KC9AEG	444.95000	+	110.9 Hz	WAR
Chilton	DMR/BM	KC9HYC	442.77500	+	CC6	
	FM	KD9TZ	444.80000	+	107.2 Hz	WAR
Chippewa Falls	DSTAR	KD9ICN	146.94000	-		WAR
	FM	AA9JL	145.23000	-	110.9 Hz	WAR
	FM	W9CVA	147.37500	+	110.9 Hz	WAR
Clam Lake	FM	K9JWM	145.21000	-	110.9 Hz	WAR
Clinton	FM	WB9SHS	146.71500	-	123.0 Hz	WAR
	FM	W9MUP	224.48000	-	123.0 Hz	
	FM	WB9SHS	443.17500	+		WAR
Colfax	FM	W9RMA	444.35000	+	110.9 Hz	WAR
Coloma	FM	W9LTA	146.70000	-	123.0 Hz	WAR
	FM	WE9COM	147.10500	-	123.0 Hz	WAR
	FM	W9LTA	442.67500	+	123.0 Hz	WAR
Cumberland	DMR/BM	KD9EJA	443.50000	+	CC4	
	FM	KD9EJA	145.39000	-	110.9 Hz	
Delafield	FM	K9ABC	146.82000	-	127.3 Hz	WAR
	FM	K9ABC	444.12500	+	127.3 Hz	WAR
Delafield, HWY C And I-94						
	FM	N9GMT	440.30000	+	123.0 Hz	WAR
Dodgeville	FM	WE9COM	145.23000	-	123.0 Hz	WAR
Dunnville	DSTAR	WW9RS	146.88000	-		WAR
Durand	FM	WB9NTO	145.35000	-	110.9 Hz	WAR
	FM	WW9RS	443.40000	+	110.9 Hz	WAR
Eagle River	FM	W9VRC	145.15000	-		
East Farmington	FM	KC9NVV	144.39000			
East Troy	FM	N9WMN	440.77500	+	127.3 Hz	WAR
Eau Claire	FM	W9EAU	146.91000	-	110.9 Hz	WAR
	FM	WI9WIN	442.80000	+	110.9 Hz	WAR
	FM	KB9R	443.30000	+	110.9 Hz	WAR
	FM	N9QWH	927.60000	902.60000	110.9 Hz	WAR
Eau Galle	DMR/BM	WW9RS	442.62500	+	CC13	WAR
	FM	WW9RS	145.21000	-	173.8 Hz	WAR
Edgerton	FM	WI9WIN	442.30000	+	123.0 Hz	WAR
Egg Harbor	FM	W9AIQ	146.73000	-	107.2 Hz	WAR
Elkhorn	DMR/MARC	WB9COW	442.84375	+	CC9	
	FM	W9ELK	146.86500	-	127.3 Hz	WAR

WISCONSIN 401

Location	Mode	Call sign	Output	Input	Access	Coordinator
Elkhorn	FM	N9LOH	443.70000	+	123.0 Hz	WAR
Evansville	FM	WB9RSQ	442.32500	+		WAR
Fitchburg	FM	KA9VDU	53.23000	51.53000	123.0 Hz	WAR
	FM	KA9VDU	444.00000	+	123.0 Hz	WAR
Fond Du Lac	DMR/MARC	KB9LQC	443.19375	+	CC12	
	DSTAR	KD9GXT	145.34500	-		
	FM	K9FDL	145.43000	-		
	FM	K9DJB	147.09000	+	107.2 Hz	WAR
	FM	KC9RUE	147.18000	+	146.2 Hz	WAR
	FM	KC9RUE	223.90000	-	107.2 Hz	
	FM	WI9WIN	442.40000	+	100.0 Hz	WAR
	FM	N9WQ	443.87500	+		WAR
	FM	N9GMT	444.60000	+	123.0 Hz	WAR
Franklin	DMR/MARC	N9OIG	443.43125	+	CC9	
Galesville	FM	N9TUU	147.00000	+	131.8 Hz	WAR
	FM	WI9WIN	442.50000	+	131.8 Hz	WAR
Germantown	FM	W9CQ	442.87500	+	127.3 Hz	WAR
	FM	WD9IEV	444.52500	+	114.8 Hz	WAR
Gillett, Jct 22-32	FM	W0LFE	444.22500	+	107.2 Hz	WAR
Gilmanton	FM	WE9COM	145.43000	-	131.8 Hz	WAR
Glenwood City	FM	N9LIE	145.27000	-	110.9 Hz	WAR
	FM	N9LIE	444.67500	+	110.9 Hz	WAR
Grafton	DSTAR	W9FRG	145.22500	-		
	DSTAR	W9FRG	442.81875	+		
	DSTAR	W9FRG	1282.15000	1262.15000		
Grand Chute	DMR/MARC	WX9KVH	434.98750	+	CC9	
Granton	FM	N9RRF	146.77500	-	114.8 Hz	WAR
Green Bay	DMR/MARC	N9PAY	443.50000	+	CC6	
	DSTAR	K9EAM	444.20625	+		
	FM	KB9GKC	146.68500	-	107.2 Hz	
	FM	N9DKH	147.07500	+	107.2 Hz	
	FM	K9GB	147.12000	+	107.2 Hz	WAR
	FM	K9EAM	147.36000	+	100.0 Hz	WAR
	FM	K9JQE	223.94000	-	107.2 Hz	WAR
	FM	N9GMT	442.80000	+	123.0 Hz	
	FM	WI9WIN	443.40000	+	100.0 Hz	WAR
	FM	KB9AMM	443.50000	+	107.2 Hz	WAR
	FM	KB9GKC	444.55000	+	107.2 Hz	WAR
	FM	W9OSL	444.75000	+	100.0 Hz	WAR
	FM	K9EAM	444.77500	+	107.2 Hz	WAR
Hales Corners	DMR/MARC	W9EMP	430.95000	438.55000	CC10	
	FM	W9JOL	443.42500	+	127.3 Hz	WAR
Hancock	FM	WI9WIN	442.72500	+	123.0 Hz	WAR
Hayward	FM	N9UPC	147.25500	+	110.9 Hz	WAR
High Bridge	FM	KC9GSK	147.21000	+	110.9 Hz	WAR
Holcombe	FM	N9LIE	52.81000	51.11000	110.9 Hz	WAR
	FM	N9LIE	145.47000	-	110.9 Hz	WAR
	FM	N9LIE	444.52500	+	110.9 Hz	WAR
Holcombe, Holcomb Hill						
	FM	AA9JL	147.34500	+	136.5 Hz	WAR
Hollandale	DSTAR	WI9WIN	147.28500	+		WAR
	FM	WI9WIN	444.55000	+	123.0 Hz	WAR
Hudson	FM	N9UPC	145.13000	-		WAR
Iola	FM	AC9F	443.60000	+	114.8 Hz	WAR
Irma	FM	KB9QJN	146.64000	-		WAR
	FM	WE9COM	146.89500	-	114.8 Hz	WAR
	FM	KC9NW	146.97000	-	71.9 Hz	WAR
	FM	KB9QJN	442.77500	+	114.8 Hz	WAR
Janesville	FM	WB9SHS	145.45000	-	123.0 Hz	
	FM	K9FRY	147.07500	+	123.0 Hz	WAR
	FM	KC9KUM	443.22500	+	123.0 Hz	WAR
Jefferson	FM	W9MQB	145.49000	-	123.0 Hz	WAR

402   WISCONSIN

Location	Mode	Call sign	Output	Input	Access	Coordinator
Juneau	FM	W9TCH	146.64000	-	123.0 Hz	WAR
Kaukauna	FM	ND9Z	444.45000	+	100.0 Hz	WAR
Kenosha	DMR/BM	KD9LUJ	434.35000	+	CC9	
	DMR/MARC	W2WAY	927.88750	902.88750	CC9	
Kewaskum	FM	N9NLU	146.79000	-	100.0 Hz	WAR
	FM	N9NLU	444.27500	+	127.3 Hz	WAR
Knapp/Menomonie						
	DMR/MARC	W8JWW	442.27500	+	CC1	
La Crosse	FM	AB9TS	444.47500	+	131.8 Hz	WAR
La Crosse, Downtown						
	FM	N0EXE	444.75000	+	131.8 Hz	
La Pointe	P25	KB9QJN	146.82000	-		WAR
Lac Du Flambeau	FM	W9BTN	146.70000	-	114.8 Hz	WAR
LaCrosse	FM	N9ETD	147.09000	+	131.8 Hz	
Lake Delton	FM	N9ROY	223.86000	-	123.0 Hz	WAR
	FM	N9ROY	443.85000	+	123.0 Hz	WAR
Lake Geneva	DMR/BM	KB9LTE	442.12500	+	CC9	WAR
	FM	N9GMT	441.52500	+	123.0 Hz	WAR
Lake Tomahawk	DSTAR	KC9ZJF	147.59000			
	DSTAR	KC9ZJF	446.42500	-		
Lampson	FM	N9PHS	146.97000	-	110.9 Hz	WAR
Lisbon	FM	KC9HBO	444.22500	+	151.4 Hz	WAR
Little Chute	DMR/MARC	KC9UHI	447.80000	-	CC6	
Madison	DMR/BM	W9YT	443.60000	+	CC9	WAR
	DMR/MARC	W9RCG	442.57500	+	CC9	WAR
	DSTAR	W9HSY	145.30500	-		WAR
	FM	N9KAN	53.07000	51.37000	103.5 Hz	WAR
	FM	WD8DAS	53.15000	51.45000	123.0 Hz	WAR
	FM	KC9FNM	145.37000	-	123.0 Hz	WAR
	FM	W9YT	146.68500	-	123.0 Hz	WAR
	FM	WR9ABE	146.94000	-	123.0 Hz	WAR
	FM	W9HSY	147.15000	+	123.0 Hz	WAR
	FM	WD8DAS	147.18000	+	107.2 Hz	WAR
	FM	WB9RSQ	224.16000	-	123.0 Hz	WAR
	FM	WD8DAS	224.18000	-		
	FM	WI9WIN	441.40000	+	123.0 Hz	WAR
	FM	N9KAN	443.40000	+	123.0 Hz	WAR
	FM	N9BDR	444.37500	+	123.0 Hz	WAR
	FM	KB9DRZ	444.57500	+	123.0 Hz	WAR
	FM	NG9V	444.77500	+	123.0 Hz	WAR
Madison, Beltline	FM	N9GMT	444.47500	+	123.0 Hz	WAR
Madison, Capital Square						
	FM	N9EM	442.55000	+		WAR
	IDAS	N9EM	145.15000	-	123.0 Hz	WAR
Madison, University Wisconsin						
	FM	WI9HF	443.77500	+	123.0 Hz	WAR
Madison, UW Hospital						
	FM	W9HSY	146.76000	-	123.0 Hz	WAR
Manitowish Waters						
	FM	KB9WCK	145.39000	-	114.8 Hz	WAR
Manitowoc	FM	W9DK	146.61000	-	107.2 Hz	WAR
	FM	W9RES	146.89500	-	146.2 Hz	
	FM	W9RES	443.15000	+	100.0 Hz	WAR
Marinette	FM	AB9PJ	444.50000	+	146.2 Hz	WAR
Markesan	FM	WB9RBC	146.95500	-	123.0 Hz	WAR
Marshfield	DSTAR	KD9FUR	147.04500	+		WAR
	FM	AA9US	147.18000	+		WAR
	FM	WI9WIN	444.85000	+	114.8 Hz	WAR
Mauston	FM	KB9WQF	146.85000	-	123.0 Hz	WAR
Medford	DSTAR	WI9WIN	444.15000	+		WAR
	FM	N9LIE	145.49000	-	114.8 Hz	WAR
	FM	KB9OBX	147.15000	+	114.8 Hz	WAR

WISCONSIN 403

Location	Mode	Call sign	Output	Input	Access	Coordinator
Medford	FM	N9LIE	444.82500	+	114.8 Hz	WAR
Menomonee Falls	DMR/MARC	W9RCG	442.01875	+	CC9	WAR
Menomonie	FM	K9KGB	146.61000	-	110.9 Hz	
	FM	N9QKK	146.68500	-	110.9 Hz	WAR
Merton	FM	W9JPE	444.62500	+	127.3 Hz	WAR
Meteor	FM	WE9COM	147.07500	+	110.9 Hz	WAR
Meteor, Meteor Hill						
	FM	N9MMU	145.11000	-	110.9 Hz	
Milwaukee	DMR/BM	K9MAR	442.05000	+	CC1	
	DMR/MARC	KD9FGF	442.20625	+	CC9	
	DMR/MARC	N9PAY	444.53125	+	CC1	
	DSTAR	KC9LKZ	145.24500	-		WAR
	DSTAR	KC9LKZ	442.46875	+		WAR
	DSTAR	KC9LKZ	1290.05000	1270.05000		WAR
	FM	W9DHI	53.03000	52.03000	103.5 Hz	WAR
	FM	N9LKH	145.13000	-	127.3 Hz	WAR
	FM	KA9WXN	145.25000	-		
	FM	W9HHX	145.27000	-	127.3 Hz	WAR
	FM	W9RH	145.39000	-	127.3 Hz	WAR
	FM	N9BMH	146.62500	-	127.3 Hz	WAR
	FM	N9PAY	146.94000	-		WAR
	FM	WB0AFB	147.04500	+	127.3 Hz	WAR
	FM	K9IFF	147.10500	+	127.3 Hz	WAR
	FM	W9WK	147.16500	+	127.3 Hz	WAR
	FM	N9UUR	442.42500	+	127.3 Hz	
	FM	W9HHX	443.02500	+	114.8 Hz	WAR
	FM	WI9WIN	443.27500	+	127.3 Hz	WAR
	FM	N9PAY	443.32500	+	127.3 Hz	WAR
	FM	N9LKH	443.55000	+	127.3 Hz	WAR
	FM	N9GMT	443.80000	+	123.0 Hz	WAR
	FM	W9EFJ	444.45000	+	114.8 Hz	WAR
	FM	WB9HKE	444.75000	+		
	FM	W9DHI	444.85000	+	127.3 Hz	WAR
	FM	N9PAY	927.51250	952.51250	127.3 Hz	WAR
Monroe	FM	W9MUP	52.97000	51.27000		WAR
	FM	KO9LR	145.11000	-	123.0 Hz	WAR
	FM	K9TSU	147.34500	+	123.0 Hz	
	FM	W9MUP	443.52500	+		WAR
Montello	FM	KC9ASQ	146.74500	-	123.0 Hz	WAR
Mount Sterling	FM	W9DMH	147.36000	+	131.8 Hz	WAR
Necedah	FM	KC9IPY	147.21000	+	123.0 Hz	WAR
	FM	K9UJH	444.12500	+	123.0 Hz	WAR
New Berlin	DMR/MARC	N9PAY	441.43750	+	CC1	WAR
	FM	W9DHI	53.03000	52.03000	127.3 Hz	
	FM	WI9MRC	146.91000	-	127.3 Hz	
	FM	KB9SIF	442.07500	+		WAR
	FM	WA9AOL	442.67500	+	127.3 Hz	WAR
	FM	W9LR	443.30000	+	127.3 Hz	WAR
	FM	W9DHI	444.85000	+	127.3 Hz	
	FM	KC9FTE	927.01250	902.01250	DCS 152	
New Holstein	FM	KA9OJN	147.30000	+	107.2 Hz	WAR
Niagara	FM	W9MB	223.82000	-	114.8 Hz	WAR
North Freedom	FM	KD9UU	443.67500	+	123.0 Hz	WAR
Oconto Falls	FM	KB9DSV	146.83500	-	107.2 Hz	WAR
Oshkosh	DMR/BM	KC9LYF	444.47500	+	CC1	
	DMR/MARC	W9RCG	442.21875	+	CC6	
	FM	N9GDY	442.07500	+	107.2 Hz	WAR
	FM	N9GMT	443.62500	+	123.0 Hz	WAR
Oshkosh, Sandpit Road Tower						
	FM	KC9SDK	147.24000	+	100.0 Hz	
Park Falls	DMR/MARC	KD9IPR	146.71500	-	CC2	
	DMR/MARC	W9PFP	444.45000	+	CC1	

404　WISCONSIN

Location	Mode	Call sign	Output	Input	Access	Coordinator
Park Falls	DSTAR	W9PFP	442.48125	+		WAR
	FM	W9PFP	444.75000	+	110.9 Hz	WAR
Parkfalls	DMR/BM	KD9IPR	444.75000	+	CC5	
Platteville	FM	WI9WIN	442.20000	+	123.0 Hz	
	FM	W9UWP	444.17500	+	131.8 Hz	
	FM	KC9KQ	444.32500	+	131.8 Hz	WAR
Pleasant Prairie, St. Catherin						
	FM	K9KEA	927.61250	902.61250	127.3 Hz	WAR
Plover	FM	W9SM	442.05000	+	114.8 Hz	WAR
Plymouth	FM	WE9COM	146.85000	-	100.0 Hz	WAR
	FM	WE9R	147.06000	+	107.2 Hz	WAR
	FM	KD9TZ	443.22500	+	107.2 Hz	WAR
	FM	WE9R	444.35000	+	114.8 Hz	WAR
Port Washington	FM	AC9CD	147.33000	+	127.3 Hz	WAR
	FM	AC9CD	443.52500	+	114.8 Hz	WAR
	FM	W9CQO	443.75000	+	127.3 Hz	WAR
Portage	DMR/MARC	N3IVK	444.07500	+	CC8	
Porterfield	FM	W4IJR	444.40000	+	100.0 Hz	WAR
Pound	FM	WI9WIN	442.00000	+	114.8 Hz	WAR
Racine	DMR/MARC	KR9RK	440.00625	+	CC9	
	FM	KR9RK	147.27000	+	127.3 Hz	WAR
	FM	KC9QKJ	224.80000	-	127.3 Hz	WAR
	FM	KR9RK	442.00000	+	127.3 Hz	
	FM	KA9LOK	444.05000	+	114.8 Hz	WAR
Radisson	FM	N9UPC	444.22500	+		
Rhinelander	FM	KC9HBX	146.94000	-	114.8 Hz	WAR
Rice Lake	FM	WI9WIN	442.10000	+	110.9 Hz	WAR
	P25	KD9EJA	444.45000	+	DCS 351	WAR
Richland Center	DSTAR	KC9WDW	147.19500	+		WAR
	DSTAR	KC9WDW	442.48750	+		
	FM	W9PVR	146.91000	-	131.8 Hz	WAR
	FM	WI9WIN	442.70000	+	131.8 Hz	WAR
Ripon	FM	N9GMT	444.95000	+	123.0 Hz	WAR
River Falls	FM	WI9WIN	443.02500	+	110.9 Hz	WAR
Roberts	FM	N9UPC	147.33000	+	110.9 Hz	WAR
Rosendale	FM	KB9YET	147.37500	+	107.2 Hz	
Rubicon	FM	WB9KPG	145.35000	-	123.0 Hz	WAR
Rudolph	FM	WD9GFY	444.32500	+	114.8 Hz	WAR
Sayner	DSTAR	KD9JHE	444.40000	+		WAR
	FM	WE9COM	145.13000	-	114.8 Hz	WAR
	FM	KD9JHE	147.36000	+		WAR
Shawano	FM	KA9NWY	145.35000	-	114.8 Hz	WAR
Sheboygan	DSTAR	KC9WUS	146.98500	-		
	DSTAR	KC9SJY	147.25500	+		WAR
	DSTAR	KC9SJY	442.48125	+		WAR
	FM	KC9AXZ	443.45000	+	107.2 Hz	
	FM	WI9WIN	444.30000	+	146.2 Hz	
Sheboygan Falls	FM	KB5ZJU	224.94000	-	100.0 Hz	WAR
Siren	FM	N9PHS	146.62500	-	110.9 Hz	WAR
Sister Bay	FM	W9AIQ	147.18000	+	100.0 Hz	WAR
Slinger	FM	KC9PVD	147.21000	+	127.3 Hz	WAR
	FM	WB9BVB	442.65000	+	127.3 Hz	
	FM	KC9PVD	443.82500	+	127.3 Hz	WAR
Solon Springs	FM	AA9JL	145.49000	-	110.9 Hz	WAR
Spooner	FM	KB9OHN	147.30000	+	110.9 Hz	WAR
	FM	KB9OHN	443.50000	+	110.9 Hz	WAR
Spooner/Hertel	FM	KB9OHN	145.19000	-	110.9 Hz	WAR
St. Lawrence	FM	WB9BVB	146.73000	-	127.3 Hz	
Stevens Point	DSTAR	N9NMH	146.50000			
	FM	WB9QFW	146.67000	-	114.8 Hz	WAR
	FM	WB9QFW	146.98500	-	114.8 Hz	WAR
	FM	KC9NW	444.70000	+	114.8 Hz	WAR

WISCONSIN 405

Location	Mode	Call sign	Output	Input	Access	Coordinator
Sturgeon Bay	FM	W9DOR	147.21000	+	107.2 Hz	WAR
Sturtevant	DMR/BM	W2WAY	927.72500	902.72500	CC9	
Superior	FM	WA9KLM	145.17000	-		WAR
	FM	K9UWS	146.76000	-	110.9 Hz	WAR
Suring	FM	WE9COM	145.29000	-	114.8 Hz	WAR
	FM	AB9PJ	145.47000	-	114.8 Hz	WAR
	FM	AB9PJ	442.55000	+	146.2 Hz	WAR
Sussex	FM	W9CQ	147.39000	+	127.3 Hz	WAR
Three Lakes	FM	N9GHE	147.19500	+	114.8 Hz	WAR
	FM	N9GHE	224.54000	-	114.8 Hz	WAR
Tomah	FM	KC9KVE	146.80500	-	131.8 Hz	WAR
	FM	WI9WIN	444.80000	+	131.8 Hz	WAR
Tomahawk	FM	N9MEA	52.83000	51.13000	114.8 Hz	WAR
	FM	N9CLE	145.43000	-	114.8 Hz	WAR
	FM	N9CLE	223.76000	-	114.8 Hz	WAR
	FM	N9CLE	444.57500	+	114.8 Hz	WAR
Town Of Weston	FM	WW9RS	146.80500	-	110.9 Hz	WAR
Trevor	FM	KA9VZD	442.60000	+	123.0 Hz	WAR
Tripoli	FM	KC9HBX	147.12000	+	114.8 Hz	WAR
Union Grove	DMR/BM	N9OIG	146.74500	-	CC9	
	DMR/BM	N9OIG	442.25000	+	CC9	WAR
	DMR/MARC	N9OIG	442.24375	+	CC9	
Unity, Brighton Tower						
	FM	W9BCC	145.41000	-	114.8 Hz	
Viroqua	FM	N9TUU	145.17000	-	131.8 Hz	WAR
Walworth	DMR/MARC	N9OIG	442.25625	+	CC9	
Washburn	FM	KB9JX	145.15000	-	103.5 Hz	WAR
Washington Island						
	FM	WI9DX	145.49000	-	100.0 Hz	
Waterford	DMR/MARC	N9OIG	440.76875	+	CC4	
Watertown	DSTAR	W9TTN	440.15000	+		
	FM	K9LUK	145.19000	-	123.0 Hz	WAR
Waukesha	FM	WE9COM	145.47000	-	127.3 Hz	WAR
	FM	W9RIX	224.82000	-	127.3 Hz	
	FM	WQ9A	444.20000	+		
Waupaca	FM	W9GAP	146.92500	-	118.8 Hz	WAR
	FM	W9KL	147.39000	+	118.8 Hz	
	FM	WI9WIN	444.67500	+	114.8 Hz	WAR
	FM	N5IIA	444.90000	+	114.8 Hz	WAR
Wausau	DMR/BM	W9KFD	442.12500	+	CC1	
	DMR/MARC	W9RCG	442.13125	+	CC9	
	DSTAR	W9BCC	145.24500	-		WAR
	DSTAR	W9BCC	442.46875	+		WAR
	DSTAR	W9BCC	1282.10000	1262.10000		WAR
	FM	W9SM	29.64000	-		WAR
	FM	W9MEA	52.89000	51.19000	114.8 Hz	WAR
	FM	KB9KST	145.37000	-	114.8 Hz	WAR
	FM	W9SM	146.86500	-	114.8 Hz	WAR
	FM	W9SM	147.13500	+	114.8 Hz	WAR
	FM	W9SM	224.64000	-	114.8 Hz	WAR
	FM	WI9WIN	442.20000	+	114.8 Hz	WAR
	FM	KA9HQE	443.32500	+	100.0 Hz	WAR
	FM	W9SM	443.52500	+	114.8 Hz	WAR
	FM	KC9NW	443.75000	+	71.9 Hz	WAR
	FM	W9SM	444.10000	+	114.8 Hz	WAR
	FM	W9BCC	444.30000	+	114.8 Hz	WAR
	FM	W9SM	444.42500	+	114.8 Hz	WAR
Wausau, Rib Mountain						
	FM	W9BCC	146.73000	-		
	FM	W9BCC	146.82000	-	114.8 Hz	WAR
Wausaukee	FM	WA8WG	146.88000	-	136.5 Hz	WAR
Webster	DMR/BM	N0DZQ	443.15000	+	CC3	

406 WISCONSIN

Location	Mode	Call sign	Output	Input	Access	Coordinator
West Allis	FM	KA9JCP	224.52000	-	127.3 Hz	WAR
	FM	N9MKX	444.42500	+	127.3 Hz	WAR
West Milwaukee	FM	N9FSE	147.13500	+	141.3 Hz	WAR
Willard	FM	N9UWX	147.27000	+	114.8 Hz	WAR
Wisconsin Rapids	FM	W9MRA	146.79000	-	114.8 Hz	WAR
	FM	W9MRA	147.33000	+	114.8 Hz	WAR
Wisconsin Rapids, East Side Wa						
	FM	W9MRA	442.42500	+	114.8 Hz	
Wonewoc	DSTAR	KD9BLN	146.71500	-		WAR
	DSTAR	KD9BLN	443.69375	+		WAR
Woodruff	DSTAR	KA9SRO	445.97500	-		

WYOMING

Location	Mode	Call sign	Output	Input	Access	Coordinator
Afton, The Narrows						
	FM	KD7LVE	146.97000	-	100.0 Hz	
Big Piney	FM	KC7BJY	145.14500	-	100.0 Hz	WRCG
	FM	KC7BJY	146.88000	-	100.0 Hz	WRCG
Big Piney, The Hogsback						
	FM	KC7BJY	146.88000	-	100.0 Hz	WRCG
Boysen, Boysen Peak						
	DSTAR	N7HYF	147.06000	+		WRCG
Buffalo	FM	NX7Z	146.88000	-	100.0 Hz	WRCG
	FM	WY7BRK	147.18000	+	110.9 Hz	WRCG
Buffalo, Windy Ridge						
	FM	NX7Z	146.88000	-	100.0 Hz	WRCG
Burlington, Tatman Mtn						
	DSTAR	KG7PRH	147.21000	+	123.0 Hz	WRCG
Casper	DMR/MARC	WY7EOC	449.98750	-	CC11	WRCG
	FM	W7VNJ	52.98000	-	131.8 Hz	WRCG
	FM	K7PLA	145.46000	-	110.9 Hz	WRCG
	FM	KD7AGA	146.64000	-	173.8 Hz	WRCG
	FM	W7VNJ	146.94000	-	123.0 Hz	WRCG
	FM	NG7T	449.10000	-		
	FM	W7VNJ	449.57500	-	173.8 Hz	WRCG
Casper, Casper Mountain						
	FM	W7VNJ	145.23500	-	100.0 Hz	
	FM	K7PLA	145.46000	-	110.9 Hz	WRCG
	FM	W7VNJ	146.94000	-	123.0 Hz	WRCG
	FUSION	N7RRB	449.90000	-		WRCG
Casper, Casper Mtn.						
	FM	NB7I	449.50000	-	100.0 Hz	
Casper, Wyoming Med Center						
	FM	W7VNJ	449.57500	-	173.8 Hz	WRCG
Cedar Mtn	FM	KE7UJB	444.77500	-	103.5 Hz	
Cheyenne	DMR/BM	N7JJY	448.87500	-	CC1	
	DMR/BM	KC7DHF	448.95000	-	CC15	WRCG
	DMR/MARC	K7PFJ	449.93750	-	CC7	WRCG
	FM	KC7SNO	146.77500	-	114.8 Hz	WRCG
	FM	KC7SNO	147.10500	+	114.8 Hz	WRCG
	FM	KB7SWR	448.15000	-	100.0 Hz	WRCG
	FM	N7JJY	449.90000	-	100.0 Hz	
Cheyenne, Chalk Bluffs						
	FM	N7JJY	449.30000	-		WRCG
Cheyenne, Denver Hill						
	FM	KC7SNO	146.77500	-	114.8 Hz	WRCG
Cheyenne, North Park						
	FM	KC7SNO	147.10500	+	91.5 Hz	WRCG
Cody	FM	KC7NP	29.68000	-		
	FM	KI7W	146.85000	-	103.5 Hz	WRCG
	FM	KE7UJB	147.12000	+		WRCG
	FM	KE7UJB	444.50000	+	103.5 Hz	
Cody, Cedar Mtn	FM	KE7UJB	147.12000	+		WRCG

WYOMING   407

Location	Mode	Call sign	Output	Input	Access	Coordinator
Cody, McCullough Peaks						
	FM	KI7W	146.85000	-	103.5 Hz	WRCG
Douglas	FM	KK7BA	147.15000	+		
Dubois	FM	KD7BN	146.82000	-	100.0 Hz	WRCG
Evanston	FM	K7JL	449.15000	-	100.0 Hz	UVHFS
Evanston, Medicine Butte						
	FM	K7JL	146.86000	-	100.0 Hz	WRCG
Gilette	FM	W7WBW	145.33000	-	123.0 Hz	
Gillette	DSTAR	NE7WY	146.96000	-		WRCG
	FM	NE7WY	147.36000	+	123.0 Hz	WRCG
	FM	NE7WY	448.75000	-		WRCG
Gillette, Antelope Butte						
	FM	NE7WY	147.36000	+	123.0 Hz	WRCG
Gillette, Bliss Ranch						
	FM	NE7WY	147.27000	+	123.0 Hz	WRCG
Green River, Green River High						
	FM	AD0BN	444.77500	+	123.0 Hz	WRCG
Jackson	FM	W7TAR	146.73000	-	123.0 Hz	WRCG
	FM	W7TAR	146.91000	-	123.0 Hz	WRCG
	FM	W7TAR	447.70000	-	123.0 Hz	WRCG
Jackson, Snow King Mountain						
	FM	W7TAR	146.91000	-		WRCG
	FM	W7TAR	447.70000	-	123.0 Hz	WRCG
Kaycee	FM	W7QQA	145.43000	-	110.9 Hz	WRCG
Kaycee, Pack Saddle						
	FM	W7QQA	145.43000	-	110.9 Hz	WRCG
Kemmerer	FM	KA7SHX	449.82500	-	100.0 Hz	WRCG
Kemmerer, Dempsey Ridge						
	FM	KF7EHE	449.07500	-	123.0 Hz	
Kemmerer, Qualey Ridge						
	FM	N7ERH	449.30000	-	127.3 Hz	
Kemmerer, Quealy Peak						
	FM	KG7VVQ	147.09000	+	100.0 Hz	WRCG
	FM	KA7SHX	449.82500	-	100.0 Hz	WRCG
Lander	FM	N7HYF	53.03000	52.03000	DCS 261	WRCG
	FM	KD7PPP	145.44500	-	110.9 Hz	WRCG
	FM	N7HYF	449.97500	-	100.0 Hz	WRCG
Lander, Airport	FM	KD7PPP	145.44500	-	110.9 Hz	WRCG
Lander, Limestone Mountain						
	FM	N7HYF	449.90000	-	DCS 261	WRCG
Laramie	DMR/MARC	WY7EOC	447.22500	-	CC11	WRCG
	FM	N7UW	146.61000	-		WRCG
	FM	KC7SNO	146.82000	-	114.8 Hz	WRCG
	FM	N7UW	147.01500	+	146.2 Hz	WRCG
Laramie, Beacon Hill						
	FM	KC7SNO	146.82000	-	114.8 Hz	WRCG
Laramie, Jelm Mountain						
	FM	N7UW	147.01500	+	146.2 Hz	
Laramie, University						
	FM	N7UW	146.61000	-		WRCG
Lovell	DMR/MARC	W7BEQ	446.92500	-	CC11	
	FM	W7BEQ	147.16500	+	103.5 Hz	WRCG
Lovell, Medicine Mountain						
	FM	W7BEQ	147.16500	+	103.5 Hz	WRCG
Lusk, 77 Hill	FM	KG7OMT	147.33000	+	103.5 Hz	
Marbleton	FM	KC7BJY	145.14500	-	100.0 Hz	WRCG
Meeteetse	FM	KI7W	147.33000	+	103.5 Hz	WRCG
Meeteetse, 3 Mile Hill						
	FM	KI7W	147.33000	+	103.5 Hz	WRCG
Mountain View, Hickey Mtn						
	FM	WY7BV	144.63500	+	100.0 Hz	
Newcastle	FM	NE7WY	147.30000	+	162.2 Hz	WRCG

408 WYOMING

Location	Mode	Call sign	Output	Input	Access	Coordinator
Newcastle, Mt Pisgah						
	FM	NE7WY	147.30000	+	162.2 Hz	WRCG
Pinedale	FM	KC7BJY	448.10000	-	100.0 Hz	WRCG
Powell	DMR/BM	KG7PRH	444.72500	+	CC3	
Rawlins	FM	N7RON	146.70000	-	162.2 Hz	WRCG
	FM	KC7OZU	146.76000	-	100.0 Hz	
	FM	N7GCR	147.24000	+	100.0 Hz	WRCG
	FM	KD7BN	147.39000	+	100.0 Hz	
Rawlins, Elk Mountain						
	FM	N7GCR	147.24000	+	100.0 Hz	WRCG
Riverton	DMR/BM	KA0NDS	147.94000	-	CC1	
Riverton, Griffey Hill						
	FM	K0FOP	145.11500	-	100.0 Hz	WRCG
Riverton, Griffy Hill						
	DSTAR	N7DMO	441.67000	+		
	FM	N7HYF	449.97500	-	127.3 Hz	WRCG
Rock River	FM	K7UWR	53.03000	52.03000		
Rock Springs	DMR/BM	KE7UUJ	444.77500	+	CC12	
	DMR/MARC	KF7OBL	147.30000	+	CC1	
	DMR/MARC	K7DRA	447.11250	-	CC11	
	DMR/MARC	KF7OBL	448.87500	-	CC1	
	FM	KE7UUJ	146.65500	-	107.2 Hz	WRCG
	FM	N7ABC	444.50000	+	123.0 Hz	
Rock Springs, Aspen Mountain						
	FM	KE7FGD	146.61000	-	100.0 Hz	WRCG
	FM	KE7UUJ	146.65500	-	107.2 Hz	
	FM	KE7UUJ	146.94000	-	100.0 Hz	WRCG
Rocky Point	FM	NE7WY	147.27000	+	123.0 Hz	WRCG
Sheridan	DMR/MARC	W7BEQ	446.72500	-	CC11	
	FM	W7GUX	449.70000	-	100.0 Hz	WRCG
Sheridan, Big Horn Mountain						
	FM	W7GUX	449.70000	-	100.0 Hz	WRCG
Shoshoni	DMR/MARC	W7BEQ	446.82500	-	CC11	
	FM	W7BEQ	146.80500	-	100.0 Hz	WRCG
Shoshoni, Copper Mtn						
	FM	W7BEQ	146.80500	-	100.0 Hz	WRCG
Statewide	DMR/MARC	WY7EOC	445.01250	-	CC11	
Sundance	FM	NE7WY	146.79000	-	100.0 Hz	WRCG
Sundance, Warren Peak						
	FM	NE7WY	146.79000	-	100.0 Hz	WRCG
Ten Sleep	DMR/MARC	W7BEQ	446.62500	-	CC11	
	FM	W7BEQ	145.14500	-	103.5 Hz	WRCG
Ten Sleep, Meadowlark Mountain						
	FM	KG7KBJ	147.37500	+	110.9 Hz	
Tensleep	FM	W7BEQ	145.14500	-	103.5 Hz	WRCG
	FM	KG7KBJ	147.37500	+	110.9 Hz	WRCG
Teton Village, Rendezvous Moun						
	FM	W7TAR	146.73000	-	123.0 Hz	WRCG
Thayne	FUSION	KN6LL	448.20000	-	123.0 Hz	WRCG
Torrington	FM	KD7JNQ	448.32500	-	151.4 Hz	
Wheatland	FM	WA7SNU	146.88000	-		
	FM	WA7SNU	449.62500	-	151.4 Hz	
Worland	FM	W7BEQ	444.82500	-	103.5 Hz	
Wright	FM	NX7Z	146.98500	-	100.0 Hz	WRCG
	FM	NE7WY	147.06000	+	100.0 Hz	WRCG
Wright, Pumpkin Butte						
	FM	NX7Z	146.98500	-	100.0 Hz	WRCG

ALBERTA 409

Location	Mode	Call sign	Output	Input	Access	Coordinator

CANADA

ALBERTA

Location	Mode	Call sign	Output	Input	Access	Coordinator
Airdrie	FM	VE6AA	145.31000	-	100.0 Hz	ARLA
	FM	VE6JBJ	147.54000			
Alberta	FM	VE6HUB	443.57500	+		
Aldersyde	FM	VE6HRA	147.00000	+	110.9 Hz	ARLA
Aldersyde, Gladys Ridge						
	FM	VE6RPX	145.47000	-	100.0 Hz	
Alix	FM	VE6PAT	147.21000	+		ARLA
	FM	VE6PAT	448.97500	-		ARLA
Andrew	FM	VE6JET	146.70000	-		ARLA
Athabasca	FM	VE6BOX	146.73000	-		ARLA
Balzac	FM	VE6EDS	444.85000	+		ARLA
Banff	FM	VE6FAA	147.57000		100.0 Hz	
	FM	VE6MPR	444.78750	+		ARLA
Banff National P	DSTAR	VE6WRO C	147.03000	+		
	DSTAR	VE6WRO B	444.82500	+		
Banff National Park						
	DSTAR	VE6WRO A	1248.50000			
Banff, Tunnel Mtn	FM	VE6BNF	146.67000	-	131.8 Hz	
Barons	FM	VE6CAM	146.88000	-		ARLA
Beaver Lodge	FM	VE6BL	146.85000	-		ARLA
Big Valley	FM	VE6UK	145.25000	-		ARLA
Bonneville	FM	VE6ADI	146.71500	-		ARLA
Borradaile	FM	VE6BDL	449.07500	-		ARLA
Bragg Creek	FM	VE6RAY	444.87500	+	110.9 Hz	ARLA
Brooks	FM	VE6HBR	145.35000	-		ARLA
	FM	VE6SPK	446.30000	-		
Burmis	FM	VE6HRP	145.39000	-		ARLA
Calgary	DMR/MARC	VE6RYC	444.00000	+	CC1	ARLA
	DSTAR	VA6ACW C	145.65000			
	DSTAR	VE6WRN C	146.80500	145.60500		
	DSTAR	VE6GHZ C	147.09000	+		
	DSTAR	VE6IPG C	147.28500	+		
	DSTAR	VA6MEO B	433.87500	-		
	DSTAR	VE6WRE B	444.82500	+		
	DSTAR	VE6WRN B	444.92500	+		
	DSTAR	VE6GHZ B	444.95000	+		
	DSTAR	VE6IPG B	444.96250	+		
	DSTAR	VA6ACW B	445.85000	-		
	DSTAR	VA3URU B	448.65000	+		
	DSTAR	VE6WRN A	1247.50000			
	DSTAR	VE6IPG A	1248.05000			
	DSTAR	VE6GHZ A	1253.00000			
	DSTAR	VE6IPG A	1275.95000	+		
	DSTAR	VE6WRN A	1287.50000	1267.50000		
	DSTAR	VE6GHZ A	1287.97500	1267.97500		
	FM	VE6RYC	53.03000	52.03000	110.9 Hz	ARLA
	FM	VA6TWO	53.39000	52.39000	131.8 Hz	
	FM	VE6ARA	145.15000	-	110.9 Hz	
	FM	VE6ZV	145.23000	-	110.9 Hz	
	FM	VA6CTV	145.29000	-		ARLA
	FM	VE6OIL	146.61000	-	114.8 Hz	ARLA
	FM	VE6CID	146.68500	-		ARLA
	FM	VE6MX	146.73000	-		ARLA
	FM	VE6RYC	146.85000	-	110.9 Hz	ARLA
	FM	VE6RPT	146.94000	-		ARLA
	FM	VE6REC	147.18000	+		ARLA
	FM	VE6RPC	147.21000	+	110.9 Hz	ARLA

410 ALBERTA

Location	Mode	Call sign	Output	Input	Access	Coordinator
Calgary	FM	VE6QCW	147.24000	+		ARLA
	FM	VE6RY	147.27000	+		ARLA
	FM	VA6TWO	147.39000	+	100.0 Hz	
	FM	VE6RYC	224.85000	223.15000	110.9 Hz	ARLA
	FM	VA6TWO	442.20000	+	131.8 Hz	
	FM	VE6OIL	442.90000	+	131.8 Hz	ARLA
	FM	VE6NZ	443.15000	+		ARLA
	FM	VE6ZV	444.27500	+	110.9 Hz	ARLA
	FM	VE6EHX	444.35000	+	110.9 Hz	ARLA
	FM	VE6RY	444.57500	+	110.9 Hz	ARLA
	FM	VE6FIL	444.67500	+		ARLA
	FM	VE6DDC	444.80000	+		ARLA
	FM	VA6TRE	444.87500	+	131.8 Hz	
	FM	VA6TWO	927.01250	902.01250		ARLA
	P25	VE6WRO	927.05000	902.05000	131.8 Hz	
	P25	VA6TRE	927.72500	902.72500	131.8 Hz	
Calgary Alberta	FM	VE6AZX	147.50000		100.0 Hz	
Camrose	FM	VE6UU	146.76000	-		ARLA
	FM	VE6UU	444.02500	+		ARLA
Canmore	FM	VE6RJZ	146.58000		100.0 Hz	
	FM	VE6XRP	147.30000	+		ARLA
	FM	VE6RMT	147.36000	+	110.9 Hz	ARLA
Carbon	FM	VE6RCB	146.71500	-	110.9 Hz	ARLA
Cheadle	FM	VE6GLR	147.01500	+		ARLA
Chipman	FM	VE6TNC	146.61000	-	100.0 Hz	ARLA
Claresholm	FM	VE6AAH	145.21000	-	103.5 Hz	ARLA
	FM	VE6ROT	146.79000	-		ARLA
Claresholm, Burton Creek						
	FM	VE6HRK	145.43000	-	110.9 Hz	
Cochrane	FM	VE6PR	147.37500	+		ARLA
	FM	VE6PR	449.05000	-		ARLA
	P25	VE6RPT	444.90000	+		ARLA
Cold Lake	FM	VE6ADI	147.09000	+		ARLA
Coleman	FM	VE6FRC	145.49000	-		
College Heights	FM	VA6REB	146.42000	+		
Crossfield	FM	VE6TPA	147.13500	+		
	FM	VE6YXR	448.75000	-	107.2 Hz	ARLA
Crossfield, Moneys Mushrooms						
	FM	VE6HRF	145.35000	-	110.9 Hz	
Crowsnest Pass	FM	VE6CNP	145.49000	-		ARLA
Delia	FM	VE6GWR	444.42500	+	114.8 Hz	
	FM	VE6HB	448.12500	-		ARLA
Drayton Valley	FM	VE6HUB	442.90000	+	131.8 Hz	ARLA
Edmonton	DMR/BM	VE6EMS	442.25000	+	CC1	
	DMR/MARC	VE6VPR	440.80000	+	CC1	
	DSTAR	VE6KM C	145.47000	-		
	DSTAR	VA6XG C	145.71000			
	DSTAR	VA6XG C	147.42000			
	DSTAR	VE6BHX B	440.75000	+		
	DSTAR	VA6KGA B	440.80000	+		
	DSTAR	VE6KM B	444.90000	+		
	DSTAR	VA6XG B	446.47500	-		
	DSTAR	VA6XG A	1282.00000	-		
	FM	VA6WY	145.01000			
	FM	VE6RPA	145.19000	-		ARLA
	FM	VE6NHB	145.41000	-	114.8 Hz	ARLA
	FM	VE6QCR	146.64000	-	100.0 Hz	ARLA
	FM	VE6FDX	146.83500	-		
	FM	VE6OG	146.85000	-	100.0 Hz	ARLA
	FM	VE6HM	147.06000	+	100.0 Hz	ARLA
	FM	VE6EDM	147.12000	+	100.0 Hz	
	FM	VE6PAW	147.18000	+	100.0 Hz	ARLA

ALBERTA 411

Location	Mode	Call sign	Output	Input	Access	Coordinator
Edmonton	FM	VE6UV	147.24000	+	100.0 Hz	ARLA
	FM	VE6TOP	147.30000	+	100.0 Hz	ARLA
	FM	VE6TOP	147.39000	+	100.0 Hz	
	FM	VA6XG	147.42000		110.9 Hz	
	FM	VA6RS	224.56000	-	123.0 Hz	ARLA
	FM	VE6HM	224.76000	-	100.0 Hz	ARLA
	FM	VE6AEC	442.15000	+	146.2 Hz	ARLA
	FM	VE6DBD	444.02500	+	146.2 Hz	
	FM	VE6HM	444.10000	+	100.0 Hz	ARLA
	FM	VE6GPS	444.40000	+		ARLA
	FM	VE6EHR	444.67500	+		
	FM	VE6TOP	444.70000	+	123.0 Hz	ARLA
	FM	VE6EDM	444.75000	+	136.5 Hz	
	FM	VE6NHB	444.95000	+	114.8 Hz	ARLA
	FM	VE6FDX	449.02500	-	100.0 Hz	
	FUSION	VE6SCA	444.72500	+	146.2 Hz	
Edmonton, River Cree Casino						
	FM	VE6JN	147.33000	+	100.0 Hz	
Edmonton, Sherwood Park						
	FM	VE6HM	53.43000	52.43000	100.0 Hz	
Edmonton, South West						
	FM	VA6KGA	440.80000	+		
Edson	FM	VE6YFR	146.68500	-		
	FM	VE6RDF	147.37500	+		ARLA
	FM	VA6JAC	147.55500			
	FM	VE6MBX	444.62500	+	156.7 Hz	ARLA
Evanston	FM	VE6BUL	145.57000	+		
Falun	FM	VE6PLP	147.09000	+	100.0 Hz	ARLA
Fort McMurray	FM	VE6TBC	146.94000	-	100.0 Hz	
	FM	VE6TRC	147.00000	+	100.0 Hz	ARLA
	FM	VA6CYR	147.15000	+		
Fortress Mountain	FM	VE6AQA	147.12000	+	110.9 Hz	ARLA
Ft. Saskatchewan	FM	VE6CWW	147.27000	+		ARLA
Furman	FM	VE6ARS	145.21000	+		
Glendon	FM	VE6HOG	145.45000	-		ARLA
	FM	VE6COW	444.97500	+		ARLA
Grande Cache	FM	VE6YGR	147.39000	+		ARLA
Grande Prairie	FM	VE6OL	147.06000	+		ARLA
	FM	VE6XN	147.15000	+		ARLA
	FM	VE6AAV	444.77500	+	146.2 Hz	ARLA
	FM	VE6MDK	449.10000	-	100.0 Hz	ARLA
Grande Prairie EchoIRLP 61640						
	FM	VE6HIM	146.58000		146.2 Hz	
Grimshaw	FM	VE6AAA	448.51500	-		ARLA
Hailstone Butte	DSTAR	VE6WRT C	147.39000	+		
Hanna	FM	VE6HB	146.82000	-		ARLA
Hardisty	FM	VE6HDY	145.17000	-		ARLA
High Prairie	FM	VE6PRR	146.64000	-		ARLA
High River	FM	VE6HRB	145.17000	-		ARLA
	FM	VA6HRH	443.55000	+	110.9 Hz	ARLA
High River, Hospital						
	FM	VA6PF	146.70000	-	110.9 Hz	
Hinton	FM	VE6YAR	146.76000	-		ARLA
Innisfree	FM	VE6INN	147.34500	+		ARLA
Jasper	FM	VE6YPR	147.15000	+		
Kathyrn	FM	VE6OTR	145.39000	-		
Kingman	FM	VE6MTR	444.17500	+	123.0 Hz	ARLA
Kitscoty	FM	VE6YHB	444.22500	+	141.3 Hz	ARLA
Lake Eliza	FM	VE6BGB	146.67000	+		
Lake Louise	FM	VE6BNP	146.88000	-		ARLA
	FM	VE6HWY	147.33000	+		ARLA
Lethbridge	FM	VA6IRL	146.97000	-		

412   ALBERTA

Location	Mode	Call sign	Output	Input	Access	Coordinator
Lethbridge	FM	VE6UP	147.15000	+		ARLA
	FM	VE6XA	444.85000	+		ARLA
	FM	VE6LRH	448.38000	-		
	FUSION	VE6CV	442.07500	+		
Lethbridge EchoIRLP 2722						
	FM	VE6COM	446.27500	-	131.8 Hz	
Lime Stone Mtn	FM	VE6MTR	145.27000	-	250.3 Hz	ARLA
Little Smoky	FM	VE6MBX	147.07500	+	156.7 Hz	ARLA
Lloydminster	FM	VE5FN	146.94000	-		
	FM	VE5YLL	444.72500	+	100.0 Hz	ARLA
Longview	FM	VE6HRL	145.37000	-		ARLA
	FM	VE6WRT	224.94000	-		ARLA
Medicine Hat	DSTAR	VE6MHD C	147.09000	+		
	DSTAR	VE6VOA B	433.10000			
	DSTAR	VA6SRG B	444.80000	+		
	DSTAR	VE6MHD B	445.90000	-		
	DSTAR	VE6MHD A	1287.50000	-		
	FM	VE6VVR	145.41000	-	DCS 100	
	FM	VE6HHO	146.70000	-	100.0 Hz	ARLA
	FM	VE6HAT	147.06000	+	100.0 Hz	ARLA
	FM	VE6RCM	147.57000		100.0 Hz	
	FM	VE6VOA	449.92500	-		
Medicine Hat, Bowell						
	FM	VE6BWL	449.92500	-		
Meeting Creek	FM	VE6REP	449.17500	-	114.8 Hz	
Millarville	FM	VE6HRC	145.19000	-		ARLA
Morinville	FM	VA6CYR	146.55000			
	FM	VE6TOP	447.75000	-	100.0 Hz	
Nanton	FM	VE6HRB	449.15000	-		ARLA
New Brigden	FM	VE6NBR	146.89500	-		
Nordegg	FM	VE6PZ	145.21000	-		ARLA
North Red Deer	FM	VE6YXR	449.55000	-		
Olds	FM	VE6OLS	145.49000	-		ARLA
	FM	VA6SVM	147.42500			
Peace River	FM	VE6AAA	145.49000	-		ARLA
	FM	VE6PRR	146.82000	-		ARLA
Pigeon Lake	FM	VE6SS	146.88000	-	100.0 Hz	ARLA
Pincher Creek	FM	VE6PAS	145.45000	-	110.9 Hz	ARLA
Pine Lake	FM	VE6REP	224.80000	-		ARLA
Poe	FM	VE6POE	145.49000	-		ARLA
Raymond	FM	VE6EVY	146.67000	-		ARLA
Red Deer	FM	VE6REP	145.33000	-		ARLA
	FM	VE6QE	147.15000	+		ARLA
	FM	VE6QE	444.75000	+		ARLA
Red Deer, Pine Lake						
	FM	VE6REP	443.57500	+		
Riverbend	FM	VE6RGB	147.42000	+		
Rocky Mtn Hse	FM	VE6VHF	146.91000	-		ARLA
Sangudo	FM	VE6TOP	147.30000	+	100.0 Hz	
Sherwood Park	FM	VE6VPR	145.29000	-		ARLA
Slave Lake	FM	VE6SLR	147.03000	+		ARLA
Smoky Lake	FM	VE6TOP	146.68500	-	100.0 Hz	
	FM	VE6SSM	446.97500	-	114.8 Hz	ARLA
St. Albert	FM	VE6LAW	144.94000	-		
St. Paul	FM	VE6SB	146.67000	-		ARLA
St.Albert	DSTAR	VE6JKB B	444.25000	+		
Sundre	FM	VE6AMP	147.03000	+	131.8 Hz	ARLA
	FM	VE6GAB	147.07500	+	100.0 Hz	ARLA
	FM	VE6MTR	443.57500	+	114.8 Hz	ARLA
Swan Hills	FM	VE6SHR	145.35000	-		ARLA
Valley Ridge	FM	VE6CIZ	223.52000	-	88.5 Hz	
Valley View	FM	VE6YK	147.24000	+		ARLA

ALBERTA 413

Location	Mode	Call sign	Output	Input	Access	Coordinator
Vulcan	FM	VE6AAP	444.97500	+	114.8 Hz	ARLA
Wabamun	FM	VE6PLP	444.27500	+	446.0 Hz	ARLA
Weatheradio-Bassano						
	FM	VFU885	162.52500			
Weatheradio-Brooks						
	FM	VDC816	162.40000			
Weatheradio-Burmis						
	FM	VBX254	162.55000			
Weatheradio-Calgary						
	FM	XLF339	162.40000			
Weatheradio-Cold Lake						
	FM	VFZ535	162.52500			
Weatheradio-Cooking Lake						
	FM	XOF962	162.47500			
Weatheradio-Drumheller						
	FM	VBX367	162.55000			
Weatheradio-Edmonton						
	FM	XLM572	162.40000			
Weatheradio-Edson						
	FM	VBU827	162.40000			
Weatheradio-Fort Chipewyan						
	FM	VFR368	162.55000			
Weatheradio-Grande Prairie						
	FM	VBA557	162.40000			
Weatheradio-Highvale						
	FM	VBU829	162.47500			
Weatheradio-Limestone Mountain						
	FM	VDA280	162.40000			
Weatheradio-Long Lake						
	FM	VFS310	162.55000			
Weatheradio-Medicine Hat						
	FM	VBK616	162.55000			
Weatheradio-Milk River						
	FM	XKA598	162.40000			
Weatheradio-Peace River						
	FM	VBU374	162.47500			
Weatheradio-Red Deer						
	FM	VBC336	162.55000			
Weatheradio-Two Hills						
	FM	VXF723	162.52500			
Weatheradio-Whitecourt						
	FM	VBU828	162.55000			
Wetaskiwin	FM	VE6WCR	145.37000	-		ARLA
	FM	VE6MTR	449.32500	-	123.0 Hz	ARLA
White Court	FM	VE6PP	146.82000	-	100.0 Hz	ARLA
	FM	VE6MTR	449.30000	-	156.7 Hz	ARLA
Wild Cat Hills	FM	VE6AUY	147.06000	+	110.9 Hz	ARLA
Wimborne	FM	VE6BT	146.97000	-		ARLA

BRITISH COLUMBIA

Location	Mode	Call sign	Output	Input	Access	Coordinator
100 Mile House	FM	VE7SCQ	146.74000	-		BCARCC
Abbotsford	FM	VE7RVA	52.52500	51.52500		
	FM	VE7RVA	52.85000	51.15000	100.0 Hz	BCARCC
	FM	VE7PKV	145.03000	+		BCARCC
	FM	VE7RVA	146.61000	-	110.9 Hz	BCARCC
	FM	VE7ASM	147.28000	+	110.9 Hz	BCARCC
	FM	VE7RVA	442.02500	+	110.9 Hz	BCARCC
Anvil Island	FM	VE7QRO	52.91000	51.21000		BCARCC
Apex Mtn	FM	VE7OKN	146.92000	-		BCARCC
Atlin Mountain	FM	VA7ATN	146.34000	+		BCARCC
Atlin, Atlin Mountain						
	FM	VA7ATN	147.36000	+	100.0 Hz	

414   BRITISH COLUMBIA

Location	Mode	Call sign	Output	Input	Access	Coordinator
Barriere	FM	VE7RTN	147.24000	+		BCARCC
	FM	VE7RTN	147.30000	+		BCARCC
	FM	VA7RTN	442.65000	+		BCARCC
	FM	VE7RTN	442.87500	+		BCARCC
Barriere, Garrison Mtn						
	FM	VE7LMR	147.38000	+		
Bennett Lake	FM	VE7RFT	147.24000	+		BCARCC
Blackpool	FM	VE7RBP	146.90000	-	100.0 Hz	BCARCC
	FM	VE7RBP	444.00000	+		BCARCC
British Columbia	FM	VE7PQU	147.08000	+	141.3 Hz	
	FM	VE7RTN	147.32000	+		
Burnaby	FM	VE7TEL	145.17000	-		BCARCC
	FM	VE7RBY	145.35000	-	127.3 Hz	BCARCC
	FM	VE7RBY	224.80000	-	127.3 Hz	BCARCC
	FM	VE7VYL	224.96000	-		BCARCC
	FM	VE7REM	442.05000	+	156.7 Hz	BCARCC
	FM	VA7LNK	442.20000	+	110.9 Hz	
	FM	VE7RBY	442.85000	+		BCARCC
	FM	VE7TEL	442.87500	+		BCARCC
	FM	VE7TEL	443.42500	+		
	FM	VE7CBN	443.67500	+	114.8 Hz	BCARCC
	FM	VE7ROX	444.75000	+	123.0 Hz	BCARCC
Burnaby, TELUS	FM	VE7TEL	145.09000	+		
Burns Lake	FM	VE7LRB	146.94000	-	100.0 Hz	BCARCC
Campbell River	FM	VE7CRC	146.55000			
	FM	VE7XJR	146.76000	-		BCARCC
	FM	VE7RVR	146.82000	-		BCARCC
	FM	VE7CRC	146.96000	-		BCARCC
	FM	VE7NVI	442.45000	+		BCARCC
	FM	VE7CRC	443.65000	+	100.0 Hz	BCARCC
Carmacks	FM	VY1RMB	146.82000	-		BCARCC
Castlegar	FM	VE7FL	147.44000		100.0 Hz	
Chemainus	FM	VE7RNA	146.68000	-	141.3 Hz	BCARCC
	FM	VE7RNA	224.94000	-	141.3 Hz	BCARCC
	FM	VE7RNA	442.60000	+	141.3 Hz	BCARCC
Chetwynd	FM	VA7XX	146.91000	-	100.0 Hz	
Chilliwack	DMR/BM	VA7CRC	443.00000	+	CC1	BCARCC
	FM	VA7RSH	145.11000	-		BCARCC
	FM	VE7VCR	146.86000	-	88.5 Hz	BCARCC
	FM	VA7CRC	146.96000	-	110.9 Hz	BCARCC
	FM	VE7VCR	147.00000	+	88.5 Hz	
	FM	VE7RCK	147.10000	+	110.9 Hz	BCARCC
	FM	VE7VCR	147.22000	+	88.5 Hz	
	FM	VE7TMQ	147.52500			
	FM	VE7RVA	224.26000	-	110.9 Hz	BCARCC
	FM	VA7RSH	442.80000	+	110.9 Hz	BCARCC
	FM	VE7TPC	443.37500	+		
	FM	VE7RAD	444.70000	+		BCARCC
Clearwater	FM	VE7RWG	146.92000	-		BCARCC
Clinton	FM	VE7RKL	146.68000	-		BCARCC
	FM	VE7LMR	147.36000	+		BCARCC
	FM	VE7LMR	442.65000	+		BCARCC
	FM	VE7LMR	442.82500	+		BCARCC
Clinton Village	FM	VA7MWR	446.20000	-	67.0 Hz	
Cobble Hill	FM	VE7BH	144.68000	+		
Comox Valley	FM	VE7RCV	146.78000	-	141.3 Hz	BCARCC
	FM	VE7RAP	447.57500	-		BCARCC
Copper Creek	FM	VE7JMN	146.46000	+		
	FM	VE7EHP	147.55500	+		
Coquihalla	FM	VE7TYN	146.98000	-	123.0 Hz	BCARCC
	FM	VE7LGN	147.10000	+		BCARCC
Coquitlam	FM	VE7MFS	145.31000	-	127.3 Hz	BCARCC

BRITISH COLUMBIA   415

Location	Mode	Call sign	Output	Input	Access	Coordinator
Coquitlam	FM	VE7NZ	223.42000	-	156.7 Hz	
	FM	VE7VFB	223.92000	-		BCARCC
	FM	VE7MFS	224.92000	-		BCARCC
	FM	VE7MFS	224.94000	-		
	FM	VE7KHZ	443.60000	+	100.0 Hz	
Courtenay	DMR/MARC	VE7RAP	442.57500	+	CC1	
	FM	VE7NIR	146.62000	-	141.3 Hz	BCARCC
	FM	VE7RAP	146.91000	-	141.3 Hz	BCARCC
	FM	VE7NIR	443.70000	+	141.3 Hz	BCARCC
Cowichan Valley	FM	VE7RNA	224.90000	-	141.3 Hz	BCARCC
Cranbrook	FM	VE7CAP	146.94000	-		BCARCC
	FM	VE7CAP	443.62500	+		BCARCC
Creston	FM	VE7RCA	146.80000	-		BCARCC
Dawson Creek	FM	VE7RMS	146.76000	-	100.0 Hz	BCARCC
	FM	VE7RDC	146.94000	-		BCARCC
Delta	FM	VE7SUN	147.34000	+	107.2 Hz	BCARCC
	FM	VE7EPP	442.35000	+		BCARCC
	FM	VE7EPP	443.35000	+	127.3 Hz	BCARCC
	FM	VA7RPA	443.55000	+		BCARCC
	FUSION	VA7PI	448.55000	-		
Duncan	FM	VE7RVC	145.47000	-	127.3 Hz	BCARCC
	FM	VA7CDH	442.15000	+	141.3 Hz	BCARCC
East Sooke, Mount Matheson						
	FM	VE7RAH	145.43000	-	100.0 Hz	
Edgewood/Nakusp						
	FM	VE7SMT	449.25000	-		BCARCC
Esquimalt / Victoria						
	FM	VE7RRU	446.02500	-	123.0 Hz	
FairmontHotSpring						
	FM	VE7RIN	146.85000	-		BCARCC
Faro	FM	VY1RRH	147.06000	+		BCARCC
Fishpot - Nazko	FM	VE7MBM	147.15000	+		BCARCC
Fort Nelson	FM	VE7VFN	146.94000	-		BCARCC
Fort St. James	FM	VE7RFF	147.24000	+	100.0 Hz	BCARCC
	FM	VE7DPG	147.33000	+	100.0 Hz	BCARCC
Fort St. John	FM	VA7XX	146.64000	-	100.0 Hz	
	FM	VE7RUC	147.21000	+	100.0 Hz	BCARCC
Fraser Lake	FM	VE7RES	146.84000	-	100.0 Hz	BCARCC
Fraser Mountain	FM	VE7RFT	146.94000	-		BCARCC
Gabriola Island	FM	VE7GEC	443.00000	+	141.3 Hz	BCARCC
Grand Forks	FM	VE7RGF	146.94000	-	100.0 Hz	BCARCC
	FM	VE7RGF	147.28000	+		BCARCC
	FM	VE7KGF	147.33000	+	67.0 Hz	
	FM	VA7KT	147.52500		67.0 Hz	
Granite Peak	FM	VE7RNH	146.76000	-		BCARCC
Haines Junction	FM	VY1RHJ	146.82000	-		BCARCC
	FM	VY1RPM	146.88000	-		BCARCC
Haney	FM	VE7HNY	443.07500	+		
Hayes Peak	FM	VY1RHP	147.06000	+		BCARCC
Hazelton	FM	VE7RHD	146.80000	-	100.0 Hz	BCARCC
Hope	FM	VE7UVR	146.70000	-	77.0 Hz	BCARCC
	FM	VE7RVB	147.08000	+	110.9 Hz	BCARCC
Horsefly	FM	VE7WLP	147.18000	+	162.2 Hz	BCARCC
Houston	FM	VE7RHN	147.06000	+	100.0 Hz	BCARCC
Houston B.C.	FM	VE7CUP	147.31000	+	100.0 Hz	
Hudson's Hope	FM	VA7RHH	146.88000	-	100.0 Hz	BCARCC
Hudsons Hope	FM	VA7RHH	146.54000		100.0 Hz	
ICOM Canada	DSTAR	VA7ICM AD	1293.15000	-		
Kamloops	DMR/MARC	VE7RLO	442.65000	+	CC1	
	FM	VE7DUF	146.94000	-		BCARCC
	FM	VE7TSI	146.96000	-		
	FM	VE7RLD	147.00000	+		

416   BRITISH COLUMBIA

Location	Mode	Call sign	Output	Input	Access	Coordinator
Kamloops	FM	VE7KEG	147.18000	+		BCARCC
	FM	VE7RLO	147.32000	+		BCARCC
	FM	VE7CRW	442.05000	+	103.5 Hz	BCARCC
	FM	VE7RLO	442.12500	+		BCARCC
	FM	VE7RLO	442.15000	147.95000		BCARCC
	FM	VE7RLO	442.17500	147.87500		BCARCC
	FM	VE7TPK	442.55000	+		
	FM	VE7JFB	446.27500	-		
	FM	VE7RXD	447.50000	-	146.2 Hz	
	FM	VE7RLO	449.25000	-		BCARCC
	FM	VE7RHM	449.30000	-		BCARCC
	FM	VE7KIG	449.50000	-		BCARCC
	FUSION	VE7UT	448.52500	-		
Kelowna	DMR/BM	VA7NBC	147.10000	+	CC1	
	DMR/BM	VA7UN	147.26000	+	CC1	BCARCC
	DMR/BM	VA7CNN	444.00000	+	CC1	
	DMR/BM	VA7NBC	444.80000	+	CC1	
	FM	VE7HWY	145.49000	-	88.5 Hz	BCARCC
	FM	VE7OGO	146.62000	-		
	FM	VE7OGO	146.68000	-		BCARCC
	FM	VE7EJP	146.72000	-		BCARCC
	FM	VE7SFX	146.78000	-	88.5 Hz	BCARCC
	FM	VE7ROC	146.82000	-		BCARCC
	FM	VE7RBG	146.86000	-	88.5 Hz	BCARCC
	FM	VE7OGO	147.00000	+	88.5 Hz	
	FM	VE7VTC	147.14000	+		BCARCC
	FM	VE7RIM	147.24000	+		BCARCC
	FM	VE7KTV	147.30000	+		BCARCC
	FM	VA7JPL	147.36000	+	88.5 Hz	BCARCC
	FM	VA7YLW	147.42000			
	FM	VE7OGO	147.57000		100.0 Hz	
	FM	VA7SPY	147.60000	-		
	FM	VA7KRG	444.10000	+		
	FM	VE7KTV	444.30000	+		BCARCC
	FM	VA7KEL	444.82500	+	88.5 Hz	
	FM	VA7UN	447.30000	-	88.5 Hz	BCARCC
	FM	VE7KEL	447.77500	-		BCARCC
Kelowna, Mount Dilworth						
	FM	VA7OGO	146.68000	-	88.5 Hz	
Kelowna, Mount Last						
	FM	VA7KEL	449.12500	-		
Keno	FM	VY1RBT	146.94000	-		BCARCC
Kimberley	FM	VE7REK	145.19000	-		
Kitimat	FUSION	VE7SNO	146.82000	-	100.0 Hz	BCARCC
	FUSION	VE7RAF	147.06000	+	100.0 Hz	BCARCC
Kootenay Nat. Park						
	FM	VE7KNP	146.70000	-	131.8 Hz	BCARCC
Ladysmith	FM	VA7DXH	224.04000	-	141.3 Hz	BCARCC
	FM	VE7RNX	444.80000	+	156.7 Hz	BCARCC
Langford BC	FM	VE7LEP	442.72500	+	123.0 Hz	
Langley	FM	VE7LGY	146.78000	-		BCARCC
	FM	VE7RMH	441.37500	+		BCARCC
	FM	VE7RLY	443.97500	+		BCARCC
	FM	VE7NPN	444.12500	+	127.3 Hz	BCARCC
	FM	VE7ICA	446.75000	-		
Langley East	FM	VE7RLY	147.38000	+	203.5 Hz	BCARCC
Lillooett	FM	VE7TJS	147.38000	+		
Logan Lake	FM	VE7CPQ	146.58000			
Lone Butte	FM	VE7AZQ	147.22000	+	88.5 Hz	BCARCC
Loos	FM	VE7RES	146.88000	-	100.0 Hz	BCARCC
Lund	FM	VA7LND	147.00000	+	100.0 Hz	BCARCC
	FM	VA7LND	444.35000	+	100.0 Hz	BCARCC

BRITISH COLUMBIA 417

Location	Mode	Call sign	Output	Input	Access	Coordinator
Lytton	FM	VE7HGR	147.06000	+		BCARCC
Mackenzie	FM	VE7MKR	146.82000	-	100.0 Hz	
	FM	VE7ZBK	147.33000	+		BCARCC
Malahat	FM	VE7XMR	146.98000	-	123.0 Hz	BCARCC
	FM	VA7XMR	443.02500	+		
Maple Ridge	FM	VE7RMR	146.80000	-	156.7 Hz	BCARCC
	FM	VE7RMR	224.88000	-		BCARCC
	FM	VE7RMR	443.62500	+	156.7 Hz	BCARCC
McBride	FM	VE7RMB	146.76000	-	100.0 Hz	BCARCC
Merritt	FM	VE7IRN	146.66000	-		BCARCC
	FM	VE7RIZ	147.08000	+	110.9 Hz	BCARCC
Monashee Pass	FM	VE7SMT	146.74000	-	123.0 Hz	BCARCC
Mt. Avola	FM	VE7RBP	145.35000	-		BCARCC
Nakusp	FM	VE7EDA	146.94000	-		
Nanaimo	FM	VA7DJA	145.43000	-	141.3 Hz	BCARCC
	FM	VA7ANI	146.98000	-	141.3 Hz	BCARCC
	FM	VE7RBB	147.18000	+	100.0 Hz	BCARCC
	FM	VA7SZU	442.52500	+		
	FM	VE7DJA	443.90000	+	141.3 Hz	BCARCC
	FM	VE7ITS	444.72500	+	141.3 Hz	BCARCC
	FM	VA7ZSU	444.80000	+		
Nanoose Bay	FM	VA7LPG	444.30000	+	141.3 Hz	BCARCC
Nelson	DMR/MARC	VE7BTU	147.06000	+	CC1	BCARCC
	FM	VE7RCT	146.64000	-		BCARCC
	FM	VE7RCT	444.55000	+		BCARCC
New Denver	FM	VE7FL	146.56000	+		
New Westminster	DMR/MARC	VE7NWR	444.60000	+	CC1	BCARCC
	FM	VE7WCC	145.15000	-	123.0 Hz	
	FM	VE7NWR	145.39000	-	100.0 Hz	BCARCC
	FM	VA7HPS	442.37500	+	110.9 Hz	BCARCC
Nimpo Lake	FM	VA7SPY	444.82500	+	88.5 Hz	BCARCC
North Vancouver	DMR/MARC	VE7RAG	443.40000	+	CC1	
North Vancouver, Mount Seymour						
	FM	VE7LAN	145.07000	+		
Oak Bay	FM	VE7XIC	146.84000	-	107.2 Hz	BCARCC
Okanagan Falls	FM	VE7DTT	144.56000	+	123.0 Hz	
	FM	VE7DTT	147.33000	+	100.0 Hz	
	FM	VE7DTT	147.56000		123.0 Hz	
Okanagan/Shuswap						
	FM	VE7RNR	147.06000	+		BCARCC
Oliver	FM	VE7RBD	147.16000	+		BCARCC
	FM	VE7ROR	147.38000	+		BCARCC
	FM	VE7RSO	444.60000	+	100.0 Hz	
Osoyoos	FM	VE7OSY	145.27000	-	107.2 Hz	BCARCC
	FM	VE7EHF	145.29000	-	107.2 Hz	BCARCC
	FM	VE7OSY	146.66000	-	156.7 Hz	
	FM	VE7STA	146.94000	-		BCARCC
	FM	VE7OJP	147.18000	+	88.5 Hz	BCARCC
	FM	VE7OSY	222.60000	+	156.7 Hz	BCARCC
Osoyoos, Mount Kobau						
	FM	VE7EHF	147.34000	+	107.2 Hz	
Parksville	FM	VE7RPQ	145.37000	-	100.0 Hz	BCARCC
	FM	VA7RFR	147.08000	+	141.3 Hz	BCARCC
	FM	VE7PQA	147.28000	+	141.3 Hz	BCARCC
	FM	VE7MIR	147.34000	+	141.3 Hz	
	FM	VE7MIR	440.85000	+	100.0 Hz	
	FM	VE7JPS	442.27500	+	136.5 Hz	
	FM	VE7PQD	444.20000	+		BCARCC
Peachland	FM	VA7OKV	447.22500	-	88.5 Hz	BCARCC
Pemberton	FM	VE7PVR	146.98000	-		BCARCC
Penticton	DMR/BM	VA7PTV	447.70000	-	CC1	
	FM	VE7PEN	146.58000		123.0 Hz	

418 BRITISH COLUMBIA

Location	Mode	Call sign	Output	Input	Access	Coordinator
Penticton	FM	VE7RCP	146.64000	-	131.8 Hz	BCARCC
	FM	VE7RPC	147.12000	+		BCARCC
	FM	VE7RPC	444.50000	+		BCARCC
	FM	VE7RCP	444.77500	+	136.5 Hz	
Pine Pass	FM	VE7RES	146.64000	-	100.0 Hz	BCARCC
Pitt Meadows	FM	VE7MTY	443.62500	+		
Port Alberni	FM	VE7RPA	147.15000	+	141.3 Hz	BCARCC
	FM	VE7KU	147.24000	+	141.3 Hz	BCARCC
	FM	VE7KU	444.45000	+		BCARCC
Port Alberni, Mount Cokely						
	FM	VE7RTU	444.75000	+	100.0 Hz	
Port Coquitlam	FM	VA7RPC	145.49000	-	94.8 Hz	BCARCC
	FM	VE7UDX	443.10000	+	94.8 Hz	BCARCC
Port McNeil	FM	VA7RNI	146.92000	-		BCARCC
	FM	VE7RNI	146.94000	-		BCARCC
Port McNeill	FM	VE7KJA	146.44500			
Pouce Coupe	FM	VE7AGJ	146.25000	+		BCARCC
Powell River	FM	VE7PRR	147.20000	+	141.3 Hz	BCARCC
	FM	VE7PRR	444.50000	+		BCARCC
Prince George	FM	VE7RES	145.43000	-	100.0 Hz	BCARCC
	FM	VE7RWT	146.91000	-	100.0 Hz	BCARCC
	FM	VE7RPM	146.94000	-	100.0 Hz	BCARCC
	FM	VE7RES	147.30000	+	100.0 Hz	BCARCC
	FM	VE7RQU	442.86200	+		
	FM	VE7RUN	444.00000	+		
Prince George, Tabor Mountain						
	FM	VE7FFF	146.70000	-	100.0 Hz	
	FM	VE7RUN	147.00000	+		
Prince Rupert	FM	VE7RPR	146.88000	-		BCARCC
	FM	VE7RKI	146.94000	-		BCARCC
	FM	VE7RMM	147.28000	+		BCARCC
	FM	VE7DQC	147.33000	+		
Qualicum Beach	FM	VE7RPQ	442.25000	144.75000	141.3 Hz	BCARCC
	FM	VE7RQR	445.00000	144.80000		BCARCC
Queen Charlotte	FM	VE7RQI	146.68000	-		BCARCC
Quesnel	FM	VE7RQL	147.06000	+		BCARCC
	FM	VE7RES	147.21000	+	100.0 Hz	BCARCC
	FM	VE7RQM	444.30000	+		
Quesnel, Airport	FM	VE7YQZ	146.97000	-		
Radium, Mount Sinclair						
	FM	VE7PNR	146.88000	-	131.8 Hz	
Revelstoke	FM	VA7AZG	146.72000	-		BCARCC
	FM	VE7RJP	147.20000	+		BCARCC
Richmond	FM	VE7RMD	147.14000	+	79.7 Hz	BCARCC
	FM	VE7BAS	147.51000		156.7 Hz	
	FM	VE7RMD	442.37500	+	203.5 Hz	BCARCC
	FUSION	VA7REF	442.50000	+		
River Springs	FM	VE7SVG	445.55000	-		
Saanich	FM	VE7SER	145.29000	-	167.9 Hz	BCARCC
	FM	VA7XMR	443.07500	+		BCARCC
	FM	VE7SLC	449.45000	-	100.0 Hz	BCARCC
Salmo	FM	VE7KNL	144.39000	+		
Salmon Arm	FM	VE7CAL	146.16000	+		BCARCC
	FM	VE7RAM	146.64000	-		BCARCC
	FM	VE7RSA	147.02000	+		BCARCC
	FM	VE7RAM	442.45000	+		BCARCC
Salt Spring Is	FM	VA7VIC	146.66000	-	100.0 Hz	
Salt Spring Island	FM	VE7GDH	147.57000		100.0 Hz	
Saltspring Is	FM	VE7RSI	147.32000	+	88.5 Hz	BCARCC
Santa Rosa, Christina Lake						
	FM	VE7RCL	146.70000	-		
Saturna Island	FM	VA7RMI	444.55000	+	97.4 Hz	

BRITISH COLUMBIA   419

Location	Mode	Call sign	Output	Input	Access	Coordinator
Sayward	FM	VE7RNC	146.70000	-		BCARCC
	FM	VE7RNC	224.62000	-		BCARCC
	FM	VE7RNC	443.70000	+		BCARCC
Sechelt	FM	VE7SSC	444.62500	+		
Sherwood Park	DSTAR	VE6DXH B	444.30000	+		
Shirley/Jordon River						
	FM	VE7RSK	147.22000	+	123.0 Hz	BCARCC
Shirley/Otter Point	FM	VE7RYF	145.41000	-	100.0 Hz	BCARCC
Shuswap	FM	VE7LIM	147.08000	+		BCARCC
Sicamous	FM	VE7QMR	145.47000	-		BCARCC
	FM	VE7BYN	147.54000			
	FM	VE7BYN	147.57000			
Sidney	FM	VE7XMT	222.54000	146.04000	100.0 Hz	BCARCC
Silver Star	FM	VE7RHW	146.90000	-	123.0 Hz	BCARCC
Smithers	FM	VE7RBH	146.88000	-	100.0 Hz	BCARCC
	FM	VE7RBH	147.33000	+	100.0 Hz	
Sooke	FM	VE7RWS	145.41000	-	103.5 Hz	
	FM	VE7XSK	146.84000	-	123.0 Hz	BCARCC
	FUSION	VE7RYF	145.43000	-		
Sorrento	FM	VE7RXX	146.64000	-		BCARCC
	FM	VA7AHR	147.14000	+		BCARCC
	FM	VE7SPG	444.10000	+		BCARCC
South Okanagan	FM	VE7RSO	147.34000	+	107.2 Hz	BCARCC
Sparwood	FM	VE7RSQ	147.30000	+	100.0 Hz	BCARCC
Squamish	FM	VE7SQR	147.00000	+	77.0 Hz	BCARCC
Squilax	FM	VE7FPG	442.52500	+		
Stewart Crosng	FM	VY1RFH	147.06000	+		BCARCC
Summerland	FM	VE7NUT	145.35000	-		
Sunshine Coast	FM	VE7RXZ	147.22000	+	100.0 Hz	BCARCC
	FM	VE7RXZ	442.65000	+	123.0 Hz	BCARCC
Sunshine Hills	FM	VA7DEP	444.42500	+	107.2 Hz	
Surrey	FM	VE7DQ	53.53000	52.53000		
	FM	VE7RSC	147.36000	+	110.9 Hz	BCARCC
	FM	VE7IKB	223.40000	-		
	FM	VE7MAN	441.07500	+		
	FM	VE7MAN	443.60000	+		BCARCC
	FM	VE7RSC	443.77500	+	110.9 Hz	BCARCC
Survey Mountain	FM	VE7RYF	444.92500	+	100.0 Hz	BCARCC
Tappen	FM	VE7RAM	146.48500			
Tatlayoko Lake	FM	VE7SML	146.82000	-		
Tatlayoko Valley	FM	VA7TKR	147.28000	+	162.2 Hz	BCARCC
Terrace	FM	VE7RTK	146.60000	-	100.0 Hz	
	FM	VE7FFU	146.80000	-		BCARCC
	FM	VE7RDD	146.94000	-		BCARCC
	FM	VE7FFU	147.33000	+		
	FM	VE7RDD	444.97500	+		BCARCC
Texada Island	FM	VE7TIR	444.02500	+	141.3 Hz	BCARCC
Tofino	FM	VE7TOF	146.88000	-	141.3 Hz	BCARCC
Triangle Mtn	FM	VE7RMT	146.84000	-	131.8 Hz	BCARCC
Tumbler Ridge	FM	VE7RTR	147.27000	+	100.0 Hz	
Ucluelet	FM	VE7RWC	147.00000	+	100.0 Hz	BCARCC
Valemont	FM	VE7YCR	146.60000	-		BCARCC
	FM	VE7RES	147.00000	+	100.0 Hz	BCARCC
Vancouver	DMR/MARC	VA7XPR	443.50000	+	CC1	
	FM	VE7HCP	52.89000	51.19000		BCARCC
	FM	VE7ROX	145.15000	-	123.0 Hz	BCARCC
	FM	VE7RTY	145.21000	-		BCARCC
	FM	VE7RHS	145.27000	-	100.0 Hz	BCARCC
	FM	VA7IP	145.29000	-	100.0 Hz	BCARCC
	FM	VE7RBI	146.72000	-		BCARCC
	FM	VE7RPT	146.94000	-		BCARCC
	FM	VE7RCH	147.04000	+		BCARCC

420   BRITISH COLUMBIA

Location	Mode	Call sign	Output	Input	Access	Coordinator
Vancouver	FM	VE7VAN	147.12000	+	156.7 Hz	BCARCC
	FM	VE7RNS	147.26000	+		BCARCC
	FM	VE7RDX	147.30000	+		BCARCC
	FM	VE7RPT	224.30000	-		BCARCC
	FM	VE7NYE	224.60000	-	127.3 Hz	BCARCC
	FM	VE7RVK	224.64000	-		BCARCC
	FM	VE7RHS	224.70000	-		BCARCC
	FM	VE7RHS	441.97500	+		BCARCC
	FM	VE7RPS	442.22500	+	88.5 Hz	BCARCC
	FM	VE7VHF	442.32500	+	100.0 Hz	BCARCC
	FM	VE7UBC	442.45000	+		
	FM	VE7ZIT	442.57500	+	114.8 Hz	BCARCC
	FM	VE7AAU	442.95000	+	114.8 Hz	BCARCC
	FM	VE7YV	443.05000	+	110.9 Hz	BCARCC
	FM	VE7NSR	443.20000	+		BCARCC
	FM	VE7RCH	443.25000	+		BCARCC
	FM	VE7RCI	443.27500	+		BCARCC
	FM	VE7RPT	443.52500	+		BCARCC
	FM	VE7UHF	443.80000	+	100.0 Hz	BCARCC
	FM	VE7URG	444.00000	+	156.7 Hz	BCARCC
	FM	VE7TOK	444.07500	+		BCARCC
	FM	VE7ROY	444.10000	+	100.0 Hz	BCARCC
	FM	VE7RIO	444.17500	+	156.7 Hz	BCARCC
	FM	VE7PRA	444.47500	+		BCARCC
	FM	VE7VYL	444.82500	+	156.7 Hz	BCARCC
	FM	VE7WAR	444.92500	+		
	FM	VE7RNV	444.95000	+		BCARCC
	FM	VE7SKY	444.97500	+		BCARCC
	NXDN	VE7NYE	443.15000	+	127.3 Hz	BCARCC
	FUSION	VA7CAB	442.30000	+		
Vancouver WIN System Affiliate						
	FM	VA7SCA	444.40000	+	100.0 Hz	
Vancouver, Anvil Island						
	FM	VE7QRO	444.40000	+	100.0 Hz	
Vancouver, Seymour Mtn						
	FM	VE7RVF	145.45000	-	100.0 Hz	
Vanderhoof	FM	VE7RSM	146.80000	-	100.0 Hz	BCARCC
	FM	VE7RON	146.88000	-		BCARCC
Vanway	FM	VE7CKZ	446.10000	-		
Verdun	FM	VE7LRB	146.76000	-	100.0 Hz	BCARCC
Vernon	DMR/BM	VA7VTV	447.77500	-	CC1	
	DMR/MARC	VE7KHZ	145.05000	+	CC1	
	FM	VA7VMR	52.01000	-	110.9 Hz	
	FM	VE7EGO	145.45000	-		
	FM	VE7KI	146.56500	+		
	FM	VE7EGO	146.80000	-		BCARCC
	FM	VE7RSS	146.88000	-		BCARCC
	FM	VE7RIP	147.04000	+		BCARCC
	FM	VA7VMR	147.22000	+		BCARCC
	FM	VE7DIR	147.49500		151.4 Hz	
	FM	VA7NWS	441.15000	147.55000		BCARCC
	FM	VE7PE	442.42500	+		
	FM	VE7RFM	444.35000	+	100.0 Hz	BCARCC
	FM	VE7RVP	447.42500	-		
	FM	VE7RVP	447.50000	-		
	FUSION	VE7EGO	147.38000	+		
Vernon, Silver Star Mtn						
	FM	VE7RVN	444.27500	+	110.9 Hz	
Victoria	FM	VE7DAT	51.27000	-		
	FM	VE7RSX	52.83000	51.13000	100.0 Hz	BCARCC
	FM	VE7RFR	52.97000	51.27000		BCARCC
	FM	VE7US	145.13000	-	114.8 Hz	BCARCC

BRITISH COLUMBIA   421

Location	Mode	Call sign	Output	Input	Access	Coordinator
Victoria	FM	VE7VIC	146.84000	-	100.0 Hz	BCARCC
	FM	VE7RYF	146.98000	-	103.5 Hz	BCARCC
	FM	VE7RBA	147.12000	+	100.0 Hz	BCARCC
	FM	VE7RFR	147.24000	+		BCARCC
	FM	VE7VIC	224.14000	-	100.0 Hz	BCARCC
	FM	VE7RGP	224.50000	-		BCARCC
	FM	VE7OVY	441.50000	+	100.0 Hz	
	FM	VE7RFR	442.70000	+		BCARCC
	FM	VA7CRT	442.77500	+		BCARCC
	FM	VE7RYF	443.02500	+	100.0 Hz	BCARCC
	FM	VE7RAA	443.57500	+		BCARCC
	FM	VE7VOP	443.82500	+	100.0 Hz	BCARCC
	FM	VE7RFR	443.90000	+	141.3 Hz	BCARCC
	FM	VE7RTC	443.95000	+	123.0 Hz	BCARCC
	FM	VE7US	444.15000	+	103.5 Hz	BCARCC
	FM	VE7FNI	444.25000	+		
	FM	VE7SLC	444.45000	+	100.0 Hz	
	FM	VE7XIC	449.87500	146.27500		BCARCC
Victoria, BC, Canada						
	DSTAR	VE7VIC B	447.00000	+		
	DSTAR	VE7VIC AD	1261.75000			
Victoria, Fire Hall 1						
	FM	VE7GHO	443.82500	+	100.0 Hz	
Victoria, Mount McDonald						
	FM	VE7BEL	224.90000	-		
	FM	VE7USA	444.87500	+	107.2 Hz	
Weatheradio-Alert Bay						
	FM	VAF	162.55000			
Weatheradio-Barry Inlet						
	FM	XLK897	162.40000			
Weatheradio-Bowen Island						
	FM	XLK672	162.47500			
Weatheradio-Calvert Island						
	FM	VGL24	162.40000			
Weatheradio-Castlegar						
	FM	XMD482	162.55000			
Weatheradio-Chilliwack						
	FM	VFV785	162.40000			
Weatheradio-Cranbrook						
	FM	VBI853	162.40000			
Weatheradio-Crawford Bay						
	FM	VFD904	162.42500			
Weatheradio-Cumshewa						
	FM	XLK894	162.47500			
Weatheradio-Eliza Dome						
	FM	VGI57	162.55000			
Weatheradio-Fort Nelson						
	FM	VXB567	162.55000			
Weatheradio-Fort St John						
	FM	VXL336	162.47500			
Weatheradio-Kelowna						
	FM	XMD480	162.55000			
Weatheradio-Klemtu						
	FM	XLK899	162.55000			
Weatheradio-Mount Gil						
	FM	XLK898	162.40000			
Weatheradio-Mount Helmcken-Vic						
	FM	XLA726	162.47500			
Weatheradio-Naden Harbour						
	FM	XLK895	162.47500			
Weatheradio-Penticton						
	FM	XMD481	162.47500			

422   BRITISH COLUMBIA

Location	Mode	Call sign	Output	Input	Access	Coordinator
Weatheradio-Port Alberni						
	FM	XLA823	162.40000			
	FM	VFM825	162.52500			
Weatheradio-Port Hardy						
	FM	VBH444	162.47500			
	FM	VFM839	162.52500			
Weatheradio-Prince George						
	FM	VGB723	162.40000			
Weatheradio-Prince Rupert						
	FM	VXB571	162.42500			
Weatheradio-Texada Island						
	FM	VBG969	162.55000			
Weatheradio-Van Inlet						
	FM	XLK896	162.55000			
Weatheradio-Vancouver						
	FM	VXL665	162.55000			
Weatheradio-Vernon						
	FM	VFM608	162.47500			
Weatheradio-Victoria						
	FM	XKK506	162.40000			
Wells/Barkerville	FM	VE7RLS	147.38000	+	100.0 Hz	BCARCC
West Vancouver	DMR/MARC	VE7SLV	443.85000	+	CC1	BCARCC
Westbank	FM	VE7CJU	147.20000	+		
Westview	FM	VA7UQ	146.56500	+		
Whistler	FM	VE7WHR	147.06000	+		BCARCC
White Rock	FM	VE7RWR	146.90000	-	91.5 Hz	BCARCC
Whitehorse	FM	VY1IRL	146.88000	-	100.0 Hz	BCARCC
	FM	VY1RPT	146.94000	-		BCARCC
	FM	VY1RM	147.18000	+		BCARCC
	FM	VY1ECH	147.28000	+		BCARCC
Williams Lake	FM	VE7RTI	146.62000	-	100.0 Hz	BCARCC
	FM	VE7RWL	147.12000	+	100.0 Hz	BCARCC
	FM	VE7ZIG	444.10000	+		BCARCC
Woss Lake	FM	VE7RWV	146.88000	-		BCARCC
Yennadon	FM	VE7NLY	146.50000	+		
MANITOBA						
Adam Lake	FM	VE4IHF	146.25000	+		NDFC
	FM	VE4IHF	146.85000	-	127.3 Hz	NDFC
Austin	FM	VE4MTR	146.91000	-	127.3 Hz	MARCC
	FM	VE4MTR	444.27500	+		MARCC
Baldy Mountain	FM	VE4BMR	147.03000	+		MARCC
	FM	VE4BMR	448.40000	-		
Basswood	FM	VE4BAS	145.15000	-	127.3 Hz	MARCC
Beausejour	FM	VE4BRC	147.54000			MARCC
Brandon	FM	VE4CTY	146.64000	-	127.3 Hz	MARCC
	FM	VE4TED	146.73000	-		MARCC
	FM	VE4BDN	146.94000	-		MARCC
	FM	VE4CTY	443.70000	+	127.3 Hz	MARCC
Bruxelles	FM	VE4MRS	145.31000	-		MARCC
	FM	VE4HS	146.88000	-		MARCC
Channing	FM	VA4BG	145.00000	+		
Chatfield	FM	VE4TGN	443.80000	+		
East Selkirk	FM	VE4SLK	146.73000	-	127.3 Hz	MARCC
Elie	FM	VE4RAG	147.24000	+	127.3 Hz	MARCC
Falcon Lake	FM	VE4FAL	146.64000	-		MARCC
Flin Flon	FM	VA4BOB	145.00000		100.0 Hz	
	FM	VE4FFR	146.94000	-		MARCC
	FM	VA4BOB	147.00000	+	100.0 Hz	
Flin Flon, Smoke Stack						
	FM	VE5ROD	146.94000	-		
Gimli	FM	VE4GIM	146.85000	-		MARCC

MANITOBA 423

Location	Mode	Call sign	Output	Input	Access	Coordinator
Hadashville	FM	VE4EMB	147.36000	+		MARCC
Headingley	FM	VE4AGA	444.50000	+	127.3 Hz	MARCC
Killarney	FM	VE4KIL	444.50000	+	123.0 Hz	MARCC
Lundar	FM	VE4LDR	146.97000	-		MARCC
Milner Ridge	FM	VE4MIL	145.21000	-		MARCC
Morris	FM	VE4CDN	145.27000	-		MARCC
Notre Dame	FM	VE4HJ	444.32500	+		MARCC
Pinawa	FM	VE4PIN	146.49000			MARCC
Portage	FM	VE4PLP	147.16500	+		MARCC
Rice Creek	FM	VE4JAR	147.36000	+		MARCC
Russell	FM	VE4BVR	147.24000	+		MARCC
Selkirk	DMR/MARC	VE4SLK	444.15000	+	CC1	MARCC
Spearhill	FM	VE4SHR	146.70000	-	127.3 Hz	MARCC
St. Andrews	FUSION	VE4COR	146.82000	-		
Starbuck	FM	VE4MAN	146.61000	-		MARCC
Steinbach	DMR/BM	VE4UHF	444.65000	+	CC4	
Swan River	FM	VE4SRR	146.94000	-		MARCC
	FM	VE4SRR	443.40000	+		MARCC
Teulon	FM	VE4TEU	145.41000	-		MARCC
The Pas	FM	VE4PAS	145.35000	-		MARCC
	FM	VE4PAS	145.40000	-	127.3 Hz	
Thompson	FM	VE4TPN	146.94000	-		MARCC
Weatheradio-Altona						
	FM	VFN684	162.42500			
Weatheradio-Brandon						
	FM	VAO302	162.55000			
Weatheradio-Dauphin						
	FM	VBA814	162.55000			
Weatheradio-Falcon Lake						
	FM	VXE212	162.42500			
Weatheradio-Haywood						
	FM	VBL854	162.40000			
Weatheradio-Long Point						
	FM	VCI386	162.55000			
Weatheradio-Pointe Du Bois						
	FM	VXG567	162.45000			
Weatheradio-Reston						
	FM	VXK206	162.42500			
Weatheradio-Riverton						
	FM	XLF471	162.40000			
Weatheradio-Steinbach						
	FM	VFN683	162.47500			
Weatheradio-Thompson						
	FM	VXI858	162.40000			
Weatheradio-Winnipeg						
	FM	XLM538	162.55000			
Windy Hill	FM	VE4WHR	145.45000	-		MARCC
Winkler	FM	VE4VRG	145.19000	-		MARCC
	FM	VE4BBS	146.91000	-		
	FM	VE4TOM	147.33000	+	127.3 Hz	
Winnipeg	ATV	VE4EDU	1289.25000	914.95000		MARCC
	DMR/MARC	VE4DHR	442.00000	+	CC1	
	DSTAR	VE4WDR C	145.49000	-		
	DSTAR	VE4WDR B	444.57500	+		
	FM	VE4PAR	145.17000	-	141.3 Hz	MARCC
	FM	VE4ARC	145.23000	-	127.3 Hz	MARCC
	FM	VE4ARS	145.35000	-	127.3 Hz	MARCC
	FM	VE4PNO	146.46000		100.0 Hz	MARCC
	FM	VE4JRA	146.67000	-		MARCC
	FM	VE4UMR	147.27000	+	127.3 Hz	MARCC
	FM	VE4EDU	147.30000	+		MARCC
	FM	VE4WPG	147.39000	+	127.3 Hz	MARCC

424 MANITOBA

Location	Mode	Call sign	Output	Input	Access	Coordinator
Winnipeg	FM	VE4WSC	147.57000			MARCC
	FM	VA4ARS	443.22500	+	127.3 Hz	MARCC
	FM	VE4VJ	443.50000	+	127.3 Hz	MARCC
	FM	VE4UHF	444.00000	+	DCS 432	MARCC
	FM	VA4ART	444.10000	+		MARCC
	FM	VA4UHF	444.75000	+		MARCC
	FM	VE4KEY	444.93750	+	88.5 Hz	
	P25	VE4KEY	147.01500	+	127.3 Hz	MARCC
	P25	VE4KEY	927.01250	902.01250	DCS 664	MARCC
	FUSION	VE4UMR	147.27000	+		
	FUSION	VE4RNP	444.10000	+		
Woodlands	FM	VE4SIX	145.43000	-	127.3 Hz	MARCC

NEBRASKA

Location	Mode	Call sign	Output	Input	Access	Coordinator
Birchy Lake	FM	VO1BLR	146.64000	-		

NEW BRUNSWICK

Location	Mode	Call sign	Output	Input	Access	Coordinator
Acadieville	FM	VE9ACD	145.43000	-		
Ajax	DMR/BM	VE3SBX	442.36250	+	CC3	
Alberta	FM	VE3KPT	444.55000	+	136.5 Hz	
Amherst, Nova Scotia Provincia						
	FM	VE1WRC	147.28500	+		
Antigonish	DMR/BM	VE1JSR	441.80000	+	CC1	
	DSTAR	VE1JSR B	441.80000	+		
Aquith	DSTAR	VE5KEV B	449.95000	-		
Argentia	FM	VO1ARG	146.82000	-		
Bathurst	FM	VE1BRD	147.31500	+		
	FUSION	VE9BAT	147.24000	+		
Borden	FM	KP4HD	147.43500	+		
Bridgewater	DSTAR	VE1DSR C	145.29000	-		
British Columbia	FM	VE7DJA	145.43000	-		
	FM	VE7RVA	146.60000	-	110.9 Hz	
	FM	VE7RSL	147.36000	+		
	FM	VE7PQE	442.52500	449.52500		
	FM	VE7RSL	443.77500	+		
	FM	VE7PQA	444.20000	+		
Campbellton	FM	VE1CTN	146.65500	-		
	FM	VE9SMR	146.95500	-		
	FM	VE1LES	147.00000	+		
Campbellton, Seven Mile Ridge						
	FM	VE9VDR	147.39000	+		
Cape Pine	FM	VO1CPR	147.12000	+		
Charlottetown	DSTAR	VE1UHF C	146.71500	-		
	DSTAR	VE1UHF B	443.30000	+		
Chicoutimi	DMR/BM	VA2RFI	146.65500	-	CC7	
	DMR/BM	VE2RRY	448.12500	-	CC2	
	DMR/MARC	VE2RRV	448.92500	-	CC2	
Chilliwack	DSTAR	VE7RCK C	145.06000	+		
	DSTAR	VE7RCK B	444.62500	+		
Chipman, Bronson						
	FM	VE9GLA	145.19000	-	123.0 Hz	
Coley's South Point						
	FM	VO1IC	146.73000	-	94.8 Hz	
Collines Poudrier	FM	VE2GPA	449.72500	444.73000	100.0 Hz	
Covey Hill	DSTAR	VA2REX A	1248.50000			
Crabbe Mountain	FM	VE1PD	146.76000	-		
Dalhousie	FM	VE9DNB	145.49000	-		
Delta	DSTAR	VE7SUN B	440.72500	+		
	DSTAR	VA7GQ B	445.77500	+		
Dezedeash, Klukshu Mountain						
	FM	VY1RDP	147.06000	+	100.0 Hz	
Doaktown	FM	VE1XI	146.91000	-		

NEW BRUNSWICK   425

Location	Mode	Call sign	Output	Input	Access	Coordinator
Edmunston	FM	VE9RCV	145.13000	-		
Elmtree	FM	VE9ELM	145.41000	-		
Essex County	DMR/MARC	VA3LLL	444.71250	+	CC7	
Fairfield	FM	VE9SKV	145.23000	-		
Faro, Rose Hill	FM	VY1RRH	146.82000	-	100.0 Hz	
Ferryland	FM	VO1CQD	147.28000	+		
Fournier	FM	VE9EDM	146.78000	-		
Fredericton	DMR/MARC	VE9FTN	147.16500	+	CC1	
	FM	VE1BM	147.12000	+		
	FM	VE9FNB	147.30000	+		
	FM	VE9FTN	927.68750	902.68750	100.0 Hz	
Fredericton, CBC Fredericton S						
	FM	VE9ARZ	448.70000	-	141.3 Hz	
Fredericton, Nashwaaksis						
	FM	VE9CWM	146.65500	-	123.0 Hz	
	FM	VE9HAM	147.25500	+	123.0 Hz	
Fundy National Park						
	FM	VE9TCF	145.17000	-		
Gatineau	DMR/BM	VE2RAO	441.95000	+	CC1	
Grand Falls	FM	VE9GFL	146.94000	-		
Grand Manan Island						
	FM	VE9GMI	146.95500	-		
Hamilton	DMR/MARC	VE3UHM	444.03750	+	CC1	
Hammond	DMR/BM	VE3PRV	442.85000	+	CC1	
Hartland	FM	VE9DKS	147.18000	+		
Jersey Side	FM	VO1PFR	147.01000	+		
Jonquiere-nord	DMR/BM	VA2NA	145.35000	-	CC7	
Kanata-Katimavik	DSTAR	VA3AIT C	145.51000	-		
	DSTAR	VA3AIT B	445.90000	-		
	DSTAR	VA3AIT A	1283.50000	-		
Kelowna	DSTAR	VA7DIG C	145.03000	+		
	DSTAR	VA7DIG B	440.95000	+		
	DSTAR	VA7DIG A	1247.00000	+		
	FM	VA7JPL	444.00000	5444.00000	88.5 Hz	
	FM	VA7JPL	444.50000	5444.50000	88.5 Hz	
	FM	VA7UN	444.70000	1044.70000	88.5 Hz	
Kensington	DSTAR	VY2DSR C	147.31500	+		
Kirkfield	DMR/BM	VE3NYY	444.58750	+	CC1	
La Pocatiere	FM	VE2RDJ	147.36000	+	100.0 Hz	
Lac Cayamant	FM	VE2RBL	147.19500	+		
Lac Ouachishmana						
	FM	VA2RLL	147.09000	+		
Lavenir	DMR/MARC	VA2RHP	443.75000	+	CC1	
Lewisporte	FM	VO1LJR	147.32000	+		
Maces Bay	FM	VE9MBY	444.87500	+		
Martensville	DSTAR	VE5MBX C	146.44000	-		
	DSTAR	VE5MBX B	449.50000	-		
Mckendrick	FM	VE9LRC	146.95500	140.95500		
Miramichi	FM	VE9MIR	147.15000	+		
Moncton	DMR/MARC	VE9DMR	146.92500	-	CC1	
	FM	VE9TCR	147.34500	+		
	FM	VE9SHM	449.32500	-		
Moncton, Indian Mountain						
	FM	VE1MTN	147.09000	+		
Moncton, Lutz Mountain						
	FM	VE1RPT	146.88000	-		
Mont-Fournier	FM	VE2RHA	447.37500	-	110.9 Hz	
Mont-Saint-Gregoire						
	DSTAR	VE2RVR A	1247.00000			
Mont-Yamaska - UHF						
	DMR/MARC	VE2RAU	444.52500	+	CC1	
Montreal	DSTAR	VA2RKA C	144.91000	+		

426   NEW BRUNSWICK

Location	Mode	Call sign	Output	Input	Access	Coordinator
Montreal	DSTAR	VE2RIO C	144.95000	-		
	DSTAR	VE2PUK B	443.81250	+		
	DSTAR	VE2RIO A	1248.00000			
	DSTAR	VE2RIO A	1283.00000	-		
Montreal West	DSTAR	VE1FKB B	442.25000	+		
Moosonee	FM	KM4OOD	147.33000	+		
Mt Seymour	DSTAR	VE7RAG C	147.02000	+		
	DSTAR	VE7RAG B	443.40000	+		
	DSTAR	VE7RAG A	1271.94000			
	DSTAR	VE7RAG A	1291.94000	1271.94000		
New Harbour	FM	VE1HMY	146.55000	+		
North Battleford	DSTAR	VE5RAD C	147.12000	+		
North Campbell River						
	FM	VA7UW	147.57000	+		
Northwest Territories						
	FM	VE3KRG	146.97000	-	118.8 Hz	
Ontario	FM	VE3NFM	145.11000	-		
	FM	VE3DPL	146.66500	-	131.8 Hz	
	FM	VE3BIC	147.22500	+	131.8 Hz	
	FM	VE3KAR	147.69000	-	131.8 Hz	
	FM	VE3BGA	442.82500	+	100.0 Hz	
	FM	VE3OBN	443.90000	+	127.3 Hz	
	FM	VA3DJJ	448.07500	-	123.0 Hz	
Oromocto	FM	VE9OPH	145.31000	-		
	FM	VE1BAS	145.74000			
Parry Sound, McKellar						
	FM	VE3AAY	53.11000	52.11000		
Perce	FM	VE2RLC	146.79000	-		
Perth-Andover, Kintore Mountai						
	FM	VE9KMT	147.06000	+		
Pleasant Ridge	FM	VE1BI	146.70000	-		
Prince George	DSTAR	VE7RES C	147.02000	+		
	DSTAR	VE7RES B	442.10000	+		
	DSTAR	VE7RES A	1291.94000	1271.94000		
Princeville	DMR/MARC	VE2TXD	443.45000	+	CC2	
Quebec	DSTAR	VE2RMF A	1243.00000			
	FM	VE2RXD	144.81000	+		
	FM	VE3CRA	146.94000	-	100.0 Hz	
	FM	VE2RXD	441.17000	+		
	FM	VA2EZ	442.90000	447.20000	103.5 Hz	
	FM	VA2MB	444.00000	+	100.0 Hz	
Raglan	DMR/BM	VE3OBI	442.13750	+	CC1	
	DMR/BM	VE3LBN	443.98750	+	CC3	
Repentigny	DMR/MARC	VE2MRC	448.47500	442.47500	CC1	
Richmond	DSTAR	VA7REF C	144.93000			
Roxboro	DMR/BM	VE2RRC	443.50000	+	CC1	
Saint George, Poor House Hill						
	FM	VE9STG	147.22500	+		
Saint John	DMR/MARC	VE9SJN	145.49000	-	CC1	
	DSTAR	VE9SJN C	145.29000	-		
	FM	VE9STJ	146.82000	-		
	FM	VE9SJW	146.89500	-		
	FM	VE9SJW	443.60000	+		
Saint John, Baxters Mountain						
	FM	VE9PSA	147.39000	+		
Saint John, Dickie Mountain						
	FM	VE9HPN	145.13000	-		
Saint John, Grove Hill						
	FM	VE9STM	145.33000	-		
Saint Leonard	FM	VE9STL	145.35000	-		
Saint Quentin	FM	VE9SQN	145.23000	-		
Saint-Calixte	DSTAR	VA2RKB C	144.91000	+		

NEW BRUNSWICK 427

Location	Mode	Call sign	Output	Input	Access	Coordinator
Saint-Jean-sur-Riche						
	DMR/BM	VA2DGR	449.22500	-	CC2	
Saint-Pascal-Baylon						
	FM	VE3PRV	147.33000	+	110.9 Hz	
Salt Spring Isl	DSTAR	VE7MDN B	440.95000	+		
Salt Spring Island	DSTAR	VE7XNR B	444.85000	+		
Sandy Point Road	FM	VE9SJN	147.27000	+		
Saquenay	DSTAR	VE2RVI C	145.21000	-		
Saskatoon	DSTAR	VA5DR B	448.12500	-		
	FM	VE5WY	447.00000	-		
Shediac	DMR/MARC	VE9SBR	147.22500	+	CC1	
	FM	VE9DRB	147.37500	+		
Sherbrooke	DSTAR	VE2RQF A	1248.00000			
Shoal Harbour	FM	VO1SHR	146.66000	-		
Skiff Lake, Canterbury						
	FM	VE9IRG	145.37000	-		
Sorel-Tracy	DSTAR	VE2FCT C	146.61000	-		
	DSTAR	VE2FCT A	1249.00000			
Springhill	DMR/MARC	VE1XPR	145.29000	-	CC1	
St Johns	DSTAR	VO1TZ C	145.09000	+		
	DSTAR	VO1TZ B	443.40000	+		
St. Alban's	FM	VO1BDR	147.38000	+		
St. John's	FM	VO1GT	146.94000	-		
St. Stephen	FM	VE1IE	146.64000	-		
Stanley	FM	VE9NRV	147.03000	+		
Stewart Crossing, Ferry Hill						
	FM	VY1RFH	146.82000	-	100.0 Hz	
Sussex, Scotch Mountain						
	FM	VE9SMT	146.61000	-		
Toronto	DMR/BM	VA3WIK	442.18750	+	CC3	
	DMR/BM	VA3DVN	444.73750	+	CC1	
	DMR/MARC	VA3XFT	441.95000	+	CC1	
Tracadie	FM	VE9SID	146.70000	-		
	FM	VE1AZU	147.03000	+		
Tracadie, Academie Ste Familli						
	FM	VE9CR	145.47000	-		
Trenton	FM	VA3RUZ	443.07500	+	118.8 Hz	
Trois-Rivieres	DMR/MARC	VA2LX	440.10000	-	CC0	
	DSTAR	VE2LKL B	449.17500	-		
	DSTAR	VE2LKL A	1246.00000			
Truro	DMR/BM	VA1DIG	442.65000	+	CC1	
	FM	VE1HAR	147.13500	+		
Turtle Lake	DSTAR	VE5TLK C	145.45000	-		
Val Cartier	DMR/BM	VE2RAG	442.50000	+	CC1	
Val-Alain	DMR/MARC	VE2DXI	447.47500	-	CC1	
Vancouver	DMR/MARC	VE7RHS	442.45000	+	CC1	
	DSTAR	VA7ICM C	145.04000	+		
	DSTAR	VA7ICM B	442.00000	+		
	DSTAR	VA7ICM A	1247.00000	+		
	DSTAR	VA7ICM A	1293.15000	-		
Victoria	DSTAR	VE7VIC C	145.08000	+		
	DSTAR	VE7VIC A	1291.50000	-		
Wallaceburg	DMR/MARC	VA3YFU	147.27000	+	CC5	
Weatheradio-Dalhousie						
	FM	XLK418	162.55000			
Weatheradio-Eskimo Point						
	FM	CKO583	162.40000			
Weatheradio-Fort McMurray						
	FM	CFA340	162.40000			
Weatheradio-Fredericton						
	FM	VCF757	162.47500			

428   NEW BRUNSWICK

Location	Mode	Call sign	Output	Input	Access	Coordinator
Weatheradio-Hay River						
	FM	CIE211	162.55000			
Weatheradio-Holden						
	FM	CFB635	162.55000			
Weatheradio-Kamloops						
	FM	CIT768	162.40000			
Weatheradio-La Ronge						
	FM	CFJ262	162.40000			
Weatheradio-Lougheed						
	FM	CFB636	162.40000			
Weatheradio-Millville						
	FM	XLM404	162.55000			
Weatheradio-Miscou Island						
	FM	XMQ533	162.55000			
Weatheradio-Moncton						
	FM	XLM467	162.55000			
Weatheradio-Perth-Andover						
	FM	VFH526	162.50000			
Weatheradio-Revelstoke						
	FM	CIT386	162.40000			
Weatheradio-Saint-Isidore						
	FM	XLK417	162.40000			
Weatheradio-Scotch Mountain						
	FM	XLM403	162.40000			
Weatheradio-Sicamous						
	FM	CIQ882	162.40000			
Weatheradio-St Margarets						
	FM	VFQ820	162.45000			
Weatheradio-St Paul						
	FM	CIM235	162.40000			
Weatheradio-St Stephen						
	FM	XLM490	162.47500			
Weatheradio-Tagish						
	FM	VFS369	162.55000			
Weatheradio-Temagami						
	FM	CFE261	162.40000			
Weatheradio-Ucluelet						
	FM	CIZ319	162.52500			
Weatheradio-Whitehorse						
	FM	CIY270	162.40000			
Whistler	DSTAR	VA7WHI C	146.22000	+		
Whitehorse	DSTAR	VY1RDS C	146.84000	-		
	DSTAR	VY1RDS B	443.97500	+		
	DSTAR	VY1RDS A	1247.00000	1267.00000		
Whitehorse, 6m Beacon						
	FM	VY1DX	50.03000	-		
Whitehorse, Haeckel Hill						
	FM	VY1IRL	444.25000	+	100.0 Hz	
Yorkville	FM	VA3ATL	444.77500	+	110.9 Hz	

NEWFOUNDLAND AND LABRADOR

Location	Mode	Call sign	Output	Input	Access	Coordinator
Bay De Verde	FM	VO1TBR	147.30000	+		
Big Pond	FM	VO1ISR	147.22000	+		
Botwood	FM	VO1BOT	147.38500	+		
Cape Norman	FM	VO1STA	147.96500	-		
Carbonear, Freshwater Ridge						
	FM	VO1FRR	147.39000	+		
	FM	VO1FRR	444.97500	+		
Corner Brook	FM	VO1MO	147.36000	+		
Gander	FM	VO1ADE	146.88000	-		
	FM	VO1GLR	147.18000	+		
Goose Bay	FM	VO2GB	146.34000	+		

NEWFOUNDLAND AND LABRADOR   429

Location	Mode	Call sign	Output	Input	Access	Coordinator
Grand Falls, Red Cliff						
	FM	VO1GFR	146.91000	-		
Grand Falls-Windsor, Water Tow						
	FM	VO1JY	146.76000	-		
Grand-Falls Windsor, Hodges Hi						
	FM	VO1HHR	146.60000	-		
Holyrood, Hawke Hill						
	FM	VO1BT	146.76000	-		
Hopedale	FM	MB7IPT	145.21200	+		
Labrador City, Round Hill						
	FM	VO2LMC	146.76000	-		
Marystown	FM	VO1MST	146.85000	-		
	FM	VO1AWR	147.22000	+		
New Harbour NL Canada						
	FM	VO1PCR	147.09000	+		
Placentia	FM	VO1SEP	447.12500	-		
Portugal Cove South						
	FM	VO1ILR	147.03000	+		
Ramea	FM	VO1RIR	147.28000	+		
Shearstown	FM	VO1EHC	447.50000	-		
St. Anthony	FM	VO1GNP	147.96500	-		
St. John S NL	FM	VO1RCR	147.34500	+	100.0 Hz	
St. John's	FM	VO1EHC	147.57000			
St. John's, NTV / OZFM Tower						
	FM	VO1NTV	147.06000	+		
St. Lawrence	FM	VO1AIR	147.19000	+		
Swift Current	FM	VO1PBR	147.22000	+		
Torbay	FM	VO1ZA	147.00000	+		
Weatheradio-Birchy Lake						
	FM	XLM665	162.40000			
Weatheradio-Brent S Cove						
	FM	XLW297	162.40000			
Weatheradio-Codroy Pond						
	FM	XLW201	162.40000			
Weatheradio-Conche						
	FM	XLW296	162.55000			
Weatheradio-Corner Brook						
	FM	XLW200	162.55000			
Weatheradio-Gander						
	FM	XLM616	162.40000			
Weatheradio-Grand Falls						
	FM	XLM664	162.55000			
Weatheradio-Hermitage						
	FM	XLW204	162.55000			
Weatheradio-Marystown						
	FM	XLM663	162.40000			
Weatheradio-Mount St Margaret						
	FM	XLW295	162.55000			
Weatheradio-Port Rexton						
	FM	XLM615	162.55000			
Weatheradio-Portland Creek						
	FM	XLW298	162.40000			
Weatheradio-Red Rocks						
	FM	XLW202	162.55000			
Weatheradio-St Anthony						
	FM	XLW299	162.40000			
Weatheradio-St John S						
	FM	XLM614	162.40000			
Weatheradio-Trepassey						
	FM	XLM662	162.55000			

430   NORTHWEST TERRITORIES

Location	Mode	Call sign	Output	Input	Access	Coordinator

NORTHWEST TERRITORIES

Location	Mode	Call sign	Output	Input	Access	Coordinator
Weatheradio-Inner Whaleback Ro						
	FM	XKI403	162.55000			
Weatheradio-Inuvik						
	FM	VBU996	162.40000			
Weatheradio-Pine Point						
	FM	XJS786	162.47500			
Weatheradio-Yellowknife						
	FM	VBC200	162.40000			
Yellowknife, Jackfish Lake						
	FM	VE8YK	146.94000	-	100.0 Hz	

NOVA SCOTIA

Location	Mode	Call sign	Output	Input	Access	Coordinator
Antigonish	FM	VE1RTI	146.82000	-		
	FUSION	VE1JCS	441.80000	+		
Antigonish NS	DSTAR	VE1JSR C	145.67000			
	DSTAR	VE1JSR D	445.67000	-		
Barrington	FM	VE1OPK	147.25500	+		
	FM	VE1KDE	443.80000	+		
Barrington Passage						
	FM	VE1JNR	146.88000	-		
Biblehill	DSTAR	VE1DNR C	144.95000	+		
Boisdale	FM	VE1HAM	146.88000	-		
Bridgetown	FM	VE1BO	147.06000	+		
Bridgewater	FM	VE1KIN	147.12000	+		
Cape Smokey	FM	VE1CBI	147.24000	+		
Chester	FM	VE1LUN	147.33000	+		
Dartmouth	FM	VE1DAR	147.15000	+		
	FM	VE1DAR	444.60000	+		
Digby	FM	VE1AAR	147.01500	+		
East Kemptville	FM	VE1EKV	147.10500	147.75000		
Gore	FM	VE1OM	51.62000	-		
	FM	VE1OM	146.64000	-		
Granite Village	FM	VE1BBY	147.36000	+		
Greenwood	FM	VE1WN	147.24000	+		
Halifax	FM	VE1PSR	53.55000	52.55000	151.4 Hz	
	FM	VE1HNS	146.94000	-	82.5 Hz	
	FM	VE1PSR	147.27000	+		
	FM	VE1CDN	444.00000	+		
	FM	VE1PSR	444.35000	+		
	FM	VE1PS	449.25000	-		
Halifax, Upper Sackville						
	FM	VE1CDN	146.97000	-		
Hammonds Plains	FM	VE1PKT	146.68500	-		
Hebron	FUSION	VE1LN	146.86500	-		
Italy Cross	FM	VE1VL	147.09000	+		
Kejimkujik National Park						
	FM	VE1KEJ	147.19500	+		
Kentville	FM	VE1WRG	144.43000	+		
Kiltarlity	FM	VE1KIL	146.73000	-		
Kingston, Stronach Mountain						
	FM	VE1VLY	444.05000	+		
Liverpool	FM	VE1QW	147.06000	+		
	FM	VE1VO	147.30000	+		
Lower Sackville	FM	VA1AA	147.58500			
Lundy	FM	VE1GYS	146.70000	-		
Middlefield	FM	VE1AVA	147.39000	+		
New Glasgow	FM	VE1HR	146.76000	-		
Oban	FM	VE1OBN	147.10500	+		
Parrsboro	FM	VE1PAR	145.47000	-		
	FM	VE1NET	146.74500	-		

NOVA SCOTIA 431

Location	Mode	Call sign	Output	Input	Access	Coordinator
Plympton Station	FM	VE1JSO	147.01500	+		
Sand River	FM	VE1SDR	145.27000	-		
Shelburne	FM	VE1SCR	146.61000	-		
Sherbrooke	FM	VE1SAB	145.39000	-		
Southampton	FUSION	VE1BFB	146.77500	-		
Southampton, Nova Scotia Provi						
	FM	VE1EWS	443.45000	+		
Springfield	FM	VE1LCA	146.83500	-		
Springhill	FM	VE1SPR	444.20000	+		
Springhill, All Saints Hospita						
	FM	VE1SPH	146.80500	-		
Springhill, Lynn Mountain						
	FM	VE1SPR	147.00000	+		
Stronach Mountain						
	FM	VE1VAL	145.21000	-		
	FM	VE1VLY	449.05000	-		
Sugarloaf	FM	VE1BHS	145.35000	-		
	FM	VE1BHS	448.92500	-		
Sydney	FM	VE1HK	146.94000	-		
Truro	FM	VE1XK	146.79000	-		
Truro, Nuttby Mountain						
	FM	VE1TRO	147.21000	+		
Weatheradio-Aspen						
	FM	XLK499	162.40000			
Weatheradio-Ben Eoin						
	FM	XLW262	162.47500			
Weatheradio-Bridgewater						
	FM	XLK409	162.40000			
Weatheradio-Cheticamp						
	FM	XLW263	162.47500			
Weatheradio-Halifax						
	FM	XLK473	162.55000			
Weatheradio-Middleton						
	FM	XLK497	162.55000			
Weatheradio-New Tusket						
	FM	XLK496	162.55000			
Weatheradio-Oak Park						
	FM	XLW502	162.47500			
Weatheradio-River Denys						
	FM	XLK445	162.55000			
Weatheradio-Shelburne						
	FM	XLK410	162.55000			
Weatheradio-Sydney						
	FM	XLK444	162.40000			
Weatheradio-Truro						
	FM	XLK498	162.40000			
Weatheradio-Yarmouth						
	FM	XLW573	162.47500			
West Chezzetcook						
	FM	VE1ESC	147.03000	+		
Windsor	FM	VE1HCA	146.91000	-		
Yarmouth	FM	VE1YAR	146.73000	-		

NUNAVUT

Location	Mode	Call sign	Output	Input	Access	Coordinator
Baker Lake	FM	VY0MBK	146.76000	+		
Iqaluit	FM	VY0SNO	146.94000	-		
Resolute	FM	VY0FG	448.92500	-		
Weatheradio-Cape Dorset						
	FM	XJS717	162.55000			
Weatheradio-Iqaluit						
	FM	VEV284	162.55000			

432   NUNAVUT

Location	Mode	Call sign	Output	Input	Access	Coordinator
Weatheradio-Rankin Inlet						
	FM	XJS716	162.40000			
ONTARIO						
Acton	FM	VE3RSS	147.03000	+		
	FM	VE3PAQ	442.12500	+	131.8 Hz	
	FM	VA3GTU	442.82500	+	103.5 Hz	
Agincourt	FM	VE3WOO	29.64000	-		
Ajax	FM	VE3SPA	147.37500	+	103.5 Hz	
	FM	VE3DAX	444.60000	+	103.5 Hz	
Alban	FM	VE3BLZ	146.70000	-	100.0 Hz	
Alfred	FM	VA3TLO	443.50000	+	110.9 Hz	
Alfred, Water Tower						
	FM	VA3PRA	145.47000	-	110.9 Hz	
Almonte	DSTAR	VA3AAR C	145.55000	-		
	DSTAR	VA3AAR B	444.10000	+		
	DSTAR	VA3AAR A	1281.00000	-		
	FM	VA3AAR	147.27000	+		
Almonte, Almonte Fire Hall						
	FM	VA3AAR	444.30000	+	100.0 Hz	
Almonte, Union Hall						
	FM	VA3ARE	147.24000	+	100.0 Hz	
	FM	VA3ARE	444.10000	+		
Alvinston	FM	VE3TTP	147.21000	+	114.8 Hz	
	FM	VE3TTP	442.85000	+	114.8 Hz	
Ancaster	FM	VE3RDM	145.27000	-	136.5 Hz	
	FM	VE3RTJ	442.50000	+		
	FM	VE3DJ	446.43700	-		
Arnprior	DSTAR	VA3JJA C	145.63000			
	DSTAR	VA3JJA B	445.84000	-		
	DSTAR	VA3JJA A	1286.00000	-		
Arnprior, Police / Fire Bldg						
	FM	VE3YYX	443.20000	+	114.8 Hz	
Atikokan	FM	VE3RIB	147.12000	+		
Aurora	FM	VE3YRA	145.35000	-	103.5 Hz	
	FM	VE3YRC	147.22500	+	103.5 Hz	
	FM	VE3ULR	224.88000	-		
Baden, Baden Sand Hill						
	FM	VE3KSR	146.97000	-	131.8 Hz	
Ballantrae	DMR/MARC	VE3URU	443.38750	+	CC1	
	FM	VA3PWR	53.13000	52.13000	103.5 Hz	
	FM	VE3ULR	145.47000	-	103.5 Hz	
	FM	VA3BAL	147.33000	147.83000	103.5 Hz	
	FM	VE3ULR	442.02500	+	103.5 Hz	
	FM	VA3URU	442.47500	+	103.5 Hz	
	FM	VE3SNM	442.85000	+	136.5 Hz	
	FM	VA3BAL	443.70000	+	103.5 Hz	
Ballinafad	FM	VA3LNK	443.42500	+		
Bancroft	FM	VE3BNI	146.61000	-	100.0 Hz	
Banks	FM	VA3ROG	145.37000	-	156.7 Hz	
Barrie	FM	VE3RAG	147.00000	+	156.7 Hz	
Barrys Bay	FM	VE3RKA	146.97000	-		
Belleville	FM	VE3QAR	146.98500	-	118.8 Hz	
	FM	VE3ALC	147.51000	-		
	FM	VE3QAR	444.47500	+	118.8 Hz	
Berkeley	FM	VA3CAX	145.29000	-	156.7 Hz	
Birkendale	FM	VE3MUS	146.77500	-		
Black Hawk	FM	VE3RBK	147.04500	+		
Blenheim	DSTAR	VA3IBA C	145.79000			
Bolton	FM	VA3EDE	146.43000		67.0 Hz	
	FM	VA3OPG	146.83500	-	103.5 Hz	
	FM	VA3OPR	444.95000	+	103.5 Hz	

ONTARIO 433

Location	Mode	Call sign	Output	Input	Access	Coordinator
Bradford	FM	VE3ZXN	224.48000	-		
	FM	VE3ZXN	927.90000	902.90000		
Brampton	FM	VE3PRC	146.88000	-	103.5 Hz	
	FM	VA3AGC	147.51000	+		
	FM	VE3WSA	224.24000	-		
	FM	VE3PRC	443.55000	+	103.5 Hz	
Brantford	FM	VE3MBX	145.27000	-	131.8 Hz	
	FM	VE3TCR	147.15000	+	131.8 Hz	
	FM	VE3TCR	443.02500	+	131.8 Hz	
	FM	VE3DTE	921.00000	-		
Bridgenorth	FM	VE3BTE	147.30000	+		
Brockville	FM	VE3IWJ	146.82000	-		
Buckhorn	FM	VE3KLR	443.15000	+	103.5 Hz	
Burks Falls	FM	VA3BFR	145.17000	-	156.7 Hz	
Burlington	DSTAR	VE3RSB B	442.03750	+		
	FM	VE3DUO	53.59000	52.59000	131.8 Hz	
	FM	VE3RSB	147.21000	+		
	FM	VE3WIK	224.96000	-	131.8 Hz	
	FM	VE3BUR	443.15000	+		
	FM	VE3RSB	444.82500	+	131.8 Hz	
Burlington ON	DSTAR	VE3OBP B	443.20000	+		
Cachet	FM	VA3CTR	442.27500	+		
Caledon	FM	VE3SKV	146.89500	-		
	FM	VE3UPR	443.60000	+		
	FM	VE3WOO	444.17500	+	103.5 Hz	
Cambridge	FM	VE3SWR	146.79000	-	131.8 Hz	
Campbellford	FM	VE3KFR	145.33000	-	162.2 Hz	
Campbellville	FM	VE3RJS	446.95000	-		
Campden	FM	VE3ALS	443.57500	+	107.2 Hz	
Cannington	FM	VA3TVE	444.45000	+		
	FM	VA3TVE	1288.00000	-		
Carlisle	FM	VE3WIK	53.11000	52.11000	131.8 Hz	
	FM	VE3WIK	146.71500	-	131.8 Hz	
	FM	VE3WIK	443.67500	+	131.8 Hz	
Carp	FM	VA3WJC	444.05000	+	123.0 Hz	
	FM	VE3IEV	445.07500	-	100.0 Hz	
Carson Grove	FM	VE3JGL	444.50000	+		
Chatham	DMR/MARC	VA3XLT	443.98750	+	CC7	
	FM	VE3KCR	144.39000	+		
	FM	VE3COZ	145.19000	-		
	FM	VA3KCR	146.68000	-		
	FM	VE3KCR	147.12000	+	100.0 Hz	
	FM	VE3CBS	224.12000	-	131.8 Hz	
	FM	VE3MGK	442.20000	+		
	FM	VE3COZ	444.32500	+	250.3 Hz	
Clarendon Station	FM	VE3KAR	147.09000	+	151.4 Hz	
Cobourg / Peterb	DSTAR	VE3RTR C	146.89500	-		
Cobourg, Rice Lake						
	FM	VE3RTR	145.15000	-	162.2 Hz	
	FM	VE3MXR	444.97500	+	162.2 Hz	
Colemans	FM	VE3NUU	446.80000	-	107.2 Hz	
Collingwood	DMR/BM	VA3BMR	442.38750	+	CC1	
	FM	VE3RMT	53.15000	52.15000	156.7 Hz	
	FM	VE3BMR	146.79000	-	156.7 Hz	
	FM	VE3BMR	442.60000	+	156.7 Hz	
	FM	VE3QCR	443.27500	+		
	FM	VE3RMT	921.50000	-	156.7 Hz	
Cookstown	FM	VA3TWO	444.68750	+	156.7 Hz	
Cornwall	DMR/MARC	VA3EDG	442.10000	+	CC1	
	DSTAR	VA3SDG C	145.57000	-		
	DSTAR	VA3SDG B	444.45000	+		
	FM	VE3YGM	145.17000	-		

434   ONTARIO

Location	Mode	Call sign	Output	Input	Access	Coordinator
Cornwall	FM	VE3XID	146.47500	+		
	FM	VE3PGC	443.00000	+	110.9 Hz	
	FM	VE3MTA	443.65000	+	110.9 Hz	
Cornwall, Hotel Dieu Hosp						
	FM	VE3SVC	147.18000	+	110.9 Hz	
Courtland	DMR/BM	VE3DPL	146.65500	-	CC1	
Coventry	FM	VA3JMF	146.83500	-		
Cumberland	DSTAR	VA3ODH C	145.61000	-		
	DSTAR	VA3ODH B	445.85000	+		
	DSTAR	VA3ODH A	1282.00000	-		
	FM	VA3RCB	444.35000	+	100.0 Hz	
Devlin	FM	VE3BIK	146.57000		123.0 Hz	
Distillery District	FM	VE3BGD	147.27000	+		
Dorchester	FM	VE3NDT	147.24000	+		
Dryden	FM	VE3DRY	147.25500	+		
	FM	VA3DIS	147.37500	+		
	FUSION	VE3YHD	146.94000	-	123.0 Hz	
Dryden ON	FM	VA3ANE	147.34500	+	123.0 Hz	
Dundalk	FM	VA3WWM	442.92500	+	131.8 Hz	
Dunnville	FM	VE3HNR	147.07500	+	123.0 Hz	
	FM	VE3KYO	444.70000	+		
Dutton	FM	VE3ISR	147.36000	+	114.8 Hz	
Dwight	FM	VE3MUS	146.82000	-	156.7 Hz	
Dysart	FM	VA3LTX	147.10500	+		
Eagle Lake	FM	VA3HAL	442.00000	+	162.2 Hz	
	FM	VA3HAL	444.80000	+	162.2 Hz	
Edgar	DMR/BM	VE3UHF	442.88750	+	CC1	
	DSTAR	VE3LSR AD	1248.50000			
	FM	VE3LSR	146.85000	-	156.7 Hz	
	FM	VE3KES	147.15000	+	156.7 Hz	
	FM	VA3IMB	147.28500	+	156.7 Hz	
	FM	VE3LSR	147.31500	+	156.7 Hz	
	FM	VE3LSR	442.57500	+	103.5 Hz	
	FM	VA3BNI	444.27500	+	103.5 Hz	
Eldorado	FM	VA3SDR	145.41000	-	118.8 Hz	
Elizabeth Park	FM	VE3RIX	145.45000	-		
Elliot Lake	FM	VE3TOP	147.00000	+		
Elmira	FM	VE3ERC	147.39000	+	123.0 Hz	
	FM	VE3ERC	444.70000	+	131.8 Hz	
Embrun	FM	VE3EYV	147.19500	+	110.9 Hz	
Englehart	FM	VE3TAR	146.97000	-		
ERA System Southern Ontario						
	FM	VA3ERA	443.67500	+	131.8 Hz	
Essex	FM	VE3SMR	53.03000	52.03000	118.8 Hz	
Essonville	FM	VE3TBF	224.84000	-		
Etobicoke	FM	VA3GTU	442.77500	+	103.5 Hz	
	FM	VA3GTU	442.80000	+	103.5 Hz	
Exeter	FM	VE3JEZ	146.76000	+		
Fauquier	FM	VA3FOK	443.80000	+	156.7 Hz	
Fergus	FM	VA3EHI	146.56500	+		
Fisherville	FM	VA3WJO	444.70000	+		
Fonthill	DMR/MARC	VA3XPR	442.71250	+	CC1	
	FM	VE3PLF	53.29000	52.29000	107.2 Hz	
	FM	VE3WCD	147.30000	+	107.2 Hz	
	FM	VE3UCS	224.58000	-	107.2 Hz	
	FM	VE3EI	224.80000	-	107.2 Hz	
	FM	VE3EI	444.72500	+	107.2 Hz	
Fonthill, Lookout Village						
	FM	VE3RNR	443.17500	+	107.2 Hz	
Fonthill- ON	DSTAR	VE3PLF B	444.72500	+		
Forest Hill	FM	VE3NOR	443.65000	+		
Fort Frances	FM	VE3RLC	146.82000	-		

ONTARIO 435

Location	Mode	Call sign	Output	Input	Access	Coordinator
Fort Frances	FUSION	VE3BVC	145.11000	-		
Foymount	FM	VE3UCR	145.43000	-	114.8 Hz	
Frankford	FM	VE3TRR	147.01500	+		
Franktown	FM	VE3WCC	444.30000	+	100.0 Hz	
Frontenac County	FM	VE3FRG	146.20500	+	151.4 Hz	
Gatineau	FM	VA2XAD	430.05000	434.95000		
Glasgow	FM	VA3TE	146.43000		67.0 Hz	
	FM	VA3OPG	444.95000	+		
Goderich	DMR/MARC	VE3OBC	442.07500	+	CC1	
	FM	VE3OBC	146.91000	-	123.0 Hz	
Goodwood	FM	VE3GTU	442.07500	+	103.5 Hz	
Grand Bend	FM	VE3SRT	442.05000	+	114.8 Hz	
	FM	VE3SRT	442.07500	+	114.8 Hz	
Grasshill	FM	VE3LNZ	147.19500	+		
Grassie	FM	VE3BQQ	442.72500	+	131.8 Hz	
Grassmere	FM	VA3BFR	145.27000	-		
	FM	VE3RAK	444.70000	+		
Grimsby	FM	VE3IUW	1283.60000	-		
Guelph	FM	VE3ZMG	145.21000	-	131.8 Hz	
	FM	VE3OVQ	147.54000		131.8 Hz	
Guelph Ontario	FM	VA3SLD	147.36000	+	131.8 Hz	
Guelph, Guelph General Hospita						
	FM	VE3GEG	147.00000	+	131.8 Hz	
Haliburton	FM	VE3SRU	53.05000	52.05000		
Haliburton County	FM	VE3ZHR	146.65500	-	162.2 Hz	
Hamilton	DSTAR	VE3WIK C	145.33000	-		
	DSTAR	VA3FS C	145.51000	-		
	DSTAR	VE3WIK C	146.71500	-		
	DSTAR	VE3WIK B	443.63750	+		
	DSTAR	VA3FS B	445.82500	-		
	DSTAR	VA3FS A	1286.00000	-		
	FM	VA3TVW	146.58000			
	FM	VE3NCF	146.76000	-	131.8 Hz	
	FM	VE3ISX	147.58500		136.5 Hz	
	FM	VE3NCF	444.07500	+	131.8 Hz	
	FM	VE3TTO	446.57500	-	131.8 Hz	
Hamilton ON	FM	VE3TTO	146.46000		131.8 Hz	
Hamilton, Gore Park						
	FM	VE3ESM	442.55000	+	131.8 Hz	
Hampton Heights	FM	VE3RFI	443.25000	442.65000		
Hearst, Ontario	FM	VA3YHF	146.70000	-		
Hillsburgh	FM	VE3ZAP	443.87500	+	88.5 Hz	
Horton	FM	VE3ZRR	146.91000	-		
Huntsville	FM	VE3URU	146.95500	-	100.0 Hz	
Ignace, Ignace School						
	FM	VA3IGN	147.18000	+	123.0 Hz	
Ingersoll	FM	VE3NRJ	145.14000	+		
	FM	VA3PLL	145.17000	-	114.8 Hz	
	FM	VE3OHR	147.27000	÷	114.8 Hz	
	FM	VE3OHR	443.45000	÷	114.8 Hz	
Inglewood	FM	VE3RDP	146.70000	-	103.5 Hz	
Ipperwash Beach	FM	VE3TCB	146.94000	-	114.8 Hz	
Iroquois	FM	VE3IRO	145.29000	-		
Kagawong	FM	VE3LTR	146.67000	-		
Kapuskasing	FM	VE3YYU	442.00000	+		
Kenora	FM	VE3YQK	146.56000		123.0 Hz	
	FM	VE3LWR	147.03000	+		
	FUSION	VE3XTI	146.91000	-		
Keswick	FM	VA3PTX	147.28500	+	156.7 Hz	
Kincardine	FM	VE3TIV	146.61000	-		
King City	FM	VE3GSR	145.31000	-	103.5 Hz	
	FM	VE3WAS	146.61000	-	103.5 Hz	

436   ONTARIO

Location	Mode	Call sign	Output	Input	Access	Coordinator
King City	FM	VE3UKC	444.30000	+	103.5 Hz	
Kingston	P25	VE3KTO	443.30000	+		
Kingston, Kingston ARC						
	FM	VE3KBR	146.94000	-	151.4 Hz	
Kitchener	FM	VE3SED	53.37000	52.37000	131.8 Hz	
	FM	VA3XTO	146.58000			
	FM	VE3RCK	146.86500	-	131.8 Hz	
	FM	VE3IXY	224.34000	-	131.8 Hz	
	FM	VE3SED	442.20000	+	131.8 Hz	
	FM	VE3BAY	442.35000	447.37000	131.8 Hz	
	FM	VE3RBM	444.87500	+	131.8 Hz	
Lanark	DSTAR	VE3ENH B	442.00000	+		
	DSTAR	VE3ENH A	1282.50000	-		
Lancaster	FM	VE2REH	147.52500	145.67000	131.8 Hz	
Lansdowne	FM	VA3LGA	146.62500	-	100.0 Hz	
Lavant	FM	VE3KJG	146.64000	-		
Lavant Station, Mountain Top						
	FM	VA3LGP	53.23000	52.23000	141.3 Hz	
Leamington	DMR/MARC	VE3TOM	443.26250	+	CC7	
	DSTAR	VE3LNK C	145.69000			
	FM	VA3LLL	146.97000	-	118.8 Hz	
	FM	VE3ZZZ	147.10000	+	118.8 Hz	
	FM	VE3TOM	147.30000	+	118.8 Hz	
	FM	VA3LLL	442.05000	+	118.8 Hz	
Lindsay	FM	VE3RWN	145.11000	-	118.8 Hz	
Lindsay, City Of Kawartha Lake						
	FM	VE3CKL	145.45000	-	162.2 Hz	
	FM	VE3CKL	444.65000	+	162.2 Hz	
Lionshead	FM	VE3CAX	146.71500	-	156.7 Hz	
Listowel	FM	VA3LIS	147.12000	+	114.8 Hz	
Little Current	DSTAR	VE3RXR C	145.31000	-		
	DSTAR	VE3RXR B	442.05000	+		
	FM	VE3RQQ	146.55000		156.7 Hz	
	FM	VE3RMI	147.27000	+		
	FM	VE3RQQ	444.30000	+	100.0 Hz	
London	DMR/BM	VE3TTT	147.18000	+	CC1	
	DMR/MARC	VE3RGM	444.61250	+	CC7	
	DSTAR	VE3TTT B	442.30000	+		
	FM	VE3GYQ	145.35000	-	114.8 Hz	
	FM	VA3MGI	145.39000	-	114.8 Hz	
	FM	VE3OME	145.45000	-	114.8 Hz	
	FM	VE3OES	146.49000		114.8 Hz	
	FM	VA3LON	147.06000	+	114.8 Hz	
	FM	VA3FEZ	444.10000	+	114.8 Hz	
	FM	VE3SUE	444.40000	+	114.8 Hz	
	FM	VE3LSG	446.15000	-		
	FM	VA3CCC	922.00000	-	114.8 Hz	
London, Ontario	DSTAR	VE3TTT A	1285.50000	-		
Mallorytown	FM	VE3IGE	146.97000	-		
	FM	VE3IGE	443.90000	+		
Manitowaning	FM	VE3RII	444.17500	+	156.7 Hz	
Maple	DSTAR	VE3LEO C	144.30000			
	DSTAR	VA3ITL B	442.21250	+		
	DSTAR	VE3LEO B	446.00000	-		
	DSTAR	VE3LEO A	1282.50000	-		
	FM	VA3ITL	442.41250	-		
Marathon	FUSION	VE3JTD	147.21000	+		
Marmora	FM	VE3OUR	443.47500	+		
Mattawa	FM	VE3NBR	147.15000	+		
Maynooth	FM	VE3WPR	147.00000	+		
Mcarthur Mills	FM	VA3PLA	147.18000	+		
McGregor	FM	VE3KUC	145.39000	-	118.8 Hz	

ONTARIO   437

Location	Mode	Call sign	Output	Input	Access	Coordinator
Mcgregor	FM	VE3RRR	145.47000	-	118.8 Hz	
	FM	VE3RRR	224.70000	-	118.8 Hz	
	FM	VE3SOT	1282.50000	-		
Midland	DMR/BM	VE3UGB	443.88750	+	CC1	
	FM	VE3UGB	146.91000	-		
Milton	FM	VE3HAL	442.30000	+	131.8 Hz	
	FM	VE3ADT	444.12500	+	131.8 Hz	
Milton ON	DSTAR	VE3ELF B	446.07500	-		
Milverton	DSTAR	VE3NMN C	147.16500	+		
	DSTAR	VE3NMN C	147.41500			
	DSTAR	VE3NMN B	438.10000			
	DSTAR	VE3NMN A	1282.45000	-		
	FM	VE3NMN	444.92500	+		
Mindemoya	FM	VE3WVU	444.45000	+	156.7 Hz	
Minden	FM	VE3VHH	147.07500	+	162.2 Hz	
Mississauga	DSTAR	VE3RSD C	147.54000			
	DSTAR	VE3RSD B	443.81250	+		
	DSTAR	VA3PMO B	444.25000	+		
	FM	VE3RSD	224.62000	-	103.5 Hz	
	FM	VE3RSD	224.72000	-	103.5 Hz	
	FM	VE3RSD	443.81250	+	103.5 Hz	
	FM	VE3RSD	444.25000	+	103.5 Hz	
Mississauga ON	DSTAR	W8ORG C	146.50000			
	DSTAR	VE2YUU B	442.05000	+		
Mississauga ON Canada						
	FM	VE3MIS	145.43000	-	103.5 Hz	
Mitchell	FM	VE3XMM	147.28500	+	114.8 Hz	
Monkland	FM	VE2NUU	445.92500	-		
	FM	VE2NUU	445.97500	-		
Mont-Laurier	DMR/BM	VE2REH	444.15000	+	CC1	
Moose Creek	DMR/BM	VE3OJE	145.37000	-	CC1	
	DMR/BM	VE3TYF	443.05000	+	CC1	
Moosonee	FM	VE3CIJ	446.25000	-	103.5 Hz	
Morrisburg	DMR/BM	VE3SVR	146.76000	-	CC1	
Mount Forest	FM	VA3CRV	147.16500	+	156.7 Hz	
Mt.St. Patrick	DSTAR	VE3STP C	147.54000			
	DSTAR	VE3STP B	443.60000	+		
	DSTAR	VE3STP B	443.81250	+		
	DSTAR	VE3STP A	1282.45000	-		
Nestor Falls	FUSION	VE3JJA	146.71500	-		
New Dundee	FM	VE3RND	145.33000	-	131.8 Hz	
New Liskeard	FM	VE3CIJ	145.19000	-		
Newmarket	DSTAR	VE3YRK C	147.18000	+		
	DSTAR	VE3YRK B	444.51250	+		
	FM	VA3PWR	443.27500	+	103.5 Hz	
Niagara Falls	DSTAR	VA3WAJ B	442.42500	+		
	FM	VA3WAJ	224.18000	-	107.2 Hz	
	FM	VA3WAJ	442.42500	+	107.2 Hz	
	FM	VE3GRW	442.90000	+	107.2 Hz	
Niagara On The L	DSTAR	VA3YYZ C	144.93000	+		
	DSTAR	VA3YYZ B	442.68750	+		
	DSTAR	VA3NAG A	1282.50000	-		
Niagara On The Lake						
	DSTAR	VA3NAG A	1299.15000			
Norfolk	DSTAR	VE3HJ B	444.55000	+		
North Bay	FM	VE3NFM	147.30000	+	107.2 Hz	
North York	FM	VE3TNC	147.27000	+	103.5 Hz	
	FM	VA3OBN	442.75000	+	103.5 Hz	
Oak Ridges	FM	VE3MPI	146.49000	+		
Oakville	DSTAR	VE3OBP B	445.01250	-		
	FM	VE3MIJ	147.01500	+	131.8 Hz	
	FM	VE3OKR	442.45000	+		

438   ONTARIO

Location	Mode	Call sign	Output	Input	Access	Coordinator
Oakville	FM	VE3OAK	444.32500	+	131.8 Hz	
	FM	VE3XCN	445.17500	-	131.8 Hz	
Oakville ON	FM	VA3WIK	438.17500		131.8 Hz	
Omemee	FM	VA3OME	147.09000	+	162.2 Hz	
	FM	VA3MME	444.90000	+		
Ontario	FM	VE3OBC	146.31000	+	97.4 Hz	
	FM	VA3YOS	146.64000	-	156.7 Hz	
	FM	VA3JFD	442.07500	+	123.0 Hz	
	FM	VE3AA	444.90000	+		
	FM	VA3JFD	447.07500	-	123.0 Hz	
	FM	VE3WWD	448.07500	-	97.4 Hz	
Orangeville	FM	VA3FYI	145.23000	-		
	FM	VE3ORX	444.02500	+	156.7 Hz	
	FM	VA3FYI	444.62500	+		
Orchard	FM	VA3MFD	446.12500	-		
Orillia	DMR/BM	VE3ORC	444.56250	+	CC1	
	DMR/MARC	VA3URU	444.68250	+	CC1	
	FM	VE3SYY	146.50500		156.7 Hz	
	FM	VE3URG	146.56500		156.7 Hz	
	FM	VA3OPS	146.65500	-		
	FM	VE3ORR	147.21000	+	156.7 Hz	
	FM	VE3ORC	444.55000	+		
Oshawa	DMR/BM	VE3OUR	442.87500	+	CC1	
	FM	VE3OSH	147.12000	+	156.7 Hz	
	FM	VE3NAA	443.00000	+	136.5 Hz	
Ottawa	DMR/MARC	VA3RFT	444.47500	+	CC1	
	DSTAR	VA3ODG C	145.53000	-		
	DSTAR	VA3ODG B	444.85000	+		
	DSTAR	VE3FSR B	445.86000	-		
	DSTAR	VA3ODG A	1282.00000	-		
	DSTAR	VA3ODG A	1299.20000			
	FM	VE3TST	29.62000	-	136.5 Hz	
	FM	VE3RVI	53.03000	52.03000		
	FM	VE3TST	53.09000	52.09000	136.5 Hz	
	FM	VE3OTW	145.19000	-	136.5 Hz	
	FM	VA3OFS	146.67000	-	136.5 Hz	
	FM	VE3TST	146.70000	-	136.5 Hz	
	FM	VA3LCC	146.79000	-		
	FM	VE3ORF	146.85000	-	136.5 Hz	
	FM	VE3OCE	146.88000	-	136.5 Hz	
	FM	VA3EMV	146.98500	-	100.0 Hz	
	FM	VE3TWO	147.30000	+	100.0 Hz	
	FM	VE3PNO	147.57000		103.5 Hz	
	FM	VA3OTW	224.68000	-		
	FM	VE3ORF	224.72000	-		
	FM	VE3OCE	443.80000	+	136.5 Hz	
	FM	VE3TST	444.12500	+	136.5 Hz	
	FM	VE3TWO	444.20000	+	110.9 Hz	
	FM	VE3ORF	444.55000	+	136.5 Hz	
	FM	VA3DSP	444.82500	+		
	FM	VA3OFS	444.95000	+	136.5 Hz	
	FM	VA3OFS	449.95000	-	136.5 Hz	
	FUSION	VE3IGN	145.45000	-		
	FUSION	VE3DRE	146.80500	-		
	FUSION	VE3ORF	444.55000	+		
Oungah	FM	VE3UGG	443.00000	442.40000		
Owen Sound	FM	VE3OSR	146.94000	-		
	FM	VE3OSR	442.35000	+	156.7 Hz	
Paisley	FM	VE3RTE	146.73000	-	156.7 Hz	
Pakenham	FM	VE2REH	145.33000	-	110.9 Hz	
Paris	FM	VE3DIB	145.49000	-	131.8 Hz	
Parry Sound	DMR/BM	VE3UPS	444.80000	+	CC1	

ONTARIO 439

Location	Mode	Call sign	Output	Input	Access	Coordinator
Parry Sound	FM	VE3RPL	145.49000	-		
	FM	VE3RPL	443.57500	+		
Pefferlaw	FM	VA3PTX	53.09000	52.09000		
Pembroke	FM	VE3KKO	144.50000	+		
	FM	VE3NRR	146.76000	-	100.0 Hz	
	FM	VE3NRR	448.02500	-	100.0 Hz	
Penetang	FM	VE3SGB	146.76000	-	156.7 Hz	
Perth	DSTAR	VA3RDD C	147.24000	+		
	DSTAR	VA3RDD B	444.80000	+		
	FM	VE3LCA	146.95500	-		
Perth, Christie Lake						
	FM	VA3TEL	145.23000	-		
Perth, Otty Lake	FM	VE3IEV	444.45000	+	100.0 Hz	
Petawawa	FM	VA3AUL	446.00000	-		
Peterborough	DMR/MARC	VE3BTE	443.13750	+	CC1	
	DSTAR	VE3SSF C	147.36000	+		
	FM	VE3ACD	146.62500	-		
	FM	VA3RZS	147.30000	+	162.2 Hz	
	FM	VE3SSF	443.37500	+	162.2 Hz	
Peterborough, Kawarthas						
	FM	VE3PBO	146.62500	-	162.2 Hz	
Picton	FM	VE3TJU	146.73000	-	118.8 Hz	
Pine Grove	FM	VE3VTG	147.51000	+		
Point Alexander, Laurentian Hi						
	FM	VA3RBW	146.79000	-	114.8 Hz	
Port Crewe	FM	AC8LR	146.44000	+		
Port Elgin	FM	VE3PER	146.82000	-		
Pottageville	FM	VA3ATL	444.77500	+		
Powassan, Powassan Hill						
	FM	VE3ERX	147.03000	+		
Queenswood Village						
	FM	VE3MPC	147.15000	+		
Raglan	DMR/MARC	VA3BMI	442.13750	+	CC0	
	FM	VE3OSH	53.27000	52.27000	DCS 010	
	FM	VE3OBI	444.52500	+	103.5 Hz	
Ramsayville	FM	VE3YRR	224.94000	-		
Red Lake	FM	VE3RLD	147.00000	+		
Renfrew, Mt St Patrick						
	FM	VE3STP	147.06000	+	114.8 Hz	
Richmond Hill	FM	VE3YRC	53.49000	52.49000		
	FM	VE3YRC	444.22500	+		
Rideau Ferry, Bell Tower						
	FM	VE3REX	442.20000	+		
Ridgeway	FM	VE3RAC	147.16500	+	107.2 Hz	
Sarnia	FM	VE3WHO	146.95500	-		
	FM	VE3RCA	224.96000	-		
	FM	VE3WHO	442.35000	+		
	FM	VA3SAR	444.55000	+	123.0 Hz	
Sarnia, BlueWater Health Hospi						
	FM	VE3SAR	145.37000	-	123.0 Hz	
Sault Ste. Marie	DSTAR	VA3SNR C	145.21000	-		
	FM	VE3SSM	146.94000	-	107.2 Hz	
	FM	VE3MHL	147.06000	+		
	FM	VE3SSM	442.65000	+		
Scarborough	DSTAR	VE3VXZ C	147.54000			
	DSTAR	VE3VXZ B	443.14000	+		
	DSTAR	VE3VXZ A	1282.50000	-		
	FM	VE3CTV	145.37000	-		
	FM	VE3RAL	224.78000	-		
	FM	VE3RTC	443.35000	+	131.8 Hz	
	FM	VA3GTU	443.75000	+	103.5 Hz	
Scotland	FM	VE3PPO	53.23000	52.23000		

440   ONTARIO

Location	Mode	Call sign	Output	Input	Access	Coordinator
Scotland	FM	VE3PPO	224.44000	-		
Shelburne	FM	VE3ZAP	146.68500	-	88.5 Hz	
Simcoe	FM	VE3TCO	147.50000		131.8 Hz	
Simcoe County	DMR/BM	VE3LSR	443.56250	+	CC1	
	DSTAR	VE3LSR C	145.19000	-		
	DSTAR	VE3LSR B	444.35000	+		
Simcoe, Norfolk County						
	DMR/BM	VE3SME	146.92500	-	CC1	
Singhampton	FM	VA3WIK	444.90000	+	156.7 Hz	
Sioux Lookout	FM	VA3SLT	147.31500	+	127.3 Hz	
	FUSION	VA3SLT	147.19500	+		
Sioux Lookout, Sioux Mountain						
	FM	VE3YXL	146.85000	-		
Sioux Narrows	FM	VE3RSN	145.17000	-		
	FM	VE3JJA	146.58000		123.0 Hz	
	FUSION	VE3JJA	147.33000	+		
Smiths Falls	FM	VA3WDP	444.75000	+	136.5 Hz	
	FUSION	VE3UIL	147.21000	+		
Smooth Rock Falls						
	FM	VA3UHY	147.36000	+		
South Frontenac	FM	VE3FRG	146.80500	-	203.5 Hz	
	FM	VE3FRG	445.13750	-	203.5 Hz	
South River	FM	VA3URU	146.97000	-	100.0 Hz	
South Western Ontario						
	FM	VE3XXL	145.23000	-	131.8 Hz	
Southampton	FM	VA3ITG	445.20000	-	103.5 Hz	
St Catharines	DSTAR	VE3NUU B	443.83750	+		
St Thomas	FM	VE3STR	147.33000	+	114.8 Hz	
	FM	VE3STR	224.78000	-	114.8 Hz	
	FM	VE3STR	443.82500	+		
St. Marys	FM	VA3SMX	444.37500	+	114.8 Hz	
Stewartville	DSTAR	VE3BFH C	145.64000			
	DSTAR	VE3BFH B	442.00000	+		
	DSTAR	VE3BFH B	445.83000	-		
	DSTAR	VE3BFH A	1282.50000	-		
Stittsville	DSTAR	VA3HOA C	145.51000	-		
	DSTAR	VA3HOA B	443.60000	+		
	DSTAR	VA3HOA B	445.84000	+		
	DSTAR	VA3HOA A	1286.00000	-		
Stoney Creek	FM	VE3OBP	147.34500	+	131.8 Hz	
Stratford	FM	VE3RFC	145.15000	-	114.8 Hz	
	FM	VE3FCG	444.97500	+	114.8 Hz	
Stratford, Ont.CA.	DSTAR	VE3FCD C	145.65000			
Sudbury	DSTAR	VA3SRG C	147.09000	+		
	FM	VE3YGR	146.92500	-	156.7 Hz	
	FM	VE3SRG	147.06000	+	100.0 Hz	
	FM	VE3RVE	147.21000	+	100.0 Hz	
	FM	VE3RKN	147.39000	+		
	FM	VE3YGR	444.20000	+	100.0 Hz	
Tarts / Era	DSTAR	VE3YYZ AD	1250.00000			
Tavistock	DSTAR	VE3DSL B	441.00000	+		
Thornhill	DSTAR	VE3EBX B	444.00000	+		
Thornill	DSTAR	VE3QSB C	144.00000			
	DSTAR	VE3QSB B	442.00000	+		
	DSTAR	VE3QSB A	1282.50000	-		
Thorold	FM	VE3RAF	145.19000	-	107.2 Hz	
	FM	VE3NRS	147.24000	+	107.2 Hz	
	FM	VA3RFM	443.72500	+	107.2 Hz	
Thunder Bay	FM	VA3LU	145.45000	-	123.0 Hz	
	FM	VA3LU	146.82000	-	107.2 Hz	
	FM	VE3WNJ	146.94000	-	100.0 Hz	
	FM	VE3FW	147.06000	+		

ONTARIO 441

Location	Mode	Call sign	Output	Input	Access	Coordinator
Tillsonburg	DMR/BM	VE3WHR	443.92500	+	CC1	
	DMR/BM	VE3PPO	444.57500	+	CC1	
Timmins	DSTAR	VE3YTS C	145.67000	-		
	DSTAR	VE3TIR C	147.21000	+		
	DSTAR	VE3KKA B	443.92500	+		
	DSTAR	VE3YTS B	444.40000	+		
	FM	VE3OPO	146.61000	-	110.9 Hz	
Timmins On. Canada						
	FM	VE3AA	147.06000	+	156.7 Hz	
Toledo	FM	VE3HTN	146.86500	-		
Toronto	DMR/MARC	VA3XPR	442.33750	+	CC1	
	DSTAR	VA3MCU C	144.91000			
	DSTAR	VE3YYZ C	144.93000	+		
	DSTAR	VE3YYZ C	144.94000	+		
	DSTAR	VA3SLU B	442.41250	+		
	DSTAR	VE3YYZ B	442.70000	+		
	DSTAR	VA3MCU B	446.65000	+		
	DSTAR	VE3YYZ A	1250.00000			
	DSTAR	VE3YYZ A	1287.50000	-		
	FM	VA3GTU	53.35000	52.35000	103.5 Hz	
	FM	VA3ECT	53.39000	52.39000		
	FM	VE3WOO	145.11000	-	82.5 Hz	
	FM	VE3OBN	145.23000	-	103.5 Hz	
	FM	VE3YYZ	147.18000	+		
	FM	VA3WHQ	224.30000	-	103.5 Hz	
	FM	VE3KRC	224.40000	-	103.5 Hz	
	FM	VA3GTU	442.37500	+	103.5 Hz	
	FM	VE3URU	442.47500	+	103.5 Hz	
	FM	VE3CAY	442.60000	+	103.5 Hz	
	FM	VA3GTU	442.65000	+	100.0 Hz	
	FM	VE3WOO	442.75000	+	103.5 Hz	
	FM	VA3GTU	442.97500	+	103.5 Hz	
	FM	VA3SCR	443.02500	+	103.5 Hz	
	FM	VE3YYZ	443.05000	+	103.5 Hz	
	FM	VE3SKI	443.10000	+		
	FM	VE3RAK	443.12500	+	103.5 Hz	
	FM	VE3VOP	443.32500	+	103.5 Hz	
	FM	VE3WOO	443.90000	+	127.3 Hz	
	FM	VA3BMI	444.45000	+	103.5 Hz	
	FM	VE3URU	444.47500	+	103.5 Hz	
	FM	VE3UKW	444.85000	+	136.5 Hz	
	FM	VA3DHJ	444.92500	+	100.0 Hz	
	FM	VA3IHX	445.95000	-		
	FM	VE3YYZ	1250.00000			
	FM	VA3GTU	1284.00000	-	103.5 Hz	
	FUSION	VE3SKY	146.98500	-		
Toronto, 1st Canadian Pl						
	FM	VE3NIB	443.50000	+	103.5 Hz	
Toronto, Canada	DSTAR	VA3SLU C	145.53500			
Toronto, CN Tower						
	FM	VE3TWR	145.41000	-	103.5 Hz	
Toronto, First Canadian Pl						
	FM	VE3WOO	443.90000	+	127.3 Hz	
Toronto, St. James Town						
	FM	VA3XPR	441.95000	+		
Tottenham	FM	VE3VGA	1285.00000	-		
Tweed	FM	VE3RNU	145.37000	-		
Unionville	FM	VE3OC	444.10000	+	103.5 Hz	
Upsala	FM	VE3UPP	145.47000	-		
Uxbridge	DSTAR	VE3RPT C	145.25000	-		
	DSTAR	VE3RPT B	443.22500	+		
	FM	VE3TFM	29.62000	-	103.5 Hz	

442   ONTARIO

Location	Mode	Call sign	Output	Input	Access	Coordinator
Uxbridge	FM	VE3PIC	146.67000	-	67.0 Hz	
	FM	VE3RPT	224.86000	-	103.5 Hz	
Uxbridge, Skyloft	FM	VE3SIX	53.03000	52.03000	103.5 Hz	
	FM	VE3RPT	147.06000	+	103.5 Hz	
	FM	VE3RPT	442.10000	+	103.5 Hz	
	FM	VE3RPT	1286.00000	-	103.5 Hz	
Val Albert	FM	VA3NKP	146.93000	-		
Vinemount	FM	VE3VSC	444.65000	+	131.8 Hz	
	FM	VE3VSC	1283.50000	-	131.8 Hz	
Walker Lk Huntsville						
	FM	VE3KR	446.55000	-	103.5 Hz	
Wallaceburg	FM	VE3OEN	446.10000	-		
Wasaga Beach	DSTAR	VE3UHF B	444.82500	+		
Waterloo	FM	VE3WFM	147.09000	+	131.8 Hz	
Weatheradio-Algonquin Park						
	FM	VEF956	162.40000			
Weatheradio-Atikokan						
	FM	VFI331	162.40000			
Weatheradio-Barry S Bay						
	FM	VFK722	162.52500			
Weatheradio-Beardmore						
	FM	XLJ892	162.47500			
Weatheradio-Belleville						
	FM	VFK720	162.42500			
Weatheradio-Britt	FM	VFJ213	162.47500			
Weatheradio-Brockville						
	FM	VFK721	162.42500			
Weatheradio-Collingwood						
	FM	XMJ316	162.47500			
Weatheradio-Fort Frances						
	FM	VDB224	162.40000			
Weatheradio-Goderich						
	FM	XLT839	162.40000			
Weatheradio-Greater Sudbury						
	FM	XLJ898	162.40000			
Weatheradio-Kawartha Lakes						
	FM	VAW217	162.40000			
Weatheradio-Kenora						
	FM	XLJ890	162.47500			
Weatheradio-Kingston						
	FM	XJV363	162.40000			
Weatheradio-Kitchener						
	FM	XMJ330	162.55000			
Weatheradio-Lavant						
	FM	VBE716	162.55000			
Weatheradio-Little Current						
	FM	XMJ375	162.47500			
Weatheradio-London						
	FM	XLN470	162.47500			
Weatheradio-Marathon						
	FM	VAT341	162.55000			
Weatheradio-Montreal River						
	FM	VAT404	162.47500			
Weatheradio-Moose Creek						
	FM	VBE718	162.45000			
Weatheradio-Mount Forest						
	FM	XLN600	162.40000			
Weatheradio-Nipigon						
	FM	XLJ891	162.55000			
Weatheradio-Normandale						
	FM	VFI621	162.45000			

ONTARIO 443

Location	Mode	Call sign	Output	Input	Access	Coordinator
Weatheradio-North Bay						
	FM	XLJ893	162.47500			
Weatheradio-Orillia						
	FM	VBV562	162.40000			
Weatheradio-Ottawa-Gatineau						
	FM	VBE719	162.55000			
Weatheradio-Paisley						
	FM	XMJ320	162.55000			
Weatheradio-Pembroke						
	FM	VAV559	162.47500			
Weatheradio-Peterborough						
	FM	VEU671	162.55000			
Weatheradio-Ramore						
	FM	VDB885	162.40000			
Weatheradio-Renfrew						
	FM	VEA549	162.42500			
Weatheradio-Rosseau						
	FM	VBT629	162.55000			
Weatheradio-Sarnia-Oil Springs						
	FM	XJV492	162.40000			
Weatheradio-Sault Ste Marie						
	FM	XMJ373	162.40000			
Weatheradio-Shelburne						
	FM	VXB212	162.52500			
Weatheradio-St Catharines						
	FM	VAD320	162.47500			
Weatheradio-Thunder Bay						
	FM	XMJ374	162.47500			
Weatheradio-Timmins						
	FM	VDB886	162.47500			
Weatheradio-Toronto						
	FM	XMJ225	162.40000			
Weatheradio-Windsor						
	FM	VAZ533	162.47500			
Whitby	DMR/BM	VE3WOM	443.47500	+	CC3	
	FM	VA3SUP	146.97000	-	156.7 Hz	
	FM	VA3UYP	224.66000	-	103.5 Hz	
	FM	VE3WOQ	444.37500	+	103.5 Hz	
Whitechurch	DMR/MARC	VE3WWD	443.07500	+	CC1	
	DSTAR	VE3WWD C	147.19500	+		
Wilberforce	FM	VE3BDJ	147.24000	+		
Winchester	FM	VA3NDC	146.97000	-	100.0 Hz	
Windsor	DMR/MARC	VE3RRR	444.30000	+	CC7	
	DMR/MARC	VE3UUU	444.40000	+	CC7	
	DSTAR	VA3WDG C	145.61000			
	DSTAR	VE3ZIN C	147.42000			
	DSTAR	VA3WDG B	445.82500	-		
	DSTAR	VA3WDG A	1282.00000	-		
	FM	VE3RRR	53.05000	52.05000	118.8 Hz	
	FM	VE3SXC	145.39000	-		
	FM	VE3EOW	145.41000	-	118.8 Hz	
	FM	VE3WHT	146.88000	-	118.8 Hz	
	FM	VE3WIN	147.00000	+	118.8 Hz	
	FM	VE3III	147.06000	+	118.8 Hz	
	FM	VA3BBB	147.39000	+	118.8 Hz	
	FM	VE3OOO	224.66000	-		
	FM	VA3ARK	444.11250	+	118.8 Hz	
	FM	VE3III	444.50000	+	118.8 Hz	
	FM	VE3WIN	444.60000	+	118.8 Hz	
Windsor Downtown						
	DMR/MARC	VE3WIN	442.06250	+	CC7	
Woodstock	FM	VA3OHR	442.87500	+	131.8 Hz	

444 PRINCE EDWARD ISLAND

Location	Mode	Call sign	Output	Input	Access	Coordinator
PRINCE EDWARD ISLAND						
Bucktown	DSTAR	W4IHS A	1297.80000			
Cavendish	FM	VY2PEI	145.15000	-		
Charlottetown	DMR/MARC	VE1CRA	443.30000	+	CC1	
	FM	VE1CRA	146.67000	-		
	FM	VY2CS	147.39000	+		
	FM	VE1CRA	444.40000	+		
	FM	VY2UHF	448.35000	-		
Glen Valley	FM	VY2SIX	53.59000	52.59000		
Glenfinnan	FM	VY2WU	147.57000	+		
Hazel Grove, Fredricton PEI						
	FM	VE1HI	146.94000	-	77.0 Hz	
O'Leary, Aliant Cellular Tower						
	FM	VY2CFB	147.12000	+		
Summerside, Water Tank						
	FM	VE1CFR	146.85000	-		
Weatheradio-Charlottetown						
	FM	XLM647	162.40000			
Weatheradio-O Leary						
	FM	XLK645	162.47500			
Weatheradio-Souris						
	FM	XLK644	162.40000			
QUEBEC						
Acton Vale	FM	VE2RBY	443.95000	+	110.9 Hz	
Albanel	FM	VA2TFL	147.37500	147.98000		
Alma	DSTAR	VE2RVI B	442.65000	+		
	FM	VE2RYK	53.25000	52.25000		
	FM	VE2RCA	146.67000	-	131.8 Hz	
	FM	VE2CVT	147.00000	+		
	FM	VA2RIT	147.04500	147.65000		
	FM	VE2RVX	147.27000	+		
	FM	VE2RPJ	147.28500	147.89000		
	FM	VE2LPO	147.36000	+		
	FM	VE2RIU	449.42500	444.43000		
	FM	VA2RIU	449.62500	444.63000		
Amos	FUSION	VE2MBT	147.15000	+		
Anjou	FM	VA2GGR	440.44000	+		
Arthabaska	FM	VE2RBF	147.14000	+		
Baie-Comeau	FM	VA2RRB	145.23000	-		
	FM	VA2RSP	146.82000	-		
	FM	VE2RMH	146.97000	-		
	FM	VA2LMH	147.04500	147.65000		
	FM	VA2RGV	147.16500	147.77000	151.4 Hz	
	FM	VE2RBC	147.30000	+		
	FM	VE2RDE	147.39000	+		
	FM	VE2RBG	442.35000	+	131.8 Hz	
	FM	VE2RUU	442.62500	447.63000		
	FM	VE2RMH	443.85000	+		
	FM	VA2RLP	444.00000	+	123.0 Hz	
	FM	VE2RBC	447.62500	442.63000		
	FM	VE2RBD	449.60000	-		
Baie-Johan-Beetz	FM	VE2RJI	146.94000	-		
Baie-Trinite	FM	VA2RBT	145.47000	-		
Beaconsfield	FM	VE2RNC	224.60000	-	100.0 Hz	
	FM	VE2RNC	224.80000	-	100.0 Hz	
Beauport	FM	VE2RIG	146.88000	-		
Beloeil	FM	VE2RGB	147.16500	147.77000		
Blainville	DSTAR	VE2YUU C	144.00000			
	DSTAR	VE2YUU B	442.00000	+		
	DSTAR	VE2YUU A	1282.50000	-		

QUEBEC 445

Location	Mode	Call sign	Output	Input	Access	Coordinator
Blainville	FM	VE2RNO	53.31000	52.31000		
	FM	VE2THE	146.82000	-	136.5 Hz	
	FM	VE2RVV	449.72500	444.73000	103.5 Hz	
	FUSION	VE2RMP	146.76000	-		
	FUSION	VE2RZY	443.55000	+		
Blainville Nord	FM	VE2RNO	146.86000	146.22000	103.5 Hz	
Bois-Franc	FM	VA2REH	145.17000	-	100.0 Hz	
Boucherville	FM	VE2MRQ	53.13000	52.31000	103.5 Hz	
	FM	VE2MRQ	145.25000	-	103.5 Hz	
	FM	VE2MRQ	449.82500	444.83000	103.5 Hz	
Cantley	FM	VA2CMB	444.70000	+	123.0 Hz	
Canton-de-Melbourne						
	DMR/MARC	VE2RHK	443.60000	+	CC1	
Cap-a-L'aigle	FM	VA2RKT	145.29000	-		
Cap-de-la-Madeleine						
	FM	VE2RBN	145.74500		136.5 Hz	
Carleton	FM	VE2RXT	147.06000	+		
Causapscal	FM	VE2RTF	145.13000	-		
Chambord	FM	VE2RVP	146.64000	-		
Chandler	FM	VE2CGR	146.85000	-		
Charlesbourg	FM	VA2UX	147.18000	+	100.0 Hz	
Charny	FM	VE2RDB	444.60000	+		
Chelsea	FM	VE2KPG	147.36000	+	203.5 Hz	
Chibougamau-Chapais						
	FM	VA2RRC	147.39000	+		
Chicoutimi	FM	VE2RHS	145.23000	-	127.3 Hz	
	FM	VE2RMI	145.43000	-	85.4 Hz	
	FM	VE2RCI	147.07500	+	127.3 Hz	
	FM	VE2RCC	147.12000	+	127.3 Hz	
	FM	VE2RPA	147.30000	+		
	FM	VE2RKA	449.02500	444.03000		
	FM	VE2RDH	449.70000	-	100.0 Hz	
Chisasibi	DMR/BM	VA2RJB	448.62500	-	CC7	
Chute-Des-Passes						
	FM	VE2RFN	147.27000	+		
Coaticook	FM	VE2RDM	147.36000	+	118.8 Hz	
	FM	VE2RJV	449.27500	444.28000	118.8 Hz	
	FUSION	VE2RDM	147.36000	+		
Contrecoeur	FM	VE2CKC	145.35000	-	141.3 Hz	
Contrecouer	FM	VE2CKC	443.65000	+	141.3 Hz	
Coupe Du Ciel	FM	VE2RKJ	147.03000	+	127.3 Hz	
Covey Hill	DMR/MARC	VE2REX	448.52500	-	CC1	
	DSTAR	VE2REX C	145.59000	-		
	DSTAR	VA2REX B	448.32500	-		
	DSTAR	VA2REX A	1266.30000	1246.30000		
	DSTAR	VA2REX A	1283.50000	-		
	FM	VE2REX	146.68500	-	100.0 Hz	
	FM	VA2CYH	444.82500	+		
	FM	VA2REX	448.52500	-	100.0 Hz	
Cowansville	FM	VE2RCZ	447.67500	442.68000	118.8 Hz	
Crabtree	DMR/MARC	VA2RIU	442.55000	+	CC2	
	DSTAR	VA2RIU C	145.00000			
	DSTAR	VA2RIU B	446.30000	-		
	DSTAR	VA2RIU A	1283.50000	-		
Davidson Corner	FM	VA2LOJ	446.12500	-		
Delson	FM	VE2LHF	442.15000	+		
Dolbeau	FM	VE2RCD	146.70000	-		
Donnacona	FM	VE2RBJ	145.33000	-	100.0 Hz	
	FM	VE2LKL	147.22500	147.83000		
Drummondville	FM	VE2ROC	53.27000	52.27000		
	FM	VE2RDL	146.62500	146.03000	110.9 Hz	
	FM	VA2RCQ	146.83500	146.24000	110.9 Hz	

446 QUEBEC

Location	Mode	Call sign	Output	Input	Access	Coordinator
Drummondville	FM	VE2RDV	147.09000	+	110.9 Hz	
	FM	VE2RBU	442.95000	+	110.9 Hz	
	FM	VE2RDL	444.15000	449.10000	110.9 Hz	
Drummondvlle	FM	VE2RBZ	444.05000	+	110.9 Hz	
Dstar-rptr(B)	DSTAR	VE3AZX B	445.85000	-		
Eastmain	FM	VE2LRE	147.53500		85.4 Hz	
	FM	VE2LRE	147.55500		85.4 Hz	
Eastman, Mont-Orford						
	FM	VA2CAV	448.37500	-	123.0 Hz	
Faribault	FM	VA2RSJ	147.09000	+		
Farnham	FM	VE2RDH	224.84000	-	103.5 Hz	
Fermont	FM	VE2RGA	146.82000	-		
Fire Lake	FM	VE2RGF	147.06000	+		
Fleurimont	FM	VA2LGX	146.86500	146.30000		
	FM	VE2RLX	442.92500	447.93000		
Forestville	FM	VE2RLI	146.70000	-	151.4 Hz	
	FM	VE2RFG	146.91000	-	151.4 Hz	
	FM	VE2REE	147.25500	+	100.0 Hz	
	FM	VE2REJ	147.28500	147.89000		
Frampton	FM	VA2III	145.31000	-		
Gagnon	FM	VE2RGH	146.69000	-		
GaspÃ©	FM	VE2OK	146.55000	+		
Gaspe	FM	VE2RLE	146.86500	146.30000		
	FM	VE2RLE	146.89500	146.30000		
Gatineau	DSTAR	VE2REG C	147.18000	+		
	DSTAR	VE2REG B	442.65000	+		
	FM	VE2REH	29.68000	-	173.8 Hz	
	FM	VE2REH	53.11000	52.11000	110.9 Hz	
	FM	VA2XAD	145.35000	-	162.2 Hz	
	FM	VE2REH	145.49000	-	110.9 Hz	
	FM	VE2REH	224.76000	-	110.9 Hz	
	FM	VA2XAD	444.90000	+	162.2 Hz	
	FM	VE2NTM	445.17500	-	110.9 Hz	
Gatineau, Hull Hospital						
	FM	VE2RAO	443.95000	+	118.8 Hz	
Granby	FM	VE2RVM	146.79000	-		
	FM	VE2RTA	147.18000	+	118.8 Hz	
	FM	VE2RGJ	448.62500	443.63000	118.8 Hz	
Granby, Mont Yamaska						
	FM	VA2RMY	442.35000	+	103.5 Hz	
Grand Fonds, Mont-Noir						
	FM	VE2CTT	147.00000	+		
Grand-Mere	DSTAR	VE2SKG C	144.00000			
	DSTAR	VE2SKG C	145.55000			
	DSTAR	VE2SKG B	442.00000	+		
	DSTAR	VE2SKG A	1282.50000	-		
	FM	VA2RTI	53.11000	52.11000	141.3 Hz	
	FM	VE2RLM	449.52500	444.53000		
	FM	VE2RGM	449.67500	444.68000	110.9 Hz	
	FM	VA2RTI	449.92500	444.93000		
	FM	VA2RDX	449.97500	444.98000	110.9 Hz	
Grand-Mere, Mauricie						
	FM	VE2RGM	146.92500	146.33000	110.9 Hz	
Grande-Anse	FM	VE2RLT	147.00000	+		
Grande-Rivi	FM	VE2RDI	145.17000	-		
Grenville	FM	VE2RCS	53.01000	52.01000		
	FM	VE2RCS	146.71500	-	123.0 Hz	
	FM	VE2RWC	146.80500	146.21000		
Grnade-Riviere	FM	VE2RBM	146.73000	-		
Grosses-Roches	FM	VA2RLJ	145.37000	-		
Havelock	DMR/MARC	VE2OCZ	449.95000	-	CC1	
	FUSION	VA2CLM	449.07500	-		

QUEBEC 447

Location	Mode	Call sign	Output	Input	Access	Coordinator
Havelock	FUSION	VA2SPB	449.72500	-		
Havre-Saint-Pierre						
	FM	VE2TIO	145.49000	-		
	FM	VE2RFD	146.97000	-		
Henrysburg	FM	VA2IPX	440.00000	+		
Herbertville	FM	VE2RCV	146.79000	-	127.3 Hz	
Hull	FM	VE2CRA	443.30000	+	100.0 Hz	
Iberville	FM	VE2RJE	449.75000	-	103.5 Hz	
Joliette	FM	VE2RHO	147.03000	+	103.5 Hz	
	FM	VE2RLJ	147.30000	+		
	FM	VA2ATV	439.25000	910.00000		
	FM	VE2RLJ	444.62500	449.63000	103.5 Hz	
	FM	VE2RIA	444.80000	+	103.5 Hz	
	FM	VE2RHO	449.12500	444.13000		
Jonquiere	FM	VE2REY	146.41500		100.0 Hz	
	FM	VE2VP	146.82000	-		
	FM	VE2DHC	147.06000	+		
	FM	VE2RVG	147.24000	+		
	FM	VE2RLG	147.39000	+	100.0 Hz	
	FM	VE2RNU	444.07500	+		
	FM	VE2RPA	449.00000	-		
	FM	VE2RFL	449.10000	-		
L'Aanse Saint-Jean						
	FM	VA2RUA	146.77500	-	85.4 Hz	
L'Aascension	FM	VA2RGP	147.51000			
L'anse-Saint-Jean	FM	VE2RME	449.90000	-		
L'Anse-St-Jean	FM	VE2RME	145.15000	-		
L'Avenir	FM	VA2RHP	147.25500	147.86000	123.0 Hz	
La Baie	DMR/BM	VE2RRX	146.76000	-	CC7	
	FM	VE2RCZ	145.27000	-	127.3 Hz	
	FM	VE2RCX	146.61000	-	85.4 Hz	
	FM	VE2RCE	146.73000	-	179.9 Hz	
	FM	VE2RCX	444.95000	+		
	FM	VE2RRZ	448.12500	-		
La Pocatiere	FM	VE2RDJ	146.62500	146.03000	151.4 Hz	
	FM	VE2RIP	448.97500	443.98000		
La Sarre	FM	VE2RSL	146.70000	-		
La Tuque	FM	VE2RTL	146.79000	-		
	FM	VA2RVD	146.83500	146.24000		
	FM	VE2RLF	146.94000	-		
Labrieville	FM	VE2ROA	147.10500	147.71000		
Lac Aux Sables	FM	VE2RSA	147.21000	+		
Lac Brassard	FM	VE2RIT	145.21000	-		
Lac Canot Mont-Valin						
	FM	VE2RGU	145.45000	-		
Lac Castor	FM	VA2RLC	145.39000	-		
	FM	VE2RUR	444.95000	+		
Lac Daran	FM	VE2RLD	145.29000	-		
	FM	VE2RLD	442.10000	447.00000		
Lac Des Commissaires						
	FM	VE2RHC	146.97000	-		
Lac Edouard	FM	VE2RCL	147.22500	147.83000		
Lac Etchemin	FM	VE2RKM	147.24000	+	100.0 Hz	
Lac Ha! Ha!	FM	VE2RCK	147.33000	+	127.3 Hz	
Lac Larouche	FM	VE2RPV	449.62500	444.63000		
Lac Larouche, Parc De La Veren						
	FM	VE2RPV	145.49000	-		
Lac P	FM	VA2JAC	146.85000	-	127.3 Hz	
Lac Pail	FM	VA2ZGB	147.19500	+	103.5 Hz	
Lac Paul	FM	VA2RHS	147.69000	-	127.3 Hz	
Lac St-Arnault	FM	VA2ZGB	147.24000	+	103.5 Hz	
Lac-a-la-Tortue	FM	VE2RBR	146.86500	146.30000		

448   QUEBEC

Location	Mode	Call sign	Output	Input	Access	Coordinator
Lac-a-la-Tortue	FM	VE2RBR	443.20000	+		
Lac-Castor	FM	VA2RUR	146.68500	-	85.4 Hz	
Lac-des-Commissaires						
	FM	VE2RHC	146.73000	-		
Lac-Echo	DSTAR	VA2RMP C	145.51000	-		
	DSTAR	VA2RMP B	443.55000	+		
	DSTAR	VA2RMP B	448.62500	+		
	DSTAR	VA2RMP A	1286.00000	-		
Lac-Jacques-Cartier						
	FM	VE2RPL	145.49000	-		
Lac-Pratte	FM	VE2REY	444.40000	445.00000		
Lac-Sainte-Marie	FM	VE2REH	53.21000	52.21000	136.5 Hz	
	FM	VE2REH	146.61000	-	110.9 Hz	
Lachute	FM	VE2RCS	224.58000	-		
	FM	VE2RCS	443.85000	+	123.0 Hz	
Laterriere	FM	VE2RGT	146.76000	-	85.4 Hz	
Lauzon	FM	VE2RYC	145.11000	-	100.0 Hz	
Laval	FM	VA2RTO	146.76000	146.19000		
	FM	VE2REZ	442.90000	+	103.5 Hz	
	FM	VE2JKA	448.17500	443.18000	107.2 Hz	
	FM	VE2CSA	449.97500	-	74.4 Hz	
Le Bic	FM	VE2BQA	147.19500	147.80000		
Le Gardeur	FM	VE2CZX	442.30000	+	103.5 Hz	
Legardeur	FM	VE2CZX	442.00000	+	103.0 Hz	
Les Coteaux	FM	VA2BDL	443.35000	+	103.5 Hz	
Les Escoumins	FM	VE2REB	146.67000	-		
Les M	FM	VE2RNM	147.27000	+		
Levis-Lauzon	FM	VE2RCT	147.15000	+	100.0 Hz	
Longueuil	DSTAR	VE2QE C	147.54000			
	DSTAR	VE2QE B	446.25000	-		
	DSTAR	VE2QE A	1241.00000	1261.00000		
	FM	VE2RSM	145.39000	-	103.5 Hz	
	FM	VE2RVC	146.73000	-	100.0 Hz	
	FM	VE2RXN	146.83500	146.24000	100.0 Hz	
	FM	VE2RSM	445.22500	-		
Longueuil, Hospital Pierre-Bou						
	FM	VE2HPB	442.40000	+	103.5 Hz	
Magog	FM	VE2RZZ	444.25000	+	118.8 Hz	
Maliotenam	FM	VA2RUM	147.18000	+		
Mascouche	FM	VE2RHL	147.34500	147.95000	103.5 Hz	
	FM	VE2RHL	443.80000	+	141.3 Hz	
Maskinonge	FM	VA2MLP	147.09000	+		
Matagami	FM	VE2RBO	146.91000	-		
Matane	FM	VA2RAM	146.88000	-		
	FM	VE2RAS	147.12000	+		
Mercier	FM	VE2RTF	442.10000	447.00000		
	FM	VE2RTS	444.50000	+		
Mille-Isles	FM	VA2RJZ	447.72500	-	103.5 Hz	
Mont Rougemont	FM	VE2RAW	144.39000			
	FM	VE2RAW	444.32500	-	103.5 Hz	
	FM	VA2RDG	444.90000	+	107.2 Hz	
Mont Saint-Bruno	FM	VE2RST	449.10000	-		
Mont St-Gregoire	DSTAR	VA2RKA C	145.53000	-		
	DSTAR	VA2RKA B	446.15000	-		
	DSTAR	VA2RKA A	1266.20000	1246.20000		
	FM	VE2RKL	147.39000	+	100.0 Hz	
Mont Ste-Marie	FM	VA2REH	146.61000	-	110.9 Hz	
Mont Sutton	FM	VE2RTC	442.20000	447.50000		
Mont Tremblant	FM	VE2RNF	145.43000	-	136.5 Hz	
Mont Victor Trembley						
	FM	VE2RTV	147.34500	147.95000		
Mont Yamaska	FM	VA2WDH	449.85000	-	100.0 Hz	

QUEBEC 449

Location	Mode	Call sign	Output	Input	Access	Coordinator
Mont-Apica	FM	VE2RHX	145.35000	-		
	FM	VE2RCP	146.91000	-	127.3 Hz	
Mont-Ate-Anne	FM	VE2RAA	447.37500	442.38000	110.9 Hz	
Mont-Belair	FM	VE2OM	146.00000	+	100.0 Hz	
Mont-Carmel	FM	VE2RTR	146.67000	-		
	FM	VE2RIR	447.07500	442.08000	110.9 Hz	
	FM	VA2RES	447.67500	-	100.0 Hz	
Mont-Cosmos	FM	VA2III	449.87500	444.88000	156.7 Hz	
Mont-Fournier	FM	VE2RMQ	147.27000	+		
Mont-gladys	FM	VE2RMG	442.40000	+		
Mont-Laurier	FM	VE2REH	147.10500	+	131.8 Hz	
Mont-Laurier, Mont Sir Wilfrid						
	FM	VE2RMC	146.97000	-		
Mont-Laurier, Montagne Du Diab						
	FM	VE2RMC	444.62500	449.63000		
Mont-ONeil	DSTAR	VE2RMF A	1283.00000	1263.00000		
	FM	VE2RMF	147.39000	+		
Mont-Orford	DMR/MARC	VA2CAV	443.25000	+	CC11	
	DSTAR	VE2RTO B	442.00000	+		
	DSTAR	VE2RTO A	1266.10000	1246.10000		
	FM	VE2RTO	145.27000	-	103.5 Hz	
Mont-Saint-Grego	DSTAR	VE2RVR B	444.20000	+		
Mont-Saint-Gregoire						
	FM	VE2RKL	145.51000			
	FM	VE2RKL	147.39000	+	100.0 Hz	
	FM	VE2RKL	444.00000	+	DCS 103	
Mont-Saint-Hilaire	FM	VE2RMR	147.19500	+	103.5 Hz	
Mont-Sainte-Anne	FM	VE2RAA	146.82000	-	100.0 Hz	
Mont-Sainte-Marguerite						
	FM	VE2LRE	146.85000	-		
Mont-Tanguay - Dixville						
	DMR/MARC	VE2RWE	448.12500	-	CC1	
Mont-Valin	FM	VE2RES	146.88000	-		
Mont-Wright	FM	VE2RGW	145.13000	-	77.0 Hz	
Mont-Yamaska - VHF						
	DMR/MARC	VE2RAU	146.77500	-	CC1	
Montmagny	FM	VE2RAB	146.97000	-		
Montreal	DMR/MARC	VE2RCM	447.62500	-	CC1	
	DSTAR	VE2RHH C	147.54000			
	DSTAR	VE2RHH B	443.81250	+		
	DSTAR	VE2RHH A	1247.00000	+		
	FM	VE2PSL	146.67000	-	103.5 Hz	
	FM	VE2BG	147.06000	+		
	FM	VE2RED	147.27000	+	103.5 Hz	
	FM	VA2OZ	147.37500	147.98000	107.2 Hz	
	FM	VE2RXM	442.25000	+	141.3 Hz	
MontrÃ©a!	FM	VE2MRC	442.45000	+	103.5 Hz	
Montreal	FM	VE2RNO	442.60000	+	103.5 Hz	
	FM	VE2ETS	442.65000	+	103.5 Hz	
	FM	VE2JGA	442.80000	+		
	FM	VE2RVL	443.10000	+	141.3 Hz	
	FM	VA2KWG	443.20000	+	127.3 Hz	
	FM	VA2KWG	444.05000	+	127.3 Hz	
	FM	VE2RVH	444.25000	+	88.5 Hz	
	FM	VE2REM	444.40000	+		
	FM	VE2REH	444.60000	+	114.8 Hz	
	FM	VE2RJX	447.02500	442.03000	103.5 Hz	
	FM	VA2RTO	447.37500	442.38000	107.2 Hz	
	FM	VE2TLM	447.77500	-	141.3 Hz	
	FM	VA2RJX	447.97500	442.98000	103.5 Hz	
	FM	VA2US	448.17500	443.18000	107.2 Hz	
	FM	VA2CME	448.72500	443.73000	103.5 Hz	

450   QUEBEC

Location	Mode	Call sign	Output	Input	Access	Coordinator
Montreal	FM	VE2AIF	449.30000	-		
	FM	VE2RGN	449.42500	444.43000	103.5 Hz	
	FM	VE2WM	449.47500	444.48000		
	FM	VE2TLM	449.77500	444.78000	103.5 Hz	
	FM	VE2RJS	449.97500	444.98000	77.0 Hz	
Montreal, Ecole Polytechnique						
	FM	VE2RHH	224.90000	-	100.0 Hz	
Montreal, Mont-Royal						
	FM	VE2RWI	146.91000	-	88.5 Hz	
	FM	VE2RVL	147.07500	+	103.5 Hz	
	FM	VE2RWI	443.05000	+	141.3 Hz	
Montreal, Pineridge						
	FM	VE2RMP	146.76000	-	103.5 Hz	
Montreal-Nord	FM	VE2RPT	448.65000	-		
Montreal-PVM	DMR/MARC	VA2OZ	442.37500	+	CC1	
Mount Rigaud	FM	VE2RM	224.98000	-		
Mt Rougemont	DSTAR	VE2RIO AD	1297.47500			
Mt-Oorford	FM	VE2DCR	446.50000	-	71.9 Hz	
Mt-Shefford	FM	VE2RWQ	443.00000	+		
Mt-St-Gregoire	DSTAR	VA2RKA AD	1248.00000			
	DSTAR	VA2RKA A	1286.00000	-		
New Richmond	FM	VE2RPG	147.00000	+		
	FM	VE2RPG	147.18000	+		
Notre Dame De L'ile Perrot						
	FM	VE2USL	443.50000	+	97.4 Hz	
Notre-Dame-d-Mont-Carmel						
	FM	VA2LX	145.10000	-	110.9 Hz	
Old Chelsea, Camp Fortune						
	FM	VE2CRA	146.94000	-	100.0 Hz	
Orford	DMR/MARC	VA2TB	442.70000	+	CC13	
Parc Chibougamau						
	FM	VE2RTG	147.03000	+		
Parc De Chibougamau						
	FM	VA2RRH	145.11000	-		
Parc Des Laurentides						
	FM	VE2RPE	147.15000	+	127.3 Hz	
Parc Des Laurentides, Mont-Gla						
	FM	VE2RMG	147.09000	+		
Parent	FM	VE2RPC	145.19000	-		
	FM	VE2LVJ	444.00000	+		
Passes Dangereuses						
	FM	VA2ADW	145.25000	-		
Petite-Rivi	FM	VA2RAT	147.39000	+		
Pierrefonds	DMR/MARC	VE2JKA	448.37500	-	CC1	
	FM	VE2RKE	145.49000	-		
Pohénégamook						
	FM	VE2CKN	145.17000	+		
Pointe-Claire	DMR/MARC	VE2ROR	443.17500	+	CC1	
	FM	VE2RHI	448.65000	-		
Princeville	FM	VA2AD	146.58000	+		
Projet-Laplante	FM	VE2DPF	53.21000	52.21000		
	FM	VE2RBV	147.21000	+		
Quebec	DMR/BM	VA2MB	444.45000	+	CC1	
	DMR/BM	VE2RIX	448.62500	-	CC7	
	FM	VE2RHT	145.17000	-	100.0 Hz	
Québec	FM	VE2RDS	146.55000	-		
Quebec	FM	VE2RQR	146.61000	-	100.0 Hz	
	FM	VE2RUK	147.01500	147.62000		
	FM	VA2PRC	147.10500	+	100.0 Hz	
	FM	VE2RHD	444.10000	+		
	FM	VE2REA	444.50000	+	100.0 Hz	
	FUSION	VE2MSW	145.11000	-		

QUEBEC 451

Location	Mode	Call sign	Output	Input	Access	Coordinator
Quebec	FUSION	VE2RFU	147.12000	+		
Quebec Centre Ville, Hotel Des						
	FM	VE2RRS	145.35000	-	85.4 Hz	
Quebec City	DSTAR	VE2RQT C	144.95000	-		
	DSTAR	VE2RQT B	449.92500	-		
	FM	VE2RAX	146.79000	-	100.0 Hz	
	FM	VE2RUR	147.01500	147.62000	100.0 Hz	
	FM	VA2SHO	147.10500	147.71000	100.0 Hz	
	FM	VE2RCQ	147.30000	+	100.0 Hz	
	FM	VE2RXR	444.30000	+		
	FM	VE2RTB	444.50000	+	100.0 Hz	
	FM	VA2MD	445.10000	-	100.0 Hz	
Quebec City, CEGEP Limoilou						
	FM	VE2RQE	147.25500	+	100.0 Hz	
Quebec City, Mont-Belair						
	FM	VE2OM	146.94000	-		
	FM	VA2TEL	147.07500	147.68000	100.0 Hz	
	FM	VE2RAJ	147.43500	147.74000		
	FM	VE2RAX	444.20000	+		
	FM	VA2TEL	444.90000	+	100.0 Hz	
Quebec City, Parc Des Laurenti						
	FM	VE2RMG	442.37500	447.38000	110.9 Hz	
	FM	VE2RMG	442.50000	+		
Quebec, Mont St-Castin						
	FM	VE2UCD	146.65500	146.06000	100.0 Hz	
Rapide Blanc	FM	VE2RRB	146.61000	-		
Rigaud	FM	VA2OZ	145.21000	-	110.9 Hz	
	FM	VA2OZ	146.76000	146.19000	107.2 Hz	
	FM	VE2RM	442.65000	+	103.5 Hz	
	FM	VE2PCQ	443.45000	+	103.5 Hz	
Rigaud Mountain	DSTAR	VE2RM B	442.25000	+		
Rigaud, Mount Rigaud						
	FM	VE2RM	147.00000	+	103.5 Hz	
Rimouski	DSTAR	VE2RKI C	144.95000	+		
	DSTAR	VE2RKI C	145.31000	-		
	FM	VE2RNJ	145.17000	-	141.3 Hz	
	FM	VE2RKI	145.31000	-		
	FM	VE2LAM	146.58000	+		
	FM	VE2RWM	146.61000	-		
	FM	VE2CSL	146.94000	-		
	FM	VE2RIM	147.34000	+	141.3 Hz	
	FM	VE2ROE	147.36000	+		
	FM	VE2RWM	442.62500	447.63000	123.0 Hz	
	FUSION	VE2EBR	444.90000	+		
Rimouski, College De Rimouski						
	FM	VE2RXA	147.24000	+		
Rimouski, QC, Canada						
	DSTAR	VA2BQ C	147.53500			
	DSTAR	VA2BQ B	445.85000	-		
Ripon	FM	VE2REH	53.31000	52.31000	123.0 Hz	
	FM	VE2RBH	145.41000	-	123.0 Hz	
	FM	VE2REH	147.39000	+	146.2 Hz	
Riviere-au-Tonnerre						
	FM	VE2RET	147.03000	+		
Riviere-des-Prairies						
	FM	VE2RMK	146.82000	-	103.5 Hz	
	FM	VE2FXD	448.22500	443.23000	107.2 Hz	
Riviere-du-Loup	FM	VE2RAY	147.15000	+	107.2 Hz	
	FM	VE2RYE	444.40000	+	103.5 Hz	
Roberval	DMR/MARC	VE2RWJ	449.95000	-	CC1	
	FM	VA2RRE	145.49000	-		
	FM	VA2NA	145.53000	-	100.0 Hz	

452   QUEBEC

Location	Mode	Call sign	Output	Input	Access	Coordinator
Roberval	FM	VE2RRE	146.74500	146.15000	136.5 Hz	
	FM	VE2RSF	147.01500	147.62000	136.5 Hz	
Rosemere	FM	VE2RXZ	445.15000	-		
Rougemont	DSTAR	VE2RIO C	145.53000	-		
	DSTAR	VE2RIO B	449.92500	-		
	DSTAR	VE2RIO A	1266.40000	1246.40000		
	FM	VE2RXW	146.70000	-	103.5 Hz	
Rougemont, Mont Rougemont						
	FM	VE2RAW	145.31000	-	103.5 Hz	
Rouyn-Noranda	FM	VE2RNR	146.64000	-		
	FM	VE2RON	146.82000	-		
	FM	VE2RYN	147.09000	+		
Saguenay	FM	VE2ADW	147.09000	+	100.0 Hz	
	FM	VA2SYD	147.16500	147.77000	127.3 Hz	
	FM	VA2RUR	444.95000	+		
	FM	VE2XEN	446.12500	-		
	FM	VE2XEN	448.02500	-		
	FM	VE2XZM	448.22500	-		
	FM	VA2RU	448.52500	443.53000	85.4 Hz	
Sagunay, Monts-Valin						
	FM	VE2RJZ	147.21000	+	127.3 Hz	
Saint Francois D'Assise						
	FM	VA2RDP	146.62000	-		
Saint-ad	FM	VE2PCQ	444.98800	449.99000		
Saint-Adolphe D'Howard						
	FM	VE2RUN	146.89500	146.30000	141.3 Hz	
Saint-Adolphe-d'Howard						
	FM	VE2RYV	146.65500	-		
Saint-Aime	FM	VE2RJO	147.13500	147.74000		
Saint-Armand	FM	VE2RSN	443.80000	+	103.5 Hz	
Saint-Calixte	DSTAR	VA2RKB C	145.51000	-		
	FM	VE2RVK	53.07000	52.07000	141.3 Hz	
	FM	VE2PCQ	146.73000	-	103.5 Hz	
	FM	VE2RVK	146.86500	-	141.3 Hz	
	FM	VE2REM	147.01500	147.62000	103.5 Hz	
	FM	VA2RLD	442.60000	+	103.5 Hz	
	FM	VA2RLD	442.72500	447.73000	103.5 Hz	
	FM	VE2RVK	443.60000	+	141.3 Hz	
	FM	VA2RLD	444.00000	+	103.5 Hz	
	FM	VE2PCQ	449.98800	444.99000		
	FM	VE2RVK	1283.60000	1263.60000	141.3 Hz	
Saint-Calixte, North Of Montre						
	FM	VA2RLD	145.19000	-	141.3 Hz	
Saint-Calixte-Nord						
	FM	VE2PCQ	447.12500	-		
Saint-Charles-de-Bourget						
	FM	VE2RCR	146.94000	-	127.3 Hz	
	FM	VE2RCR	444.20000	+		
Saint-Constant	FM	VA2RSC	442.10000	447.00000		
	FM	VE2APO	442.20000	+		
Saint-Damien	FM	VE2RGC	145.29000	-	103.5 Hz	
Saint-Denis-sur-Richelieu						
	FM	VE2RSO	447.12500	442.13000		
Saint-Donat	FM	VA2RIA	147.00000	+	103.5 Hz	
	FM	VE2RRA	147.09000	+	103.5 Hz	
	FM	VE2RRA	444.60000	+	103.5 Hz	
	FM	VA2RIA	444.80000	+	103.5 Hz	
Saint-Donat De Rimouski						
	FM	VE2RAC	146.73000	-	123.0 Hz	
Saint-Donat-de-Montcalm						
	FM	VA2RSD	147.37500	147.98000	141.3 Hz	
Saint-Eleuthere	FM	VE2NY	442.37500	447.38000	110.9 Hz	

QUEBEC   453

Location	Mode	Call sign	Output	Input	Access	Coordinator
Saint-Eleuthere	FM	VE2NY	442.62500	447.63000	123.0 Hz	
Saint-Elzear De Beauce						
	FM	VE2RVD	146.76000	-	100.0 Hz	
Saint-Etienne-des-Gres						
	FM	VE2RZX	147.19000	+		
Saint-Fracois-De-Sales						
	FM	VE2RRR	145.47000	-		
Saint-Frederic	FM	VE2LFO	147.24000	+	136.5 Hz	
Saint-Georges	FM	VA2RME	146.89500	+		
	FM	VE2RSG	147.28500	147.89000		
Saint-Georges-de-Windsor						
	DMR/MARC	VA2RSG	443.55000	+	CC1	
Saint-Honore	FM	VA2RCH	145.33000	-		
	FM	VA2RCH	145.39000	-		
	FM	VE2RKT	147.18000	+		
	FM	VA2RMV	147.22500	147.83000		
	FM	VA2RCH	147.48000			
	FM	VA2RCR	442.60000	+	127.3 Hz	
Saint-Hubert	FM	VA2CSA	147.30000	+		
	FM	VA2ASC	449.02500	444.03000		
Saint-Jean De Matha						
	FM	VE2RMM	145.41000	-	103.5 Hz	
Saint-Jean-de-Matha, Montagne						
	FM	VE2RHR	447.82500	442.83000	103.5 Hz	
Saint-Jean-Port-Joli						
	FM	VA2RWW	147.31500	+	100.0 Hz	
Saint-Jean-sur-Richelieu						
	FM	VE2RVR	442.85000	+	141.3 Hz	
Saint-Jerome	FM	VE2RVS	146.85000	-		
Saint-Jogues	FM	VE2RIN	146.82000	-		
Saint-Joseph Du Lac						
	FM	VA2RSD	145.33000	-	141.3 Hz	
Saint-Joseph-de-Beauce						
	FM	VE2RSJ	146.98500	146.39000	100.0 Hz	
	FM	VE2RSJ	445.05000	-		
Saint-Joseph-de-Sorel						
	FM	VE2CBS	446.25200	144.77000	103.5 Hz	
Saint-Joseph-du-Lac						
	FM	VE2RST	53.05000	52.05000		
	FM	VE2REL	147.31500	+		
	FM	VE2RST	449.87500	-	103.5 Hz	
Saint-Jospeh-du-Lac						
	FM	VE2RST	29.68000	-		
Saint-Laurent, College Saint-M						
	FM	VE2RVQ	439.25000	910.00000		
	FM	VE2RMS	444.70000	+	103.5 Hz	
	FM	VE2RVQ	923.25000	439.25000		
Saint-Lin-Laurentides						
	FM	VE2RFO	147.09000	+	103.5 Hz	
	FM	VE2RFO	147.36000	+	103.5 Hz	
	FM	VE2RFO	439.25000	1255.00000		
	FM	VE2RFO	444.65000	+	103.5 Hz	
Saint-Marguerite	FM	VE2RIX	53.09000	52.09000		
Saint-Michel Des Saints, Mont						
	FM	VA2HMC	145.47000	-	141.3 Hz	
Saint-Michel-des-Saints						
	FM	VE2RLP	145.33000	-		
	FM	VE2ESN	443.85000	+	103.5 Hz	
Saint-Michel-des-Saints, Mont-						
	FM	VA2HMC	444.55000	+	103.5 Hz	
Saint-Nazaire	FM	VE2DCR	145.15000	-	136.5 Hz	
	FM	VA2RAU	145.19000	-		

454   QUEBEC

Location	Mode	Call sign	Output	Input	Access	Coordinator
Saint-Nazaire	FM	VE2DCR	145.29000	-	136.5 Hz	
	FM	VE2RUB	145.37000	-	85.4 Hz	
	FM	VE2DCR	145.49000	-	136.5 Hz	
Saint-Onesime	FM	VE2RAF	147.21000	+		
Saint-Pacome	FM	VE2RAK	146.70000	-		
Saint-Pascal	FM	VE2RGP	147.06000	+		
	FUSION	VE2MEL	449.57500	-		
Saint-Paul-d'Abbotsford, Mont						
	FM	VE2RMV	224.30000	-	103.5 Hz	
	FM	VE2RMV	443.30000	+	103.5 Hz	
SAINT-Philippe	FM	VA2RMS	145.37000	-	100.0 Hz	
Saint-Raymond	FM	VE2RCJ	147.28500	147.89000	100.0 Hz	
Saint-Simon-les-Mines						
	FM	VE2RSG	146.64000	-		
	FM	VE2RSG	449.97500	444.98000	100.0 Hz	
Saint-Sophie D'Halifax						
	FM	VE2CTM	146.73000	-		
Saint-Therese	FM	VE2RWW	448.42500	443.43000	103.5 Hz	
Saint-Tite	FM	VE2RJA	147.04500	147.65000		
Saint-Tite-des-Caps						
	FM	VE2RTI	145.47000	-		
	FM	VA2RSL	147.04500	147.65000	100.0 Hz	
	FM	VE2RHM	147.34500	147.95000		
	FM	VA2RSL	147.37000	150.37000		
	FM	VE2RSB	447.20000	-		
SAINT-Ubalde	FM	VE2RBT	145.39000	-	100.0 Hz	
	FM	VE2RZT	442.30000	+		
Saint-Urbain	FM	VE2RAT	146.91000	-		
Sainte-Apolline-de-Patton						
	FM	VE2RIX	147.24000	+	127.3 Hz	
Sainte-Foy	FM	VE2SRC	147.12000	+		
	FM	VE2RCH	442.70000	+		
	FM	VE2RSX	444.70000	+		
	FM	VA2ROY	444.80000	+		
Sainte-Foy-Sillery-Cap-Rouge						
	FM	VE2RQR	146.61000	-	100.0 Hz	
Sainte-Hedwidge-de-Roberval						
	FM	VA2RRA	146.74500	+		
Sainte-Marie	FM	VA2ABA	147.49500	+		
Sainte-Sophie D'Halifax						
	FM	VE2RNB	448.92500	443.93000		
Sainte-VIictoire De Sorel						
	FM	VE2RBS	446.50000	144.77000	103.5 Hz	
Salaberry-de-Valleyfield						
	FM	VE2RVF	449.67500	-		
Sarsfield	FM	VA2UHF	444.25000	+	141.3 Hz	
Scotstown, Mont Megantic						
	FM	VE2RJC	147.10500	147.71000	118.8 Hz	
Senneterre	FM	VE2RSZ	145.11000	-		
Sept-Iles	FM	VE2RDO	145.19000	-	88.5 Hz	
	FM	VE2RNN	146.64000	-	88.5 Hz	
	FM	VE2RRU	146.79000	-	88.5 Hz	
	FM	VA2RCJ	146.88000	-		
	FM	VE2RSI	146.94000	-		
Shawinigan	FM	VE2REY	53.39000	52.39000	103.5 Hz	
Sherbrooke	DMR/BM	VE2RVO	442.50000	+	CC7	
	DSTAR	VE2RQF C	147.06000	+		
	DSTAR	VE2RQF A	1266.50000	1246.50000		
	FM	VE2RCO	29.63000	-		
	FM	VE2PAK	144.99000			
	FM	VE2RGX	145.23000	-	123.0 Hz	
	FM	VE2PAK	145.51000			

QUEBEC 455

Location	Mode	Call sign	Output	Input	Access	Coordinator
Sherbrooke	FM	VE2PAK	145.61000			
	FM	VE2RSH	146.97000	-	118.8 Hz	
	FM	VE2RQM	444.75000	+		
Sherbrooke - Belvedere						
	DMR/MARC	VE2RHE	448.37500	-	CC7	
Sillery	FM	VE2REA	146.68500	146.09000		
Sorel	DSTAR	VE2CST C	146.98500	-		
	DSTAR	VE2FCT B	446.25000	-		
	DSTAR	VE2CST A	1241.00000	1261.00000		
	FM	VE2FCT	146.61000	-	103.5 Hz	
	FM	VE2CBS	446.25000	-	103.5 Hz	
Sorel-Tracy	DSTAR	VE2CST D	1299.00000			
	FUSION	VE2RBS	145.37000	-		
St-Ambroise	FM	VE2HOM	144.99000			
ST-Calixte	FM	VA2RLD	145.43000	-	103.5 Hz	
St-Felix-de-Valo	DSTAR	VA2RVB B	446.27500	-		
St-Felix-de-Valois	DMR/BM	VA2RVB	442.80000	+	CC2	
St-Hippolite	DMR/MARC	VE2RST	145.13000	-	CC1	
St-Hyacinthe	FM	VE2RBE	146.95500	146.36000		
St-j	FM	VE2RFR	145.29000	-	141.3 Hz	
	FM	VE2RFR	146.82000	-	103.5 Hz	
St-Joseph Du Lac	DMR/BM	VE2RCW	442.50000	+	CC1	
St-Medard	FM	VE2RWO	147.14000	+	107.2 Hz	
St-Medard EQC	FM	VA2RXY	147.03000	+	141.3 Hz	
St-Ubalde	FM	VE2RPW	146.85000	-	100.0 Hz	
St-z	FM	VA2KIK	147.36000	+	103.5 Hz	
St. Calixte	DMR/MARC	VA2RLD	443.15000	+	CC2	
St. Jerome	FM	VA2RMP	449.97500	-	74.4 Hz	
Ste Sophie D'Halifax						
	FUSION	VE2NBE	448.92500	-		
Sully	FM	VE2RXY	145.45000	-		
Sutton	FM	VE2RTC	146.64000	-		
Tach	FM	VE2RTX	449.02500	444.03000		
Tadossac	FM	VE2REY	444.55000	+	103.5 Hz	
Tadoussac	FM	VE2RSB	145.35000	-		
Talon	FM	VE2RVR	147.24000	+		
Thetford Mines	FM	VE2RVA	145.13000	-	100.0 Hz	
	FM	VE2RDT	147.16500	147.77000		
	FM	VE2CVA	448.17500	443.18000		
Thetford Mines, Pontbriand						
	FM	VE2RSQ	147.37500	147.98000		
Tour-Val-Marie	FM	VE2RNI	147.07500	147.68000	141.3 Hz	
Tracy Sorel	DMR/BM	VE2RBS	448.87500	-	CC2	
Trois Pistoles	FM	VE2SLJ	147.10500	+		
Trois-Rivieres	DSTAR	VA2LX B	448.67500	-		
	FM	VE2ROX	146.98500	146.39000	110.9 Hz	
	FM	VE2CTR	147.06000	+		
	FM	VE2LKL	147.27000	+		
	FM	VE2REY	442.50000	+	103.5 Hz	
	FM	VE2VIP	442.75000	+		
	FM	VE2RBN	448.67500	443.68000	136.5 Hz	
	FM	VE2RTZ	449.17500	444.18000	110.9 Hz	
	FM	VE2VIP	772.75000	447.75000		
Universite De Sherbrooke						
	FM	VE2PAK	441.14500	441.74500		
Vad-D'Irene	FM	VE2RDD	147.37500	147.98000		
Val-Alain	DMR/MARC	VE2DXI	447.47500	-	CC1	
Val-Belair	FM	VE2RGG	447.62500	442.63000		
Val-Brillant	FM	VE2ROL	147.00000	+		
Val-d'Or	FM	VE2RYL	146.76000	-	114.8 Hz	
Val-des-Monts	FM	VE2GFV	444.02500	+		

456   QUEBEC

Location	Mode	Call sign	Output	Input	Access	Coordinator
Valcartier-Village, Mont Triqu						
	FM	VE2RAG	145.45000	-		
Varennes	FM	VE2REQ	145.17000	-		
	FM	VE2REQ	448.27500	443.28000		
Variable	FM	VE2VK	447.52500	-	141.3 Hz	
Victoriarille	DSTAR	VA2RVO C	145.57000	-		
	DSTAR	VA2RVO C	147.39000	+		
	DSTAR	VA2RVO B	443.35000	+		
Victoriaville	FM	VE2RMD	144.81000	+	88.5 Hz	
	FM	VE2RMD	442.85000	+		
	FM	VE2RQC	443.00000	+	DCS 103	
	FM	VE2RQC	443.35000	+	103.5 Hz	
	FM	VE2RBF	443.50000	+		
	FM	VE2RHY	444.60000	+	110.9 Hz	
	FM	VE2RMD	448.82500	443.83000	97.4 Hz	
Ville De Saguenay, Holiday Inn						
	FM	VE2RKA	145.41000	-	100.0 Hz	
Ville-Marie	FM	VE2RTE	146.73000	-		
Waterloo	FM	VE2ESM	443.90000	+		
Weatheradio-Amqui						
	FM	XLR528	162.40000			
Weatheradio-Baie-TrinitÃ©						
	FM	VDD596	162.47500			
Weatheradio-Blanc-SablonÂ						
	FM	XLR526	162.40000			
Weatheradio-Carleton-sur-Mer						
	FM	VDD598	162.50000			
Weatheradio-Chibougamau						
	FM	XLR749	162.55000			
Weatheradio-DÃ©gelis						
	FM	VDD225	162.55000			
Weatheradio-GaspÃ©						
	FM	VDD597	162.55000			
Weatheradio-La Malbaie						
	FM	XLR611	162.47500			
Weatheradio-La Tuque						
	FM	VBB499	162.47500			
Weatheradio-Longue-Pointe-de-M						
	FM	VOR669	162.40000			
Weatheradio-Magdalen Islands						
	FM	VOR668	162.55000			
Weatheradio-Mont-Fournier						
	FM	VDD464	162.40000			
Weatheradio-Mont-Laurier						
	FM	XLR969	162.55000			
Weatheradio-Mont-MÃ©gantic						
	FM	XLR420	162.55000			
Weatheradio-Mont-Tremblant						
	FM	VAF367	162.47500			
Weatheradio-Montreal						
	FM	XLM300	162.55000			
Weatheradio-Quebec						
	FM	XLM369	162.55000			
Weatheradio-Rimouski						
	FM	XLR617	162.55000			
Weatheradio-RiviÃ¨re-au-Renard						
	FM	XLR525	162.47500			
Weatheradio-Rouyn-Noranda						
	FM	XLR748	162.40000			
Weatheradio-Saguenay						
	FM	XLR285	162.55000			

Location	Mode	Call sign	Output	Input	Access	Coordinator
Weatheradio-Saint-FÃ©licien						
	FM	VBS906	162.47500			
Weatheradio-Sainte-Marie						
	FM	XLR527	162.52500			
Weatheradio-Sept-ÃŽles						
	FM	XLR519	162.55000			
Weatheradio-Sherbrooke-Magog						
	FM	XLR412	162.47500			
Weatheradio-Trois-RiviÃ¨res						
	FM	XLR411	162.40000			
Weatheradio-Val-d Or-Amos						
	FM	XLR747	162.47500			
Weatheradio-Ville-Marie						
	FM	XLR750	162.55000			
Wendake	FM	VA2RZ	147.57000	+		
Wheelock Mill	FM	VE2FSK	146.58000	+		
Zec Du Gros Brochet						
	FM	VA2ZGB	147.39000	+	103.5 Hz	
	FM	VA2ZGB	442.27500	+	103.5 Hz	

SASKATCHEWAN

Location	Mode	Call sign	Output	Input	Access	Coordinator
Arcola, Moose Mountain						
	FM	VE5MMR	146.82000	-		
Asquith	FM	VE5KEV	146.70000	-	100.0 Hz	
Avonlea	FM	VE5ARG	147.06000	+		
	FM	VE5ARG	444.15000	+		
Battleford	FM	VE5BRC	147.24000	+		
Cactus Lake	FM	VE5IPL	146.91000	-		
Canora	FM	VE5RJM	147.30000	+		
	FM	VE5RJM	445.57500	-		
Christopher Lake	FM	VE5LAK	146.61000	-		
Davidson	FM	VE5RPD	145.19000	-		
Deschambault Lake						
	FM	VA5DES	145.30000	-		
Emma Lake	FM	VE5QU	145.35000	-		
Endeavour	FM	VA5INV	147.08000	+		
Engelfeld	FM	VE5NJR	146.73000	-		
Estevan	FM	VA5EST	147.03000	+		
	FM	VE5EST	147.18000	+		
	FM	VA5EST	224.70000	-		
	FM	VA5EST	444.80000	+		
Eyebrow	FM	VE5YMJ	147.36000	+		
La Ronge	FM	VE5LAR	146.97000	-		
Leroy	FM	VE5HVR	146.91000	-		
	FM	VE5HVR	445.57500	-		
Little Bear Lake	FM	VE5NLR	146.85000	-		
Lizard Lake	FM	VA5LLR	145.39000	-		
Lucky Lake	FM	VE5XW	146.73000	-		
Martensville	FUSION	VE5MBX	449.50000	-		
Meadow Lake	FM	VE5MLR	147.33000	+		
Melfort	FM	VE5MFT	146.88000	-		
Melville Sk. Canada						
	FM	VE5MDM	147.00000	+		
Minatinas Hills	FM	VE5RPA	147.15000	+		
Moose Jaw	FM	VE5CI	146.94000	-		
	FM	VE5PSC	147.52500	-	100.0 Hz	
Moosomin Sk Canada						
	FM	VE5MRC	146.79000	-		
Nipawin	FM	VE5NIR	443.75000	+		
North Battleford	FM	VE5BRC	146.88000	-		
Preeceville	FM	VE5SS	146.61000	-		
	FM	VE5SEE	147.39000	+		

458   SASKATCHEWAN

Location	Mode	Call sign	Output	Input	Access	Coordinator
Prince Albert	FM	VE5PA	147.06000	+		
Regina	FM	VE5REC	146.64000	-		
	FM	VE5WM	146.88000	-	103.5 Hz	
	FM	VE5YQR	147.12000	+	100.0 Hz	
	FM	VE5UHF	444.25000	+		
	FM	VE5EIS	446.25000	-	100.0 Hz	
	FM	VE5BBZ	447.25000	-		
	FUSION	VE5BBZ	447.25000	-		
Rocanville	FM	VE5LCM	146.31000	+		
Saskatoon	FM	VA5SV	145.33000	-	100.0 Hz	
	FM	VE5SK	146.64000	-		
	FM	VE5CC	146.97000	-	100.0 Hz	
	FM	VE5HRF	147.39000	+		
	FM	VE5FUN	441.65000	+	100.0 Hz	
	FM	VE5CC	449.97500	449.97000		
Saskatoon WIN System Affiliate						
	FM	VE5FUN	147.52500		100.0 Hz	
Snowden	FM	VE5NDR	147.09000	+		
Strasbourg, Last Mountain						
	FM	VE5AT	146.85000	-		
Swift Current	DMR/MARC	VE5DMR	443.00000	+	CC1	
	FM	VE5SCR	146.79000	-		
	FM	VE5SCC	146.88000	-		
Tisdale	FM	VE5FXR	146.70000	-		
Turtle Lake	FM	VE5TLK	145.45000	-		
Unity	FM	VE5URC	147.00000	+		
Va5eis B 145 670	DSTAR	VA5EIS C	144.95000	+		
Watrous	FM	VE5IM	146.70000	-		
Weatheradio-Broadview						
	FM	VCB462	162.47500			
Weatheradio-Elbow						
	FM	VBP687	162.47500			
Weatheradio-Estevan						
	FM	VAM595	162.40000			
Weatheradio-Lanigan						
	FM	VBU746	162.40000			
Weatheradio-North Battleford						
	FM	VAR552	162.47500			
Weatheradio-Prince Albert						
	FM	VAR551	162.40000			
Weatheradio-Regina						
	FM	XLM537	162.55000			
Weatheradio-Regina Beach						
	FM	VBC936	162.40000			
Weatheradio-Saskatoon						
	FM	XLF322	162.55000			
Weatheradio-Stranraer						
	FM	VAR554	162.40000			
Weatheradio-Swift Current						
	FM	XLF524	162.55000			
Weatheradio-Waseca						
	FM	VDI204	162.40000			
Weatheradio-Yorkton						
	FM	VAM594	162.55000			
Weyburn	FM	VE5WEY	146.70000	-		
Wolseley	FM	VE5WRG	146.67000	-		
Yellow Creek	FM	VE5AG	147.18000	+		

NOTES

NOTES

NOTES

NOTES

NOTES